Petroleum Refinery Process Modeling

Petroleum Refinery Process Modeling

Integrated Optimization Tools and Applications

Y. A. Liu, Ai-Fu Chang, and Kiran Pashikanti

WILEY-VCH

Authors

Prof. Y. A. Liu
Virginia Polytechnic Institute
Department of Chemical Engineering
and State University
245 Goodwin Hall
735 Prices Fork Road
Blacksburg 24061, VA
United States

Dr. Ai-Fu Chang
Chevron Phillips Chemical CO
1862 Kingwood Drive
Kingwood 77339, TX
United States

Dr. Kiran Pashikanti
Chevron Phillips Chemical CO
1862 Kingwood Drive
Kingwood 77339, TX
United States

Cover Credits: photodisc.com

Library of Congress Card No.: applied for

British Library Cataloguing-in-Publication Data
A catalogue record for this book is available from the British Library.

Bibliographic information published by the Deutsche Nationalbibliothek
The Deutsche Nationalbibliothek lists this publication in the Deutsche Nationalbibliografie; detailed bibliographic data are available on the Internet at <http://dnb.d-nb.de>.

Print ISBN: 978-3-527-34423-9
ePDF ISBN: 978-3-527-81336-0
ePub ISBN: 978-3-527-81338-4
Mobi ISBN: 978-3-527-81337-7
oBook ISBN: 978-3-527-81339-1

Cover Design Schulz Grafik-Design, Fußgönheim, Germany
Typesetting SPi Global, Chennai, India
Printing and Binding C.O.S. Printers Pte Ltd, Singapore

Printed on acid-free paper

10 9 8 7 6 5 4 3 2 1

Contents

About the Authors

Y.A. Liu, the Alumni Distinguished Professor and the Frank C. Vilbrandt Endowed Professor of Chemical Engineering at Virginia Tech, received his BS, MS, and PhD degrees from National Taiwan University, Tufts University and Princeton University, respectively.

He taught at Auburn University from 1974 to 1981, where his last position was Alumni Associate Professor endowed by the Auburn Alumni Association. He joined Virginia Tech as a Professor of Chemical Engineering in 1982. In 1983, he was appointed the Vilbrandt Professor. In 2012, he was appointed the Alumni Distinguished Professor. He has published numerous papers and ten books. These include six pioneering chemical engineering textbooks on *Artificial Intelligence in Chemical Engineering* (1991) and on *Neural Networks in Bioprocessing and Chemical Engineering* (1995), both by Academic Press; on *Industrial Water Reuse and Wastewater Minimization* (1999) by McGraw-Hill; on *Step-Growth Polymerization Process Modeling and Product Design* (2008), by Wiley; on *Refinery Engineering – Integrated Process Modeling and Optimization* (2012), by Wiley-VCH; and on *Design, Simulation and Optimization of Adsorptive and Chromatographic Separations: A Hands-on Approach* (2018), by Wiley-VCH.

His contributions to chemical engineering teaching, research, and industrial outreach have been recognized by over 30 university, national, and international awards. He is a Fellow of the American Institute of Chemical Engineers, a member of Virginia Tech's Academy of Teaching Excellence, and a recipient of the 1996 AspenTech International Award for University Teaching Excellence in computer-aided design. He has received three awards from the American Society of Engineering Education (ASEE): the Fred Merryfield Design Award (1993) for creativity and excellence in teaching and research of engineering design; the George Westinghouse Award (1990), ASEE's highest honor for an engineering educator under the age of 45 years for outstanding achievements in both teaching and scholarship; and the Western Electric Award (1984) for excellence in instruction of engineering students. In 1986, he received the National Catalyst Award for excellence in chemical education from the Chemical Manufacturers Association. He received the Distinguished Alumni Award in 1990 and the Outstanding Career Achievement Award in 2010, both from Tufts University. In 2015, he received the Virginia Professor of the Year Award from the US. Professors of the Year Awards Program, the only US national awards that recognize and celebrate extraordinary dedication to undergraduate teaching, sponsored by the Carnegie Foundation for the Advancement of Teaching and the Council for Advancement and Support of Education.

Over the past 25 years, he devoted his school breaks helping petrochemical industries in developing countries and chemical industries in Virginia with technology development and engineering training. He has taught intensive training courses on computer-aided design, advanced process control, energy and water savings, and refinery and polymerization process modeling to over 7000 practicing engineers in China, Taiwan, and in the USA. For his contributions to teaching, research, and industrial outreach, he received the Virginia Outstanding Faculty Award from Virginia's Governor Jim Gilmore and the National Friendship Award from China's Premier Zhu Ronjie in 2000.

Ai-Fu Chang received his PhD in the Department of Chemical Engineering at Virginia Polytechnic Institute and State University ("Virginia Tech") in September, 2011. He received his BS in Chemical Engineering from National Taiwan University in 2001. He completed his doctoral dissertation on integrated process modeling and product design of biodiesel manufacturing and refinery reaction and fraction systems. The latter was the basis of this textbook. He has worked on several industrial modeling projects, including poly (acrylonitrile–vinyl acetate), hydrocracking, and biodiesel. These projects were collaborative efforts between Virginia Tech, Aspen Technology, and industrial manufacturers. He is currently employed by Chevron Phillips Chemical Company.

Kiran Pashikanti was a PhD student in the Department of Chemical Engineering at Virginia Tech. He received his BS in Chemical Engineering from Virginia Commonwealth University in 2005 and his PhD in Chemical Engineering from Virginia Tech in September, 2011. He has worked on several industrial modeling projects on integrated modeling of refinery reaction and fraction systems and of carbon-dioxide capture processes. This textbook grows out of his doctoral dissertation on the predictive modeling of fluid catalytic cracking (FCC) and catalytic reforming processes. He is currently a Senior Olefins and NGL Technical Support Engineer at Chevron Phillips Chemical Company.

Foreword by Lawrence B. Evans

Petroleum refining is a huge industry. Everyday the industry produces more than $8 billion of refined products worldwide. Small improvements in the design and operation of a refinery can deliver large economic value. Crude petroleum is a natural material containing thousands of chemical compounds. The refinery converts the crude into a wide range of products from transportation fuels and petrochemical feedstocks to asphalt and coke. All of these products must meet demanding specifications while the refinery stays within tight environmental constraints.

Computer models are routinely used today to model petroleum refining processes. Engineers use them to design new refineries, to improve the operation of existing refineries, to make decisions on purchasing crude, and to optimize the planning of production. The ability to accurately model each step in the refining process is the key to optimizing the performance of the integrated refinery. Modeling a refinery is challenging because crude petroleum consists of thousands of chemical compounds. The refinery takes the large molecules in crude oil and cracks them into the smaller molecules of transportation fuels. It must also carry out chemical reactions to tailor the composition of products to meet specifications. These reactions take place through a complex set of reaction pathways.

For most of my career, I have worked on the development of computer models of chemical processes. Today, very good commercial software systems exist that enable engineers to build and use sophisticated models for refinery simulation and optimization. But these tools are mainly used by experts. This book by Professor Liu and his colleagues represents a major advance by enabling engineers who are not experts to develop and use state-of-the-art computer models for the simulation and optimization of integrated refinery reaction and fractionation processes.

The book is very well organized and systematic. It starts in the first chapter by showing how to represent the thermodynamic and physical properties of crude petroleum and the complex materials that comprise the intermediate streams in a refinery. The next two chapters cover the major separation units in a refinery: the atmospheric distillation unit (ADU) and the vacuum distillation unit (VDU). The final four chapters cover the most important chemical conversion units together with their product fractionation systems. These include the fluid catalytic cracking (FCC) process, the continuous catalytic reforming process, the hydroprocessing units, the alkylation and delayed coking units, and refinery-wide process simulation. Each chapter follows the same pattern starting first with a description of the unit, methods to organize and use the pertinent data from the refinery, and then the workflows to construct a rigorous model using existing commercial software. Finally, the chapter concludes with strategies to tune the models to match performance followed by case study examples, and the discussion of other applications of the models such as for refinery production planning. The book uses Aspen HYSYS for modeling, but most of the concepts are also applicable to other systems. A supplement to the book provides relevant spreadsheets and simulation files for all the models and examples presented in the book.

One of the unique strengths of the book is that it does not stop with the theory, or even the case study examples and hands-on workshops. It covers very practical problems of how to work with real data, how to construct the right level of detail for the problem you are solving and the data you have, and how to tune the model to plant data because, as the authors note, no model is perfect. The book contains numerous up-to-date references to the literature. The field of refinery process modeling is constantly developing and evolving. Individuals who need to contribute to this development or explore new directions will find the review of existing work valuable. The current edition of this book will be valuable to industrial practitioners and to the academic chemical engineers by exposing them to the problem of modeling and optimization of refinery processes and enabling them to solve realistic problems. It will take this work from a technology used mostly by experts to a tool that engineers in a refinery can use in their everyday work.

Lawrence B. Evans

Professor Emeritus of Chemical Engineering
Massachusetts Institute of Technology
Founder, Aspen Technology, Inc.
Member, National Academy of Engineering
Past President, American Institute of Chemical Engineers

Foreword by Steven R. Cope

Petroleum refining is one of the most important, exciting, and challenging industries on the face of the earth. It has been in existence for about 100 years, and during that time, it has evolved and advanced to the point where today's modern refinery is full of complex, cutting-edge technologies. Examples include state-of-the-art catalyst systems, complex reactor designs, sophisticated computer control hardware and software, and advanced safety and environmental controls.

A typical medium-sized refinery has hundreds of pumps, heat exchangers, and drums; dozens of furnaces, compressors, and high-temperature/high-pressure reactors; and thousands of control loops and associated advanced computer control technologies. This same typical refinery has dozens of different crudes and other feedstocks to choose from and dozens of products to maximize or minimize based on consumer demands and global marketplace economics. In addition to daily decisions about feedstocks and products, there are also hundreds of decisions to be made each day about operating temperatures, pressures, unit feed rates, catalyst addition rates, cycle times, distillation cut points, product specifications, inventory levels, etc.

In this very competitive global industry, it is critical to minimize overall operating costs while achieving the maximum possible "upgrade" for each hydrocarbon molecule. This is commonly referred to as "molecule management." All of these decisions and options require complex computer modeling to aid in the selection of feedstocks and product slates and in troubleshooting and optimizing the performance of individual refinery "processes" such as distillation units, fluidized catalytic cracking units, continuous catalyst regeneration reforming units, hydrocracking and hydrotreating units, and alkylation and delayed coking units. And eventually, all of these individual parts have to be pulled together into a refinery-wide process simulation model in order to feed a linear program (LP) model capable of optimizing the overall refinery. This integrated process modeling and optimization is the subject of the present book by Y. A. Liu, Ai-Fu Chang, and Kiran Pashikanti.

Based on my review of this book, I believe that it provides a very solid introduction to integrated refinery process modeling and optimization, using many of the tools and techniques currently employed in modern refineries. I believe that this book and associated hands-on workshops would be a highly desirable investment by any engineering student considering a career in the petroleum refining industry.

Refining Director – North America
ExxonMobil Refining & Supply Company

Steven R. Cope

Preface

This book is a substantial revision of our book, *Refinery Engineering: Integrated Process Modeling and Optimization*, published in 2012. We rename our book as *Petroleum Refinery Process Modeling: Integrated Optimization Tools and Applications*, to better reflect the focus and contents of the new edition. Since the publication of our previous edition, there are three primary changes to the petroleum refining industry and simulation software technology that provide a strong motivation for us to update, revise, and expand the coverage of our textbook.

The first primary change to the petroleum refining industry is the declining crude oil price and its impact on the petrochemical industries. The senior author has worked as an advisor to FORTUNE's 2017 global top two oil companies (SINOPEC and PetroChina) since 1993 and has developed good knowledge and experience of the worldwide petrochemical industries. For many petroleum companies with aged oil fields that rely on old technology and extensive labor, it is more profitable to purchase cheaper crude oil from the world market rather than continue with expensive crude oil extraction and production. Thus, many petrochemical corporations are under tremendous pressure to cut their losses with upstream oil extraction and production and to increase their downstream profits in refining and chemical production. As a result, integrated process modeling and optimization of petroleum refining to improve the profit margins, as covered in our book, has become vitally important.

Next, petroleum refining has traditionally relied heavily on the knowledge and experience of engineers and operators to make good estimates of refining unit operating conditions and process performance. However, the growing tide of retiring professionals and increasing loss of experience throughout the industry is making this experience-based task difficult or impossible. Therefore, the use of user-friendly simulation tools and techniques becomes invaluable. Additionally, with a growing focus on intelligence manufacturing in the refining industries, the development of operator training simulators to aid in the training of new operators and engineers requires modeling tools that are user-friendly, efficient, and accurate in the prediction of plant data.

Third, since 2012, there has been a number of significant advances in the user interface and simulation capability for advanced software tools for refinery process modeling and optimization. In particular, Aspen HYSYS Petroleum Refining has included the new Petroleum Assay Manager that represents a major improvement over the old Oil Manager in defining the hypothetical components (hypos or pseudocomponents) based on boiling point ranges of petroleum assay and in quantifying the physical properties of the hypos. It has added the powerful new tool of column hydraulic analysis for the sizing and rating (performance evaluation) of distillation columns and refinery fractionators. It has provided new modeling tools for refining reaction processes such as alkylation, delayed coking, and isomerization.

Additionally, Aspen HYSYS as well as other Aspen engineering suites of software tools have developed a new "common" user interface that is vastly different from and more user-friendly than the old user interface shown in the hundreds of figures of our first edition published in 2012, making it essential for us to update and revise our textbook. The new user interface is easy to use and it combines simulation, integration, and optimization into a single framework through activation tools. Once a user has developed a process simulation model, the new model interface enables the user to "activate": (1) the rigorous energy-saving analysis based on pinch technology; (2) the rigorous heat exchanger design and rating; and (3) process economic evaluation for profitability analysis. Additionally, the new user interface "integrates" process simulation with new tools for safety analysis based on process relief devices and safety valves. By doing so, the new Aspen HYSYS interface greatly accelerates collaboration among engineers and specialists involved in process, production, equipment design, cost, safety, and so on into a common software platform. Other commercial simulation tools do not have this unique and significant capability in activation and integration.

Our specific revisions include the following:

1) We have replaced the screen images of our illustrative examples and hands-on workshop steps in the first six chapters based on the new user interface. This amounts to over 600 figures.
2) We have included new workshops and application case studies in each chapter and have expanded the discussion of our hands-on workshops with additional step-by-step illustrations. Sections 1.7 (petroleum assay manager), 1.8 (from oil manager to petroleum assay manager), 2.4 (model development and recommendation to correctly handle column efficiency), 2.10, 2.12, and 2.13 (new case studies), 2.14 (new column hydraulic analysis tool), 2.15 (petroleum distillation column), 4.13 (complete simulation of integrated FCC reactor, main fractionator and gas plant system), and 6.12 (fractionation system for hydrocracking reactor) are all new. We cover the new petroleum assay manager and its improvement over the old oil manager, column hydraulic analysis, and so on. We teach the reader how to use the new column hydraulic analysis tool for rating and retrofit of existing columns. Workshop 4.3 is significant, as it is the first detailed, step-by-step demonstration of how to build a simulation model for the FCC main fractionator and gas plant system that has ever appeared in a textbook or in the literature. Refinery engineers can apply the same procedure to build the simulation models for the fractionation systems for reformers, hydrocrackers, delayed cokers, and so on.

3) We have written a new Chapter 7 that covers the simulation and optimization of alkylation, delayed coking, and refinery-wide simulation and profit margin analysis.
4) We have updated our reference sections to include new literature published since 2012 and additional references for further reading.

Lastly, we should mention that we have not found any new competing textbooks and reference volumes published since 2012, and we have taught the materials in this revision to engineers and scientists in global top two oil companies. Our trainees find our materials, particularly the hands-on workshops and case studies, easy to learn and very useful to their simulation and optimization of refining processes from plant data and to significantly increase their refinery profit margins.

Virginia Polytechnic Institute
and State University
Blacksburg

Y. A. Liu

Acknowledgments

It is a pleasure to thank a number of very special persons and organizations that contributed to the preparation of this book.

We would like to express our sincere appreciation to the following senior leaders of Aspen Technology, Inc., for their strong support of the Center of Excellence in Process System Engineering at Virginia Tech since 2002: Antonio Pietri, President and CEO; Willie Chan, Senior Vice President and Chief Technology Officer; Filipe Soares-Pinto, Senior Vice President, Asia-Pacific; Andy Lui, Vice President, R & D; Vikas Dhole, Vice President, Engineering Product Management; and Steven Qi, Vice President, Customer Support and Training; Fran Royer, Senior Manager, University Programs; and Theresa Foley, Senior Customer Care Specialist. We thank the following refinery modeling experts at Aspen Technology for teaching us the art and practice of refinery process modeling: Sandeep Mohan, Dinu Ajikutira, Stephen Dziuk, Hiren Shethna, Dave Dhaval, Darin Campbell, Maurice Jett, and John Adams.

We would like to thank the China Petroleum and Chemical Corporation (SINOPEC) and Formosa Petrochemical Corporation (FPCC) for challenging us to enter the field of refinery process modeling in 2007.

We thank BAE Systems, Aspen Technology, SINOPEC, China National Petroleum Company (PetroChina), Novozymes Biologicals, Universal Fibers, Eastman Chemical, and Mid-Atlantic Technology, Research and Innovation Center for supporting our educational programs in computer-aided design and process system engineering at Virginia Tech. We are very grateful to Mr Cao Xianghong for his strong support of this work during his tenure as Senior Vice President and Chief Technology Officer of SINOPEC and to Mr He Shengbao, Mr Du Jizhou, Mr Xu Yingjun and Mr Chen Yuanpeng, senior executives for research and development at PetroChina for their strong support in refinery technology development and engineering training.

We thank Professor Lawrence B. Evans of Massachusetts Institute of Technology and Mr Steve Cope, ExxonMobil, who kindly took time to write the FOREWORD for our text.

The senior author would like to thank his wife, Hing-Har Liu, for her support through the laborious process of this book writing and revision.

Scope of Textbook

The purpose of this text to guide senior-level undergraduates, graduate students, and industrial practitioners how to quantitatively model key refinery reaction and fractionation processes. In addition, this text contains advanced modeling topics (such as kinetic network calibration) that will prove useful to researchers and practitioners alike. After following the procedures in this text, the reader will be able to (1) identify key data required for building reaction and fractionation models with commercial software; (2) filter extensive data available at the refinery and use plant data to begin calibrating available models; (3) extend model to include key fractionation submodels; (4) provide a sound and informed basis to understand and exploit plant phenomena to improve yield, consistency, and performance of a given unit; and (5) apply models in an overall refinery context through refinery planning based on linear programming (LP).

We present the topics in a logical progression from basic refinery thermodynamic and physical property predictions to detailed guides for modeling complex reaction and fractionation units. Chapter 1 introduces the reader to the basics of dealing with the thermodynamics and physical property predictions of hydrocarbon components in the context of process modeling. Chapters 2 and 3 use the key concepts of fractionation lumps and physical properties to develop detailed models and workflows for atmospheric or crude (CDU) and vacuum (VDU) distillation units. Chapters 4–7 are largely self-contained that discuss modeling FCC, catalytic reforming, hydroprocessing units, alkylation and delayed coking. In general, we discuss each unit in the following order.

- Process description
- Modeling and literature review
- Key modeling details
  - Kinetic models
  - Fractionation models
- Model calibration
- Model validation with industrial data collected by the authors
- Industrially relevant case studies
- Model application in LP refinery planning context
- Workshops and step-by-step guides for building models using commercial software

In addition, we provide significant supporting materials alongside the text (accessible through Wiley-VCH website, wiley-vch.de/en/→shop→bookfinder→ petroleum refinery process modeling→student materials). These materials include relevant spreadsheets, guides, and sample simulation files for all models developed in the workshops throughout this text.

We hope that this text allows both academia and industrial practitioners to understand, model, and optimize complex refinery reaction and fractionation systems. The goal of all modeling exercises presented is to improve yield, consistency, profitability, and performance of a given unit and the refinery as a whole.

Software Selection and Copyright Notice

Aspen HYSYS and Aspen HYSYS Petroleum Refining (Version 8.8 or new version) are available from Aspen Technology, Cambridge, MA (http://www.aspentech.com/).

Microsoft Excel and Visual Basic for Applications (VBA) are available as part of Microsoft's Office software package (http://office.microsoft.com/en-us/default.aspx).

Aspen HYSYS® and Aspen HYSYS Petroleum Refining® are available from Aspen Technology, Bedford, MA (http://www.aspentech.com/).

1

Characterization and Physical and Thermodynamic Properties of Oil Fractions

This chapter introduces the common methods for characterizing crude oils and petroleum fractions (i.e., oil fractions) and estimating their thermophysical properties. We begin by defining the essential bulk and fractional properties of oil fractions and by explaining the various types of distillation curves and their interconversion (Section 1.1.1). Next, we explain the generation of hypothetical components ("hypos") or pseudocomponents of oil fractions based on boiling point ranges and the estimation of density and molecular weight distributions of the resulting hypos (Section 1.3). Sections 1.4–1.9 present six hands-on workshops using Excel spreadsheets and Aspen HYSYS Petroleum Refining for (1) the interconversion of distillation curve data; (2) the extrapolation of an incomplete distillation curve data; (3) the calculation of the mean average boiling point (MeABP) of a given oil fraction; (4) specifying an oil fraction in the *old* oil manager; (5) representing an oil fraction in the *new* petroleum assay manager; and (6) conversion from the oil manager to petroleum assay manager and improvements of the petroleum assay manager over the oil manager.

Section 1.10 introduces the essential thermophysical properties for developing refinery reaction and fractionation process models. Section 1.10.1 presents the useful methods for estimating the thermophysical properties (e.g., molecular weight, liquid density, critical properties, ideal gas heat capacity, and heat of vaporization) of pseudocomponents of oil fractions. Section 1.11 describes the important thermodynamic models for refinery reaction and fractionation processes. Section 1.12 discusses the estimation methods for other physical properties such as flash point, freeze point, and PNA (paraffin, naphthalene, and aromatic) content of a refinery feed. Section 1.13 summarizes the conclusion of this chapter. Finally, we present the nomenclature and bibliography.

1.1 Crude Assay

Crude oils and petroleum fractions are the most important feedstocks for refining processes. To properly simulate the refining processes, we must have good understanding of the compositional information and thermophysical properties of crude oils and petroleum fractions. However, the complexity of molecular composition of crude oils and petroleum fractions makes it hardly possible to

Petroleum Refinery Process Modeling: Integrated Optimization Tools and Applications,
First Edition. Y. A. Liu, Ai-Fu Chang, and Kiran Pashikanti.

identify individual molecules. Instead, modern refiners use assay to characterize crude oils and petroleum fractions.

A typical crude assay includes two types of information for an oil sample: (1) *bulk properties* and (2) *fractional properties*. Table 1.1 provides examples of both types of properties of a crude assay. For design and modeling purposes, it is always the best practice to have process data obtained in the same period as assay data, as the properties and composition of crude change over time as it is produced from a given well. Kaes [1] suggested that the assay data should not be 2 years older than the process data used to build process simulation. We explain both bulk and fractional properties in the following sections.

1.1.1 Bulk Properties

Bulk properties include specific gravity, sulfur content, nitrogen content, metal (Ni, V, Fe, etc.) content, asphaltene content, C:H ratio, pour point, flash point, freeze point, smoke point, aniline point, cloud point, viscosity, carbon residue, light hydrocarbon yields (C1–C4), acid number, refractive index, and boiling point curve. We generally use the *API (American Petroleum Institute) gravity* to specify the specific gravity (SG) of the crude oil as

$$\text{API} = (141.5/\text{SG}) - 131.5 \tag{1.1}$$

or

$$\text{SG} = 141.5/(\text{API} + 131.5) \tag{1.2}$$

SG is the specific gravity defined as the ratio of the density of the crude oil to the density of water both at 15.6 °C (60 °F). The API gravity varies from less than 10 for very heavy crudes to between 10 and 30 for heavy crudes, to between 30 and 40 for medium crudes, and to above 40 for light crudes.

The *sulfur content* is expressed as a percentage of sulfur by weight and varies from less than 0.1% to greater than 5%. Crude oils with less than 1 wt% sulfur are called *low sulfur or sweet* and those with more than 1 wt% sulfur are called *high sulfur or sour*. Sulfur-containing constituents of the crude oil include simple mercaptans (also known as thiols), sulfides, and polycyclic sulfides. *Mercaptan sulfur* is simply an alkyl chain (R–) with –SH group attached to it at the end. The simplest form of R–SH is *methyl mercaptan*, CH_3SH.

The *pour point* is a measure of how easy or difficult it is to pump the crude oil, especially in cold weather. Specifically, the pour point is the lowest temperature at which a crude oil will flow or pour when it is chilled without disturbance at a controlled rate. The pour point of the whole crude or oil fractions boiling above 232 °C (450 °F) is determined by the ASTM test method D97.

The *flash point* of a liquid hydrocarbon or an oil fraction indicates its fire and explosion potential, and it is the lowest temperature at which sufficient vapor is produced above the liquid to form a mixture with air that a spontaneous ignition can occur if a spark is present. One of the standard ASTM test methods for flash point is D3278.

The *freeze point* is the temperature at which the hydrocarbon liquid solidifies at atmospheric pressure. It is an important property of kerosene and jet fuels

Table 1.1 A typical crude assay.

	Whole crude	C4 and C4−	C5–74 °C	74–166 °C	166–480 °C	480–249 °C	249–537 °C	537 °C+
Cut volume, %	100	1.57	8.26	20.96	17.11	17.52	24.71	9.87
API gravity	38.6	117.9	80.6	55.7	42.82	34.7	25.5	10.9
Carbon, wt%		82.5	83.9	86.0	86.1	86.4	86.4	
Hydrogen, wt%		17.5	16.1	14.0	13.9	13.2	12.8	
Pour point, °C	−12.2				−53.9	−10.6	38.9	56.7
Sulfur, wt%	0.3675			0.0137	0.058	0.2606	0.6393	1.1302
Nitrogen, ppm	970	0	0	0	2.4	94.6	1346	4553
Viscosity at 20 °C/68 °/, cSt	4.59	0.41	0.46	0.73	1.74	6.76	118.4	1 789 683
Viscosity at 100 °C/212 °/, cSt	1.35	0.24	0.26	0.38	0.68	1.43	5.91	372
Mercaptan sulfur, ppm	25			22.8	35.3			
CCR, wt%	1.71					0	0.11	14.21
Nickel, ppm	1.7					0	0.1	12.8
Vanadium, ppm	5.2					0	0.1	41.5
Heat of combustion (gross), BTU/lb	19 701							
Heat of combustion (net), BTU/lb	18 496	19 078	18 729	18 561	18 546			
Salt content, ptb	1.7							
Paraffins, vol%		100	84.77	46.64	48.83	39.42	30.18	
Naphthenes, vol%		0	13.85	36.56	31.54	37.44	31.83	
Aromatics, vol%				16.8	15.15			
Freeze point, °C					−43.9	−0.6		
Smoke point, mm					23.3			
Cetane index 1990 (D4737)	37	131	44	30	43	55	59	43
Cloud point, °C					−47.8	−3.9		

(Continued)

Table 1.1 (Continued)

	Whole crude	C4 and C4−	C5–74 °C	74–166 °C	166–480 °C	480–249 °C	249–537 °C	537 °C+
Aniline point, °C					57.7	69.5	87.1	
Distillation type	D1160	D86	D86	D86	D86	D86	D1160	D1160
ASTM IBP, °C	0.2	−70.9	−57.2	206.9	97.2	263.1	365.2	559.1
5 vol%, °C	51.9	−27.3	−32.9	212.1	100.1	265.6	367.8	561.7
10 vol%, °C	79.7	13.8	−10.1	214.8	101.6	266.7	373.1	565.7
20 vol%, °C	119.9	30.2	−1.0	220.8	104.9	269.7	384.1	575.1
30 vol%, °C	160.7	36.8	2.7	227.6	108.7	273.7	396.7	585.8
40 vol%, °C	205.6	38.2	3.4	235.8	113.2	278.4	410.8	598.2
50 vol%, °C	254.3	38.3	3.5	244.1	117.8	283.2	426.3	612.4
60 vol%, °C	308.7	42.7	5.9	254.1	123.4	288.7	442.8	631.2
70 vol%, °C	364.0	46.5	8.1	265	129.4	294.8	459.5	653.1
80 vol%,°C	425.6	49.3	9.6	276.8	136.0	301.4	477.6	681.3
90 vol%, °C	502.9	47.5	8.6	289.4	143.0	308.3	496.0	718.7
95 vol%, °C	570.9	47.1	8.4	296.4	146.9	312.2	507.4	751.0
ASTM EBP, °C	730.7	47.9	8.8	307.7	153.2	318.2	520.7	791.6

because of the very low temperatures encountered at high altitudes in jet planes. A standard test method for the freeze point is ASTM D4790.

The *smoke point* refers to the height of a smokeless flame of fuel in millimeters beyond which smoking takes place. It reflects the burning quality of kerosene and jet fuels and is determined by ASTM D1322.

The *aniline point* represents the minimum temperature for complete miscibility of equal volumes of aniline and petroleum oil. It is an important property of diesel fuels and is measured by ASTM D611.

The *cloud point* refers to the temperature at which solidifiable components (waxes) present in the oil sample begin to crystallize or separate from solution under a method of prescribed chilling. It is an important specification of middle distillate fuels, as determined by ASTM D2500.

The *Conradson carbon residue (CCR)* results from ASTM D189. It measures the coke-forming tendencies of oil. It is determined by destructive distillation of a sample to elemental carbon (coke residue) in the absence of air, expressed as the weight percentage of the original sample. A related measure of the carbon residue is called *Ramsbottom carbon residue* , as determined by ASTM D524-15. A crude oil with a high CCR has a low value as a refinery feedstock.

The *acid number* results from ASTM D3339-11 that determines the organic acidity of a refinery stream.

The *refractive index* represents the ratio of the velocity of light in a vacuum to that in the oil. It is determined by ASTM D1218.

The *gross heat of combustion or high heating value (HHV)* is the amount of heat produced by the complete combustion of a unit quantity of fuel. We obtain the gross heat of combustion by cooling down all products of the combustion to the temperature before the combustion and by condensing all the water vapor formed during combustion.

The *net heat of combustion or lower heating value (LHE)* is obtained by subtracting the latent heat of vaporization of the water vapor formed by combustion from the gross heat of combustion or higher heating value.

The *true boiling point (TBP) distillation* [1] of a crude oil or petroleum fractions results from using the US Bureau of Mines Hempel method and ASTM D285. Neither of these methods specifies the number of theoretical stages or the molar reflux ratio used in the distillation. Consequently, there is a trend toward applying a 15.5 distillation according to ASTM D2892, instead of the TBP. The 15.5 distillation uses 15 theoretical stages and a molar reflux ratio of 5.

A key result from a distillation test is *the boiling point curve,* that is, the boiling point of the oil fraction versus the fraction of oil vaporized. The *initial boiling point (IBP)* is the temperature at which the first drop of liquid leaves the condenser tube of the distillation apparatus. The final boiling point or the *end point (EP)* is the highest temperature recorded in the test.

In addition, oil fractions tend to decompose or crack at a temperature of approximately 650 °F (344 °C) at 1 atm. Thus, the pressure of TBP distillation is gradually reduced to as low as 40 mmHg, as this temperature is approached to avoid cracking of the sample and distorting the measurements of true components in the oil.

The TBP distillation typically takes much time and labor. In practice, we carry out the distillation test of oil fractions using other less costly ASTM methods and convert the resulting boiling point curve into TBP curve using correlations, as given in the *API Technical Data Book – Petroleum Refining* [2]. We have implemented these correlations in an Excel spreadsheet, *ASTMConvert.xls*, for the interconversion of boiling point curves from typical ASTM distillation methods in a hands-on workshop in Section 1.4.

The *ASTM D86 distillation* of an oil fraction takes place at laboratory room temperature and pressure. Note that the D86 distillation will end below an approximate temperature of 650 °F (344 °C), at which petroleum oils begin to crack at 1 atm.

The *ASTM D1160 distillation* of an oil fraction is applicable to high-boiling oil samples (e.g., heavy heating oil, cracker gas oil feed, and residual oil) for which there is a significant cracking at atmospheric pressures. The sample is distilled at a reduced pressure, typically at 10 mmHg, to inhibit cracking. In fact, at 10 mmHg, we can distill an oil fraction up to temperatures of 950–1000 °F (510–538 °C), as reported on a 760 mmHg basis. The reduced pressure used for D1160 distillation produces a separation of components that is more ideal than that for D86 distillation.

The *ASTM D2887 distillation* of an oil fraction is a popular chromatographic procedure to "simulate" or predict the boiling point curve of an oil fraction. We determine the boiling point distribution by injecting the oil sample into a gas chromatograph that separates the hydrocarbons in a boiling point order. We then relate the retention time inside the chromatograph to the boiling point through a calibration curve.

1.1.2 Fractional Properties

Bulk properties provide a quick understanding of the type of the oil sample such as sweet and sour, and light and heavy. However, refineries require *fractional properties* of the oil sample that reflects the property and composition for a specific boiling point range to properly refine it into different end products such as gasoline, diesel, and raw materials for chemical process. Fractional properties usually contain PNA contents, sulfur content, and nitrogen content for each boiling point range; octane number for gasoline; freezing point; cetane index; and smoke point for kerosene and diesel fuels.

The *octane number* is a measure of the knocking characteristics of a fuel in a laboratory gasoline engine according to ASTM D2700 [1]. We determine the octane number of a fuel by measuring its knocking value compared to the knocking of a mixture of *n*-heptane and isooctane or 2-2-4-trimethylpentane (224TMP). By definition, we assign an octane number of 0 to pure *n*-heptane and of 100–224TMP. Therefore, a mixture of 30% heptanes and 70% isooctane has an octane number of 70.

There are two specific octane numbers in use. The *motor octane number (MON)* reflects the engine performance at highway conditions with high speeds (900 rpm), whereas *the research octane number (RON)* corresponds to the low-speed city driving (600 rpm). RON is typically higher than MON because of engine test efficiencies. The posted octane number is an average of MON and RON.

The *cetane number* measures the ease for self-ignition of a diesel fuel sample and is essentially an opposite of the octane number. It represents the percentage of pure cetane (*n*-hexadecane) in a blend of cetane and alpha-methylnaphthalene that matches the ignition quality of a diesel fuel sample. This quality is important for middle distillate fuels.

The *cetane index* is a substitute for the cetane number of diesel fuel. It is calculated based on the fuel's specific gravity and distillation range using ASTM methods D976 and D4737.

1.1.3 Interconversion of Distillation Curves

While building a refining process simulation, the distillation curve of the oil sample is the most confusing information among assay data, as different methods are used to obtain volatility characteristics of an oil sample. The most widely used tests of distillation curve are ASTM D86, ASTM D1160 (atmospheric distillation), ASTM D1160 (vacuum distillation), ASTM D2887 (chromatographic simulation), and TBP. API Technical Databook [2] presents the characteristics of each test and gives the correlations to perform interconversion among these ASTM distillation types. Most commercial process simulators include the capability to convert one type of distillation curve into the other. We develop an MS Excel spreadsheet, which automates the API conversion between any two of the ASTM distillation types (see Figure 1.1). Section 1.4 presents a hands-on workshop for this interconversion of distillation curve data.

760 mmHg ASTM-D86 (C)	Vol %	760 mmHg ASTM-D86 (F)	760 mmHg TBP (F)	760 mmHg TBP (C)	760 mmHg TBP (C)	760 mmHg TBP (F)	760 mmHg ASTM-D86 (F)	760 mmHg ASTM-D86 (C)
160.0	0%	320	259.1	126.2	126.2	259.1	320	160.0
176.7	10%	350	316.5	158.1	158.1	316.5	350	176.7
193.3	30%	380	372.6	189.2	189.2	372.6	380	193.3
206.7	50%	404	411.2	210.7	210.7	411.2	404	206.7
222.8	70%	433	451.2	232.9	232.9	451.2	433	222.8
242.8	90%	469	496.7	258.2	258.2	496.7	469	242.8
248.9	100%	480	503.0	261.7	261.7	503.0	480	248.9

ASTM-D2887(C)	Wt%/Vol%	ASTM-D2887(F)	760 mmHg TBP (F)	760 mmHg TBP (C)	760 mmHg TBP (C)	760 mmHg TBP (F)	ASTM-D2887 (F)	ASTM-D2887(C)
145.0	5%	293	322.2	161.2	348.0	658.4	639.1711023	337.3
151.7	10%	305	327.7	164.3	369.0	696.2	685.3443333	363.0
162.2	30%	324	332.4	166.9	406.0	762.8	756.2204757	402.3
168.9	50%	336	336.0	168.9	433.0	811.4	811.4	433.0
173.3	70%	344	339.6	170.9	459.0	858.2	861.2301007	460.7
181.7	90%	359	350.1	176.7	495.0	923.0	922.5542047	494.8
187.2	95%	369	357.4	180.8	512.0	953.6	974.5478925	523.6
198.9	100%	390	366.2	185.7	556.0	1032.8	1038.378625	559.1

ASTM-D2287 (C)	Wt%/Vol. %	ASTM-D2287 (F)	760 mmHg ASTM-D86 (F)	760 mmHg ASTM-D86 (C)	760 mmHg ASTM-D86 (C)	760 mmHg ASTM-D86 (F)	ASTM-D2887 (F)	ASTM-D2287 (C)
25.0	0%	77	121.3	49.6	298.8	569.9	446.4892018	230.3
33.9	10%	93	128.2	53.5	349.7	661.5	605.3731877	318.5
64.4	30%	148	154.8	68.2	392.0	737.5	715.3377437	379.6
101.7	50%	215	206.3	96.8	424.2	795.5	787.7262099	419.8
140.6	70%	285	270.6	132.5	459.0	858.2	856.5296061	458.1
182.2	90%	360	334.0	167.8	514.5	958.0	964.7774337	518.2
208.9	100%	408	367.5	186.4	577.9	1072.2	1273.441992	689.7

760 mmHg ASTM-D1160 (C)	Vol%	760 mmHg ASTM-D1160 (F)	760 mmHg TBP (F)	760 mmHg TBP (C)	760 mmHg TBP (C)	760 mmHg TBP (F)	760 mmHg ASTM-D1160 (F)	760 mmHg ASTM-D1160 (C)
369.0	10%	696.2	686.2	363.4	143.1	289.5	300.1	149.0
406.0	30%	762.8	757.9	403.3	201.5	394.7	400.1	204.5
433.0	50%	811.4	811.4	433.0	246.1	475.0	475.0	246.1
459.0	70%	858.2	857.9	458.8	287.7	549.9	550.0	287.8
495.0	90%	923	922.5	494.7	343.3	650.0	650.0	343.4

Figure 1.1 Conversion spreadsheet for distillation curves.

1.2 Boiling Point-Based Hypothetical or Pseudocomponent Generation

To simulate refining processes, the first task is to construct a hypothetical (hypo) or pseudocomponent scheme to characterize the feedstock. Data requirement and definition of the hypos or pseudocomponents depend on the type of the refining process to be modeled. There are different issues to consider when specifying hypos for fractionation and reaction units.

The hypos for fractionation units have to accurately characterize volatilities of the hydrocarbons in the feedstock in order to calculate the vapor–liquid equilibrium in distillation columns. Therefore, refiners use hypos based on boiling point ranges to represent the feedstock and model fractionation units. For modeling of reaction units, refiners partition the hydrocarbons into multiple lumps (or model compounds) based on molecular structure or/and boiling point ranges and assume the hydrocarbons of each lump to have an identical reactivity in order to develop the reaction kinetics for reaction units.

This section deals with hypo or pseudocomponent generation based on boiling point ranges for fractionation units. Chapters 4–7 will represent the hypo schemes for the major reaction units in modern refinery – fluid catalytic cracking (FCC) unit, catalytic reformer, catalytic hydrocracker, delayed coker, and alkylation unit.

Most commercial process simulators include the capability to generate hypos based on boiling point ranges representing the oil fractions. Workshop 1.4 in Section 1.7 demonstrates how to use Aspen HYSYS to generate hypos based on boiling point ranges for a given oil fraction with required analysis data.

Specifically, there are four steps to develop pseudocomponents based on boiling point ranges to represent petroleum fraction.

1) Convert ASTM D86/ASTM D1160/ASTMD2887 into TBP curve if TBP curve is not available.
   - We develop a spreadsheet, *ASTMConvert.xls*, that allows interconversion between different ASTM distillation types based on the correlations from [2] (see Figure 1.1).
2) Cut the entire boiling point range into a number of cut point ranges, which are used to define pseudocomponents (see Figure 1.2).
   - The determination of number of cuts is arbitrary. Table 1.2 provides the typical boiling point widths for pseudocomponents in commercial process simulators.
3) Estimate the density distribution of pseudocomponents if only the bulk density is available.
   - Assume the UOP or Watson–Murphy "characterization factor" or *K* factor to be constant throughout the entire boiling point range and calculate the mean average boiling point (MeABP). Dissimilar to weight average boiling point (WABP), MeABP is defined as the average of molal average boiling

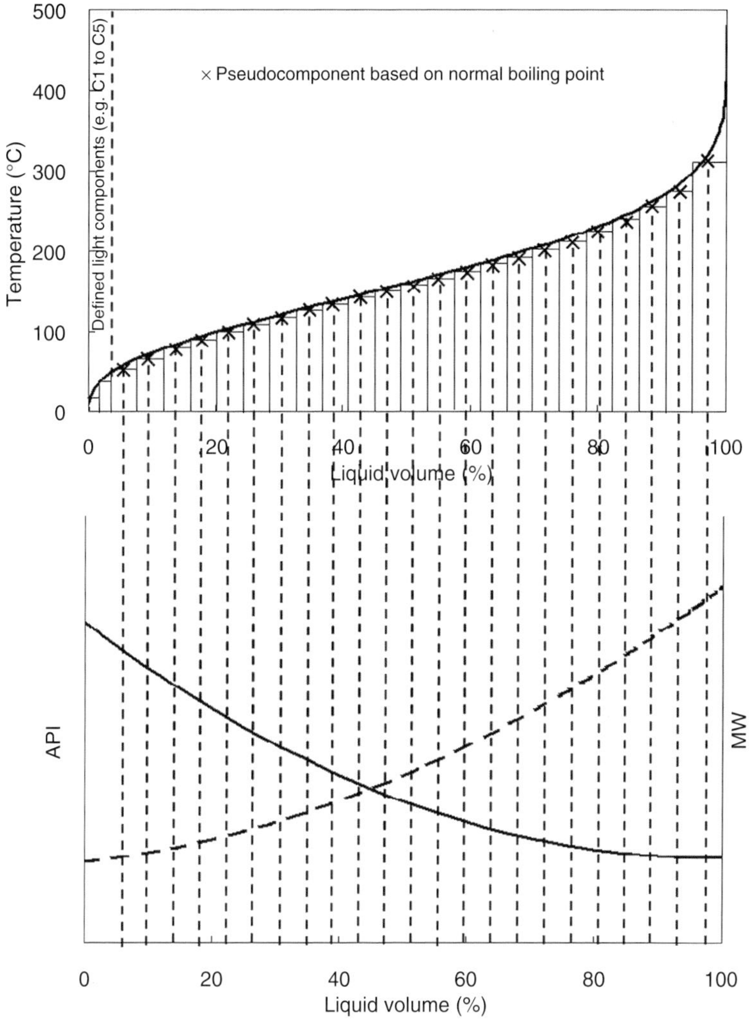

Figure 1.2 Relationship between pseudocomponent properties and the TBP curve. (Redraw from [1].)

Table 1.2 Typical boiling point widths for pseudocomponents in commercial process simulators.

Boiling point range	**Suggested number of pseudocomponents**
IBP–800 °F (425 °C)	30
800–1200 °F (650 °C)	10
1200–1650 °F (900 °C)	8

point (MABP) and cubic average boiling point (CABP). The following equations define these four boiling point indicators:

$$\mathrm{WABP} = \sum_{i=1}^{n} x_i T_{\mathrm{b}i} \tag{1.3}$$

$$\mathrm{MABP} = \sum_{i=1}^{n} x_i T_{\mathrm{b}i} \tag{1.4}$$

$$\mathrm{CABP} = \left(\sum_{i=1}^{n} x_i T_{\mathrm{b}i}^{1/3} \right)^3 \tag{1.5}$$

$$\mathrm{MeABP} = \frac{\mathrm{MABP} + \mathrm{CABP}}{2} \tag{1.6}$$

where $T_{\mathrm{b}i}$ indicates the boiling point of component i and x_i in Eqs. (1.3)–(1.5) indicates weight fraction, molar fraction, and volume fraction of component i, respectively. Here, we create a spreadsheet tool (see Figure 1.3) to perform the iteration of estimating MeABP based on the methods presented by Bollas *et al.* [3] (see Section 1.5)

$$K_{\mathrm{avg}} = [\mathrm{MeABP}]^{0.333} / \mathrm{SG}_{\mathrm{avg}} \tag{1.7}$$

where K_{avg} is the Watson K factor and $\mathrm{SG}_{\mathrm{avg}}$ is the bulk specific gravity 60 °F/60 °F.

- Calculate the density distribution of the entire boiling point range.

$$\mathrm{SG}_i = [T_{i,\mathrm{b}}]^{0.333} / K_{\mathrm{avg}} \tag{1.8}$$

where SG_i is the specific gravity 60 °F/60 °F of pseudocomponent i and $T_{i,\mathrm{b}}$ is the normal boiling point of pseudocomponent i.

4) Estimate molecular weight distribution of the entire boiling point range if not available and required properties for modeling purpose (see Section 1.4 for details).

Lacking the analysis data of high boiling point range (>570 °C) is a common problem while building pseudocomponents based on boiling point ranges. Therefore, we need to extrapolate the incomplete distillation curve in order to cover the entire boiling point range. Least squares and probability distribution functions are most widely used to perform the extrapolation of distillation curve in most commercial process simulators. Sanchez *et al.* [5] presented a comprehensive review of using probability distribution functions to fit distillation curves of petroleum fraction. They conclude that the cumulative beta function (with four parameters) can represent a wide range of petroleum products. The beta cumulative density function is

$$f(x, \alpha, \beta, A, B) = \int_{A}^{x \leq B} \left(\frac{1}{B - A} \right) \frac{\Gamma(\alpha + \beta)}{\Gamma(\alpha)\Gamma(\beta)} \left(\frac{x - A}{B - A} \right)^{\alpha - 1} \left(\frac{B - x}{B - A} \right)^{\beta - 1} \tag{1.9}$$

where α and β refer to the positive-valued parameters that control the shape of the distribution and Γ refers to the standard gamma function, which is an extension of the factorial function, with its argument shifted down by 1 to real and

	A	B	C	D
4	TBP Curve @ 760 mmHg			
5	Vol%	Temperature (F)		Initial
6	0	256.8		0
7	10	368.2		5
8	30	447.2		10
9	50	516.9		15
10	70	583.9		20
11	90	633.4		25
12	100	722.2		30
13				35
14	Specific gravity	0.8505		40
15	Refractive index @ 20 C			45
16	Oxygen content (wt%)	0.00		50
17	Initial MeABP (F) [Enter as first guess in yellow cell]	506.76		55
18				60
19	Trial MeABP (F)	497.46		65
20	Trial MeABP (R)	957.13		70
21	Watson-K	11.59		75
22				80
23	Calc. VABP (R)	969.22		85
24	Calc. WABP (R)	972.98		90
25	Calc. MABP (R)	948.85		95
26	Calc. CABP (R)	965.42		
27				
28	Calc. MeABP (R)	957.13		
29				
30	Error (Trial MeABP - Calc. MeABP)	0.00000	(Use goalseek to drive	
31				
32	Correlation for refractive index	A	B	C
33	Naphthas	1.028	0.53	
34	Straight or hydrosulfurized gas oils	0.9734	0.59	
35	Deeply hydrogenated fractions	0.9713	0.59	
36	Short residues	0.9345	0.63	0.006
37	FCC feeds	Naphthas		0.006
38	Coal liquids	Straight or hydrosulfurized gas oils		0.006
39	Stream cracker residue	Deeply hydrogenated fractions		
40		Short residues FCC feeds		
41	Selected correlation	Coal liquids		
42		Stream cracker residue		
43		FCC feeds		

Figure 1.3 Iteration spreadsheet for MeABP calculation.

complex numbers. That is, if v is a positive integer, then $\Gamma(v) = (v-1)\,!$ A and B parameters set lower and upper bounds on the distribution and x represents normalized recovery. We develop an MS Excel spreadsheet, *Beta.xls*, to perform the extrapolation of distillation curve by using the cumulative beta distribution function (see Figure 1.4).

Section 1.5 presents Workshop 1.2 for applying our spreadsheet to extrapolate an incomplete distillation curve. We note that we should use the density distribution together with the boiling point whenever the density distribution is available (in step 3), because the assumption of constant Watson K factor always fails in low and high boiling point ranges of the distillation curve. Figure 1.5 compares the pseudocomponents generated from constant Watson K factor and from density distribution. Using a constant Watson K factor shows significant deviations from assay data on estimating the densities of pseudocomponents, particularly in both

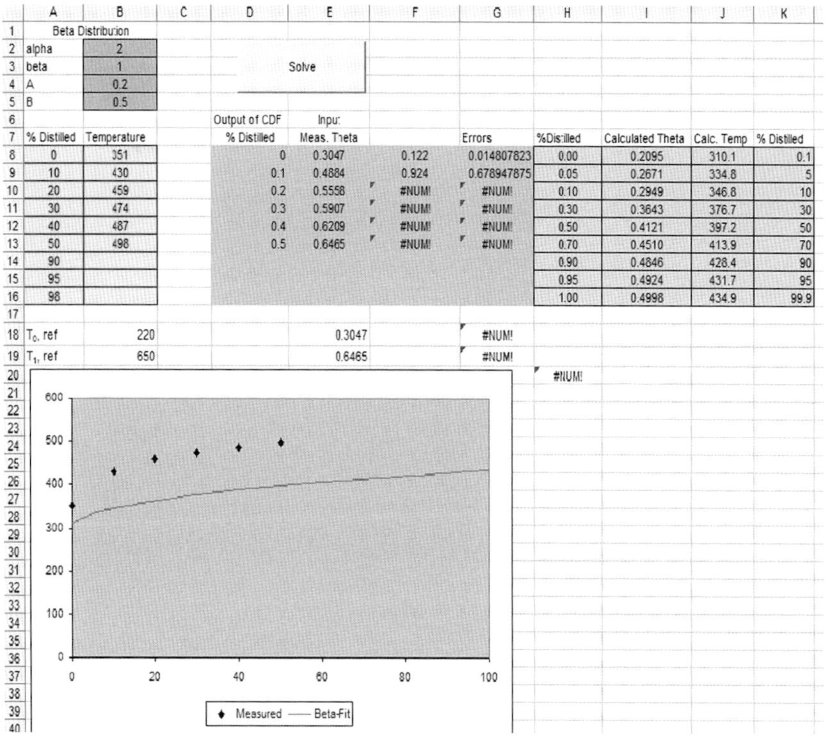

Figure 1.4 Spreadsheet for extrapolating distillation curve.

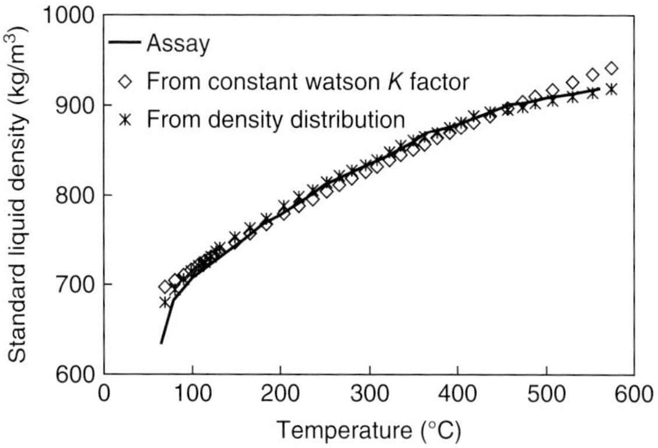

Figure 1.5 Comparison of the pseudocomponents generated from constant Watson *K* factor and density distribution. (Adapted from Kaes 2000 [1].)

light and heavy ends of the distillation curve. On the other hand, using the density distribution is able to provide good estimation of the densities of pseudocomponents. Estimating the densities of pseudocomponents is the most important part when developing pseudocomponents because density is required for most physical property estimations.

1.3 Workshop 1.1 – Interconvert Distillation Curves

There are two situations that we may encounter when the distillation curve available is not a TBP curve and needs to be converted: (1) it is another ASTM type, and (2) it is ASTM D1160 at vacuum pressure. The spreadsheet we have developed is able to solve these two cases. The following steps demonstrate how to convert an ASTM D1160 curve (at 10 mmHg) into a TBP curve.

Step 1. Open *WS1.1 ASTMConvert.xls* (Figure 1.6).
Step 2. Copy and paste the ASTM D1160 curve into the sheet for interconversion among different testing pressures of ASTM D1160 (Figure 1.7).
Step 3. Input the testing pressure, which is 10 mmHg in this case (Figure 1.8).
Step 4. The blue cells will show the converted results, which correspond to ASTM D1160 at 1 atm (Figure 1.9).
Step 5. Copy the values of ASTM D1160 (at 1 atm) to the sheets for converting ASTM D1160 at 1 atm into TBP (Figure 1.10).
Step 6. The blue cells reveal the converted TBP curve (Figure 1.11).

1.4 Workshop 1.2 – Extrapolate an Incomplete Distillation Curve

Step 1. Open *WS1.2 Beta.xls.* Purple cells show the adjustable parameters in beta distribution function, yellow cells require the input of the distillation curve, tan cells and the graph indicate the fitted results (Figure 1.12).
Step 2. Input the incomplete distillation curve into yellow cells. The user is allowed to add/remove the cells of "% Distilled" and "Temperature" according to the number of points in distillation curve (Figure 1.13).
Step 3. Click "solve" to run the fitting program (Figure 1.14).
Step 4. The purple cells show the fitted parameters. The tan cells and the graph represent the extrapolated distillation curve (Figure 1.15).

1.5 Workshop 1.3 – Calculate MeABP of a Given Assay

Step 1. Open *WS1.3 MeABP Iteration.xls* (Figure 1.16).
Step 2. Select type of the oil fraction. We choose naphtha in this case (Figure 1.17).
Step 3. Input TBP curve and specific gravity in blue cells (Figure 1.18).
Step 4. Go to Tool/Goal Seek (for new version of Excel, Data → What–If Analysis → Goal Seek) (Figure 1.19).
Step 5. Assign yellow cell to "By changing cell" and green cell to "Set cell" and input "0" in "To value." And then, click "OK" (Figure 1.20).
Step 6. The yellow cell reveals the calculated MeABP for the given oil fraction (Figure 1.21).

760 mmHg		760 mmHg	760 mmHg	760 mmHg	760 mmHg	760 mmHg	760 mmHg	760 mmHg
ASTM-D86 (C)	Vol. %	ASTM-D86 (F)	TBP (F)	TBP (C)	TBP (C)	TBP (F)	ASTM-D86 (F)	ASTM-D86 (C)
160.0	0%	320	259.1	126.2	126.2	259.1	320	160.0
176.7	10%	350	316.5	158.1	158.1	316.5	350	176.7
193.3	30%	380	372.6	189.2	189.2	372.6	380	193.3
206.7	50%	404	411.2	210.7	210.7	411.2	404	206.7
222.8	70%	433	451.2	232.9	232.9	451.2	433	222.8
242.8	90%	469	496.7	258.2	258.2	496.7	469	242.8
248.9	100%	480	503.0	261.7	261.7	503.0	480	248.9

			760 mmHg	760 mmHg	760 mmHg	760 mmHg		
ASTM-D2887(C)	Wt%/Vol%	ASTM-D2887(F)	TBP (F)	TBP (C)	TBP (C)	TBP (F)	ASTM-D2887 (F)	ASTM-D2887(C)
145.0	5%	293	322.2	161.2	348.0	658.4	639.1711023	337.3
151.7	10%	305	327.7	164.3	369.0	696.2	685.3443333	363.0
162.2	30%	324	332.4	166.9	406.0	762.8	756.2204757	402.3
168.9	50%	336	336.0	168.9	433.0	811.4	811.4	433.0
173.3	70%	344	339.6	170.9	459.0	858.2	861.2301007	460.7
181.7	90%	359	350.1	176.7	495.0	923.0	922.5542047	494.8
187.2	95%	369	357.4	180.8	512.0	953.6	974.5478925	523.6
198.9	100%	390	366.2	185.7	556.0	1032.8	1038.378625	559.1

			760 mmHg	760 mmHg	760 mmHg	760 mmHg		
ASTM-D2287 (C)	Wt%/Vol. %	ASTM-D2287 (F)	ASTM-D86 (F)	ASTM-D86 (C)	ASTM-D86 (C)	ASTM-D86 (F)	ASTM-D2887 (F)	ASTM-D2287 (C)
25.0	0%	77	121.3	49.6	298.8	569.9	446.4892018	230.3
33.9	10%	93	128.2	53.5	349.7	661.5	605.3731877	318.5
64.4	30%	148	154.8	68.2	392.0	737.5	715.3377437	379.6
101.7	50%	215	206.3	96.8	424.2	795.5	787.7262099	419.8
140.6	70%	285	270.6	132.5	459.0	858.2	856.5298061	458.1
182.2	90%	360	334.0	167.8	514.5	958.0	964.7774337	518.2
208.9	100%	408	367.5	186.4	577.9	1072.2	1273.441992	689.7

760 mmHg		760 mmHg	760 mmHg	760 mmHg	760 mmHg	760 mmHg	760 mmHg	760 mmHg
ASTM-D1160 (C)	Vol%	ASTM-D1160 (F)	TBP (F)	TBP (C)	TBP (C)	TBP (F)	ASTM-D1160 (F)	ASTM-D1160 (C)
280.8	10%	537.3541391	527.3	275.2	143.1	289.5	300.1	149.0
350.6	30%	663.1131895	657.8	347.7	201.5	394.7	400.1	204.5
402.7	50%	756.9327522	756.9	402.7	246.1	475.0	475.0	246.1
450.5	70%	842.8909373	842.9	450.5	287.7	549.9	550.0	287.8
513.0	90%	955.4507826	955.6	513.1	343.3	650.0	650.0	343.4

Figure 1.6 *WS1.1 ASTMConvert.xls.*

Pressure =	30	mmHg	2 =< P =< 760			
X	0.00180742			760 mmHg	760 mmHg	760 mmHg
TBP/D1160 (C)	Vol%	TBP/D1160 (F)	TBP/D1160 (R)	TBP/D1160 (R)	TBP/D1160 (F)	TBP/D1160 (C)
143.1	10%	289.5	749.2	941.7	482.1	250.0
201.5	30%	394.7	854.4	1063.5	603.8	317.7
246.1	50%	475.0	934.7	1154.8	695.1	368.4
287.7	70%	549.9	1009.6	1238.8	779.1	415.1
343.3	90%	650.0	1109.7	1349.2	889.5	476.4

Figure 1.7 Input cells of ASTM D1160 interconversion in ASTMConvert.xls.

47	Pressure =	10	mmHg	2 =< P =< 760			
48	X	0.00195599			760 mmHg	760 mmHg	760 mmHg
49	TBP/D1160 (C)	Vol%	TBP/D1160 (F)	TBP/D1160 (R)	TBP/D1160 (R)	TBP/D1160 (F)	TBP/D1160 (C)
50	143.1	10%	289.5	749.2	997.0	537.4	280.8
51	201.5	30%	394.7	854.4	1122.8	663.1	350.6
52	246.1	50%	475.0	934.7	1216.6	756.9	402.7
53	287.7	70%	549.9	1009.6	1302.6	842.9	450.5
54	343.3	90%	650.0	1109.7	1415.1	955.5	513.0

Figure 1.8 Input pressure for ASTM D1160 interconversion.

47	Pressure =	10	mmHg	2 =< P =< 760			
48	X	0.00195599			760 mmHg	760 mmHg	760 mmHg
49	TBP/D1160 (C)	Vol%	TBP/D1160 (F)	TBP/D1160 (R)	TBP/D1160 (R)	TBP/D1160 (F	TBP/D1160 (C)
50	143.1	10%	289.5	749.2	997.0	537.4	280.8
51	201.5	30%	394.7	854.4	1122.8	663.1	350.6
52	246.1	50%	475.0	934.7	1216.6	756.9	402.7
53	287.7	70%	549.9	1009.6	1302.6	842.9	450.5
54	343.3	90%	650.0	1109.7	1415.1	955.5	513.0

Figure 1.9 Results of ASTM D1160 interconversion.

37	760 mmHg		760 mmHg	760 mmHg	760 mmHg
38	ASTM-D1160 (C)	Vol%	ASTM-D1160 (F)	TBP (F)	TBP (C)
39	280.8	10%	537.3541391	527.3	275.2
40	350.6	30%	663.1131895	657.8	347.7
41	402.7	50%	756.9327522	756.9	402.7
42	450.5	70%	842.8909373	842.9	450.5
43	513.0	90%	955.4507826	955.6	513.1

Figure 1.10 Input cells for other ASTM interconversion in ASTMConvert.xls.

37	760 mmHg		760 mmHg	760 mmHg	760 mmHg
38	ASTM-D1160 (C)	Vol%	ASTM-D1160 (F)	TBP (F)	TBP (C)
39	280.8	10%	537.3541391	527.3	275.2
40	350.6	30%	663.1131895	657.8	347.7
41	402.7	50%	756.9327522	756.9	402.7
42	450.5	70%	842.8909373	842.9	450.5
43	513.0	90%	955.4507826	955.6	513.1

Figure 1.11 Result cells for other ASTM interconversion in ASTMConvert.xls.

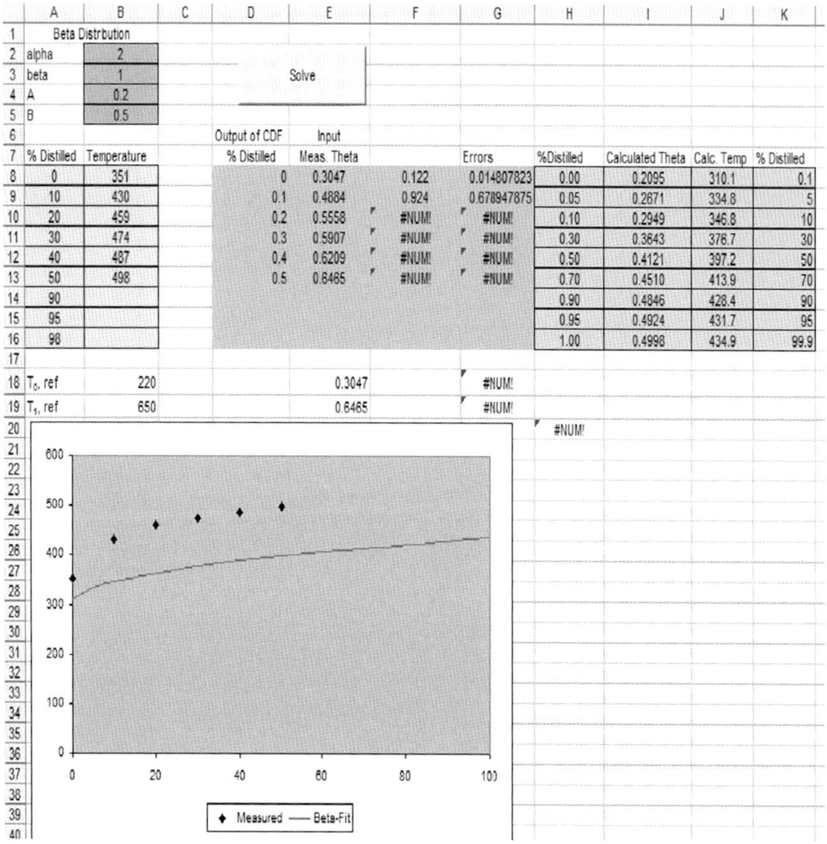

Figure 1.12 *WS1.2 Beta.xls.*

1.6 Workshop 1.4 – Represent an Oil Fraction by the Old Oil Manager in Aspen HYSYS Petroleum Refining

Step 1. Start a new case in Aspen HYSYS Petroleum Refining and save as *WS1.4 Oil Manager.hsc* (Figure 1.22).

Step 2. Click "add" to add a new component list (Figure 1.23).

Step 3. Click "view" to edit the component list. Add light components, which are shown in assay data (Figure 1.24).

Step 4. Click "add" in "fluid pkgs" tab to add the thermodynamic model (Figure 1.25).

Step 5. Select the Peng–Robinson method (Figure 1.26).

Step 6. Click "Input Assay" in "Oil Manager" environment (Figure 1.27).

Step 7. Add an assay by inputting the TBP curve, bulk density, and light end composition (Figure 1.28).

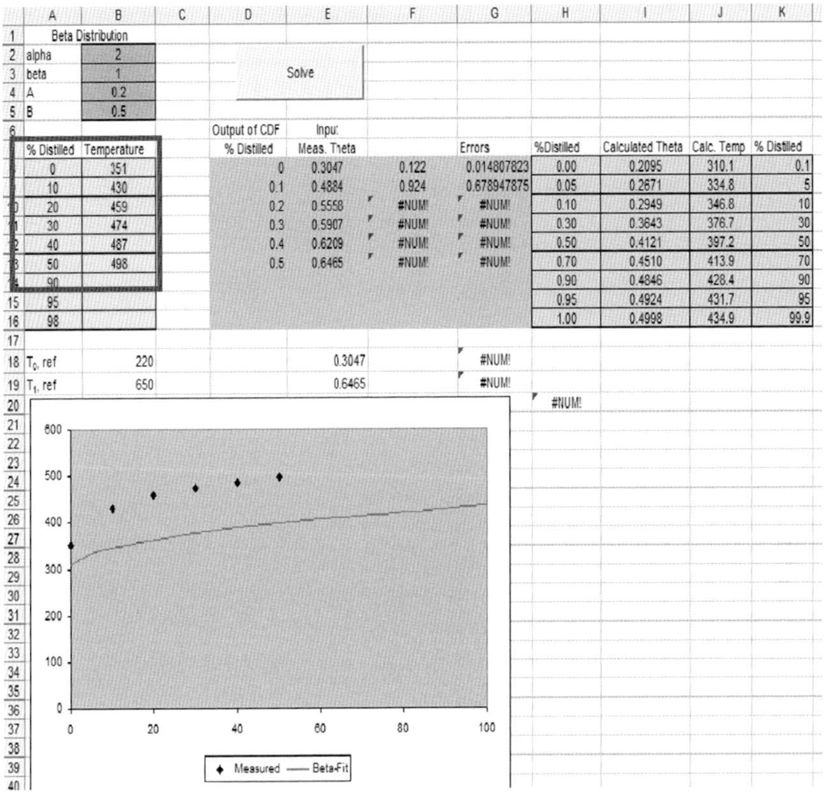

Figure 1.13 Input cells in *WS1.2 Beta.xls*.

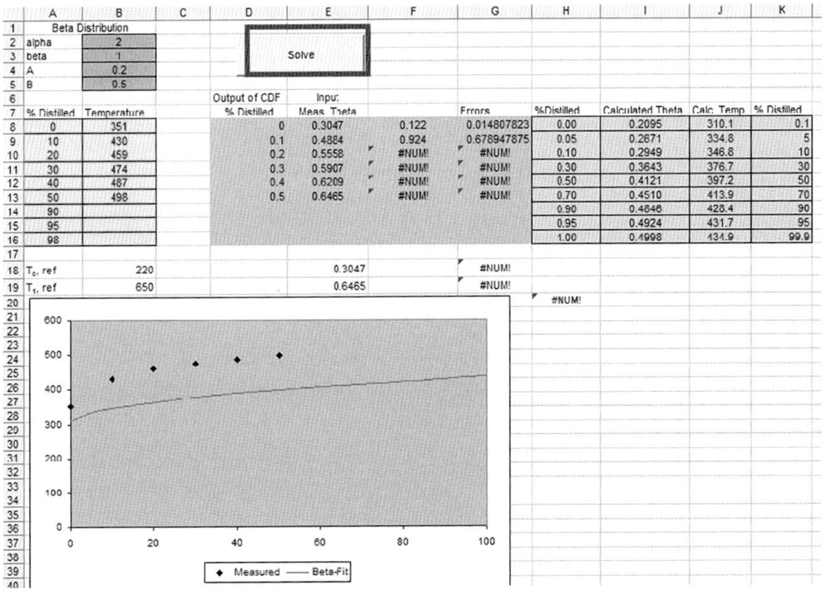

Figure 1.14 Activation button in *WS1.2 Beta.xls*.

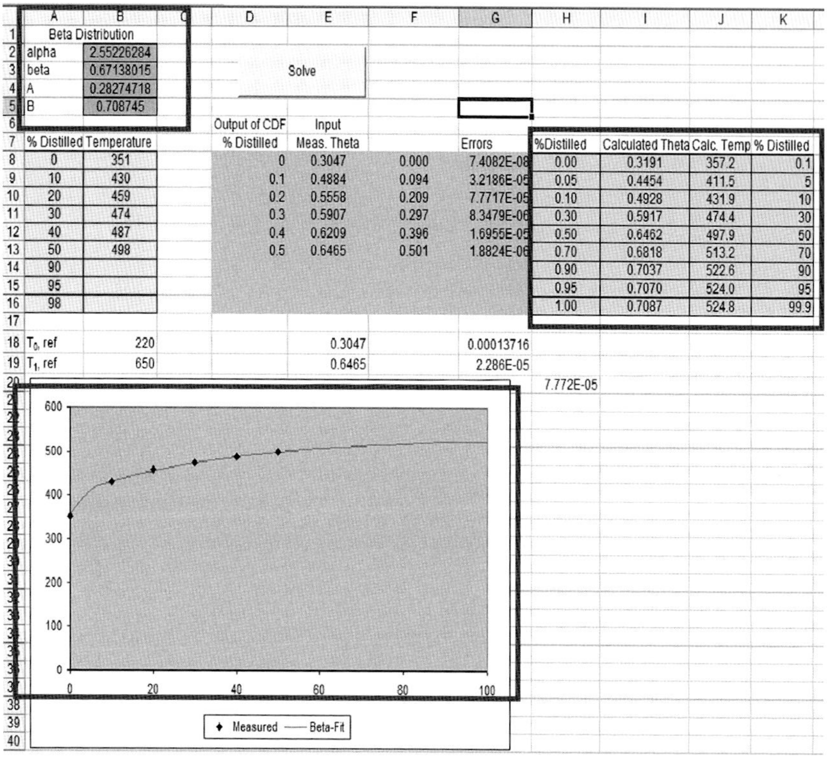

	A	B	C	D	E	F	G	H	I	J	K
1	Beta Distribution										
2	alpha	2.55226284									
3	beta	0.67138015			Solve						
4	A	0.28274718									
5	B	0.708745									
6				Output of CDF	Input						
7	% Distilled	Temperature		% Distilled	Meas. Theta		Errors	%Distilled	Calculated Theta	Calc. Temp	% Distilled
8	0	351		0	0.3047	0.000	7.4082E-08	0.00	0.3191	357.2	0.1
9	10	430		0.1	0.4884	0.094	3.2186E-05	0.05	0.4454	411.5	5
10	20	459		0.2	0.5558	0.209	7.7717E-05	0.10	0.4928	431.9	10
11	30	474		0.3	0.5907	0.297	8.3479E-06	0.30	0.5917	474.4	30
12	40	487		0.4	0.6209	0.396	1.6955E-05	0.50	0.6462	497.9	50
13	50	498		0.5	0.6465	0.501	1.8824E-06	0.70	0.6818	513.2	70
14	90							0.90	0.7037	522.6	90
15	95							0.95	0.7070	524.0	95
16	98							1.00	0.7087	524.8	99.9
17											
18	T_0, ref	220			0.3047		0.00013716				
19	T_1, ref	650			0.6465		2.286E-05				
20								7.772E-05			

Figure 1.15 Fitted results in *WS1.2 Beta.xls*.

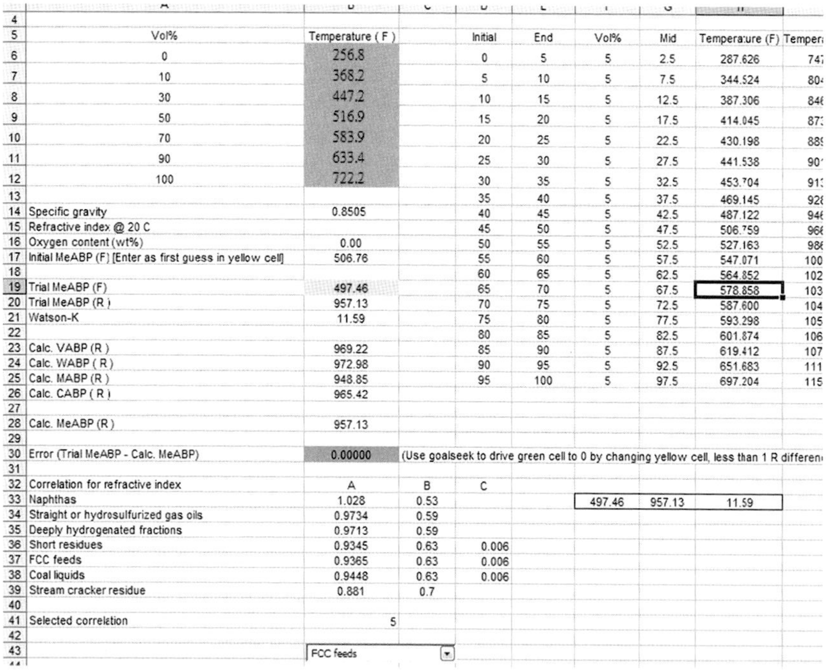

	A	B	C	D	E	F	G	H	I
4									
5	Vol%	Temperature (F)		Initial	End	Vol%	Mid	Temperature (F)	Temper
6	0	256.8		0	5	5	2.5	287.626	74
7	10	368.2		5	10	5	7.5	344.524	80
8	30	447.2		10	15	5	12.5	387.306	84
9	50	516.9		15	20	5	17.5	414.045	87
10	70	583.9		20	25	5	22.5	430.198	88
11	90	633.4		25	30	5	27.5	441.538	90
12	100	722.2		30	35	5	32.5	453.704	91
13				35	40	5	37.5	469.145	92
14	Specific gravity	0.8505		40	45	5	42.5	487.122	94
15	Refractive index @ 20 C			45	50	5	47.5	506.759	96
16	Oxygen content (wt%)	0.00		50	55	5	52.5	527.163	98
17	Initial MeABP (F) [Enter as first guess in yellow cell]	506.76		55	60	5	57.5	547.071	100
18				60	65	5	62.5	564.852	102
19	Trial MeABP (F)	497.46		65	70	5	67.5	578.858	103
20	Trial MeABP (R)	957.13		70	75	5	72.5	587.600	104
21	Watson-K	11.59		75	80	5	77.5	593.298	105
22				80	85	5	82.5	601.874	106
23	Calc. VABP (R)	969.22		85	90	5	87.5	619.412	107
24	Calc. WABP (R)	972.98		90	95	5	92.5	651.683	111
25	Calc. MABP (R)	948.85		95	100	5	97.5	697.204	115
26	Calc. CABP (R)	965.42							
27									
28	Calc. MeABP (R)	957.13							
29									
30	Error (Trial MeABP - Calc. MeABP)	0.00000	(Use goalseek to drive green cell to 0 by changing yellow cell, less than 1 R differen						
31									
32	Correlation for refractive index	A	B	C					
33	Naphthas	1.028	0.53				497.46	957.13	11.59
34	Straight or hydrosulfurized gas oils	0.9734	0.59						
35	Deeply hydrogenated fractions	0.9713	0.59						
36	Short residues	0.9345	0.63	0.006					
37	FCC feeds	0.9365	0.63	0.006					
38	Coal liquids	0.9448	0.63	0.006					
39	Stream cracker residue	0.881	0.7						
40									
41	Selected correlation	5							
42									
43		FCC feeds							

Figure 1.16 *WS1.3 MeABP Iteration.xls*.

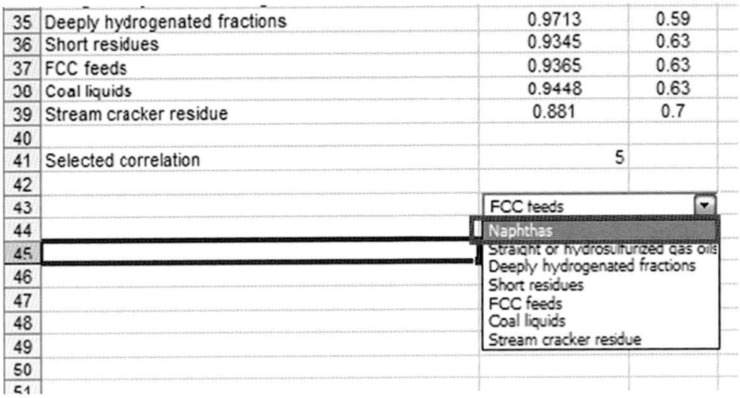

Figure 1.17 Select oil type.

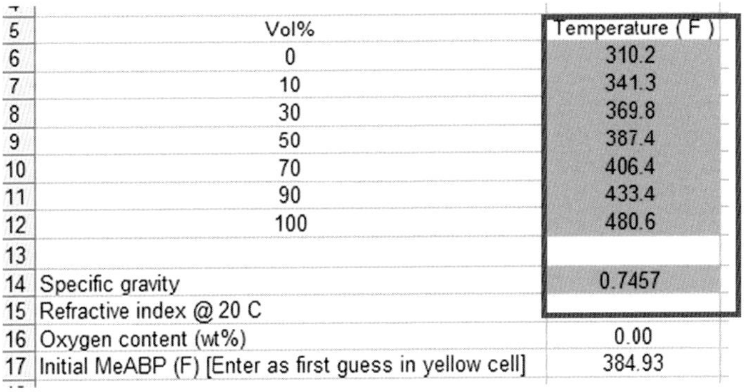

Figure 1.18 Input distillation curve and specific gravity.

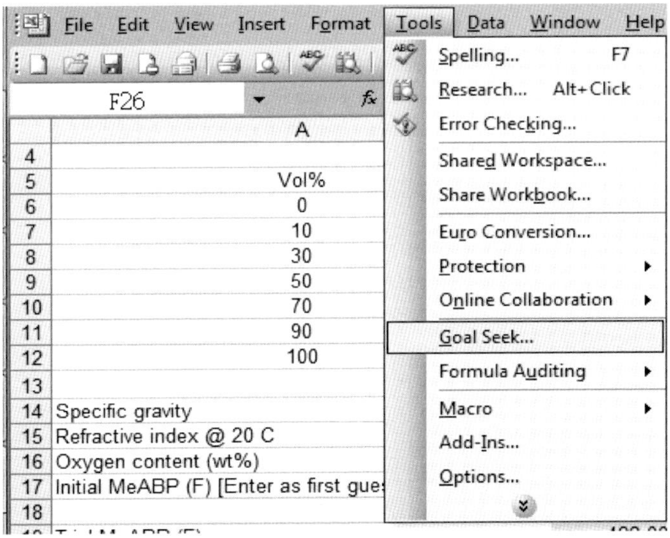

Figure 1.19 Activate "goal seek" in *WS1.3 MeABP Iteration.xls*.

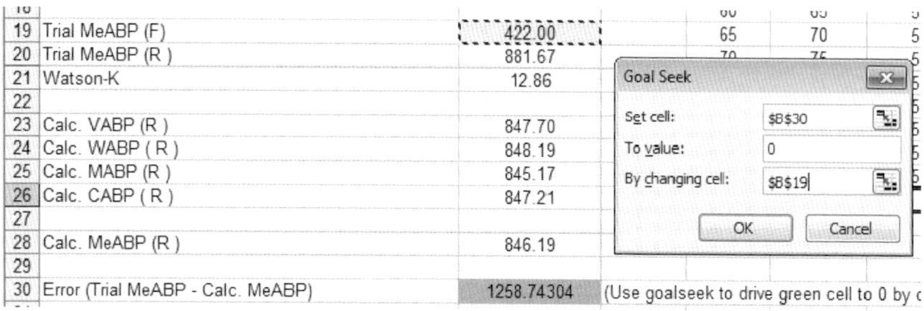

Figure 1.20 Assign tuning and objective cells.

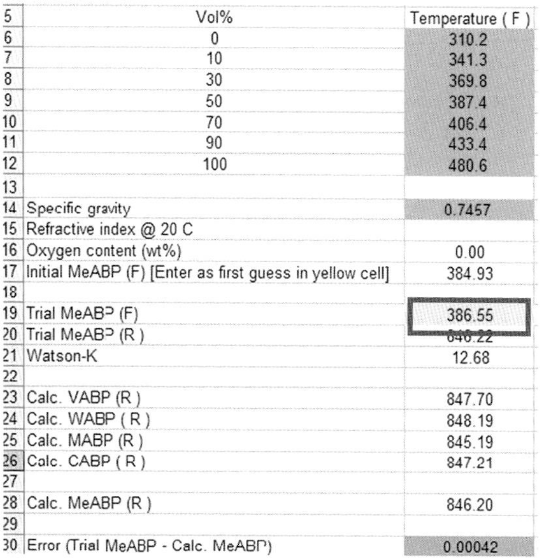

Figure 1.21 Iterative MeABP in *WS1.3 MeABP Iteration.xls*.

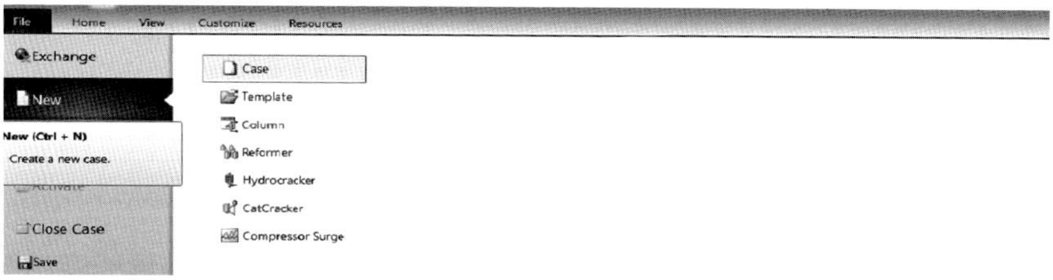

Figure 1.22 Start a new case in Aspen HYSYS Petroleum Refining.

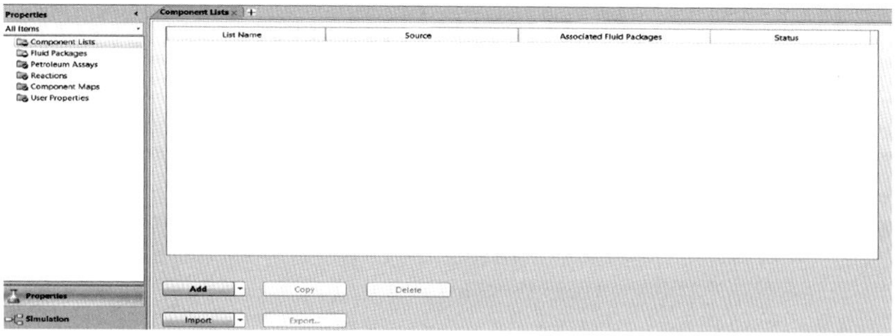

Figure 1.23 Add a new component list.

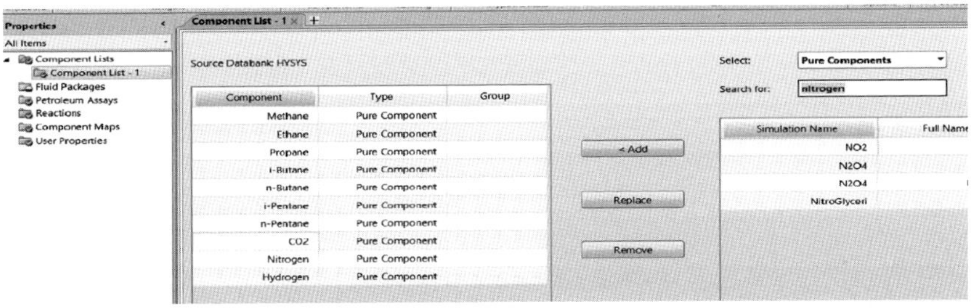

Figure 1.24 Add light components.

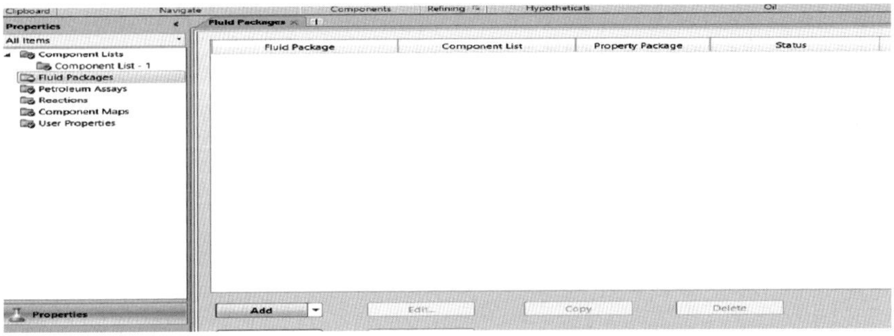

Figure 1.25 Click "add" to enter the list of thermodynamic models.

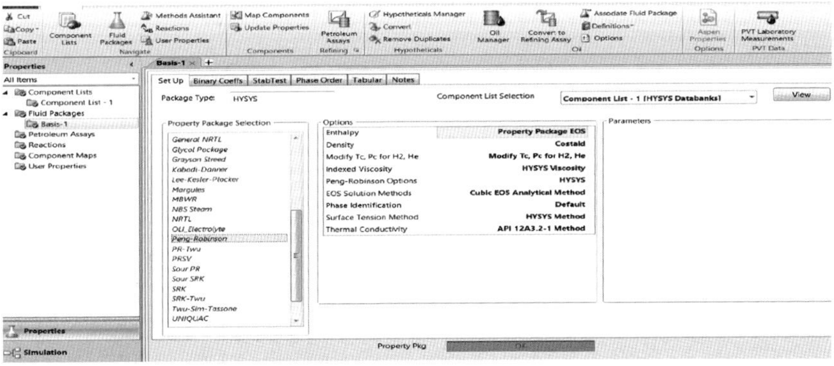

Figure 1.26 Select the Peng–Robinson thermodynamic model and click on "Oil Manager" tab.

Step 8. Check "distillation" and click "edit assay" to input the distillation curve. Refer to the data in the spreadsheet, *WS1.4 Distillation Curve and Light End Composition.xlsx*. Note that the temperature unit in Figure 1.28 is degree Fahrenheit. To change this to degree Centigrade, go to the File menu and click Options. This will open the Simulation Options window. On the Variables tab, click Units. Choose SI units and then the temperature unit becomes degree Centigrade (Figure 1.29).

Step 9. Check "bulk props" to input the bulk density and other bulk properties if available (Figure 1.30).

Step 10. Check "light ends" to input the light end composition (Figure 1.31).

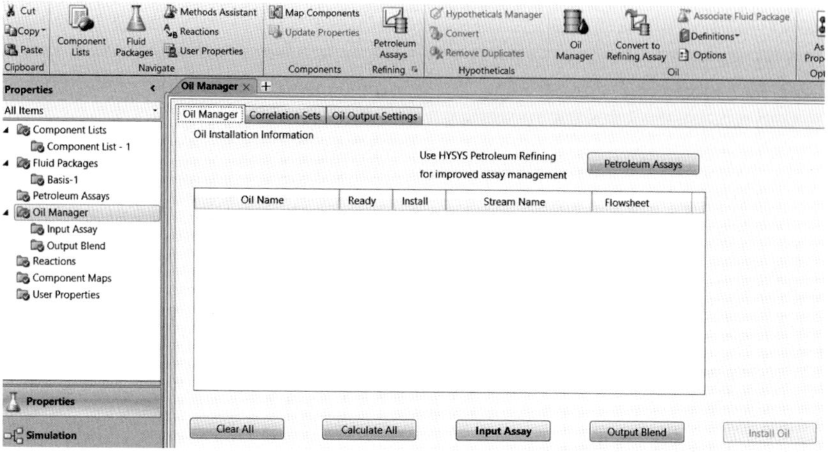

Figure 1.27 "Input Assay" to define a new assay.

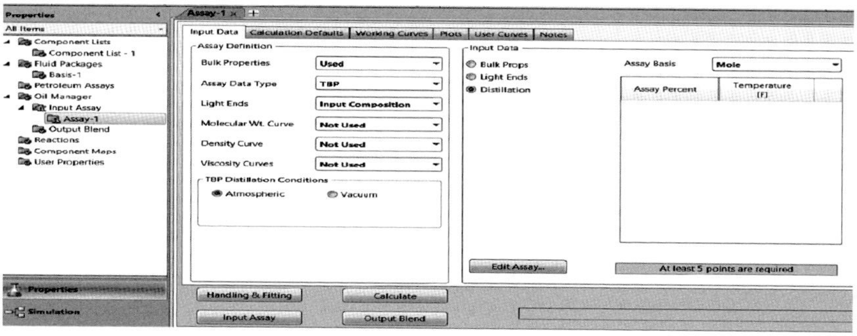

Figure 1.28 Select the data to be used to define an assay.

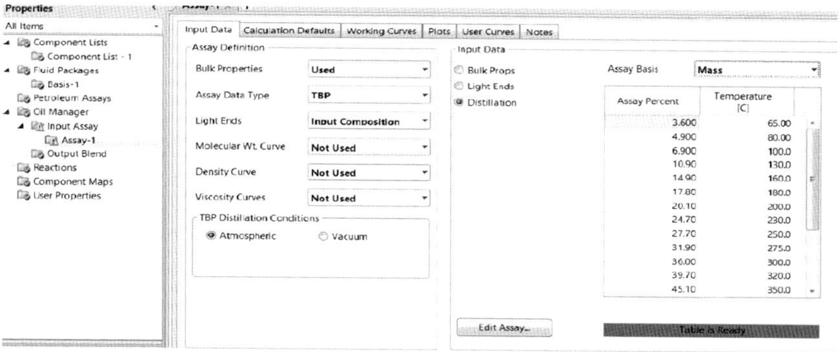

Figure 1.29 Enter the distillation curve.

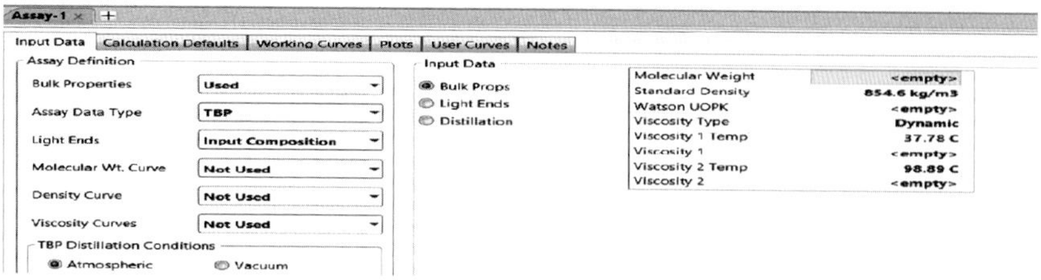

Figure 1.30 Enter the bulk density.

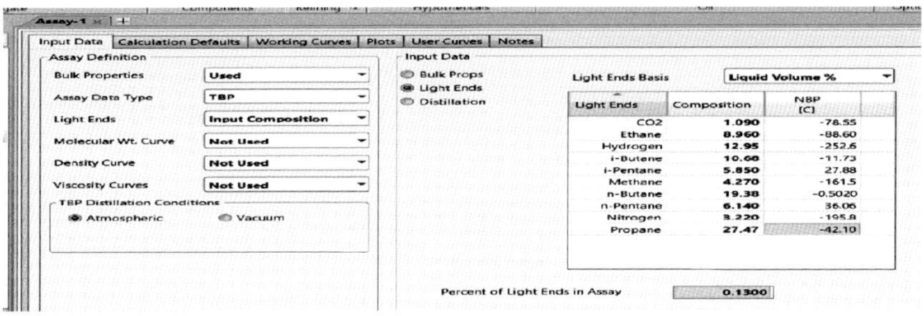

Figure 1.31 Enter the composition of light components.

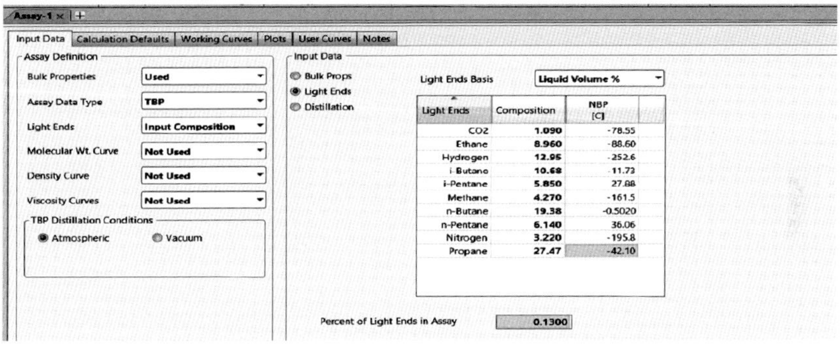

Figure 1.32 Click "calculate" for calculation and generate the pseudocomponents.

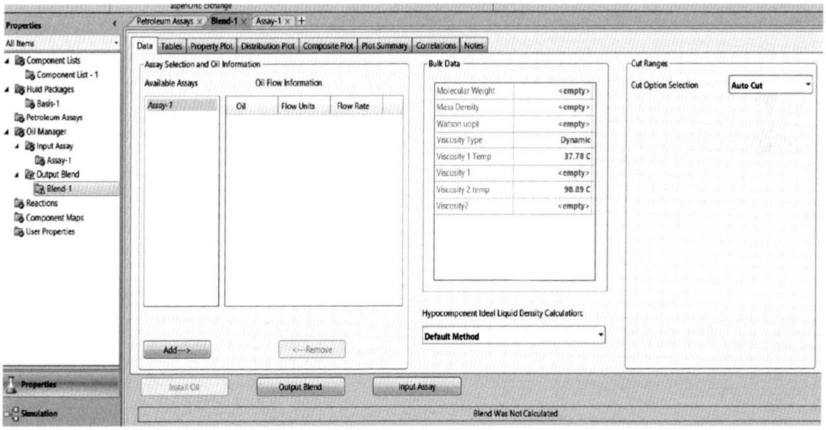

Figure 1.33 Create a new blend, Blend-1. See our previously defined assay, Assay-1.

Step 11. Click "calculate" to enable the calculations by Aspen HYSYS Petroleum Refining to generate pseudocomponents (Figure 1.32).

Step 12. Click on "Output Blend" and click "Add" to create a new blend, Blend-1 (Figure 1.33).

Step 13. Select "Assay-1" and click add to generate the corresponding pseudo-components (Figure 1.34).

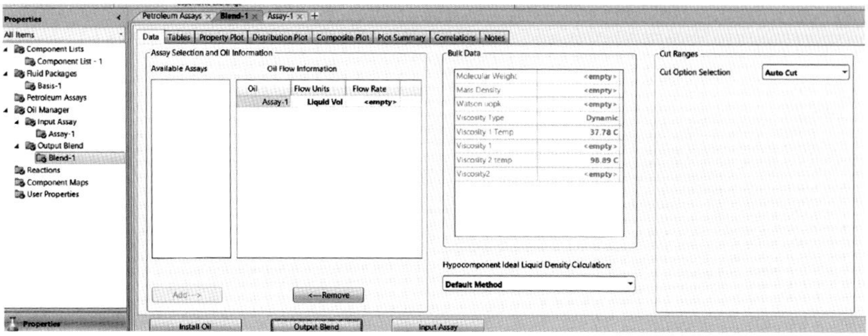

Figure 1.34 Select Assay-1 used to be cut or blended and enable the blend calculation.

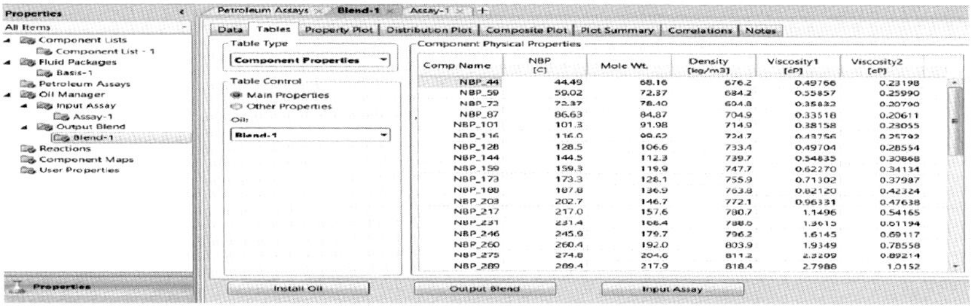

Figure 1.35 The pseudocomponents used to represent the cut or blend.

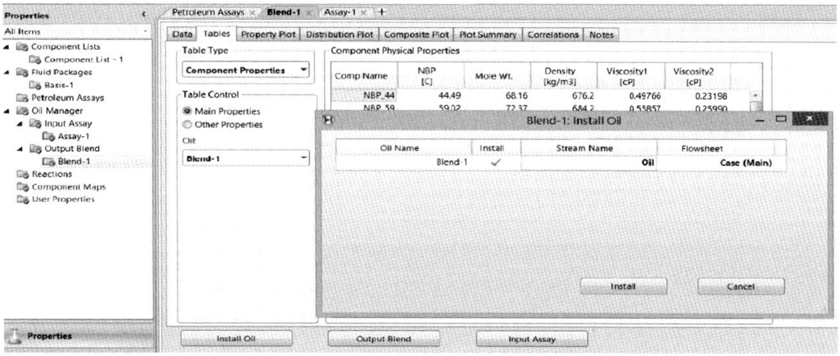

Figure 1.36 Install the cut/blend into simulation.

Step 14. Go to "Tables" tab to check the generated pseudocomponents (Figure 1.35).

Step 15. Click on "Install Oil" tab, enter "Oil" as the stream name, and click the "Install" box (Figure 1.36).

Step 16. Go to the simulation environment. The stream "Oil" represents the created oil fraction. Click on the stream to see the Composition under Worksheet. We have duplicated the oil fraction within the Oil Manager within Aspen HYSYS Petroleum Refining (Figure 1.37).

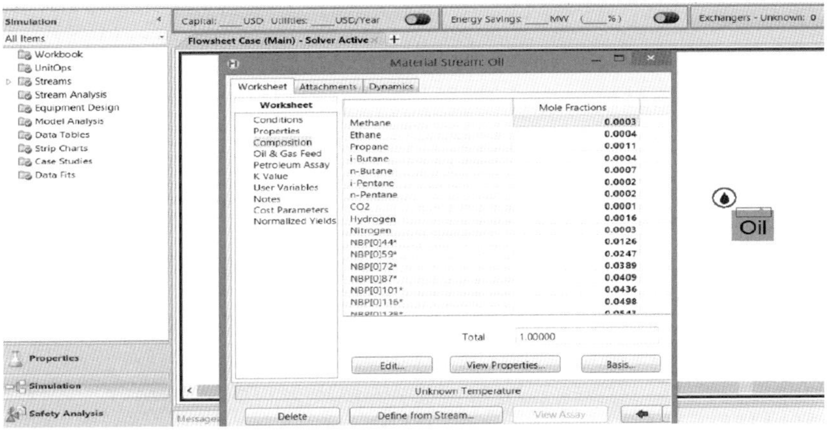

Figure 1.37 The stream in the simulation environment represents the created oil fraction.

1.7 Workshop 1.5- Represent an Oil Fraction by the New Petroleum Assay Manager in Aspen HYSYS Petroleum Refining

Step 1. Start a new case in Aspen HYSYS Petroleum Refining (Figure 1.38).

Step 2. Right-click "Petroleum Assays" and select "Add new essays" to add a new assay. Choose "Manually enter" option. For "Assay Component Selection," choose "Assay Component Celsius to 850 °C." Click OK (Figure 1.39).

Step 3. This generates the "New Assay" form of Figure 1.40a. Choose "Single Steam Properties." Copy and paste the TBP distillation curve from

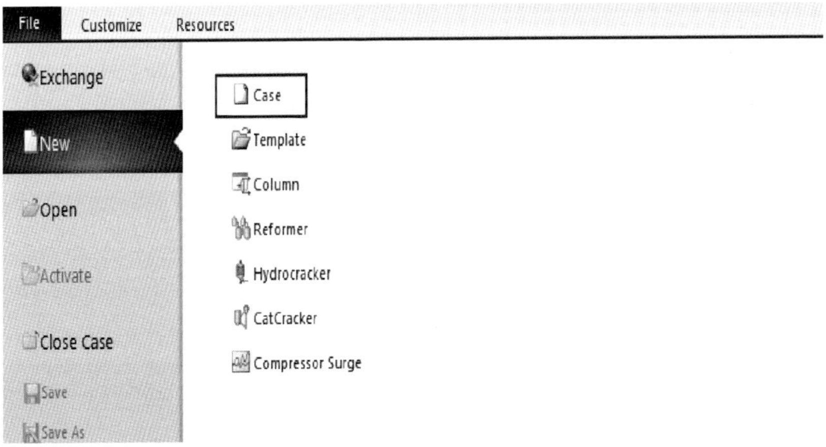

Figure 1.38 Start new case in Aspen HYSYS Petroleum Refining and save as *WS1.5 Petroleum Assay Manager.hsc*. Add the same components (C1, C2, C3, *i*C4, *n*C4, *i*C5, *n*C5, CO_2, H_2, and N_2) and fluid package (Peng–Rob) as shown in Figures 1.24 and 1.25 in *WS1.4*.

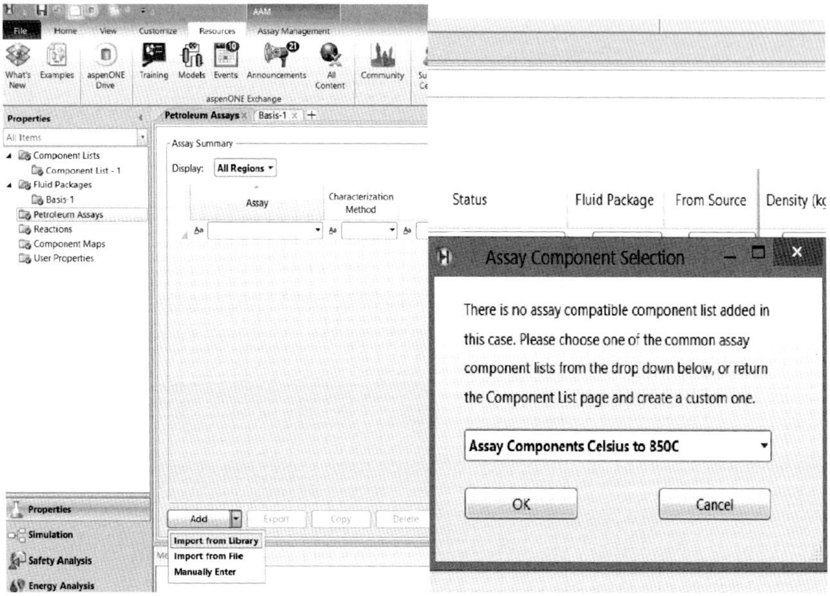

Figure 1.39 Right-click "Petroleum Assays" to add a new assay and choose "Assay Components Celsius to 850 °C" for "Assay Component Selection" and click "OK."

WS1.4 Distillation Curve and Light End Compositions.xls into the New Assay form. This results in an input summary of Figure 1.40b.

Step 4. Input the bulk density and other bulk properties if available (Figure 1.41).

Step 5. In "Pure Component," add a new cut named "LightEnd" and set the IBP as its initial temperature and final boiling point (FBP) as its final temperature. Then, input the light end compositions following the data in *WS1.4 Distillation Curve and Light End Compositions.xls* (Figure 1.42).

Step 6. In "Input Summary" form, click on "Characterize Assay" to enable the Aspen HYSYS Petroleum Refining to do crude characterization (Figure 1.43).

Step 7. After characterizing the assay, we can create plots of cut yields, distillations, crude properties, cut viscosities, and PNA (Figures 1.44 and 1.45).

Step 8. Click "Simulation" to enter the simulation environment (Figure 1.46).

Step 9. Click "Model Palette" to open the window of unit models (Figure 1.47).

Step 10. Click " Refining > Petroleum Feeder" to add a petroleum feed (Figure 1.48).

Step 11. Add a feed stream (Figure 1.49).

Step 12. Click the feeder and select feed assays and the product stream (Figure 1.50).

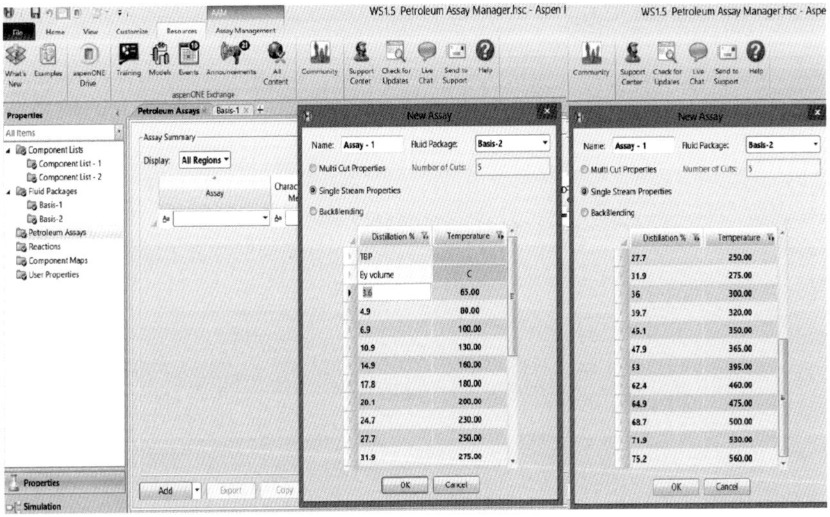

(a)

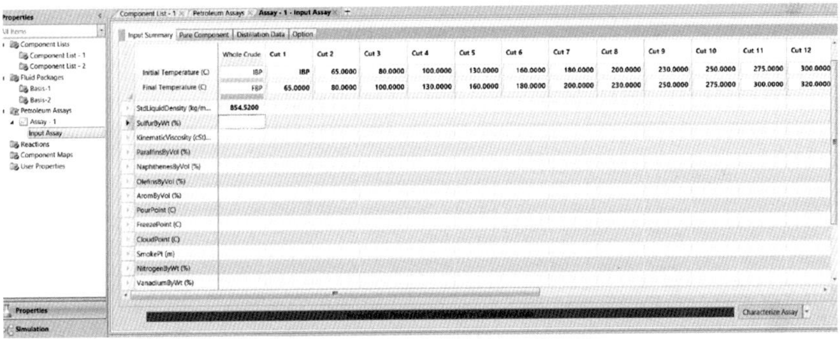

(b)

Figure 1.40 (a) Enter the TBP distillation curve into "New Assay" form. (b) The resulting input summary form.

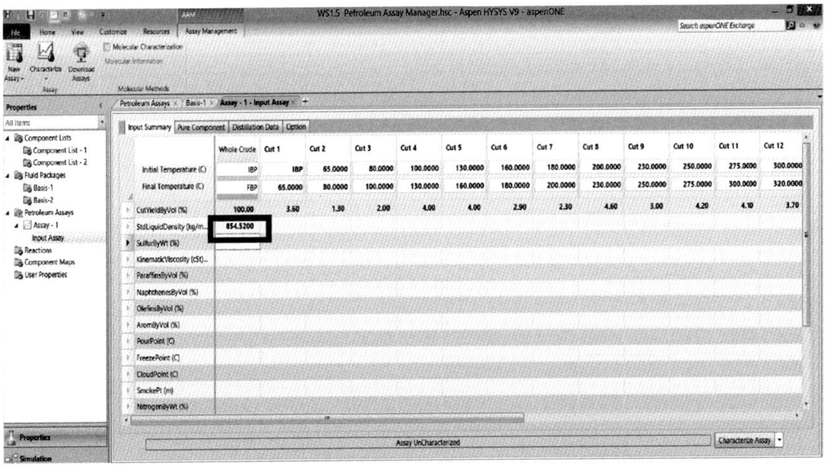

Figure 1.41 Enter the bulk density of 854.62 kg/m^3.

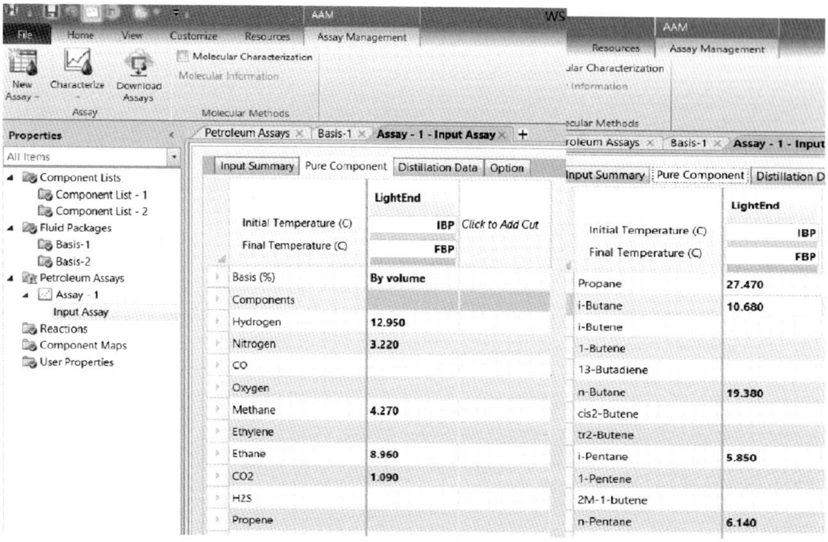

Figure 1.42 Enter the compositions of light components.

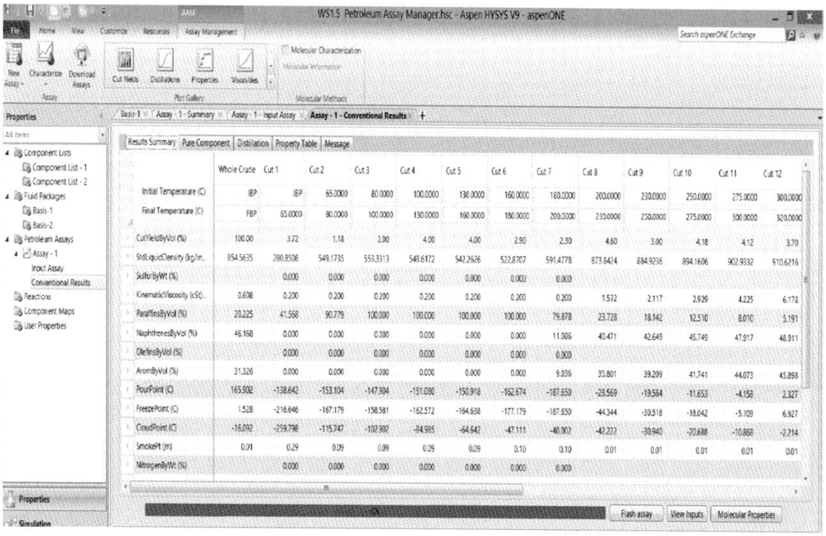

Figure 1.43 Characterize the assay.

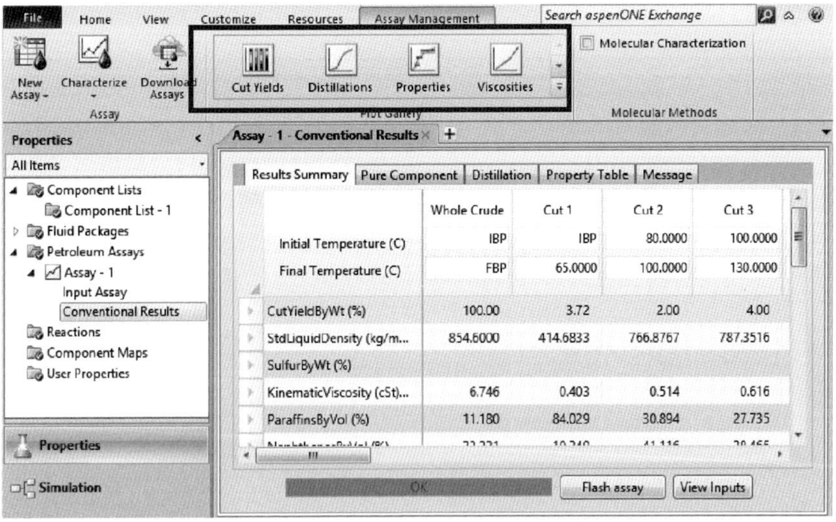

Figure 1.44 Add and edit assay data.

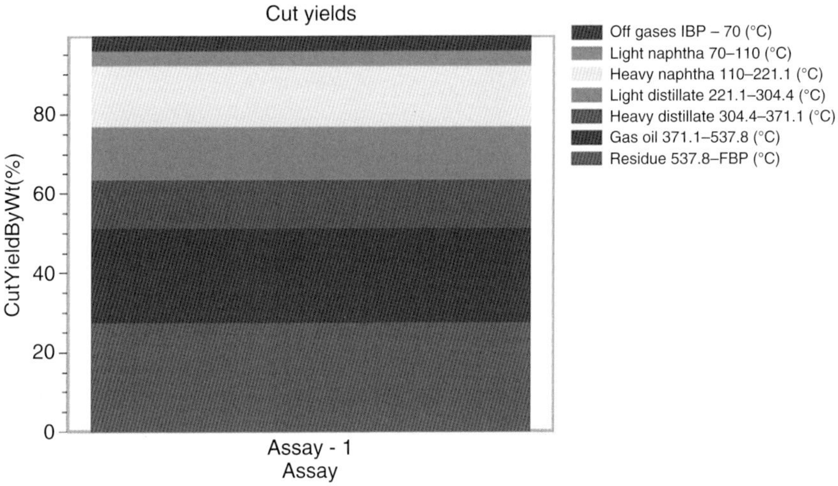

Figure 1.45 Plot of cut yields.

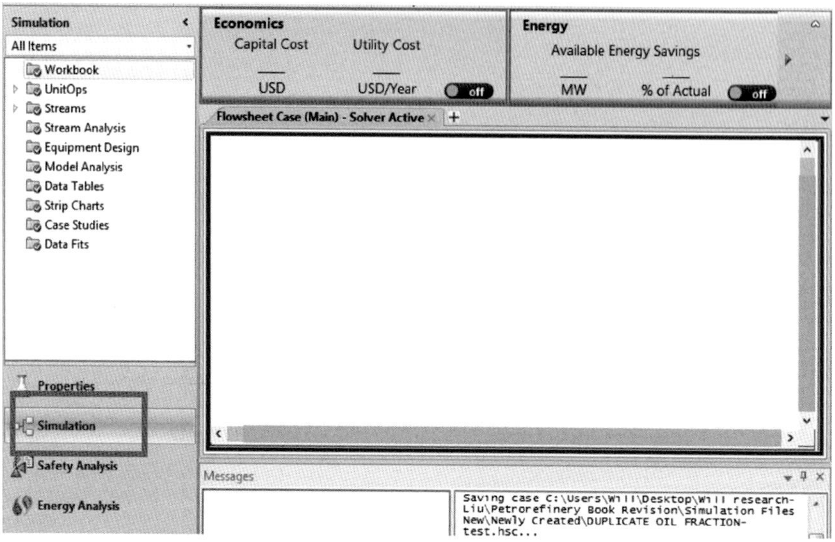

Figure 1.46 Enter the simulation environment.

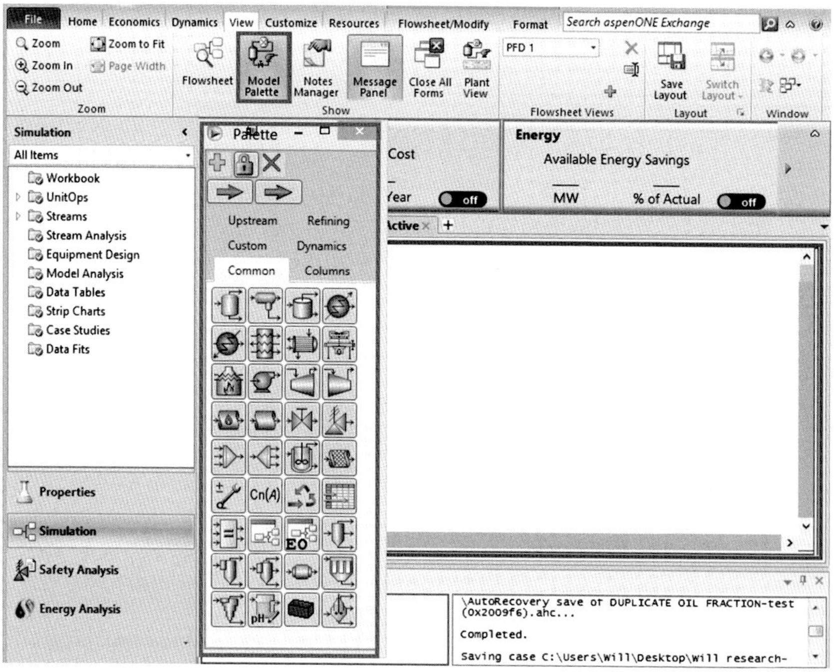

Figure 1.47 Open the window of unit models.

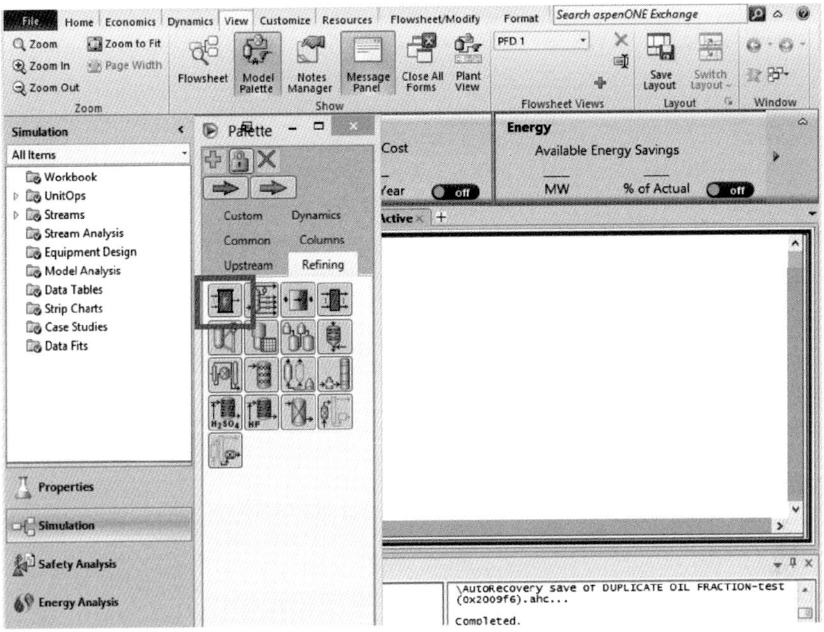

Figure 1.48 Add a petroleum feeder.

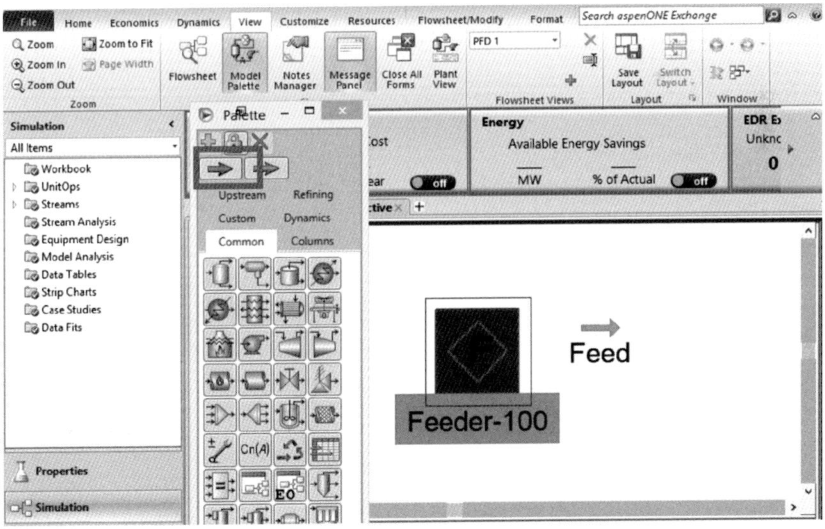

Figure 1.49 Add a feed stream.

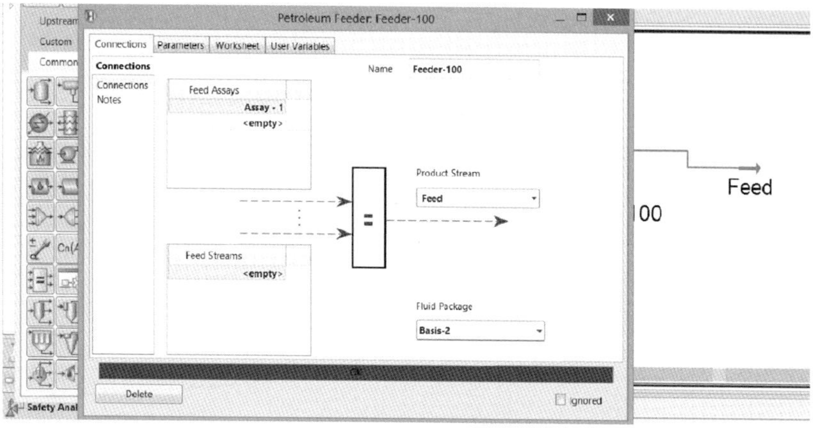

Figure 1.50 Specify feed assays in the petroleum feeder.

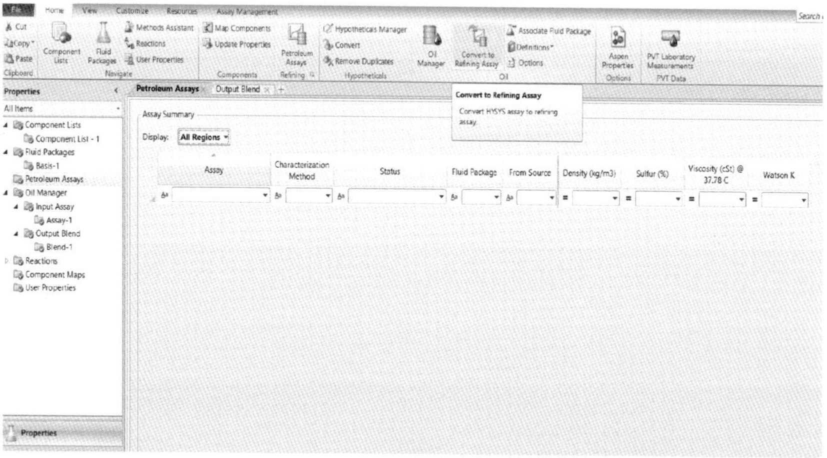

Figure 1.51 Convert the representation from oil manager to petroleum assay manager.

1.8 Workshop 1.6 – Conversion from the Oil Manager to Petroleum Assay Manager and Improvements of the Petroleum Assay Manager over the Oil Manager

We open the file, *WS1.4 Oil Manager.hsc*, and save as *WS1.6 Conversion from Oil Manager to Petroleum Assay Manager.hsc*. Figure 1.51 shows where we highlight the Petroleum Assay within the Properties Environment and then click on the button, Convert to Refinery Assay, to make the conversion.

This is given in Figure 1.52, in which we choose to use the existing fluid package and then click on Convert.

The conversion results in Figure 1.53, which is identical to the representation in the petroleum assay manager in Figure 1.42.

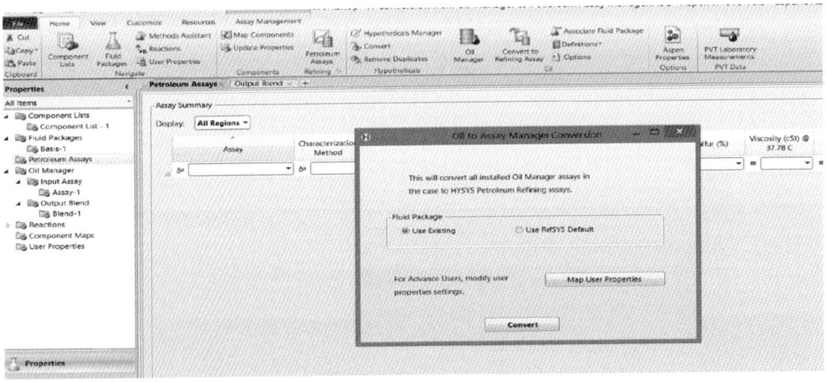

Figure 1.52 Oil to petroleum assay manager conversion.

Table 1.3 summarizes the improvements of the new petroleum assay manager over the old oil manager.

We strongly recommend the use of the petroleum assay manager to represent oil assays.

1.9 Property Requirements for Refinery Process Models

We classify the processes in modern refinery into two categories: separation units and reaction units. To develop a process model for any unit, we need to check the mass and energy balances of the flowsheet and perform calculations to describe the performance of the target unit. Therefore, the essential properties (physical and chemical) used to simulate these processes depend on the target unit, the chosen pseudocomponent scheme, and the selected kinetic model for reaction unit. Chapters 4 through 6 will represent the relevant issues for the three major reaction units in a modern refinery – FCC, catalytic reformer, and hydrocracker – and Chapter 7 covers additional refinery reaction units such as alkylation and delayed coking. While this chapter focuses primarily on the thermophysical properties required for modeling fractionation processes, the general framework for developing these properties for different kinds of pseudocomponents (i.e., those generated by kinetic lumping networks) is the same.

The previous sections in this chapter address the creation of pseudocomponents by cutting an assay curve into a set of discrete components based on boiling point ranges. We also briefly consider physical properties and process thermodynamics selection in the earlier workshops of this chapter. In this section, we discuss, in detail, the problem of how to represent these components in process modeling software. There are two major concerns in this area: physical properties of pseudocomponents and selection of a thermodynamic model that can deal with these hydrocarbon pseudocomponents in the context of refinery modeling. An accurate selection of physical properties and process thermodynamics results in a process model that can accurately account for material and energy flows in both vapor and liquid process streams.

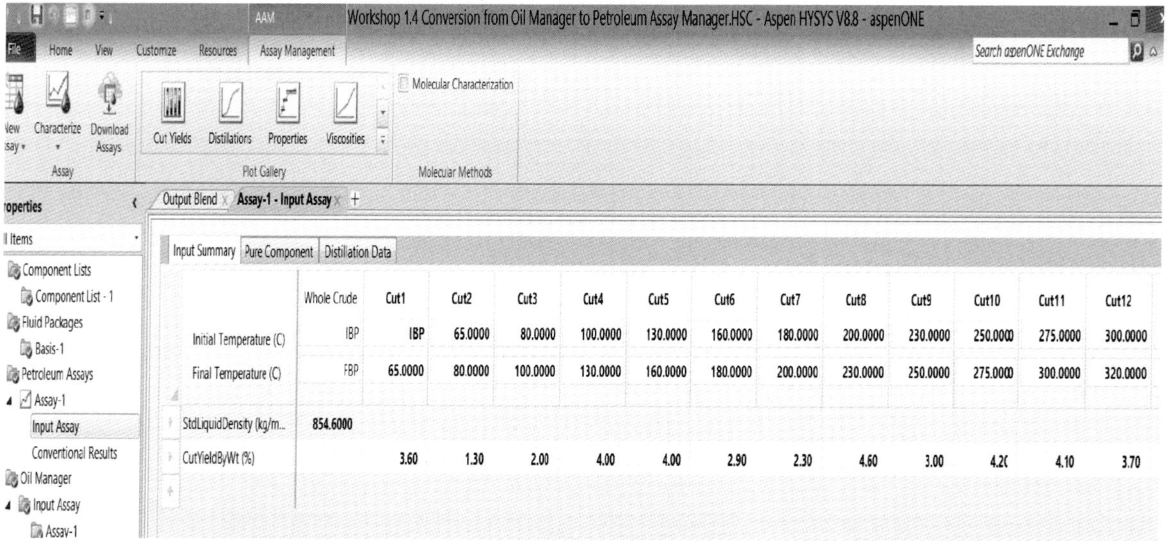

Figure 1.53 The petroleum assay resulting from the conversion from the oil manager.

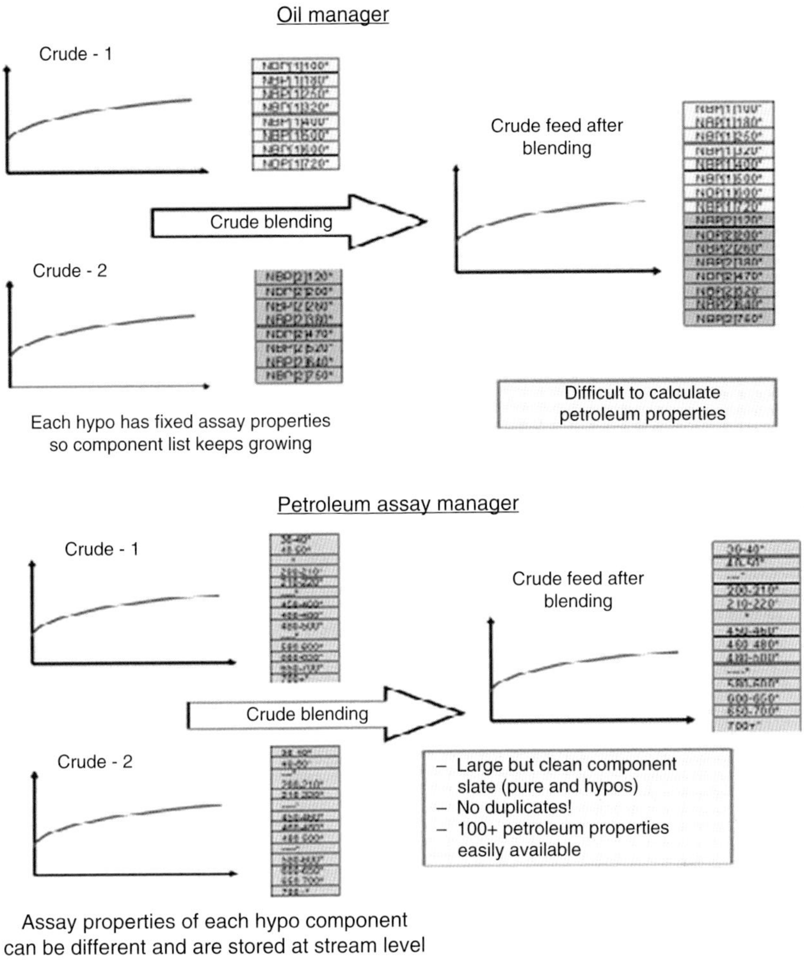

Figure 1.54 Comparison of assay representation in the oil manager and in the petroleum assay manager. (Courtesy of Aspen Technology, Inc.)

Table 1.3 Improvements of the petroleum assay manager over the oil manager.

Aspen HYSYS oil manager	Aspen HYSYS petroleum refining
Each petroleum assay blend has its own set of component lists (illustrated in Figure 1.54 and in Section 2.11.2 and Figure 2.82)	Multiple assay blend shares the same component list (illustrated in Figure 1.54 and in Section 2.11.2 and Figure 2.83)
Use less accurate blending rules, as each assay blend has its own component list	Calculate property values based on accurate blending rules, as all assays share the same component list
Allows to change very few petroleum properties	Allows the user to change more petroleum properties
Use a simplified option to characterize a petroleum assay	Use advanced options to characterize a petroleum assay

Table 1.4 Required properties for each phase.

Phase	Required properties
Vapor	Ideal gas heat capacity (CP_{IG})
Liquid	Liquid heat capacity (CP_L), liquid density (ρ_L), latent heat of vaporization (ΔH_{VAP}), vapor pressure (P_{VAP})
Both	Molecular weight (MW)

1.10 Physical Properties

For any process simulation that involves only vapor–liquid phases, certain key physical and thermodynamic properties must be available for each phase. Table 1.4 lists these properties for all phases. We can typically obtain these properties for pure components (i.e., *n*-hexane and *n*-heptane) from widely available databases such as DIPPR [2]. Commercial process simulation software (including Aspen HYSYS) also provides access to a large set of physical and thermodynamic properties for thousands of pure components. However, using these databases requires us to identify a component by name and molecular structure first and use experimentally measured or estimated values from the same databases. Given the complexity of the crude feed, it is not possible to completely analyze the crude feed in terms of pure components. Therefore, we must be able to estimate these properties for each pseudocomponent based on certain measured descriptors.

It is important to note that the properties given in Table 1.4 are the minimal physical properties required for rigorous accounting of the material and energy flows in the process. As we discuss in the subsequent sections, process models may require additional properties (especially vapor pressure) depending on the type of thermodynamic models being considered.

1.10.1 Estimating Minimal Physical Properties for Pseudocomponents

We have shown in the previous sections that the minimal amount of information to create pseudocomponents is a distillation curve and a specific gravity or density distribution. If only the bulk density is available, we can use constant Watson K factor assumption to estimate the density distribution. If only a partial density distribution is available, we can use the beta function to extrapolate an incomplete distillation curve. Note that it is usually better to incorporate as much experimentally measured information about the density curve as possible when building the process model. Once the distillation and density curve are available, we can cut the curve into a set of discrete pseudocomponents, each with its own boiling point and density. We can then use these two measured properties to estimate a variety of different types of physical properties (i.e., molecular weight, critical temperature, critical pressure, and acentric factor). Using these estimated physical properties, we can derive additional estimates for minimal physical properties required for process simulation. We have also provided a Microsoft Excel spreadsheet, *Critical_Property_Correlations.xls*, in the material that accompanies this text, which includes many of the correlations given in this section.

1.10.2 Molecular Weight

The molecular weight is the most basic information for a given pseudocomponent. Molecular weight is a required property to ensure a material balance throughout the process flowsheet. Researchers have extensively studied the trends of molecular weight for a variety of pure hydrocarbons and oil fractions. Several correlations are available to estimate the molecular weight as a function of boiling point, density, and viscosity. In general, correlations that only require the boiling point are the least accurate and correlations that require values of boiling point, density, and viscosity tend to be the most accurate. We use viscosity as a parameter in these correlations because it correlates well with molecular type – which can further refine the molecular weight estimate. In most cases, we use correlations that require the boiling point and density of a given component. Two popular correlations are the Lee–Kesler correlation [9, 10], Eq. (1.10), and the Twu correlation [11], Eqs. (1.11)–(1.13), respectively.

$$\begin{aligned}\mathrm{MW} = &-12272.6 + 9486.4(\mathrm{SG}) + (8.3741 - 5.99175 \cdot \mathrm{SG})T_\mathrm{b}\\ &+(1 - 0.77084 \cdot \mathrm{SG} - 0.02058 \cdot \mathrm{SG}^2)\\ &\times\left(0.7465 - \frac{222.466}{T_\mathrm{b}}\right)\cdot\frac{10^7}{T_\mathrm{b}} + (1 - 0.80882\cdot\mathrm{SG} - 0.02226\cdot\mathrm{SG}^2)\\ &\times\left(0.3228 - \frac{17.335}{T_\mathrm{b}}\right)\cdot\frac{10^{12}}{T_\mathrm{b}^3}\end{aligned} \tag{1.10}$$

$$\mathrm{MW}^\mathrm{o} = \frac{T_\mathrm{b}}{5.8 - 0.0052T_\mathrm{b}} \tag{1.11}$$

$$\mathrm{SG}^\mathrm{o} = 0.843593 - 0.128624\alpha - 3.36159\alpha^3 - 13749.5\alpha^{12} \tag{1.12}$$

$$\begin{aligned}T_\mathrm{c}{}^\mathrm{o} = &T_\mathrm{b}(0.533272 + 0.343838 \times 10^{-3} \times T_\mathrm{b} + 2.52617 \times 10^{-7} \times T_\mathrm{b}{}^2\\ &-1.654881 \times 10^{-10} \times T_\mathrm{b}{}^3 + 4.60773 \times 10^{-24} \times T_\mathrm{b}{}^{-13})^{-1}\end{aligned} \tag{1.13}$$

$$\alpha = 1 - \frac{T_\mathrm{b}}{T_\mathrm{c}{}^\mathrm{o}} \tag{1.14}$$

$$\ln(\mathrm{MW}) = \ln(\mathrm{MW}^\mathrm{o})\left[\frac{(1 + 2f_M)}{(1 - 2f_M)^2}\right] \tag{1.15}$$

$$f_M = \Delta\mathrm{SG}_M\left[\chi + \left(-0.0175691 + \frac{0.143979}{T_\mathrm{b}{}^{0.5}}\right)\right]\Delta\mathrm{SG}_M \tag{1.16}$$

$$\chi = \left|0.012342 - \frac{0.244515}{T_\mathrm{b}{}^{0.5}}\right| \tag{1.17}$$

$$\Delta\mathrm{SG}_M = \exp[5(\mathrm{SG}^\mathrm{o} - \mathrm{SG})] - 1 \tag{1.18}$$

Riazi [4] listed several other correlations such as Cavett and Goosens for molecular weight, but they generally do not have significant advantage over the Lee–Kesler or Twu correlations. The Lee–Kesler correlation was developed by

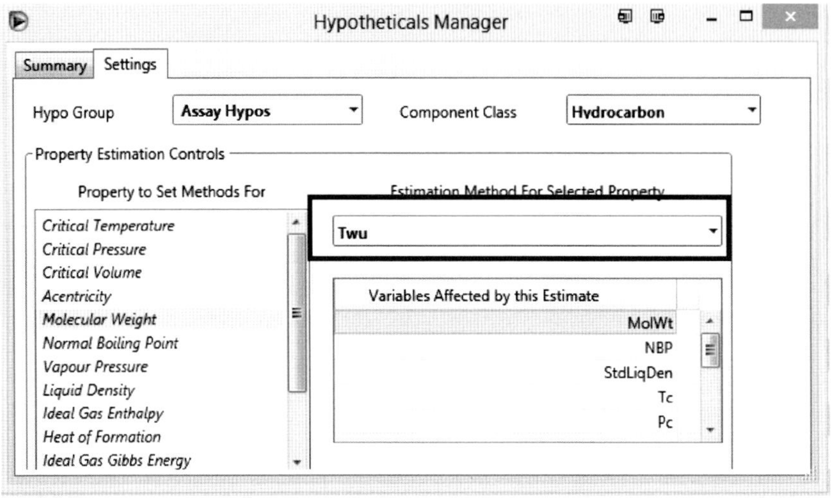

Figure 1.55 Modify the molecular weight correlation in Aspen HYSYS Hypotheticals Manager.

correlating light oil fractions (<850 °F or 454 °C) from a variety of sources. As a result, the Lee–Kesler correlation tends to be less accurate for pseudocomponents with high boiling point temperatures. The Twu correlation includes a significant number of data points to account for heavier components. Aspen HYSYS uses the Twu correlation to calculate the molecular weight. Figure 1.55 shows how to select the molecular weight correlation for a particular blend (shown in earlier workshops) in Aspen HYSYS Hypotheticals Manager.

1.10.3 Critical Properties

Many properties that are required for rigorous accounting of material and energy flows (Table 1.4) in process models are not well defined for pseudocomponents. Fortunately, researchers have found that these required properties correlate well with critical temperature (T_c), critical pressure (P_c), and acentric factor (ω) for different types of hydrocarbons from many sources. Therefore, when we use pseudocomponents of any kind, we must also estimate these critical properties. Just as with molecular weight, many critical property estimation methods are available in the literature. These correlations differ on the basis of the parameters required and underlying data used to create the correlation. We note that as the components get heavier and boil at higher temperatures, the associated change in critical pressure tends to diminish. Hence, correlations for critical pressure tend to be logarithmic formulas. A modeling consequence is that particularly accurate measures of these critical pressures are not required for good modeling results. In addition, most refinery process conditions do not approach the critical properties of these pseudocomponents.

Lee–Kesler [9, 10] and Twu [11] have also produced correlations for critical properties. In our work, we have used the Lee–Kesler correlations extensively. Equations (1.19) and (1.20) give the correlations for critical temperature (T_c) and critical pressure (P_c) using the Lee–Kesler correlations. We recommend using

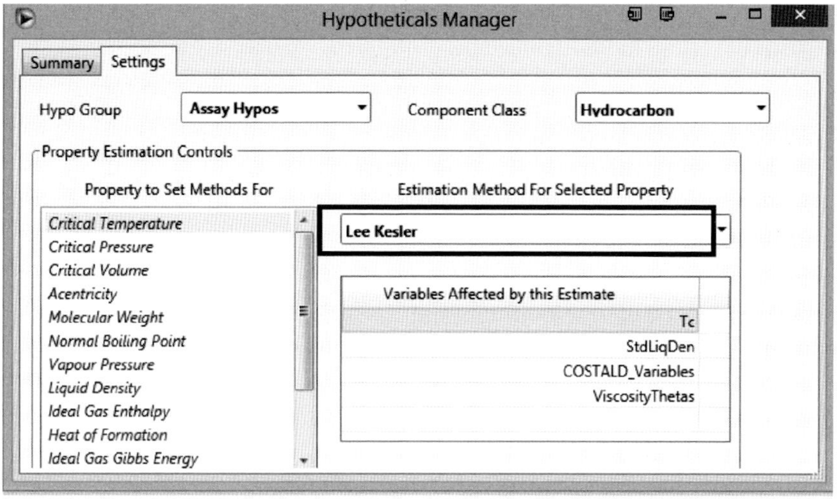

Figure 1.56 Modify T_c correlation in Aspen HYSYS Hypotheticals Manager.

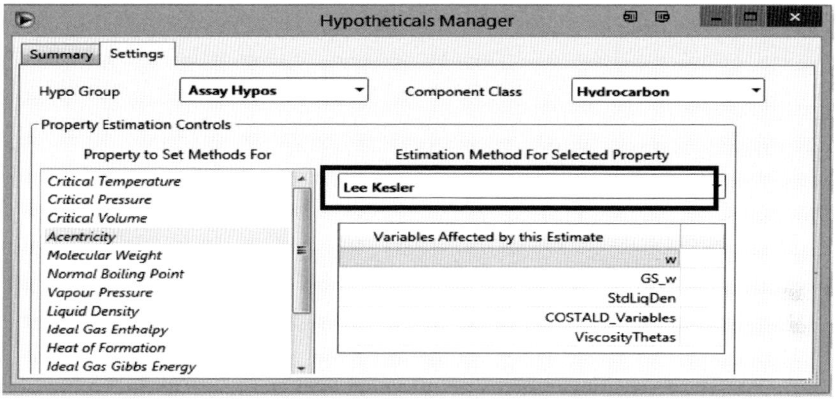

Figure 1.57 Modify acentric factor correlation in Aspen HYSYS Hypotheticals Manager.

these correlations for all boiling point ranges as the differences that arise from using other correlations are often minor. Figures 1.56 and 1.57 show how we can change the correlation for each blend in Aspen HYSYS Hypotheticals Manager.

$$T_c = 189.8 + 450.6\text{SG} + (0.4244 + 0.1174\text{SG})T_b + (0.1441 - 1.0069\text{SG})10^5/T_b \tag{1.19}$$

$$P_c = 5.689 - \frac{0.0566}{\text{SG}} - \left(0.43639 + \frac{4.1216}{\text{SG}} + \frac{0.21343}{\text{SG}^2}\right) \times 10^{-3} T_b + \left(0.47579 + \frac{1.182}{\text{SG}} + \frac{0.15302}{\text{SG}^2}\right) \times 10^{-6} T_b^{\,2} - \left(2.4505 + \frac{9.9099}{\text{SG}^2}\right) \times 10^{-10} T_b^{\,3} \tag{1.20}$$

A related property is the acentric factor. The acentric factor accounts for the size and shape of various kinds of molecules. Simple molecules have an acentric factor close to 0, whereas large or complex hydrocarbon molecules may have values approaching 0.5–0.6 [6]. The acentric factor is not measured but defined as an explicit function of the ratio of vapor pressure at the normal boiling point to the measured or estimated critical pressure. We show the definition of the acentric factor in Eq. (1.21).

$$\omega = -\log_{10}(P_r^{VAP}) - 1.0 \tag{1.21}$$

where P_r^{VAP} represents the reduced vapor pressure, that is, the pseudocomponent vapor pressure divided by its critical pressure, when the reduced temperature, T_r, that is, the temperature divided by the critical temperature, is equal to 0.7.

Given the small range of values for the acentric factor, most correlations can provide useful results. The accuracy of the acentric correlation depends largely on the accuracy of the critical temperature and pressure correlations. However, even large relative errors do not result in significant deviation of derived properties such as ideal gas heat capacity. We again choose the Lee–Kesler [9, 10] correlation for the acentric factor. This correlation, given by Eq. (1.22), relies on extensive vapor pressure data collected by Lee and Kesler for the critical temperature and pressure correlations. The correlation is technically limited to the reduced boiling point temperature (T_{br}) of less than 0.8 but has been successfully used at high T_{br} values. Figure 1.57 shows how we can modify the acentric factor estimation method for oil blends in Aspen HYSYS Hypotheticals Manager.

$$\omega = \frac{\begin{array}{c} -\ln(P_C/1.01325) - 5.92714 + \dfrac{6.09648}{T_{br}} \\ +1.28862\ \ln(T_{br}) - 0.169347\ T_{br}^{\ 6} \end{array}}{15.2518 - \dfrac{15.6875}{T_{br}} - 13.4721 \ln(T_{br}) + 0.43577\ T_{br}^{\ 6}} \tag{1.22}$$

In this equation, T_{br} represents the reduced boiling point, that is, the normal boiling point divided by the critical temperature T_c.

1.10.4 Liquid Density

The liquid density of hydrocarbons is essential for modeling purposes to convert molar and mass flows into volumetric flows. Many processes in the refinery operate on the basis of volumetric flow. In addition, the density of the products is an important constraint while marketing the refinery's products for sale. In the context of process modeling, liquid density is also a property parameter that must be correlated as many of the equation-of-state (EOS) thermodynamic models cannot accurately predict liquid densities. Even when a given process modeling software uses an EOS approach for refinery modeling, liquid density is often calculated independently to ensure accurate results. Figure 1.58 shows how Aspen HYSYS calculates the liquid density independently even when we use an EOS (in this case, Peng–Robinson) as the thermodynamic model.

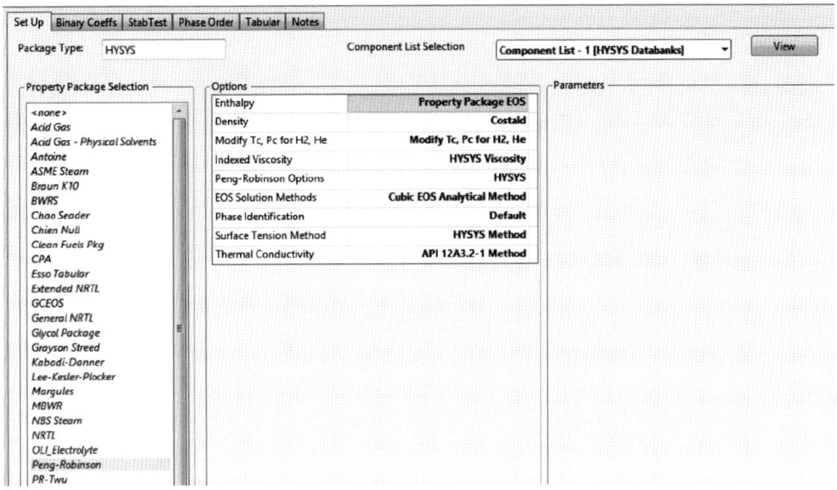

Figure 1.58 Options for Peng–Robinson equation of state in Aspen HYSYS.

Several correlations are available in the literature for liquid mass density or liquid molar volume as functions of various critical properties. It is possible to convert from liquid mass density to liquid molar volume using the molecular weight of the component in question. This also means that errors in the molecular weight or critical property predictions can introduce additional error in the liquid density or molar volume correlations. Popular correlations for liquid density include Yen-Woods [12], Gunn-Yamada [13], and Lee–Kesler [9, 10]. An accurate correlation (when the reduced temperature is less than 1) of liquid density is the Spencer–Danner (modified Rackett) method [14] with COSTALD (Corresponding States Liquid Density) [15] correction for pressure. Equation (1.23) gives the standard Spencer–Danner equation. This equation actually predicts the molar volume at saturated liquid conditions. We can convert this molar volume into liquid density using the molecular weight.

$$V^{\mathrm{SAT}} = \left(\frac{\mathrm{R}T_{\mathrm{C}}}{P_{\mathrm{c}}}\right) Z_{\mathrm{RA}}{}^{n} \quad \text{with } n = 1.0 + (1.0 - T_{\mathrm{r}})^{2/7} \tag{1.23}$$

$$Z_{\mathrm{RA}} = 0.29056 - 0.08775\omega \tag{1.24}$$

Z_{RA} is a special parameter to account for the critical compressibility of the component. Tables of Z_{RA} for many pure components are part of the pure component databases in Aspen HYSYS. We may estimate Z_{RA} for pseudocomponents from Eq. (1.24) as a function of the correlated acentric factor. The liquid density from Spencer–Danner equation is a function of temperature only. Refinery processing conditions can be severe enough where the liquid density is also a function of pressure. To correct the liquid density for high pressure, we can introduce the COSTALD correction given by Eq. (1.25). This equation requires the liquid density, $\rho_{P^{\mathrm{o}}}$, at a certain reference pressure, P^{o}, obtained from

Eq. (1.25) and predicts the density, ρ_P, at an elevated pressure, P, as a function of two parameters, C and B.

$$\rho_P = \rho_{P^o}\left[1 - C\ \ln\left(\frac{B+P}{B+P^o}\right)\right]^{-1} \tag{1.25}$$

$$e = \exp(4.79594 + 0.250047\omega + 1.14188\omega^2) \tag{1.26}$$

$$B = P_c(-1 - 9.0702(1.0 - T_r)^{\frac{1}{3}} + 62.45326(1.0 - T_r)^{\frac{2}{3}} - 135.1102(1.0 - T_r) + e(1.0 - T_r)^{\frac{4}{3}}) \tag{1.27}$$

$$C = 0.0861488 + 0.0344483\omega \tag{1.28}$$

The COSTALD correlation is quite accurate even at high reduced temperatures and pressures. Predicted liquid densities generally agree with measured values within 1–2%, provided the errors in the critical property predictions are low. A potential problem can occur if the reduced temperature is greater than 1. There can be discontinuity from the Spencer–Danner equation in the density prediction, which may cause some process models to fail. However, at a reduced temperature greater than 1, the EOS becomes more accurate and can be used directly. Aspen HYSYS includes a smoothing approach (using the Chueh and Prausnitz correlation [16]) to ensure a smooth transition from the COSTALD densities to EOS-based densities.

1.10.5 Ideal Gas Heat Capacity

The last property that is often directly correlated is the ideal gas heat capacity of pseudocomponents. The ideal gas heat capacity represents the vapor heat capacity of the pseudocomponent at a given standard condition. The standard conditions typically refer to 25 °C and 1 atm or 77 °F and 14.696 psia. It is well known that the heat capacity of hydrocarbons can be modeled with a simple polynomial expression as a function of temperature. Lee and Kesler [9, 10] presented a popular correlation, Eq. (1.29) to Eq. (1.36), where M is molecular weight, T in Kelvin, and K_w is Watson K factor. These parameters may be estimated from other correlations, including Lee–Kesler equation for MW in Section 1.10.3, Eq. (1.10). The heat capacities of hydrocarbons do not vary significantly over a wide range of temperatures, so very accurate heat capacities are not necessary for good modeling results. We present this correlation in Eq. (1.29). Figure 1.59 shows how we can modify the ideal gas heat capacity estimation method for oil blends in Aspen HYSYS Hypotheticals Manager.

$$CP_{IG} = \text{MW}[A_0 + A_1T + A_2T^2 - C(B_0 + B_1T + B_2T^2)] \tag{1.29}$$

$$A_0 = -1.41779 + 0.11828K_w \tag{1.30}$$

$$A_1 = -(6.99724 - 8.69326K_w + 0.27715K_w{}^2) \times 10^{-4} \tag{1.31}$$

$$A_2 = -2.2582 \times 10^{-6} \tag{1.32}$$

$$B_0 = 1.09223 - 2.48245\omega \tag{1.33}$$

$$B_1 = -(3.434 - 7.14\omega) \times 10^{-3} \tag{1.34}$$

$$B_2 = -(7.2661 - 9.2561\omega) \times 10^{-7} \tag{1.35}$$

$$C = \left[\frac{(12.8 - K_w) \times (10 - K_w)}{10\omega}\right]^2 \tag{1.36}$$

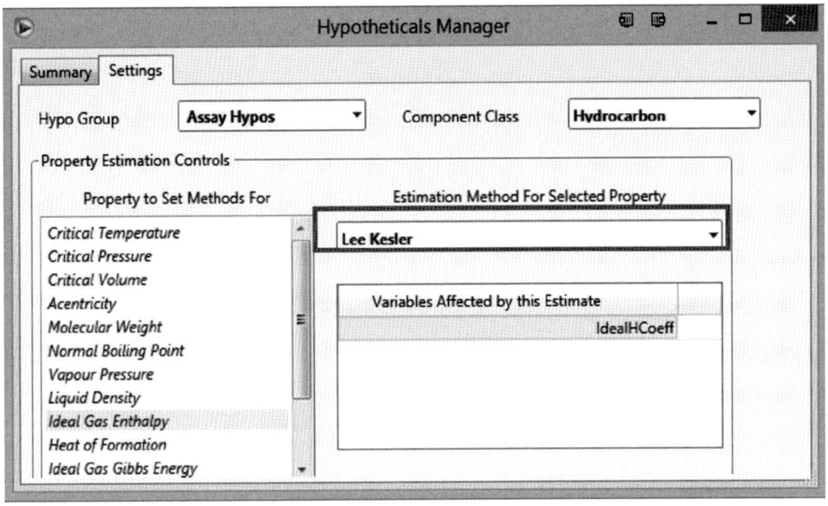

Figure 1.59 Modify ideal gas heat capacity correlation in Aspen HYSYS Hypotheticals Manager.

1.10.6 Other Derived Physical Properties

Once we have obtained the boiling point, density or specific gravity, molecular weight, and critical properties of a particular pseudocomponent, we can also generate estimates for other required properties for process simulation provided in Table 1.4. The accuracy of these predictions is largely a function of the accuracy of the molecular weight and critical property predictions. In addition, depending on the thermodynamic method chosen, we may not require any correlations for certain properties. For example, if we choose an EOS, we do not require any additional correlations for the vapor pressure (P_{VAP}) or heat of vaporization (ΔH_{VAP}), as these values will be calculated directly by the EOS. We discuss such features of the EOS in the following section. In this section, we present correlations for all required properties so that model developers are aware of the model limitations and additional data requirements when we do not use an EOS for modeling process thermodynamics.

The liquid heat capacity of pseudocomponents in refinery modeling is largely constant. Walas [6] noted that as the boiling point and density of the pseudocomponent increase, the heat capacity of hydrocarbons tends to approach a value of 1.8–2.2 kJ/kg K near the normal boiling point. Consequently, rough estimates of heat liquid capacities do not affect model results significantly. Two correlations are available for liquid heat capacities of hydrocarbons that are in general use. Equation (1.37) is a correlation by Kesler and Lee [9, 10] and Eq. (1.41) is a correlation recommended by API. Either correlation may be used with equally acceptable results. We generally do not encounter these temperature limits prescribed for both of these correlations. We also note that these correlations are weak functions of temperature. Process modeling software programs have a variety of models to estimate liquid heat capacity, but these methods are only marginally better when compared to the simple correlations given here.

When $145\ \mathrm{K} < T < 0.8T_\mathrm{c}$

$$CP_L = a(b + cT) \tag{1.37}$$

$$a = 1.4651 + 0.2302K_\mathrm{w} \tag{1.38}$$

$$b = 0.306469 - 0.16734\mathrm{SG} \tag{1.39}$$

$$c = 0.001467 - 0.000551\mathrm{SG} \tag{1.40}$$

When $T_\mathrm{r} < 0.85$

$$CP_L = A_1 + A_2T + A_3T^2 \tag{1.41}$$

$$A_1 = -4.90383 + (0.099319 + 0.104281\mathrm{SG})K_\mathrm{w} + \left(\frac{4.81407 - 0.194833K_\mathrm{w}}{\mathrm{SG}}\right) \tag{1.42}$$

$$A_2 = (7.53624 + 6.214610K_\mathrm{w}) \times \left(1.12172 - \frac{0.27634}{\mathrm{SG}}\right) \times 10^{-4} \tag{1.43}$$

$$A_3 = -(1.35652 + 1.11863K_\mathrm{w}) \times \left(2.9027 - \frac{0.70958}{\mathrm{SG}}\right) \times 10^{-7} \tag{1.44}$$

Another property related to the heat capacity is the heat of vaporization of pseudocomponent as a liquid. The heat of vaporization represents the heat required to vaporize a given mass (or volume) of liquid into vapor. Similar to heat capacities, there are several correlations to calculate the heat of vaporization in the literature. We present two popular correlations here. Equation (1.45) is the Riedel correlation [17] and Eq. (1.46) is the Chen and Vettere correlation [17]. We note that both correlations rely on critical temperatures and pressure and give the heat of vaporization at the normal boiling point. We can obtain the heat of vaporization at a different temperature by using the Watson relation [1] in Eq. (1.47). Either of the correlations can provide very good results for hydrocarbons (<2% average relative deviation, ARD). We recommend the use of either correlation if the process modeling software does not already include a correlation. In addition to these correlations, Aspen HYSYS offers a more advanced proprietary correlation using two reference state liquids.

$$\Delta H_\mathrm{NBP}^\mathrm{VAP} = 1.093RT_CT_\mathrm{br}\frac{\ln P_\mathrm{c} - 1.013}{0.93 - T_\mathrm{br}} \tag{1.45}$$

$$\Delta H_\mathrm{NBP}^\mathrm{VAP} = RT_CT_\mathrm{br}\frac{3.978T_\mathrm{br} - 3.958 + 1.555\ln P_\mathrm{c}}{1.07 - T_\mathrm{br}} \tag{1.46}$$

$$\Delta H^\mathrm{VAP} = \Delta H_\mathrm{NBP}^\mathrm{VAP}\left(\frac{1 - T_\mathrm{r}}{1 - T_\mathrm{br}}\right)^{0.38} \tag{1.47}$$

The vapor pressure of pseudocomponents is also an important property when an EOS is not used. All other approaches to process thermodynamics require some form of vapor pressure correlation. The vapor pressure for pure hydrocarbons has been extensively tabulated in many component databases such as DIPPR (Design Institute for Physical Property Research, American Institute of Chemical Engineers) and significant libraries are available in modern process modeling software. Several correlations for the vapor pressure of pseudocomponents are available in the literature. It is important to recall that the vapor pressure and heat vaporization are related through the Clausius–Clapeyron equation (Eq. 1.48) [17]. This relationship imposes a constraint if we wish the model to be thermodynamically consistent. In general, most of the popular correlations for vapor pressure such as the Lee–Kesler [9, 10] agree well with heat of vaporization correlations and maintain thermodynamic consistency. We present the Lee–Kesler vapor pressure correlation in Eq. (1.49).

$$\frac{\mathrm{d}\ln P}{\mathrm{d}T} = \frac{\Delta H_{\mathrm{VAP}}}{RT^2} \tag{1.48}$$

$$\ln P_{\mathrm{r}}^{\mathrm{VAP}} = 5.92714 - \frac{6.096648}{T_{\mathrm{r}}} - 1.28862 \ln T_{\mathrm{r}} + 0.169347 T_{\mathrm{r}}^{6} + \omega\left(15.2518 - \frac{15.6875}{T_{\mathrm{r}}} - 13.4721 \ln T_{\mathrm{r}} + 0.43577 T_{\mathrm{r}}^{6}\right) \tag{1.49}$$

The Lee–Kesler correlation for vapor pressure is quite accurate for low-to-medium boiling pseudocomponents. For very light components, we recommend using pure component properties directly. In the case of heavy components, Ambrose [17] has presented an additional term for the Lee–Kesler correlation. In practice, however, the additional term is not necessary for refinery modeling purposes.

1.11 Process Thermodynamics

After we have fully characterized the pseudocomponents and any true components in the process model, we must choose a thermodynamic model. The thermodynamic model here refers to a framework that allows us to describe whether a particular mixture of components forms one phase or two phases, the distribution of components within these phases, and material and energy flows of these phases given a set of process conditions. Process thermodynamics also set material and energy transfer limits on various fractionation and reaction units in the model and in the actual plant itself.

Modern refineries deal with a multitude of complex systems that may require different thermodynamic models for each refinery plant and its associated process model. For example, we cannot model the sour gas units that deal with acid gases and water with the same thermodynamic model that we use for the crude fractionation system. In fact, reasonable thermodynamic models form the heart of any process model. Chen and Mathias [7] have documented a variety of thermodynamic models available for frequently encountered chemical and physical systems. Agarwal *et al.* [18] presented a detailed account about the pitfalls of choosing a poor thermodynamic system for process models and the undesired consequences of using these poor models to modify plant operations. Process model developers and users must be aware of the underlying thermodynamics and its limitations.

Given that the field of thermodynamic models is vast, we choose to focus on thermodynamic models that deal with hydrocarbon–hydrocarbon interactions only and can model many units in the refinery quite accurately. The only complication (aside from the choice of an appropriate thermodynamic model) is the presence of large amounts of water in the form of steam in various fractionation and reaction units. In most cases, we can simply deal with the hydrocarbon and water phases as immiscible. This is known as *the free-water approach*. Kaes [1] discussed this approach extensively and it is a common approach in many process simulators. Some software may include *a dirty-water approach*. This approach uses correlations to model the solubility of water in the hydrocarbon and the solubility of light acid gases in water. For the purposes of refinery reaction and fractionation modeling in this text, both approaches have negligible effects on the overall process model. We give the general statement of vapor–liquid equilibrium for any thermodynamic model in Eq. (1.50).

$$y_i \varphi_i^{\ V} P = x_i \varphi_i^{\ L} P \tag{1.50}$$

where y_i refers to vapor phase molar composition of component i, φ_i^V refers to the vapor phase fugacity coefficient of component i, P is overall pressure, x_i is the liquid phase molar composition of component i, and φ_i^L refers to the liquid phase fugacity coefficient of component i.

For refinery fractionation modeling, several simplifications are possible. Each one of these simplifications represents a different thermodynamic approach. We list major approaches, required pseudocomponent properties, and our recommendation for use in Table 1.5. We discuss each of these approaches and their requirements in subsequent sections.

Table 1.5 Comparison of various thermodynamic approaches.

Approach	Required physical properties	Recommended
Simple	Molecular weight (MW) Ideal gas heat capacity (CP_{IG}) Vapor pressure (P_{VAP}) Heat of vaporization (ΔH_{VAP}) Liquid heat capacity (CP_L) Liquid density (ρ_L)	No
Mixed or activity coefficient	Molecular weight (MW) Ideal gas heat capacity (CP_{IG}) Vapor pressure (P_{VAP}) Heat of vaporization (ΔH_{VAP}) Liquid heat capacity (CP_L) Liquid density (ρ_L) Solubility parameter (δ)	Yes, however, best with heavy components that the equation-of-state (EOS) approach cannot deal with
Equation of state	Molecular weight (MW) Critical temperature (T_c) Critical pressure (P_c) Acentric factor (ω) Ideal gas heat capacity (CP_{IG}) Liquid density (ρ_L) Interaction parameter (k_{ij})	Yes, with adequate corrections of liquid density

1.11.1 Process Thermodynamics

The simple approach is the most basic and least rigorous thermodynamic approach. In the simple approach or Raoult's law approach, we assume that vapor phase and liquid phase are ideal. In this case, we may write the general statement of equilibrium equation (1.50), as Eq. (1.51), where y_i is the vapor phase molar composition of component i, P is the pressure, x_i is the liquid phase molar composition, and $P^{\mathrm{SAT}}(T)$ is the vapor pressure of component i as a function of temperature only. These properties are routinely available for pure components and we have extensively discussed how to obtain the required properties from pseudocomponents.

$$y_i P = x_i P^{\mathrm{SAT}}(T) \tag{1.51}$$

A variation of this equation is to rearrange the equation to obtain the equilibrium distribution ratio, y_i/x_i as shown in Eq. (1.52). This distribution ratio is also known as the K-value for component i. Numerous correlations for K-values exist for a variety of pure components and pseudocomponents. The Braun-K10 (BK-10) correlation is a popular correlation of this type [6].

$$K_i = \frac{y_i}{x_i} = \frac{P^{\mathrm{SAT}}(T)}{P} = f(T) \tag{1.52}$$

Once we obtain a K-value at a given temperature and pressure, we can perform mass and energy balances that include isothermal, isobaric, and isenthalpic flashes. We can also use the ideal gas heat capacity of the vapor phase, heat of vaporization, and heat of capacity of the liquid to represent the enthalpies of relevant vapor and liquid streams.

Most process simulators include these types of correlations, but they are largely of historical interest or used to maintain compatibility with old models. We do not recommend using simple methods, as they cannot adequately quantify the transition from vapor to liquid phases beyond the original correlation. In addition, these correlations tend to be thermodynamically poor (do not consider any interactions between components and thermodynamically inconsistent at higher pressures). We cannot integrate models using these correlations into new models that use an EOS or activity coefficient approach without significant efforts.

1.11.2 Mixed or Activity Coefficient-Based Approach

The mixed or activity coefficient approach uses the concept of activity coefficients to separate out the effects of nonideality because of component interactions and the effect of pressure. For the activity coefficient approach, we can rewrite the general equilibrium statement as

$$y_i \varphi_i^{V} P = x_i \gamma_i \varphi_i^{\mathrm{SAT}} P^{\mathrm{SAT}}(T) \mathrm{PF}_i \tag{1.53}$$

$$\mathrm{PF}_i = \exp\left(\int_{P^{\mathrm{SAT}}}^{P} \frac{V_i(T,\pi)}{RT} \mathrm{d}\pi\right) \tag{1.54}$$

In the equations, y_i is vapor molar composition of component i, φ_i^{V} is the vapor phase fugacity coefficient for component i, P is the system pressure, x_i is the liquid molar composition of component i, φ_i^{SAT} is the fugacity coefficient for vapor pressure of component i, $P^{\mathrm{SAT}}(T)$ is the vapor pressure of component i, and PF_i is the Poynting factor for component i at pressure P. V_i is the molar volume of component i as a function of temperature, T, and pressure, π (integrated from P^{SAT} to P). The PF_i factor is generally close to a value of 1 unless the system pressure is very high [17]. We can now rewrite the equilibrium relationship in the form of K-values as Eq. (1.55).

$$K_i = \frac{y_i}{x_i} = \frac{\gamma_i \varphi_i^{\mathrm{SAT}} P^{\mathrm{SAT}}(T)}{\varphi_i^{V} P} \tag{1.55}$$

We apply the Redlich–Kwong (RK) EOS [6] and liquid phase correlation (or an EOS) to obtain expressions for φ_i^V and φ_i^{SAT} as function of temperature, pressure, and component critical properties. This is the approach taken by the very popular Chao–Seader and Grayson–Streed methods [6]. The only factor that remains undefined is the liquid activity coefficient. The Chao–Seader and Grayson–Streed methods use regular solution theory to obtain an expression for γ_i as follows:

$$\ln \gamma_i = \frac{V_i}{RT}(\delta_i - \overline{\delta}) \tag{1.56}$$

$$\overline{\delta} = \frac{\sum x_i V_i \delta_i}{\sum x_i V_i} \tag{1.57}$$

where V_i is the liquid molar volume of component i and δ_i is the solubility parameter for component i. Molar volumes for pure components are readily available and we discussed several methods to estimate molar volumes for pseudocomponents in Section 1.10.5. We can obtain the solubility parameter for pseudocomponents using Eq. (1.56), where ΔH_{VAP} is the heat of vaporization, R is the universal gas constant, and T is system temperature. We have discussed how to calculate the heat of vaporization for pseudocomponents in Section 1.11.

$$\delta_i = \left(\frac{\Delta H_{\mathrm{VAP}} - RT}{V_i}\right)^{0.5} \tag{1.58}$$

We use the K-value expression to calculate various equilibrium properties and perform typical process modeling flashes. As with the simple thermodynamic approach, we can use the heat capacities and heats of vaporization to obtain enthalpy balances for vapor and liquid streams. In addition, as we account for vapor and liquid phase nonideality due to component interaction, and temperature and pressure effects, we can also apply standard thermodynamic relationships to compute excess properties for enthalpies, and so on. The excess properties account for deviation of ideal mixing behavior and resulting deviations in equilibrium behavior.

Using the activity coefficient approach in the form of the Chao–Seader or Grayson–Streed method for refinery modeling is a significant improvement over the simple approach. The activity coefficient approach accounts for vapor and liquid phase nonidealities accurately in both the equilibrium and the enthalpy calculations. In addition, this approach is easy to integrate with other types of activity coefficient models that we may use in refinery models (especially for sour water systems). We prefer to use activity coefficient models while dealing with heavy components that occur especially in vacuum distillation systems. A key shortcoming of this approach is that light components may require fictitious solubility parameters fitted to certain data sets and performance of this approach degrades quickly near the vicinity of the critical point. In general, however, this method is a reasonable thermodynamic model for real and pseudocomponents that we find in refinery reaction and fractionation systems.

1.11.3 Equation-of-State Approach

The most rigorous approach is the EOS approach. When we use an EOS, both vapor and liquid phases use the same model. We do not modify the general equilibrium statement from Eq. (1.50) because we can calculate the fugacity coefficients directly after we choose a particular EOS.

There are many types of EOS with a wide range of complexity. The RK model is a popular EOS that relies only on critical temperatures and critical pressures of all components to compute equilibrium properties for both liquid and vapor phases. However, the RK EOS does not represent liquid phases accurately and is not widely used, except as a method to compute vapor fugacity coefficients in activity coefficient approaches. On the other hand, the Benedict–Webb–Rubin–Starling (BWRS) EOS [6] has up to 16 constants specific for a given component. This EOS is quite complex and is generally not used to predict properties of mixture with more than few components.

For the purposes of refinery reaction and fractionation modeling, the most useful EOS models derive from either the Peng–Robinson (PR) EOS [6] or the Soave–Redlich–Kwong (SRK) EOS [6]. Both the PR and SRK are examples of cubic equations of state. Cubic EOSes are quick and easy to use for modeling work and provide a good balance between thermodynamic robustness and prediction accuracy. In our work, we have used the PR EOS with good results throughout many reaction and fractionation processes in refineries. More advanced EOS models are available in the context of refinery modeling, but we limit the scope of our discussion to the PR EOS.

We give the basic form of the PR EOS in Eq. (1.65). The PR EOS requires three main properties: critical temperature, critical pressure, and acentric factor.

$$a_i = 0.45724 R^2 \frac{T_{ci}^2}{P_{ci}} \tag{1.59}$$

$$b_i = 0.07780 R \frac{T_{ci}}{P_{ci}} \tag{1.60}$$

$$\alpha_i = [1 + (0.37464 + 1.5426\omega_i - 0.26992\omega_i^2)(1 - T_{ri}^{0.5})]^2 \tag{1.61}$$

$$a\alpha_{\mathrm{MIX}} = \sum \sum x_i x_j (a\alpha)_{ij} \tag{1.62}$$

$$b_{\mathrm{MIX}} = \sum x_i b_i \tag{1.63}$$

$$a\alpha_{ij} = \sqrt{a\alpha_{ii} a\alpha_{jj}}(1 - k_{ij}) \tag{1.64}$$

$$P = \frac{RT}{V_{\mathrm{MIX}} - b_{\mathrm{MIX}}} - \frac{a\alpha_{\mathrm{MIX}}}{V_{\mathrm{MIX}}^2 + 2b_{\mathrm{MIX}}V_{\mathrm{MIX}} + b_{\mathrm{MIX}}^2} \tag{1.65}$$

where V_{MIX} is the molar volume of the mixture and k_{ij} is an interaction parameter for each *i* and *j* pair of components. The critical properties and interaction parameters for a large number of pure components are available within most process modeling software. We discussed how to obtain the critical properties of pseudocomponents in Section 1.10.4. In general, we can set the interaction parameters for pseudocomponents to 0 without significantly changing model results. Riazi [4] discussed several correlations to estimate the interaction parameters as functions of critical volumes of the components.

The EOS approach is robust and can generate the vapor pressure, heat of vaporization, liquid density, and liquid heat capacity using standard thermodynamic relationships and basic information such as critical properties and ideal gas heat capacities for all components. We refer the reader to the excellent text by Poling *et al.* [17] where there are detailed formulas for all these derived properties from the EOS directly. In general, the PR EOS makes good predictions of equilibrium distributions for light and medium boiling components. In addition, we ensure thermodynamic consistency by design as we use the same model for the vapor and liquid phases. The PR EOS also generates mostly acceptable predictions for vapor and liquid enthalpies and displays good behavior near the critical point.

A key shortcoming in the EOS approach (specifically PR) is that predictions of liquid density are quite poor and not sufficient for process modeling. The most popular method to deal with this problem is to ignore liquid density prediction from the EOS and use COSTALD method described in Section 1.10.5 to provide accurate density predictions. With similar reasoning, some process modeling software programs replace the enthalpy methods of EOS with Lee–Kesler correlations for heat capacity and enthalpy. However, this is not entirely necessary given the inaccuracies in the pseudocomponent physical property predictions themselves. Finally, the presence of very light components such as hydrogen and helium can sometimes provide spurious results. Aspen HYSYS includes several modifications (shown in Figure 1.58) for light components to prevent undesired behavior of light components. In general, we recommend using the EOS approach when developing refinery reaction and fractionation process models.

1.12 Miscellaneous Physical Properties for Refinery Modeling

In addition to thermophysical properties required for modeling purposes, a complete model must also make predictions regarding several fuel properties routinely measured at the refinery. Typically, these fuel or product properties include measurements such as flash point, freeze point, cloud point, and PNA content. These properties not only serve as indicators of product quality and distribution but may also be limited by government or internal refinery regulations. We can often justify the use of process modeling in the refinery by making sure that models also include predictions of these useful fuel properties. We will briefly discuss two approaches in this area and give concrete examples with flash point, freeze point, and PNA content. We choose these particular properties because they display characteristics common to many types of fuel property correlation methods. We refer the reader to API standards [2] and Riazi [4] for more detailed expositions on various types of correlations for fuel properties not discussed in this section.

1.12.1 Two Approaches for Estimating Fuel Properties

Fuel or product properties can be a complex function of feed composition, process conditions, and analysis method. It is generally not possible to take into account all of these variables when estimating fuel properties.

The simplest approach is to correlate the relevant fuel property against modeled or measured bulk properties. For example, the flash point maybe correlated with the 10% point of the ASTM-D86 curve. We can obtain the required distillation curve from the pseudocomponent stream composition. The software accomplishes this task by arranging pseudocomponents in an ascending order of boiling point and creating a running cumulative sum of the liquid fractions of these pseudocomponents. This process results in the TBP curve of a given stream. Most software programs (including Aspen HYSYS) include methods to automatically convert this TBP curve into ASTM D86 or D1160 curve. Once we obtain this distillation curve, we can use several correlations to estimate flash point, freeze point, and so on. This method is simple to use and adaptable to any process simulator. However, this method relies on the availability of good correlations. It is important to remember that such correlations may not be valid or accurate for refineries that process frequently changing feedstocks.

A second approach is to use indexes based on pseudocomponent compositions. In an index-based approach, we represent each fuel property using the following equation:

$$\mathrm{PROP}_{\mathrm{MIX}} = \sum_{i=0}^{N} \mathrm{PROP}_i w_i \tag{1.66}$$

where $\mathrm{PROP}_{\mathrm{MIX}}$ represents a given fuel property; PROP_i represents the property index for pseudocomponent i; w_i corresponds to the liquid, molar, or weight fraction; and N is the total number of pseudocomponents. Process modeling software tools and the literature have used this approach to quantify fuel properties such as octane numbers. An important advantage of this approach is that we can tune the property prediction to a particular plant by modifying the value of PROP_i. This allows the model user to track plant performance accurately. This method is also very useful while attempting to correlate the flash point of various blends of fuels. However, this approach is generally not portable across various process modeling software programs and requires a large initial data set to regress starting values for PROP_i. In addition, there is a danger of overfitting these values to match plant performance. Overfitting the property indexes renders the model less useful for predictive purposes. In our work, we have used both approaches with equal success. However, for simplicity, we recommend the first approach; especially in light of the fact that large sets of data may not be available for determining initial PROP_i values.

1.12.2 Flash Point

The flash point of a fuel typically refers to the temperature at which the fuel can ignite in the presence of an ignition source and sufficient air. A low flash point is an important consideration for gasoline engines as "sparking" or igniting the gasoline fuel is critical to optimum engine performance. In contrast, engines that use diesel and jet fuels do not rely on ignition (but on compression) and require fuels with a high flash point. The API [2] has correlated numerous data for a variety of fuels and found that the open- and closed-cup flash points (alternative measurement methods) linearly correlate well with the 10% ASTM-D86 distillation temperature.

The flash point correlation is

$$\mathrm{FP} = A(\mathrm{D86}_{10\%}) + B \tag{1.67}$$

where FP is the flash point measured in degree Fahrenheit and $\mathrm{D86}_{10\%}$ refers to the 10% distillation temperature measured in degree Fahrenheit. A and B are specific constants for various feed types. Typical values of A and B are 0.68–0.70 and 110–120, respectively. We recommend performing a simple linear regression to tune existing measurements into this correlation. API notes that this correlation may be improved using the 5% distillation temperature instead of the 10% distillation temperature. Deviations of 5–7 °F are within the tolerance of this correlation.

1.12.3 Freeze Point

The freeze point refers to the temperature at which solid crystals start to appear as a given fuel sample is being cooled. The freeze point dictates how a given fuel may be sold and if additives or blendings are required to ensure that the fuel does not clog engines at low ambient temperatures. A related concept is the cloud point, which is the temperature at which the sample takes a cloudy appearance. This is due to the presence of paraffins, which solidify at a higher temperature than other components. The freeze point and cloud point do not correlate well with each without considering the paraffin content of the stream. The API [2] has correlated freeze point as follows:

$$\mathrm{FRP} = A(\mathrm{SG}) + B(K_{\mathrm{w}}) + C(\mathrm{MeABP}) + D \tag{1.68}$$

where FRP is the freeze point in degree Fahrenheit, SG is the specific gravity, K_{w} is the Watson K factor, and MeABP refers to the mean average boiling point. A, B, C, and D refer to specific constants for a given fuel composition. Typical values for A, B, C, and D are 1830, 122.5, −0.135, and −2391.0, respectively. We can also fix the value of K_{w} to a constant (roughly 12) for narrowly distributed petroleum cuts. We can calculate the value of MeABP using the spreadsheet procedure described in Section 1.4. It is important to compare this correlation to that for the flash point. This correlation uses more bulk measurements (SG and K_{w}) to capture the effect of feed composition on the freeze point.

1.12.4 PNA Composition

The last sets of correlations we discuss are composition correlations. These correlations identify chemical composition in terms of total PNA content of a particular feed based on key bulk measurements. These correlations are useful in two respects. First, we use these correlations to screen feeds to different refinery reaction units. For example, we may wish to send a more paraffinic feed to a reforming process when we need to increase the yield of aromatic components from the refinery. Second, these types of correlations form the basis of more detailed lumping for kinetic models that we discuss in detail in subsequent chapters of this book. We use these types of correlations to build extensive component lists that we can use to model refinery reaction processes.

Compositional information is quite useful to the refiner, and many correlations are available in the literature that attempt to correlate PNA content with various bulk measurements. In general, these correlations rely on density or specific gravity, molecular weight, distillation curve, and one or more viscosity measurements. The *n–d–M* (refractive index, density, and molecular weight) [1], API/Riazi-Daubert [2, [4]], and TOTAL [19] correlations are just a few of the correlations available. The Riazi–Daubert correlation relies on the most directly observed information and we expect it to show the smallest deviation from measured values. The other correlations require parameters (aniline point, etc.) that may not be routinely measured for all feeds. The Riazi–Daubert correlation takes the form

$$\%X_{\mathrm{P}}\,or\,\%X_{\mathrm{N}}\,or\,\%X_{\mathrm{A}} = A + B\cdot R_{\mathrm{i}} + C\cdot \mathrm{VGC}' \tag{1.69}$$

where $\%X$ represents the percent molar or volumetric composition of paraffins, naphthenes, or aromatics (based on the subscript chosen); R_{i} is the refractive index; and VGC′ is the viscosity gravity constant or viscosity gravity factor defined in ASTM D2501-14. Coefficients *A*, *B*, and *C* take on different values based on whether an aromatic, naphthene, or paraffin is chosen as the subscript. This correlation can provide reasonably accurate results when we know the values of key input parameters with high accuracy. Overall, this method indicates a 6–7% absolute average deviation (AAD) from known measurement test cases.

We have extended the correlation by Riazi [1] to include the specific gravity, refractive index, and the stream viscosity. Our updated correlation is given by

$$\%X_{\mathrm{P}}\,\mathrm{or}\,\%X_{\mathrm{A}} = A + B\cdot \mathrm{SG} + C\cdot R_i + D\cdot \mathrm{VGC}' \tag{1.70}$$

$$\%X_{\mathrm{N}} = 1 - (X_{\mathrm{P}} + X_{\mathrm{A}}) \tag{1.71}$$

In the equations, $\%X$ represents the percent molar or volumetric composition of paraffins (P), naphthenes (N), or aromatics (A) (based on the subscript chosen); SG is the specific gravity; R_{i} is the refractive index; and VGC′ is the viscosity gravity constant or viscosity gravity factor. In addition, the constants *A* to *D* are given for paraffins and naphthenes and for each fuel type. We list our updated constants in Tables 1.6 and 1.7. We also group the constants in this updated correlation by boiling point ranges (light naphtha, etc.). This correlation reproduces plant data with 3–4% AAD, which is a significant improvement over the Riazi–Daubert

Table 1.6 Coefficients for paraffin content in petroleum fractions.

	Paraffin (vol%)				
Boiling point range	A	B	C	D	AAD
Light naphtha	311.146	−771.335	230.841	66.462	2.63
Heavy naphtha	364.311	−829.319	278.982	15.137	4.96
Kerosene	543.314	−1560.493	486.345	257.665	3.68
Diesel	274.530	−712.356	367.453	−14.736	4.01
VGO	237.773	−550.796	206.779	80.058	3.41

Table 1.7 Coefficients for aromatic content in petroleum fractions.

	Aromatic (vol%)				
Boiling point range	A	B	C	D	AAD
Light naphtha	−713.659	−32.391	693.799	1.822	0.51
Heavy naphtha	118.612	−447.589	66.894	185.216	3.08
Kerosene	400.103	−1500.360	313.252	515.396	1.96
Diesel	228.590	−686.828	12.262	372.209	4.27
VGO	−159.751	380.894	−150.907	11.439	2.70

correlation. We show how the grouping constants by boiling point ranges can be useful while creating kinetic lumping procedures for the FCC in Chapter 4.

1.13 Conclusion

This chapter discusses several key modeling steps regarding the characterization and the thermophysical properties of crude oil and petroleum fractions. The basic process for developing a set of pseudocomponents for modeling refinery fractionation systems is as follows:

1) The feed to the fractionation system is often poorly defined in terms of actual components. We may only have an assay and associated bulk property measurements (such as density). We use the techniques discussed in Sections 1.1.1–1.4 to produce a complete TBP distillation curve and a density or specific gravity distribution.
2) Once we obtain the TBP and density curve, we can cut the components into a number of pseudocomponents. Each of these pseudocomponents has at least a TBP and a density, by definition. The number of pseudocomponents for each cut point range can vary depending on the product range of the fractionation system. We have suggested the number of pseudocomponents for a few product ranges in Table 1.2. Subsequent chapters include more information for specific fractionation systems.

3) After obtaining the pseudocomponents, we decide how to model key physical properties (Section 1.10.1) for these components. Process modeling software often includes a large variety of correlations and estimation methods. However, for almost all cases, the Lee–Kesler correlations for critical properties and ideal gas heat capacities are sufficient. We have used the extended Twu correlation for molecular weight in our work. After obtaining the critical properties and molecular weight for a given pseudocomponent, we may estimate all other required properties (heat capacities, etc.) with correlations given by Riazi [1].
4) We also select a thermodynamic model to represent vapor–liquid equilibrium for these pseudocomponents. For crude fractionation columns, an EOS approach yields good results. However, an EOS approach does not predict liquid densities accurately and tends to give poor equilibrium predictions of heavy pseudocomponents. We can improve the EOS density predictions with more accurate density correlations such as COSTALD, Eq. (1.25). If the feed and products contain significant amounts of heavy products, it may be better to rely on empirical thermodynamic models such as Grayson–Streed or BK-10.
5) Lastly, we must make sure to use the product pseudocomponent information to verify measured product properties. In this chapter, we discuss the flash point, freeze point, and chemical composition properties of the products. The reader may find additional correlations for other fuel properties from the API handbook [2] and work by Riazi [1].

Although this chapter has focused extensively on the requirements for modeling fractionation systems, we can use the same techniques in the context of modeling refinery reaction process as well. We illustrate this process in Chapters 4–7. It is possible to obtain good predictive results for fractionation systems provided we make reasonable choices for the thermodynamics and physical properties of the pseudocomponents involved.

Nomenclature

A, B, α, β	Fitting parameters for cumulative beta distribution
CP_{IG}	Ideal gas heat capacity, J/mol K
CP_L	Liquid heat capacity, J/mol K
δ	Solubility parameter, $(J/cc)^{0.5}$
$\overline{\delta}$	Mean weighted solution solubility parameter, $(J/cc)^{0.5}$
$D86_{10\%}$	10% ASTMD86 distillation point, °F
FP	Flash point, °F
FRP	Freeze point, °F
γ	Activity coefficient, unitless
ΔH_{VAP}	Heat of vaporization, J/mol
ΔH_{VAP}^{NBP}	Heat of vaporization at normal boiling point temperature, J/mol
K_i	K-value, ratio of y_i/x_i, unitless
K_w	Watson K factor, unitless

K_{avg}	Watson K factor, unitless
k_{ij}	Interaction parameter for component i and component j in PR EOS, unitless
MeABP	Mean average boiling point temperature, K
MW	Molecular weight, g/mol
P	Pressure, bar
P_c	Critical pressure, bar
P_r	Reduced pressure = P/P_c, unitless
P^{SAT}	Saturation or vapor pressure, bar
PF_i	Poynting correction factor, unitless
$PROP_{MIX}$	Mixture of indexed fuel properties
$PROP_i$	Fuel property index for a given component
φ_i^V	Vapor phase fugacity coefficient for component i
φ_i^{SAT}	Liquid phase fugacity coefficient corrected to saturation pressure for component i
φ_i^L	Liquid phase fugacity coefficient for component i
R	Universal gas constant, 8.315 J/mol K
T	Temperature, K
T_c	Critical temperature, K
T_r	Reduced temperature = T/T_c, unitless
T_b	Boiling point temperature, K
T_{br}	Reduced boiling point temperature = T_b/T_c, unitless
ρ_L	Liquid density, g/cc
ρ_P	Liquid density at pressure P, g/cc
ρ_{P^o}	Liquid density at reference pressure P^o, g/cc
R_i	Refractive index, unitless
SG	Specific gravity, unitless
V^{SAT}	Molar volume of saturated liquid, cc/mol
V_i	Molar volume of component i as a function of temperature and pressure, cc/mol
VGC′	Viscosity gravity constant or viscosity gravity factor, unitless
w_i	Weighting factor for property index mixing
$\%X_P$	Molar or volumetric composition of paraffins
$\%X_N$	Molar or volumetric composition of naphthenes
$\%X_A$	Molar or volumetric composition of aromatics
x_i	Liquid phase composition of component i
y_i	Vapor phase composition of component i
Z_{RA}	Rackett parameter, unitless
ω	Acentric factor, unitless

Bibliography

1 Kaes, G.L. (2000) *Refinery Process Modeling. A Practical Guide to Steady State Modeling of Petroleum Processes*, The Athens Printing Company, Athens, GA.

2 Daubert, T.E. and Danner, R.P. (1997) *API Technical Data Book – Petroleum Refining*, 6th edn, American Petroleum Institute, Washington, DC.

3 Bollas, G.M., Vasalos, I.A., Lappas, A.A., Iatridis, D.K., and Tsioni, G.K. (2004) Bulk molecular characterization approach for the simulation of FCC feedstocks. *Industrial and Engineering Chemistry Research*, **43**, 3270.
4 Riazi, M.R. (2005) *Characterization and Properties of Petroleum Fractions*, 1st edn, American Society for Testing and Materials, West Conshohocken, PA.
5 Sanchez, S., Ancheyta, J., and McCaffrey, W.C. (2007) Comparison of probability distribution functions for fitting distillation curves of petroleum. *Energy & Fuels*, **21**, 2955.
6 Walas, S.M. (1985) *Phase Equilibria in Chemical Engineering*, Butterworth-Heinemann, Burlington, MA.
7 Chen, C.C. and Mathias, P.M. (2002) Applied thermodynamics for process modeling. *AIChE Journal*, **48**, 194.
8 de Hemptinne, J.C. and Behar, E. (2006) Thermodynamic modeling of petroleum fluids. *Oil and Gas Science and Technology*, **61**, 303.
9 Lee, B.I. and Kesler, M.G. (1975) A generalized thermodynamic correlation based on three-parameter corresponding states. *AIChE Journal*, **21**, 510.
10 Kesler, M.G. and Lee, B.I. (1976) Improve prediction of enthalpy of fractions. *Hydrocarbon Processing*, **55**, 153.
11 Twu, C.H. (1984) An internally consistent correlation for predicting the critical properties and molecular weights of petroleum and coal-tar liquids. *Fluid Phase Equilibria*, **16**, 137.
12 Rackett, H.G. (1970) Equation of state for saturated liquids. *Journal of Chemical and Engineering Data*, **15**, 514.
13 Yamada, T.G. (1973) Saturated liquid molar volume. The Rackett equation. *Journal of Chemical and Engineering Data*, **18**, 234.
14 Spencer, C.F. and Danner, R.P. (1972) Improved equation for the prediction of saturated liquid density. *Journal of Chemical and Engineering Data*, **2**, 236.
15 Thomson, G.H., Brobst, K.R., and Hankinson, R.W. (1982) An improved correlation for densities of compressed liquids and liquid mixtures. *AIChE Journal*, **28**, 671.
16 Cheuh, P.L. and Prausnitz, J.M. (1969) A generalized correlation for the compressibilities of normal liquids. *AIChE Journal*, **15**, 471.
17 Poling, B.E., Prausnitz, J.M., and O'Connell, J.P. (2000) *Properties of Gas and Liquids*, 5th edn, McGraw-Hill, New York.
18 Agarwal, R., Li, Y.K., Santollani, O., Satyro, M.A., and Vieler, A. (2001, May) Uncovering the realities of simulation. *Chemical Engineering Progress*, **42**.
19 Sadeghbeigi, R. (2000) *Fluid Catalytic Cracking Handbook. Design, Operation and Troubleshooting of FCC Facilities*, Gulf Publishing Company, Houston, TX.
20 Mohan, S.R. (2016) *Five Best Practices for Refineries: Maximizing Profit Margins through Process Engineering*, https://www.aspentech.com/White-Paper-Five-Best-Practices-Refineries.pdf.
21 Niederberger, N. (2009) *Modeling, Simulation and Optimization of Refining Processes*, https://www.slideshare.net/Bioetanol/wks-biorefinery-jacques-niederbergerpetrobras-refino.
22 Wu, Y. (2010) Molecular modeling of refinery operations. Ph.D. dissertation. University of Manchester, Manchester, United Kingdom, https://www.escholar

.manchester.ac.uk/api/datastream?publicationPid=uk-ac-man-scw:93706&datastreamId=FULL-TEXT.PDF.

23 Mullick, S., Dooley, K., Dziuk, S., and Ajikutira, D. (2012) *Benefits of Integrating Process Models with Planning and Scheduling in Refining Operations*, https://www.scribd.com/document/269573466/Integrating-Process-Models-With-Refining-PandS.

24 Aspen, Technology, Inc. (2012) *Molecule-Based Characterization Methodology for Correlation and Prediction of Properties for Crude Oil and Petroleum Fractions* (Molecular Characterization White Paper), https://origin-www.aspentech.com/Molecular_Characterization_White_Paper.pdf.

2

Atmospheric or Crude Distillation Unit (CDU)

2.1 Introduction

Crude distillation is the oldest and most important part of any refinery. The distillation of crude provides refined products such as gasoline and diesel for direct sale and feedstocks to other units in the refinery. With the advent of large, highly integrated refineries, it is critical to understand the operation of major units and predict desirable and undesirable changes in unit performance as a function of key operating conditions. Recent emphasis on nontraditional crude feeds, reduced energy consumption, operational optimization, production planning modeling, and CO_2 release provides additional impetus to clearly understand the relationship among process variables, feed qualities, and product profiles [1].

Crude distillation has a long history and refiners have developed hundreds of empirical correlations for feed conditions and process variables. It is very difficult to develop general correlations applying a variety of operating scenarios that modern refineries face. Many of these correlations are proprietary or not widely available. These correlations are often not very useful when current unit operations and feed conditions vary significantly from the conditions used for their development. Often, experienced engineers and operators can make good estimates of unit performance. However, the growing tide of retiring professionals and loss of experience throughout the industry make this difficult or impossible.

Given the above issues, the use of simulation tools and techniques becomes invaluable. In fact, refineries were one of the first users of computational models to improve process operation. The rapid pace of advance of computer hardware and software has enabled an engineer to develop a multitude of models for most processes in the refinery. Although the task of building a model is not difficult now, the ability to build a model that accurately reflects the plant operation and has predictive capability remains elusive. We must always remember a fundamental modeling premise "GARBAGE IN = GARBAGE OUT."

It is in this context that we present the bulk of this chapter. We discuss how to model existing crude units, including relevant data collection and validation, estimation of missing data, model development and validation, and model

Petroleum Refinery Process Modeling: Integrated Optimization Tools and Applications,
First Edition. Y. A. Liu, Ai-Fu Chang, and Kiran Pashikanti.

applications in the form of case studies. This chapter summarizes our own experiences in refinery modeling from plant data and related work presented in the open literature.

2.2 Scope of the Chapter

In this chapter, we address several important issues relevant to the simulation of atmospheric or crude distillation units (CDUs).

- Overview of CDU (Section 2.3.1)
  Description of a modeler's view of atmospheric crude distillation and recommendation of techniques for column efficiency and calculation convergence for solving refinery distillation models (Section 2.4.1).
- Characterization of the feed to a CDU (Section 2.6)
- Discussion of key data required and estimates for missing or incomplete data (Section 2.7)
- Illustration of representative CDU data for modeling purposes (Section 2.8)
- Building a model based on plant data using Aspen HYSYS (Sections 2.8.2–2.8.4)
- Initializing and converging models successfully (Sections 2.8.5–2.8.8)
- Validating the model predictions with plant data (Section 2.9)
- Industrially relevant case studies that focus on improving the profitability, yields, and predictability (Section 2.10.1)
- Hands-on workshops on model building using backblending procedure (Section 2.11.1), investigating new product profiles with new product demands (Section 2.12.1), studying the effects of process variables on product qualities (Section 2.14), and application of column internal tools (column hydraulic analysis) (Section 2.15)
- Conclusion (Section 2.16), Nomenclature and Bibliography.

2.3 Process Overview

Figure 2.1 shows the general process flow in the initial distillation and product recovery of a refinery. The solid lines refer to primary *material flows* and the dashed lines refer to *energy flows*. Crude from multiple sources (storage, pipeline, etc.) enters the refinery after some initial treatment to remove impurities and sediments. This crude enters an initial heat recovery section to raise its temperature and recover heat from downstream units. The heated crude then enters a desalting section, which removes dissolved salts and associated impurities. Once the salts (and the water associated with the salt removal process) have been sufficiently treated, the crude enters the preheat train, consisting of heat exchangers associated with various downstream equipment throughout the refinery. The preheat train typically raises the temperature of crude significantly and reduces the overall energy consumption in the refinery. Following the preheat train, the crude enters the primary crude furnace. The crude furnace vaporizes a major portion of the crude and feeds this vapor–liquid mixture into

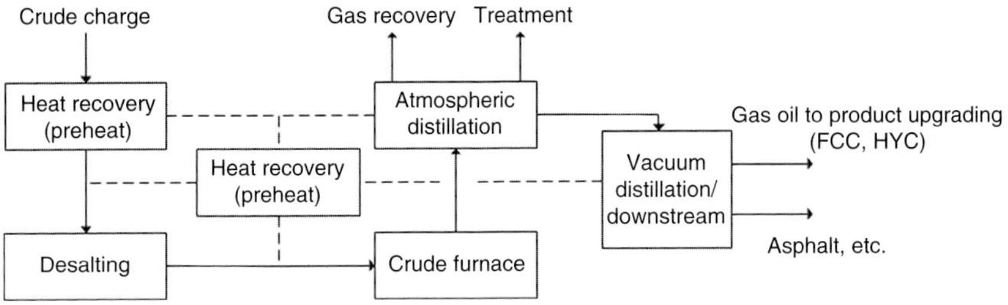

Figure 2.1 General process flow of initial crude processing.

the atmospheric distillation tower. Most refineries today recover the products and send them for further processing into vacuum distillation section and product upgrading sections (catalytic cracking, hydrocracking, reforming, etc.).

Each of the processes shown in Figure 2.1 is quite extensive and can be surprisingly complex in an integrated refinery. In this work, we limit the scope by presenting brief summaries of the units and how to quantitatively simulate the performance of each unit in a modeling context. Our specific focus in this chapter is how to model the atmospheric distillation section.

2.3.1 Desalting

Figure 2.1 shows the basic flow of primary distillation process in modern refineries. Before the crude enters the actual crude distillation column, it must go through several steps to ensure reliable operation. The main steps are as follows:

- Desalting
- Dewatering
- Solid removal.

Most crude contains appreciable levels of salts (20–500 ppm) [9]. It is critical to remove these salts to prevent fouling and scaling of heat transfer surfaces. Loss of heat transfer efficiency can significantly increase the energy required for distillation.

Figure 2.2 shows a typical crude desalter. The charge crude from the storage is heated to a particular salt removal temperature (around 80–150 °C) [9]. Large quantities of water come in contact with the crude. The salts will dissolve preferentially in the water until its saturation point. Next, the water coalesces to large water droplets in the presence of a strong electric field. The droplets begin to settle out from the oil due to gravity. Refiners processing a variety of crudes may include several desalting stages to ensure that the process effectively reduces salts to a minimal level. Due to the presence of other impurities in the crude, refiners may also add de-emulsification agents to prevent forming of oil–water emulsions.

The desalting process is mostly driven by thermodynamic and hydrodynamic constraints. For the purpose of modeling the CDU, we consider the desalting operation as a simple component splitter that removes any water present in the feed crude. Desalting and dewatering processes are very effective and do not consume significant resources compared to other units, so this simple model representation is justified.

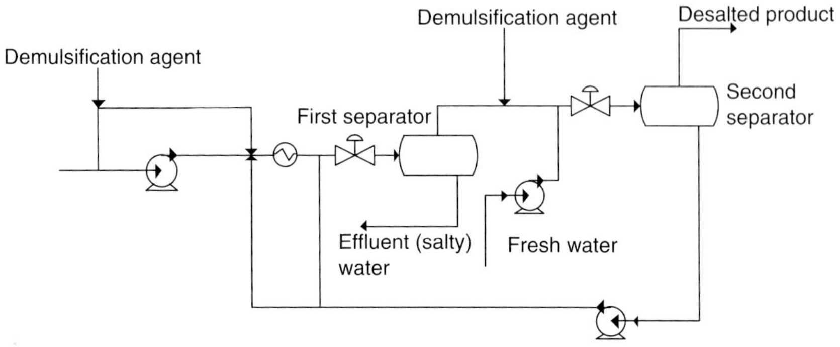

Figure 2.2 Simplified desalting and dewatering process.

2.3.2 Preheat Train and Heat Recovery

The CDU consumes 20–30% [13] of the total energy required to distill a given crude into products. Therefore, it is critical to optimize and recover as much heat as possible from heat streams throughout the refinery to optimally heat and vaporize the crude for fractionation. The preheat train consists of heat exchangers that incrementally heat up the crude feed using hot streams from the crude distillation and other downstream units in the refinery. Crude exits in the preheat train at around 250 °C [13] (Figure 2.3).

Modeling and optimizing the preheat train can be a significant undertaking. Although process simulators can handle the complexity of the network, we require additional tools to optimize this heat exchanger network (e.g., Aspen Technology's Aspen Energy Analyzer). There has been significant work in this area, and impressive savings are possible using pinch technology [11] and mathematical optimization methods [12]. The methods are not the focus of our book; hence, in our simulations, we simplify the preheat train and model it as a simple heater with a variable heat duty.

2.3.3 Atmospheric Distillation

After leaving the preheat train, the heated crude enters the atmospheric (or topping) crude distillation furnace (Figure 2.4). The main purpose of the furnace is to vaporize the portion of the crude that we recover as products from the column. Typically, we set the heat (and outlet temperature) of the furnace so that the amount of crude vaporized equals the sum of the products recovered from the column plus a small percentage. This small percentage excess (typically in the range of 2–10%) [8, 9, 14] is called *overflash*. The overflash indicates the amount of the heavy residue that will be distributed into the lighter products, increasing the D86 95% point of these products. This vapor–liquid crude mixture enters the column around 380–410 °C [14] and immediately flashes in the bottom few trays of the column (flash zone). Most column configurations also include a significant amount of steam at the bottom. This steam serves to strip any residue and prevent excessive thermal cracking of crude due to high temperatures.

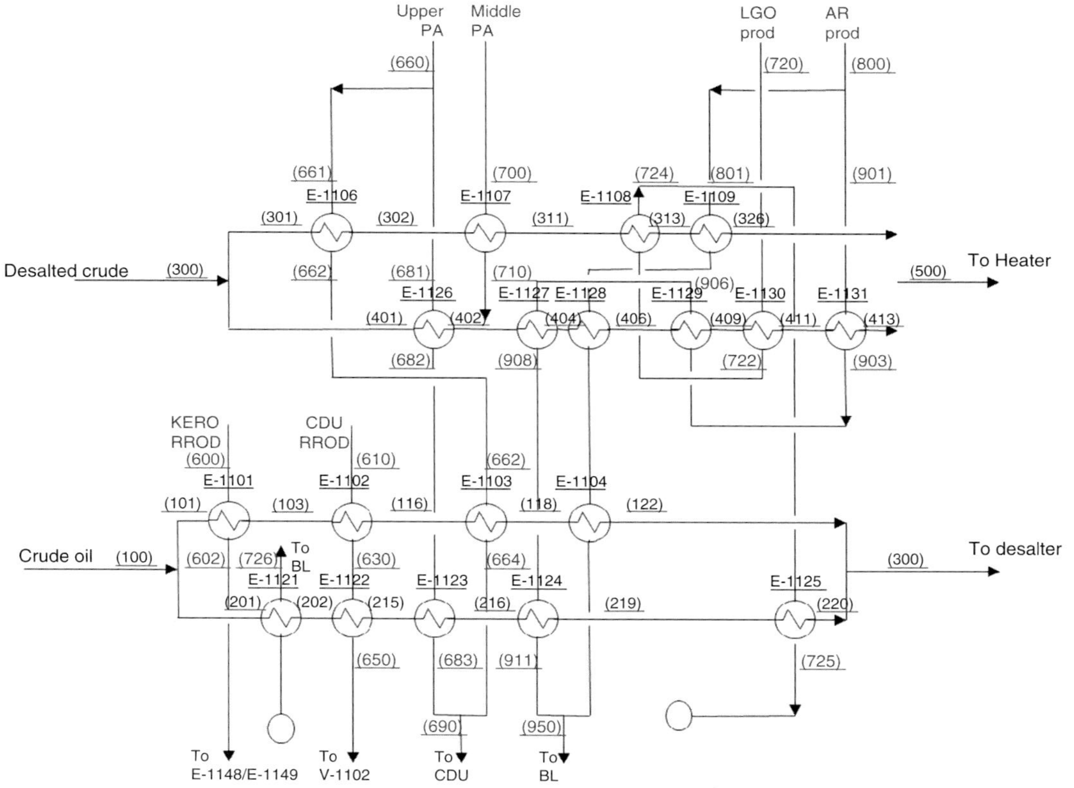

Figure 2.3 Typical preheat train in refineries.

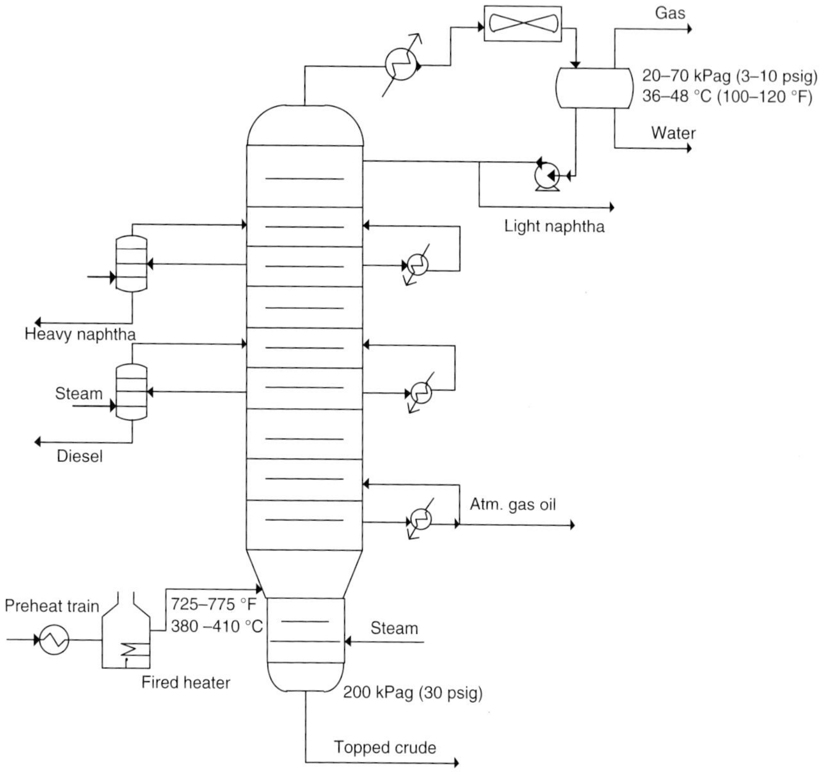

Figure 2.4 Typical flow of a refinery atmospheric distillation unit.

The column typically has 50–60 [3, 14] physical trays or separation stages. These do not correspond to ideal separation stages (discussed in Section 2.4.2), which is an entirely different concept. In general, there are about five stages for each side product and about 10–12 stages are required for the column bottoms and flash zone [9].

As lighter components of the crude move up the column, we draw various side streams at different locations. The draw locations represent the temperature range of the liquid products that we can collect from the given draw location. Table 2.1 shows the typical products recovered from the atmospheric column. There are many possible side draw locations and configurations given the product demand and refinery economics. The side draws typically have a low D86 5% point, indicating the presence of many light components.

Light components (i.e., pentanes and lighter) travel up the column and leave from the off gas and column condenser liquid output. The temperature of the condenser depends on the column operating pressure and other variables. The typical range is between 30 and 65 °C [14]. An additional feature in most columns is the presence of side coolers or pumparounds. These units reduce the vapor flow in the column (by lower temperature) and allow for heat recovery. Many of the exchangers in the preheat train use pumparound oils as the hot side fluid.

For simulation purpose, we represent the furnace as a simple heater with a variable duty to match the overflash specification. There are many detailed models

Table 2.1 Major products from atmospheric crude distillation unit [13].

Cut	ASTM distillation range (°C)
Light straight-run naphtha (LSR)	32–104
Heavy straight-run naphtha (HSR)	82–204
Kerosene	166–282
Light gas oil (LGO)	216–338
Atmospheric gas oil (AGO)	288–443

for the fired heaters in process simulators. Using these fired heater models can be sophisticated and is not the focus of this text. We model the crude column rigorously and include all side operations.

2.4 Model Development

The theory of modeling distillation columns is quite extensive, and many authors have written on a variety of approaches in modeling distillation columns. In general, the two major approaches for modeling columns are the rate-based approach and equilibrium-stage approach. In the rate-based approach, multistage operations are defined on the basis of rigorous heat and mass transfer rates between the vapor and liquid phases, and thermodynamic equilibrium between the vapor and liquid phases occurs at the vapor–liquid interface. This approach can be highly accurate and account for numerous phenomena including the physical layout of the column. However, this approach is also very demanding in terms of the parameter values required to produce a reasonable model.

The more traditional approach for modeling multistage operations is the equilibrium-stage approach. In this approach, we consider each stage to separate the vapor–liquid mixture based on thermodynamics and on heat and mass balance constraints alone. Figure 2.5 shows a general schematic of the

Figure 2.5 On-stage convention.

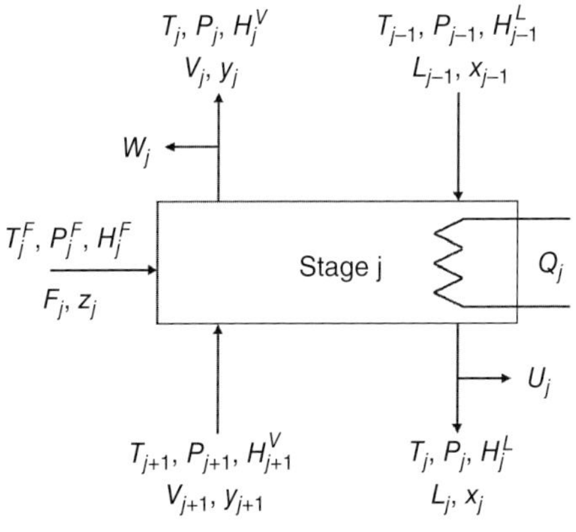

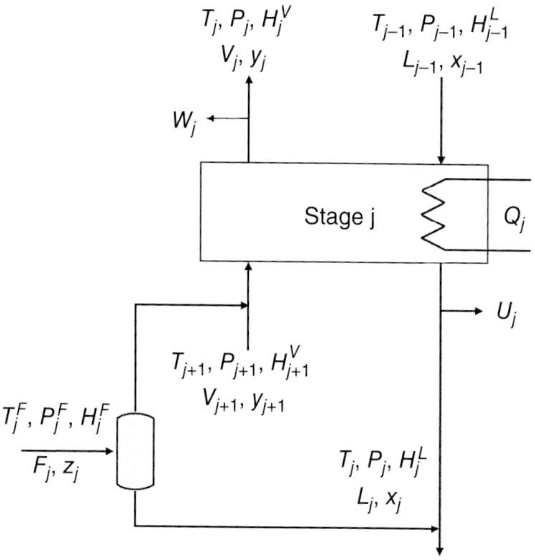

Figure 2.6 Feed flash variation.

stage-by-stage equilibrium process and a variation (Figure 2.6) that includes a flash tank to separate vapor and liquid flows (see Nomenclature).

2.4.1 MESH Equations

$$L_{j-1}x_{i,j-1} + V_{j+1}y_{i,j+1} + F_j z_{i,j} - (L_j + U_j)x_{i,j} - (V_j + W_j)y_{i,j} = 0 \quad \text{Material (M)} \tag{2.1}$$

$$y_{i,j} - K_{i,j}x_{i,j} = 0 \quad \text{Equilibrium (E)} \tag{2.2}$$

$$\sum_{i=1}^{N} x_i - \sum_{i=1}^{N} y_i = 0 \quad \text{Summation (S)} \tag{2.3}$$

$$L_{j-1}H_{L,j-1} + V_{j+1}H_{V,j+1} + F_jH_{F,j} - (L_j + U_j)H_{L,j} - (V_j + W_j)H_{V,j} - Q_j = 0 \quad \text{Heat/Energy(H)} \tag{2.4}$$

$$K_i = f(T, P, x, y) \quad K\text{-Value} \tag{2.5}$$

$$H_L = f(T, P, x, y) \quad \text{Liquid Enthalpy} \tag{2.6}$$

$$H_F = f(T, P, x, y) \quad \text{Feed Enthalpy} \tag{2.7}$$

$$H_V = f(T, P, x, y) \quad \text{Vapor Enthalpy} \tag{2.8}$$

We collectively refer to Eqs. (2.1)–(2.5) as material, equilibrium, summation, and heat equations or *MESH equations* [5, 7]. Given functions for the K-values and enthalpies and values for all the material and energy feeds, we can solve these equations with a variety of techniques. We will discuss the solution techniques in Section 2.5.

2.4.2 Overall Column Efficiency and Murphree Stage Efficiency

A key assumption in the simulation of a multistage distillation column is the equilibrium-stage assumption. For the vapor stream V_j and liquid stream L_j leaving stage j shown in Figure 2.5, the equilibrium-stage assumption implies that both streams have identical temperature T_j and pressure P_j. In addition, the mole fractions of component i leaving stage j in the vapor stream and in the liquid stream, $y_{i,j}$ and $x_{i,j}$, are in thermodynamic equilibrium and are related by the equilibrium ratio or K factor according to Eq. (2.2).

To simulate an actual column, we first convert the actual number of stages to an equivalent number of equilibrium stages by using an *overall stage efficiency*. Overall stage efficiency refers to the ratio of ideal (or theoretical) stages contained in the column (excluding the condenser and the reboiler) to the actual physical trays. If a distillation column has 20 physical trays and an overall efficiency of 0.5, we model it as a column with 10 ideal stages. *Every stage now remains in thermodynamic equilibrium.* For hydrocarbon distillation, the overall stage efficiencies are 50–90%. For absorption processes, the range is 10–50%.

Table 2.2 lists the number of theoretical stages required for each zone of the crude column. In general, the overall efficiency is roughly 0.5 and the column model contains around 28 theoretical stages (excluding the side strippers) [3].

Figures 2.14 and 2.15 compare a CDU with 56 actual physical trays (excluding side strippers) and its representation by 28 equivalent theoretical stages.

A popular method in dealing with the nonideal stage behavior in process simulators is to use an individual stage efficiency, called Murphree vapor stage efficiency.

$$E_{MVij} = (y_{i,j}-y_{i,j+1})/(y^*_{i,j}-y_{i,j+1}) \tag{2.9}$$

In the equation, subscript i refers to the component and j denotes the stage number. $y^*_{i,j}$ is the vapor mole fraction of ith component leaving stage j that would be in thermodynamic equilibrium with the liquid mole fraction of ith component leaving stage j, that is, $x_{i,j}$. $y_{i,j}$ and $y_{i,j+1}$ are actual vapor mole fractions of ith component leaving stage j and $j+1$, respectively. Further, in implementing the Murphree vapor stage efficiency in process simulators, we commonly assume that this efficiency is independent of the components involved. This simplifies Eq. (2.9) to

$$E_{MVn} = (y_n-y_{n+1})/(y^*_n-y_{n+1}) \tag{2.10}$$

where subscript n refers to stage number.

Table 2.2 Theoretical stages for each fractionation zone of a CDU [3].

Zone/location	Theoretical stages	Overall zone efficiency
Column top to naphtha	6–8	0.6
Naphtha to kerosene	4–5	0.5
Kerosene to diesel	3–4	0.5
Diesel to gas oil	4–5	0.4
Gas oil to flash zone	3–4	0.3
Flash zone to column bottoms	1–2	0.2
Total CDU, excluding side strippers	*21–28*	
Steam stripped side columns	2–3	0.3
Reboiled side stripper columns	3–4	0.5

Note that after applying the Murphree vapor stage efficiency, the vapor leaving a stage is no longer in vapor–liquid equilibrium with the liquid leaving the same stage; that is, the stage no longer corresponds to a theoretical stage in the distillation model.

2.4.3 Recommendation for Correctly Handling the Efficiency

We summarize the fundamental insights into the use of individual stage efficiency and the overall stage efficiency by Kaes [3] and Kister [5] in the following paragraphs.

Users of process simulators often enter the actual number of stages or trays from the column process flow diagram (PFD) to the simulator and enter individual stage efficiency values independent of components, Eq. (2.10), for different stages in the column such that the actual column operation is replicated. In fact, commercial simulators support this line of thinking by providing a variety of stage efficiency models. Unfortunately, many engineers do not understand the serious limitations of the individual stage efficiency models.

The nonequilibrium liquid and vapor leaving the stages cause uncertainties in heat and mass transfer calculations. In addition, individual stage efficiency models such as Murphree are too simplistic for petroleum columns, in which there are a large number of components and widely varying process conditions. As a result, there are no reports of typical individual stage efficiency values for petroleum columns. The user must then manipulate individual stage efficiency values in the column in order for the simulation results to match the plant data. This manipulation of individual stage efficiencies is similar to the regression fitting of data to an equation form. The rigorous distillation theory of predicting results other than the current column operating conditions is thoroughly masked and violated by the individual stage efficiency. This follows because the principle of equilibrium vapor and liquid leaving each stage has been violated. Therefore, the worst consequence of using individual stage efficiencies is that the simulation model would likely become useless for predicting the performance of the column at operating conditions different from those used to develop the column model.

By contrast, models based on the overall stage efficiency always correspond to rigorous distillation, as the liquid and vapor leaving a stage are in thermodynamic equilibrium. These models tend to be more effective and accurate in the prediction mode because distillation theory will always predict an answer that is directionally correct for a new set of operating conditions.

To summarize, we agree with Kaes [3] and Kister [5] and recommend against the use of individual stage efficiency models such as Murphree vapor stage efficiency. When modeling existing columns, we recommend to use a typical overall stage efficiency value to convert the actual number of trays (stages) to the equivalent number of theoretical stages.

Table 2.3 summarizes the overall stage efficiencies in refinery distillation that we have adapted from Kaes [3]. This table does not include the overall stage efficiencies for CDU given previously in Table 2.2 and those for vacuum distillation unit (VDU) given in Chapter 3 and for FCC main fractionation column given in Chapter 4.

Table 2.3 Overall stage efficiencies for refinery distillation (Adapted from [3]).

Column	Typical number of actual stages	Typical overall stage efficiency (%)	Typical number of theoretical stages
Simple absorber/stripper	20–30	20–30	4–9
Steam side stripper	5–7	30–40	2–3
Reboiled side stripper	7–10	30–40	3–4
Reboiled absorber	20–40	40–50	10–20
De-ethanizer	25–35	65–70	16–24
Depropanizer	35–40	70–80	25–32
Debutanizer	38–45	85–90	32–40
Alkylation De-*i*C4 (reflux)	75–90	85–90	64–81
Alkylation De-*i*C4 (no reflux)	55–70	55–60	30–42
Naphtha splitter	25–35	70–75	18–26
C2 splitter	110–130	95–100	104–130
C3 splitter	200–250	95–100	190–250
C4 splitter	70–80	85–90	60–72
Amine contactor	20–24	20–30	4–7
Amine stripper	20–24	45–55	9–13

2.4.4 Inside-Out Algorithm for Distillation Column Calculation Convergence

For refinery distillation column calculation convergence, we recommend the inside-out method first proposed by Boston [7] and further developed later by many others, as reviewed by Kister [5] and Seader *et al.* [8]. The inside-out method does not require significant estimates and converges robustly. This method provides quick convergence and allows for multiple subunit operations.

The inside-out algorithm starts by approximating the K-value and liquid and vapor enthalpies as simple functions of temperature and parameters that are to be fitted. For a reference component or a base component B, we write the relative volatility of component i relative to the base component B as

$$\alpha_i = \alpha_{i,B} = (y_i/x_i)/(y_B/x_B) = K_i/K_B \tag{2.11}$$

We rewrite the equation as

$$K_i = K_B \alpha_i \tag{2.12}$$

The inside-out algorithm expresses K_B as a function of temperature T and a reference temperature T_{ref}.

$$\ln K_B = A + B\,(1/T - 1/T_{\text{ref}}) \tag{2.13}$$

Next, the inside-out algorithm defines the vapor and liquid phase enthalpies, H^V and H^L, as the sum of the enthalpy of an ideal gas, H^{IG}, and the corresponding vapor and liquid phase excess enthalpies, ΔH^V and ΔH^L, and then expresses each excess enthalpy as a function of temperature and the reference temperature.

$$H^V = H^{IG} + \Delta H^V \tag{2.14}$$

$$\Delta H^V = C + D\,(T - T_{ref}) \tag{2.15}$$

$$H^L = H^{IG} + \Delta H^L \tag{2.16}$$

$$\Delta H^L = E + F(T - T_{ref}) \tag{2.17}$$

Figure 2.7 illustrates the steps involved in the inside-out algorithm. The *outer loop* does the thermodynamic calculations. These involve finding the equilibrium ratios and relative volatilities of all components on various stages of the column, based on the initial temperature and pressure profiles, and fitting the parameters *A*, *B*, *C*, *D*, *E*, and *F* in Eqs. (2.13)–(2.17) based on the predicted values of equilibrium ratios and vapor and liquid enthalpies from the selected thermodynamic model, such as Peng–Robinson equation of state. The *inside loop* does the phase equilibrium calculations and solves the equations for component and overall mass balances, and for energy balance. When the inside loop calculations fail to convergence with the specified error tolerance, the method returns to the outer loop to redo the thermodynamic calculations and update the parameters *A*, *B*, *C*, *D*, *E*, and *F*.

Commercial simulators, such as Aspen HYSYS Petroleum Refining, include the inside-out algorithm and its improved version, modified inside-out algorithm, for refinery distillation column simulations. In general, if the inside-out algorithm fails to converge, we can choose *the modified HYSYS inside-out algorithm* with an adaptive damping factor to facilitate the convergence. Figure 2.8 shows this option within Aspen HYSYS Petroleum Refining.

The term "modified" refers to the fact that the solution procedure uses a full Newton–Raphson method to converge the inner loop (i.e., the stage-by-stage mass and energy balances of the column at fixed stage temperature and pressure specified by the outer loop that focuses on phase-equilibrium calculations) of the

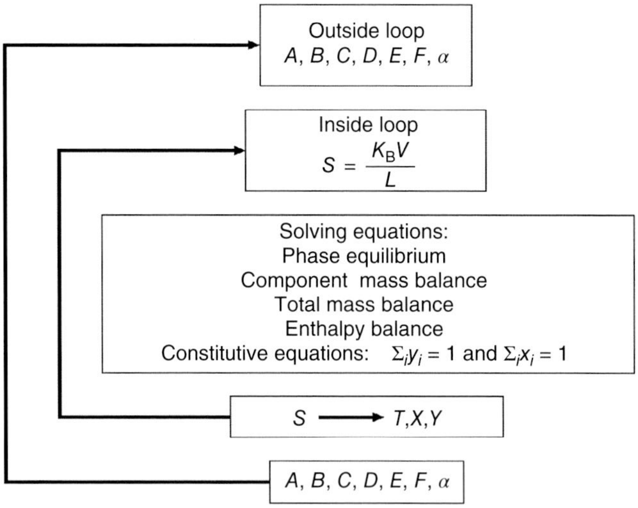

Figure 2.7 The inside-out convergence algorithm for distillation column calculations.

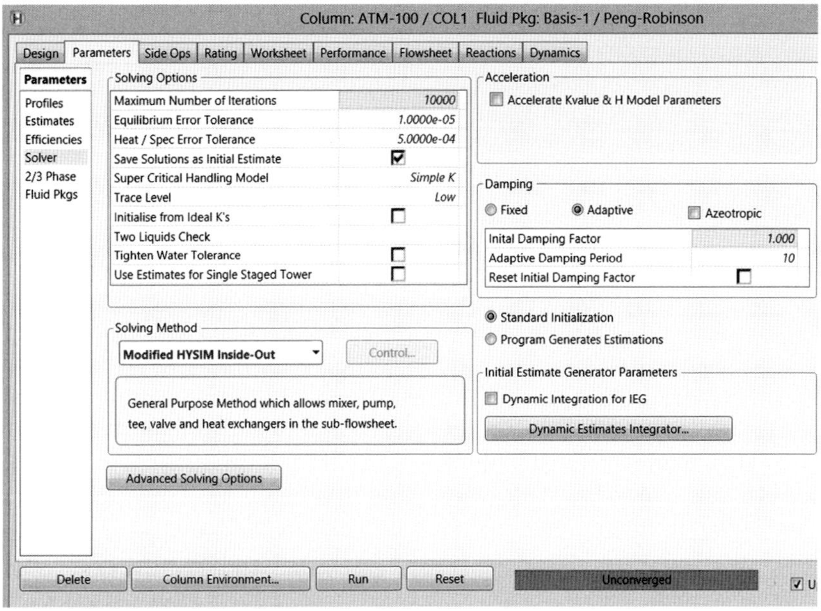

Figure 2.8 Choosing the "modified HYSYS inside-out algorithm" with an adaptive damping factor to improve distillation column convergence calculations: Parameters ->Solver ->Solver Method and Damping.

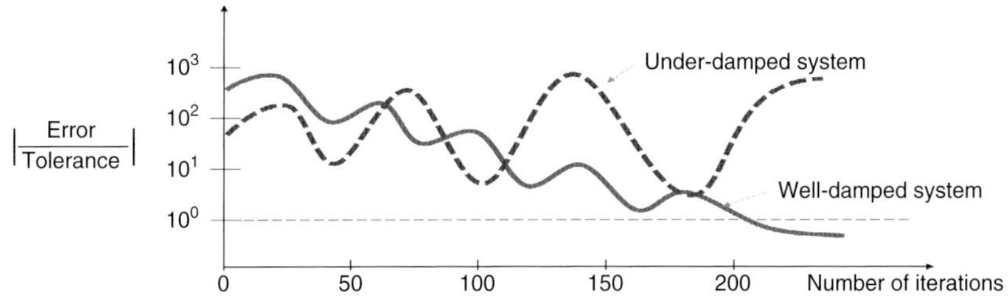

Figure 2.9 An illustration of how applying a damping factor reduces the ratio of simulation error to error tolerance to less than 1 and thus achieves convergence within a finite number of iterations in a well-damped system.

solution algorithm. Applying the damping factor often reduces the ratio of the root-mean-squared error of process variables to the error tolerance to less than 1 within a finite number of iterations and thus achieves convergence. Figure 2.9 illustrates how this ratio reduces to less than 1 within a finite number of iterations in a well-damped system after applying an adaptive damping factor, and how this ratio grows in an under-damped system.

Many software packages include additional options to speed up and improve convergence, and developers of software packages have tuned these algorithms for the best performance. We emphasize that failure of algorithms to converge is often a result of poor column specifications and not the underlying algorithm itself. We discuss valid specifications and required estimates in the subsequent sections.

2.5 Feed Characterization

Crude oil is a mixture of enormous variety of hydrocarbons derived from multiple sources and it contains hundreds of thousands of different molecules. As a result, we generally do not deal with crude in terms of molecular composition, especially in the case of crude fractionation. We indicate the composition of crude (and refined hydrocarbon products) in terms of bulk properties and distillation-based properties.

Bulk properties refer to properties measured while considering the whole crude. These properties are typically density, viscosity, refractive index, and so on and are useful (but are not sufficient) to define the crude or a cut from this crude. Distillation-based properties refer to the bulk properties measured for small amounts of crude based on that small amount's boiling point. Typically, we present these properties as a function of these small amounts as density distributions, boiling point distributions (TBP, D-2887, SimDist.), and so on. When a refiner considers a particular crude for use, the collection of bulk and distillation-based properties forms the assay of a particular crude. This assay indicates how much of a given cut (or product) we can produce for a selected boiling point range from a given crude. Tables 2.6–2.9 show crude assays for Arab Heavy and Arab Light crude.

When we work with the crude in the process simulator, we deal with specific cuts based on the boiling point distribution of a particular crude feed as shown in Figure 2.10. Each individual bar represents a hypothetical component with pseudo properties (such as critical points, heat of vaporization, and heat capacity) calculated from a correlation. These correlations typically rely on the boiling point and specific gravity or density. The goal is to find a minimum number of pseudocomponents such that the combination of these pseudocomponents approximates the properties and behavior of the entire crude.

We generally need to minimize the number of pseudocomponents to reduce the complexity of the process flowsheet. We show our recommendations for the number of pseudocomponents as a function of boiling point range in Table 2.4. The number of components in the table is greater than that recommended by

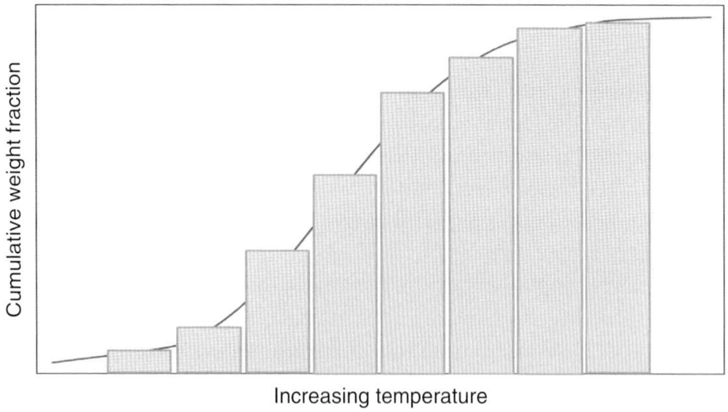

Figure 2.10 Creating pseudocomponents from boiling point distillation curve.

Table 2.4 Pseudocomponents for each boiling point range.

Boiling point range (°F)	Suggested number of hypothetical components
100–800	30
800–1200	10
1200–1600	8

Kaes [3]. We suggest increasing the number of pseudocomponents in the heavier part of the crude to account for heavy crude that most refiners process today. In addition, an increased number of cuts in the higher boiling point range will allow us to model atmospheric and vacuum distillation columns with the same component slate. The added computational requirements are not significant with today's modern computer hardware.

2.6 Data Requirements and Validation

Any modeling exercise requires a reasonable set of input data to ensure that models remain valid and predictive over a wide range of operating scenarios. The complex nature and composition of crude feeds present additional modeling complications when compared to a process that uses a well-defined component slate. Collecting the maximal possible process information is the best way to ensure that a model is valid for a variety of operating scenarios. However, it is frequently too expensive or simply infeasible to collect detailed information during regular operation of the crude tower. Consequently, we must work toward building models that do not require detailed information but also remain valid and predictive over a variety of operating scenarios.

The most important factor in the success of any model of crude columns is an accurate representation of the crude feed. There are two ways to quantify the crude feed to the unit. The first method relies on the availability of crude assays and knowledge of the ratios of the crude mixes fed to the unit. This is particularly useful when a column only processes a few types of crudes. The other method uses current column product yields and qualities to back-mix or "backblend" these products. The goal of *backblending* is to recover the composition of the crude fed to the column. This method is very useful when we have little information of the crude fed to the column or the assays are too old and unreliable. Kaes [3] provided methods to estimate missing data when backblending data for modeling.

While using the first method, it is important to recover as much information from the feed assay as possible. At a minimum, we must obtain a detailed distillation curve and density distribution. The bulk density of the crude is not sufficient to produce a reasonable set of hypothetical components. We suggest the use of the beta statistical function, Eq. (1.7), and Workshop 1.4 in Section 1.6, to fit and interpolate for missing data [3]. We can also remove irregularities in the data with this smoothing procedure. We show the results of this fitting process

	A	B	C	D	E	F	G	H	I	J	K
1	Data-fitting using the cumulative beta statistical distribution										
2	(Created by Kiran Pashikanti, Last update: 05/07/11)										
3											
4	alpha	0.325034875		θ_{XLOW}	0		Least Squares Res.	1.07E-02			
5	beta	0.40588245		θ_{XHIGH}	150		R^2	0.990155			
6	A	0.017755766		θ_{YLOW}	0.5		AAD	0.005698			
7	B	0.689926701		θ_{YHIGH}	1.2		ARD	0.7%			
8											
9	Max. X	4.9675		0.033117							
10	Min. X	100		0.666667							
11											
12	X, input var.	Y, dep. var		Norm. X	Norm. Y		Pred. Norm. Y	Pred. Y	Error	AAD	ARD
13	4.9675	0.6348		0.033117	0.192571		0.188237402	0.631766	9.20E-06	0.003034	0.004779

Figure 2.11 EXCEL spreadsheet, *Beta.xls*, for beta function data fitting.

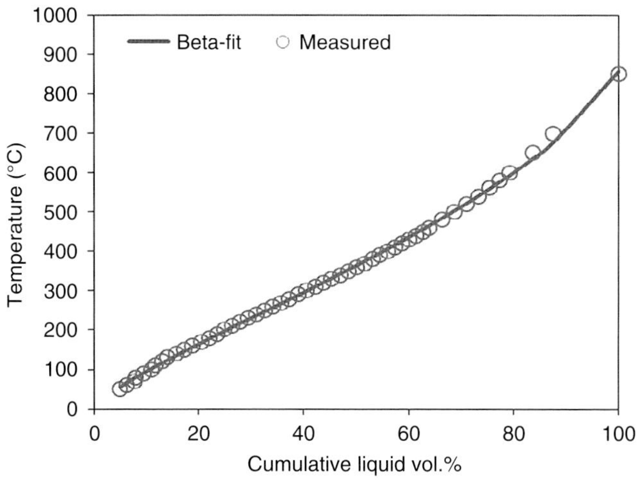

Figure 2.12 Data fit of the distillation curve as a function of liquid volume using the beta function.

in Figures 2.11–2.13. Some process simulators may provide this functionality automatically.

Once we have estimated the composition of crude feed, we must also collect data that summarize the column operating conditions and profile. Table 2.5 provides a basic list of data that we require to develop a reasonable model for the atmospheric or CDU.

One final consideration is to ensure that the collected data are consistent. This means that we must verify mass balance around the column and cannot accept

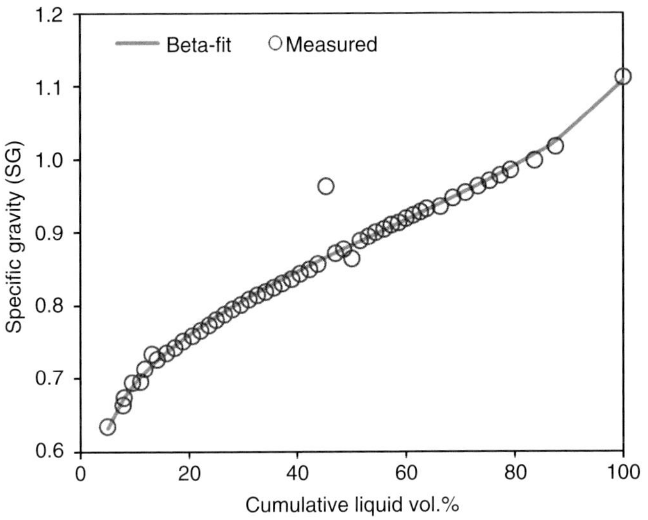

Figure 2.13 Data fit of the specific gravity as a function of liquid volume using the beta function.

Table 2.5 Basic requirements for initial column model for a CDU.

Flow rate
– Feed and product steams
– Pumparound flow rates
– Stripping steam rates
Pressure
– Flash zone
– Top of column
– Bottom column
Temperature
– Flash zone
– Top of column
– Bottom column
– Side product draw tray
– Furnace inlet and outlet temperature
– Transfer temperature
– Draw and return temperatures for all pumparounds
– Inlet and outlet temperature of all pumparound cooling streams
Analysis
– Distillation and gravity of atmospheric residue (feed)
– Distillation and gravity for all product streams
– Compositional analysis of overhead gas

yield percentages only to calculate flow rates. This may require observation of the unit over a significant period of time, in order to collect a data set that is acceptable. If this is not possible, averaging the yields and column performance over a short period may be acceptable. However, we must accept a higher threshold for error between the measured operating conditions/profiles and

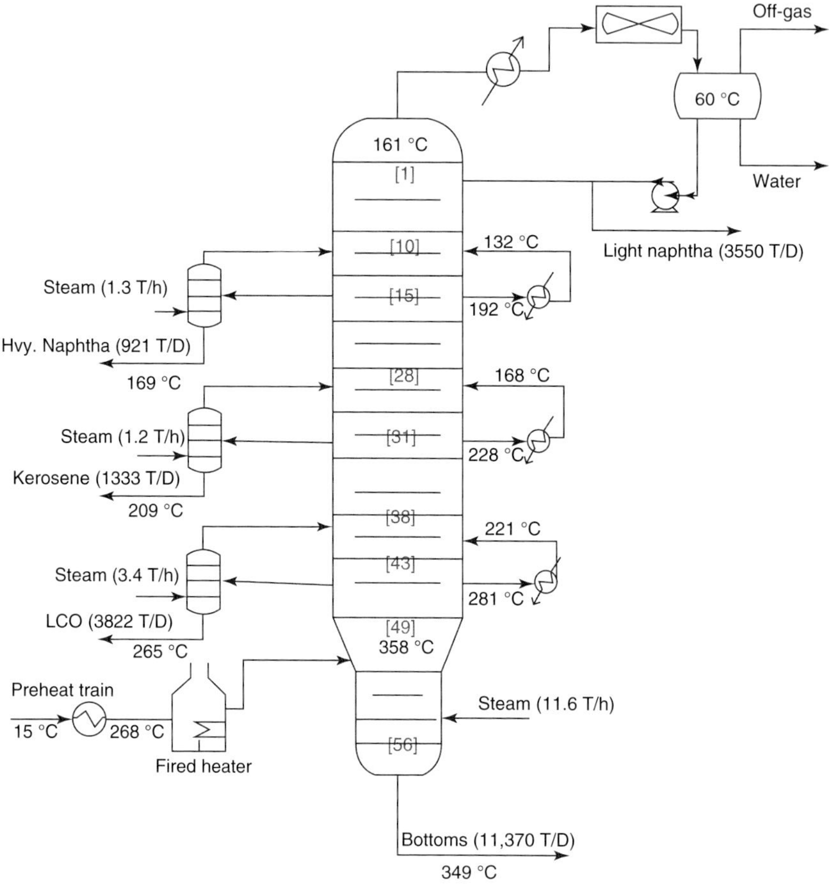

Figure 2.14 True atmospheric column.

predicted values. We can also compare the model predictions to a large databank of historical measurements (1–3 months) to help validate the model in question.

2.7 A Representative Atmospheric Distillation Unit

Figure 2.14 shows data from a typical atmospheric distillation unit that processes a variety of crudes. In subsequent sections of this chapter, we build a simulator model based on these initial data and perform several case studies. Following the previous sections, we build the model in the form of Figure 2.15 using theoretical stages. We note that the number of theoretical stages is roughly half the number of physical trays, indicating an overall stage efficiency of 50%. The locations of each zone (heavy naphtha, kerosene, etc.) reflect the zone fractionation concept from Kaes [3] described in Section 2.4.3. We also summarize key operating conditions in the following tables.

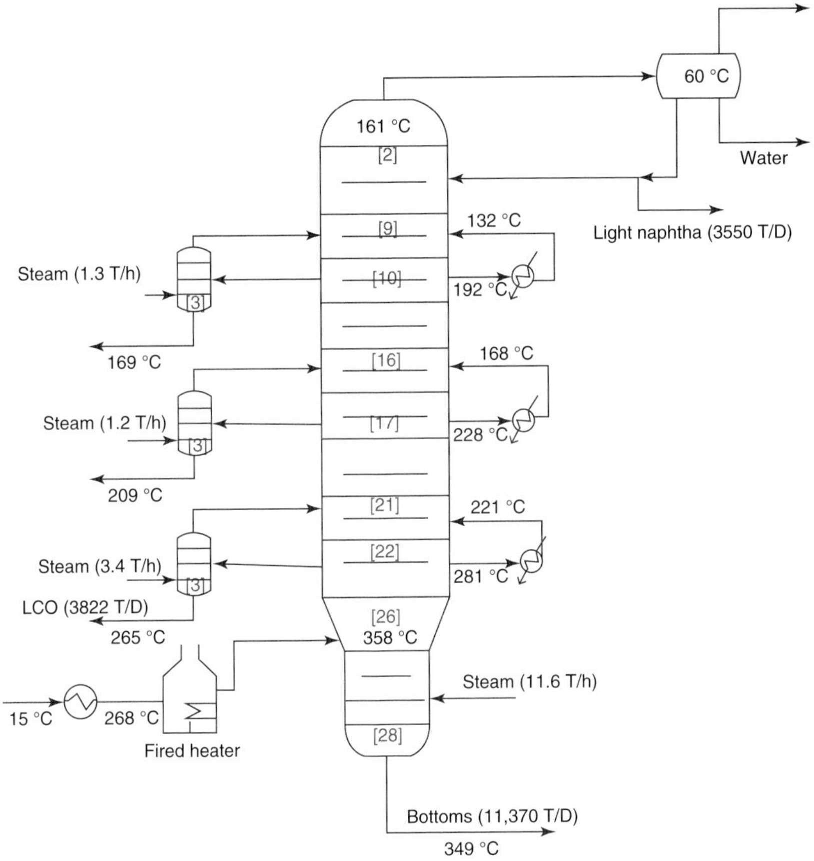

Figure 2.15 Simulation model representation.

2.8 Building the Model in Aspen HYSYS Petroleum Refining

This section documents some of the key steps required to model the representative crude unit described in the previous section. We use a 50–50% mixture of the assays presented in Tables 2.6–2.9 as the feed to the column. In the workshop examples accompanying this chapter, we will simulate this column based on backblending alone and compare the results.

Throughout this work, we have used Aspen HYSYS Petroleum Refining [13] and related software products quite extensively. Despite this fact, the techniques described in this work are applicable almost directly to many other simulation software tools. The most important considerations are the availability of a robust implementation of the inside-out method and the ability to deal with pseudocomponents and associated thermodynamics in the software chosen. Most modern process simulators meet these criteria. As discussed in Sections 1.8 and 1.9, version 8.0 and new versions of Aspen HYSYS Petroleum Refining have included the

Table 2.6 Arab Heavy TBP distillation curve.

% Distilled	Temperature (°C)	% Distilled	Temperature (°C)
4.97	50	42.18	310
6.32	60	43.78	320
7.83	70	45.38	330
8.06	80	46.97	340
9.45	90	48.54	350
11.00	100	50.09	360
11.81	110	51.61	370
13.21	120	53.10	380
14.14	130	54.56	390
15.76	140	55.99	400
17.38	150	57.39	410
18.98	160	58.76	420
20.55	170	60.10	430
22.08	180	61.41	440
23.59	190	62.70	450
25.08	200	63.96	460
26.57	210	66.42	480
28.05	220	68.79	500
29.55	230	71.07	520
31.08	240	73.27	540
32.62	250	75.36	560
34.19	260	77.37	580
35.77	270	79.28	600
37.37	280	83.67	650
38.97	290	87.53	700
40.57	300	100.00	850

new *petroleum assay manager* that uses more accurate blending rules to calculate property values. We can use the petroleum assay manager to manage multiple types of assays and blends of different assays.

Following Workshop 1.6, Section 1.7, we define two new petroleum assays, ArabLight and ArabHVY, as illustrated in Figure 2.16. We save the file as *Crude Assay Only.hsc.*

2.8.1 Entering the Crude Information

The first step in building the atmospheric distillation unit is entering the composition of the crude in order to generate the necessary pseudocomponents for the model. For the purposes of this simulation, we consider the crude assays given in Tables 2.6–2.9. It is important to remember that we may have to remove

Table 2.7 Arab Heavy density distribution.

% Distilled	SG	% Distilled	SG
4.97	0.635	43.78	0.856
7.83	0.664	45.38	0.962
8.06	0.673	46.97	0.871
9.45	0.694	48.54	0.877
11.00	0.695	50.09	0.863
11.81	0.713	51.61	0.889
13.21	0.734	53.10	0.895
14.14	0.726	54.56	0.900
15.76	0.735	55.99	0.905
17.38	0.743	57.39	0.910
18.98	0.751	58.76	0.914
20.55	0.759	60.10	0.919
22.08	0.766	61.41	0.923
23.59	0.774	62.70	0.928
25.08	0.781	63.96	0.932
26.57	0.788	66.42	0.936
28.05	0.795	68.79	0.947
29.55	0.802	71.07	0.955
31.08	0.808	73.27	0.962
32.62	0.814	75.36	0.970
34.19	0.818	77.37	0.978
35.77	0.824	79.28	0.986
37.37	0.830	83.67	0.999
38.97	0.837	87.53	1.017
40.57	0.843	100.00	1.112
42.18	0.849	Bulk	

extraneous detail from the distillation curve to avoid unusual column behavior. We use the TBP distillation, density distribution, and overall bulk density to define the CDU in Figures 2.17–2.20.

Many simulators offer the ability to build a set of pseudocomponents based on a distillation curve and bulk density. Although this method will produce a set of pseudocomponents, it is not sufficient for crude distillation. Methods that only use the bulk density fix the Watson K factor (typically to 12.0). This can lead to significant errors in predicting the equilibrium distributions of heavier components of the feed. Figures 2.18–2.20 show that we enter both a complete density distribution and the total bulk density. We recommend using the beta smoothing and correlation function described in Section 1.2 and demonstrated in Workshop 1.2 in Section 1.4 to screen out unusual density distributions or predict a density distribution based on a limited number of data points. The data given in

Table 2.8 Arab Light TBP distillation curve.

% Distilled	Temperature (°C)	% Distilled	Temperature (°C)
3.79	40	48.99	310
4.51	50	50.78	320
5.14	60	52.57	330
7.06	70	54.35	340
7.97	80	56.11	350
8.78	90	57.90	360
10.89	100	59.61	370
11.82	110	61.28	380
12.79	120	62.90	390
15.33	130	64.48	400
17.11	140	66.01	410
18.88	150	67.50	420
21.10	160	68.94	430
23.11	170	69.96	440
25.13	180	71.32	450
26.99	190	72.65	460
28.86	200	75.23	480
30.54	210	77.68	500
32.41	220	80.02	520
34.26	230	82.24	540
36.12	240	84.19	560
37.97	250	85.88	580
39.81	260	87.45	600
41.64	270	90.90	650
43.47	280	93.72	700
45.37	290	100.00	850
47.18	300		

the following screenshots come from the Arab Light and Arab Heavy assays given in Tables 2.6–2.9. Additional information such as viscosity distribution does not typically help in defining pseudocomponents.

The last step in defining a complete assay is the description of the light components of the assay (Figures 2.21–2.23). While starting with detailed assays, it is possible to obtain analysis of the feed products as well. While simulating an existing column, it is sufficient to backblend measured light gas products back into the feed of the crude. In addition, for accurate light composition, we must also consider light gas components due to thermal cracking in the column. If the light gas analysis is not available, simulators may provide an option to estimate the light gas distribution. While useful, it is unlikely that these values will be correct. As we will show in later sections, the light gas composition does not play a

Table 2.9 Arab Light density distribution.

% Distilled	SG	% Distilled	SG
3.79	0.634	48.99	0.853
4.51	0.654	50.78	0.860
5.14	0.653	52.57	0.869
7.06	0.663	54.35	0.875
7.97	0.716	56.11	0.882
8.78	0.704	57.90	0.887
10.89	0.702	59.61	0.893
11.82	0.724	61.28	0.898
12.79	0.766	62.90	0.903
15.33	0.733	64.48	0.908
17.11	0.759	66.01	0.910
18.88	0.765	67.50	0.915
21.10	0.763	68.94	0.919
23.11	0.771	69.96	0.923
25.13	0.777	71.32	0.927
26.99	0.785	72.65	0.930
28.86	0.792	75.23	0.936
30.54	0.796	77.68	0.941
32.41	0.802	80.02	0.948
34.26	0.808	82.24	0.955
36.12	0.814	84.19	0.962
37.97	0.816	85.88	0.970
39.81	0.822	87.45	0.978
41.64	0.828	90.90	0.991
43.47	0.834	93.72	1.010
45.37	0.840	100.00	1.098
47.18	0.847	Bulk	

significant role in determining the column performance. Kaes [3] gave methods to estimate the gas composition of crudes.

Depending on the available analysis, users may have to add additional light gas components into the thermodynamic basis to reflect plant measurements. In general, we do not add any components with boiling points higher than the *n*-butane or *n*-pentane series. We require the butane series if we expect the model to handle vapor pressure (Reid vapor pressure, RVP) specifications and predictions for gasoline-type cuts.

The next step in building the set of pseudocomponents is creating a blend. A blend represents a combination of two or more assays on a weight or volume basis. The combined blend is the input to the hypothetical component generator in Aspen HYSYS. For the purposes of this simulation, we use the data from

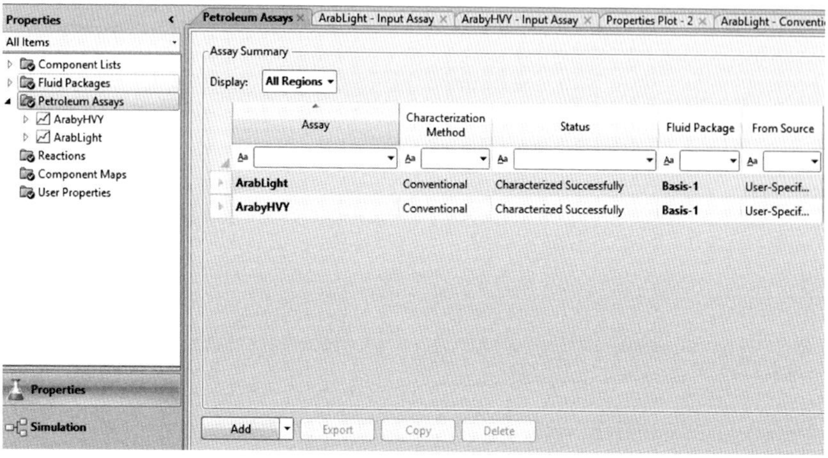

Figure 2.16 Define petroleum assays, ArabLight and ArabHVY, in Aspen HYSYS petroleum assay manager and save the file as *Crude Assays Only.hsc*.

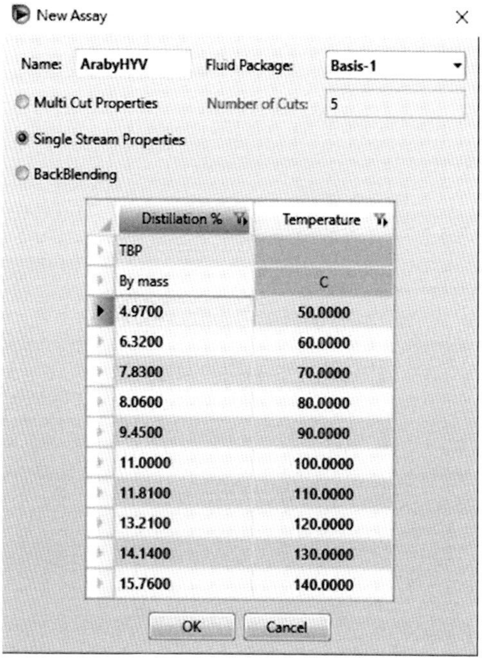

Figure 2.17 Initial assay definition-TBP distillation data (Arab Heavy).

the representative atmospheric distillation unit of Section 2.8. We can place a petroleum feeder into the flowsheet to create a simulated crude feed by backblending true column products. We show a 50–50% mixture of the assays in Figure 2.24.

Once we create the blend in the petroleum feeder, we can save the feed stream from the feeder as an assay to make several observations about the calculated properties (distillation curve, viscosity curve, etc.) in Aspen HYSYS petroleum

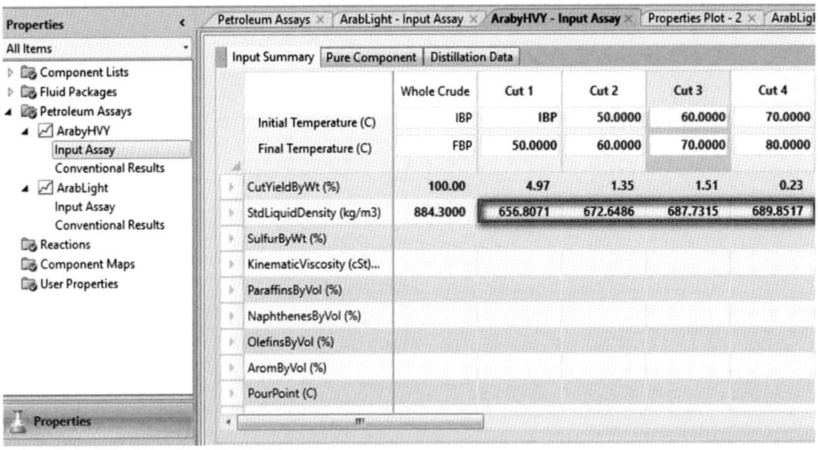

Figure 2.18 Specify density distribution.

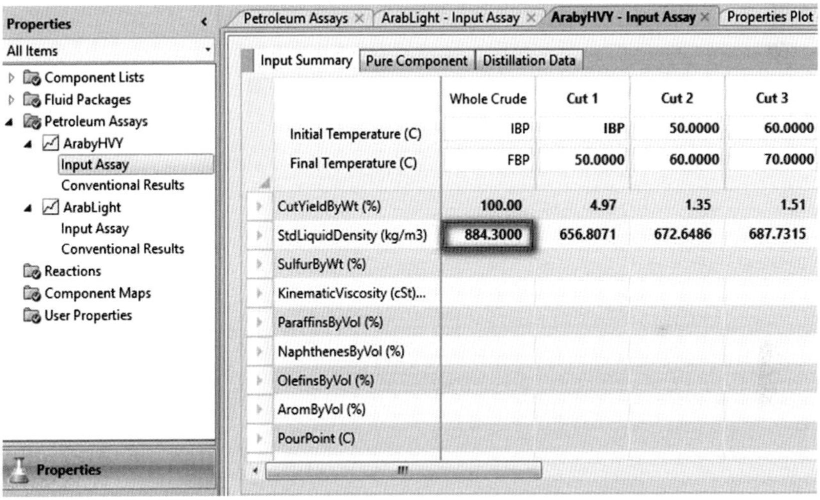

Figure 2.19 Specify bulk properties (Arab Heavy).

assay manager. Documentation from AspenTech [19] and work by Riazi [9, 11] summarize most of these correlations. We can review the generated pseudocomponent list (Figure 2.25). This pseudocomponent list shows all relevant physical property information calculated from the input distillation and density data.

The last step is to specify pressure, temperature, and flow rate of the blend stream from the petroleum feeder in the flowsheet (Figure 2.26). We must create a new blend each time the composition of the assays changes. For the purposes of a basic simulation, a blend of assays or a blend of backblended products is sufficient. In the old oil manager of Aspen HYSYS, if we need to evaluate a variety of crudes, the component list can quickly become unmanageable. By contrast, the

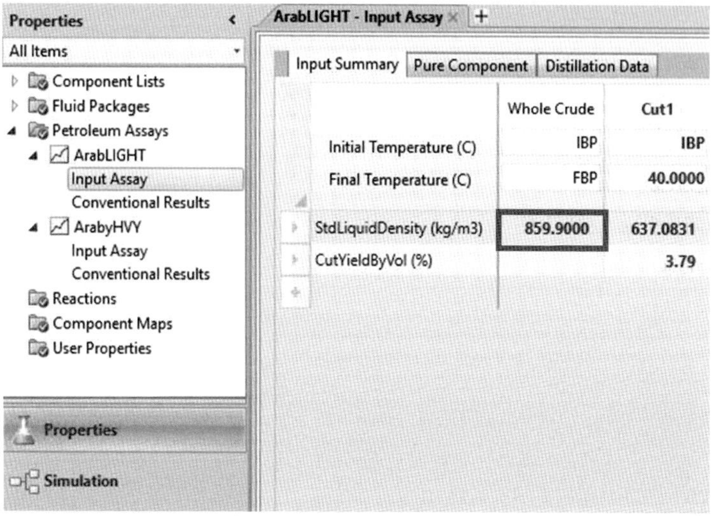

Figure 2.20 Specify bulk properties (Arab Light).

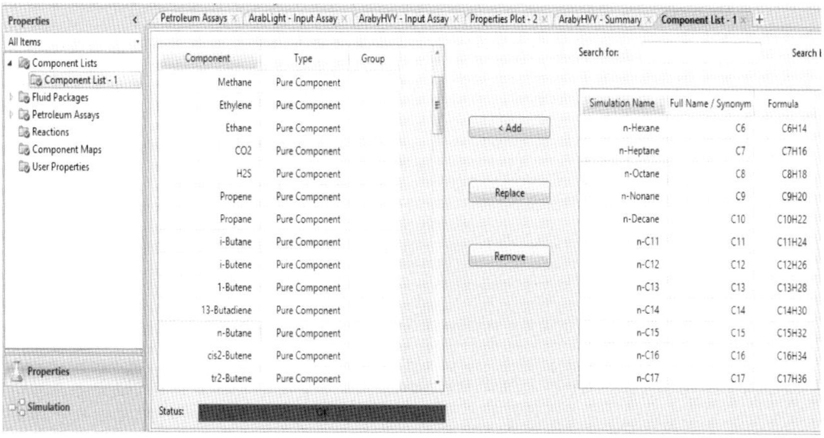

Figure 2.21 Add additional light gas components to the thermodynamic basis.

new petroleum assay manager of Aspen HYSYS generates a unified component list across all assays, which makes it more convenient to handle multiple-assay blends. See our previous comparison in Table 1.3, Section 1.8.

2.8.2 Selection of a Thermodynamic Model

The choice of a thermodynamic model can have significant impact on the results. The primary source of this error is the poor representation of K-values, especially for heavier crudes. The following options are typically available for hydrocarbon-rich streams (like crude):

- Equation-of-state (EOS) based.
  - Peng–Robinson (PR), Soave–Redlich–Kwong (SRK)

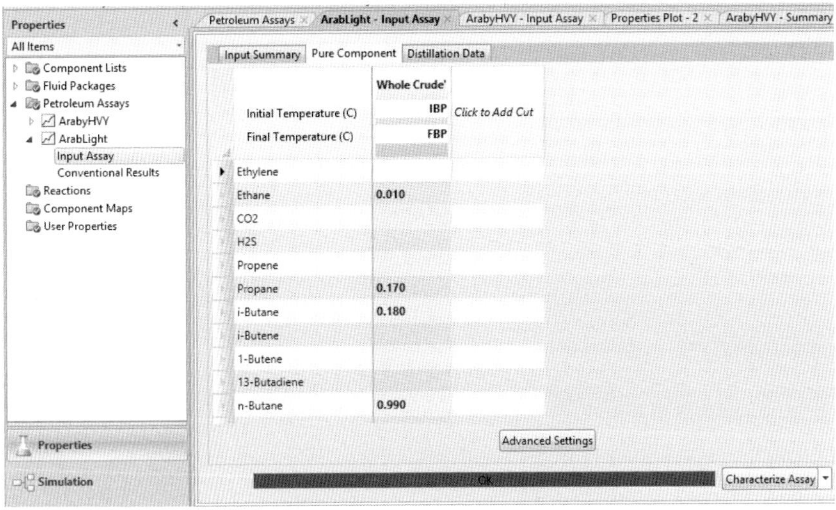

Figure 2.22 Specify light gas components of the assay (Arab Light).

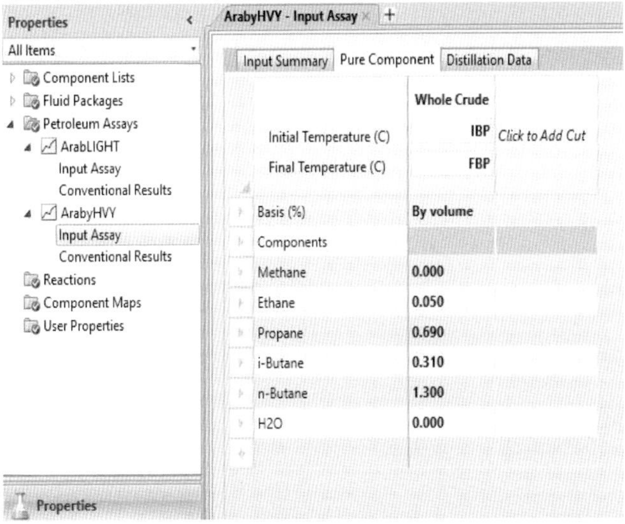

Figure 2.23 Specify light gas components of the assay (Arab Heavy).

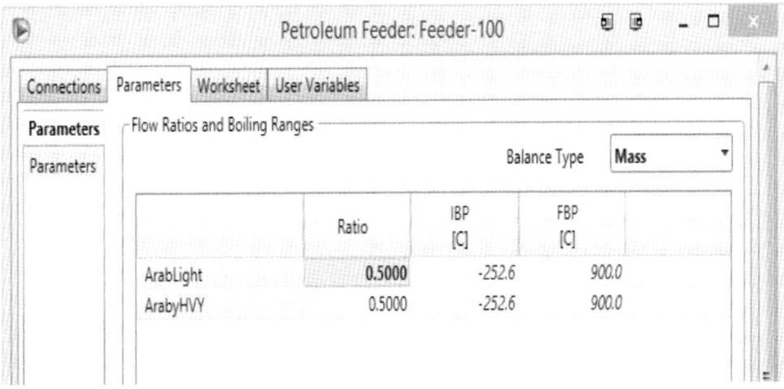

Figure 2.24 Create a blend from crude assays.

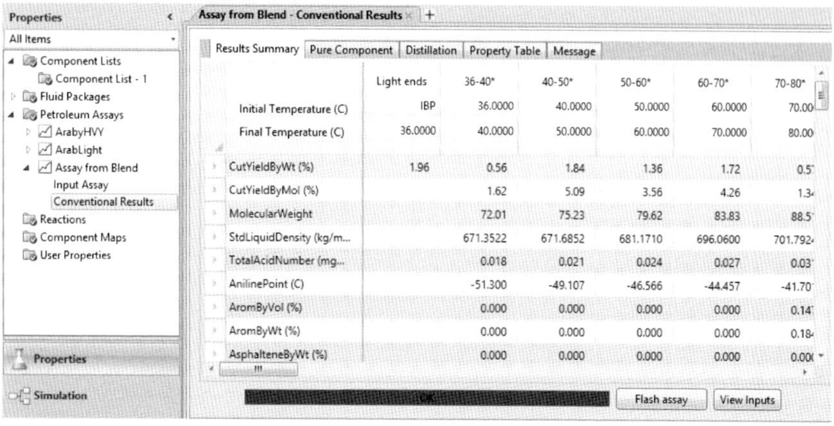

Figure 2.25 Review calculated pseudocomponent properties.

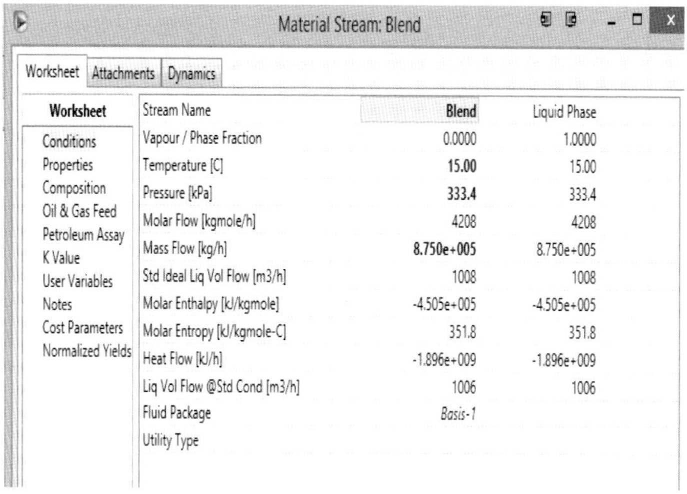

Figure 2.26 Enter specification of the blend stream.

- Fugacity correlation based.
  - Grayson–Streed, Chao–Seader
- Correlation based.
  - BK-10, ESSO, API

Equations of state generally rely on pure component properties such as critical temperature, critical pressure, and acentric factor. In addition, an interaction parameter is required to account for mixtures of components. Correlation-based approaches rely on measured vapor pressures and observed data to provide empirical correlations for various hypothetical or pseudocomponents.

We have commented extensively on thermodynamic models and various approaches in Chapter 1. In general, we recommend the use of EOS or fugacity correlation-based approaches in modern process simulators. There are minor deficiencies associated with each type of model. However, advanced options in most process simulators can counteract these problems and provide similar results. We illustrate the impact of the thermodynamic model in Section 2.10.

2.8.3 Crude Charge and Prefractionation Units

Once we have chosen an appropriate thermodynamic representation for the crude feed and associated components, we can begin building the actual simulation. The charge feed enters a simple heat exchanger that simulates the preheat train of the CDU. For a more realistic model, we should simulate the extensive preheat train. As the preheat train is not the focus of this work, we represent the preheat train with a simple heater with a variable heat duty instead (see Figure 2.27; save as *CDU EX-1.hsc*). Figure 2.28 gives the specifications for the outlet stream.

The next major unit is the crude furnace or crude heater. This is typically a very large fired heater capable of vaporizing significant quantities of crude. It is also the major energy consumer in the CDU. Aspen HYSYS includes a fired heater model if extensive information about the fired heater is available. As we do not have this information, we represent the fired heater as a simple exchanger with a given heat duty. However, we must account for the overflash specification of fired heater. The overflash specification controls how much heavy material can be recovered in the nonresidue product from the column. Figures 2.29 and 2.30 show the required setup for this furnace.

We specify the overflash specification by using an Adjust block. As a reminder, the *overflash* is the amount of liquid that is vaporized in addition to all the products recovered in the column excluding the residue. The residue is the crude that was not vaporized in the feed of the column. We specify the furnace initially by an

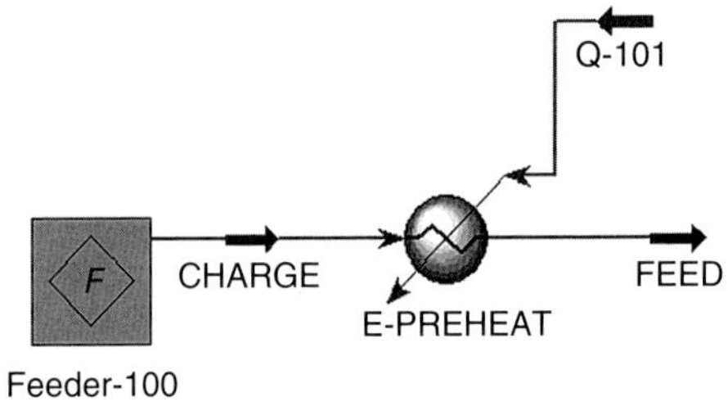

Figure 2.27 Simplified preheat train.

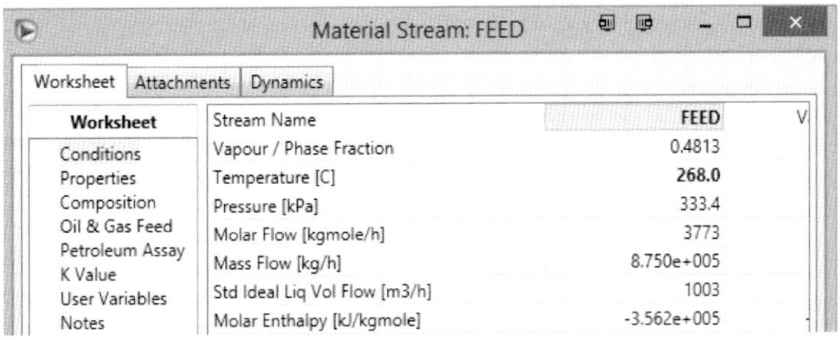

Figure 2.28 Outlet temperature of the preheat train.

estimate of the heat duty. The Adjust block will vary a manipulated variable, the furnace heat duty, until our design specification, vapor fraction (based on mass) of the furnace outlet stream, reaches a target value.

From Table 2.13, we see that the sum of the liquid products (light naphtha, heavy naphtha, kerosene, and light gas oil) is 45.84 wt%. We would like to specify an overflash of 3%. Therefore, we expect the vapor fraction (on a mass basis) of the stream leaving the furnace to be 48.84 wt% (sum of liquid products and overflash). We will adjust the heat duty to match the overflash requirement.

Figures 2.31 and 2.32 show how to select the adjust variable (in this case, the heat duty to the crude heater). We select the target variable (in this case, the mass vapor fraction of the crude heater outlet) and set the target variable value in Figure 2.33.

The Adjust block may not converge during the initial run. We can typically improve convergence by increasing the number of the iterations and the step size in the solver parameters for the Adjust block in Figure 2.34.

Figure 2.35 shows the completed flowsheet for modeling the Heated_FEED in Section 2.8.4. We save the resulting file as *CDU EX-1.hsc* and begin to configure the actual distillation column.

2.8.4 Atmospheric Distillation Column – Initial

In this section, we create and configure the representative crude unit shown in Figure 2.14 of Section 2.7. As we avoid the use of stage-by-stage efficiency factors, we will create the model based on the overall stage efficiency description from Figure 2.15. We take the data for column configuration and process data

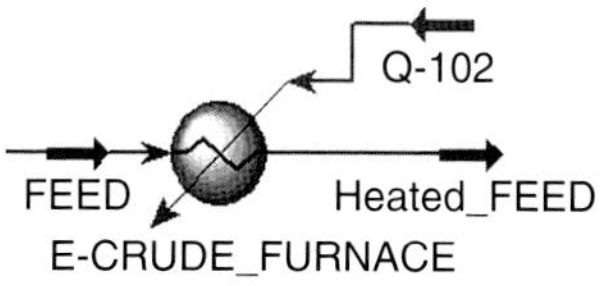

Figure 2.29 Simplified crude fired heater.

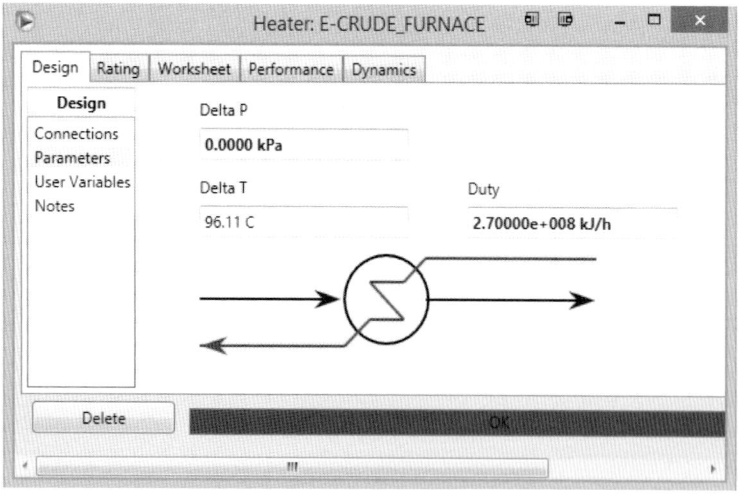

Figure 2.30 Supply initial guess for heater duty.

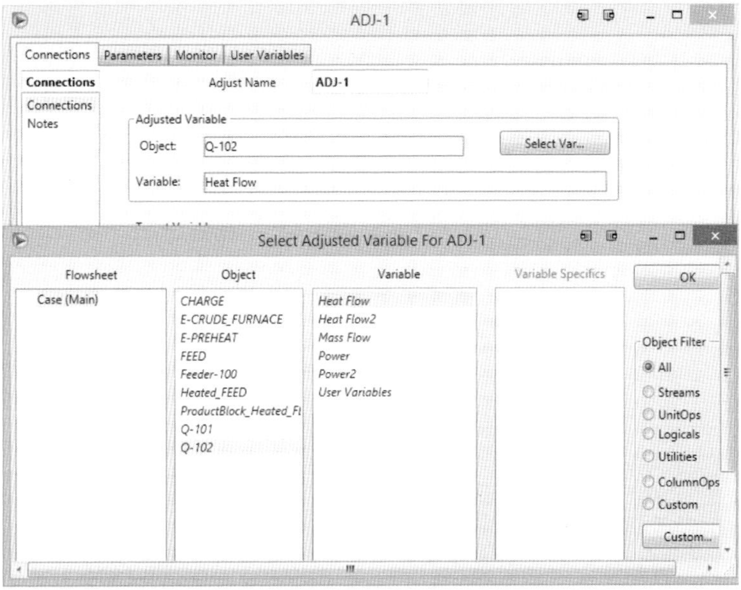

Figure 2.31 Select the heat flow as the adjusted variable.

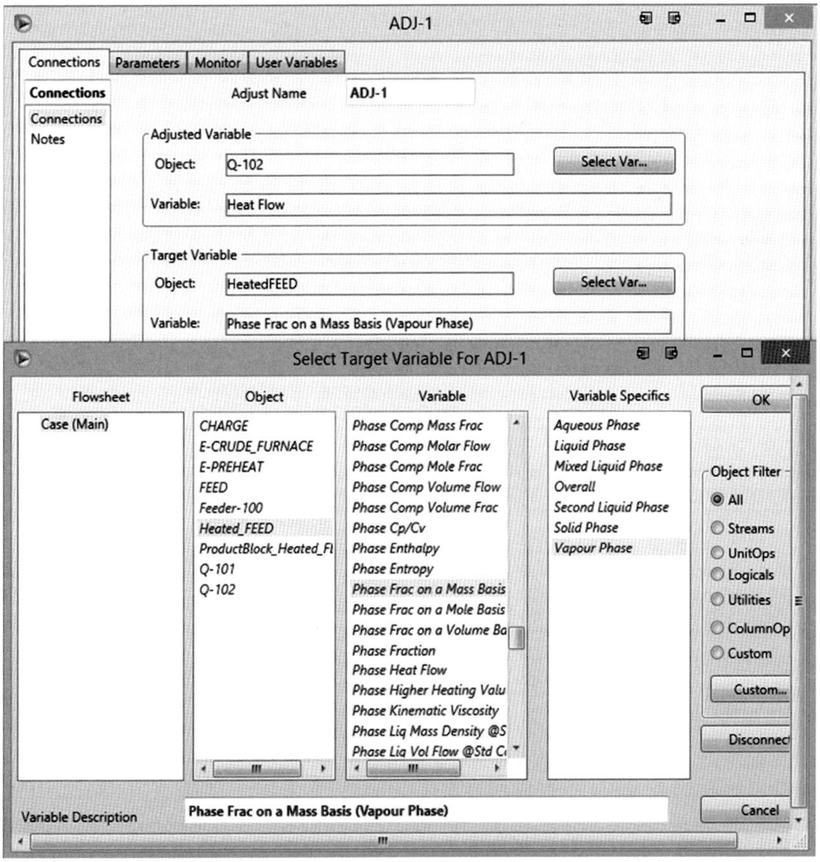

Figure 2.32 Select mass vapor fraction of heater outlet as target variable.

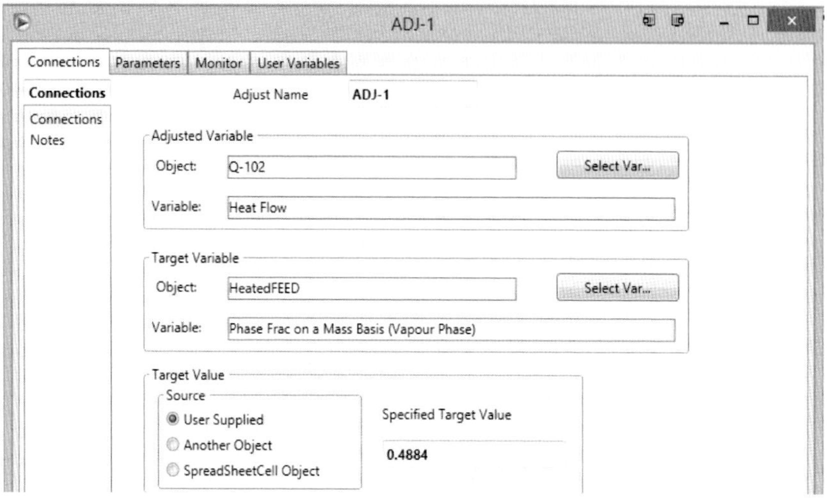

Figure 2.33 Specify the target mass vapor fraction of outlet.

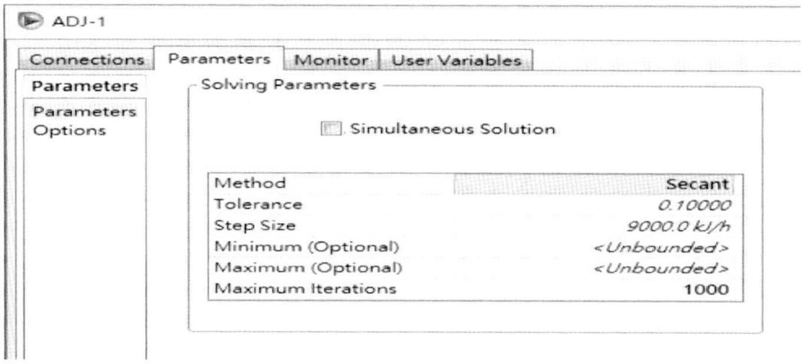

Figure 2.34 Modify solver parameters for the Adjust block.

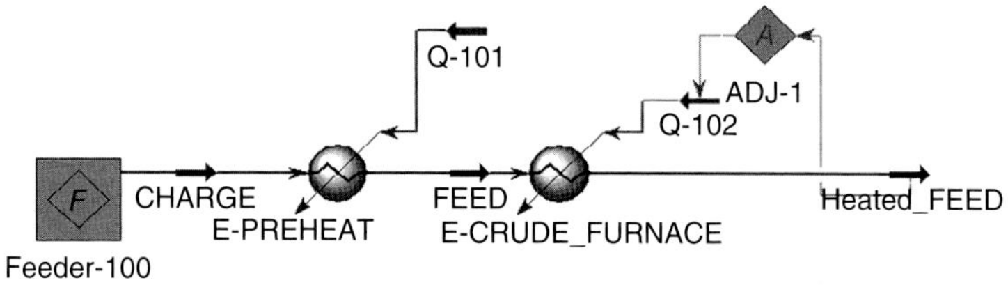

Figure 2.35 Flowsheet of the feed preheater and furnace.

from Tables 2.10 to 2.13. We follow a step-by-step procedure that develops improved initial guesses at each step. This method ensures that even relatively unsophisticated column algorithms can run with these improving column estimates. Sophisticated column algorithms (as in Aspen HYSYS) can converge quickly even without these steps; however, it is quite easy to discover problems in the data that prevent convergence with this method.

Table 2.10 Measured plant flow rates and specifications.

Feed/products	Stage	Flow rate (ton/day)	Conditions
Crude feed	49	21 000	3.0% Overflash
Bottoms steam	56	278.4	Sat'd steam @ 250 °C
Off gas	Condenser	N/A	60 °C
Light naphtha	Condenser	3549	60 °C
Residue	Bottoms	11,375	349 °C

Table 2.11 Measured pumparound flow rates and specifications.

Pumparounds	Flow rate (ton/h)	Temperature change (°C)	Duty (Gcal/h)	Draw/return
Heavy naphtha	376.1	−90	−13.9	15/10
Kerosene	234.9	−60	−9.1	31/28
LGO	298.1	−60	−12.2	43/38

Table 2.12 Measured side stripper flow rates and specifications.

Side strippers	Draw rate (ton/day)	Steam (kg/h)	Draw/return
Heavy naphtha	921	1313	15/10
Kerosene	1333	1243	31/28
LGO	3822	3418	43/38

Table 2.13 Measured product distribution and qualities.

ASTM D86 (°C)	LN	HN	Kerosene	LGO	Residue
IBP	69	137	168	218	319
5%	71	165	198	246	368
10%	74	172	203	254	381
30%	88	179	210	268	454
50%	104	183	215	283	533
70%	122	187	221	301	684
90%	146	193	229	328	874
95%	153	196	235	337	–
FBP	162	204	251	378	–
Specific gravity	0.7037	0.7826	0.8034	0.8456	0.9713
Yield (wt%)	16.9	4.39	6.35	18.2	54.16
Yield (ton/day)	3549	921	1333	3822	11,375

We continue with the file *CDU EX-1.hsc* and create a refluxed absorber: F4 → operation palette (model menu) → columns → insert "refluxed absorber T-100" (see Figure 2.36).

We then double-click the column T-100 to open the column input form, rename the column as ATM-100, and enter the relevant stream inputs. We configure the column with 27 ideal stages and attach the relevant energy and material streams (see Figure 2.37).

We specify the pressure profile in Figure 2.38.

The next step is not required by Aspen HYSYS but is *a recommended good practice* to ensure that columns converge regardless of the method used. Based on the measured plant data, we *enter estimates of the top and bottom temperatures* in Figure 2.39. For the initial run, the calculated values may differ from the given temperature estimates.

As this is a refluxed absorber, we must provide two initial specifications in Figure 2.40. We specify a vapor distillate rate (i.e., the off gas flow rate) of 1.421E4 kg/h and a liquid distillate rate (i.e., the light naphtha flow rate) of 1.479E5 kg/h. A reflux ratio of 2.0 generally ensures quick convergence. If the column does not converge with a reflux ratio of 2.0, not enough material may have vaporized in the

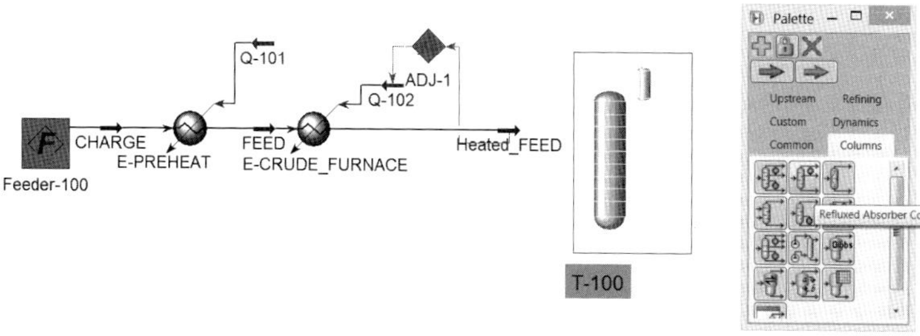

Figure 2.36 Insert a refluxed absorber model T-100 to the current flowsheet to represent the CDU.

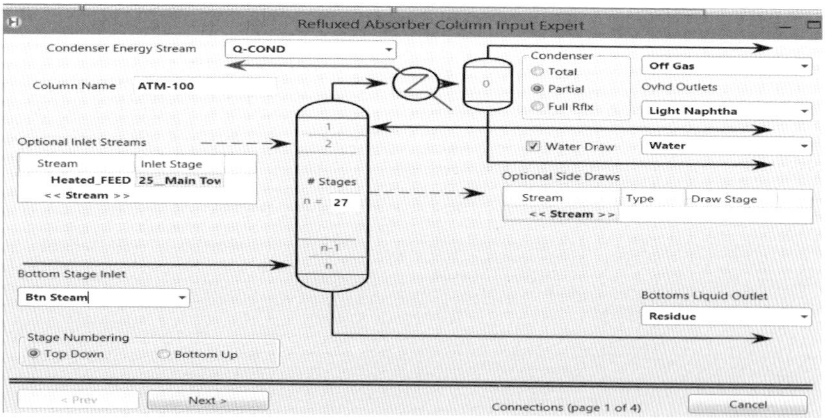

Figure 2.37 Initial stream setup for column ATM-100 with 27 ideal stages: (1) energy streams – Q-Cond and Btm Steam and (2) material streams – Heated_FEED, Off Gas, Light Naphtha, Water, and Residue.

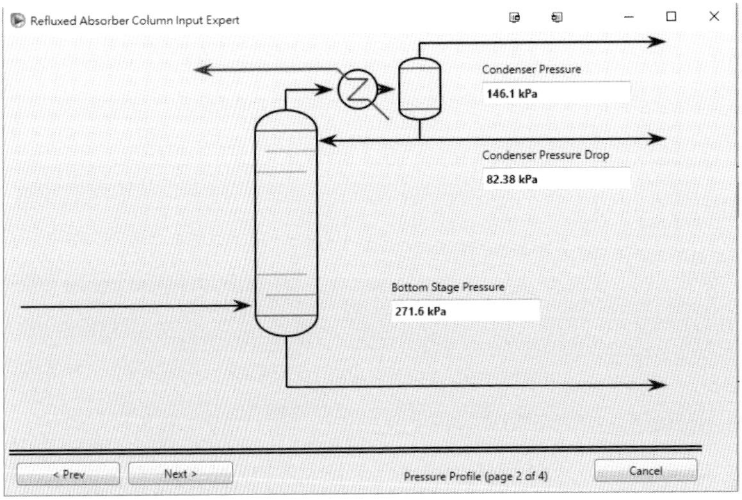

Figure 2.38 Column pressure profile.

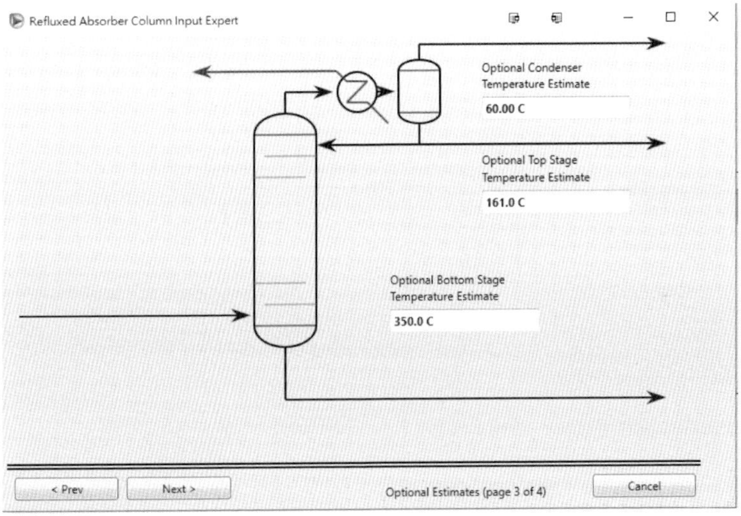

Figure 2.39 Initial estimate of temperature profile.

feed heater. We click "done" in Figure 2.40 to return to the column input summary in Figure 2.41. We see that HYSYS proceeds to do calculations on the column and shows "unconverged" in red. Why? We have not yet specified the Btm Steam input.

We close the column input window and return to the main flowsheet. Click on Btm Steam stream to enter a vapor fraction of 1, a temperature of 250 °C, a mass flow rate of 1.116E4 kg/h (under "Conditions"), and a mass fraction of 1 for water (under "Composition") (see Figure 2.42).

Next, we close the Btm Steam stream, click on the column ATM-100 again, and enter the column input summary window. We go to Column → Design → Monitor and choose to make specifications of reflux ratio and distillate rate "active"

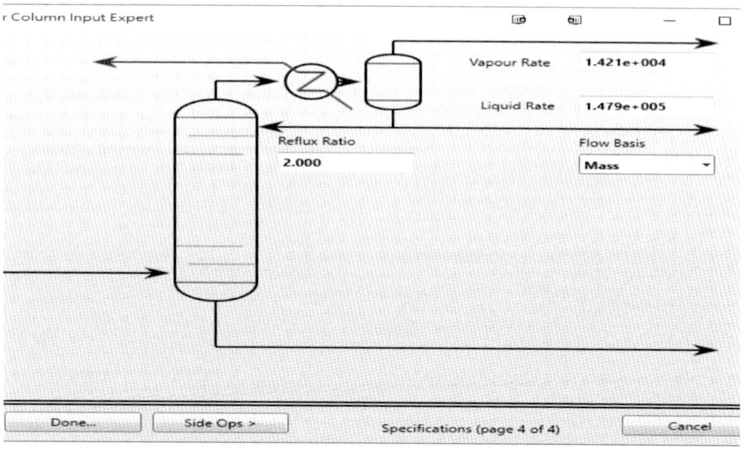

Figure 2.40 Initial flow specifications of reflux ratio and liquid and vapor distillate rates.

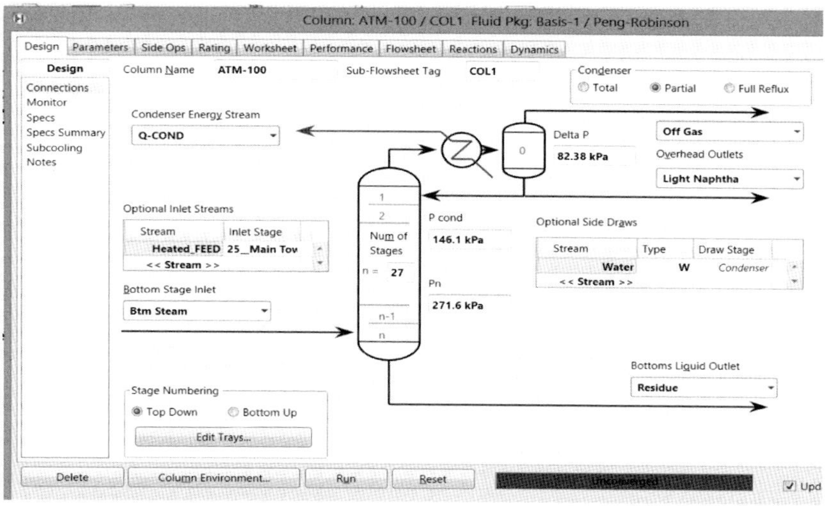

Figure 2.41 Column input summary and unconverged calculations due to a lack of specification of Btm Steam specification.

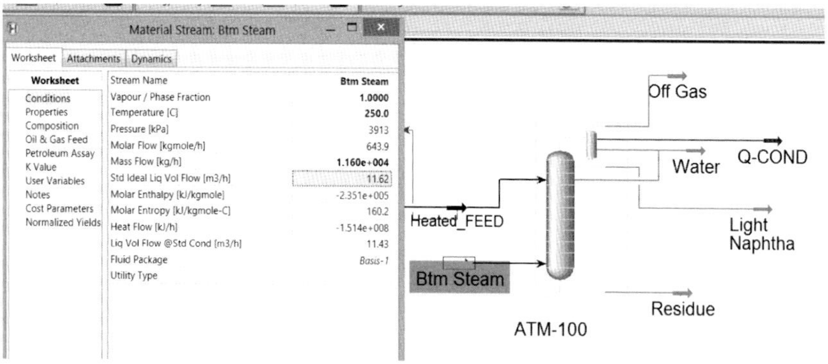

Figure 2.42 Specification of Btm Steam stream.

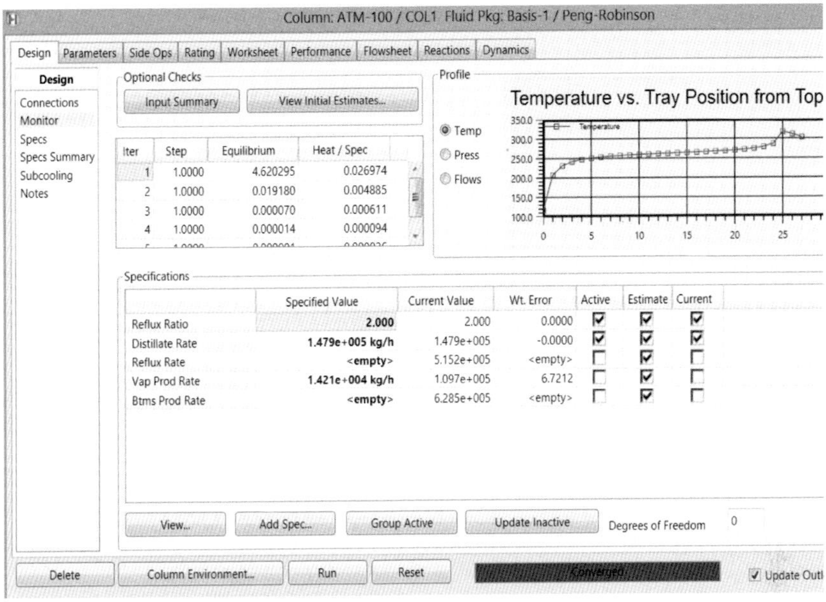

Figure 2.43 Converged initial column model.

and then click "run." The column calculations converge quickly (see Figure 2.43). We save the resulting file again as *CDU EX-2.hsc*. We may receive warnings about a potential aqueous phase in the Light Naphtha stream. We ignore these warnings until we complete building the entire column model.

2.8.5 Atmospheric Distillation Column – Side Strippers

Once we have obtained convergence only with the top and bottom products, we will add three side strippers (specified in Table 2.14) in a consecutive manner. We save file *CDU EX-2.hsc* as a new name, *CDU EX-3.hsc*, to include side strippers.

We illustrate this process only for the heavy naphtha side stripper, but it is identical for all side strippers. In Aspen HYSYS, the "Side-Ops" tab allows the user to insert side operations in the main column. By adding the side operations directly

Table 2.14 Specifications of side strippers for the CDU model.

Stripper	SS1	SS2	SS3
Draw stage	10	17	22
Return stage	9	16	21
Product name	SS heavy naphtha	SS kerosene	SS LGO
Draw spec (kg/h)	3.838E4	5.554E4	1.617E5
Stripping steam name and flow rate (kg/h)[a]	Heavy naphtha steam at 1313 kg/h	Kerosene steam at 1243 kg/h	LGO steam at 3418 kg/h

a) All stripping steam streams with a vapor fraction of 1, a temperature of 250 °C, and a composition of mass fraction of 1 for water.

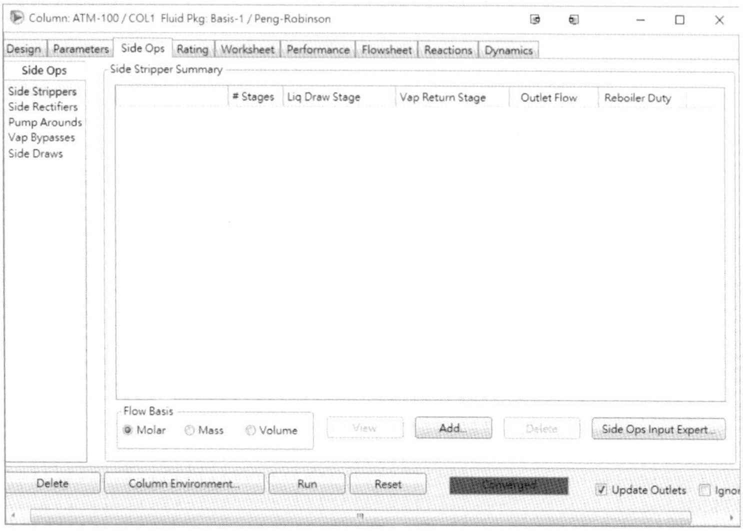

Figure 2.44 Side operations tab in Aspen HYSYS.

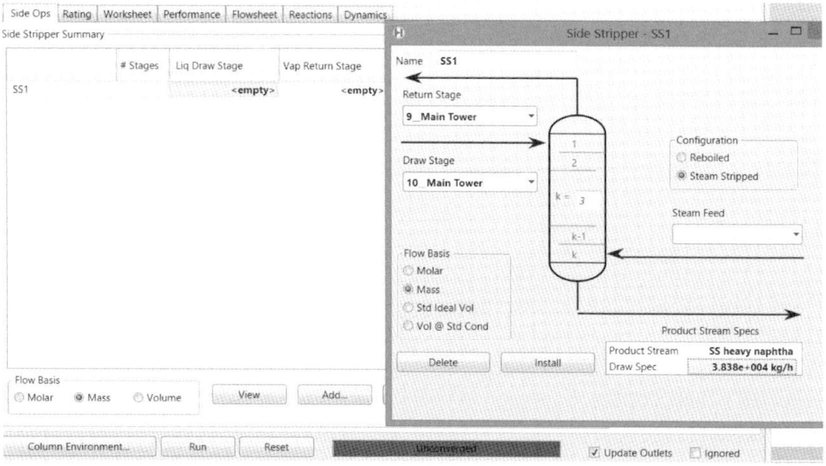

Figure 2.45 Connect and specify the product stream and the draw rate for each side stripper SS1.

to the main column, we have a great deal of flexibility when assigning column or product recovery specifications. We show the "Side-Ops" tab in Figure 2.44.

We add the heavy naphtha side stripper SS1 with three theoretical stages (suggested in Table 2.2) and specify the draw stage as 10 and the return stage as 9 in Figure 2.45. We specify the product stream, SS heavy naphtha, and its draw rate, 3.838E4 kg/h (see Figure 2.45). Then, we choose to "install" the side stripper (see Figure 2.46).

We close the side stripper window, return to the main flowsheet, and click on the "Heavy naphtha steam" stream to enter its specifications according to Table 2.14 (see Figure 2.47).

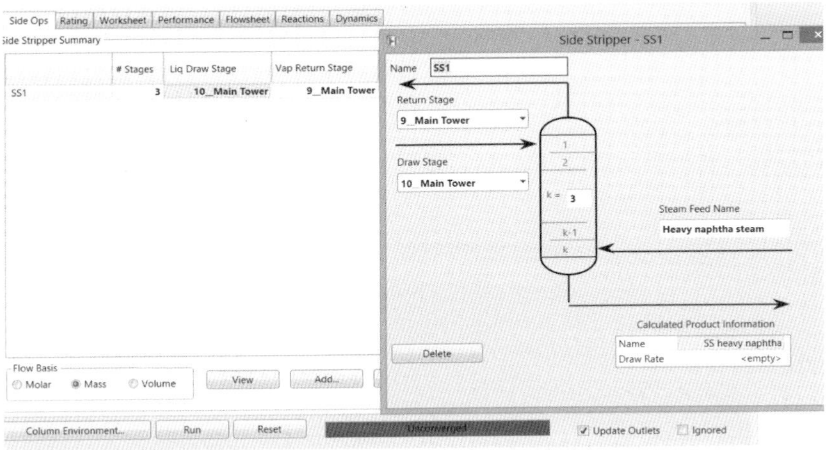

Figure 2.46 After installing the side stripper SS1, the model is missing the specifications of the "Heavy naphtha steam" stream required to calculate the draw rate.

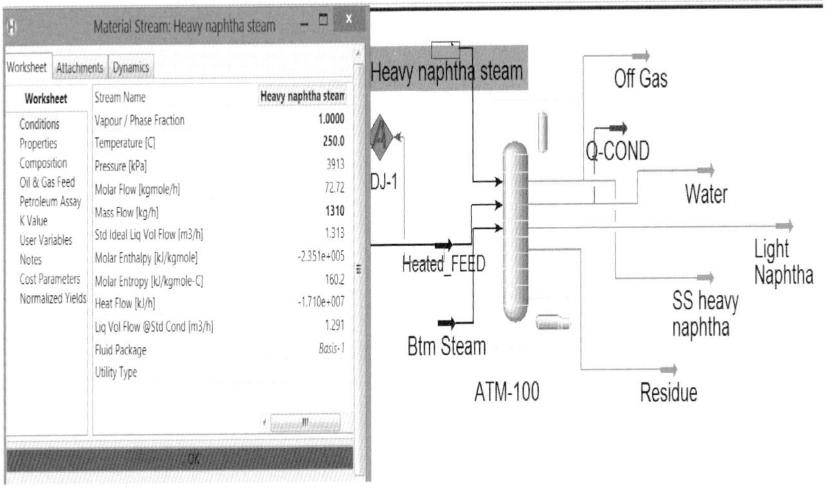

Figure 2.47 Specification of stripping stream for heavy naphtha stripper SS1.

We then close the "Heavy naphtha steam" stream window, click on column ATM-100 to enter to column input window: Column → Design → Monitor → Run → Converged (see Figure 2.48).

We recommend solving the column model after adding each side stripper. This ensures that the initial estimates are slowly refined for each step.

We continue with our simulation file, *CDU EX-3*, and save it as a new file, *CDU-EX-4*. Following the same approach from Figures 2.44 to 2.48 and the specifications in Table 2.14, we add side stripper SS2 for SS kerosene product and SS3 for SS LGO product. Figure 2.49 depicts the resulting flowsheet with three side strippers. Figure 2.50 shows the monitor displaying the active specifications of five independent variables and the column calculation convergence after adding three side strippers. Figure 2.51 displays the column profile after adding all the

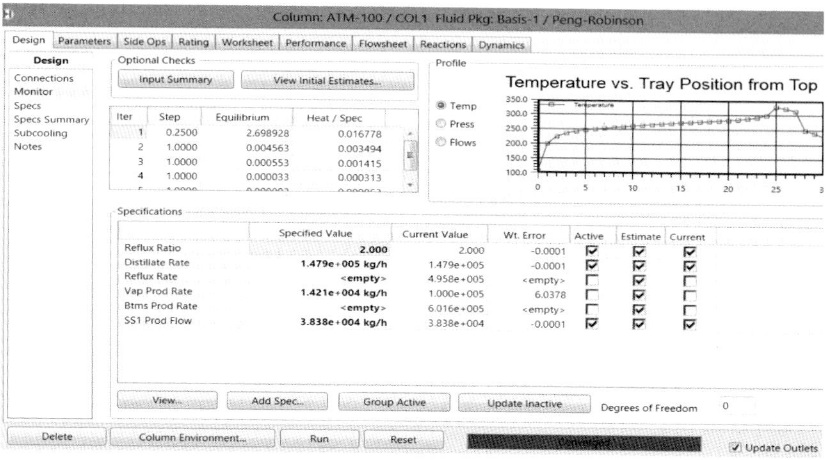

Figure 2.48 Convergence of the ATM-100 column simulation with heavy naphtha stripper SS1.

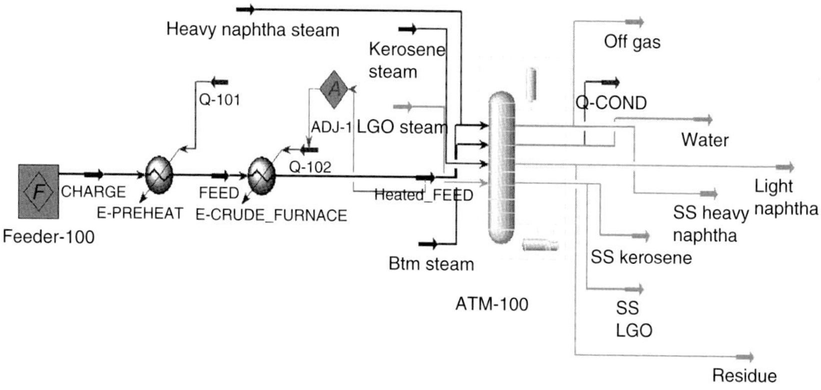

Figure 2.49 CDU flowsheet with products from three side strippers.

three side strippers. We save the resulting files after adding the three side strippers as *CD EX-4.hsc*.

We see from Figure 2.51 that the calculated condenser temperature is 90.71 °C, which is different from our initial specification of 60 °C in Figure 2.39. We shall fix the error below.

2.8.6 Atmospheric Distillation Column – Pumparounds

The last step in building the column model is adding the pumparound for each product draw. We first save the simulation file, *CDU-EX-4.hsc* as a new file, *CDU EX-5.hsc*, and reenter the "Side-Ops" tab and create the pumparound for each product. The draw and return stages are typically the same as side stripper draw and return stages. Kaes [3] documented some alternative configurations, but the difference for simulation is often small when compared to errors in other simulation assumptions. The specifications of pumparounds for the CDU model are given in Table 2.15.

It is important to specify either a duty or a temperature change across the pumparound. Specifying an absolute return temperature can often solve

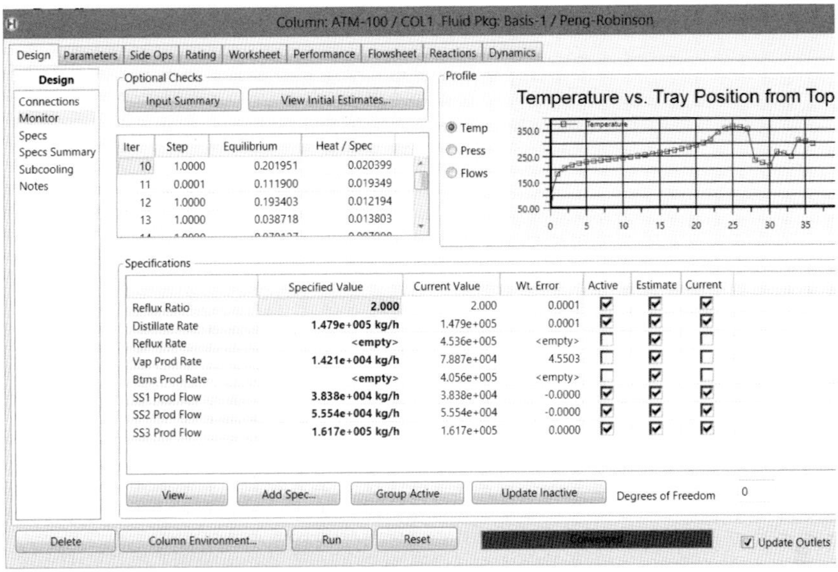

Figure 2.50 Converged CDU model after adding three side strippers.

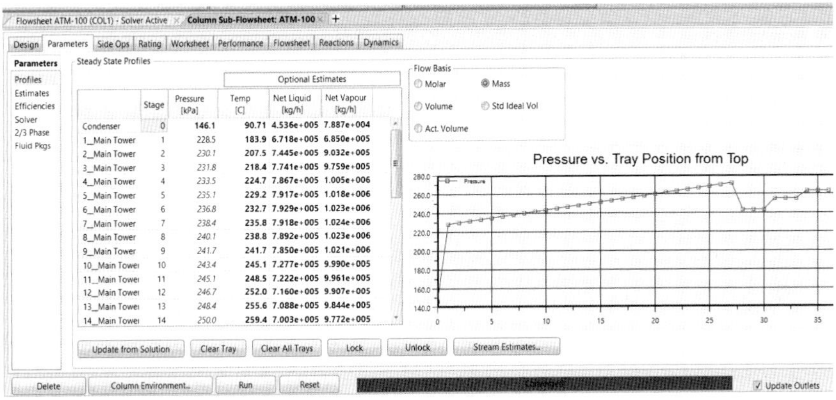

Figure 2.51 Temperature and pressure profiles of the CDU after adding three side strippers: ATM100 → Parameters → Profiles.

Table 2.15 Specifications of pumparounds for the CDU model.

Pumparound	PA1	PA2	PA3
Side stripper	SS heavy naphtha	SS kerosene	SS LGO
Draw stage	10	17	22
Return stage	9	16	21
PA rate (kg/h)	3.761E5	2.35E5	2.981E5
Temperature change, dT (°C)	90	60	60

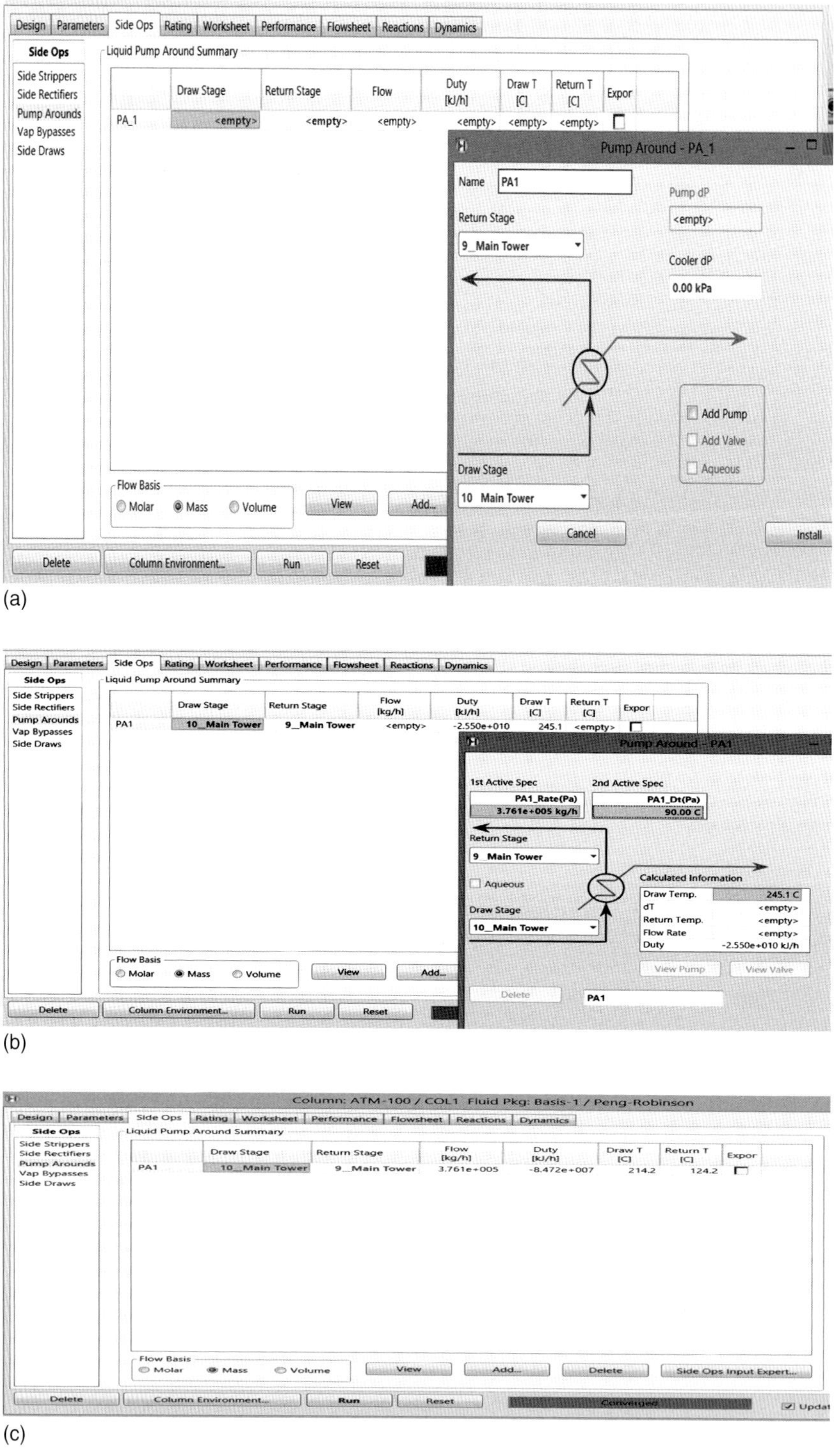

Figure 2.52 (a) Add pumparound PA1 for SS heavy naphtha. (b) Add two specifications after installing pumparound PA1. (c) Run the simulation with PA1 and achieve convergence.

significant problems in convergence, especially when the column is specified by draw temperatures instead of product yields.

To add pumparounds, we go to the same "Side-Ops" window within column ATM-100 and click "Pump Arounds." We then add PA1 by following the illustration from Figure 2.52a–c.

After achieving simulation convergence with PA1, we repeat the same process to install and run PA2 and PA3, following the specifications in Table 2.15. Our initial simulation with three PAs does not converge, as shown in Figure 2.53.

We can typically improve convergence by selecting the "Modified HYSIM Inside-Out" method (discussed in Section 2.4.4). The term "modified" refers to the fact that the solution procedure uses a full Newton–Raphson method to converge the inner loop (i.e., the stage-by-stage mass and energy balances of the column at fixed stage temperature and pressure specified by the outer loop that focuses on phase-equilibrium calculations) of the solution algorithm. The modified method can handle a wide variety of specifications just as easily as the standard inside-out algorithm (Figure 2.54).

Figure 2.51 shows that the condenser temperature is 90.71 °C, instead of our estimated 60 °C. Let us now learn how to define a design specification to correct the condenser temperature. We follow the steps: Design → Specs → Column Specifications → Add → Column Specification Types → Column Temperature → Add Spec(s) → Temp Spec: Name – Condenser Temperature, Stage – Condenser, and Spec Value – 60 °C (see Figure 2.55).

Figure 2.56 shows that we change the reflux ratio from active to estimate and set condenser temperature as estimate. The simulation converges quickly. We save the converged simulation file as *CDU EX-6.hsc.*

2.8.7 Atmospheric Distillation Column – Adding Custom Stream Properties

Figure 2.57 shows the completed distillation column simulation with all the pumparounds and side strippers. Once we add all the side operations, the

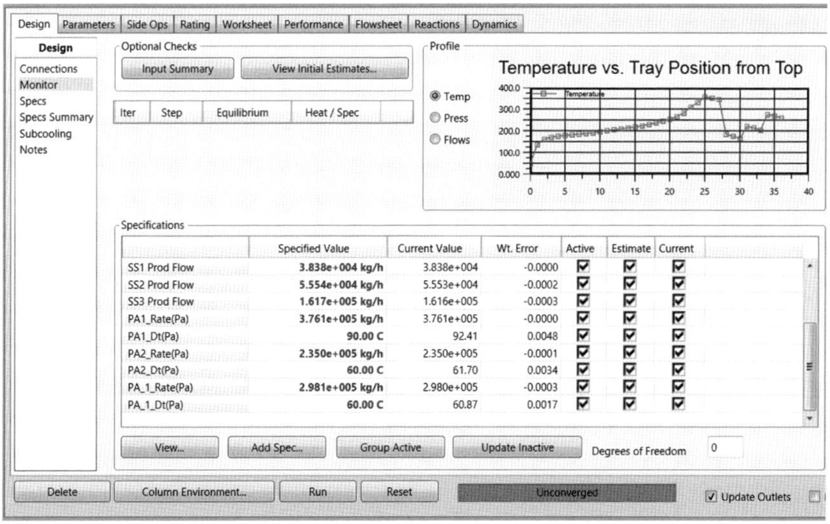

Figure 2.53 CDU model with three pumparounds not yet converged.

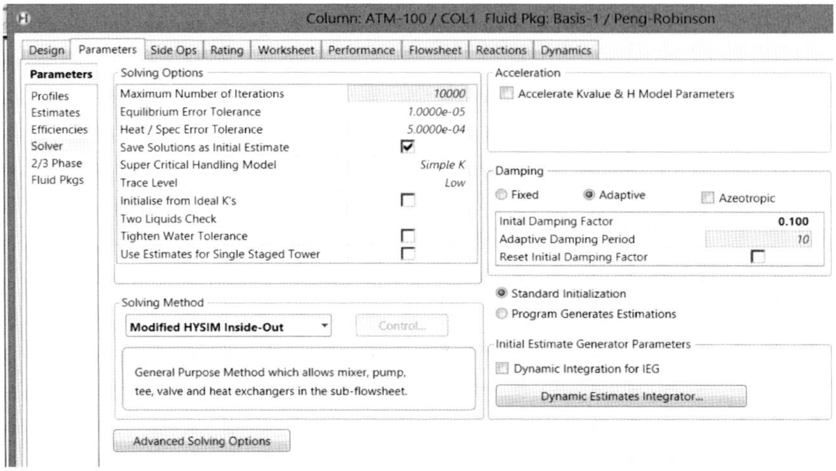

Figure 2.54 Column convergence options and the modified HYSIM inside-out algorithm: Parameters → Solver → Solving Method (modified HYSYS inside-out) and Damping (Adaptive).

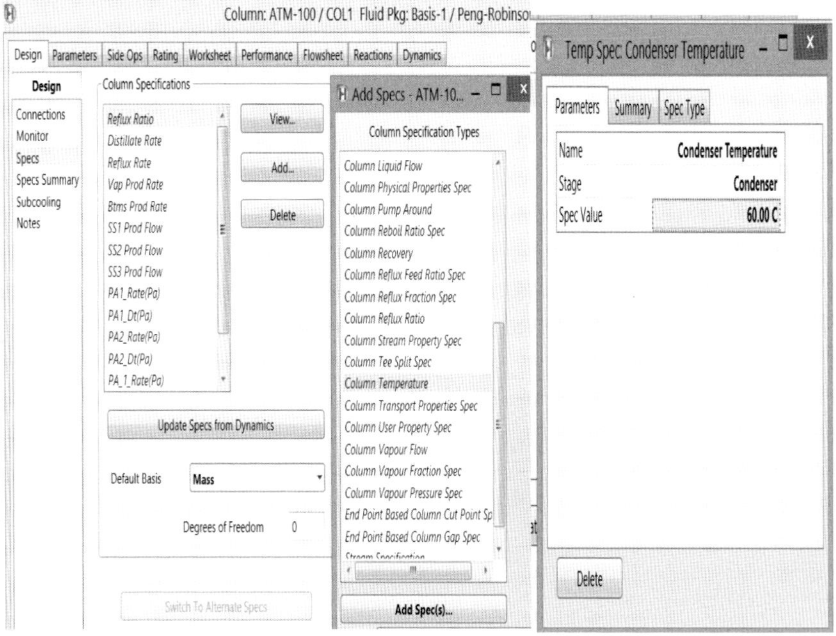

Figure 2.55 Procedure to define a design specification: an example of specifying the condenser temperature.

simulation typically converges quickly if we follow the step-by-step procedure. There may be rare cases where the simulation does not converge.

We demonstrate how to add custom petroleum properties to stream reports. As an example, we add the D86 5% and 95% temperatures to SS LGO product.

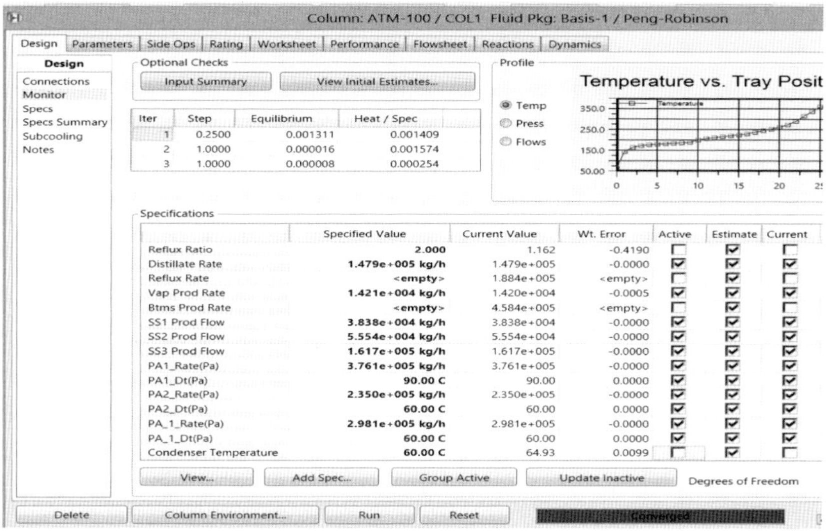

Figure 2.56 Column convergence after relaxing reflux ratio specification as an estimate and adding condenser temperature estimate.

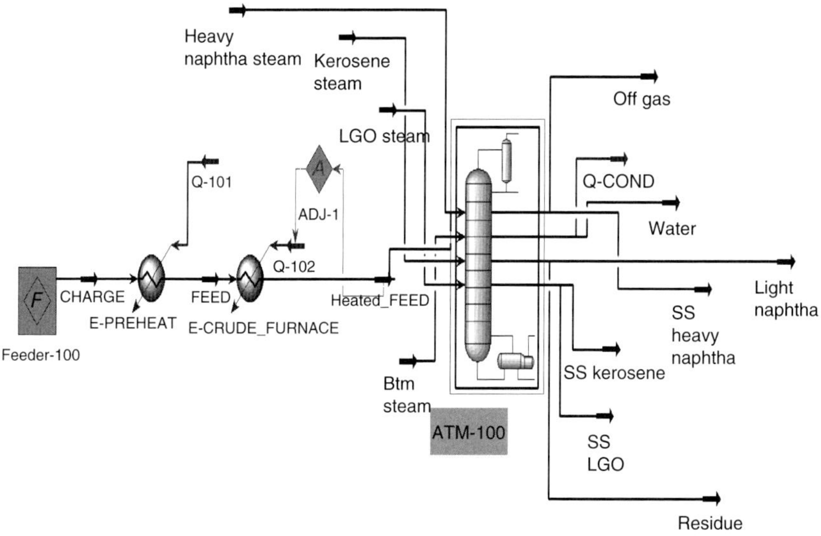

Figure 2.57 Completed atmospheric crude distillation section.

We follow the steps: Flowsheet → click on SS LGO stream → Material Stream SS LGO → Properties → click on "+" (append new correlation) → Correlation Picker → Petroleum → Choose D86 5% and apply; choose D86 95% and apply → Close (see Figure 2.58). We repeat the same steps to add the D86 5% and 95% temperatures for light naphtha, SS heavy naphtha, SS kerosene, and residue and save the resulting file as *CDU EX-7.hsc.*

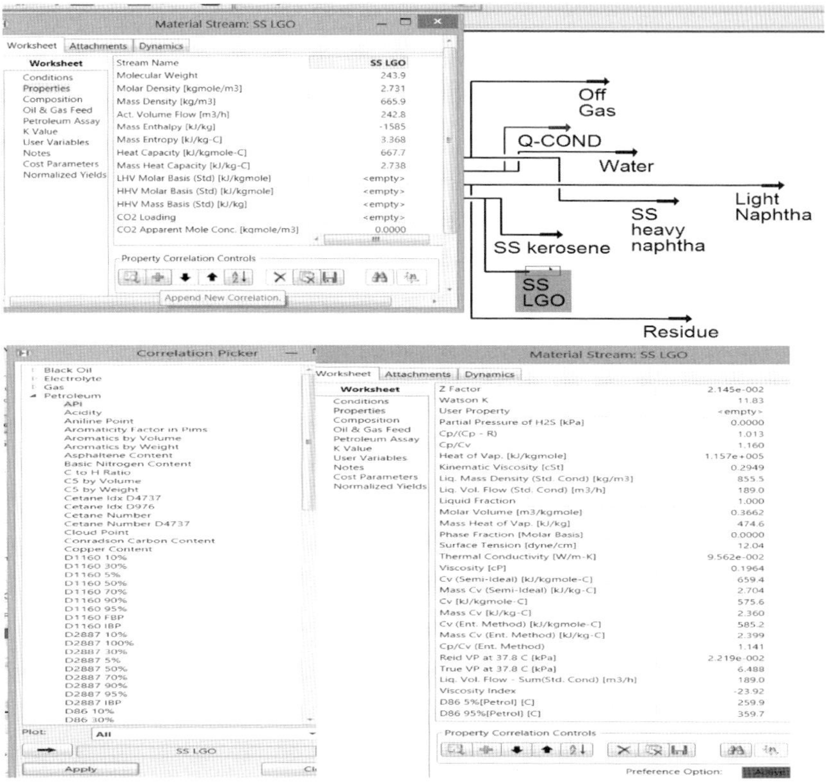

Figure 2.58 Adding custom stream correlations.

2.8.8 Post-Convergence

If the model behavior does not match plant performance even after we converge the model, several adjustments can be made to match plant performance. We suggest the following changes:

- The stripping in the column may be overoptimistic; removing stages especially above the flash zone may help predict lowered efficiency that we observe in columns.
- As steam rates will probably not be accurately measured, we may freely adjust these rates to attempt to match data. However, we must make sure that adjusted values fall within acceptable plant parameters.
- Shift the temperature change in the pumparound circuits. We may shift or lower heat duties on a given stage.
- If the column is specified with either overhead gas draw rate or bottoms rate, remove this specification or adjust this specification to match measurements of the remaining products. The overhead gas and bottoms product (topped crude) are not routinely measured.

We note that making the changes listed above will only show small changes in the yield and product quality profile. The most important contributor to yield and product quality is the feed composition. If there are significant errors, the feed composition is the most likely source.

2.9 Results

Before using the model to study different operating scenarios and perform case studies, we must ensure that the model matches the baseline column conditions and operating profiles. For the atmospheric distillation column, the important operating profile measurements are as follows:

- Column temperature profile – specifically the condenser, top dray, and bottoms temperature.
- Temperatures at the draw points of key products.
- Distillation curves of collected key products.
- Density of key products.

The ordering of the above list is significant. We expect to show good agreement with the column temperature profile first and then we should attempt to match the subsequent properties. In addition, it is very unlikely that the model will match plant behavior exactly. Kaes [3] has presented some guidelines to judge whether a model reflects the performance of a real column. We summarize these "reality checks" in Table 2.16.

With these considerations in mind, we present the results of the column simulation developed in the previous section. Figure 2.59 compares the measured column temperature with simulation results. In general, we are able to observe all the trends that Kaes [3] has described. The model stage temperature is higher than measured top stage temperature. In addition, we also note that there is a drop in temperature from the flash zone (Stage 25) to the column bottoms (Stage 27). The temperature of the bottoms stream is also lower than the temperature of the crude feed (366 °C at 3% overflash).

The next check in the model results is the prediction of the key product qualities. We typically check this by comparing the D86 (or TBP, whichever are

Table 2.16 Checks for validating column model predictions.

Model prediction or specification	Expectation/comments
Top stage temperature	Model prediction is generally higher than (7–15 °C) true column temperature
Bottom stage temperature	Model prediction must be lower than flash zone temperature due to isenthalpic cooling
	The temperature of the bottoms stream leaving the column model should be lower than feed temperature of the crude (5–7 °C)
Pumparound duties/side stripping steam rates	These values are not routinely measured during true column operation and may vary significantly. It is generally inadvisable to rely on these values to make a simulation converge
Product yields	It is not possible to match yields given a poor representation of feed. Feed crude assays may be too old or inaccurate to represent current operation. Backblending the products may be the only way to correctly represent column feed
Product quality	Adjust stripping steam rates to meet D86 5% point
	Adjust draw rate of subsequent cut to control the D86 95% point

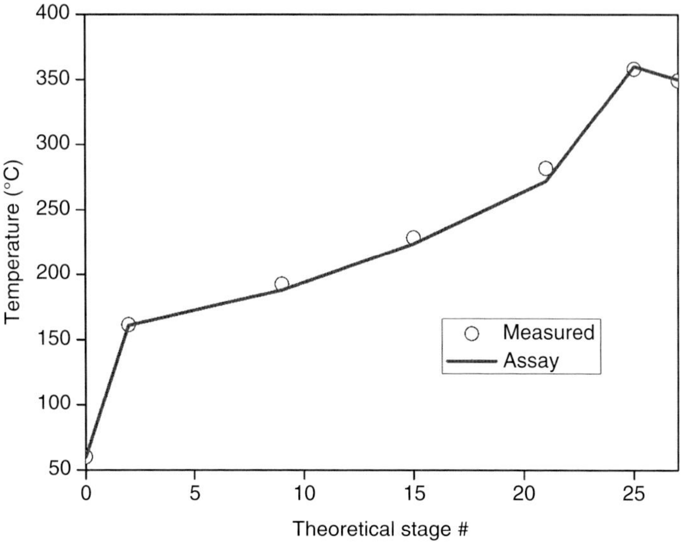

Figure 2.59 Comparison measured and predict column temperature profile.

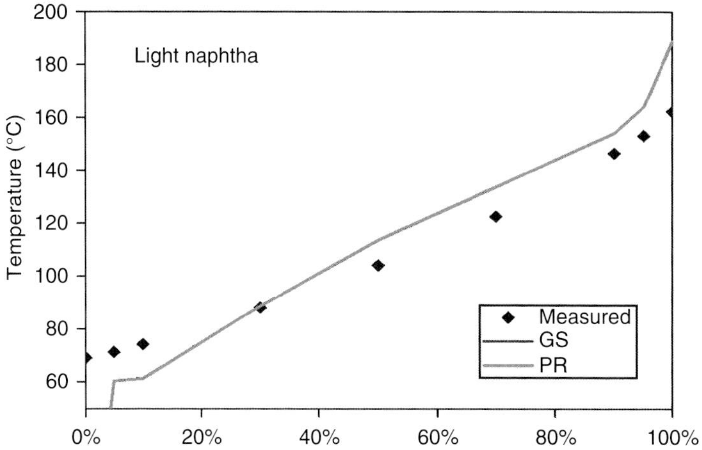

Figure 2.60 D86 comparison curve for light naphtha: GS = Grayson–Streed equation of state (EOS); PR = Peng–Robinson EOS.

available) curves for all liquids products. Figures 2.60–2.63 show the results of the model with measured values. In this case, we run the model with two different thermodynamic sets (GS –Grayson–Streed and PR – Peng–Robinson). See simulation file *CDU EX-8_GS.hsc.*

In general, we note good agreement with the D86 5% and 95% temperatures. In addition, we also correctly predict the flat distillation curve for "heartcut" draw of SS heavy naphtha and SS kerosene. Typically, more significant deviations appear in the initial and final boiling points of the distillation curves. This often results from various (and possibly) conflicting definitions of the initial and final boiling

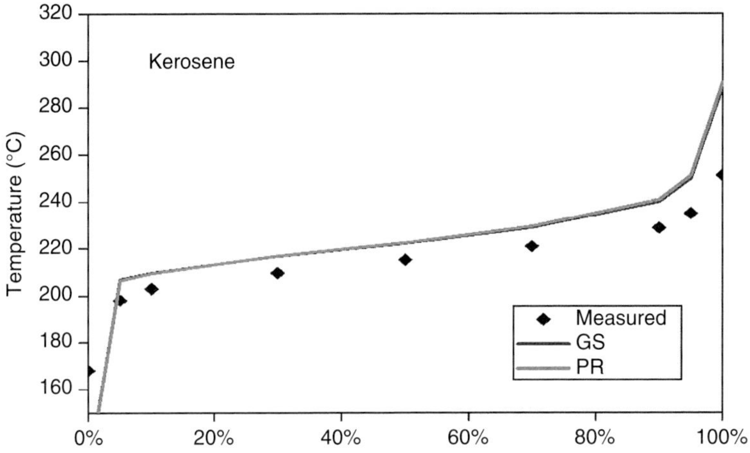

Figure 2.61 D86 comparison curve for kerosene.

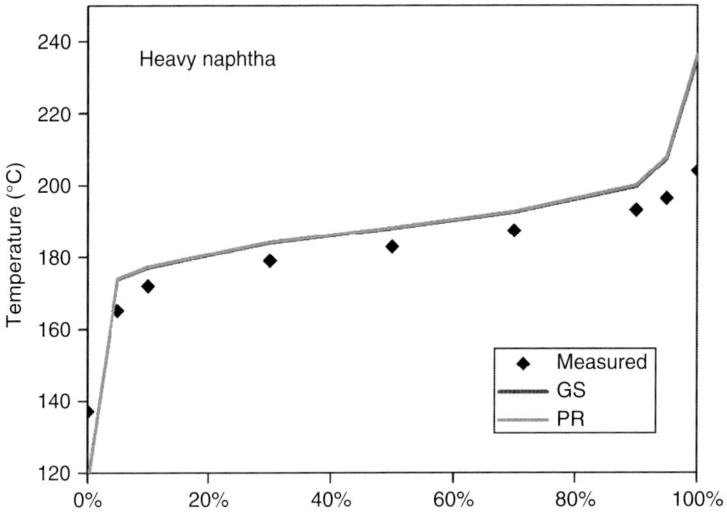

Figure 2.62 D86 comparison curve for heavy naphtha.

points in different simulators and the presence of very light components in a given cut. If accurate estimates for the light products (including the cracked light gases from the column boot) are not available, light products tend to distribute themselves throughout the column. This often leads to errors in the first few points of the distillation curve of light naphtha. We show this effect in Figure 2.60. We can typically alleviate these errors by providing better light gas estimates and specifying condenser temperature (instead of light product draw rates).

The density or specific gravity of the key products is also an important consideration when we verify the model predictions. Figure 2.64 compares the models. With most modern process simulations, it is quite easy to change the thermodynamic model and use sophisticated thermodynamic models. However,

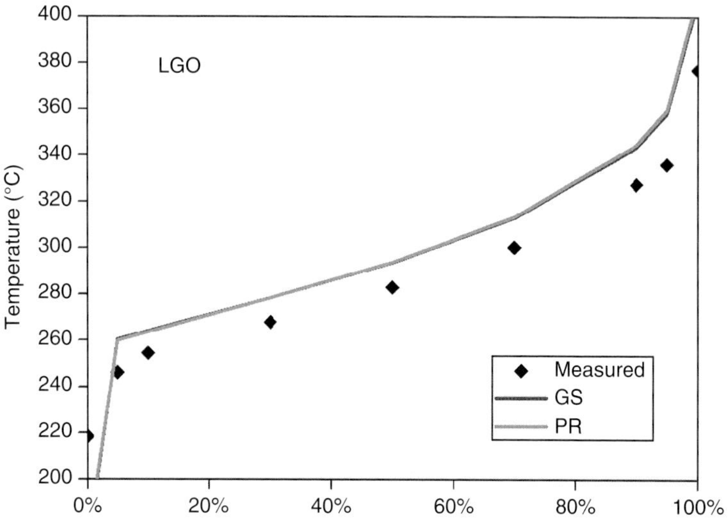

Figure 2.63 D86 comparison curve for LGO.

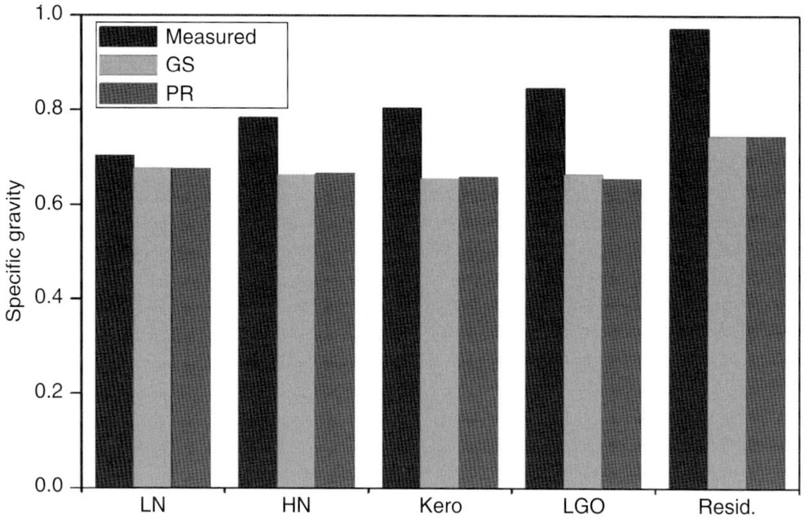

Figure 2.64 Comparison of measured and predicted product-specific gravity.

as shown in Figures 2.60–2.63, different thermodynamic models may give slightly different results. This is especially important for the crude columns, which have a large number of pseudocomponents. In general, while advanced equations of the state predict the K-values accurately, the associated predictions of density are quite poor when compared with a simpler Grayson–Streed model. Different process simulator vendors may provide individual options to keep rigorous K-value predictions using an equation of state, while supplanting other properties from simpler methods. In Aspen HYSYS, the COSTALD [13] liquid density correlation, Eq. (1.27), Section 1.10.4, gives accurate results with the flexibility of an equation of state. The model developer must be aware of these options and make sure to verify results before using the model to make predictions of column performance.

2.10 Model Applications to Process Optimization

After validating the model predictions with plant data, we would typically like to use the model to predict new operating scenarios or perform experiments that are too costly or otherwise prohibitive in actual atmospheric distillation. Refiners spend significant effort to develop models, but they are rarely used again. Often times, the users neglect these models, while the real column operation continues to change. Thus, when users actually run models, the predictions are far removed from process reality. The simple way to avoid this model stagnation is to use model to help make many different kinds of routine decisions. In this section, we consider a few common scenarios and use the model to investigate these scenarios.

2.10.1 Improve the 5% Distillation Point for an Individual Cut

As the supply and demand of global crude change, heavier crudes become more attractive to process. However, many existing columns cannot produce cuts that meet distillation product specifications. Many process changes could improve the distillation curve of a given product. However, it may be unclear what the side effects of a given change could be. In this case study, we look at how we can improve the distillation curve (5%) of the heavy naphtha and kerosene cuts. One option is to draw more or less of a particular cut to force the distillation curve to shift. However, this will affect other product draws as well.

Nelson [15] noted that the "the initial boiling point of side draw products is always low, and must be corrected by either steam stripping or reprocessing." Consequently, we perform a case study to show the effect of the stream stripping rate on both side strippers of the column. We note that as the steam rate increases, the deviation from the base case is significantly positive. We show the results of the case study in Figures 2.65 and 2.66. An important side effect, in the case of heavy naphtha, is that the distillation curves of the other products remain unaffected. However, when we increase the steam stripping rate to the kerosene side stripper, there is an appreciable loss in the D86 5% temperature of the LGO product. Depending on the subsequent processing of this LGO, this may not be a desirable situation. The use of a properly built column model can advise the refiner of such undesirable changes to the process.

2.10.2 Change Yield of a Given Cut

Modern refineries operate within strong economic, regulatory, and process constraints. Many times, the preferable operating mode for the atmospheric distillation unit may not be the operating mode that maximizes the yield of the most valuable product from the distillation unit alone. The atmospheric

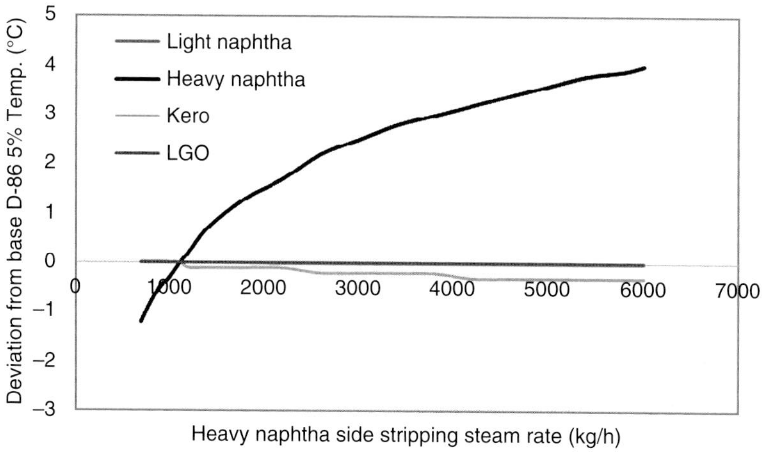

Figure 2.65 Change in distillation curves as a function of heavy naphtha stripping steam rate.

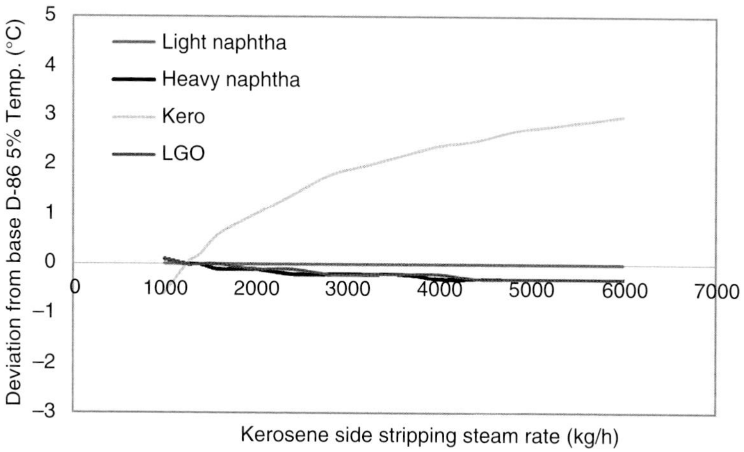

Figure 2.66 Change in distillation curves as a function of kerosene stripping steam rate.

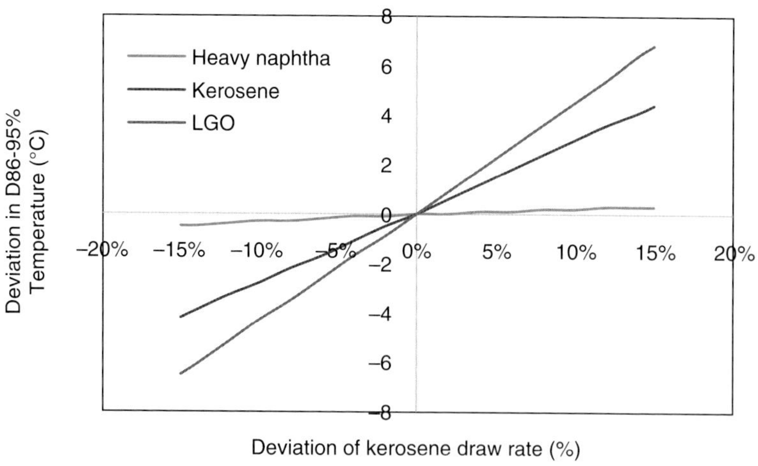

Figure 2.67 Change in D86-95% as a function of kerosene draw rate.

column operates in concert with many other units in the refinery. Therefore, it is important to understand how the product yield slate changes with different draws of a given cut.

In Figure 2.67, we use the model developed in the earlier sections to study the effect of changing the kerosene draw rate (while holding other draw rates constant), and how this affects the properties of the neighboring cuts. We note that as we increase the kerosene draw rate, the D86 95% point of the LGO increases significantly, while there is no appreciable change in the D86 profile of heavy naphtha. This indicates that if we draw more kerosene, the contribution comes mostly from the heavier portion of the crude and that heavier material is moving up the column.

Figure 2.68 shows the D86 5% point affected by the same change in kerosene yield. The interesting effect here is that the 5% point changes less significantly than the 95% point. This indicates that there is some opportunity to change the steam rate to the LGO and kerosene side strippers to manage the product profile while keeping the yield of heavy and light naphtha relatively constant. Another option that we could explore is increasing the overflash in the feed heater. The overflash is the primary control for how much heavy material is available while the stripper stream and draw rates can serve as finer controls on the heavy material shifts.

2.10.3 Workshop 2.1 – Perform Case Studies to Quantify the Effects of Stripping Steam Rate and Product Draw Rate

We open the converged simulation model, *CDU-EX-6.hsc*, and save it as *Workshop 2.1.hsc* to demonstrate how to use the "case studies" tool in HYSYS (see Figure 2.69).

We define the independent variables in Figures 2.70 and 2.71 and dependent variables in Figures 2.72 and 2.73.

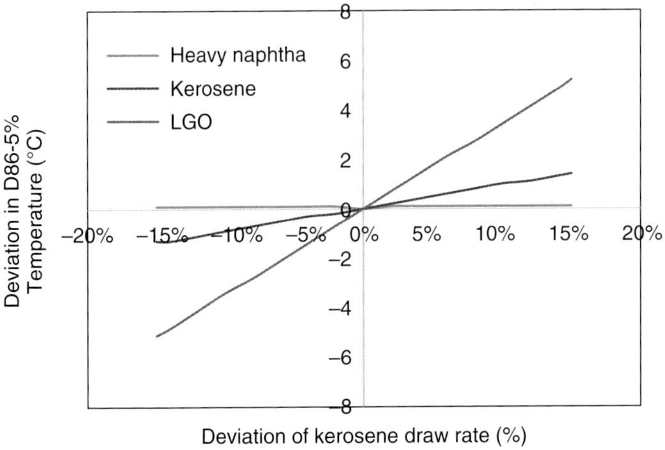

Figure 2.68 Change in D86-5% as a function of kerosene draw rate.

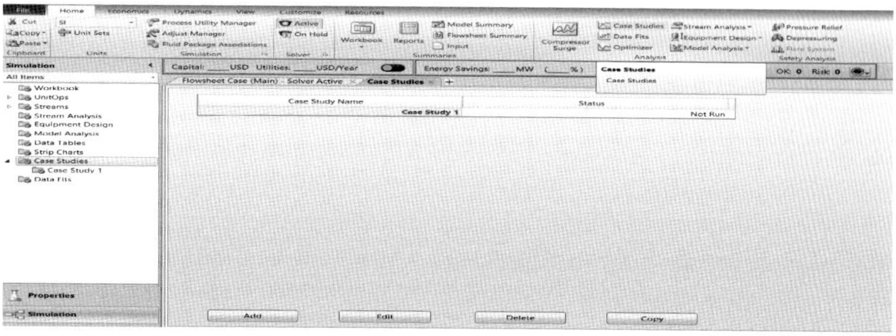

Figure 2.69 Start a new case study; Case Studies → Add → Case Study 1 → Edit.

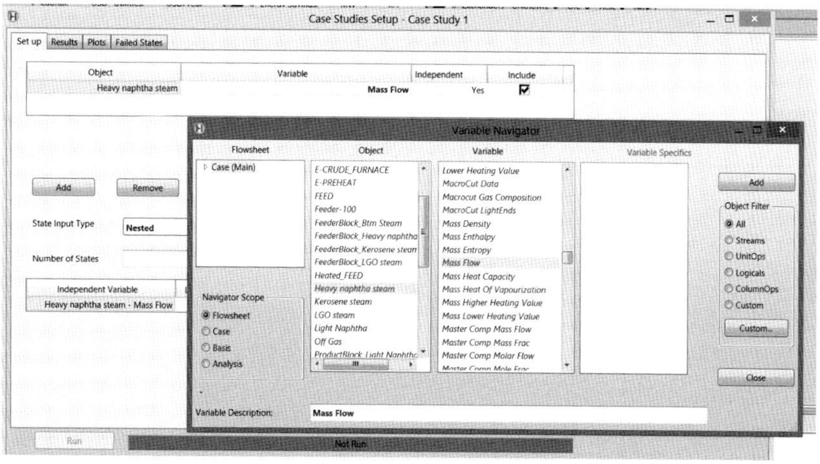

Figure 2.70 Define independent variables: (1) Flowsheet – Case(Main) → Object – Heavy naphtha steam → Variable – Mass Flow → Add. (2) Repeat the same for kerosene steam mass flow.

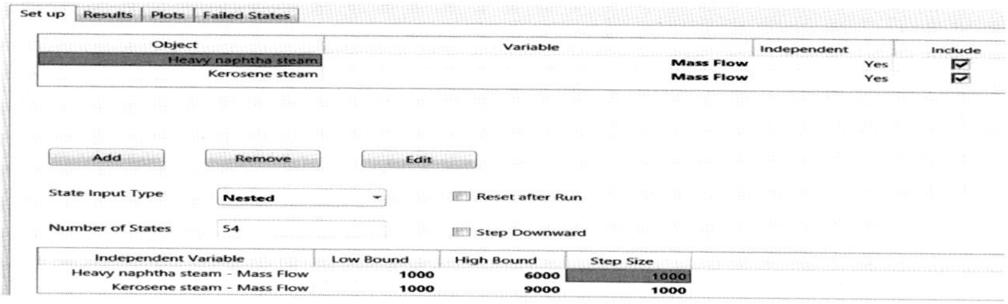

Figure 2.71 Specify lower and upper bounds of independent variable and step size.

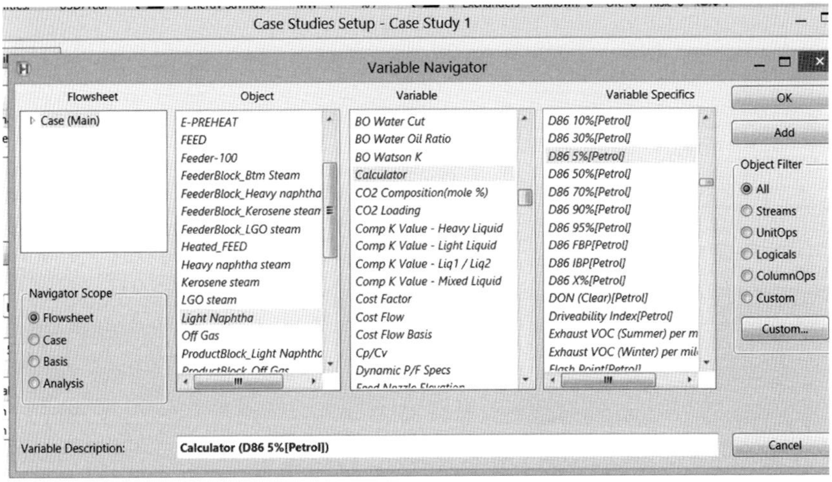

Figure 2.72 Define dependent variables: (1) Flowsheet – Case(Main) → Object – Light naphtha → Variable –Calculator → Variable Specifics – D86 5% (Petrol) → Add; (2) repeat the same for SS Heavy Naphtha, SS Kerosene, and SS LGO D86 5% (Petrol).

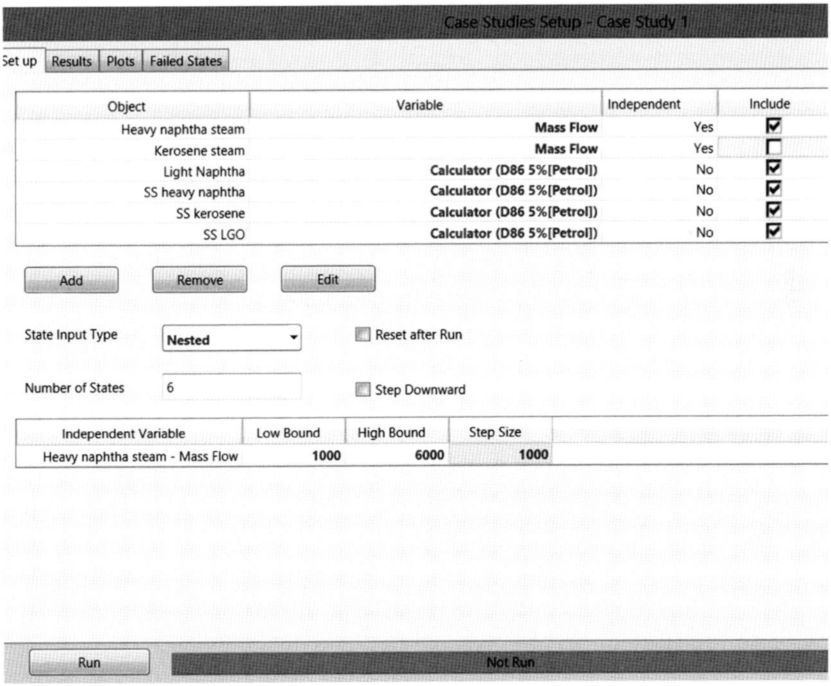

Figure 2.73 List of independent and dependent variables. Include one independent variable at a time and run simulation.

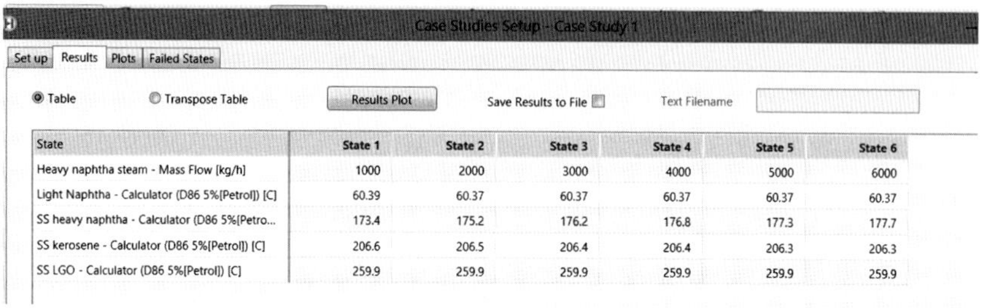

State	State 1	State 2	State 3	State 4	State 5	State 6
Heavy naphtha steam - Mass Flow [kg/h]	1000	2000	3000	4000	5000	6000
Light Naphtha - Calculator (D86 5%[Petrol]) [C]	60.39	60.37	60.37	60.37	60.37	60.37
SS heavy naphtha - Calculator (D86 5%[Petro...	173.4	175.2	176.2	176.8	177.3	177.7
SS kerosene - Calculator (D86 5%[Petrol]) [C]	206.6	206.5	206.4	206.4	206.3	206.3
SS LGO - Calculator (D86 5%[Petrol]) [C]	259.9	259.9	259.9	259.9	259.9	259.9

Figure 2.74 Effect of heavy naphtha steam mass flow on the D86 5% temperatures of CDU products.

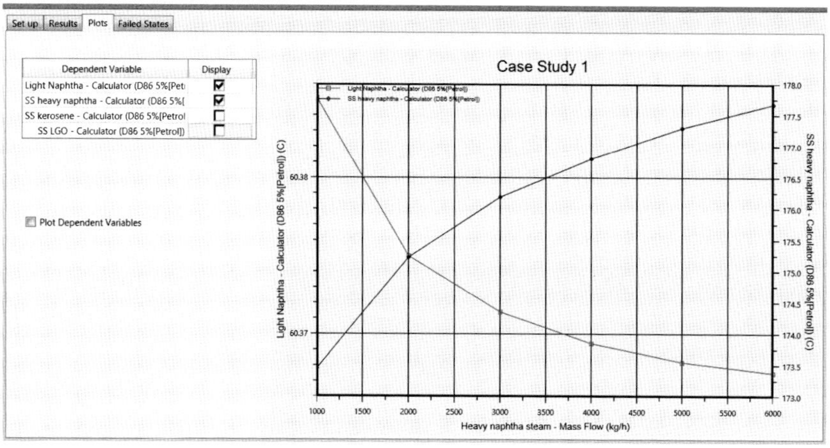

Figure 2.75 Effect of heavy naphtha mass flow and the D86 5% temperatures on light naphtha and SS heavy naphtha products.

Figures 2.74 and 2.75 illustrate the resulting table and plot from the case study. We save the simulation file as *Workshop 2.1.hsc.*

2.11 Workshop 2.2 – Rebuild Model Using "Backblending" Procedure

The procedure we used in earlier sections to build and validate the crude unit model relies on the availability of crude assays and associated density curves. Although this procedure can provide very accurate results, it can be challenging to implement directly. Often, the composition of the crude entering the atmospheric unit is ill defined and only product yield and operating measurements are available. How do we construct a model using this limited amount of information?

In this workshop, we walk through the process of building a "backblending" model. A "backblending" model refers to the process of reconstructing the feed from known product measurements and to run the crude unit model using this reconstructed feed. We start this process with an analysis of the products that includes distillation curve data and the bulk density at a minimum. Refineries routinely measure this type of data, which should be generally available for modeling

Table 2.17 Product yield and properties required for "backblending."

ASTM D86 (°C)	Heavy naphtha (HN)	Light naphtha (LN)	Kerosene	Light gas oil (LGO)	Residue
IBP	69	137	168	218	323[a]
5%	71	165	198	246	358[a]
10%	74	172	203	254	381[a]
30%	88	179	210	268	459[a]
50%	104	183	215	283	543[a]
70%	122	187	221	301	656[a]
90%	146	193	229	328	877[a]
95%	153	196	235	337	1009[a]
FBP	162	204	251	378	1178[a]
Standard liquid density (kg/m^3)	703.7	782.6	803.4	845.6	971.3
Yield (wt%)	16.9	4.39	6.35	18.2	54.16
Yield (ton/day)	3549	921	1333	3822	11375

Note: Distillation curves have been converted to D86 curves.
a) Values have been estimated.

purposes. Table 2.17 gives product yield measurements for the model developed earlier in this chapter. We will use this set of yields as a basis to reconstruct the crude feed entering the unit.

Table 2.17 contains the distillation curve and standard liquid density (or specific gravity) from each cut. If a complete distillation curve is not available, we recommend using the beta distribution fitting method to identify missing values (see Workshop 1.2, Section 1.4). The residue distillation curve may not be available routinely. We use the simple correlation outlined by Kaes [3] to identify key points on the distillation curve as a function of residue density. We can then use the same beta distribution fit to complete the entire required distillation curve. Finally, we also require the light gas composition (C1–C5) leaving the naphtha and the overhead products.

2.11.1 Import Distillation Data into Aspen HYSYS Oil Manager

We continue the development in Section 2.8.1 and open the simulation file, *Crude Assay Only.hsc,* in which we have already defined two petroleum assays, ArabianLight and ArabianHeavy within the petroleum assay manager. We rename the file as *CDU-Backblending-1.hsc* (see Figure 2.76).

Following Workshop 1.5, Section 1.7, we define the petroleum assays for the five CDU products specified in Table 2.17 within the petroleum assay manager. This involves entering the distillation curve data and the standard liquid density. Figure 2.77 shows the Assay tab in the petroleum assay manager once we have added all the product specifications. We note that the Aspen HYSYS continually updates and verifies the properties of the assays. If the calculated properties require pseudocomponents with boiling points higher than about 1100 °C, Aspen

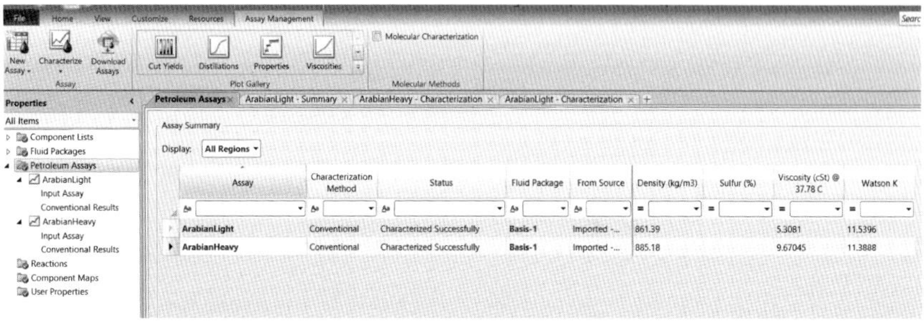

Figure 2.76 Two petroleum assays, ArabianLight and ArabianHeavy, defined within the petroleum assay manager.

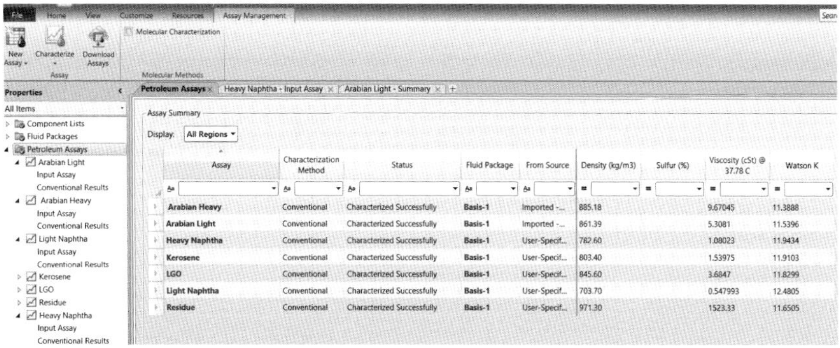

Figure 2.77 Created assays of all CDU product cuts.

HYSYS will indicate that the pseudocomponents may not yield accurate results. Although it is generally safe to ignore this warning for atmospheric units, the pseudocomponents may not be sufficient for the VDU. An alternate correlation may be required to account for these high boiling cuts.

2.11.2 Define a New Blend of the Backblended Crude Feed

We continue with the file, *CDU-Backblanding-1.hsc*, and resave it as *CDU-Backblending-2.hsc*. The next step is to create an appropriate blend of the product assays to represent the backblended crude feed. We put a petroleum feeder on the flowsheet as shown in Figure 2.78 and create a new blend stream with the name "BackBlended." We click "View" to edit the feeder and add the respective flow ratio for each product cut.

We use the flow ratio based on the flow rates given in Table 2.17 and the product assays to create a new blend, as shown in Figure 2.79.

Note that the light components are still not yet part of the reconstructed crude definition. We approximate the light gas component mole flows by copying those from the charge stream in our simulation file, *CDU EX-6.hsc*, for our converged CDU flowsheet, Figure 2.57. We show in Figure 2.80 these light gas component mole flows and their duplications to specify the BackBlended_Gas stream in the main flowsheet. We assume the BackBlended_Gas stream to

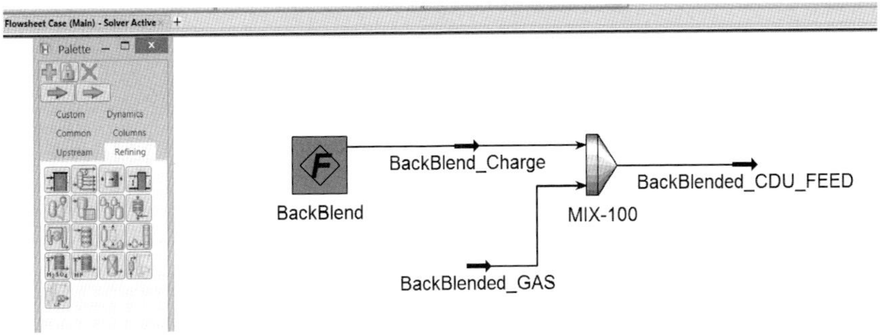

Figure 2.78 Add a new petroleum feeder to reconstruct the backblended crude charge and insert a mixer to add light gas components to the reconstructed the crude feed.

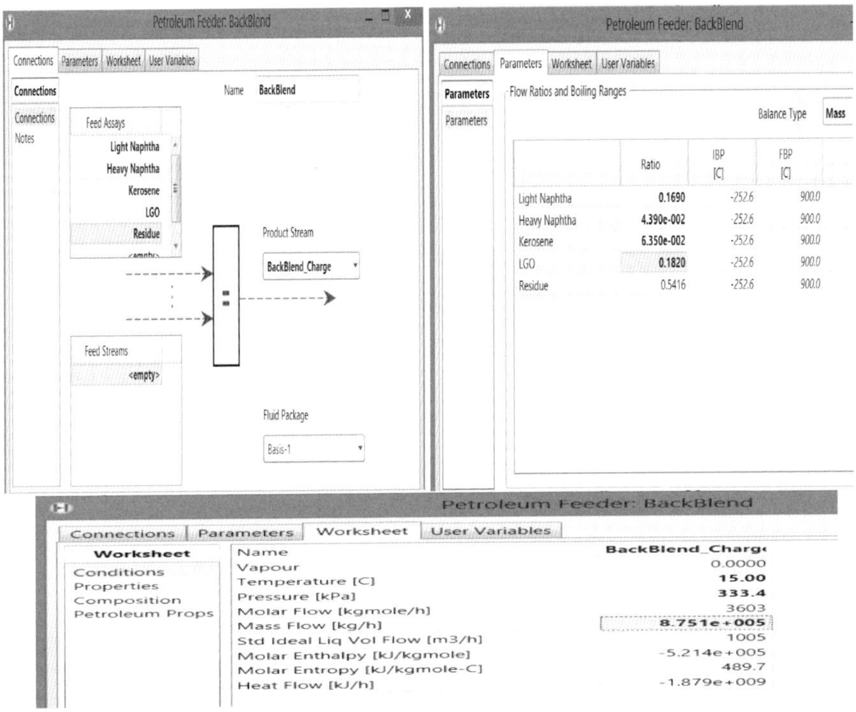

Figure 2.79 Specification of the BackBlend_Charge stream.

have the same temperature (15 °C) and pressure (333.4 kPa) as those of the BackBlended_Charge stream shown in Figure 2.79.

Figure 2.81 shows the specification of our mixer, MIX-100.

Throughout the previous steps from Figures 2.77 to 2.81, the petroleum assay manager automatically generates only *a single list of pseudocomponents* for both the assay blending and the backblending operations. We see this list of pseudocomponents from Properties → Component Lists → Component List-1, shown in Figure 2.82.

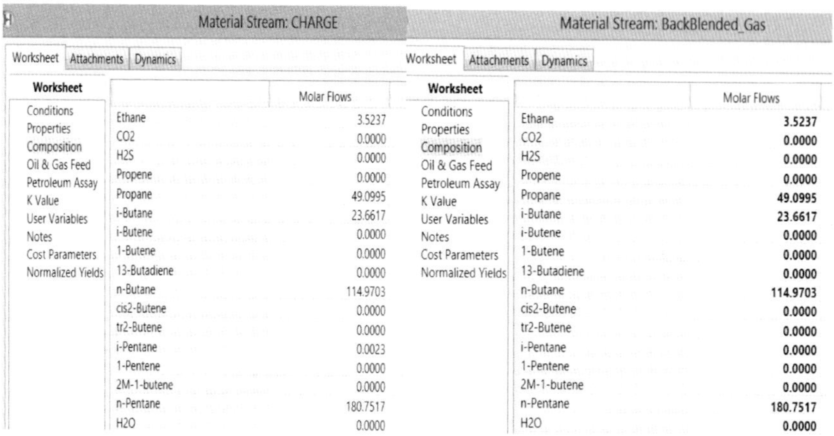

Figure 2.80 Gas component molar flows from original CHARGE stream and its duplication to BackBlended_Gas stream: total mole flow rate = 372.4 kg mol/h.

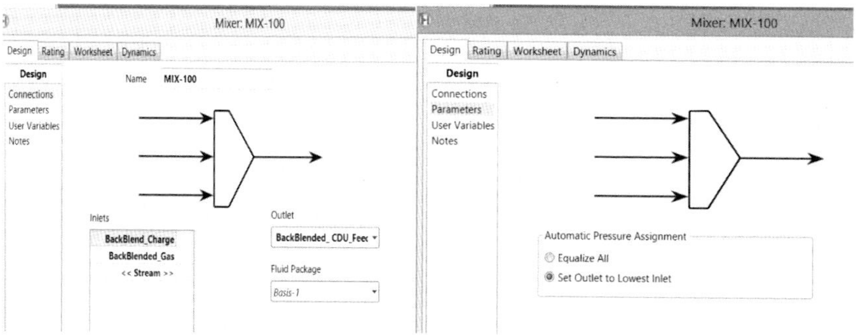

Figure 2.81 Specification of the mixer, MIX-100.

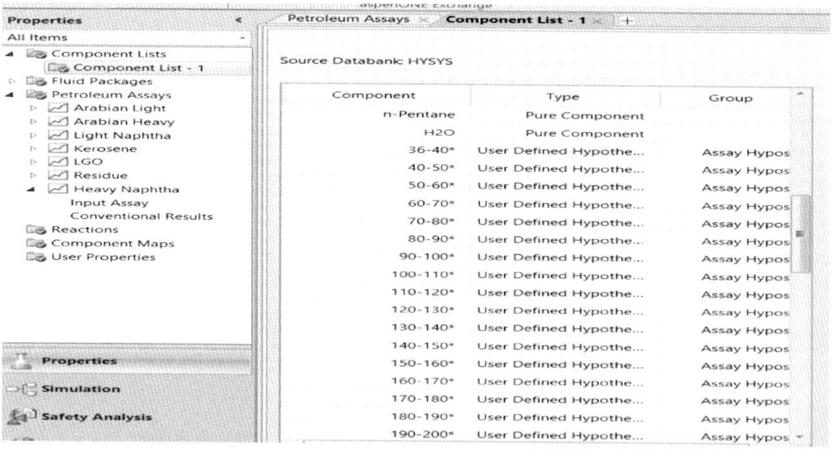

Figure 2.82 A single list of pseudocomponents generated by the petroleum assay manager.

By contrast, as we discussed previously in Table 1.3, Section 1.8, while using the old oil manager, each assay blend will generate its own set of component lists. In fact, in Figure 2.83 (which is generated by earlier versions of Aspen HYSYS Petroleum Refining when only the old oil manager was available), we see two sets of pseudocomponents, NBP[0]* and NBP[1]*. The NBP[0]* components refer to the components created from the assay blending operation, and NBP[1]* components correspond to the components created from the backblending operation. Users of *the old oil manager* must be aware that continuously adding various blends can create a very large unmanageable component list. Hence, we strongly recommend taking advantage of the significant improvements of *the new petroleum assay manager* over the old oil manager.

Figure 2.84 shows the resulting simulation flowsheet. We save the file as *CDU-Backblending-2.hsc.*

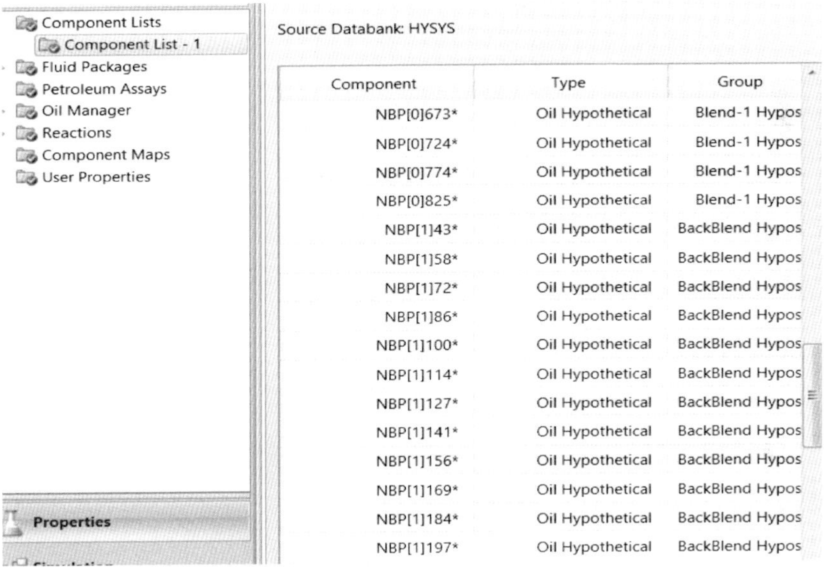

Component	Type	Group
NBP[0]673*	Oil Hypothetical	Blend-1 Hypos
NBP[0]724*	Oil Hypothetical	Blend-1 Hypos
NBP[0]774*	Oil Hypothetical	Blend-1 Hypos
NBP[0]825*	Oil Hypothetical	Blend-1 Hypos
NBP[1]43*	Oil Hypothetical	BackBlend Hypos
NBP[1]58*	Oil Hypothetical	BackBlend Hypos
NBP[1]72*	Oil Hypothetical	BackBlend Hypos
NBP[1]86*	Oil Hypothetical	BackBlend Hypos
NBP[1]100*	Oil Hypothetical	BackBlend Hypos
NBP[1]114*	Oil Hypothetical	BackBlend Hypos
NBP[1]127*	Oil Hypothetical	BackBlend Hypos
NBP[1]141*	Oil Hypothetical	BackBlend Hypos
NBP[1]156*	Oil Hypothetical	BackBlend Hypos
NBP[1]169*	Oil Hypothetical	BackBlend Hypos
NBP[1]184*	Oil Hypothetical	BackBlend Hypos
NBP[1]197*	Oil Hypothetical	BackBlend Hypos

Figure 2.83 Composite component list for multiple-assay blends generated by the old oil manager: NBP[0]* refers to the pseudocomponents created from the assay blending operation, and NBP[1]* components correspond to the pseudocomponents created from the backblending operation.

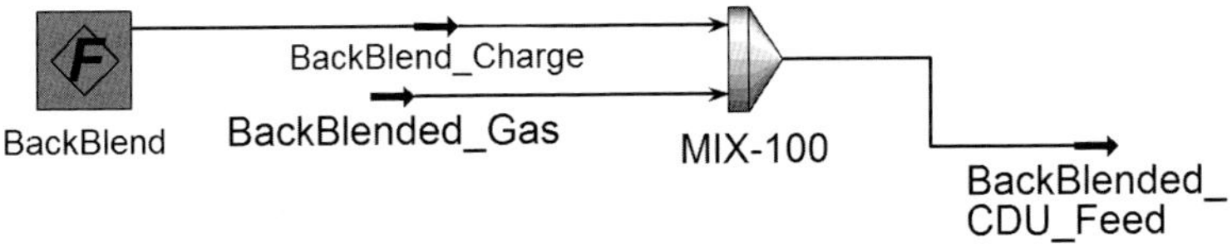

Figure 2.84 The "backblended CDU feed" replacing the "Charge" stream entering the preheat exchanger in Figure 2.27.

2.11.3 Build the CDU Model Based on the Backblended Feed

Following exactly the steps from Figures 2.29 to 2.35, we develop the simulation for the Heated_FEED to the CDU. This results in the flowsheet of Figure 2.85. We save the file as *CDU-Backblending-3.hsc.*

We then follow exactly the steps from Figures 2.36 to 2.57 to develop the complete CDU model based on the backblanded feed. Figure 2.86 shows the resulting CDU flowsheet, and we save the file as *CDU-backblending-4.hsc.*

2.11.4 Converging Column Model

When converging the updated column model, we may occasionally observe the errors shown in Figures 2.87 and 2.88. Aspen HYSYS indicates that two liquid phases may be possible on the bottom stage of the column. This is unlikely given the high temperature of the steam and stage pressure. We will modify the simulation *to include a water draw stream* to enforce that Aspen HYSYS performs a rigorous three-phase check on the bottom stage. The water draw removes all condensed liquid water from a stage to ensure that we can continue to apply the standard inside-out formulation to solve the column model.

To begin adding the water stream, we first enter the Column Environment for the CDU, as shown in Figure 2.89. We double-click on the column icon on the flowsheet and click on the "Column Environment" button. The column

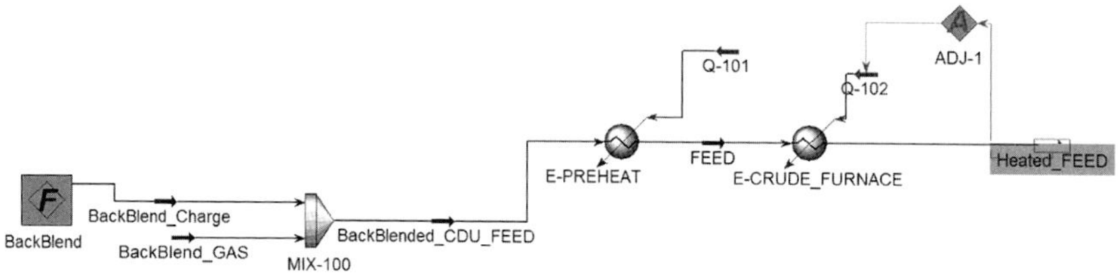

Figure 2.85 The Heated_FEED stream to the CDU developed from backblending.

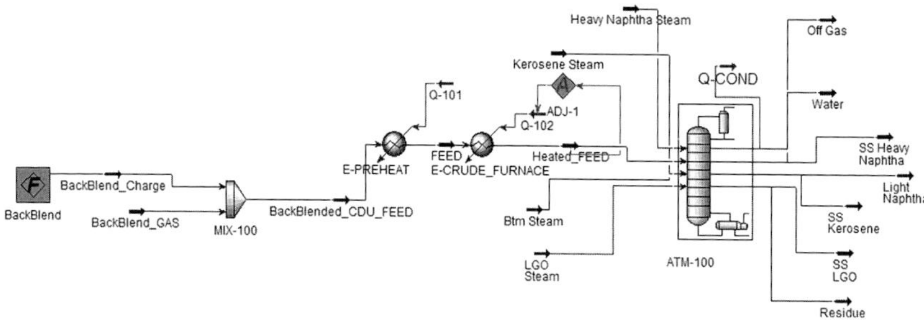

Figure 2.86 CDU flowsheet based on the backblended feed.

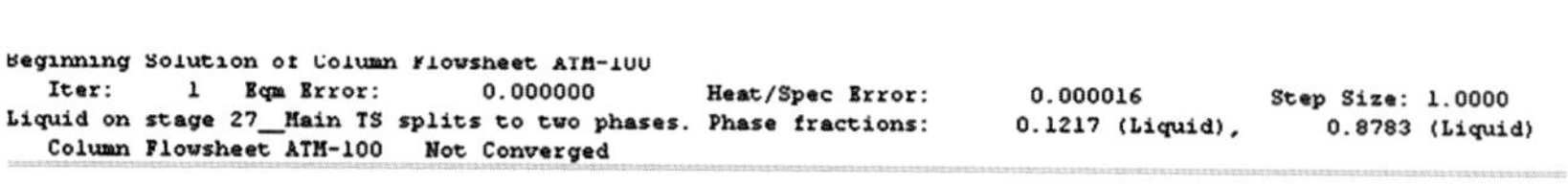

Figure 2.87 Solver output indicating two phases present.

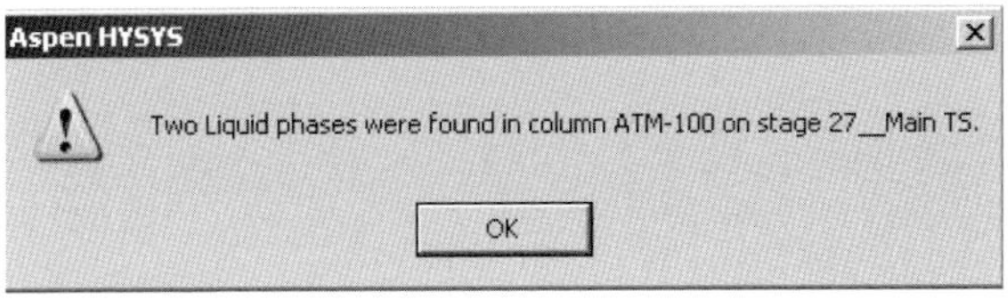

Figure 2.88 Additional warning to indicate that two liquid phases may be present.

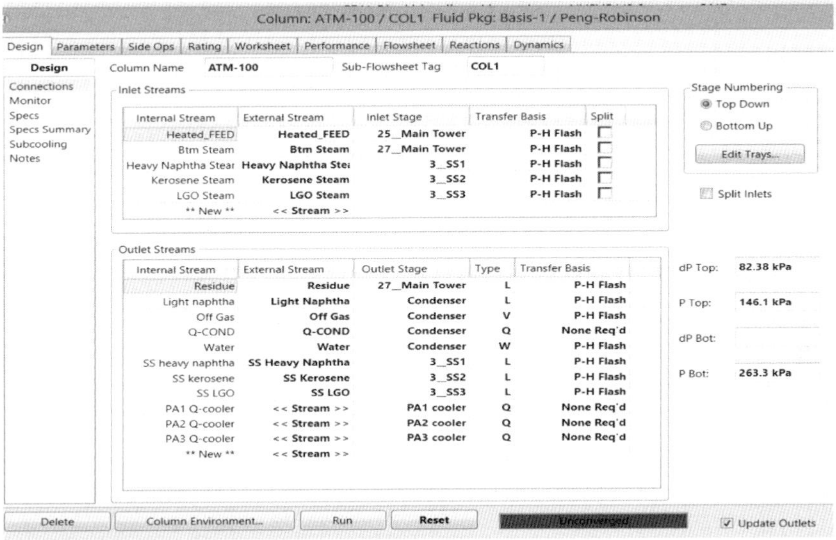

Figure 2.89 Enter Column Environment for column.

environment is essentially a subflowsheet that represents all the units internal to the column. In this environment, we can see the connections and draws for all pumparounds, side strippers, and so on for the column.

We first enter the column environment, as shown in Figure 2.90. We can then double-click on the column environment to bring up the advanced configuration for this column. From this interface, we can add nonstandard units such as thermosiphon reboilers. However, for this example, we focus on adding a water draw. We select the "Side Draws" section and create an additional "water draw" stream at tray 27 (the bottom tray), as shown in Figure 2.91. We also choose a "Total" water draw, indicating that all water will be removed from this stage. Partial water draws are not possible with the standard column solver method.

At this point, we can rerun the column to obtain the latest solution and we save the file as *CDU-backblending-5.hsc*. The column should converge quickly (<10 iterations). We must now confirm our assumption that no actual water condenses in the bottom stage. Figure 2.92 shows the results of the water draw stream. The water draw stream has a zero flow. The column solution may indicate that a small amount of water is present in the overhead light naphtha draw. This value is safe to ignore and has little bearing on the product results. In the following section, we compare the results of the "backblending" procedure and original assay method and discuss some reasons for the differences in predictions.

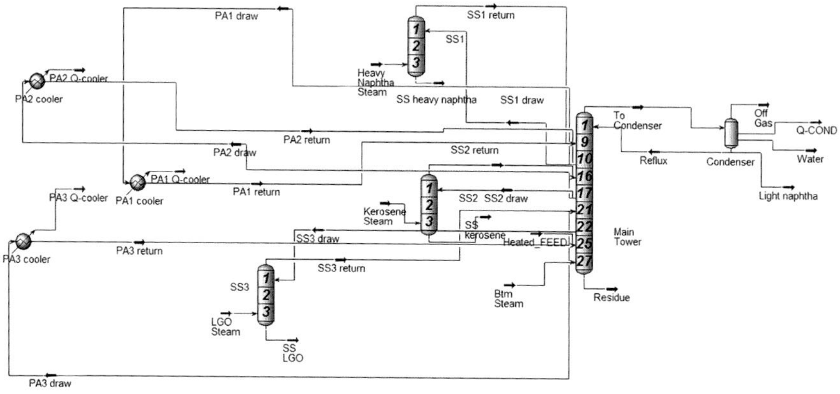

Figure 2.90 Column Environment for ATM-100.

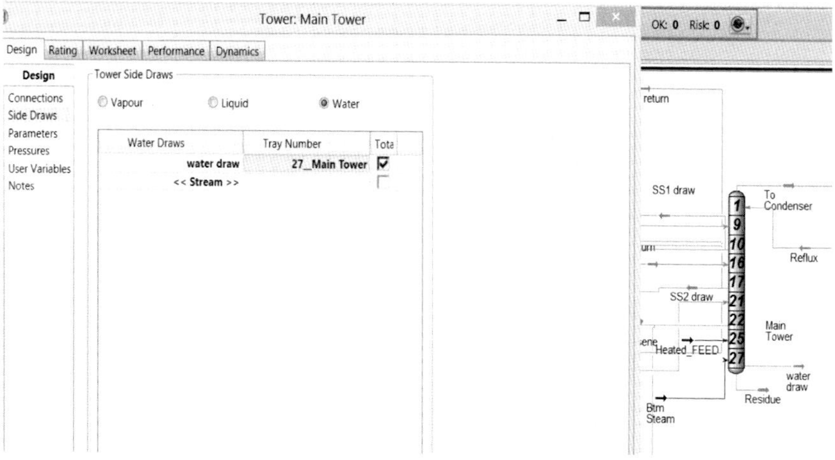

Figure 2.91 Create a "water draw" in stage 27.

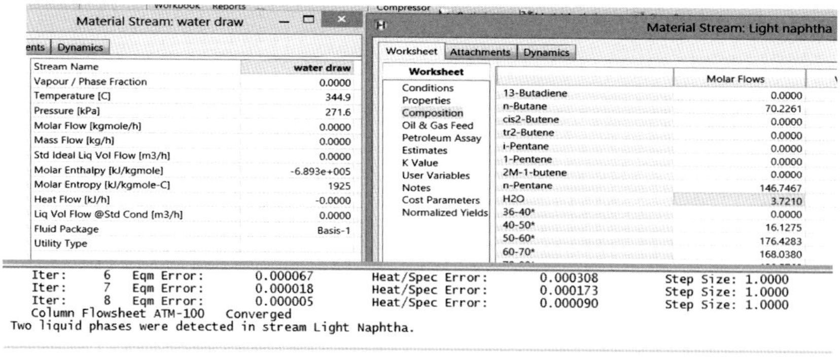

Figure 2.92 Water draw stream from column bottoms and detection of water and of two liquid phases (organic and aqueous) in stream light naphtha.

2.11.5 Comparison of Results

Figure 2.93 compares the results of the "backblending" procedure with the original assay blending procedure used for the initial column. In both cases, we used the PR EOS method to model the physical properties of the components. Both methods show good agreement with the measured temperature profile and follow the prediction guidelines given in earlier sections of this text. However, we note that the temperatures from the "backblending" case are consistently lower for the immediate stages than the assay case. This indicates that the temperatures in the distillation curves for the cuts (from the "backblending" case) will be consistently lower as well. Figures 2.94–2.97 compare these distillations curves and display the temperature deviation. This deviation tends to be most pronounced in the lighter cuts.

There are several reasons why this deviation occurs; the primary reason is that there is no detailed density distribution to model pseudocomponents for each of these cuts. The lack of a density distribution tends to create generally lighter components. As increasingly lighter components are drawn off as products in the light products, the higher boiling streams (kerosene and LGO) become lighter as well.

Matching plant results with "backblending" is generally more difficult. The most direct way to improve results is to obtain a distillation curve for the residue product. The residue product is a significant portion of the crude unit effluent and is quite heavy. These heavy components can affect the distribution of light components through all plant cuts.

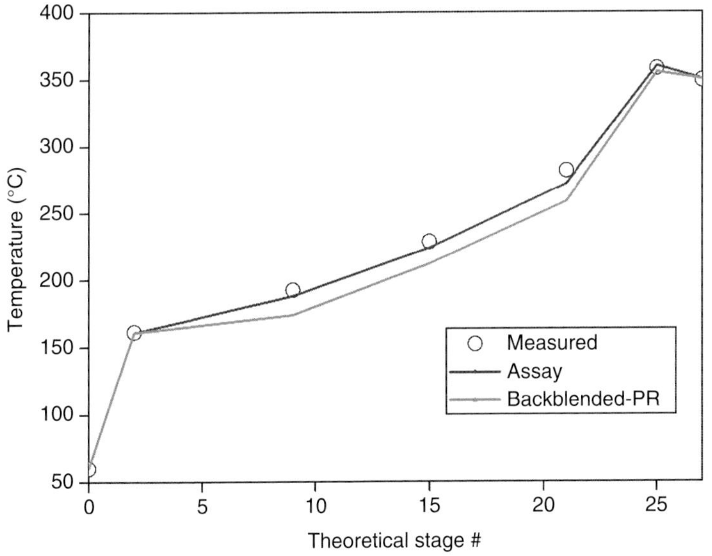

Figure 2.93 Comparison of temperature profile from "backblending" and traditional assay procedure: PR = Peng–Robinson equation of state (EOS).

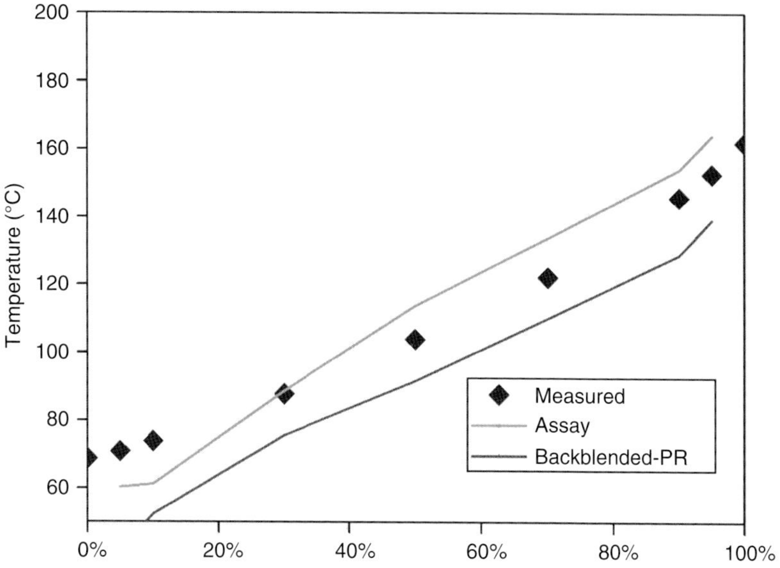

Figure 2.94 D86 comparison curve for light naphtha.

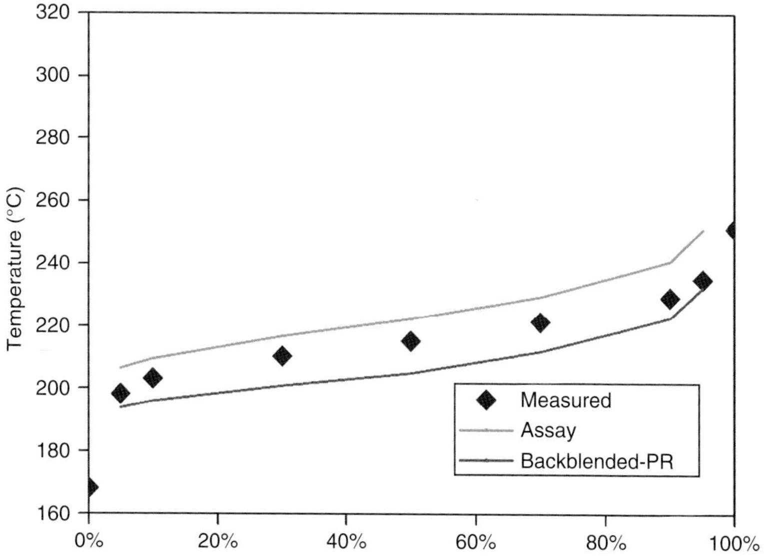

Figure 2.95 D86 comparison curve for kerosene.

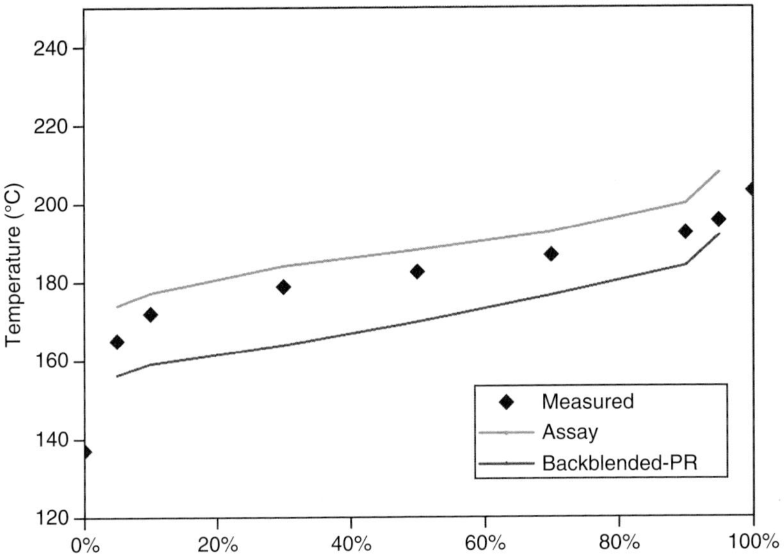

Figure 2.96 D86 comparison curve for heavy naphtha.

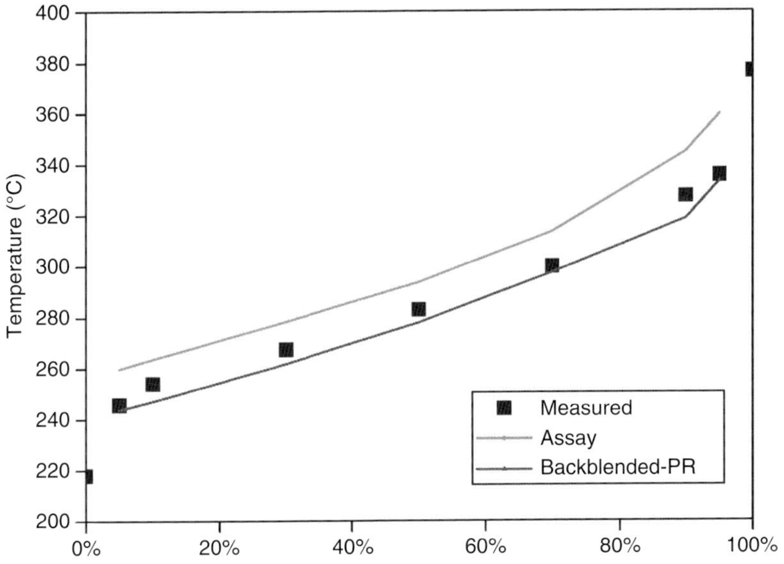

Figure 2.97 D86 comparison curve for LGO.

2.12 Workshop 2.3 – Investigate Changes in Product Profiles with New Product Demands

Seasonal demands and product quality constraints play an important role in determining the draw rates for various products from the crude column. In this workshop, we consider the effect of the draw rates on various product profiles. This type of study is particularly useful when the refiner wishes to shift product profiles in the summer or winter. An important consideration is the 10% distillation point as this point correlates well with other properties such as flash point and cloud point.

2.12.1 Update Column Specifications

We open the converged CDU simulation, *CDU EX-7.hsc*, Section 2.8.8, in which we have already added the D86 5% and 95% temperatures for all CDU products. We save the file as *Workshop 2.3.hsc.* We want to learn how to display these custom product properties while varying an independent variable using the spreadsheet tool available within Aspen HYSYS Petroleum Refining.

We must first change the specifications to allow product yields to vary, as we cannot increase the yield of one product while keeping all others constant. This violates the overall material balance of column. For this workshop, we allow the rate of the overhead vapor product to vary and keep the condenser temperature fixed. In Figure 2.98, we remove the vapor product specification and fix the condenser temperature to 65 °C (which is essentially fixing the initial boiling point of the light naphtha), the simulation converges quickly.

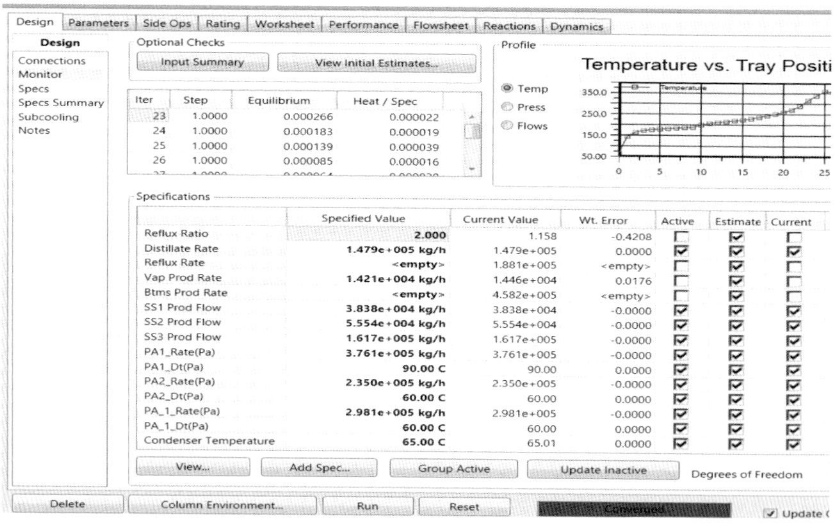

Figure 2.98 Remove vapor product rate specification and add condenser temperature specification.

2.12.2 Vary Draw Rate of LGO

We simplify the data collection process by using the spreadsheet object from the Aspen HYSYS palette. We create the flowsheet by selecting the spreadsheet, as shown in Figure 2.99.

We import the dependent variables to display in the spreadsheet as follows: Spreadsheet: SPRDSHT-1 → Parameters → Add Import → Select Import for cell → Flowsheet – Case (Main); Object – Residue; Variable – Calculator; Variable Specifics – D86 5%(Petrol) → OK. We repeat the same procedure for the D86 5% temperatures for other products (see Figures 2.100 and 2.101).

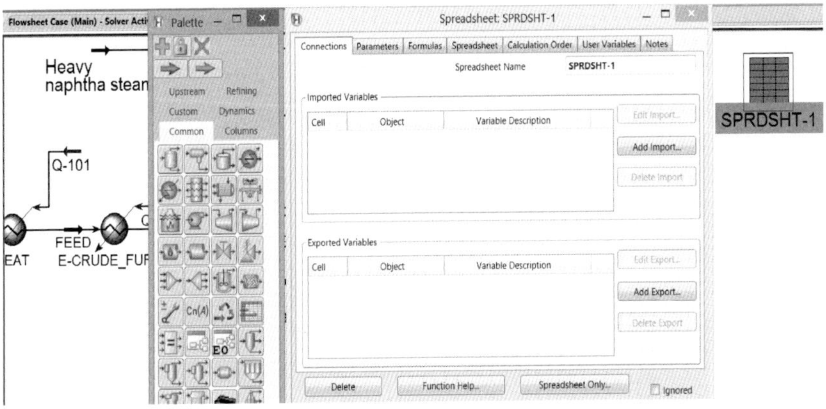

Figure 2.99 Add spreadsheet object from Aspen HYSYS palette – F4 → Common → double-click on Spreadsheet → SPRDSHT-1 and Connections interface.

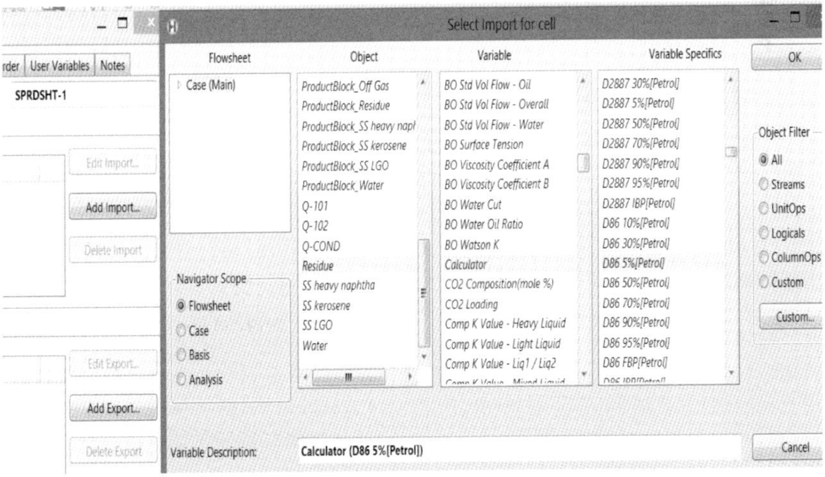

Figure 2.100 Select dependent variables to display in the spreadsheet.

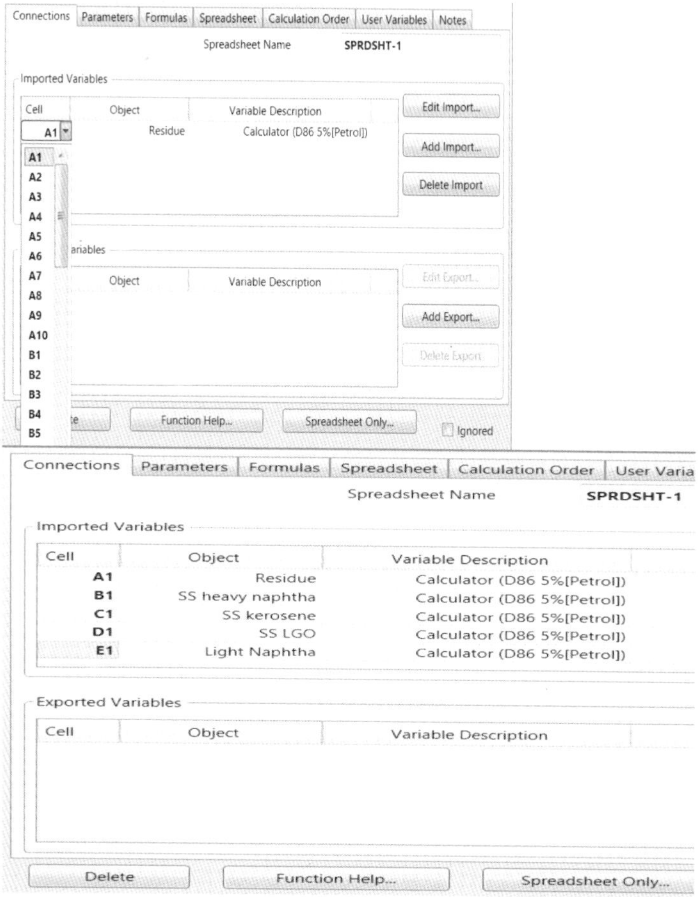

Figure 2.101 Select the variables to display in the chosen cell of the spreadsheet. To display five variables in cells A1–E1: Parameters → Spreadsheet Parameters → set the Number of Columns to 5. Click on "Spreadsheet Only" to export the values of the computed.

We open the monitor of the main column, ATM-100, and vary the SS LGO or the "SS3 product flow" from its current value of 1.617E5 to 2E5 kg/h. We click on "Spreadsheet Only" at the bottom of the spreadsheet to export the computed D86 5% temperature values to cells A1–E1 (see Figure 2.102).

We see from the resulting spreadsheets that when increasing the SS LGO draw rate, the D86 5% temperatures of the SS heavy naphtha, SS kerosene, and light naphtha essentially do not change. By contrast, the D86 5% temperatures of the SS LGO and the residue increase significantly from 259.9 to 264.3 °C and from 372.9 to 387.1 °C, respectively. This means that the LGO stream gets heavier (drawing material from residue) as the draw rate of LGO increases. However, if the refiner wishes lighter material in LGO stream, the steam stripping rate of the cut above LGO, that is, SS kerosene, should be increased.

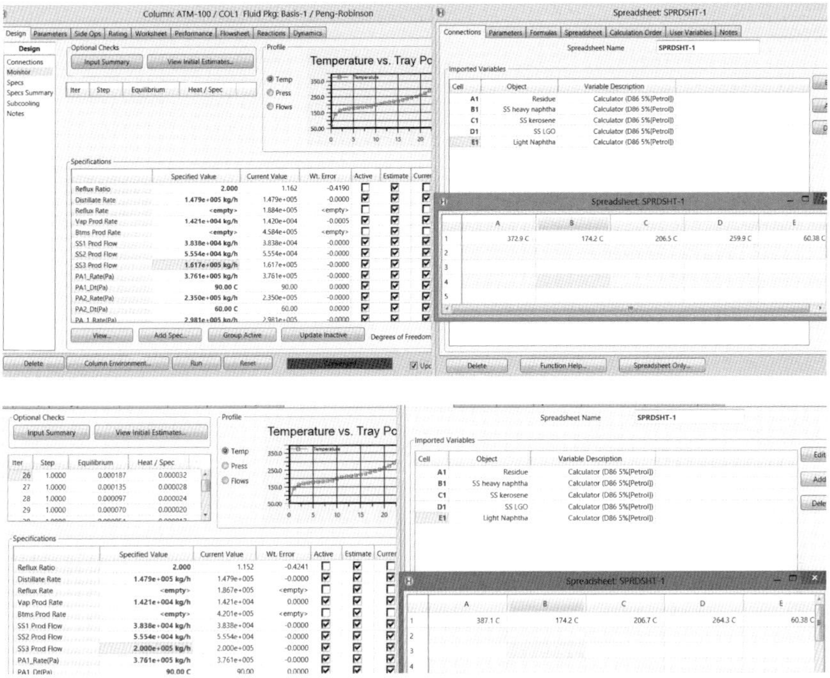

Figure 2.102 Computed D86 5% temperatures of product streams specified in spreadsheet cells A1–E1 when SS LGO or SS3 product flow varies from 1.617E5 to 2.0E5 kg/h.

2.13 Workshop 2.4 – Investigate the Effects of Process Variables on Product Qualities

Step 1. We open the converged CDU simulation, *CDU EX-7.hsc*, Section 2.8.8, in which we have already added the D86 5% and 95% temperatures for all CDU products. We save the file as *Workshop 2.4.hsc*. Click on Case Study (Figure 2.103).

Step 2. "Add" a case study (Figure 2.104).

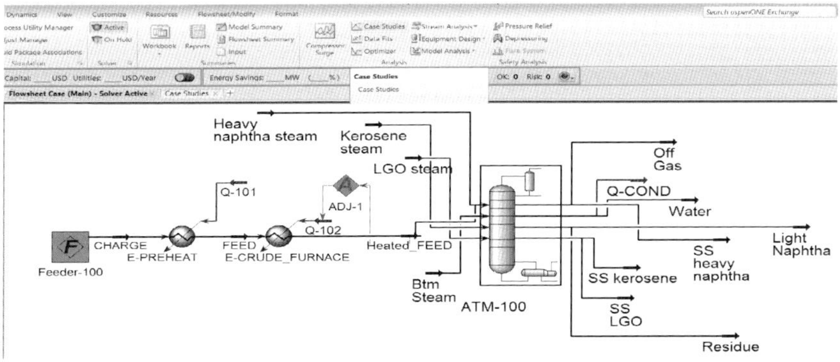

Figure 2.103 Activate "Case Studies."

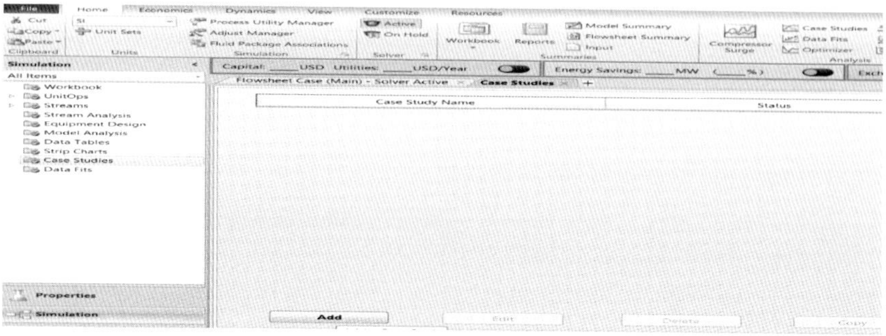

Figure 2.104 Add a case study.

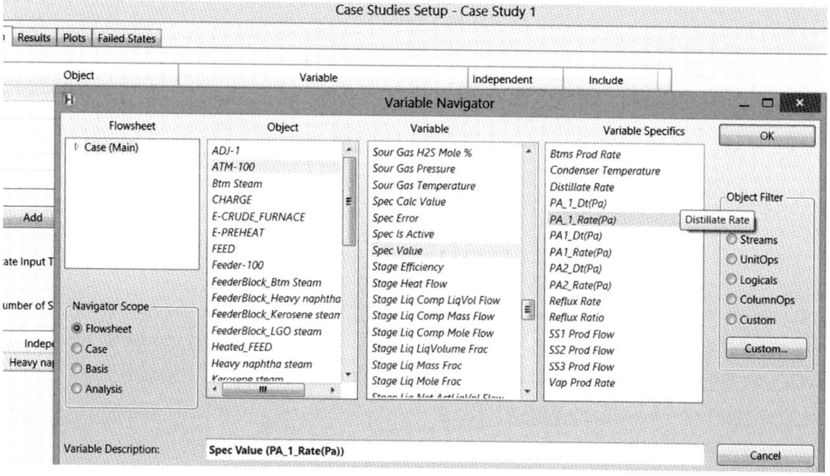

Figure 2.105 Add an independent variable, the specified heavy naphtha pumparound (PA1) flow rate.

Step 3. Add an independent variable, the specified heavy naphtha pumparound (PA1) flow rate: Variable Navigator → Flowsheet – Case(Main); Object – ATM-100;Variable – Spec Value; Variable Specifics – PA 1_Rate (Pa) → OK (Figure 2.105).

Step 4. Vary heavy naphtha pumparound (PA1) flow rate from 7.5E5 to 8.5E5 kg/h with a step size of 2E4 kg/h. Add product petroleum properties, D86 5% and 95% temperatures, for light naphtha, SS heavy naphtha, SS kerosene, SS LGO, and residue (Figure 2.106).

Step 5. Run the case study and view the result table and plot (Figures 2.107 and 2.108).

Step 6. Follow the same procedure and try to do a number of case studies yourself.

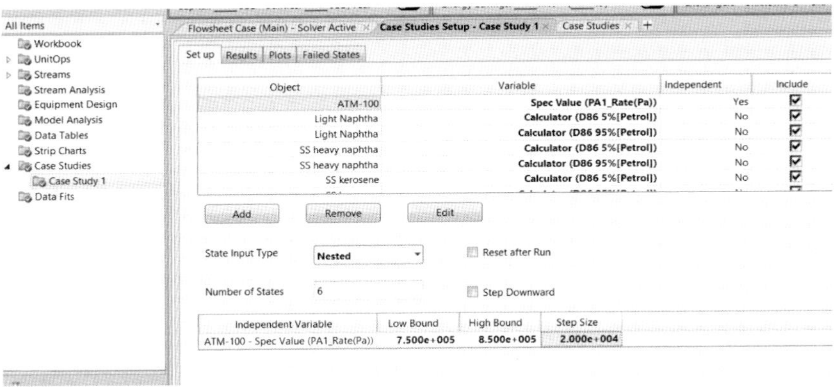

Figure 2.106 The resulting display of object, independent variable and its lower and upper bounds, and the product petroleum properties.

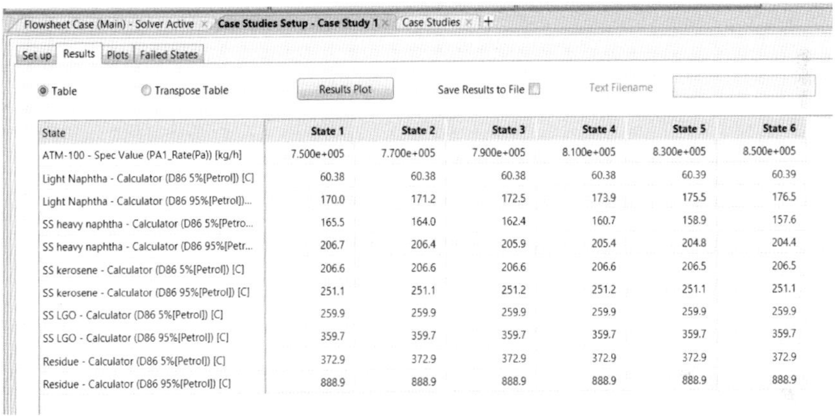

State	State 1	State 2	State 3	State 4	State 5	State 6
ATM-100 - Spec Value (PA1_Rate(Pa)) [kg/h]	7.500e+005	7.700e+005	7.900e+005	8.100e+005	8.300e+005	8.500e+005
Light Naphtha - Calculator (D86 5%[Petrol]) [C]	60.38	60.38	60.38	60.38	60.39	60.39
Light Naphtha - Calculator (D86 95%[Petrol])...	170.0	171.2	172.5	173.9	175.5	176.5
SS heavy naphtha - Calculator (D86 5%[Petro...	165.5	164.0	162.4	160.7	158.9	157.6
SS heavy naphtha - Calculator (D86 95%[Petr...	206.7	206.4	205.9	205.4	204.8	204.4
SS kerosene - Calculator (D86 5%[Petrol]) [C]	206.6	206.6	206.6	206.6	206.5	206.5
SS kerosene - Calculator (D86 95%[Petrol]) [C]	251.1	251.1	251.2	251.2	251.1	251.1
SS LGO - Calculator (D86 5%[Petrol]) [C]	259.9	259.9	259.9	259.9	259.9	259.9
SS LGO - Calculator (D86 95%[Petrol]) [C]	359.7	359.7	359.7	359.7	359.7	359.7
Residue - Calculator (D86 5%[Petrol]) [C]	372.9	372.9	372.9	372.9	372.9	372.9
Residue - Calculator (D86 95%[Petrol]) [C]	888.9	888.9	888.9	888.9	888.9	888.9

Figure 2.107 Case study result table.

2.14 Workshop 2.5 – Application of Column Internal Tools (Column Hydraulic Analysis)

We open simulation file, *CDU-blending-5*, and save it as *CDU-blending-internals*. Our goal is to become familiar with the column sizing (finding the column-section, CS, diameters), rating (performance evaluation of CSs of existing diameters), and column hydraulic analysis.

We begin with some background of column hydraulic analysis that defines the ranges of liquid and vapor flows for satisfactory column operations. Figure 2.109 illustrates the ranges of liquid and vapor loadings for satisfactory operation of sieve trays [16]. In the figure, L = liquid flow rate, lb/h*(ft^2 of empty tower cross-sectional area); G = vapor flow rate, lb/h*(ft^2 of empty tower

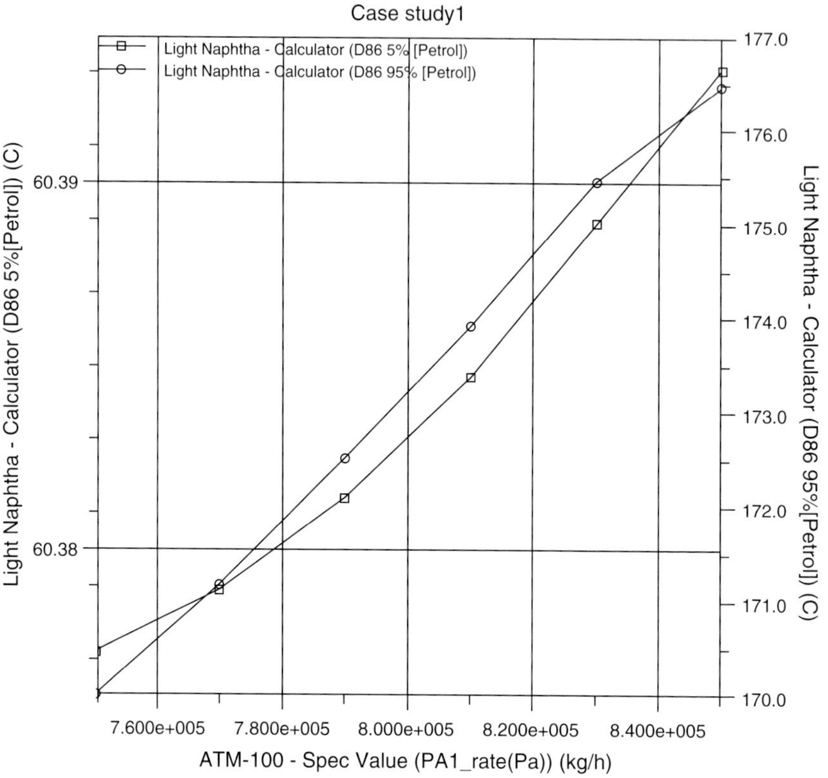

Figure 2.108 Case study result figure.

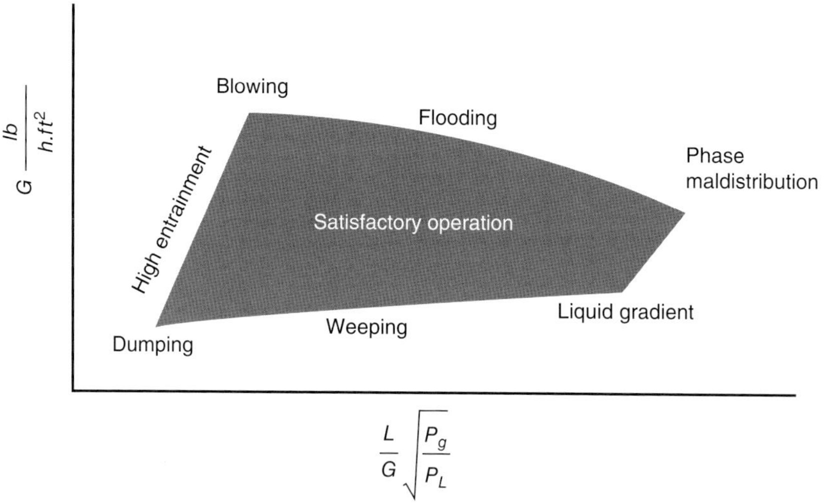

Figure 2.109 Range of satisfactory tray column operations.

cross-sectional area); ρ_G = density of gas, lb/ft^3; ρ_L = density of liquid, lb/ft^3.

The various operating limits defined in the figure include the following:

1) For most x-axis values, the capacity limit coming from too high a vapor rate will be *flooding*.
2) For low x-axis values, such as vacuum towers, the capacity limit corresponding to too high a vapor rate comes from *entrainment*.
3) At very high vapor velocities and relative low L/G, the efficiency may drop very markedly because of *blowing*, wherein the tray is blown clear of liquid in the immediate vicinity of the vapor distributors.
4) When L/G is very high, the quantity of liquid flow across the tray may require a very high *liquid or hydraulic gradient* (see Figure 2.110) in order to drive the flow.
5) Another result of too high a liquid gradient can be *phase maldistribution*, wherein the vapor flows preferentially through the perforations near the liquid outlet, and the liquid flows in part download through the perforations near the liquid inlet where the liquid depth is greatest.
6) The preferential flow of liquid downward through the perforations, rather than through the downcomer, is called, somewhat colorfully, *weeping*. This happens when vapor flow rate in the perforations is not large enough to hold the liquid out of the perforations. Massive liquid weeping is called *dumping*, resulting in severe phase maldistribution.
7) Within the shaded range of satisfactory operation, the upper portion corresponds to the *spray regime* and the lower portion corresponds to the *froth regime*.

Figure 2.110 shows a tray dynamics schematic diagram for froth regime. In the figure, *Liquid or hydraulic gradient* across tray represents the difference between

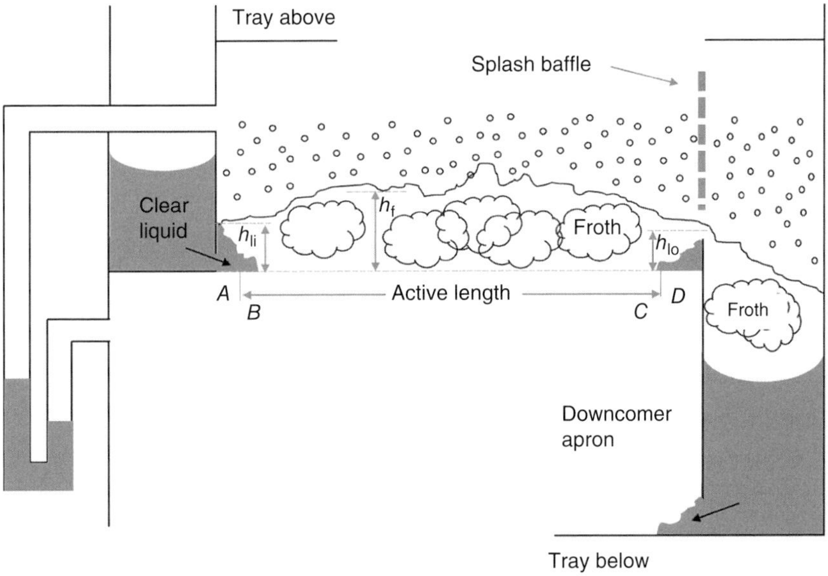

Figure 2.110 Tray dynamics schematic diagram defining the liquid or hydraulic gradient depicted in Figure 2.109.

the clear liquid heights, h_{li} (inlet location AB) and h_{lo} (outlet location CD) on the flow path above the tray.

We now open a converged CDU simulation file, *CDU-backblending-5.hsc*, and save it as *CDU-backblending-internals.hsc*. Figure 2.86 in Section 2.11.4 shows the CDU flowsheet. Double-click the column T-100 to open the CDU column. Choose the "internals" folder (see Figure 2.111).

We first explain the CS diagram below. Each CS typically includes at least a feed, a side stripper return stream, or a pumparound return stream entering the section, together with at least a product, a side stripper draw stream, or a pumparound draw stream leaving the section. For our CDU example:

1) CS-1 from stages 1 to 10 includes two input streams, SS HN return stream and PA-1 return stream, both to stage 9; two output streams, SS HN draw stream and PA-1 draw stream, both from stage 10.
2) CS-2 from stages 11 to 17 includes two input streams, SS kerosene return stream and PA-2 return stream, both to stage 16; two output streams, SS kerosene draw stream and PA-2 draw stream, both from stage 17.
3) CS-3 from stages 18 to 22 includes two input streams, SS LGO return stream and PA-3 return stream, both to stage 21; two output streams, SS LGO draw stream and PA-3 draw stream, both from stage 22.
4) CS4 from stages 23 to 27 includes one input stream, Heated_FEED to stage 25; one output stream, residue stream from stage 27.

Figure 2.112 compares the one-pass and multi-pass trays included in Figure 2.111.

Figure 2.111 Column-section diagram, specification of column-sections, and automatic sizing of column-section diameters.

In Figure 2.111, we choose to do "interactive sizing" to find the required CS diameter based on the vapor velocity at 80% of the maximum vapor velocity at the jet flooding limit. In this example, we choose sieve trays with a tray spacing of 2 ft.

We see from Figure 2.111 that the four CSs have diameters varying from 6.333 to 8.0 ft.

Next, we can change the calculation mode from "interactive sizing" to "rating" or "performance evaluation." The latter means finding the jet flooding percentage under the given feed rates and specified CS diameters (see Figure 2.113). We then click on "View Internals Summary" to see the results in Figure 2.114.

We note that column-section CS-4 (stages 23–27) has a jet flooding of 95.14%. In other words, under the given feed rates and a calculated CS diameter of 6.333 m, the linear vapor velocity is 95.1% of the maximum vapor velocity under the jet flooding limit.

The column internal analysis of Aspen HYSYS has a useful new feature, namely, to illustrate graphically if a chosen CS or a column tray is operating within the range of satisfactory column operations, using a hydraulic plot that is similar to Figure 2.109. We click on "internals" depicted in Figure 2.114 to return to the column internals window and then click on "view hydraulic plots" below the CS diagram. Pay attention to column-section CS-4, particularly stage 23 (see Figure 2.115).

To fix the problems in column-section CS-4, we return to the "Internals" form and click on the CS-4 "View" bottom to see the details (see Figure 2.116).

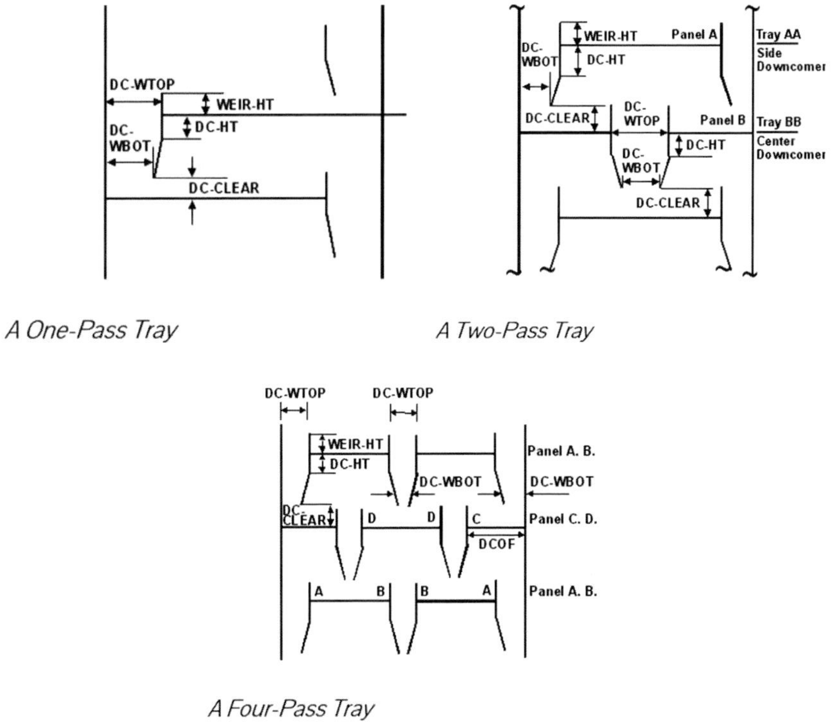

Figure 2.112 Comparison of one- and multi-pass tray configurations.

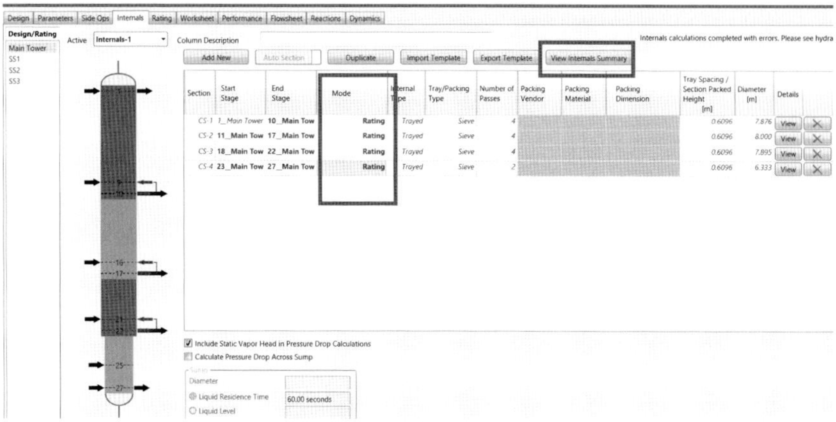

Section	Start Stage	End Stage	Mode	Internal Type	Tray/Packing Type	Number of Passes	Packing Vendor	Packing Material	Packing Dimension	Tray Spacing / Section Packed Height [m]	Diameter [m]	Details
CS-1	1_Main Tower	10_Main Tow	Rating	Trayed	Sieve	4				0.6096	7.876	View
CS-2	11_Main Tow	17_Main Tow	Rating	Trayed	Sieve	4				0.6096	8.000	View
CS-3	18_Main Tow	22_Main Tow	Rating	Trayed	Sieve	4				0.6096	7.895	View
CS-4	23_Main Tow	27_Main Tow	Rating	Trayed	Sieve	2				0.6096	6.333	View

Figure 2.113 Change from interactive sizing to rating mode.

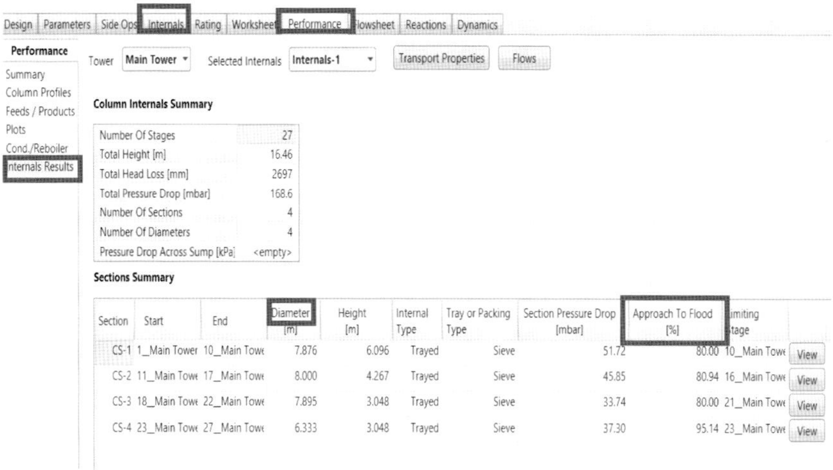

Number Of Stages	27
Total Height [m]	16.46
Total Head Loss [mm]	2697
Total Pressure Drop [mbar]	168.6
Number Of Sections	4
Number Of Diameters	4
Pressure Drop Across Sump [kPa]	<empty>

Section	Start	End	Diameter [m]	Height [m]	Internal Type	Tray or Packing Type	Section Pressure Drop [mbar]	Approach To Flood [%]	Limiting Stage	
CS-1	1_Main Tower	10_Main Towe	7.876	6.096	Trayed	Sieve	51.72	80.00	10_Main Towe	View
CS-2	11_Main Towe	17_Main Towe	8.000	4.267	Trayed	Sieve	45.85	80.94	16_Main Towe	View
CS-3	18_Main Towe	22_Main Towe	7.895	3.048	Trayed	Sieve	33.74	80.00	21_Main Towe	View
CS-4	23_Main Towe	27_Main Towe	6.333	3.048	Trayed	Sieve	37.30	95.14	23_Main Towe	View

Figure 2.114 Summary of column internal analysis.

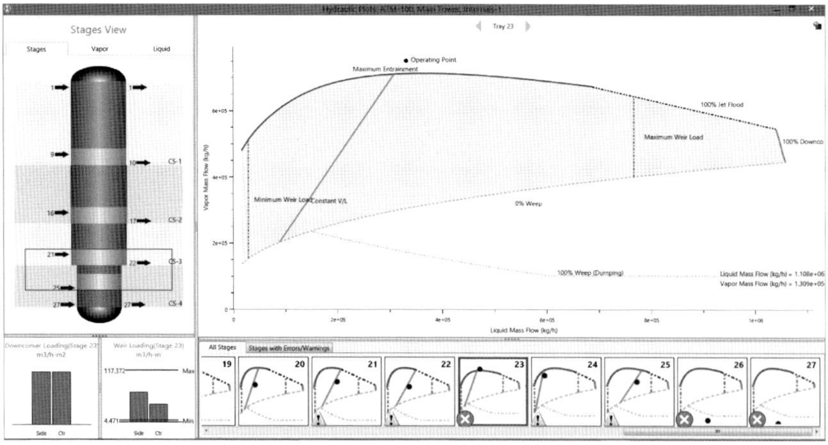

Figure 2.115 The hydraulic plot showing the operating point for stage 23 within column-section CS-4 is above the jet flooding limit, and the operating points for stage 26 and 27 are below the weeping limit.

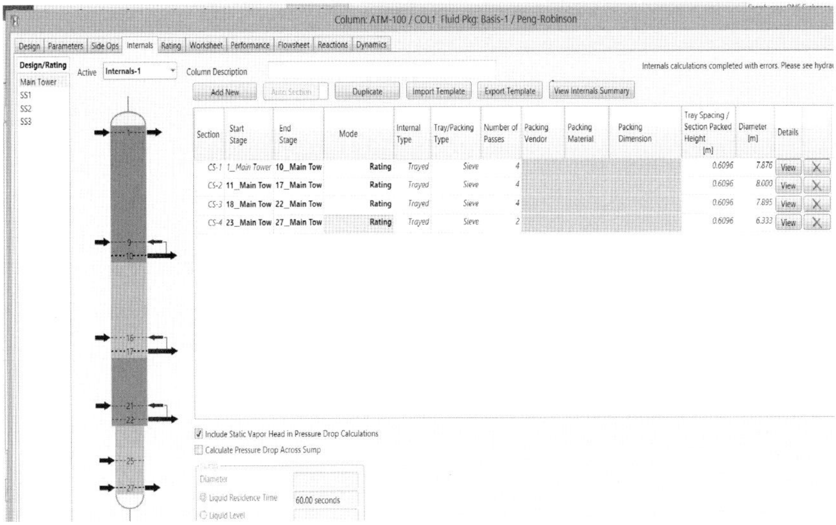

Figure 2.116 Click on "View" bottom to see details of column-section CS-4.

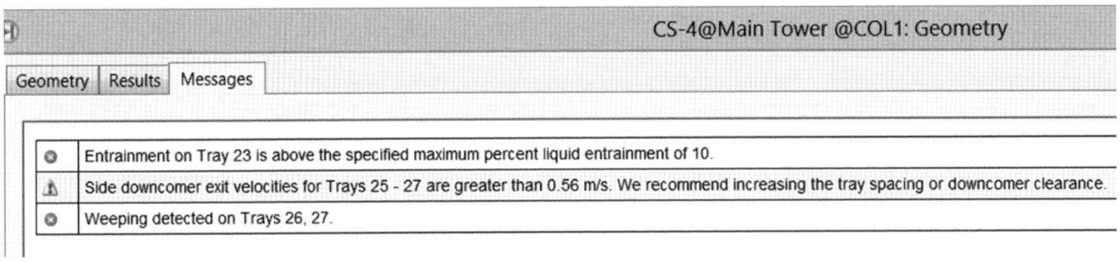

Figure 2.117 Error messages from column-section CS-4 hydraulic analysis.

When opening the "View" details, we see the red color prompt, "Errors in Hydraulics Calculations. Please Check the Messages tab," which appears on the upper left corner of the window. We see the messages shown in Figure 2.117.

We can take care of the first two errors plus the 95.14% approach to flooding shown in Figure 2.114 by increasing the CS diameter to 7.4 m and increase both the downcomer clearance and the weir height to 44 mm. To eliminate the occurrence of liquid weeping at the bottom two stages (26 and 27), *we need to change the feed stage of the Heated_FEED from stage 25 to 27* (Column ATM-100 → Design → Connections → Inlet Streams → Heated_FEED → Inlet Stage). Why? With a large liquid feed flow (Heated_FEED, entering at stage 25), and a small vapor feed flow (Btm Steam, entering at stage 27), the column-section CS4 does not have a sufficiently large vapor flow to present a massive liquid weeping (dumping) from stages 26 and 27. Figure 2.118 shows the changes, and column-section CS-4 no longer shows any error, and the resulting approach to flooding is 61.66% (Internals → View Internals Summary).

Figures 2.119 and 2.120 show the hydraulic analysis errors and the corrected column-section CS-1 specifications, in which we increase the side weir length to 4 m and the downcomer weir clearance to 48 mm.

For column-section CS-2, we follow the suggestion in the error message and make the same changes as in column-section CS-1 (Figure 2.121) to correct the specifications.

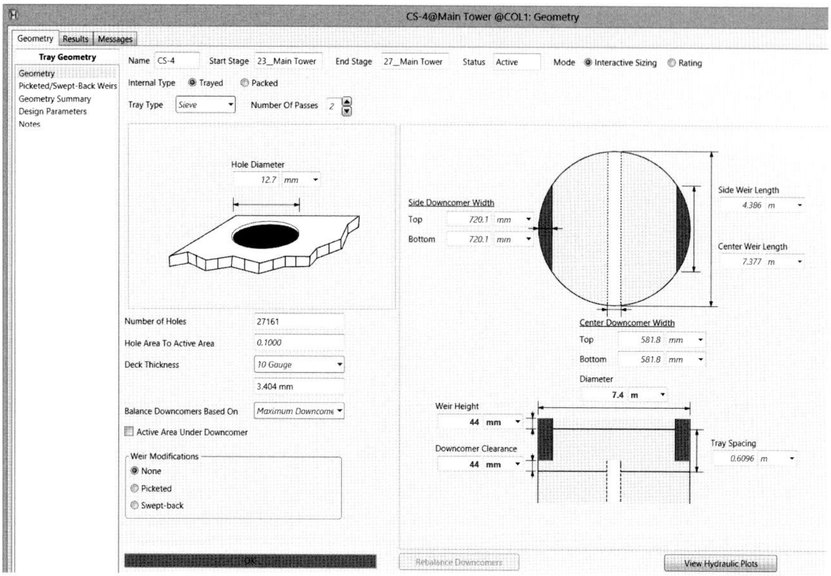

Figure 2.118 Corrected column-section C-4 specifications.

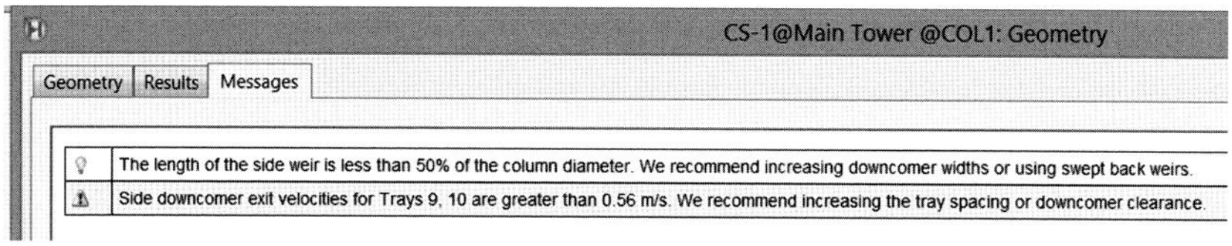

Figure 2.119 Error messages from column-section CS-1 hydraulic analysis.

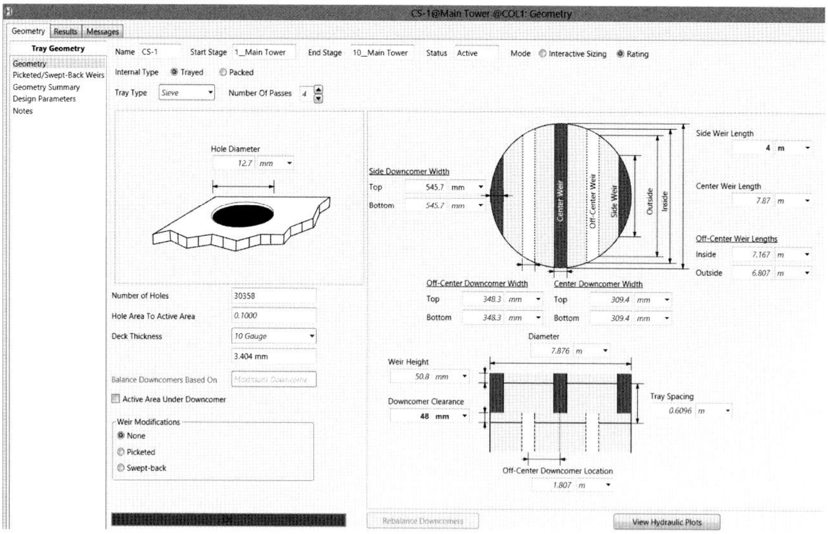

Figure 2.120 Corrected column-section CS-1 specifications.

Lastly, for column-section CS-3, we see the error messages in Figure 2.122 and make corrections by increasing the side weir length to 4 m, column diameter to 8 m, and downcomer clearance to 50 mm (see Figure 2.122).

Figure 2.123 summarizes the results of our hydraulic analysis for the column rating (performance evaluation) and minor retrofits. We encourage the reader to practice using the powerful new tool of hydraulic analysis in Aspen HYSYS V9.

Finally, as an exercise, the reader might want to increase the diameters of all column-sections, CS-1 to CS-4, to 8 m and study the resulting hydraulic analysis.

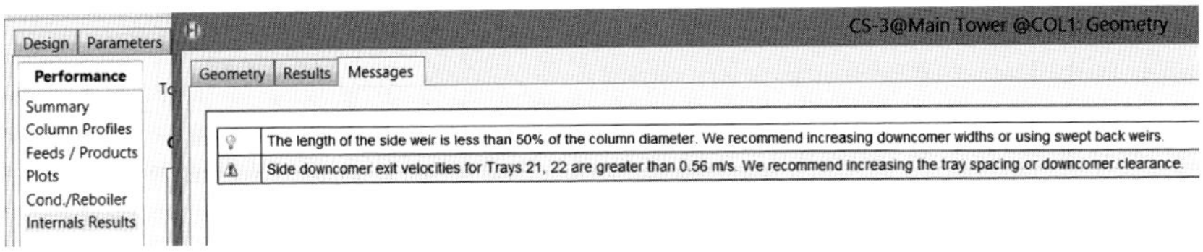

Figure 2.121 Error messages from column-section CS-3 hydraulic analysis.

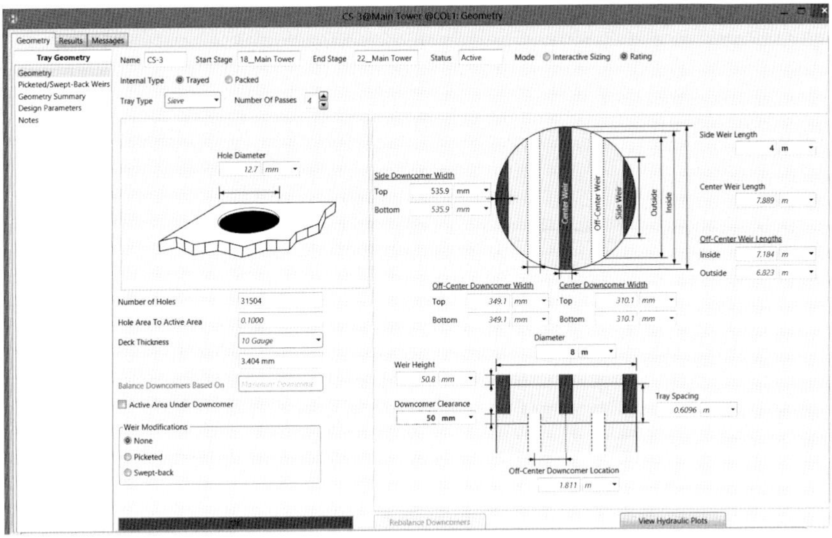

Figure 2.122 Corrected column-section CS-3 specifications.

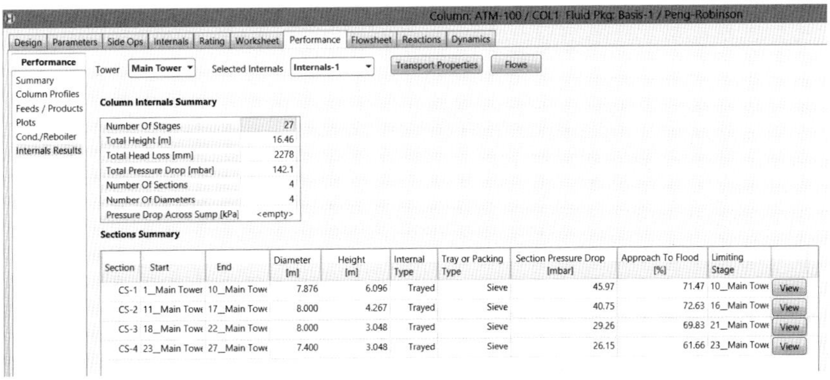

Figure 2.123 Results of column rating and minor retrofits.

2.15 Workshop 2.6 – Application of the Petroleum Distillation Column

We illustrate the use of petroleum distillation column as a simplified representation of the rigorous fractionation column.

This column model simulates a wide range of crude oils for scenarios of optimization or gradient generation (e.g., generating the delta base vector for production planning; see Sections 4.12 and Section 4.17 – Workshop 4.5 Generate Delta-Base Vectors for Linear Programming (LP)-Based Planning). In these situations, we need to simulate the column repeatedly, and the column should converge quickly and consistently in all scenarios. We also use the simplified petroleum distillation column whenever appropriate in developing the large refinery-wide simulation for profit margin analysis in Section 7.3.

This workshop uses a simulation file available in the Aspen HYSYS online example and explains the new concepts involved. We open the starting file, *Workshop 2.6_Starting.hsc* (see Figure 2.124).

We add a petroleum distillation column for the hot crude (see Figure 2.125).

We continue to specify the column input specs. Figure 2.126 shows the initial input form. We need to modify the products to include off gas, unstabilized naphtha, kerosene, LGO (light gas oil), HGO (heavy gas oil), and AR (atmospheric residue). We can do this by typing the new product name under the column "cuts" to override an existing product or to add a new product. See Figure 2.127 and the corresponding column flowsheet in Figure 2.128.

To understand the meanings of ECP (effective cut point), SI TOP (fractionation index, top section), and SI BOT (fractionation index, bottom section), let us consider a plot of Ln(Di/Bi) for all the components versus the normal boiling point (NBP). Here, Di is the mass flow rate of component *i* leaving the top section as a distillate, Bi is the mass flow rate of component *i* leaving at the bottom

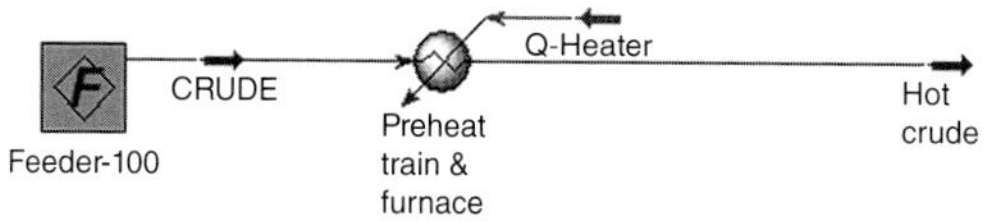

Figure 2.124 Starting flowsheet for crude distillation modeled by petroleum distillation column.

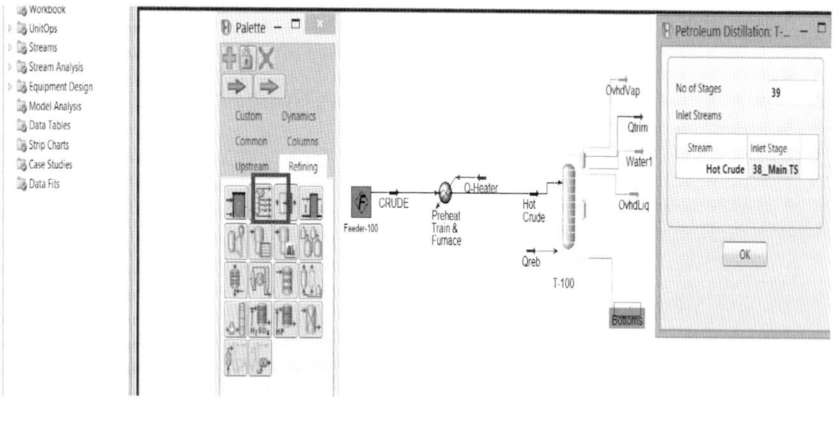

Figure 2.125 Add a petroleum distillation column.

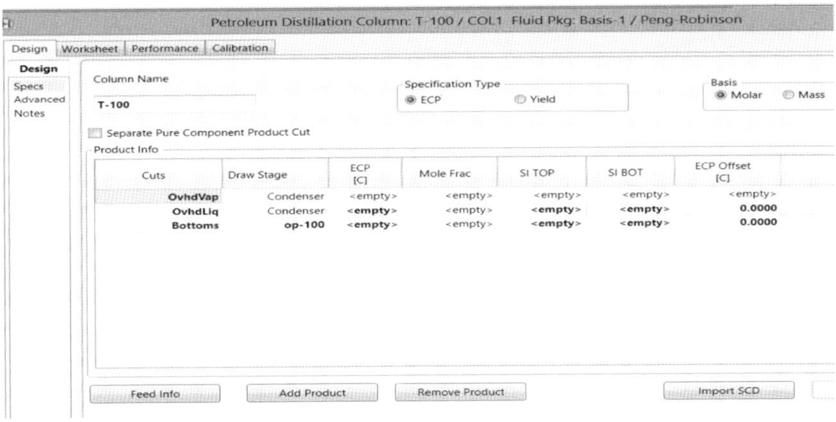

Figure 2.126 The initial input form for product specs.

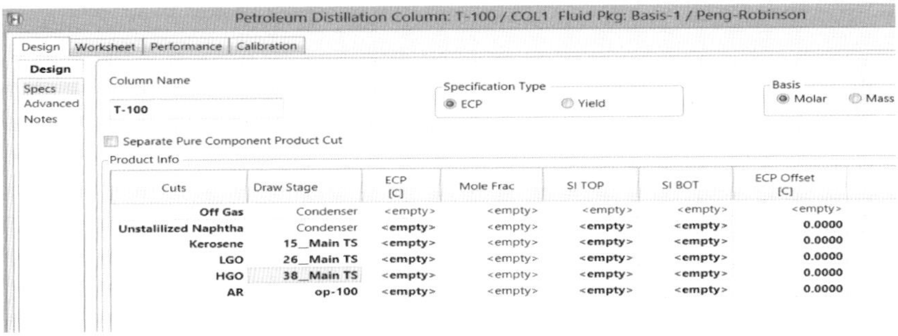

Figure 2.127 Modified input form to correspond to the crude distillation unit.

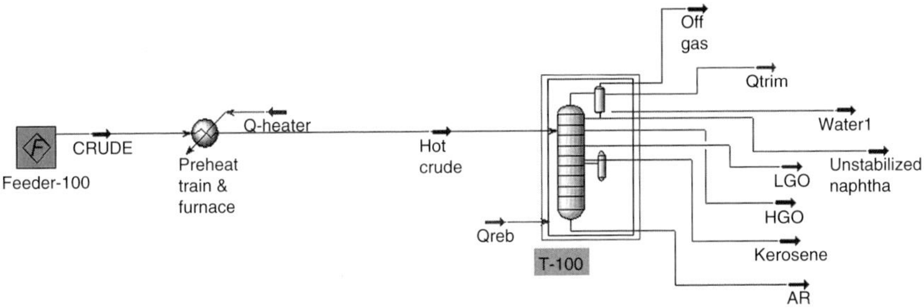

Figure 2.128 Add the petroleum distillation column.

section as a bottoms product, and $i = 1, 2, \ldots, n$. This plot is typically bilinear with two straight-line sections of different slopes in a petroleum distillation column. Figure 2.129 illustrates such a plot.

In Figure 2.129, we designate the slope of the top section as S_1 and of the bottom section as S_2. The fractionation index *SI TOP is defined as $-1/S_1$ and SI BOT is defined as $-1/S_2$.*

Both slopes S_1 and S_2 signify the extent of imperfect fractionation. For example, as S_1 tends to zero, there is virtually no separation (SI TOP = −1/0 = − infinity) and inversely, as S_1 approaches negative infinity (SI TOP = −1/(−infinity) = zero),

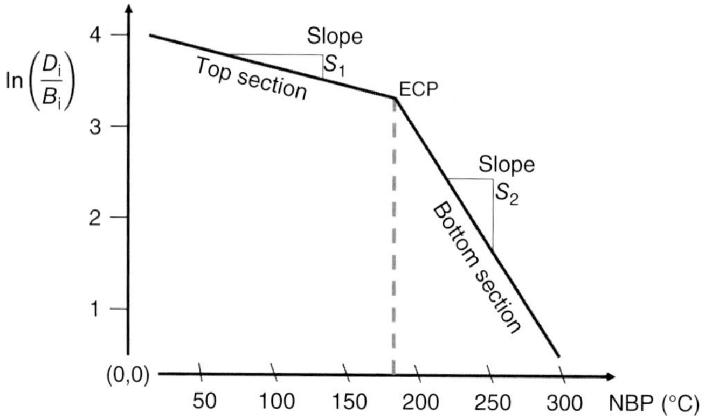

Figure 2.129 Ln(Di/Bi) versus NBP (normal boiling point) for top and bottom sections of a fractionation column.

the separation is almost perfect. The overall distillate and bottoms flow distributions decide the position of the curve horizontally. Note that a SI TOP value of zero represents perfect separation, which we do not get in reality. For practical purposes, we normally start with a value of SI TOP of 5 (i.e., with a S_1 value of −0.20). Decreasing SI TOP value from 5 to 0 implies poor split between the top and bottom sections, whereas increasing SI TOP to 10 suggests a smaller negative slope S_1 with a value of −0.10 and an improved split between the top and bottom sections. We can calibrate the column with plant data and get more realistic values of SI TOP (and SI BOT).

In Figure 2.129, the temperature or NBP at which the two straight-line sections intersect is called *the effective cut point (ECP)*, which is usually close to the TBP cut point. When the users specify the yield of each cut, the SI TOP/BOT supplied by the user is used to create the plot. The ECP is then varied to match the specified yield. With the correct ECP for a cut, users can calculate the value of Ln(Di/Bi) from the plot and then use the mass balance to get the yield for that cut.

Having explained the background of ECP, SI TOP, and SI BOT, we return to Figure 2.127. We complete the required input specs of ECP, SI TOP, and SI BOT as in Figure 2.130 and run the simulation. We can understand the input values of ECP.

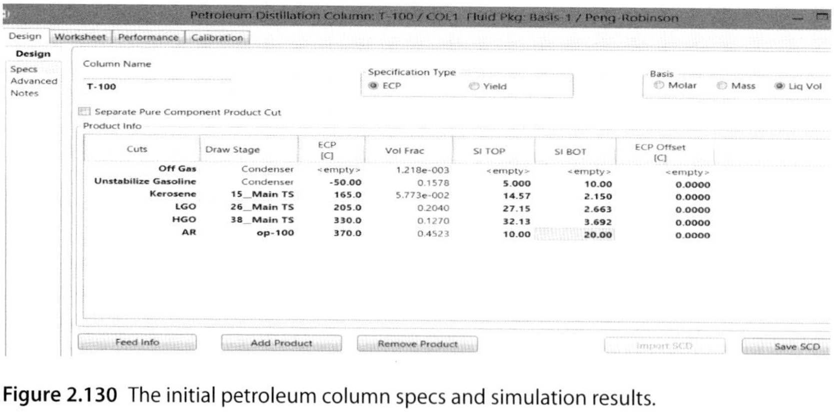

Figure 2.130 The initial petroleum column specs and simulation results.

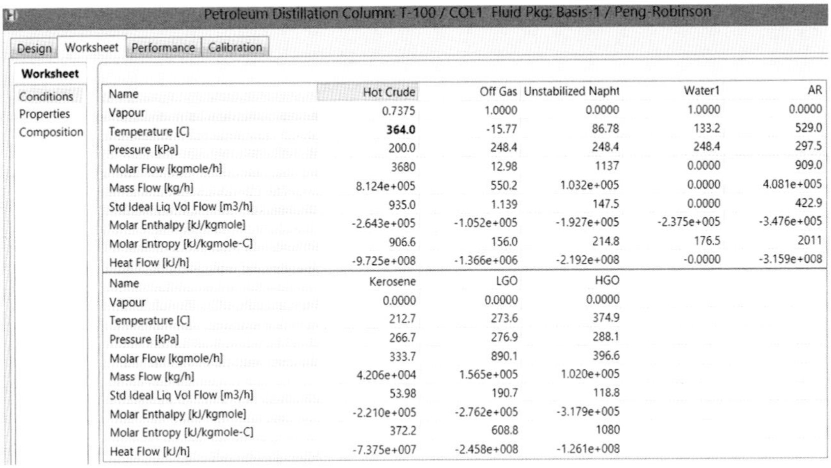
Petroleum Distillation Column: T-100 / COL1 Fluid Pkg: Basis-1 / Peng-Robinson

Design | Worksheet | Performance | Calibration

Worksheet: Conditions, Properties, Composition

Name	Hot Crude	Off Gas	Unstabilized Napht	Water1	AR
Vapour	0.7375	1.0000	0.0000	1.0000	0.0000
Temperature [C]	**364.0**	-15.77	86.78	133.2	529.0
Pressure [kPa]	200.0	248.4	248.4	248.4	297.5
Molar Flow [kgmole/h]	3680	12.98	1137	0.0000	909.0
Mass Flow [kg/h]	8.124e+005	550.2	1.032e+005	0.0000	4.081e+005
Std Ideal Liq Vol Flow [m3/h]	935.0	1.139	147.5	0.0000	422.9
Molar Enthalpy [kJ/kgmole]	-2.643e+005	-1.052e+005	-1.927e+005	-2.375e+005	-3.476e+005
Molar Entropy [kJ/kgmole-C]	906.6	156.0	214.8	176.5	2011
Heat Flow [kJ/h]	-9.725e+008	-1.366e+006	-2.192e+008	-0.0000	-3.159e+008

Name	Kerosene	LGO	HGO
Vapour	0.0000	0.0000	0.0000
Temperature [C]	212.7	273.6	374.9
Pressure [kPa]	266.7	276.9	288.1
Molar Flow [kgmole/h]	333.7	890.1	396.6
Mass Flow [kg/h]	4.206e+004	1.565e+005	1.020e+005
Std Ideal Liq Vol Flow [m3/h]	53.98	190.7	118.8
Molar Enthalpy [kJ/kgmole]	-2.210e+005	-2.762e+005	-3.179e+005
Molar Entropy [kJ/kgmole-C]	372.2	608.8	1080
Heat Flow [kJ/h]	-7.375e+007	-2.458e+008	-1.261e+008

Figure 2.131 The stream table for the petroleum distillation column.

Table 2.18 Comparison of product stream temperatures and effective cut points in the petroleum refinery distillation example.

Product stream	Temperature (°C)	Effective cut point, ECP (°C)
Off gas	−15.77	
Unstabilized naphtha	86.78	−50
Kerosene	212.7	165
LGO	273.6	205
HGO	374.9	330
AR	529	370

Figure 2.131 gives the resulting stream table for the column, which shows, among other results, the temperature of each product stream. Table 2.18 compares the product stream temperatures with the ECPs. We see that ECP values are close to the corresponding product stream temperatures. This means that if we know the product stream temperatures from plant data, we could input values close to them as initial values for ECPs.

Next, we refer to the plant data for six product streams given in the supplement to this book within Chapter 2, *Workshop 2.6_Calibration data for refinery distillation column example.xlsx*. We enter all of the plant data to the calibration folder. See an example in Figure 2.132.

We then run the calibration and see the calibrated model parameters in Figure 2.133.

We conclude this workshop by noting that if we require significant internal details of the column such as vapor–liquid traffic and temperature profiles matching closely to plant data, or if we see more flexibility in the specifications, or the topology of the column, we should use the standard rigorous column models.

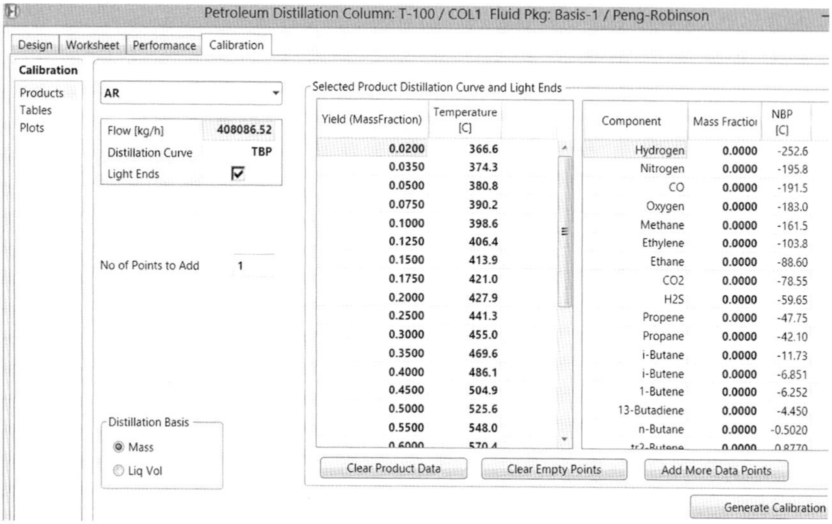

Figure 2.132 Enter the plant data for calibration of the petroleum distillation column model.

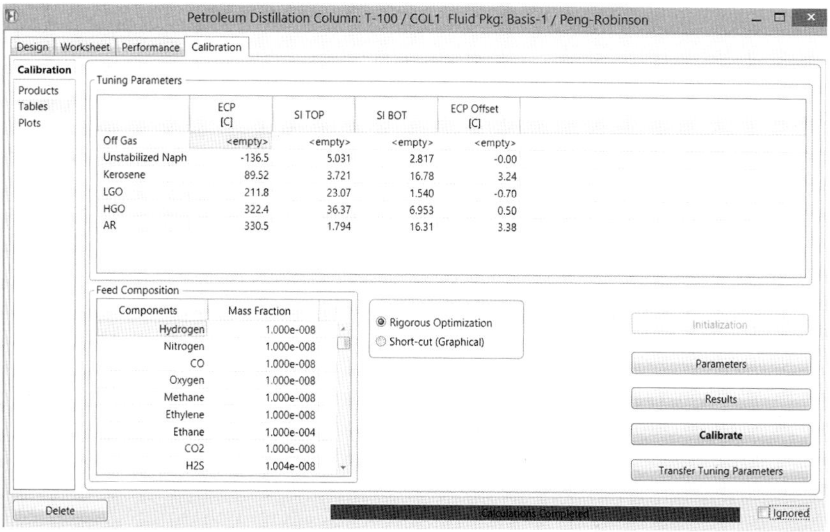

Figure 2.133 Calibrate model parameters.

2.16 Conclusions

This chapter serves as a guide to model atmospheric distillation section of the CDU. We provide relevant process and operational and modeling details to model the atmospheric column. We also discuss methods to estimate missing data for model development. We provide step-by-step instructions to model a particular column in Aspen HYSYS. We discuss how to validate the model predictions with plant data and how to use the model to perform industrially useful case studies.

Nomenclature

T	Temperature, °C
P	Pressure, kPa
F_i	Feed entering tray i, kmol/h
L_i	Liquid flow leaving tray i, kmol/h
U_i	Side draw liquid flow leaving tray i, kmol/h
V_i	Vapor flow leaving tray i, kmol/h
W_i	Side draw vapor flow leaving tray i, kmol/h
x_i	Liquid composition leaving tray
y_i	Vapor composition leaving tray
H_{F_i}	Feed molar enthalpy, kJ/kmol
H_{V_i}	Vapor molar enthalpy, kJ/kmol
H_{L_i}	Liquid molar enthalpy, kJ/kmol
K_i	Ratio of vapor to liquid composition, y_i/x_i
K_w	Watson K factor

Bibliography

1 Hsu, C.S. and Robinson, P.R. (2006) *Practical Advances in Petroleum Processing. Volume 1 &2*, Springer.
2 Daubert, T.E. and Danner, R.P. (1997) *APITechnical Data Book – Petroleum Refining*, 6th edn, American Petroleum Institute, Washington DC.
3 Kaes, G.L. (2000) *Refinery Process Modeling. A Practical Guide to Steady State Modeling of Petroleum Processes*, The Athens Printing Company, Athens, GA.
4 Riazi, M.R. (2005) *Characterization and Properties of Petroleum Fractions*, 1st edn, American Society for Testing and Materials, West Conshohocken, PA.
5 Kister, H.Z. (1992) *Distillation Design*, McGraw-Hill, Inc., New York, NY.
6 Bazaraa, M.S., Jarvis, J.J., and Sherali, H.D. (2009) *Linear Programming and Network Flows*, John Wiley and Sons.
7 Boston, J.F. (1980) *Inside-Out Algorithms for Multicomponent Separation Process Calculations*, ACS Symposium Series, vol. **124**, pp. 135–151.
8 Seader, J.D., Henley, E.J., and Roper, D.K. (2010) *Separation Process Principles*, 3rd edn, John Wiley and Sons, New York.
9 Watkins, R.N. (1979) *Petroleum Refinery Distillation*, 2nd edn, Gulf Publishing Company, Houston.
10 Gary, J.H., Handwerk, G.E., and Kaiser, M.J. (2007) *Petroleum Refining. Technology and Economics*, 5th edn, CRC Press, Boca Raton, FL.
11 Sanchez, S., Ancheyta, J., and McCaffrey, W.C. (2007) *Energy & Fuels*, **21**, 2955–2963.
12 Floudas, C.A. (1995) *Nonlinear and Mixed-Integer Programming.Fundamentals and Applications*, Oxford University Press.
13 Aspen Tech. (2017) *Aspen HYSYS User Guide.*
14 Favennec, J.P. (1998) *Fractionation Systems*, 5th edn, IFP, Paris, France.
15 Nelson, W.L. (1974) *Petroleum Refinery Engineering*, 4th edn, McGraw-Hill, New York.

16 King, C.J. (1980) *Separation Processes*, 2nd edn, McGraw-Hill, New York, pp. 591–603.

17 Shankar, N., Sivasubramanian, V., and Arunachalam, K. (2016) Steady state optimization and characterization of crude oil by Aspen HYSS. *Petroleum Science and Technology*, **34**, 1187–1194.

18 Waheed, M.A. and Oni, A.O. (2015) Performance improvement of a crude oil fractionation unit. *Applied Thermal Engineering*, **75**, 315–324.

19 Al-Mayyahi, M.A. (2014) Energy optimization of crude oil distillation using different designs of pre-flash drums. *Applied Thermal Engineering*, **73**, 1204–1210.

20 Bashir, D.M., Mohamed, S.A., and Rabah, A.A. (2014) Effect of naphtha and residue yield through different operating and design variables in atmospheric distillation column. *Journal of Petroleum Technology and Alternative Fuels*, **5**, 31–37.

21 Mittal, V., Zhang, J., Yang, X., and Xu, Q. (2011) E3 (energy, emission and economic) analysis for crude and vacuum distillation system. *Chemical Engineering and Technology*, **34**, 1854–1963.

22 Goncalves, D.D., Martins, F.G. and Azevedo, S.F.D. (2010) Dynamic Simulation and Control: Application to Atmospheric Distillation of Crude Oil Refinery. *20th European Symposium on Computer-Aided Processing Engineering- ESCAPE20*, 1–6.

23 Kim, Y.H. (2017) An energy-efficient crude distillation unit with a prefractionator. *Chemical Engineering and Technology*, **40**, 588–597.

24 Menezes, B.C., Kelly, J.D., and Grosmann, I.E. (2013) Improved swing-cut modeling for planning and scheduling of oil-refinery distillation units. *Industrial & Engineering Chemistry Research*, **52**, 18324–18333.

25 Ali, S.F. and Yusoff, N. (2012) Determination of optimal cut point temperatures at crude distillation unit using the Taguchi method. *International Journal of Engineering and Technology, IJET-IJHNS*, **12**, 36–46.

26 Ochoa-Estopier, L.M. and Jacobson, M. (2015) Optimization of heat-integrated crude oil distillation systems. Part III: Optimization framework. *Industrial & Engineering Chemistry Research*, **54**, 5018–5036.

27 Yela, S. (2009) Framework for Operability Assessment of Production Facilities: Application to a Primary Unit of a Crude Oil Refinery. M.S. thesis, Louisian State University, Chemical Engineering, https://etd.lsu.edu/docs/available/etd-11042009-012159/unrestricted/Yela_Thesis.pdf.

28 Parthlban, R., Nagarajan, N., Kumaran, V.M., and Kumar, D.S. (2013) Dynamic modelling and simulation of crude fractionation column with three side strippers using Aspen HYSYS dynamics. *Journal of Petroleum and Gas Exploration Research*, **3** (3), 31–39.

29 Fu, G. and Mahalec, V. (2015) Comparison of methods for computing crude distillation product properties in production planning and scheduling. *Industrial & Engineering Chemistry Research*, **54**, 11371–11382.

30 Lopea, D.C., Hoyos, L.J., Mahecha, C.A., Ayellano-Garcia, H., and Wozony, G. (2013) Optimization model of crude oil distillation units for optimal crude oil blending and operating conditions. *Industrial & Engineering Chemistry Research*, **52**, 12993–13005.

3

Vacuum Distillation Unit

This chapter presents the methodology for the development and applications of simulation models for vacuum distillation units (VDUs) based on plant data. We begin by describing the typical VDUs in Section 3.1 and then present the data requirements and reconciliation procedures for simulating VDUs in Section 3.2. Section 3.3.1 shows the plant data from a typical VDU, and Sections 3.3.2 and 3.3.3 demonstrate how to develop and validate a simplified model and a rigorous model for a VDU. Section 3.4 discusses the principle of applying a validated VDU model to optimize the deep-cut operation of a VDU. The goal is to process heavier crude feeds and to improve process economics by increasing the cut point of heavy vacuum gas oil (HVGO) higher than 1050 °F (565 °C) to produce more gas oil for downstream units such as fluid catalytic cracking. Section 3.5 presents a hands-on workshop for implementing the deep-cut operation of a VDU. Finally, we present bibliography at the end of the chapter.

3.1 Process Description

The distillated products of atmospheric distillation unit (also known as crude distillation unit, CDU) are limited to the boiling fractions under 350 °C such as gasoline and diesel because petroleum fractions tend to thermally degrade in high temperatures. To recover additional distillates and gas oils, the refinery uses VDU following the CDU. The reduced operating pressure of VDU allows recovery of heavy boiling fraction above 560 °C from the atmospheric residue.

There are two major types of VDU operations in a modern refinery – *feedstock preparation and lubricant production*. Feedstock preparation is the most common operation that recovers gas oil from the atmospheric residue as a feed to the downstream conversion units (e.g., FCC and hydrocracking units), which converts the gas oil into more valuable liquid products such as gasoline and diesel. Lubricant production is designed to extract petroleum fractions from the atmospheric residue to produce lubricant oil with desirable viscosity and other related properties.

Petroleum Refinery Process Modeling: Integrated Optimization Tools and Applications,
First Edition. Y. A. Liu, Ai-Fu Chang, and Kiran Pashikanti.

This chapter presents the methodology to simulate the VDU for feedstock preparation because it is the most popular operation; however, most of the guidelines in our methodology are also applicable to lubricant production units.

Figure 3.1 represents a typical process flow diagram of VDU operated under wet operations with three vacuum gas oil (VGO) side products from light (L) to medium boiling (M) to heavy (H) – LVGO, MVGO, and HVGO. The furnace outlet temperature varies from 380 to 420 °C, depending on the feedstock type. In particular, asphalt-based feedstock requires higher furnace outlet temperature than nonasphalt-based feedstock. The pressure drop across transfer line is around 20 kPa, whereas the temperature change is 10–15 °C. In the wet operation, a superheated steam is pumped into the stripping zone to enhance the vaporization of gas oil by reducing the partial pressure of hydrocarbon. Therefore, the wet operation requires a lower flash zone temperature than the dry operation for the same service. The flash zone pressure is typically controlled in the range of 2.6–13.3 kPa (20–100 mmHg). For the process shown in Figure 3.1, the wash grid exit is withdrawn from the column and routed back to the transfer line. Some VDUs recycle the wash grid exit stream back through the furnace whereas others send it to the stripping zone to mix with the vacuum residue (VR).

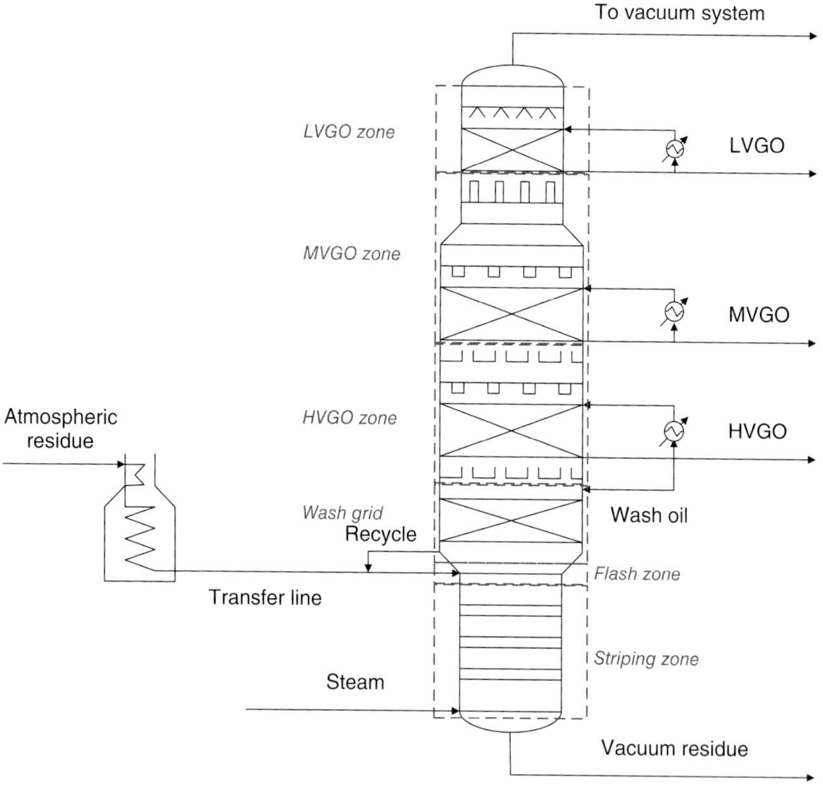

Figure 3.1 Typical process flow diagram of VDU.

3.2 Plant Data Reconciliation

3.2.1 Required Data

Simulating a VDU begins with data collection. We should collect the operating and analysis data as much as possible (see Table 3.1).

It is always helpful to collect a long period (1–3 months) of data for modeling purpose, particularly for a commercial process. As it is common to have missing data or failed meters, we need to take average of the collected data for a short period (1–3 days) or to make up the missing data by extrapolating or interpolating the data collected from adjacent time period in order to construct one complete data set for modeling. It is also important to consult plant engineers about data consistency to ensure that each complete data set does not include the data in the period of operational upset and significant operation changes. In addition, it is always helpful to revisit the original data for test run, because test run data are usually adjusted to show perfect material and heat balances [1].

3.2.2 Representation of the Atmospheric Residue

To represent the feed properly for modeling the VDU, we need to pay attention to two requirements: (1) using a sufficient number of pseudocomponents to represent the atmospheric residue, and (2) having good-quality assay data for the atmospheric residue.

The methodology used in commercial simulators for splitting a petroleum fraction into pseudocomponents typically specifies a small number of

Table 3.1 Data requirement of VDU model.

Flow rate
– Feed and product steams (overhead products are bonus)
– All pumparound streams
– All cooling streams for pumparounds
– Coil and strip steams
Pressure
– Flash zone
– Top of column
– Bottom column
Temperature
– Flash zone
– Top of column
– Bottom column
– Side product draw tray
– Furnace inlet and outlet temperature
– Transfer temperature
– Draw and return temperatures for all pumparounds
– Inlet and outlet temperature of all pumparound cooling streams
Analysis
– Distillation and gravity of atmospheric residue (feed)
– Distillation and gravity for all product streams
– Compositional analysis of overhead gas

Table 3.2 Typical boiling point widths for pseudocomponents in commercial simulator.

Boiling point range	Boiling point width of each pseudocomponent	Number of pseudocomponents per 100 °F
IBP–800°F (425 °C)	25 °F (15 °C)	4
800–1200°F (650 °C)	50 °F (30 °C)	2
1200–1650°F (900 °C)	100 °F (55 °C)	1

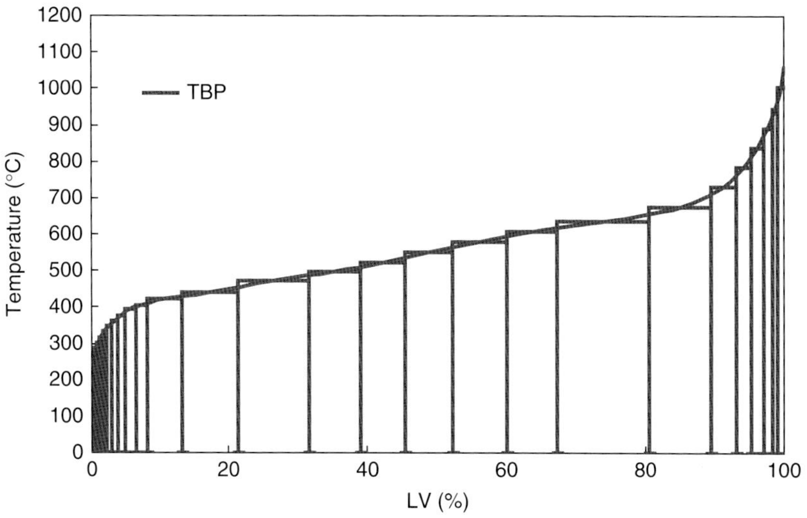

Figure 3.2 Typical pseudocomponents generated by commercial simulator.

pseudocomponents in the higher boiling point region. Table 3.2 lists the typical number of pseudocomponents specified in different boiling point ranges. Within conventional cut point between HVGO and VR (around 1000 °F), the boiling point range of each cut (50 °F) is twice as wide as that (25 °F) above 800 °F. Using a small number of pseudocomponents to define the interested boiling point range may not represent the feed accurately to reflect the real operation and production, particularly when we use the resulting model for deep-cut revamping purpose [5].

Figure 3.2 illustrates the pseudocomponent representation of an atmospheric residue generated by a commercial simulator based on its default pseudocomponent boiling point cutting scheme. By assigning a boiling point width of 25 °F to each pseudocomponent with a boiling point above 800 °F, we can more accurately represent the atmospheric residue. In general, we should do a sensitivity test, investigating the relationship between the side draw rate and the side draw temperature and associated distillation curve to ensure that the pseudocomponents based on boiling point ranges are able to provide reasonable results [1]. If the resulting relationship is stepwise rather than continuous, we should redefine the number of pseudocomponents based on boiling point ranges (Figure 3.3).

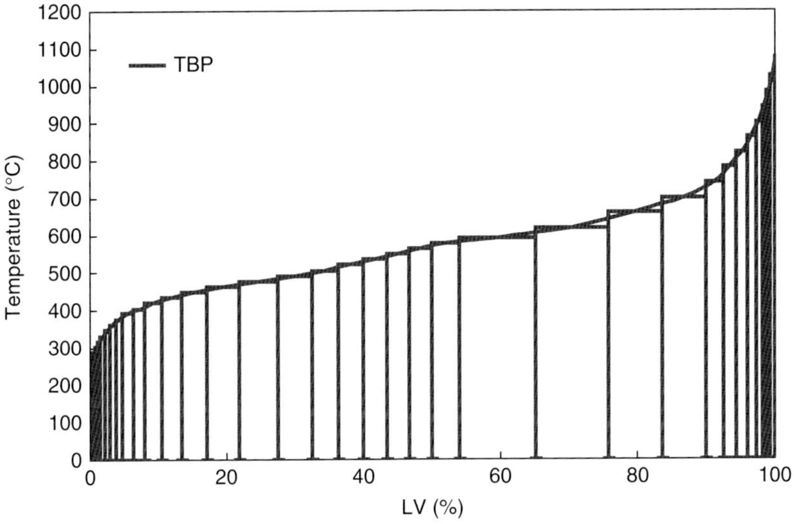

Figure 3.3 Improved pseudocomponent generation by assigning a boiling point width of 25 °F to each pseudocomponent with a boiling point above 800 °F.

High-quality assay data for the atmospheric residue are always desirable in modeling a VDU. For VDU simulation, there are three ways to obtain assay data of the atmospheric residue: (1) stream results of CDU simulation if we build the CDU and VDU together; (2) analysis of the atmospheric residue; and (3) back-blending the assay data of VDU product streams if product analyses are available.

We need to consider a number of issues when applying each approach to represent the feed for modeling purpose. Specifically, when we build the CDU and VDU models together, the representation of the atmospheric residue is reliable if a detailed feed representation of crude oil is available (by either crude assay or backblending product analysis) and the CDU model performs correctly.

However, we must pay more attention to correctly representing the atmospheric residue when using VDU operation and production data to build a VDU model. This follows because the atmospheric residue is an intermediate stream rather than a final product and a detailed stream analysis is usually not available. Most likely, we can only have the analysis results of the distillation curve below 540 °C and bulk density of the atmospheric residue. We note that when using commercial simulators to construct the atmospheric residue based on an incomplete feed analysis, the resulting pseudocomponent distribution may not represent atmospheric residue well. This is because commercial simulators typically extrapolate the distillation curve by statistical functions and assume a constant Watson K factor within the entire boiling point range. Kaes [1] suggested that it is reliable to backblend product analyses to represent the atmospheric residue for modeling purpose when detailed feed assay data are not available. Table 3.3 illustrates a data set including mass flows, specific gravities, and distillation curves of the atmospheric residue and corresponding products of a VDU [3]. Figure 3.4 shows the difference between the feed analysis and the backblend of product analyses.

Table 3.3 Analysis data of a VDU's feedstock and products [3].

	Atmospheric residue	LVGO	HVGO	Vacuum residue
Mass flow (kg/h)	234 004	35 172	103 618	94 600
Specific gravity	0.9593	0.8718	0.9321	1.0366
Liquid vaporization (LV) %	D1160 at 1 atm, °C	D1160 at 1 atm, °C	D1160 at 1 atm, °C	D1160 at 1 atm, °C
0	246.1	198.8	360.0	421.1
5	335.0	254.4	393.3	513.8
10	368.3	290.5	405.5	543.3
30	448.8	331.1	446.6	
50	506.1	351.6	475.5	
70		376.6	507.2	
90		407.2	553.3	
95		429.4		
100		475.0		

Source: Courtesy of G.L. Kaes.

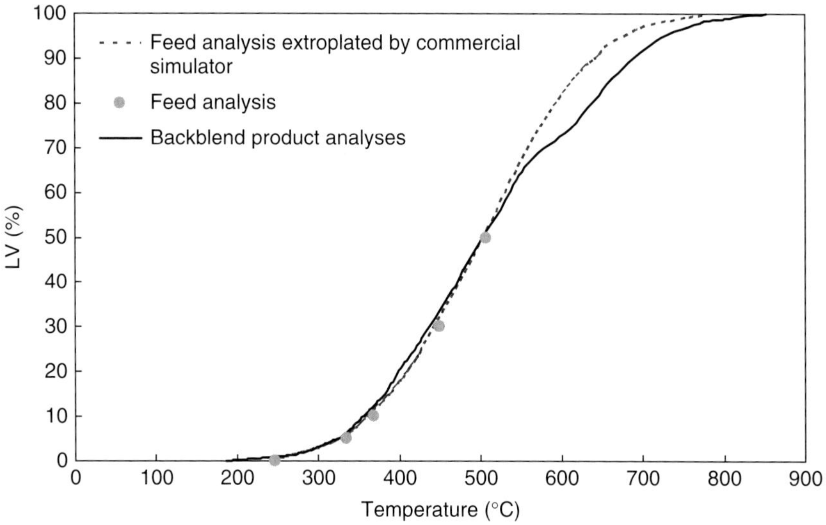

Figure 3.4 Comparison between feed analysis and backblend of product analyses.

3.2.3 Makeup of Gas Streams

Under high-temperature operations of VDU, there is always some extent of cracking of the atmospheric residue into light gases and petroleum fractions. In addition, the vacuum condition also allows some air to leak into the VDU. When the amounts of light gases and petroleum fractions are significant, we must add these materials into the atmospheric residue to represent it properly. Kaes [1]

Table 3.4 Values of parameters C_1 and C_2 for Eqs. (3.2) and (3.4).

Flash zone temperature (°F)	Flash zone temperature (°C)	C_1	C_2
800	427	1.2	67
775	413	0.6	35
750	399	0.3	20
725	385	0.15	12
700	371	0.08	5

Source: Courtesy of G.L. Kaes.

suggested that the amount of noncondensable gases that must be removed per 1000 barrels of feed charge to the VDU ranges from 15 to 50 lb (1–3.5 kg/m^3 charge stock). It consists of five different sources of light gases: (1) dissolved light gases; (2) native front-end tail; (3) cracking gas; (4) cracking front-end tail; and (5) air leaks. Kaes gave the general guideline to estimate these gas streams and corresponding compositions to adjust the atmospheric residue.

1) *Dissolved light gases.* Light gas component dissolved in the feedstock

$$\text{lb/h gas} = 11.5 * (\text{feed rate, barrels per day})/1000 \tag{3.1}$$

2) *Cracking gases.* Low-molecular weight gases resulting from thermal cracking of the feedstock in the furnace

$$\text{lb/h gas} = C_1 * (\text{feed rate, barrels per day})/24 \tag{3.2}$$

where C_1 depends on the flash zone temperature, and its value appears in Table 3.4. For simulation purposes, we may represent both the dissolved light gases and the cracking gases by a 75/25 mixture of ethane (C2) and propane (C3).

3) *Front-end tail (native).* A tail of low boiling (0.5% liquid vaporization) material inherent from the crude oil fractionation

$$\text{lb/h gas} = 50 * (\text{feed rate, barrels per day})/1000 \tag{3.3}$$

When the VDU feed rate is generated by simulating a CDU, both adjustments (1) and (3) are not necessary.

4) *Front-end tail (cracking).* Gas components resulting from thermal cracking of the CDU residue within the furnace

$$\text{lb/h gas} = C_2 * (\text{feed rate, barrels per day})/24 \tag{3.4}$$

where C_2 depends on the flash zone temperature, and Table 3.4 shows its value. For simulation purposes, we may represent both the front-end tail (native) and front-end tail (cracking) by a 50/50 mixture of normal undecane (n-C11) and normal dodecane (n-C12).

5) *Air leaks into the overhead system*

$$\text{lb/h gas} = 6.0 * [(\text{feed rate, barrels per day})/1000]^{1/2} \tag{3.5}$$

For simulation purpose, we may represent air leaks by pure nitrogen.

	A	B	C	D	E	F
1	Feed Flow Rate (bbl/day)	81003				
2	Feed Flow Rate (m3/h)	536.6				
3	Flash Zone Temperature (℃)	407				
4		Dissolved Gas	Front end tail (Native)	Front end tail (Cracking)	Cracking Gas	Air Leak
5	Flow Rate (lb/hr)	932	4050	2188	1494	54
6	Flow Rate (kg/hr)	423	1837	993	678	24
7	Molar fraction					
8	N2	0.00	0.00	0.00	0.00	1.00
9	C1	0.00	0.00	0.00	0.00	0.00
10	C2	0.75	0.00	0.00	0.75	0.00
11	C3	0.25	0.00	0.00	0.25	0.00
12	n-C11	0.00	0.50	0.50	0.00	0.00
13	n-C12	0.00	0.50	0.50	0.00	0.00
14						
15		Cracking as	FZ Temp (℃)	FZ Temp (℃)	Front end tail (Cracking)	
16		1.2	427	427	67	
17		0.6	413	413	35	
18		0.3	399	399	20	
19		0.15	385	385	12	
20		0.08	371	371	5	
21		0.44			27.01	
22						

Figure 3.5 Excel spreadsheet to estimate makeup gas streams corresponding to Figure 3.14.

We have developed an Excel spreadsheet to implement Eqs. (3.1)–(3.5), *Gases makeup.xls*. Figure 3.5 illustrates this spreadsheet. It requires the atmospheric residue (AR) flow rate and flash zone temperature (cells B2 and B3) to calculate the flow rate of gas streams. The Excel data correspond to the AR to the VDU examples in this chapter. Cells B21 and E21 represent the values of interpolated parameters C_1 (=0.44) and C_2 (=27.01) at a flash zone temperature of 407 °C according to Table 3.4.

3.3 Model Implementation

Similar to CDU simulation, it is necessary to translate a real distillation column into an equivalent configuration using theoretical stages to properly simulate the VDU. The high vapor velocities and low liquid levels of vacuum column stages make the performance of a real VDU deviating from that predicted by an ideal vapor–liquid equilibrium. Moreover, the packing section of VDU behaves as a heat transfer facility rather than a separation unit, thus making the separation performance even worse. The product distribution of VDU highly depends on the composition of atmospheric residue rather than the extent of fractionation. Therefore, simulating a VDU with two side products usually requires less than 10 theoretical stages. While building VDU simulation, it is common having trouble on converging the column model because of the low liquid flow in the column. Kaes [1] suggested *a two-step approach* to simulate VDU properly – *simplified and rigorous simulations*. A simplified model is able to produce quick and informative understanding of a real VDU, particularly for a preliminary study of revamping. In addition, the initial model from a simplified approach provides good estimates to reconcile a rigorous simulation if the column convergence is difficult. The following sections demonstrate the modeling procedures for both simplified and rigorous simulations using Aspen HYSYS and the data from operating VDUs in the Asia Pacific.

3.3.1 Plant Data and Modeling Approaches

There are two important steps to complete before building a process flowsheet in Aspen HYSYS (same as any other commercial simulator): (1) define the feed representation, and (2) select an appropriate thermodynamics model. Table 3.5 and Figure 3.6 represent the key process data and a simplified process flow diagram of a VDU in the Asia Pacific. The VDU is operated in wet conditions (with steam) and produces the VR and three valuable products – vacuum distillate (VD), LVGO, and HVGO. The D1160 analysis of "VGO" in Table 3.5 represents the distillation data of a mixture of VD, LVGO, and HVGO. Feedstock representation is always the first step for building a VDU simulation. As mentioned in Section 3.2.2, there are three approaches to obtain the assay data of the atmospheric residue. However, analysis of feedstock is the only option in this case, as the purpose is to model VDU alone and there is no analysis made for VR.

For both the simplified and rigorous simulations, Kaes [1] recommended using two to three theoretical stages to simulate each separation zone, and three theoretical stages for the wash grid. In addition, Kaes [1] suggested using a single absorber to model each separation zone in the simplified simulation. Thus, we represent the real VDU shown in Figure 3.6 by four similar absorbers with theoretical configurations of Figure 3.7 for simplified modeling. The simplified configuration divides the VDU into four absorbers with each having two theoretical stages, and it includes three theoretical stages to model the wash grid section. The four absorbers represent (1) stripping and flash zone, (2) wash grid and HVGO zone, (3) LVGO zone, and (4) VD zone, respectively. In the simplified simulation, the slop wax includes the overflash and entrained oil. Figure 3.8 shows the process flowsheet of the simplified VDU model built by Aspen HYSYS. We present a step-by-step illustration of the modeling details in the following workshop.

Table 3.5 Key operating and analysis data of VDU for modeling.

			Atmospheric residue (AR)	VGO
Flash zone temperature (°C)	407	Density (kg/m^3)	971	
Top pressure (mmHg)	76	MW	533	
Pressure of VD draw tray (mmHg)	79		D1160 at 760 mmHg	
Pressure of LVGO draw tray (mmHg)	83	IBP	319 °C	304 °C
Pressure of HVGO draw tray (mmHg)	90	5%	368 °C	341 °C
Flash zone pressure (mmHg)	100	10%	381 °C	359 °C
Bottom pressure (mmHg)	190	30%	454 °C	404 °C
		50%	533 °C	443 °C
		55%	560 °C	—
		70%		489 °C
		90%		543 °C
		95%		562 °C

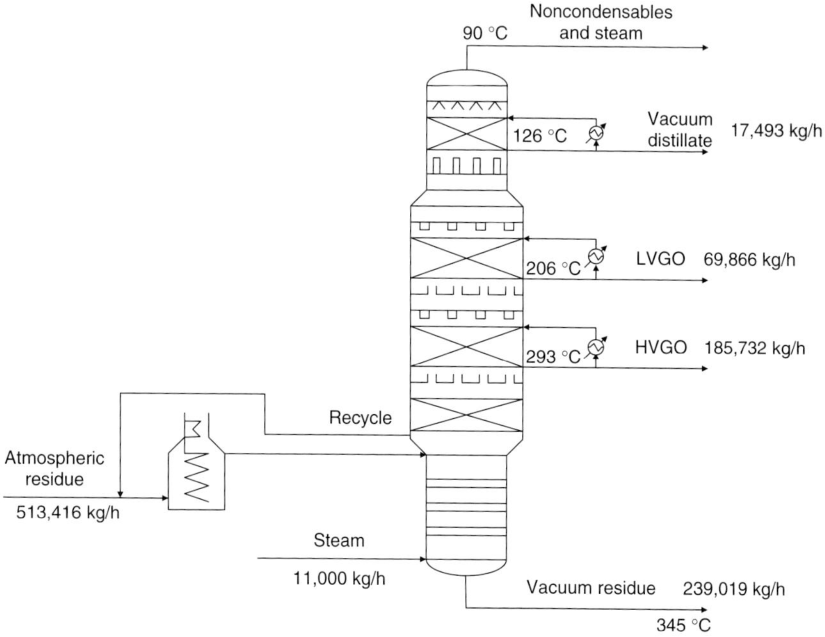

Figure 3.6 Operating and production data of a VDU in Southeast Asia.

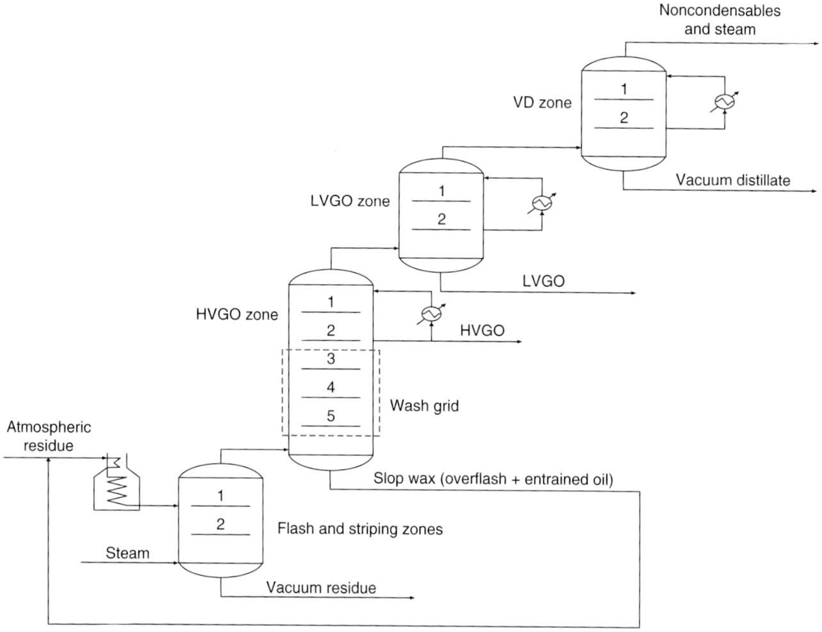

Figure 3.7 Configuration of simplified simulation.

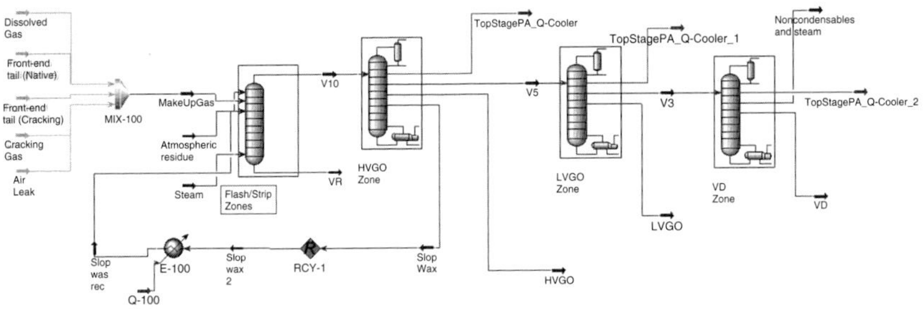

Figure 3.8 Process flowsheet of the simplified VDU model in Aspen HYSYS.

3.3.2 Workshop 3.1 – Build the Simplified VDU Model

Step 1. Define the components, properties, and petroleum assay.

Following Workshop 1.5, Section 1.7, we import the default Component List-1 (petroleumCom1.cml), and choose the default fluid package, Basis-1 (Peng–Robinson).

Make sure to add *n*-C11 and *n*-C12 to the component list. Before we define the atmospheric residue of Table 3.5, we note that the engineer often converts D1160 distillation curve obtained at reduced pressure to the corresponding atmosphere data. Therefore, the model developer should always consult with plant engineers and operators to figure out the type of distillation curve and the corresponding distillation pressure. For our problem, the D1160 distillation curve data for AR in Table 3.5 correspond to a pressure of 760 mmHg. Thus, we must set the option of "D1160 Distillation Conditions" to "Atmospheric." Currently, the Aspen HYSYS petroleum assay manager does not allow us to choose this option directly, but the oil manager does. We therefore define the AR in the old manager first and then convert it to the petroleum assay manager following Workshop 1.6, Section 1.8 (see Figures 3.9–3.13).

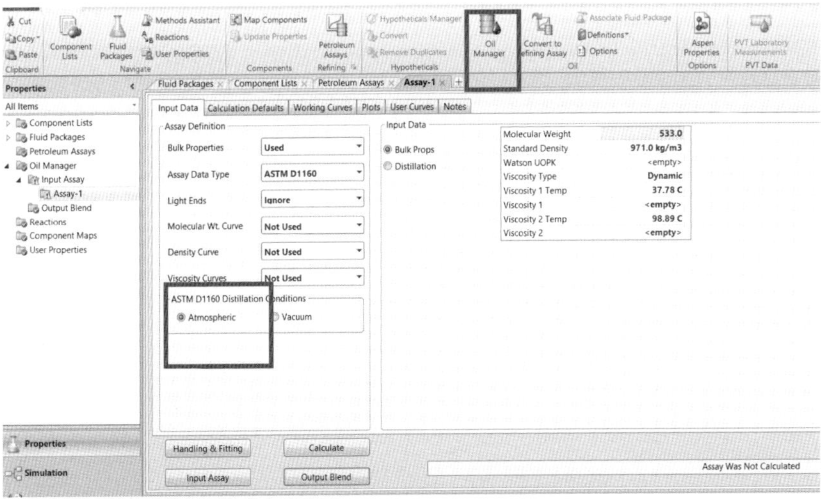

Figure 3.9 Input bulk properties and choose D1160 distillation pressure for AR.

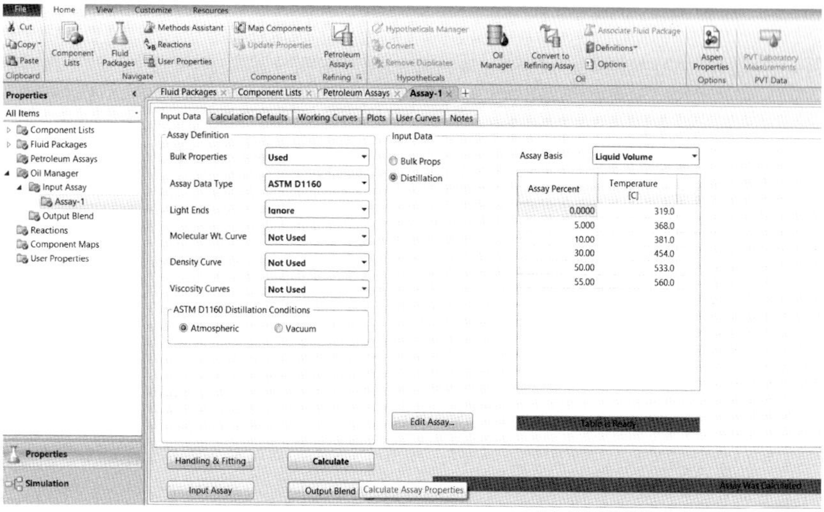

Figure 3.10 Input the D1160 distillation curve data, perform property characterization calculations, and convert to refinery from oil manager to refinery assay (petroleum assay manager).

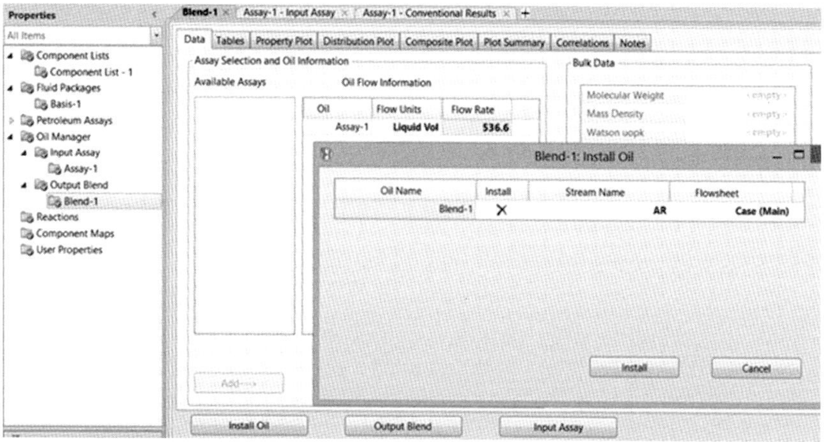

Figure 3.11 Select Assay-1 to be cut or blended as Blend-1, move it from Available Assays to Oil Flow Information, enter flow rate of 536.6 m^3/h, and install oil. Enter stream name AR within Blend-1:Install Oil, and click install Blend-1.

Step 2. Define the AR feed, makeup gas streams, and steam. Draw the flowsheet of Figure 3.14. See the stream specifications in Table 3.6. Save the resulting simulation file as *Workshop 3.1-1.hsc.*

Step 3. Simulate the flash and stripping zones by specifying a flash zone temperature of 407 °C (see Figures 3.15–3.17). Save the converged simulation as *Workshop 3.1-2.hsc.*

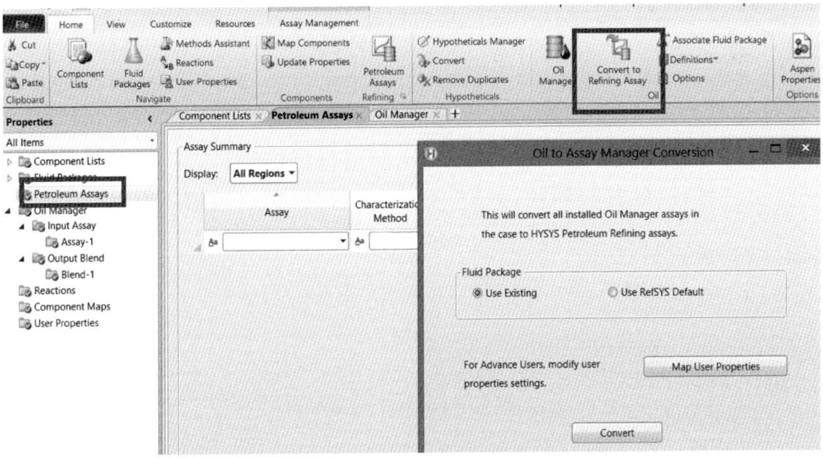

Figure 3.12 Go to Petroleum Assays and choose "Convert to Refinery Assay."

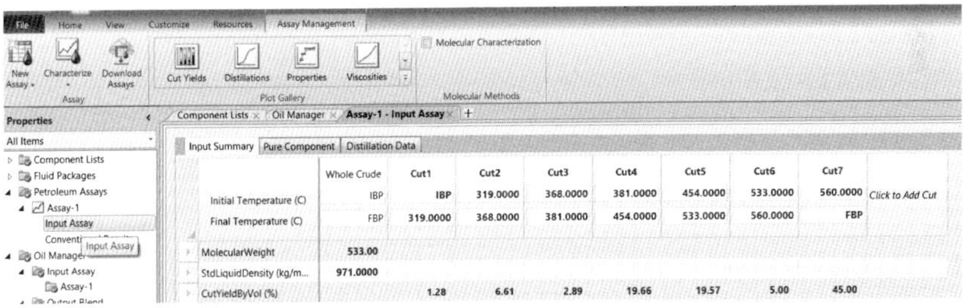

Figure 3.13 Converted Assay-1 for AR in petroleum refinery manager.

Figure 3.14 Flowsheet of the feed section of the VDU.

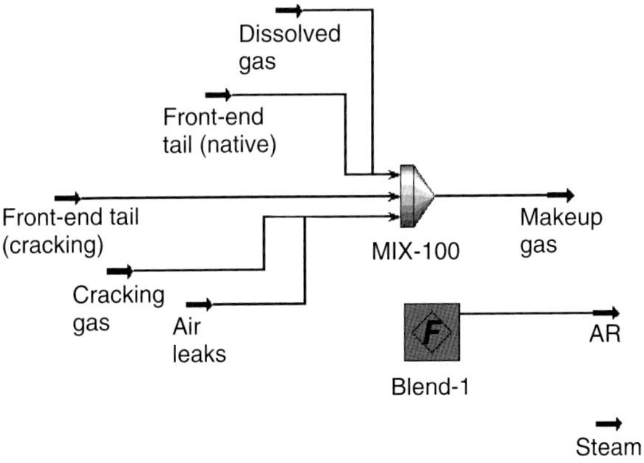

Table 3.6 Stream specifications for Figure 3.14.

Stream	Dissolved gas	Front-end tail (native)	Front-end tail (cracking)	Cracking gas	Air leaks	Steam (160 °C, 343.2 kPa)
Mass flow (kg/h)	423	1837	933	678	54	1.1E4
Mole fractions	0.75 C2 0.25 C3	0.5 *n*-C11 0.5 *n*-C12	0.5 *n*-C11 0.5 *n*-C12	0.75 C2 0.25 C3	1.0 N_2	1.0 H_2O

Note: Except for steam, all gas streams and AR feed are at 411 °C and 13.33 kPa. AR has a flow rate of 536.6 m^3/h.

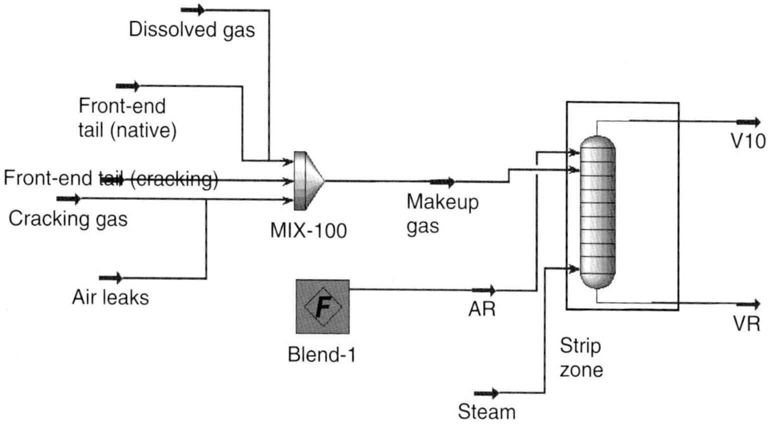

Figure 3.15 Add the flowsheet of the flash/strip zones with a simple absorber without reflux.

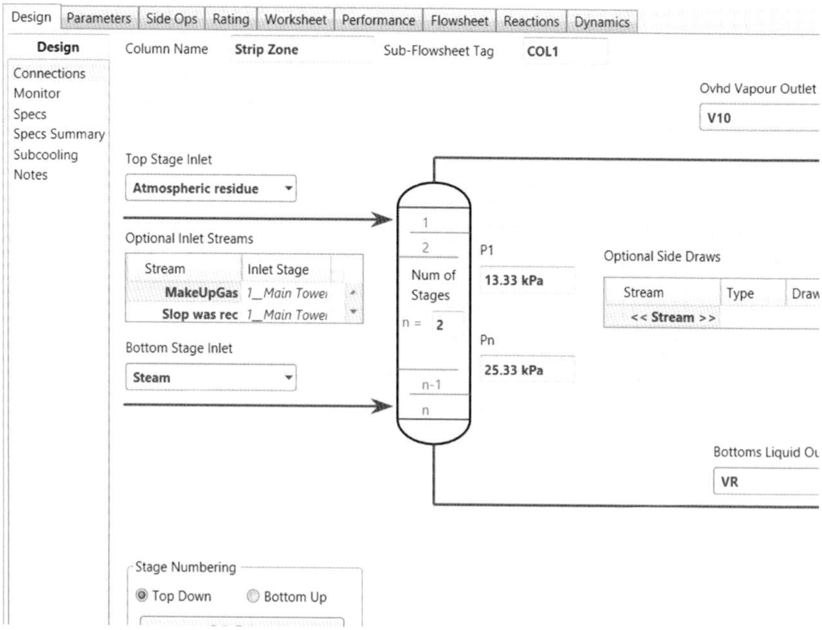

Figure 3.16 Specification of the flash/strip zones.

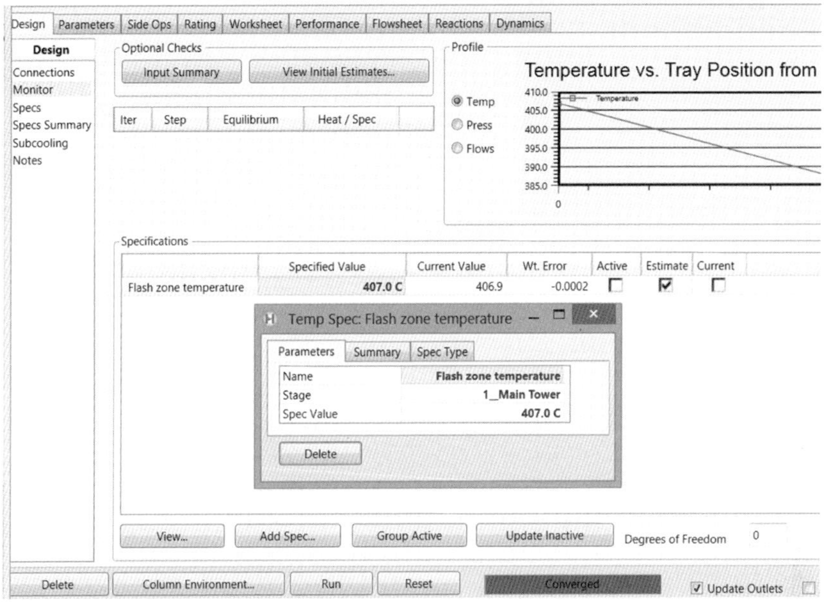

Figure 3.17 Specify the flash zone temperature.

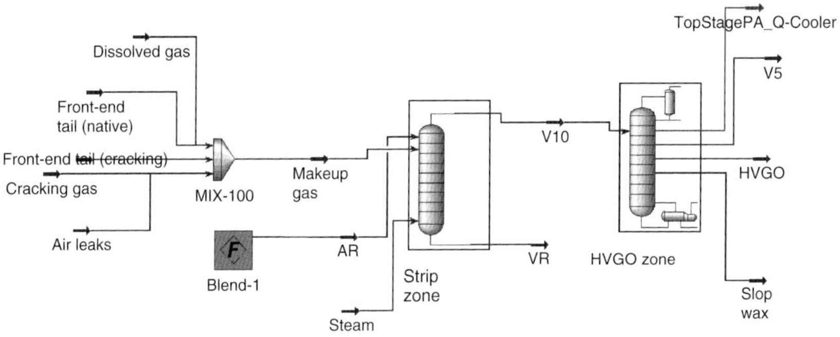

Figure 3.18 Add the flowsheet of the HVGO zone.

Step 4. Add the HVGO zone and the slop wax recycle loop by choosing an absorber with top-stage pumparound (PA) reflux. See the flowsheet and specifications in Figures 3.18–3.20. Save the converged simulation file as *Workshop 3.1-3.hsc*.

In practice, the slop wax stream has two sources, namely, overflash and entrained oil. Figures 3.19–3.21 represent the specifications used in the HVGO zone, including the circulation rate and temperature change of the pumparound stream and HVGO flow rate, following Kaes' suggestions [1]. The model predicts a flow rate of 14 m^3/h for the slop wax, which is 2.6% of the atmospheric residue flow rate of 536.6 m^3/h. Considering that the slop wax includes overflash and entrained oil, and typical volume flow ratio of overflash to feed ranges from 0.2% to 5% [4], the simplified model gives good estimates on the slop wax.

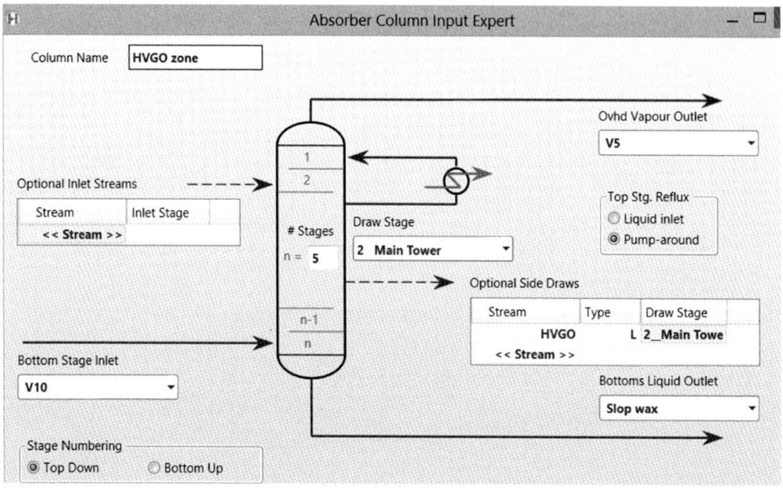

Figure 3.19 Specifications of the HVGO zone (step 1).

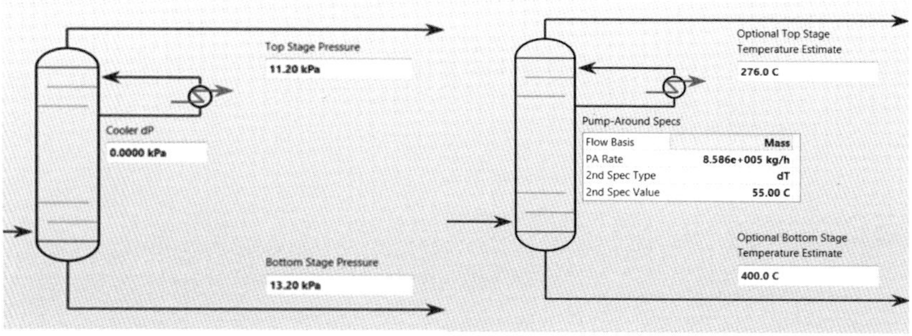

Figure 3.20 Specifications of the HVGO zone (step 2).

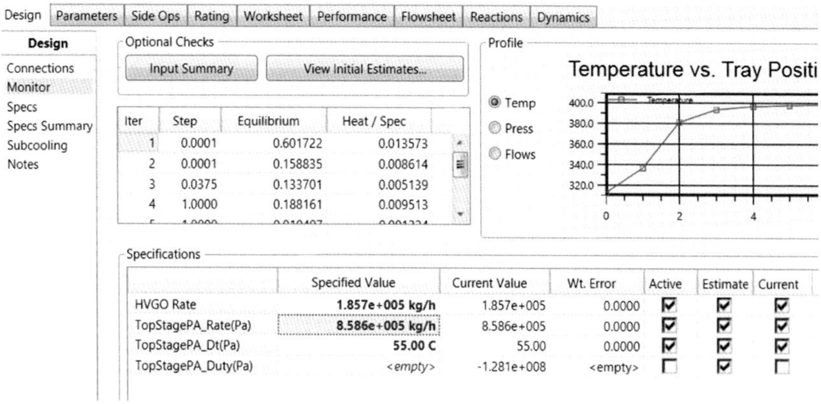

Figure 3.21 Specifications of the HVGO zone (step 3).

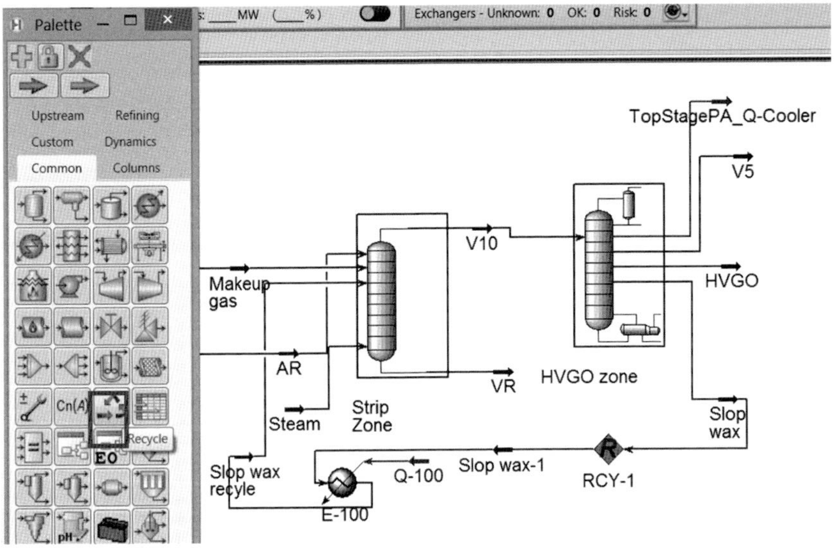

Figure 3.22 Add the flowsheet of the slop wax recycle to stage 1 of the strip zone.

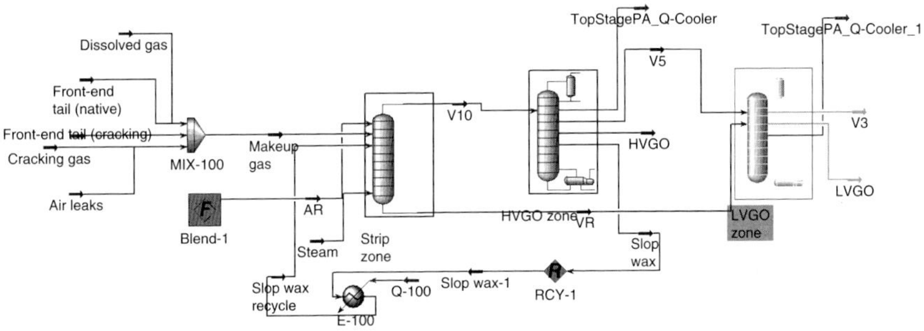

Figure 3.23 Add the flowsheet of the LVGO section.

Step 5. Complete the recycle loop for slop wax to enter the stage 1 of the strip zone. Heater E-100 heats up slop wax-1 to 411 °C. See Figure 3.22. No need for other specifications. The simulation converges quickly. Save the converged simulation as *Workshop 3.1-4.hsc.*

Step 6. Add the LVGO zone by choosing an absorber with top-stage pumparound (PA) reflux. See the flowsheet and specification in Figures 3.23–3.26. We add the specification of the LVGO draw rate of 6.987E4 kg/h as an estimate. We do this in Figure 3.26 as follows: Add Spec → Column Specification Types → Column Draw Rate → Draw Spec → Name – LVGO draw rate; Draw – LVGO; Flow Basis – mass; Spec Value – 6.987E4 kg/h. Save the converged simulation as *Workshop 3.1-5.hsc.*

Step 7. Add the vacuum distillation (VD) zone. See Figures 3.27–3.30 for the flowsheet and specifications. We add the stage 1 temperature of 90 °C. Save the converged simulation as *VDU-Simplified.hsc.*

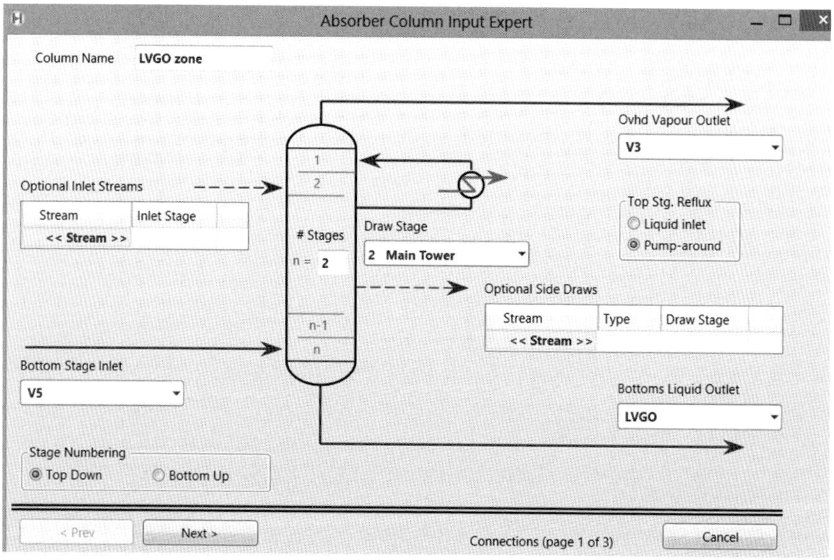

Figure 3.24 Specifications of the LVGO zone (step 1).

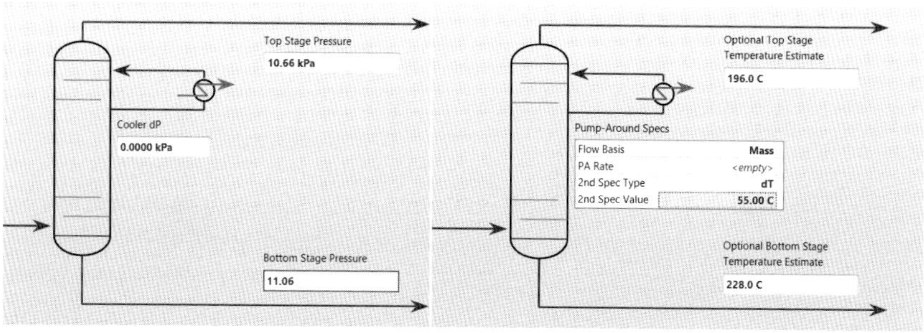

Figure 3.25 Specifications of the LVGO zone (step 2).

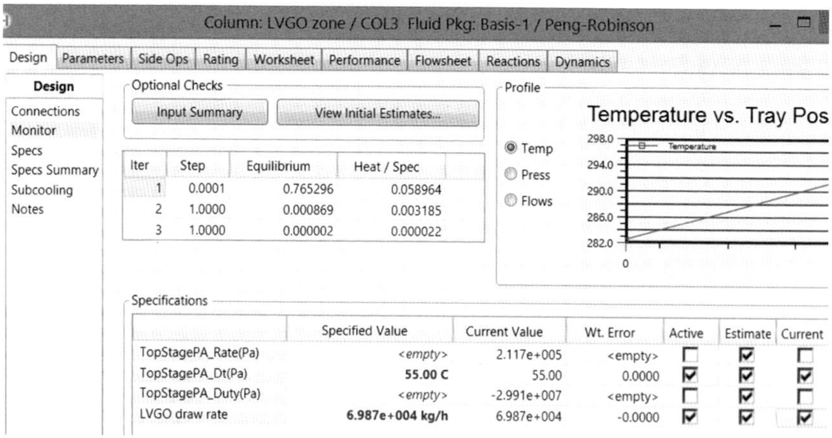

Figure 3.26 Specifications of the LVGO zone (step 3).

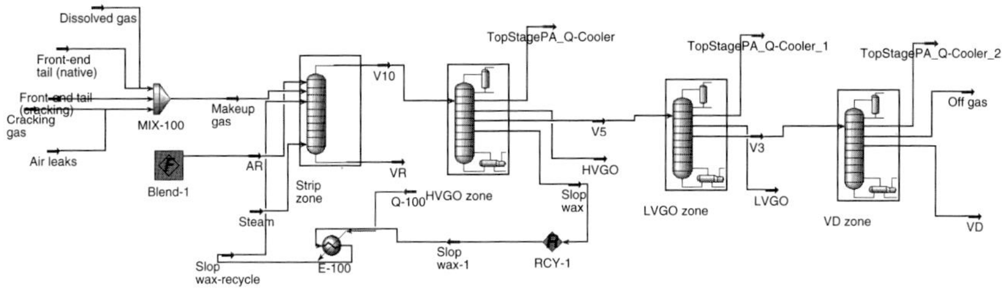

Figure 3.27 Flowsheet of the complete simplified VDU model after adding the vacuum distillation (VD) zone.

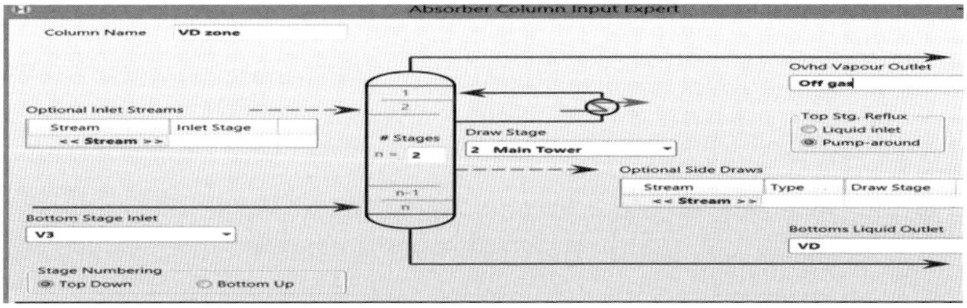

Figure 3.28 Specifications of the VD zone (step 1).

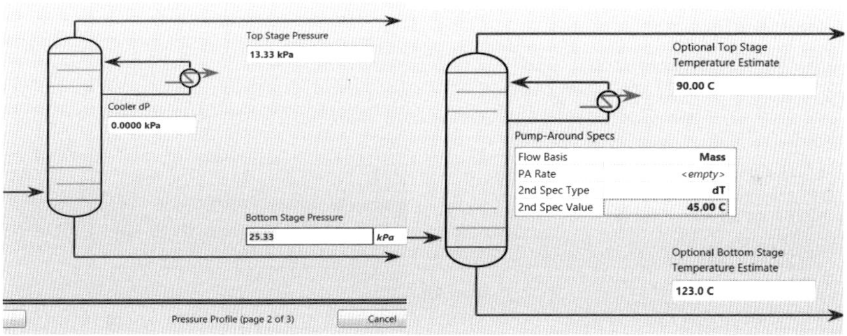

Figure 3.29 Specifications of the VD zone (step 2).

Figures 3.31–3.33 compare the simplified model predictions with plant data for the VDU temperature profile, D1160 curve of VGO, and product yields. Significantly, our simplified model is able to generate good results that not only provide a quick and informative understanding of the real VDU, but also serve as a convenient platform to investigate the consistency of plant data.

3.3.3 Workshop 3.2 – Build the Rigorous Model from a Simplified Model

Step 1. We open the simulation file for the feed section, *Workshop 3.1-1.hsc*, and save it as *Workshop 3.2-1.hsc*. See Figure 3.14 for the flowsheet.

Step 2. Add an absorber column to represent the rigorous VDU simulation model (see Figures 3.34–3.42). Specifications of rigorous VDU model are similar to

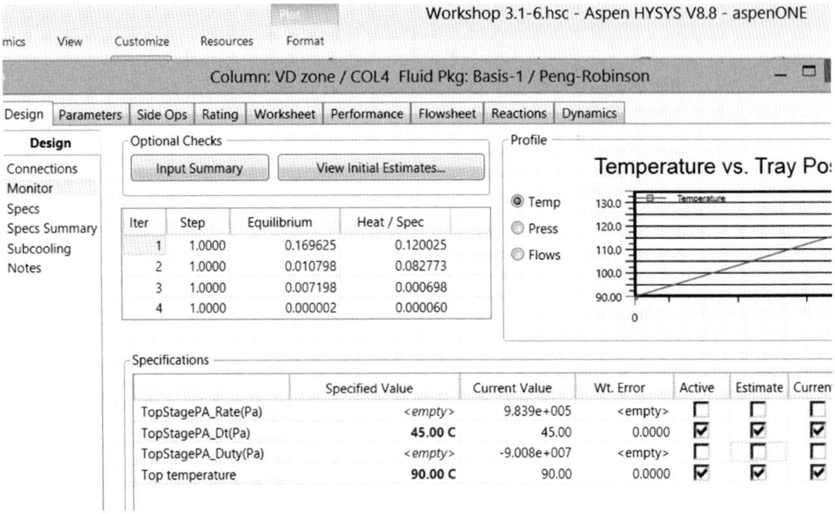

Figure 3.30 Specifications of the VD zone (step 3).

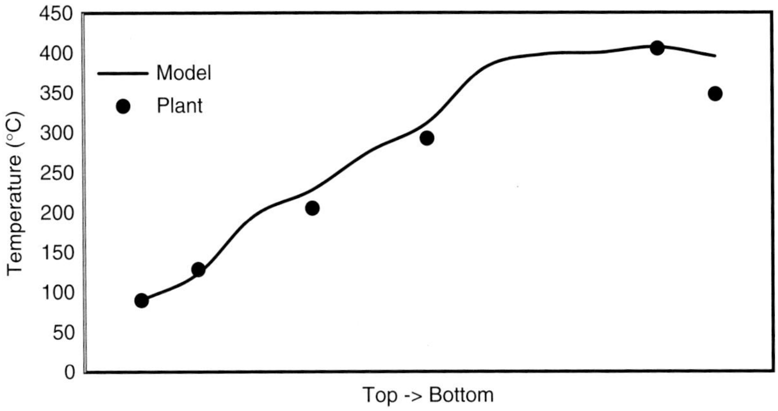

Figure 3.31 Prediction on temperature profile by the simplified model.

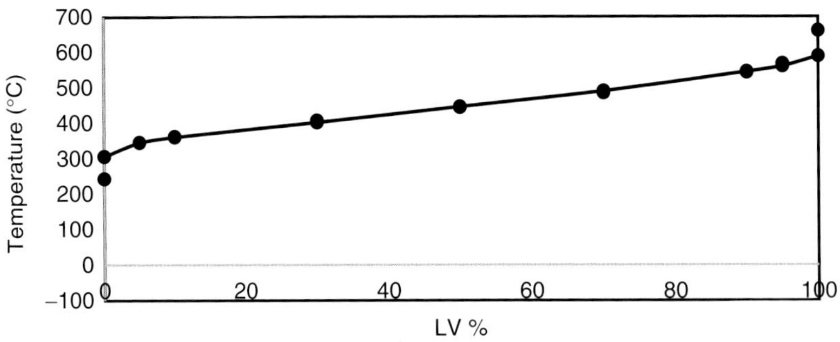

Figure 3.32 Prediction on D1160 curve of VGO by the simplified model.

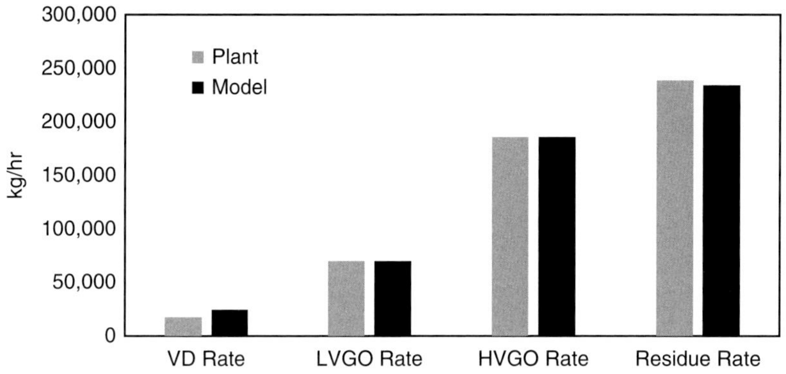

Figure 3.33 Product yields estimated by the simplified model.

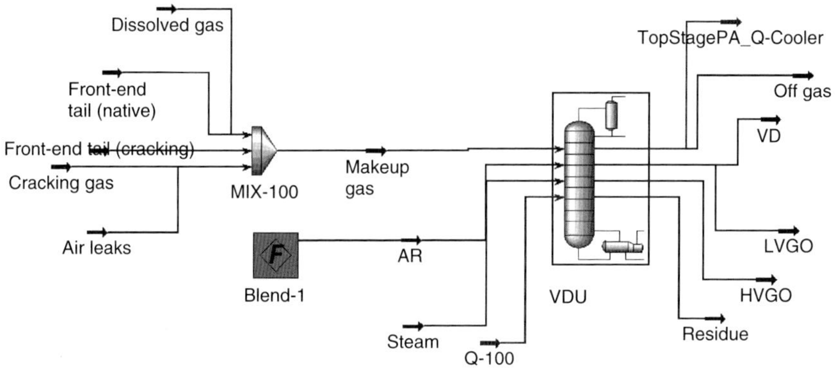

Figure 3.34 Flowsheet of the rigorous VDU simulation model.

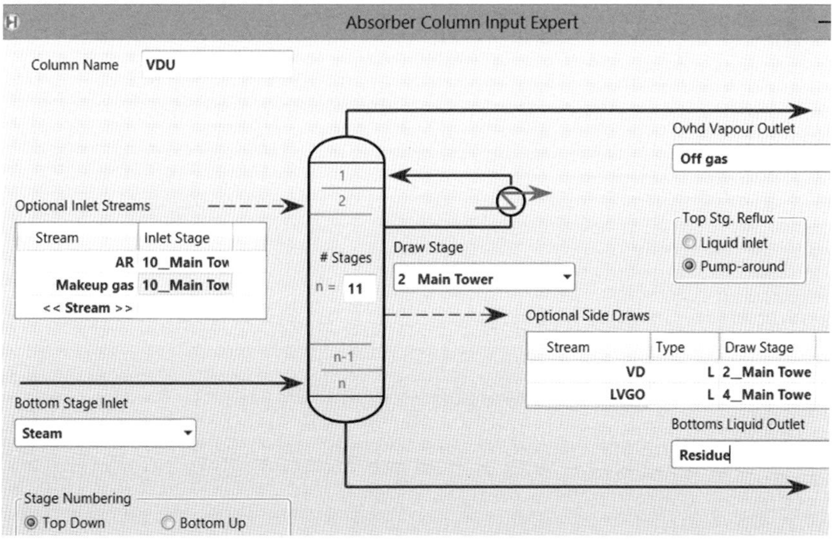

Figure 3.35 Specifications of the VDU unit (step 1): Optional Side Draws also include Stream-HVGO, Type-L, Draw Stage-6_Main Tower (not shown).

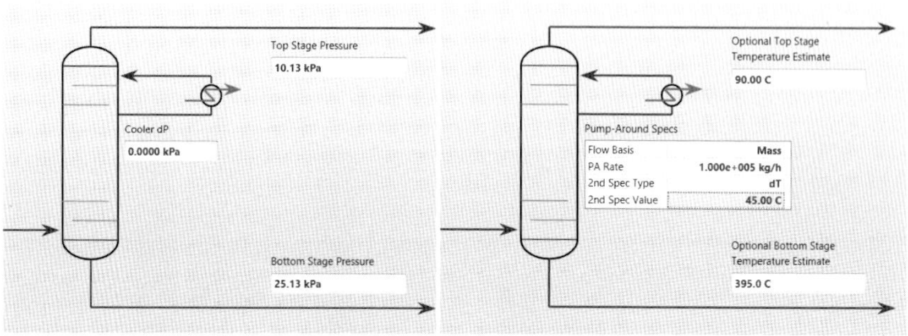

Figure 3.36 Specifications of the VDU unit (step 2).

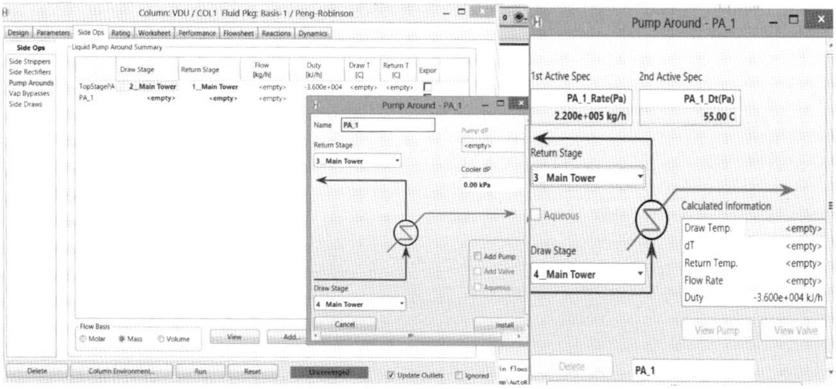

Figure 3.37 Specifications of the VDU unit (step 3): Add pumparound PA_1.

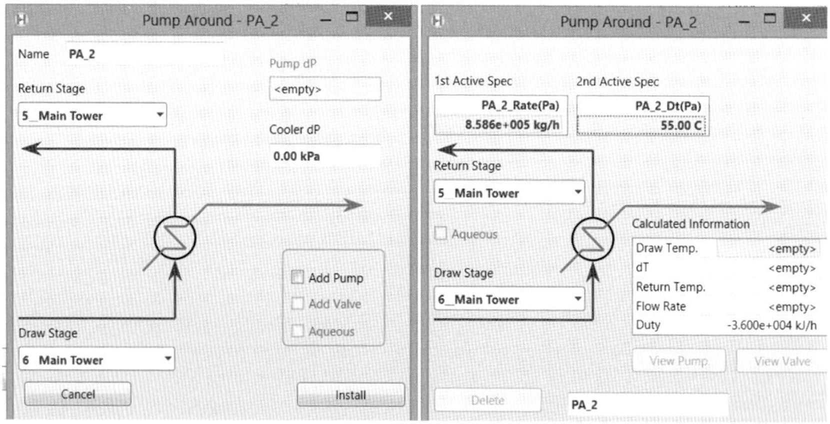

Figure 3.38 Specifications of the VDU unit (step 4): Add pumparound PA_2.

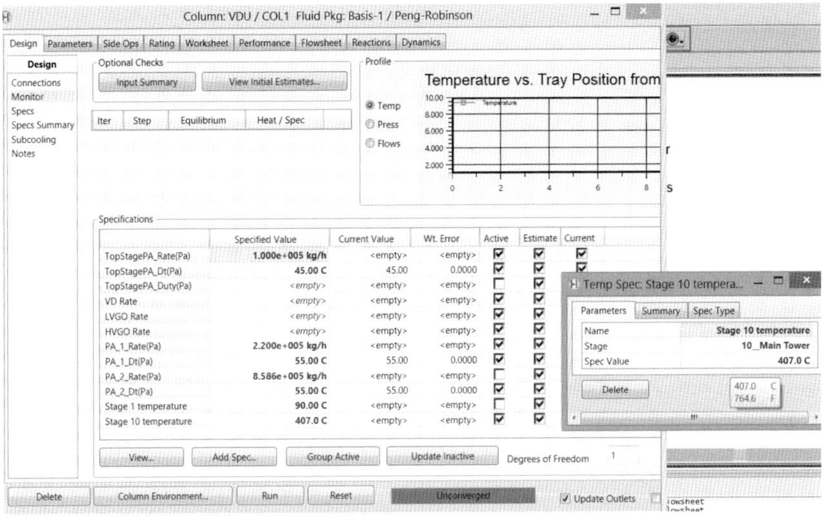

Figure 3.39 Specifications of the VDU unit (step 5): Add stage 1 temperature of 90 °C and stage 10 (flash zone) temperature of 407 °C.

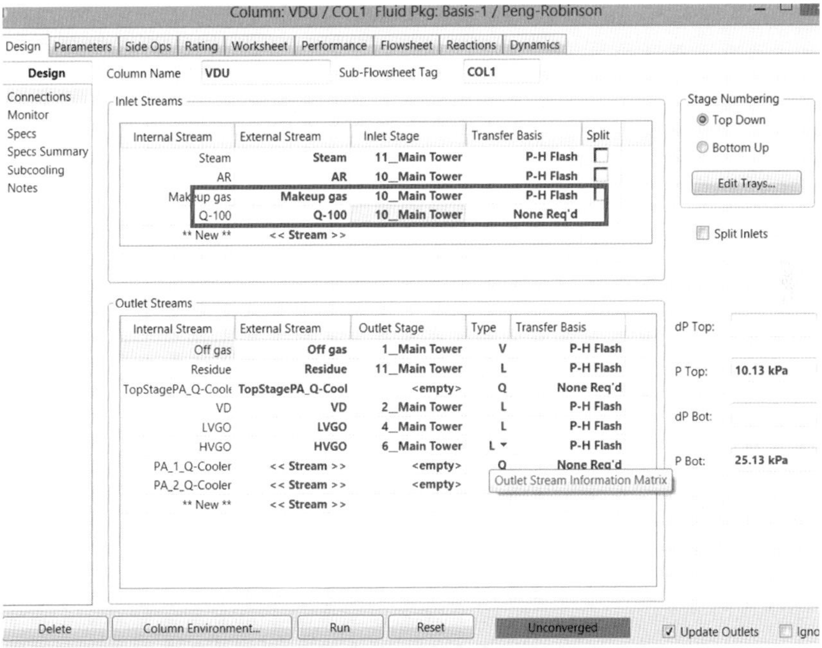

Figure 3.40 Add an energy stream Q-100 to the stage 10 to tune the flash zone temperature and to ensure the closure of the energy balance calculation. No need to specify its value.

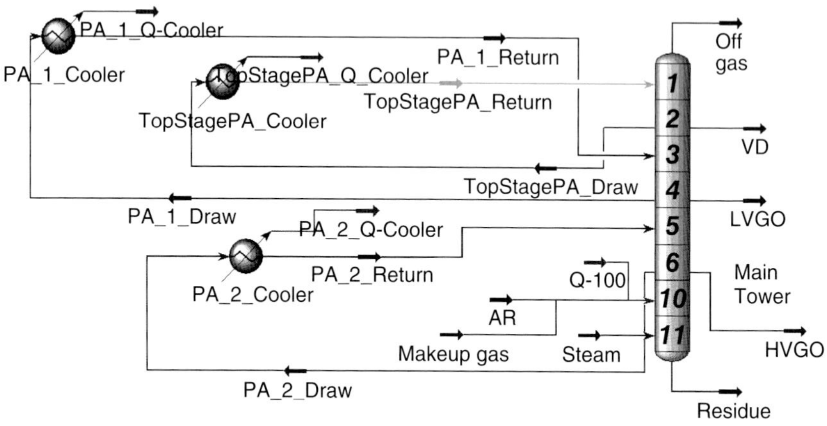

Figure 3.41 The column environment of the VDU model.

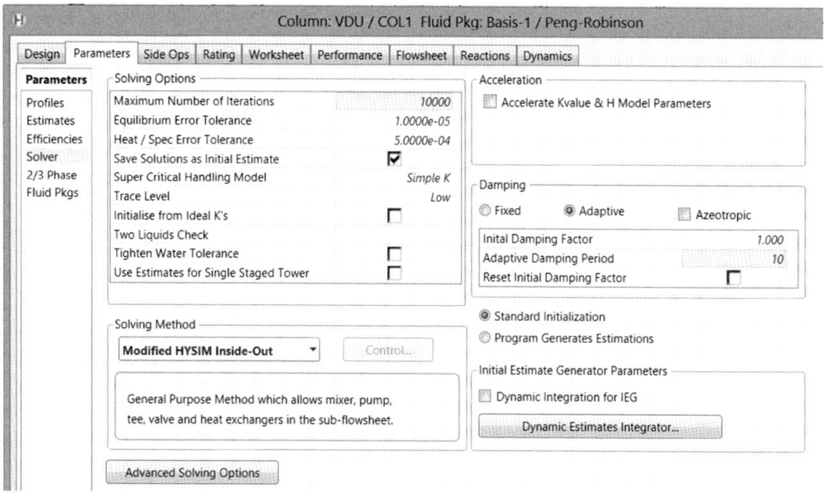

Figure 3.42 Choose Modified HYSYS Inside-Out algorithm and adaptive damping under Parameters → Solve.

the simplified CDU model, including the circulation rate and temperature change of each pumparound stream, flow rates of all liquid products except for one, top temperature, and flash zone temperature [1]. By using the results of the simplified model, the rigorous simulation converges quickly (Figure 3.43).

Figure 3.44 shows the pressure, temperature, and internal flow profiles obtained from the results of the rigorous model. We save the converged rigorous VDU model as *VDU-Rigorous.hsc*.

Figures 3.45–3.47 compare the rigorous model predictions with plant data for column temperature profile, D1160 curve of VGO, and product yields. The results demonstrate that the two-step approach of model development generates accurate predictions on key operation and production variables of VDU.

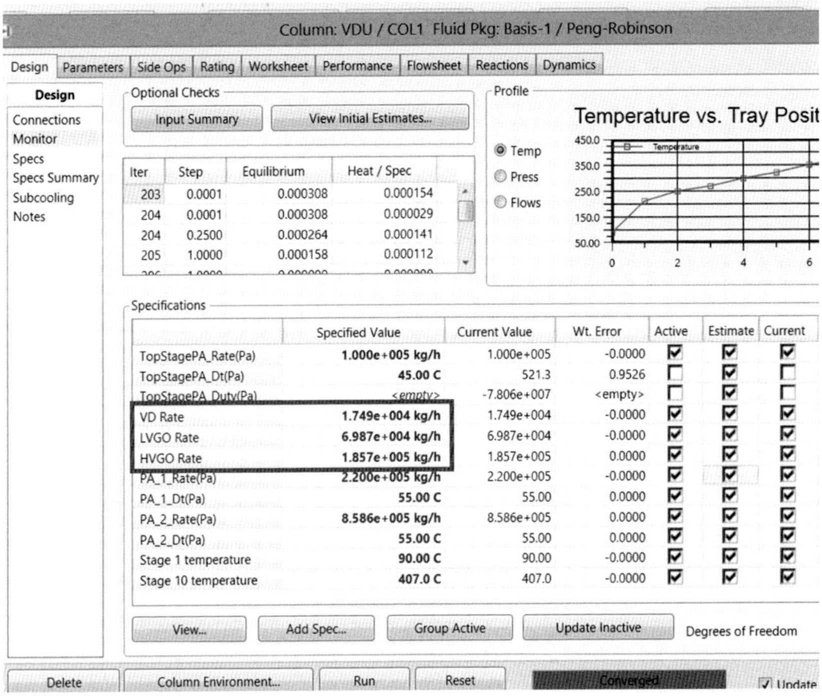

Figure 3.43 Specify rates of VD, LVGO, and HVGO to enable simulation convergence.

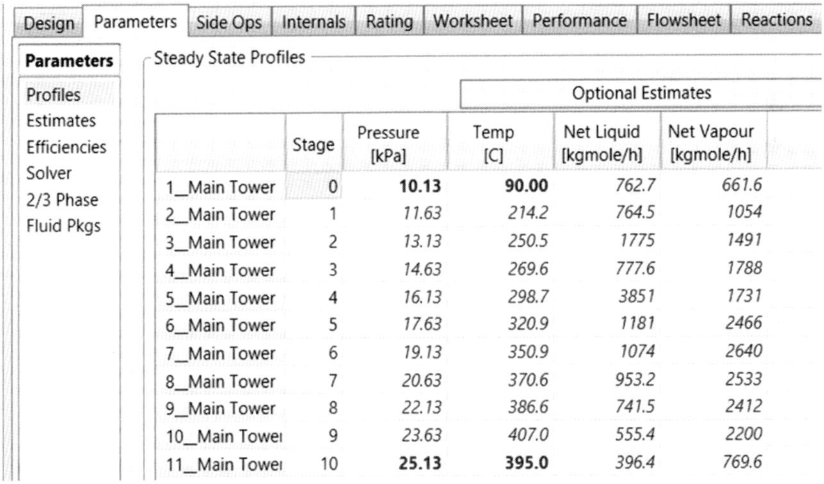

	Stage	Pressure [kPa]	Temp [C]	Net Liquid [kgmole/h]	Net Vapour [kgmole/h]
1_Main Tower	0	10.13	90.00	762.7	661.6
2_Main Tower	1	11.63	214.2	764.5	1054
3_Main Tower	2	13.13	250.5	1775	1491
4_Main Tower	3	14.63	269.6	777.6	1788
5_Main Tower	4	16.13	298.7	3851	1731
6_Main Tower	5	17.63	320.9	1181	2466
7_Main Tower	6	19.13	350.9	1074	2640
8_Main Tower	7	20.63	370.6	953.2	2533
9_Main Tower	8	22.13	386.6	741.5	2412
10_Main Towei	9	23.63	407.0	555.4	2200
11_Main Towei	10	25.13	395.0	396.4	769.6

Figure 3.44 Pressure, temperature, and internal flow profiles obtained from the rigorous simulation model.

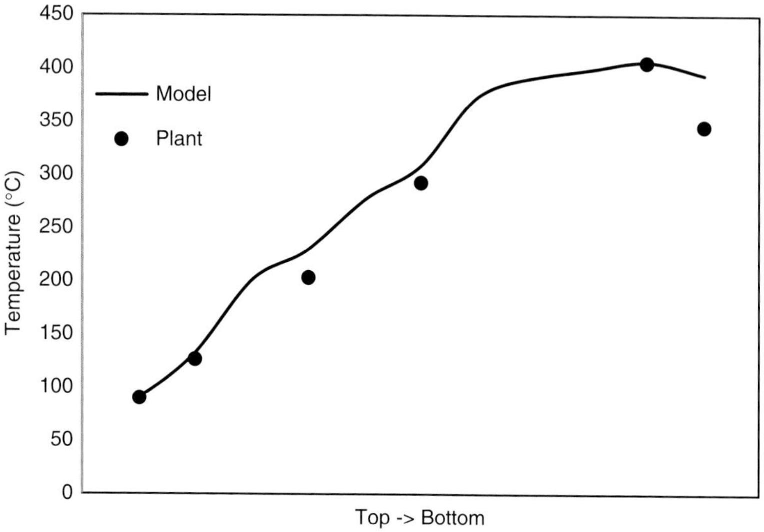

Figure 3.45 Prediction on temperature profile by rigorous model.

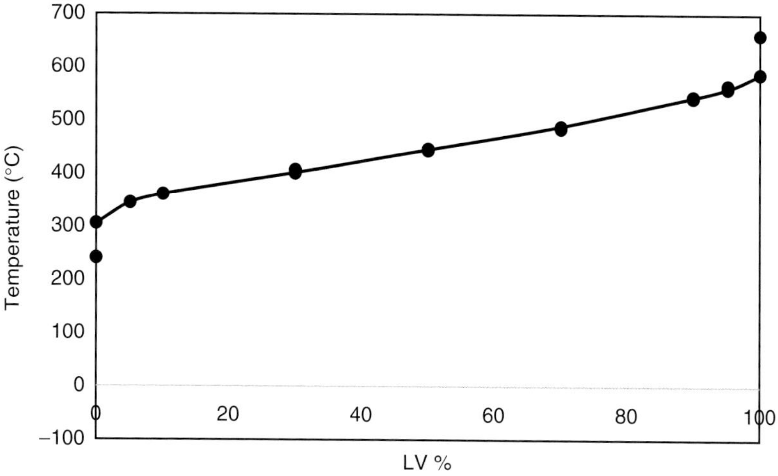

Figure 3.46 Prediction on D1160 curve of VGO by rigorous model.

3.4 Model Application – VDU Deep-Cut Operation

One of the industrially significant applications of VDU simulation is to optimize the deep-cut operation to process heavier crude because of the growing demand of the refiners to process heavier crude feeds. Deep-cut operation is expected to improve process economics by increasing the cut point of HVGO higher than 1050 °F or 565 °C to produce more gas oil for downstream units such as FCC. In our preceding example, the HVGO stream has a D86 95% point of 883 °F or

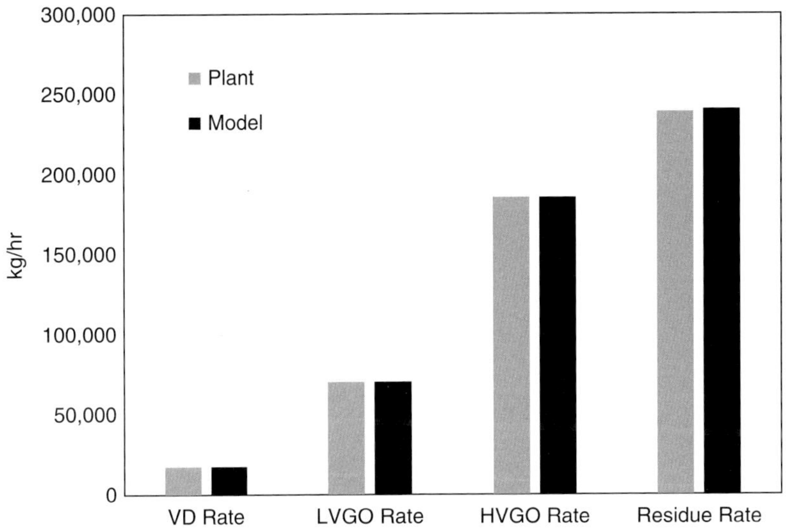

Figure 3.47 Product yields estimated by rigorous model.

489 °C only. In practice, there are four strategies and corresponding actions to perform VDU deep-cut revamping to increase the HVGO yield:

1) Increase feed vaporization.
   - Raise flash zone temperature.
   - Reduce flash zone pressure.
2) Increase distillate strip-out from residue (if column is steam-stripped).
   - Optimize stripping steam.
3) Decrease overflash while maintaining high-quality washing of the vapors rising from the flash zone.
   - Reduce wash oil.
   - Optimize wax recycle.
4) Increase vapor–liquid separation in the flash zone.
   - Optimize mechanical design of distillation column.

Except for the fourth strategy, increase vapor–liquid separation in the flash zone, we can investigate all the other three strategies by altering process operations and optimizing the process through simulations. This section presents two case studies to demonstrate the model applications to optimize VDU deep-cut operation by raising flash zone temperature and by optimizing stripping steam.

Figures 3.48 and 3.49 illustrate the effects of flash zone temperature on the mass yield and TBP 95% point of HVGO. The mass yield of HVGO increases 4% when flash zone temperature increases from 407 to 418 °C. In addition, TBP 95% point of HVGO also increases to as high as 567 °C. As higher HVGO yield and cut point result from the higher flash zone temperature, more atmospheric residue gets vaporized.

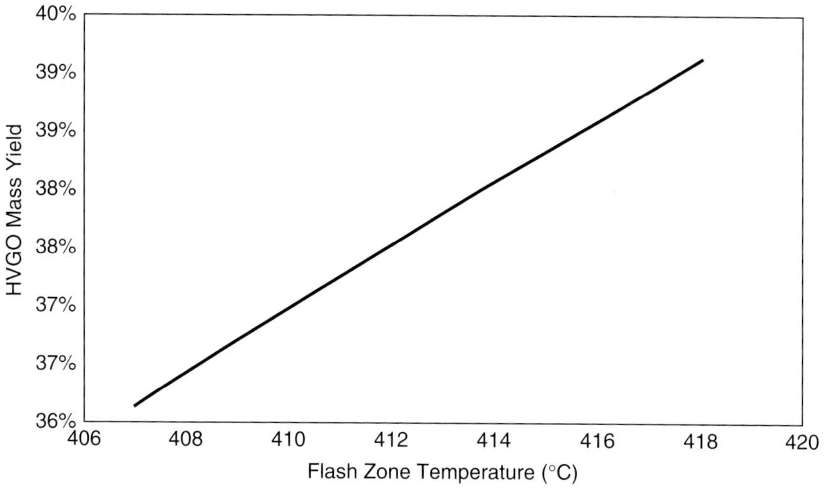

Figure 3.48 Effect of flash zone temperature on HVGO yield.

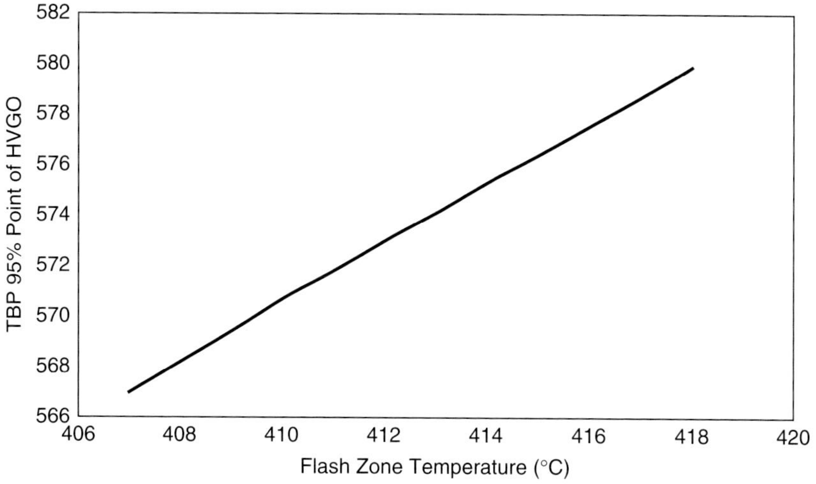

Figure 3.49 Effect of flash zone temperature on HVGO cut point.

Figures 3.50 and 3.51 illustrate the effects of stripping steam rate on the yield and TBP 95% point of HVGO. Similar to the effect of flash zone temperature, both mass yield and TBP 95% point of HVGO increase significantly with increasing stripping steam rate. We note that stripping steam has little effect on vaporizing atmospheric residue directly. Higher stripping steam rate will reduce the vapor pressure of hydrocarbon and enhance the vaporization of atmospheric residue. Both case studies show the expected operations of modern refiners when performing VDU deep-cut operation.

Although process simulation provides a guideline for deep-cut operation, it is important to keep in mind that simulation results are usually the most optimistic scenarios and we should consider other issues before making any changes to a

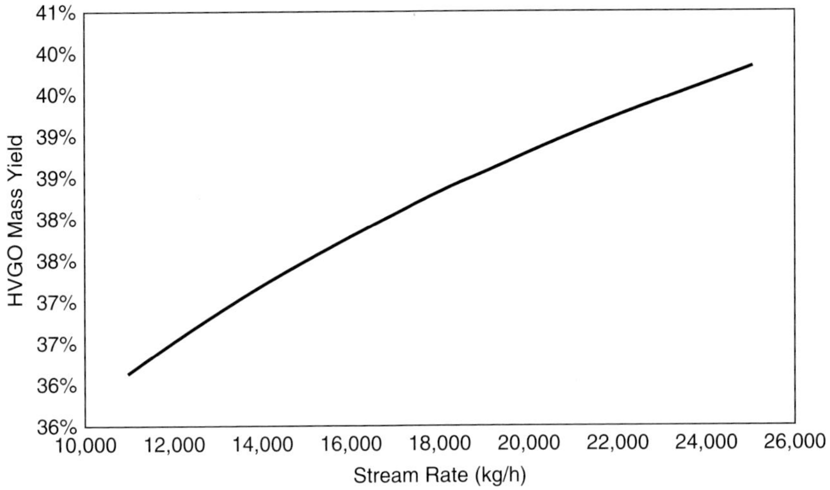

Figure 3.50 Effect of stripping steam rate on HVGO yield.

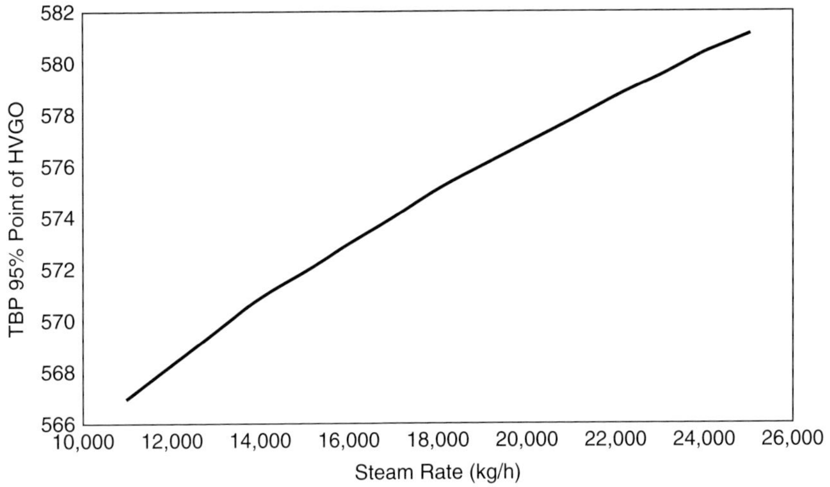

Figure 3.51 Effect of stripping steam rate on HVGO cut point.

real process. The most important issue is the feed oil quality, particularly cracking ability and contaminant distributions, such as nickel and vanadium. We need to minimize the thermal cracking and reduce coke formation in the wash grid. In addition, detailed contaminant distribution of feed stock is also important for process simulation. For example, Figures 3.52 and 3.53 illustrate the nickel and vanadium distribution of some crude oil [8]. Apparently, nickel and vanadium contents rise significantly due to metalloporphyrin components [9] over the temperature range from 550 to 600 °C, which is the target range of deep-cut operation. We must validate the VDU model simulation with a detailed analysis of contaminant distribution to determine if the produced HVGO meets the specifications.

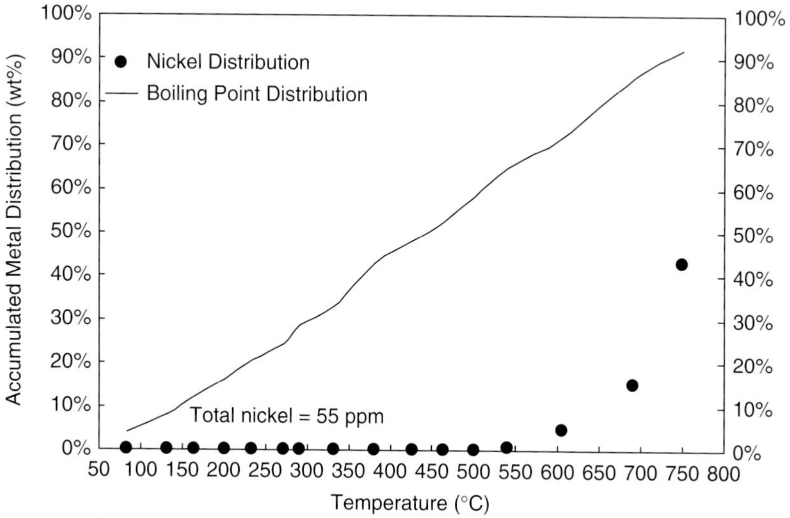

Figure 3.52 Nickel and boiling point distribution of some crude oil. (Adapted from Boduszynski [8].)

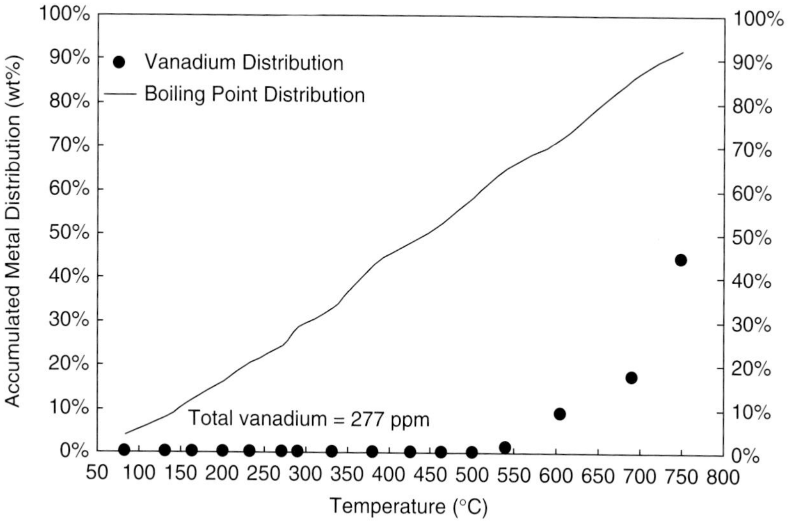

Figure 3.53 Vanadium and boiling point distribution of some crude oil. (Adapted from Boduszynski [8].)

3.5 Workshop 3.3 – Simulation of the VDU Deep-Cut Operation

This workshop provides a step-by-step guideline of how to conduct the VDU deep-cut investigation by using "Case Studies" tool that we have demonstrated previously in Workshop 2.1, Section 2.10.3, and in Workshop 2.4, Section 2.13.

Step 1. Open *VDU Deep Cut – Start.hsc*, which is a completed VDU simulation. Save the file as *Workshop 3.3-1.hsc* (Figure 3.54).

Step 2. Activate case studies.

Step 3. Add two independent variables, steam mass flow rate (1.1–2.5E4 kg/h; increment of 0.1E4 kg/h), and flash zone (stage 10) temperature (407–418 °C; increment of 1 °C) (see Figures 3.55–3.57).

Step 4. Define dependent variables: mass flow and TBP 95% point of HVGO and mass flow and TBP 95% point of LVGO (see Figures 3.58 and 3.59).

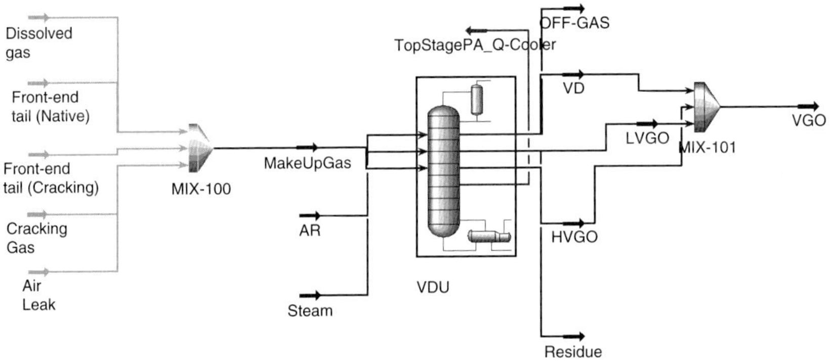

Figure 3.54 VDU model in Aspen HYSYS.

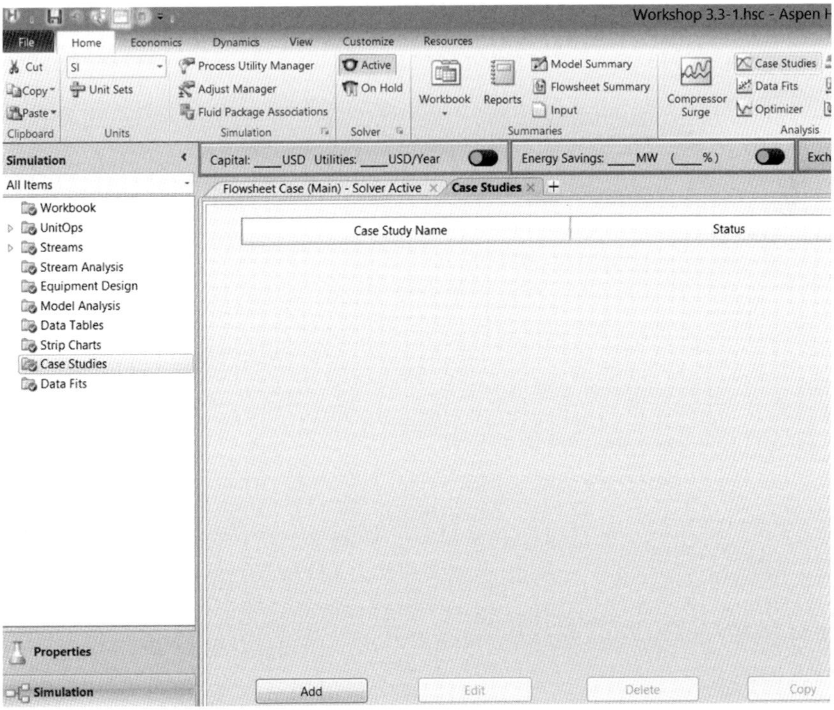

Figure 3.55 Add a case study.

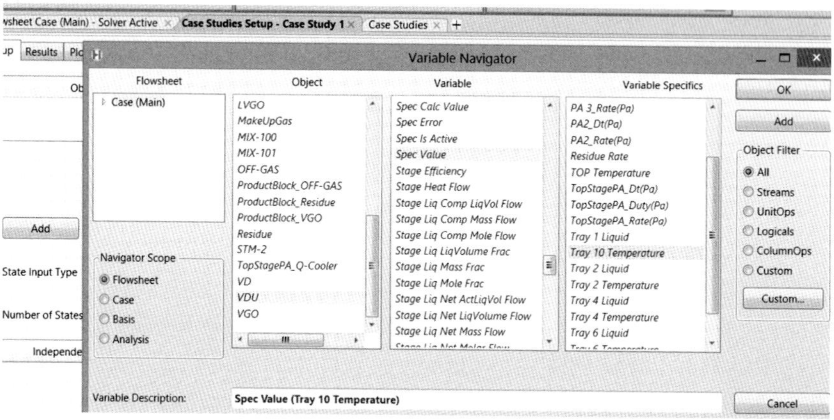

Figure 3.56 Add an independent variable, the specified stage 10 (flash zone) temperature: Variable Navigator → Flowsheet – Case (Main); Object – VDU; Variable – Spec Value; Variable Specifics – Tray 10 Temperature → OK.

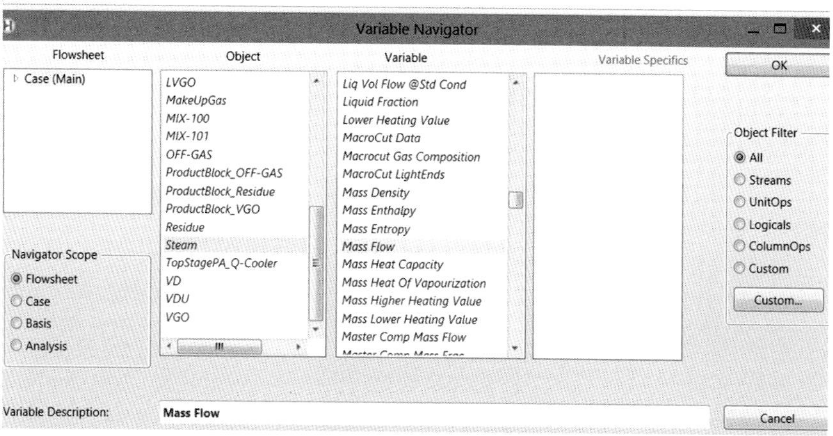

Figure 3.57 Add independent variable: steam mass flow.

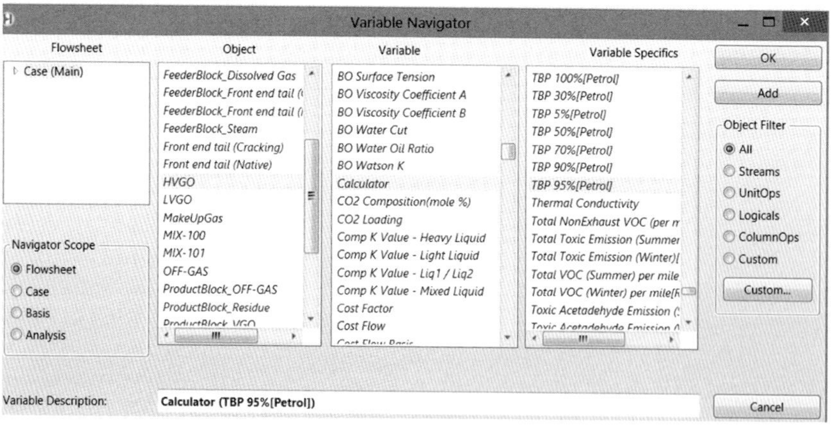

Figure 3.58 Define dependent variable: TBP 95% point of HVGO.

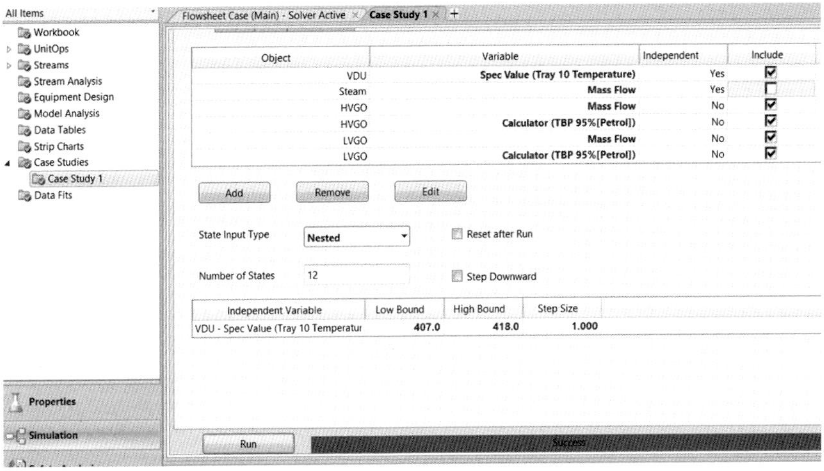

Figure 3.59 Run case study, varying the tray 10 (flash zone) temperature.

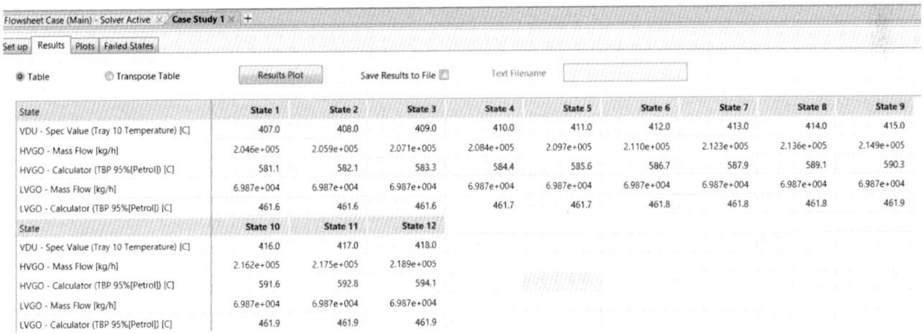

State	State 1	State 2	State 3	State 4	State 5	State 6	State 7	State 8	State 9
VDU - Spec Value (Tray 10 Temperature) [C]	407.0	408.0	409.0	410.0	411.0	412.0	413.0	414.0	415.0
HVGO - Mass Flow [kg/h]	2.046e+005	2.059e+005	2.071e+005	2.084e+005	2.097e+005	2.110e+005	2.123e+005	2.136e+005	2.149e+005
HVGO - Calculator (TBP 95%[Petrol]) [C]	581.1	582.1	583.3	584.4	585.6	586.7	587.9	589.1	590.3
LVGO - Mass Flow [kg/h]	6.987e+004	6.987e+004	6.987e+004	6.987e+004	6.987e+004	6.987e+004	6.987e+004	6.987e+004	6.987e+004
LVGO - Calculator (TBP 95%[Petrol]) [C]	461.6	461.6	461.6	461.7	461.7	461.8	461.8	461.8	461.9

State	State 10	State 11	State 12
VDU - Spec Value (Tray 10 Temperature) [C]	416.0	417.0	418.0
HVGO - Mass Flow [kg/h]	2.162e+005	2.175e+005	2.189e+005
HVGO - Calculator (TBP 95%[Petrol]) [C]	591.6	592.8	594.1
LVGO - Mass Flow [kg/h]	6.987e+004	6.987e+004	6.987e+004
LVGO - Calculator (TBP 95%[Petrol]) [C]	461.9	461.9	461.9

Figure 3.60 Increasing the flash zone temperature from 407 to 418 °C increases the mass yield of HVGO by 7.3% from 2.046E5 to 2.189E5 kg/h and increases the TBP 95% point only slightly from 461.6 to 461.9 °C.

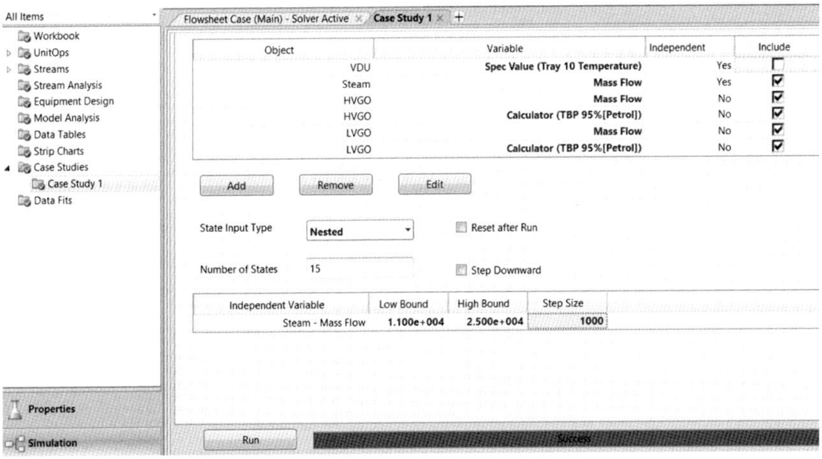

Figure 3.61 Run case study, varying the steam mass flow.

State	State 1	State 2	State 3	State 4	State 5	State 6	State 7	State 8	State 9
Steam - Mass Flow [kg/h]	1.100e+004	1.200e+004	1.300e+004	1.400e+004	1.500e+004	1.600e+004	1.700e+004	1.800e+004	1.900e+004
HVGO - Mass Flow [kg/h]	2.010e+005	2.027e+005	2.043e+005	2.058e+005	2.073e+005	2.086e+005	2.099e+005	2.112e+005	2.124e+005
HVGO - Calculator (TBP 95%[Petrol]) [C]	579.9	581.0	582.2	583.3	584.4	585.4	586.5	587.5	588.5
LVGO - Mass Flow [kg/h]	6.986e+004	6.987e+004	6.987e+004	6.987e+004	6.987e+004	6.987e+004	6.987e+004	6.987e+004	6.987e+004
LVGO - Calculator (TBP 95%[Petrol]) [C]	464.0	463.8	463.7	463.6	463.4	463.2	463.1	462.9	462.8
State	State 10	State 11	State 12	State 13	State 14	State 15			
Steam - Mass Flow [kg/h]	2.000e+004	2.100e+004	2.200e+004	2.300e+004	2.400e+004	2.500e+004			
HVGO - Mass Flow [kg/h]	2.136e+005	2.147e+005	2.158e+005	2.168e+005	2.179e+005	2.189e+005			
HVGO - Calculator (TBP 95%[Petrol]) [C]	589.4	590.5	591.4	592.3	593.2	594.1			
LVGO - Mass Flow [kg/h]	6.987e+004	6.987e+004	6.987e+004	6.987e+004	6.987e+004	6.987e+004			
LVGO - Calculator (TBP 95%[Petrol]) [C]	462.6	462.5	462.4	462.2	462.1	461.9			

Figure 3.62 Increasing the steam mass flow from 1.1E4 to 2.5E5 kg/h increases the mass yield of HVGO by 8.9% from 2.01E5 to 2.189E5 kg/h and increases the TBP 95% point significantly from 579.8 to 594.1 °C.

Step 5. Set the lower and upper bounds and step size of the flash zone (tray 10) temperature and run the case study (see Figures 3.59 and 3.60). Save the file as *Workshop 3.3-2.hsc.*

Step 6. Set the lower and upper bounds and step size of the steam mass flow and run the case study (see Figures 3.61 and 3.62). Save the file as *Workshop 3.3-3.hsc.*

Bibliography

1 Kaes, G.L. (2000) *Refinery Process Modeling A Practical Guide to Steady State Modeling of Petroleum Processes*, The Athens Printing Company, Athens, GA.

2 Nelsen, N.L. (1951) *Oil Gas Journal*, **50**, 100.

3 Kaes, G.L. (2007) *Steady State Simulation of an Oil Refinery Using Commercial Software*, Kaes Consulting, Colbert, GA.

4 Watkins, R.N. (1979) *Petroleum Refinery Distillation*, 2nd edn, Gulf Publishing, Houston, TX.

5 Remesat, D. (2008) Improving crude vacuum unit performance. *Petroleum Technology Quarterly*, **Q3**, 107.

6 Barletta, T. and Golden, S.W. (2005) Deep-cut vacuum unit design. *Petroleum Technology Quarterly*, **Q4**, 91.

7 Yahyaabadi, R. (2009 March) Consider practical conditions for vacuum unit modeling. *Hydrocarbon Processing*, **69**.

8 Boduszynski, M.M. (1987) Composition of heavy petroleum 1. Molecular weight, hydrogen deficiency, and heteroatom concentration as a function of atmospheric equivalent boiling up to 1400 °F. *Energy & Fuel*, **1**, 2.

9 Boduszynski, M.M., Grudoski, D.A., Rechsteiner, C.E., and Iwamoto, J.D. (1995, September) Deep-cut assay reveals additional yields of high-value VGO. *Oil & Gas Journal*, **11**, 39.

10 Schneider, D.F. and Musumeci, J. (1997, November) Deep cut vacuum tower processing provides major incentives. *Hydrocarbon Processing*, **76** (11), 83.

11 Al-Mutairi, E.M. (2014) Energy optimization of integrated atmospheric and vacuum crude distillation units in oil refinery with light crude. *Asia-Pacific Journal of Chemical Engineering*, **9**, 181.

12 Chen, Q. (2014) Comparative analysis and evaluation of three crude oil vacuum distillation process for process selection. *Energy*, **76**, 559.

13 Cui, Z. (2016) Application of vacuum deep-cut process in atmospheric-vacuum distillation unit and study on coking prevention. *Petroleum Refinery Engineering*, **46** (3), 14–19.

4

Predictive Modeling of the Fluid Catalytic Cracking (FCC) Process

This chapter presents the methodology to develop, validate, and apply a predictive model for an integrated fluid catalytic cracking (FCC) process. We demonstrate the methodology by using data from a commercial FCC plant in the Asia Pacific with a feed capacity of 800 000 tons per year. Our model accounts for the complex cracking kinetics in the riser–regenerator with a 21-lump kinetic model. We implement the methodology with Microsoft Excel spreadsheets and a commercial software tool, Aspen HYSYS Petroleum Refining from Aspen Technology, Inc. The methodology is equally applicable to other commercial software tools. This model gives accurate predictions of key product yields and properties, given feed qualities and operating conditions. In addition, this work presents the first lumped FCC kinetic model integrated with a gas plant model in the literature. We validate this work using 6 months of plant data. We also perform several case studies to show how refiners may apply this work to improve gasoline yield and increase unit throughput.

A key application of the integrated FCC model is to generate delta-base vectors for linear programming (LP)-based production planning to help refiners choose an optimum slate of crude feeds. Delta-base vectors quantify changes in FCC product yields and properties as functions of changes in feed and operating conditions. Traditionally, refiners generated delta-base vectors using a combination of historical data and empirical correlations. Our integrated model can eliminate guesswork by providing more robust predictions of product yields and properties.

This chapter differentiates itself from previous work in the literature through the following contributions: (1) detailed models of the entire FCC plant, including the overhead gas compressor, main fractionator, primary and sponge oil absorber, primary stripper, and debutanizer columns; (2) process to infer molecular composition required for the kinetic model using routinely collected bulk properties of feedstock; (3) predictions of key liquid product properties not published alongside previous related work (density, D86 distillation curve and flash point); (4) case studies showing industrially useful applications of the model; and (5) application of the model with an existing LP-based production planning tool.

Specifically, Section 4.1 gives the motivation of this chapter. Section 4.2 describes the typical FCC process, including both the riser–regenerator complex and downstream fractionation units. Section 4.3 summarizes the FCC

Petroleum Refinery Process Modeling: Integrated Optimization Tools and Applications,
First Edition. Y. A. Liu, Ai-Fu Chang, and Kiran Pashikanti.

process chemistry and the five main classes of reactions involved, including cracking, isomerization, hydrogen transfer, dehydrogenation and dealkylation, and aromatic ring condensation. Section 4.4 presents a literature review relevant to predictive modeling of FCC processes, covering kinetic models and unit-level models. Section 4.5 describes the features of the Aspen HYSYS Petroleum Refining FCC model, including a 21-lump kinetic model. Section 4.6 presents a step-by-step procedure of determining the parameters of the lumped kinetic model from plant data, called model calibration. Section 4.7 discusses the practical aspects of developing the simulation models of the downstream fractionation units. Section 4.8 presents the guidelines of mapping the feed information to kinetic lumps. We cover fitting distillation curves, inferring molecular compositions, and converting kinetic lumps to fractionations lumps (pseudocomponents). Section 4.9 presents the overall strategy for the model development, and Section 4.10 compares the model predictions with plant data. Section 4.11 illustrates the model applications to improve gasoline yield and increase the throughput of the FCC unit. Section 4.12 demonstrates the model applications to refinery production planning. Sections 4.13–4.17 present five hands-on workshops of development and validation of FCC reaction and fractionation systems from plant data, together with model applications to process optimization and production planning. Section 4.18 summarizes the conclusions of this chapter. Finally, we present "Nomenclature" and "Bibliography" at the end of the chapter.

4.1 Introduction

The current economic, political, and regulatory climates place significant pressures on petroleum refiners to optimize and integrate the refining process. The FCC unit is the largest producer of gasoline and light ends in the refinery [1]. It plays a critical role in the profitable operation of any refinery. Plant operators can make minor adjustments based on the experience to improve the yield and efficiency of the FCC unit. However, major improvements must come from a concerted effort that involves understanding the reaction chemistry, feed characteristics, and equipment performance. In such an endeavor, use of rigorous simulation models is critical. In particular, rigorous simulation models validated with plant data can identify key areas for process improvements.

There is a significant previous work that addresses the issues of process dynamics and control for the integrated FCC unit. We particularly note the efforts by Arbel *et al.* [2] and McFarlane *et al.* [3] in this regard. Subsequent authors [4, 5] use similar techniques and models to identify control schemes and yield behavior. However, most of the earlier work uses a simplified reaction chemistry (yield model) to represent the process kinetics. In addition, previous work in the literature (to our knowledge) does not integrate the FCC model with the complex FCC fractionation system. This work fills the gap between the development of a rigorous kinetic model and industrial application in a large-scale refinery.

4.2 Process Description

The FCC unit is the primary producer of gasoline and olefins in the refinery. Current FCC designs are based on continual improvements and advances in unit and catalyst design since 1940. There are many popular FCC designs in use today and we choose to focus on a Universal Oil Products (UOP) FCC unit. The UOP design includes many features that highlight the unique characteristics of the FCC process. Figure 4.1 shows a general schematic of the FCC unit. We discuss the process flow and unit design in the following section.

4.2.1 Riser–Regenerator Complex

Hot fluidized catalyst (1000 °F+ or 538 °C+) enters the bottom riser through a standpipe where it comes in contact with preheated gas oil feed. The gas oil feed typically consists of vacuum gas oil (VGO) from the vacuum tower, coker gas oil (CGO) from the delayed coker, and recycled products from the FCC main fractionator (Figure 4.2). The heat from the hot catalyst (and any additional

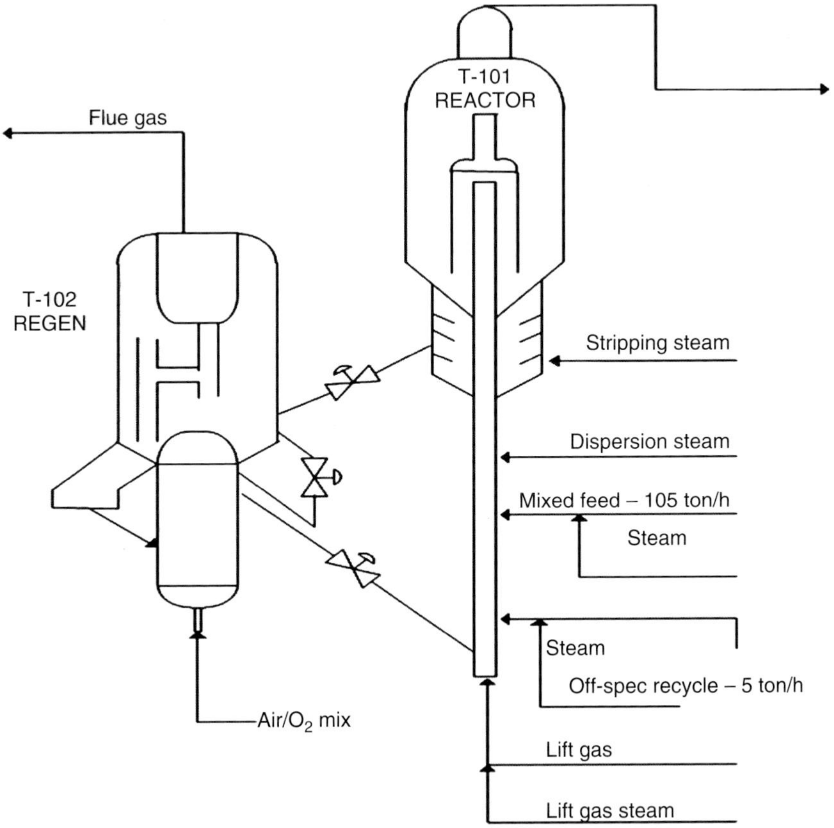

Figure 4.1 General schematic of typical FCC reactor–regenerator unit.

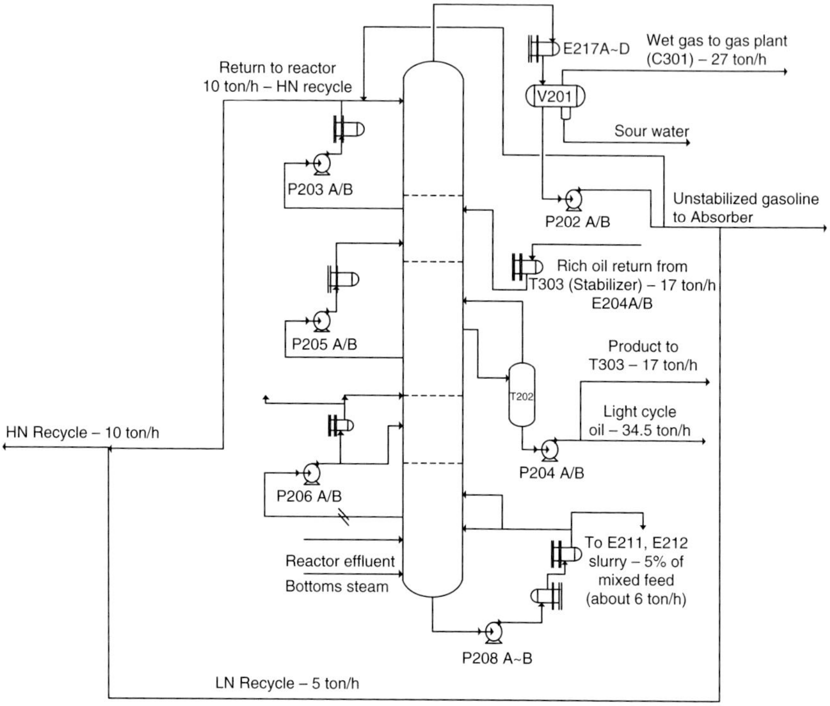

Figure 4.2 Downstream fractionation (main fractionator).

steam or fuel gas added into the standpipe) is sufficient to vaporize the gas oil feed. The components of the vaporized gas oil undergo several reactions over the catalyst surface, including hydrocracking, isomerization, hydrogenation/dehydrogenation, alkylation/dealkylation, cyclization/decyclization, and condensation. These reactions result in components that make up the product slate. The products typically present are dry gas (hydrogen, methane, and ethane), liquid petroleum gas (propanes, propylene, butanes, and butenes), gasoline (up to 430 °F or 221 °C), light cycle oil (LCO), heavy cycle oil (HCO), slurry or decant oil, and coke. Properties of the feed oil and impurities present on the catalyst significantly affect the distribution of products and the operating profile in the riser.

The catalyst travels to the top of the riser, carrying heavy components and coke deposits from the preceding reactions. The catalyst enters a stripping zone, where some steam is added to further crack and remove the heavy hydrocarbons from the catalyst surface. The catalyst then enters the reactor section, where a cyclone separates the catalyst from the product vapor. The separated product vapor is sent to the main fractionation column (Figure 4.2) that separates the product into gaseous and liquid products. The separated catalyst is piped into the regenerator where the coke on the catalyst is burned off.

The separated catalyst typically contains about 0.4–2.5% of coke by weight [1]. Air and possibly pure oxygen (depending on unit configuration) enter into the regenerator through additional ports. Fresh makeup coke also enters the FCC plant through additional ports. The coke is mostly oxidized, producing CO_2 and CO as primary products and SO_x and NO_x as secondary products. These flue gas products are typically used in heat integration loops to provide steam to the plant. The catalyst is typically oxidized to a level containing 0.05% of coke by weight [1]. This oxidization also heats the catalyst as it reenters the riser through the standpipe.

4.2.2 Downstream Fractionation

The effluent from the FCC enters the main fractionator with a significant quantity of steam, as shown in Figure 4.2. This fractionator separates the reactor effluent into five product groups: *light gases* (C1–C4), *gasoline* (C5+ to 430 °F or 221 °C), LCO, HCO (430–650 °F, or 221–343 °C), and *slurry/decant oil* (650+ °F or 343+ °C). The temperature range of these products varies in different refineries (or different operating scenarios in the same refinery) depending on product demands and current operating constraints. Several pumparounds are associated with the main fractionator, which help control the product distribution and temperature profiles. Most of the products from the main fractionator cannot be sent directly into the refinery's product blending pool. Additional fractionation and product isolation occur in the gas plant associated with the FCC unit, as shown in Figure 4.3. The overhead vapor contains some C5 components, which must be recovered in the product gasoline. A portion of the LCO product is drawn off as sponge oil to recover gasoline in a sponge oil absorber. The liquid from the overhead condenser flows to the primary absorber where C3–C4 components are recovered.

There is a significant value in separating and isolating the C3–C4 components. We may sell these components as LPG, or use them as a valuable feedstock for other petrochemical processes. The FCC gas plant is responsible for the

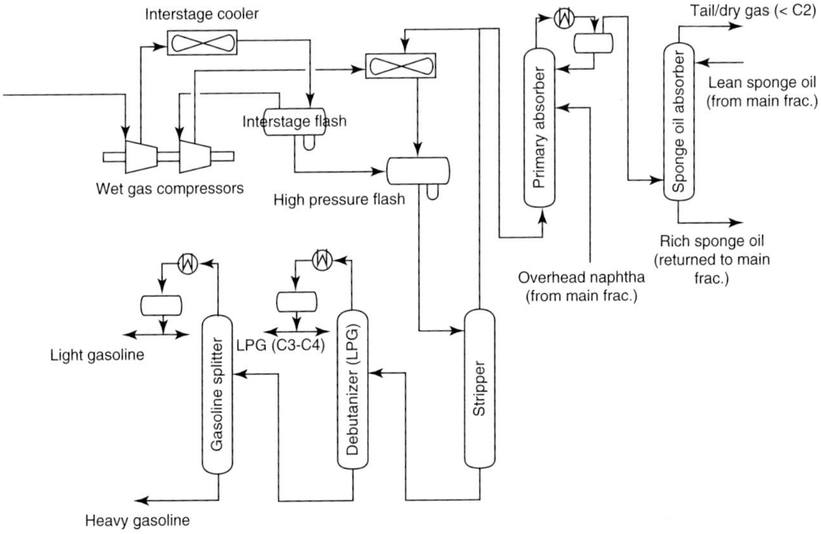

Figure 4.3 FCC gas plant section.

separation of C3–C4 components and stabilization of gasoline. The stabilization of gasoline refers to controlling the amount of C4 components present in the product gasoline.

The overhead vapor from the main fractionation column enters the wet gas compressor train. The vapor leaving the compressor train then enters a high-pressure flash system. The vapor from the high-pressure flash enters the primary absorber. The C5 components leave with the bottom product from the primary absorber. This bottom product reenters the high-pressure flash. The overhead vapor product enters a sponge oil absorber where it is contacted with LCO drawn off from the main fractionator. The overhead products of the sponge oil absorber are H2 and C1 components that can serve as fuels to meet the refinery's energy demands. The bottom product from the sponge oil absorber is recycled back to the main fractionator.

The liquid product from the high-pressure flash enters the primary stripping column. The overhead product from the stripping column mainly consists of C2 components. This product is recycled back to the high-pressure flash. The bottom product from the column mainly consists of C3–C4 components and gasoline. This product enters the primary stabilizer (sometimes called a debutanizer), which separates most of the C3–C4 components into the overhead liquid. The stabilized gasoline (containing a regulated amount of C4) leaves as the bottom product.

Some FCC gas plants further separate the gasoline product leaving the stabilizer into heavy and light gasoline. We do not include additional gasoline splitting in this work. In addition, most plants contain a water wash or injection system to control the presence of acidic compounds that lead to corrosion. This water injection typically occurs between the stages of the overhead wet gas compressor. Most of this water leaves the process flow before entering the columns of the gas plant. This water wash has little effect on the overall simulation of the process, so we do not include it in this work.

4.3 Process Chemistry

The feed to the FCC unit is a complex mixture consisting of long-chain paraffins, single- and multiple-ring cycloalkanes, and large aromatic compounds. It is impossible to list every reaction that each individual molecule undergoes in the FCC riser. However, we can place each of the reactions into five different classes based on the type of reactants and products, effect on catalyst activity, and contributions to product slate. In general, catalytic cracking occurs through formation of carbocation (from feed hydrocarbon molecule) in conjunction with a catalyst acid site. This carbocation may then undergo cracking (to produce smaller molecules), isomerization (to rearrange molecules), and hydrogen transfer (to produce aromatic compounds). Table 4.1 gives a simplified overview of key classes of reactions and the general formulas for reactants and products.

The most significant classes of reactions are cracking (reaction class 1), isomerization (reaction class 2), and hydrogen transfer (reaction class 3) [1, 6, 7].

Table 4.1 Key classes of reactions with general formulas for products and reactants.

Description	General reaction formula for reactants and products
Reaction class 1: Cracking	
Paraffin cracked to olefins and smaller paraffins	$C_{m+n}H_{2[(m+n)+2]} \rightarrow C_mH_{2m+2} + C_nH_{2n+2}$
Olefins cracked to smaller olefins	$C_{(m+n)}H_{2(m+n)} \rightarrow C_mH_{2m} + C_nH_{2n}$
Aromatic side-chain scission	$Ar{-}C_{(m+n)}H_{2(m+n)+1} \rightarrow Ar{-}C_mH_{2m-1} + C_nH_{2n+2}$
Naphthenes (cycloparaffins) cracked to olefins and smaller naphthenes	$C_{(m+n)}H_{2(m+n)}$ (naphthene) $\rightarrow C_mH_{2m}$ (naphthene) + C_nH_{2n} (Olefin)
Reaction class 2: Isomerization	
Olefin bond shift	x-$C_nH_{2n} \rightarrow y$-C_nH_{2n} (x and y are different locations of the olefin)
Normal olefin to iso-olefin	n-$C_nH_{2n} \rightarrow i$-C_nH_{2n}
Normal paraffins to isoparaffin	n-$C_nH_{2n+2} \rightarrow i$-$C_nH_{2n+2}$
Cyclohexane to cyclopentane	C_6H_{12} (naphthene) $\rightarrow C_5H_9{-}CH_3$ (naphthene)
Reaction class 3: Hydrogen transfer	
Paraffins and olefins converted to aromatics and paraffins	C_nH_{2n} (naphthene) + C_mH_{2m} (olefin) $\rightarrow$ ArC_xH_{2x+1} (aromatic) + C_pH_{2p+2} (paraffin) (where $x = m + n - 6 - p$)
Reaction class 4: Dehydrogenation and dealkylation (contaminated catalyst)	
Metal-catalyzed aromatic and light hydrocarbon production	i-$C_nH_{2n\text{-}1} + C_mH_{m\text{-}1} \rightarrow Ar + C_{(n+m\text{-}6)}H_{2(n+m\text{-}6)}$ n-$C_2H_{2n+2} \rightarrow C_nH_{2n} + H_2$
Reaction class 5: Aromatic ring condensation	
Condensation of single-ring aromatic cores to produce multiple-ring aromatic cores	$Ar{-}CHCH_2 + R_1CH{-}CHR_2 \rightarrow Ar{-}Ar + H_2$

The remaining classes are undesirable and contribute to hydrogen or coke production. The acid-catalyzed cracking reactions from reaction class 1 form the primary pathway for light gas and LPG (C3–C4) components and the long-chain paraffin components of diesel. These reactions also provide some of the lighter aromatic components present in the products. When catalytic conditions are not present (e.g., contaminated/occluded catalyst or high temperatures), a thermal cracking process takes over, promoting lower-order cracking reactions. These reactions tend to produce very large amounts of dry gas components (C1 and C2) and result in higher coke production [1, 6]. In addition, excessive thermal cracking is not an economically attractive operating scenario.

Isomerization reactions (reaction class 2) give an important pathway for high-octane components in the gasoline. This class of reactions is critical for producing high-octane components in the gasoline products. In addition, we find more valuable isobutene components due to the isomerization of butanes. The isoparaffins from the isomerization class of reactions also reduce the cloud point of the diesel product [1].

Hydrogen transfer reactions (reaction class 3) form a class of reactions that improves gasoline yield and stability (by lowering olefin content), but also lower the overall octane rating of the product. These reactions produce paraffins and aromatics that have low octane ratings. In addition, we cannot recover the olefins consumed by hydrogen transfer reactions in the LPG or the light ends of gasoline [8].

Dehydrogenation (reaction class 4) is a result of the presence of metals such as nickel and vanadium on the catalyst. The metal sites on the catalyst promote dehydrogenation and dealkylation. These reactions tend to produce large amounts of H_2 and paraffin components with low octane ratings. The coking process follows a complicated series of reactions that include olefin polymerization and aromatic ring condensation (reaction class 5). The coking reactions dominate when the unit is operating at a nonoptimal temperature (typically less than 850 °F or 454 °C, or greater than 1050 °F or 566 °C), or when the feed contains significant amounts of residue, recycled coke or olefins [8].

4.4 Literature Review

We can divide the literature on FCC modeling into two categories: kinetic and unit-level models. Kinetic models focus on chemical reactions taking place within the riser or reactor section of the FCC unit and attempt to quantify the feed as a mixture of chemical entities to describe the rate of reaction from one chemical entity to another. In contrast, unit-level models contain several submodels to take into account the integrated nature of modern FCC units. A basic unit-level model contains submodels for the riser/reactor, regenerator, and catalyst transfer sections. The riser requires a kinetic model to describe the conversion of chemical entities. The regenerator contains another kinetic model to describe the process of coke removal from the catalyst. The unit-level model also captures the heat balance between the riser and the regenerator.

4.4.1 Kinetic Models

We classify kinetic models according to the chemical entities that makeup the model. Typically, the entities or “lumps” are boiling point lumps or yield lumps, grouped chemical lumps, and full chemical lumps. Early kinetic models consist entirely of yield lumps, which represent the products that the refiner collects from the main fractionator following the FCC unit. Figure 4.4 shows a typical kinetic model based on yield lumps by Takatsuka *et al.* [9]. Many similar models have appeared in the literature. The models differentiate themselves based on their number of lumps. Models may contain as few as two [10] or three lumps [11] and as many as 50 lumps [12]. We note that models with more lumps do not necessarily have more predictive capabilities than models with fewer lumps [6].

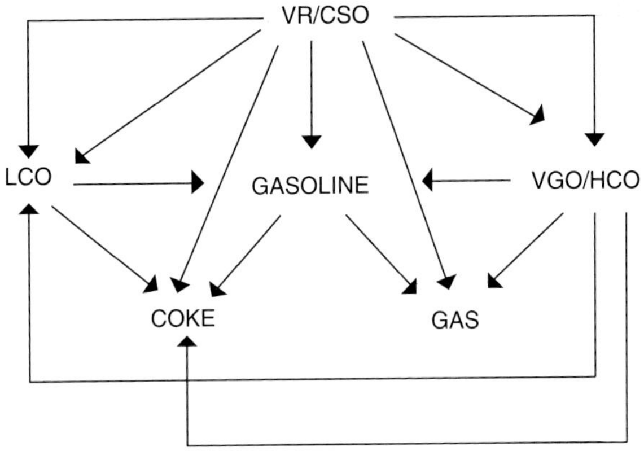

Figure 4.4 Lumped model from Takatsuka *et al.* [9]. VR= vacuum residue, CSO = coke slurry oil, VGO = vacuum gas oil, HCO = heavy cycle oil, and LCO = light cycle oil.

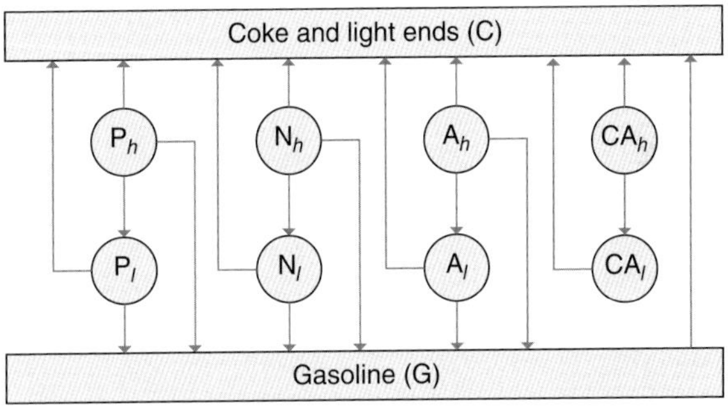

Figure 4.5 Ten-lump model from Jacob *et al.* [13].

The next class of kinetic models considers both chemical type lumps and boiling point or yield lumps. For example, Jacob *et al.* [13] present a popular 10-lump model (shown in Figure 4.5) that includes coke and light ends (C), gasoline (G, C5–221 °C), light paraffin P_l, heavy paraffin P_h, light naphthene N_l, heavy naphthene N_h, light aromatics A_l, heavy aromatics A_h, light aromatic with side chains CA_l, and heavy aromatic with side chains CA_h. The "*l*" subscript refers to "light" lumps in the boiling point range between 221 and 343 °C, whereas the "*h*" subscript refers to "heavy" lumps that have boiling point above 343 °C.

The key advantage of this lumped kinetic model is that we can measure the composition of lumps with various experimental techniques. In addition, the rate constants that arise from using this model are less sensitive to changes in feed and process conditions [14]. This model has served as the basis for models that include more chemical types. Pitault *et al.* [15, 16] have developed a 19-lump model that includes several olefin lumps. Saleh *et al.* [61] utilized a six-lump yield model to study optimization of FCC units. Xu *et al.* [62] proposed an eight-lump kinetic model to study catalytic cracking of VGO. In 2014, Xu *et al.* [63] presented a 22-lump kinetic model to investigate hydroisomerization and hydroaromatization reactions of FCC naphtha. AspenTech [17, 18] has developed a 21-lump model to address heavier and more aromatic feeds, which we will use to model reaction section of the FCC unit. We discuss this 21-lump model in Section 4.5.2.

Hsu *et al.* [6] stated that "lumped kinetic models developed by the top-down route have limited extrapolative power." To remedy this situation, many researchers have developed complex reaction schemes based on first chemical principles that involve thousands of chemical species. We can classify them into *mechanistic* models and *pathway* models. Mechanistic models track the chemical intermediates such as ions and free radicals that occur in the catalytic FCC process. Transition state theory helps in quantifying the rate constants involved in adsorption, reaction, and desorption of reactant and product species from the catalyst surface. Froment and coworkers [19] have pioneered the use of such models in a refinery context and have developed a model for catalytic VGO. Hsu *et al.* [6] claimed that using this method is challenging because of its large size and reaction complexity. *Structure-oriented lumping (SOL)* is a leading example of the pathway-based models. Quann and Jaffe [20–22] have developed a unique method for tracking molecules in the feed oil. The method tracks different compositional and structural attributes of a molecule (number of aromatic rings, number of nitrogen substituents, sulfur substituents, etc.) in a vector format. Figure 4.6 shows typical vectors for some sample molecules.

After developing these vectors for the feed oil, SOL method includes several rules to generate reaction paths that convert the feed vectors to product vectors. The rate constants and activation energies for these reactions are functions of the reaction type and the feed oil composition vector. Christensen *et al.* [23] discussed applying the SOL method to develop a FCC kinetic model, which contains over 30 000 chemical reactions and 3000 molecular species. The resulting model can accurately predict product yields, compositions, and qualities over a wide range of operating conditions. Klein and coworkers [24] have also developed similar models for FCC and catalytic reforming.

Figure 4.7 compares these kinetic models on the basis of complexity and model fidelity. The yield lump models have the lowest complexity and require the least amount of data. Typically, the feed may be treated as a single lump and there are few reaction rates to calibrate. Chemical lumps require knowledge of chemical type of the lump, namely, paraffin, naphthene, and aromatic (PNA) content of each boiling point range. Pathway and mechanistic models require the detailed analysis of the feed data to develop molecular representation. Additionally, pathway and mechanistic models require more data to calibrate the numerous kinetic parameters [6].

A6	A4	A2	N6	N5	N4	N3	N2	N1	R	br	me	IH	AA	NS	RS	AN	NN	RN	NO	RO	KO
1	0	0	0	0	0	0	0	0	0	0	0	0	0	0	0	0	0	0	0	0	0

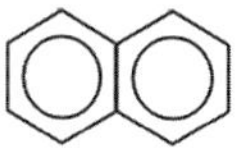

A6	A4	A2	N6	N5	N4	N3	N2	N1	R	br	me	IH	AA	NS	RS	AN	NN	RN	NO	RO	KO
1	1	0	0	0	0	0	0	0	0	0	0	0	0	0	0	0	0	0	0	0	0

Figure 4.6 Typical SOL lumping. (From [20].)

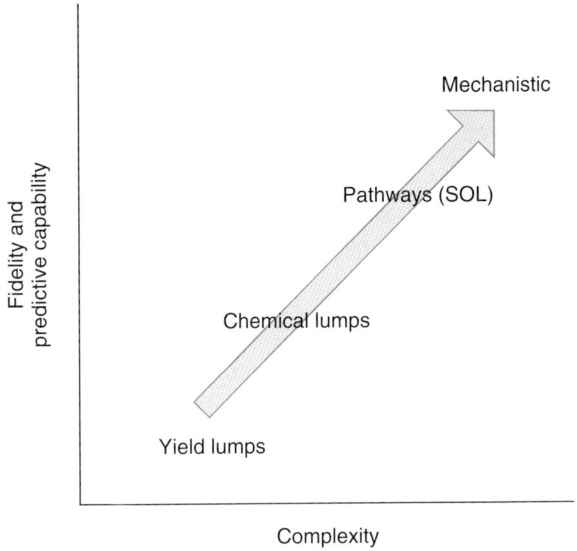

Figure 4.7 Summary of kinetic models.

4.4.2 Unit-Level Models

Table 4.2 compares a selection of published work (after 1985) regarding modeling of an entire FCC unit. This table does not include work that only compares the performance of the riser with experimental or plant data. It includes work where the authors compare the predictions of the entire FCC unit model with published data, experimental data, or plant data. The work by Lee *et al.* [10], McFarlane *et al.* [3], and Arbel *et al.* [2] provided the basis for many dynamic and process control-related models by later authors. These studies focus on optimal control strategies and the dynamic response of the FCC unit. Few papers compare the steady-state operation of the FCC unit with detailed predictions of yield and product properties with plant data. Notably, the work of Fernandes *et al.* [33] followed an industrial FCC unit over the course of 3 years and gives good predictions of the unit's performance. However, this work does not include any detailed predictions of product quality and composition. Additional work by Fernandes *et al.* [35] showed how feed and operating conditions such as coke composition, catalyst-to-oil ratio, CCR in the feed, air-to-oil ratio, and regenerator combustion modes can induce multiple steady states with implications for a general unit control strategy.

A complete unit-level model for a FCC unit includes several submodels of varying degrees of rigor. A modern FCC unit involves complex kinetic, heat management, and hydrodynamic issues. Necessarily, researchers develop models that focus on particular aspects of FCC operation. There is significant research [36] on the topic of complex hydrodynamics in the riser and regenerator sections using computational fluid dynamics (CFD). Zhang *et al.* [64] employed the CFD method to model a heavy oil riser with a bottom-lift loop mixer. These models often require detailed information about the process that is proprietary. The focus of this chapter is developing a model to predict key process output variables such as product yields, product properties, and operating profiles of the

Table 4.2 Survey of related published literature for integrated FCC modeling.

Reference	Application	Kinetics	Property predictions	Fractionation modeling	Validation data	Integration with production planning
Lee *et al.* [10]	Dynamic/ process control	3-Lump	None	None	None	None
McFarlane *et al.* [3]	Dynamic/ process control	2-Lump	None	None	None	None
Arbel *et al.* [2]	Dynamic/ process control	10-Lump	None	None	Literature	None
Khandalekar *et al.* [5]	Dynamic/ process control	3-Lump	None	None	Literature	None
Kumar *et al.* [25]	Steady state	10-Lump	None	None	Literature	None
Chitnis *et al.* [1]	Dynamic/ online optimization	4-Lump	None	None	Literature	None
Ellis *et al.* [26]	Dynamic/ process control	10-Lump	Light gas composition (C1–C4), RON/MON of gasoline products	None	Literature	None
Secchi *et al.* [27]	Dynamic	10-Lump	None	None	Industrial (dynamic)	None
Mo *et al.* [28]	Steady state/ online optimization	NA	Extensive properties of all key products	None	Industrial, pilot plant, and experimental	
Elnashaie *et al.* [29]	Steady state	3-Lump	None	None	Industrial	None
Rao *et al.* [30]	Steady state	11-Lump	None	None	Industrial	None
Arajuo-Monroy *et al.* [31]	Steady state	6-Lump	Light gas composition	None	Industrial	None
Bollas *et al.* [32]	Dynamic/pilot plant process control	2-Lump	None	None	Pilot plant	None
Fernandes *et al.* [33]	Steady state/ dynamic	6-Lump	None	None	Industrial	None
Shaikh *et al.* [34]	Steady state	4-Lump	None	None	Pilot plant	None
This work	Steady state	21-Lump	Light gas composition, flash point, density of key products, and RON/MON	Main fractionator and associated gas plant	Industrial	Export model to LP-based planning tool

Note: RON/MON = research octane number/motor octane number.

Table 4.3 Required submodels for a basic simulation of a complete FCC unit.

Submodel	Purpose	Unit operation
Riser reactor	Crack feed species to product species	Plug flow reactor (PFR) operating under pseudo-steady conditions Catalyst activity decay to due to coke formation as a result of time on stream, coke on catalyst, and catalyst type
Stripper	Removal of adsorbed hydrocarbons on the catalyst	Continuously stirred tank reactor (CSTR) with well-mixed model
Regenerator	Combust coke present on the catalyst	Stoichiometric or partial combustion of coke Bubbling bed reactor with a dense phase and a dilute phase
Feed vaporizer	Vaporize the feed species for input into the riser model	Heater with associated two-phase flash
Valves	Control the flow and pressure drop from the riser/reactor section to regenerator section	Typical valve equations based on pressure drop across the valve
Cyclones	Separate solids from the hydrocarbon and effluent vapors	Simple component splitter

FCC unit and associated gas plant. We acknowledge that the hydrodynamics and complex kinetics have significant effects on these output variables [1]. However, our goal is to develop a model that engineers can use and modify based on limited process data.

Arandes *et al.* [37] and Han *et al.* [38] summarized the key submodels required for a unit-level model that can provide necessary simulation fidelity for this work. We briefly summarize these submodels in Table 4.3 and refer readers to these two papers for detailed equations and additional references.

Modern FCC units and catalyst have very high conversions in the riser section. The conversion of feed species to product species completes within the riser; thus, we require no additional sections for feed conversion. There are units where feed conversion may occur in locations other than the riser [39, 40], but we have chosen to limit our discussion to the most common type of unit.

4.5 Aspen HYSYS Petroleum Refining FCC Model

The Aspen HYSYS Petroleum Refining FCC model relies on a series of submodels that can simulate an entire operating unit while satisfying the riser and regenerator heat balance. Note that the configuration is similar to the minimum submodels listed in Table 4.3 of the previous section. We summarize Aspen HYSYS Petroleum Refining submodels in Table 4.4 and highlight some key features in subsequent sections (Figure 4.8).

Table 4.4 Summary of Aspen HYSYS Petroleum Refining FCC submodels (adapted from [6]).

Submodel	Purpose	Unit operation	Considerations
Riser (more than one can be present)	Convert feed to product species using 21-lump kinetics	Modified plug flow reactor (PFR)	Allows any angle of inclination Pressure drop is a combination of pressure drop due to solid and vapor phases Catalyst activity decay due to kinetic and metal coke on catalyst Slip factor correlations (difference between vapor and solid velocities) to estimate specie density
Reactor/stripper	Complete feed conversion and remove adsorbed hydrocarbons	Bubbling bed reactor with two phases	Switches to fluidized-bed reactor model for units with low catalyst holdup
Regenerator	Combust coke present on catalyst	Bubbling bed reactor with two phases	Kinetic models for coke combustion with air and enriching oxygen [41]
Regenerator freeboard	Complete combustion of coke	Simple PFR	Additional kinetics to match behavior of industrial units [42]
Cyclones	Separate solids from hydrocarbon and effluent vapors	Two-phase pressure drop calculation	Pressure drop is a combination of pressure drop due to solid and vapor phases
Delumper	Converts lumped composition into set of true boiling point (TBP) pseudocomponents suitable for fractionation	–	Carries chemical information about the kinetic lumps as an attribute of the pseudocomponent Additional delumping of light gas into C1–C4 components using known kinetics [43]

4.5.1 Slip Factor and Average Voidage

An important concern in FCC riser submodels is how to calculate the slip factor, φ, and the average voidage, ε, of the riser. The slip factor is simply defined as the ratio between gas velocity and catalyst particle velocity. The slip factor plays an important part in determining the residence time of reactions and thus affects the overall conversion in the riser. Harriot described a slip factor range of 1.2–4.0 for most FCC risers, but also indicates that there is no reliable correlation available for prediction [44]. Previous authors have used a variety of approaches including constant slip factor [45], multiple slip factors [46], and correlations [47]. An alternative approach is to include additional momentum balance equations for the gas phase and catalyst phase [48]. This approach allows users to calculate velocity profiles for each phase and the overall pressure drop in the riser directly.

Aspen HYSYS uses a custom correlation based on a fully developed flow (away from the catalyst particle acceleration zone) that accounts for various angles of riser inclination. We present a similar correlation from Bolkan-Kenny *et al.* [47]

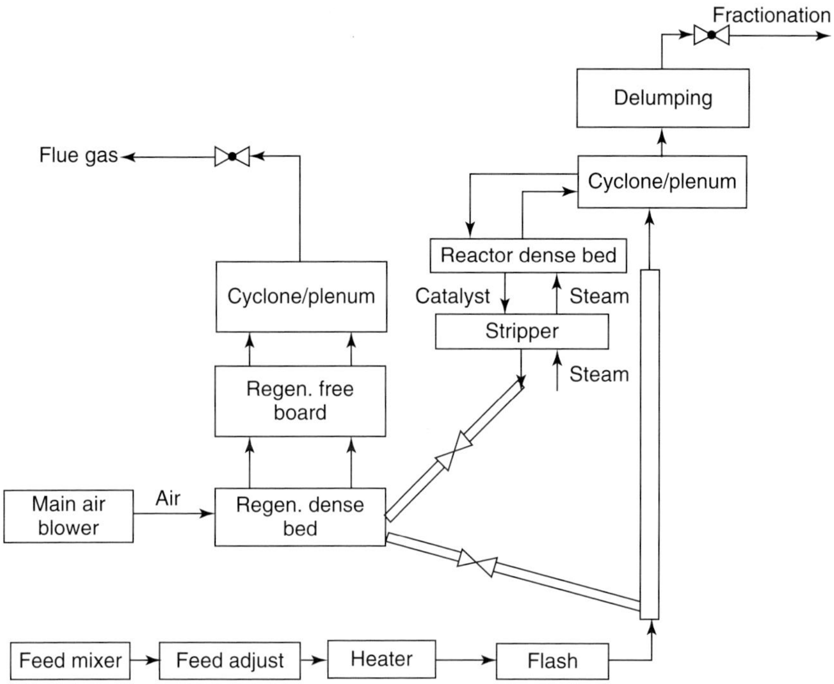

Figure 4.8 Overview of the major submodels that make up the Aspen HYSYS Petroleum Refining FCC model. (Adapted from [6].)

in Eq. (4.1) using dimensionless Froude numbers, Eqs. (4.2)–(4.3). This correlation is essentially a function of riser diameter, D; gravitational constant, g; superficial gas velocity, u_o; and u_t, terminal settling velocity of the catalyst particle.

$$\varphi = 1 + \frac{5.6}{Fr} + 0.47 Fr_t^{0.41} \quad (4.1)$$

$$Fr = \frac{u_o}{\sqrt{gD}} \quad (4.2)$$

$$Fr_t = \frac{u_t}{\sqrt{gD}} \quad (4.3)$$

4.5.2 21-Lump Kinetic Model

The 21-lump kinetic model in Aspen HYSYS Petroleum Refining is similar to the popular 10-lump model from Jacob *et al.* [13] (Figure 4.5). The 21-lump model follows the same basic structure and pathways as the 10-lump model by grouping lumps into boiling point ranges and chemical types within each boiling point range. In addition, the 21-lump model includes a boiling point range to deal with heavy feeds (boiling point greater than 510 °C) that the original 10-lump model cannot handle. To account for the differences in reactivity of various aromatic compounds, aromatic lumps are further split into lumps containing side chains and multiple rings separately. The 21-lump model also splits the original single lump for coke into two separate coke lumps. These separate lumps

Table 4.5 Summary of 21-lump kinetics (adapted from [6]).

Boiling point range	Lumps
<C5	Light gas lump
C5–221 °C	Gasoline
221–343 °C (VGO)	Light paraffin (PL)
	Light naphthene (NL)
	Light aromatics with side chains (ALs)
	One-ring light aromatics (ALr1)
	Two-ring heavy aromatics (ALr2)
343–510 °C (heavy VGO)	Heavy paraffin (PH)
	Heavy naphthene (NH)
	Heavy aromatics with side chains (AHs)
	One-ring heavy aromatics (AHr1)
	Two-ring heavy aromatics (AHr2)
	Three-ring heavy aromatics (AHr3)
510+ °C (Residue)	Residue paraffin (PR)
	Residue naphthene (NR)
	Residue aromatics with side chains (ARs)
	One-ring residue aromatics (ARr1)
	Two-ring residue aromatics (ARr2)
	Three-ring residue aromatics (ARr3)
Coke	Kinetic coke (produced by reaction scheme)
	Metal coke (produced by metal activity on catalyst)

account for coke produced from cracking reactions (called *kinetic coke*) and coke produced from metal activity (called *metal coke*) individually. We note that the rate equations in the kinetic network in Aspen HYSYS Petroleum Refining are largely similar to equations in the first-order network for 10-lump model. However, the rate equations in the 21-lump model include additional terms to account for the adsorption of the heavy hydrocarbons (due to the extended boiling point range of the lumps) and the metal activity of the catalyst. Table 4.5 lists the kinetic lumps used in the 21-lump model.

We can obtain the lump composition of the feedstock directly via GC/MS, ^{1}H NMR, ^{13}C NMR, HPLC, and ASTM methods. However, this is infeasible on a regular basis for refineries, given the changing nature of the feedstock. Aspen HYSYS Petroleum Refining includes a method that uses existing feed analysis to infer feed composition using routinely collected data. However, we have developed an alternative scheme to infer feed composition. We detail this method in Section 4.8.

4.5.3 Catalyst Deactivation

Another important consideration in the FCC unit model is the deactivation of catalyst as it circulates through the unit. Previous work has used two different approaches to model catalyst activity: time-on-stream and coke-on-catalyst [49]. As the 21-lump includes discrete lumps for the kinetic and metal coke, we use a coke-on-catalyst approach to model catalyst deactivation. In addition, we include a rate equation in the kinetic network for coke balance on the catalyst. The general deactivation function due to coke, ϕ_{COKE}, is given by Eq. (4.4).

$$\begin{aligned}\phi_{COKE} &= \phi_{KCOKE}\phi_{MCOKE}\\ &= \exp(-a_{KCOKE}C_{KCOKE})\exp(-a_{MCOKE}C_{MCOKE}f(C_{METALS}))\end{aligned} \tag{4.4}$$

In the equation, a_{KCOKE} is the activity factor for kinetic coke, a_{MCOKE} is the activity factor for metal coke, C_{KCOKE} is the concentration of kinetic coke on the catalyst, a_{MCOKE} is the activity factor for metal coke, C_{MCOKE} is the concentration of metal coke on the catalyst, and C_{METALS} represents the concentration of metals on the catalyst.

4.6 Calibrating the Aspen HYSYS Petroleum Refining FCC Model

Given the variety of feedstocks that the FCC unit processes, it is unlikely that a single set of kinetic parameters will provide accurate and industrially useful yield and property predictions. In addition, changes in catalyst may significantly alter the yield distribution. Therefore, it is necessary to calibrate the model to a base scenario. Table 4.6 lists the key calibration parameters for the FCC model. We group them by their effects on the model predictions.

Aspen HYSYS Petroleum Refining includes a base set of kinetic and calibration parameters regressed for a variety of feed oils and catalyst types. We use these as a starting point to *calibrate* the model to our specific operating scenario. Due to the chemical nature of the feed lumping, the calibration process results in only small changes in the values of calibration parameters. Significant changes from the base values may result in "overcalibration" and fix the model to a particular operating point. An "overcalibrated" model gives poor predictions even when we make small changes to input variables. It is critical to keep track of these changes

Table 4.6 Key calibration parameters for FCC model.

Parameter class	Calibration parameters
Overall reaction selectivity	Selectivity to C (coke lump) Selectivity to G (gasoline lump) Selectivity to L (VGO lump)
Distribution of light gas components (C1–C4)	Selectivities to C1–C4 light gases
Deactivation	Factors accounting for the metal contents and activity of the equilibrium catalyst (ECAT)
Equipment and process conditions	Activity for CO/CO_2 generation from coke combustion in the regenerator

in the calibration factors and make sure that they are reasonable. The key steps in the calibration process are as follows:

1) Obtain a base or reference set of operating data that fully defines the operation of the FCC unit and associated product yields. Table 4.12 lists the relevant data used for calibration in this work.
2) Use experimentally measured chemical composition of liquid products (or estimate using the methods given in Section 4.8) to calculate the expected effluent composition of kinetic lumps from FCC unit.
3) Vary the reaction selectivities for reaction paths (three parameters) that lead to *coke lumps* (kinetic coke and metal coke), *gasoline* (G lump), and *VGO* (PH, NH, AHs, AHr1, AHr2, and AHr2 lumps); deactivation activity factors (two parameters); and coke burn activity (one parameter) so that model predictions for kinetic lump compositions agree with measured (or estimated) kinetic lump compositions from step 2.
4) Vary the distribution selectivities (minimum two parameters – ratio between C1 and C2 and ratio between C3 and C4) for light gases to match the total measured light gas composition from the dry gas and LPG stream of the refinery.
5) Once calibration is complete, verify that overall material and energy balances hold.

In Aspen HYSYS, we can modify the parameters in steps 3 and 4 concurrently to simplify the calibration process. We note that if the initial kinetic parameters have been regressed from a variety of sources, small adjustments to calibration parameters are enough to match typical plant operation. In our work, the range of calibration parameters is roughly on the order of 0.5–1.5 times the initial calibration parameter values.

4.7 Fractionation

The fractionation section uses standard inside-out methods [50] implemented by many popular simulators, including Aspen HYSYS, as discussed in Section 2.4.4. This method offers robust convergence and wide flexibility in specifications. The key issue in implementing fractionation models is whether to use individual stage efficiencies, such as the Murphree stage efficiency, as defined previously in Eqs. (2.9) and (2.10) in Section 2.4.2. Readers should be careful to avoid confusion with a related concept, the overall stage efficiency. Overall stage efficiency refers to the ratio of theoretical stages used in simulations to physical stages in the actual column. For example, consider the case where we model a distillation column having 20 physical stages with simulator using only 10 theoretical stages. This column has overall efficiency of $10/20 = 0.50$. Note that each stage in the simulation operates under valid thermodynamic vapor–liquid equilibrium assumptions.

Section 2.4.3 discussed in detail that distillation column simulations using the Murphree stage efficiency violates vapor–liquid equilibrium constraints and can predict unusual and unphysical solutions for stage-by-stage simulation models. Both Kister [50] and Kaes [51] advised against the use of the stage efficiency models. They warned that simulations using these factors may lose predictive abilities and may not converge robustly. In our work, we use the rigorous stage-by-stage models for all fractionators with the overall stage efficiency concept. Kaes [51] has documented the relevant overall stage efficiencies that are reasonable for modeling columns in the FCC gas plant. Table 4.7 shows the number of theoretical stages and the overall stage efficiencies for FCC fractionation. We obtain the overall stage efficiency as the ratio of number of theoretical stages to actual physical stages in the column. For example, the main fractionator column typically has 30–40 physical stages and we find that 12–16 theoretical stages are sufficient for modeling purposes. Hence, the overall efficiency ranges from about 40% to 50%. We calculate overall efficiencies for other columns given in Table 4.7 using a typical range for the number of physical stages from various process design data.

We can actually develop the initial model for the fractionation without connecting it to the FCC model. Here, we follow the process of "backblending" (previously demonstrated in Section 2.8), as shown in Figure 4.9, to recover the reactor effluent (or fractionator feed) from a known set of product yield data [51]. This process requires that we know the yields and compositions of all the key products from the FCC plant, the feed rate to the reactor, and additional inputs (such as steam) to the reactor. We then use the composition data of the light products and the distillation curves of the liquid products to reconstruct a reactor effluent as the fractionator feed. We feed this effluent into the initial fractionation model and recover the products that are "backblended." There are two advantages to this process. First, we can verify that the fractionation model accurately reflects plant operation. We verify the fractionation model through accurate predictions of product yields, good overlap between plant and model distillation curves of liquid products, agreement of plant and model gas compositions (dry gas, LPG), and small deviations between the temperature profiles of plant and model columns. Second, this process can shorten the model development time, as we can work on modeling FCC unit and the fractionation units at the same time.

In this work, *calibrating* the fractionation section refers to the process of adjusting the number of theoretical stages in each zone (in the case of the main fractionator) or the number of theoretical stages between feed points. We use a set of a basic initialization specifications and efficiencies given in Table 4.7

Table 4.7 Theoretical stages and efficiency factors for FCC fractionation.

Fractionator	Theoretical stages	Overall efficiency (%)
Main fractionator	13–17	40–50
Primary absorber	6–10	20–30
Primary stripper	12–15	40–50
Secondary absorber	3–8	20–25
Gasoline stabilizer	25–30	75–80
LPG (C3/C4) splitter	25–30	75–80

Source: Courtesy of G. L. Kaes.

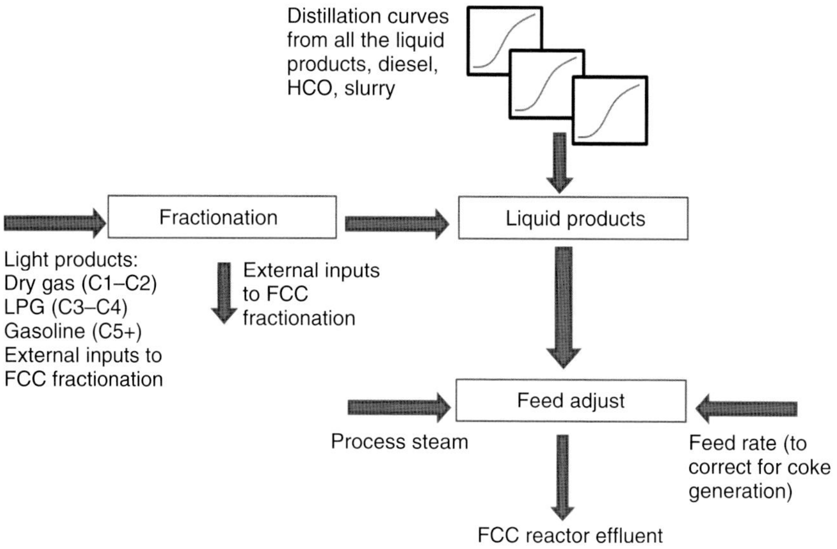

Figure 4.9 "Backblending" products to reconstitute FCC reactor effluent.

Table 4.8 Initialization and final specifications.

Column	Initial specifications	Final specifications
Main fractionator	All pumparound rates and return temperatures (or temperature changes) Draw rates for all products Bottom temperature Condenser temperature	Column overhead temperature Cut point for naphtha draws Pumparound duties Bottom temperature Condenser temperature
Primary absorber	None	None
Primary stripper	None	None
Secondary absorber	None	None
Gasoline stabilizer	Reflux ratio (around 2.0) Overhead draw rate	Gasoline *n*-butane fraction or Reid vapor pressure (RVP) in bottom Column overhead temperature or C5+ content in overhead
LPG stabilizer	Reflux ratio (around 3.0) Overhead draw rate	Reboiler temperature or bottom temperature Fraction C4 in the column overhead

to solve the column models. Typically, we only need to add or remove a few stages to calibrate the columns and achieve agreement with the plant operating profile. Once we converge the column models using the basic initialization specifications, we change (especially for the main fractionator) to specifications based on cut point and stage temperature. Kaes [51] described a similar process. We summarize the initial and final specifications in Table 4.8.

4.8 Mapping Feed Information to Kinetic Lumps

Aspen HYSYS Petroleum Refining includes a method to convert limited feed information (distillation curve, density, viscosity, refractive index, etc.) into kinetic lumps for use in the unit-level FCC model. In this section, we present an alternative method based on data and methods available in the public literature. We extend the method based on the work by Bollas *et al.* [52] to infer the kinetic lump composition from limited process data. This method uses techniques to normalize the distillation curve, cut the distillation curve into boiling point lumps, and infer the composition of each of these boiling point lumps. We have developed all of these techniques into spreadsheets using Microsoft Excel. These spreadsheets are available in the supplement to this text.

4.8.1 Fitting Distillation Curves

Distillation curves for FCC feedstock can be limited. Due to the nature of the feedstock, complete true boiling point (TBP) analysis without D-2887/SimDist methods is frequently not possible. Many refiners still use a limited D1160 distillation method to obtain some information about the distillation curve. Table 4.9 shows a typical D1160 analysis for a heavy FCC feedstock.

This curve does not contain enough information to convert into TBP curve using standard ASTM correlations. We must fit these data to a reasonable model to obtain estimates for the missing data points. We have previously demonstrated in Section 1.4, Workshop 1.2, about how to use an Excel spreadsheet, *Beta.xls*, to extrapolate incomplete distillation curve using the beta distribution function. Specifically, Sanchez *et al.* [53] have evaluated several different types of cumulative probability distribution functions to fit distillation curves of crudes and petroleum products. They conclude that the cumulative beta function with four parameters, Eq. (1.7), can represent a wide range of petroleum products [53]. We use this method to extend the measured partial distillation curve.

The beta cumulative density function is given in Eq. (1.7), renumbered as Eq. (4.5):

$$f(x,\alpha,\beta,A,B)=\int_A^{x\le B}\left(\frac{1}{B-A}\right)\frac{\Gamma(\alpha+\beta)}{\Gamma(\alpha)\Gamma(\beta)}\left(\frac{x-A}{B-A}\right)^{\alpha-1}\left(\frac{B-x}{B-A}\right)^{\beta-1} \qquad (4.5)$$

where α and β refer to the positive valued parameters that control the shape of the distribution, Γ refers to the standard gamma function, A and B parameters

Table 4.9 Typical distillation curve collected from D1160.

Recovery	Temperature (°C)
0 (initial point)	253
10	355
50	453
73 (end point)	600

set lower and upper bounds on the distribution, and x represents normalized recovery. We normalize all the temperatures between 0 and 1 using the following equation:

$$\theta_i = \frac{T_i - T_0}{T_1 - T_0} \tag{4.6}$$

where T_0 and T_1 are reference temperatures. For this work, we choose $T_0 = 250\,°\mathrm{C}$ and $T_1 = 650\,°\mathrm{C}$. Then, we apply the cumulative beta function with each normalized recovery, x_i and initial values for α, β, A, and B parameters. If we choose good estimates for parameters, then the output of the beta function must be close to the corresponding recovery for each x_i. We define the following error terms:

$$\mathrm{RSS} = \sum_{i=1}^{n} (x_{\mathrm{exp},i} - x_i)^2 \tag{4.7}$$

$$\mathrm{AAD} = \frac{1}{n} \sum_{i=1}^{n} \mathrm{abs}(x_{\mathrm{exp},i} - x_i) \tag{4.8}$$

where $x_{\mathrm{exp},\,i}$ represents the recovery measured in the distillation curve and x_i is the output of the beta function. RSS is the sum of least squares and AAD represents average absolution deviation. We now use the SOLVER method in Microsoft Excel to obtain optimized values of α, β, A, and B. Figure 4.10 shows how this fit compares with the result using a log normal distribution with two fitting parameters [53] (see *lognormal.xls* in the supplement to this text) instead of the beta

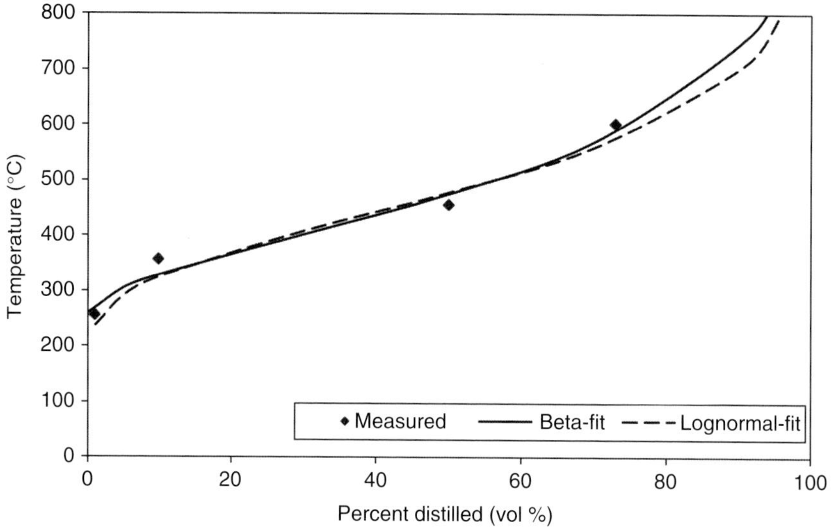

Figure 4.10 Comparison between using the beta distribution and lognormal distribution to fit the same distillation data.

function. Using the beta function, we can generate the temperatures and recoveries needed for the conversion to TBP using standard ASTM methods.

4.8.2 Inferring Molecular Composition

As mentioned earlier, we must also be able to infer the PNA composition of each boiling point range, given certain measured bulk properties in order to completely map the feed information to kinetic lumps. The API (Riazi–Daubert) [54, 55] is a popular chemical composition correlation, Eq. (1.69), renumbered as Eq. (4.9):

$$\%X_{\mathrm{P}} \text{ or } \%X_{\mathrm{N}} \text{ or } \%X_{\mathrm{A}} = a + b \cdot R_{\mathrm{i}} + c \cdot \mathrm{VGC}' \tag{4.9}$$

where X_{P}, X_{N}, and X_{A} represent the mole composition of paraffins (P), naphthenes (N), and aromatics (A); R_{i} is the refractive index and VGC′ is either the viscosity gravity constant (VGC) or the viscosity gravity factor (VGF). The parameters *a*, *b*, and *c* take on different values for each molecule type (paraffin, naphthene, or aromatic). Using the Riazi [55] correlation does not give sufficiently accurate predictions for molecular compositions for this work. We note that this correlation encompasses a wide molecular weight range of 200–600 [55].

We present an alternate correlation in Eqs. (4.10) and (4.11). Our correlation extends the original correlation from Riazi [54, 55] by including specific gravity (SG) as an additional parameter and providing different sets of correlation coefficients (*a*, *b*, *c*, and *d*) for different boiling point ranges.

$$\%X_{\mathrm{P}} \text{ or } \%X_{\mathrm{A}} = a + b \cdot \mathrm{SG} + c \cdot R_{\mathrm{i}} + d \cdot \mathrm{VGC}' \tag{4.10}$$

$$\%X_{\mathrm{N}} = 1 - (X_{\mathrm{P}} + X_{\mathrm{A}}) \tag{4.11}$$

where X_{P}, X_{N}, and X_{A} represent the mole composition of paraffins (P), naphthenes (N), and aromatics (A), respectively; R_{i} is the refractive index and VGC′ is either the VGC or the VGF. The parameters *a*, *b*, *c*, and *d* can take on different values for different molecule type and boiling point ranges.

We use a total of 233 different data points containing laboratory-measured chemical composition and bulk property information (distillation curve, density, refractive index, and viscosity) for light naphtha, heavy naphtha, kerosene, diesel, and VGO. These data points come from various plant measurements made over the six-month course of this study and a variety of light and heavy crude assay data (spanning several years) available to the refinery.

We use Microsoft Excel and the SOLVER method to fit values for the parameters *a*, *b*, *c*, and *d* that minimize the sum of squares residual between the measured $\%X_{\mathrm{P}}$ and $\%X_{\mathrm{A}}$ and calculated $\%X_{\mathrm{P}}$ and $\%X_{\mathrm{A}}$. We calculate $\%X_{\mathrm{N}}$ by difference, as shown in Eq. (4.10). We show the results of our data regression with the associated average absolute deviation (AAD) in Tables 4.10 and 4.11. Figures 4.11–4.13 compare the measured and calculated molecular compositions.

We can now use the two methods we have developed to propose a technique to use limited feed information to infer the lumped composition. This technique is similar to the one given by Bollas *et al.* [52]. However, we make several changes

Table 4.10 Coefficients for paraffin content in petroleum fractions.

	Paraffin (vol%)				
	A	B	C	D	AAD
Light naphtha	311.146	−771.335	230.841	66.462	2.63
Heavy naphtha	364.311	−829.319	278.982	15.137	4.96
Kerosene	543.314	−1560.493	486.345	257.665	3.68
Diesel	274.530	−712.356	367.453	−14.736	4.01
VGO	237.773	−550.796	206.779	80.058	3.41

Table 4.11 Coefficients for aromatic content in petroleum fractions.

	Aromatic (vol%)				
	A	B	C	D	AAD
Light naphtha	−713.659	−32.391	693.799	1.822	0.51
Heavy naphtha	118.612	−447.589	66.894	185.216	3.08
Kerosene	400.103	−1500.360	313.252	515.396	1.96
Diesel	228.590	−686.828	12.262	372.209	4.27
VGO	−159.751	380.894	−150.907	11.439	2.70

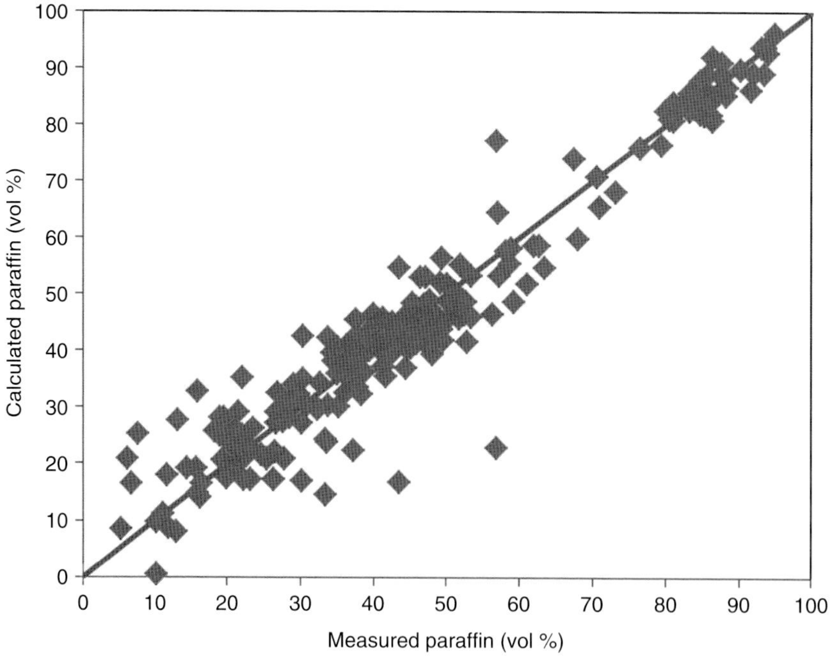

Figure 4.11 Comparison of calculated and measured paraffin content in all fractions.

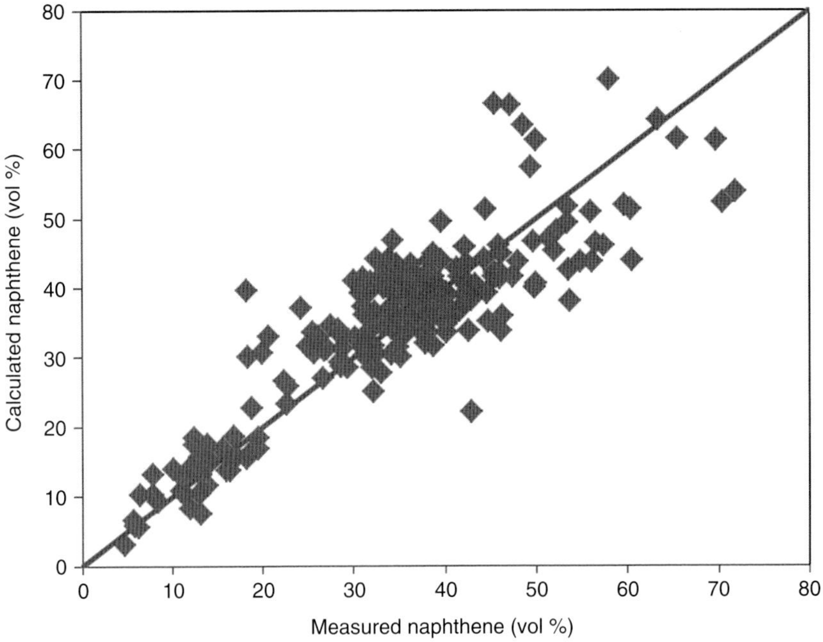

Figure 4.12 Comparison of calculated and measured naphthene content in all fractions.

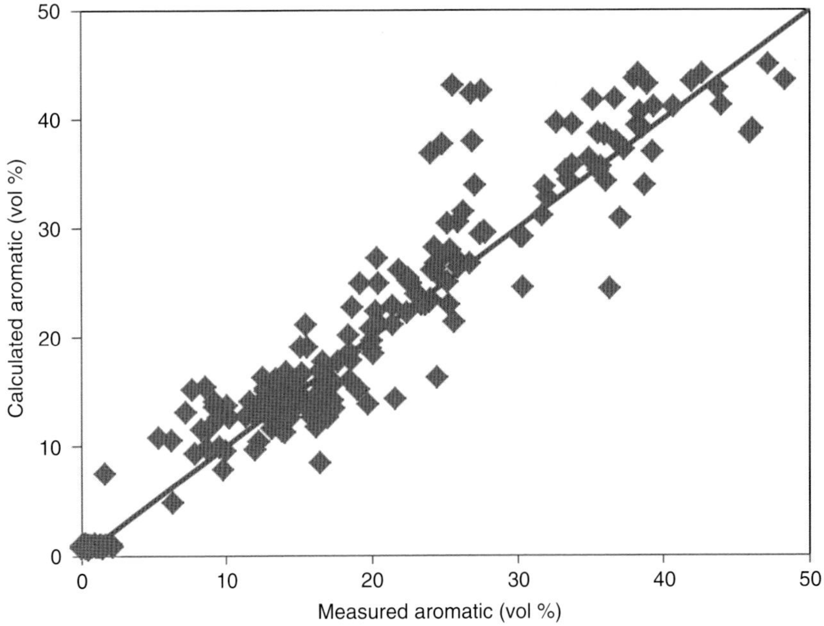

Figure 4.13 Comparison of calculated and measured aromatic content in all fractions.

to account for limited data sets. We outline the technique in the following steps (we indicate the changes from the procedure of Bollas *et al.* [52] with an *):

1) Use the beta distribution method (Use *Beta.xls*, Section 1.4) to extend partial ASTM D1160 distillation curves. (*)
2) Convert the ASTM D1160 to a TBP curve using standard API correlations [54] (use *ASTMConvert.xls*, Section 1.3). (*)
3) Using the 50% point of the TBP, estimate the Watson factor (K_w). Set the 50% TBP temperature as an initial guess for the mean average boiling point (MeABP) (use *MeABP Interation.xls*, Section1.5).
4) Use the definition of K_w to create the specific gravity distribution of the fraction.
5) Calculate pseudocomponent molecular weight using the correlation of Riazi [55].
6) Use densities and mole weights to calculate volume, cubic, molar, and MeABP of the total fraction [55].
7) If the MeABP from step 7 is close to the MeABP assumed in step 3, go to step 8. Otherwise, assume a new value for MeABP and go back to step 4.
8) Assign a lump to every boiling point range in the kinetic lumping. (*)
9) Calculate the boiling point, molecular weight, density, volume, weight, and molar concentrations of each lump.
10) Use Goosen's correlation to estimate the refractive index of each lump [56].
11) Use correlations from Riazi [55] to estimate the viscosity of the lump. (*)
12) Calculate the relevant VGF or VGC [55] for the lump. (*)
13) Use correlations (with an appropriate choice for the set of correlation coefficients) proposed in the preceding section to identify the PNA composition of the lump. (*)
14) If required, use correlations from Riazi [55] to estimate the number of aromatic rings in each aromatic fraction. (*)

We have found that this technique can provide reasonable estimates of kinetic lump composition. It is difficult to justify a more sophisticated scheme, given the limited amount of data available. Some refiners also make bulk chemical composition measurement of the feed, which includes a measurement of the total aromatic content. The sum of the aromatic kinetic lumps generated from the above technique generally agrees with the measured aromatic content.

4.8.3 Convert Kinetic Lumps to Fractionation Lumps

A related problem is the conversion of kinetic lumps back to fractionation lumps required to build rigorous fractionation models. For our models, Aspen HYSYS gives a method to transition the kinetic lumps to boiling point-based pseudocomponents typically used to model petroleum fractionation. We also propose an alternative technique that can provide similar results using methods developed earlier in this section. Essentially, we must convert the kinetic lumps back into a TBP curve. The key steps in converting the kinetic lumps to boiling point-based pseudocomponents are as follows:

1) Using the "backblending" concept from the previous section, develop a FCC effluent TBP curve from a reference set of product yields. These yields include

all liquid products, such as light and heavy naphtha, light and HCO or diesel, and slurry or decant oil.

2) Fit a cumulative beta distribution to this "backblended" reference TBP curve and obtain the best values for the cumulative beta distribution fit. We calculate this initial set of parameters only once.
3) Run the model to obtain the product distribution in terms of kinetic lumps.
4) Apply steps 3–13 of Section 4.8.2 in reverse; that is, we obtain the 50 % TBP point for each boiling point range from the known PNA distribution of the kinetic lumps involved.
5) As we know initial and final boiling points for all the kinetic lumps (by definition), use these points in conjunction with calculated 50% TBP points to generate an updated FCC effluent TBP curve.
6) Fit a new cumulative beta distribution to the updated FCC effluent TBP curve using the initial set of cumulative distribution parameters as a starting guess.
7) Cut this new TBP curve into petroleum pseudocomponents using methods commonly available in process simulations. In addition, Riazi [55] discussed several strategies to cut a TBP curve into pseudocomponents suitable for fractionation models.

4.9 Overall Modeling Strategy

This work relies primarily on data collected while the refinery is in regular operation. Related work in integrated FCC modeling often relies on pilot plant and experimental data. It is more difficult to produce a predictive model with plant operation data alone. The nature of plant operation determines that there may be abrupt changes in feed quality, operating parameters, inaccurate measurements due to poorly calibrated, failing sensors, or inconsistent data. Fernandes *et al.* [33] have encountered similar issues in the validation phase of their work. We outlined the following strategy and our specific implementation in Figure 4.14.

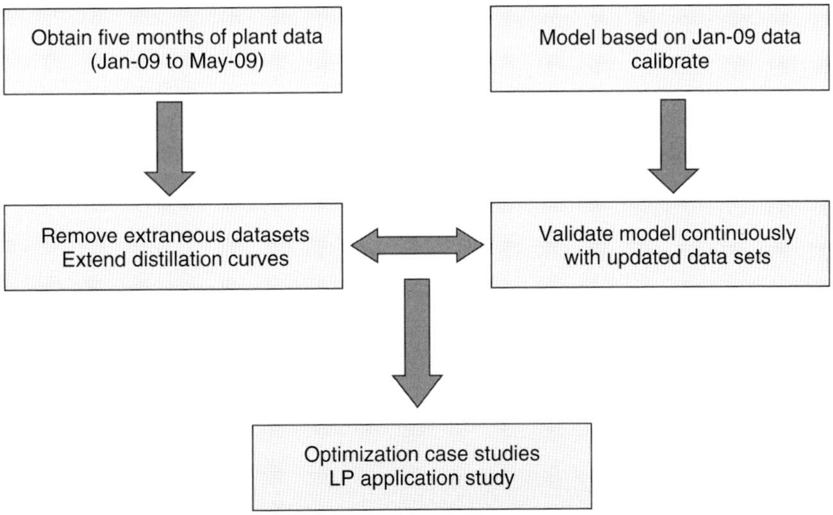

Figure 4.14 Specific implementation of overall modeling strategy.

- Obtain data on a continuous basis from the plant over a number of months:
  - Reconcile data from multiple sources (DCS, Inventory, etc.) (Table 4.12).
  - Check the consistency of the data by ensuring mass balance and enthalpy balance.
  - Accept a data set when it is consistent.
  - Track variation in the data set to ensure there are multiple operating scenarios (Figure 4.15).
- Use the first accepted data set to develop an initial model for FCC unit and fractionation section.
- Calibration:
  - The most basic calibration is to introduce a selectivity calibration factor for classes of the reactions in the kinetic network.
  - It is typically sufficient to vary the calibration selectivity factors to match plant performance during the first accepted data set.
  - The user may introduce additional factors to account for significant changes in catalyst behavior of unit profile.
  - The yield results from the initial model calibration should be within 1–2% of actual plant yield.
- Validation:
  - Use the subsequently accepted data sets to verify and track the performance of the unit and fractionation sections with the model.
  - Make sure to examine the yield of the FCC unit independently of column accuracies in the fractionation section.
  - It is typically possible to predict the mass yields of key products on a normalized feed rate with AAD of less than 2–3%.
- Case studies:
  - The model is calibrated with a finite amount of plant data, so it may not be meaningful to study changing operating parameters of the FCC over a very wide range. However, case studies on the fractionation section can take on wide ranges.
  - Recalibrate the model when significant process changes occur.

Table 4.12 Routinely monitored properties used for model development and calibration.

Feed	Products	FCC	Fractionation
Flow rate Distillation curves Specific gravity Conradson carbon residue (CCR) Sulfur content (S) Metal contents (Fe, Na, Ni, and V) Saturates, resins, aromatics, and asphaltenes (SARA)	Yield Composition (for light products) Density RON/MON Flash point Sulfur content	Temperatures (feed, riser outlet, regenerator bed, and flue gas) Pressure differential between riser/reactor and regenerator Steam usage Main air blower flow rate	Temperature profile Pressure profile Draw rates Pumparound flow rates and duties Set points (usually temperatures)

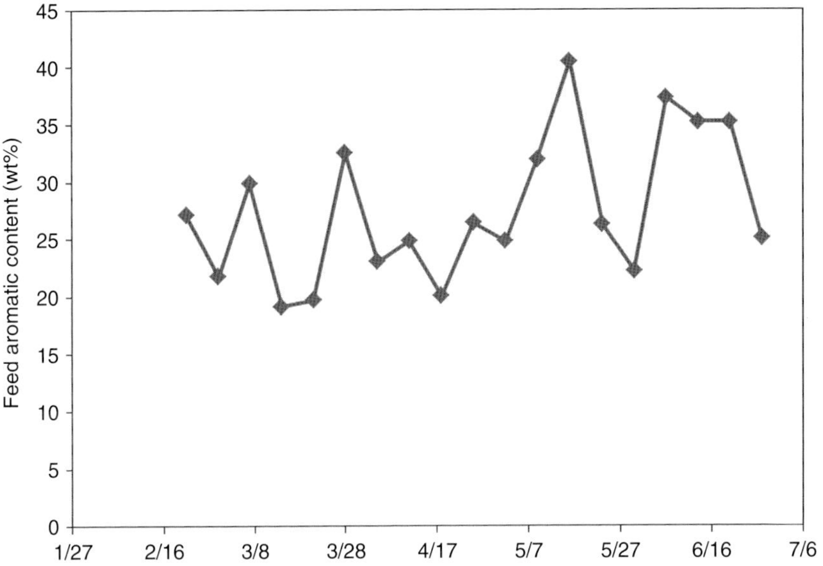

Figure 4.15 Tracking aromatic content in the feed to ensure multiple operating scenarios.

4.10 Results

We evaluate the model using over 6 months of operating data from a commercial FCC unit in the Asia Pacific with a feed capacity of 800 000 ton per year operating under a maximum diesel and gasoline plan. Figure 4.16 shows a process flow diagram (PFD) for the entire process. The evaluation of the model includes comparisons of overall reactor yield, light and heavy product compositions, and operating profiles for key equipment in the gas plant. We note that in general, the model can accurately predict the product yields and compositions over a variety of feed conditions.

The most important prediction is the overall product yields from the reactor. A validated prediction of the overall product yields allow the refiner to use the model to study different kinds of feedstock and operating conditions. Table 4.13 shows the results for product yields. The most important and valuable products are LPG, gasoline, and diesel. We use operating data from the base run to calibrate the model. In terms of overall yield, the largest errors in the base case appear with prediction of LPG and slurry. The AAD for the product over all validation cases (VALID-1 to VALID-6) is 0.96%. The AAD is much lower than the previous AAD standard of 5% for yield predictions in the plant.

Another set of key indicators is the product properties of the liquid fuel from the FCC. The properties of interest to refiners are density, flash point (volatility), RON/MON (for gasoline), sulfur content, and aromatic content. This is one of the areas where our model is different from other published work. We discussed a method of transition from kinetic lumping to fractionation lumping in Section 4.8. Not only does this method allow the user to observe the results directly, we can also see the effect of the reactor conditions on fractionated

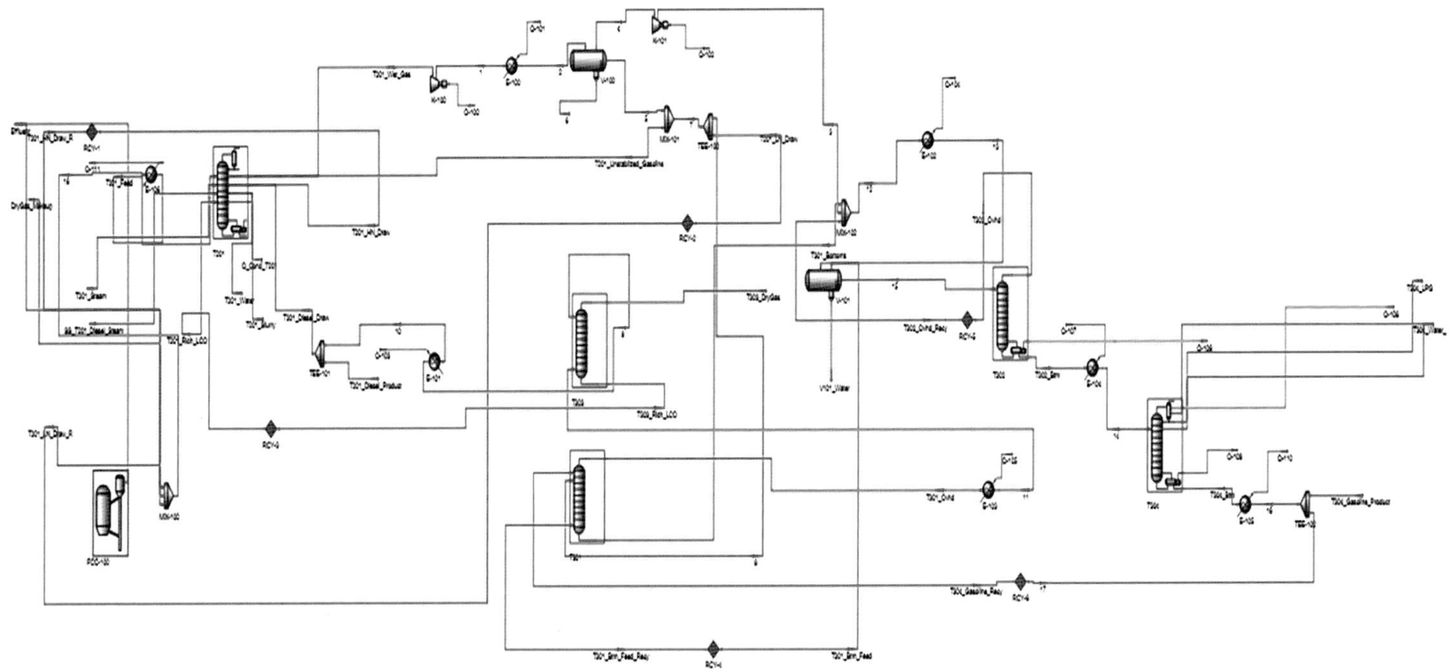

Figure 4.16 Overall Aspen HYSYS model of FCC unit and associated gas plant.

Table 4.13 Product yield results, AAD = 0.96%.

Yield	VALID-1		VALID-2		VALID-3	
Mass%	*Model*	*Plant*	*Model*	*Plant*	*Model*	*Plant*
Gasoline	43.3%	41.9%	43.3%	44.2%	40.1%	39.5%
Diesel	24.6%	23.7%	21.6%	22.0%	25.6%	25.2%
LPG	18.5%	20.1%	17.9%	19.9%	19.1%	21.1%
Dry gas	4.9%	4.4%	5.0%	4.2%	4.7%	4.1%
Slurry	1.4%	4.0%	5.5%	3.8%	4.5%	3.9%
Coke	7.3%	5.9%	6.7%	6.0%	6.0%	6.3%
Yield	**VALID-4**		**VALID-5**		**VALID-6**	
Mass%	*Model*	*Plant*	*Model*	*Plant*	*Model*	*Plant*
Gasoline	41.5%	41.2%	44.1%	44.2%	40.8%	41.2%
Diesel	24.7%	24.6%	20.8%	20.9%	24.3%	24.5%
LPG	19.3%	21.6%	17.8%	20.6%	18.6%	20.2%
Dry gas	4.8%	3.8%	4.7%	4.3%	5.3%	4.4%
Slurry	3.9%	3.9%	6.5%	3.9%	5.1%	4.0%
Coke	5.7%	4.8%	6.0%	6.2%	5.9%	5.6%

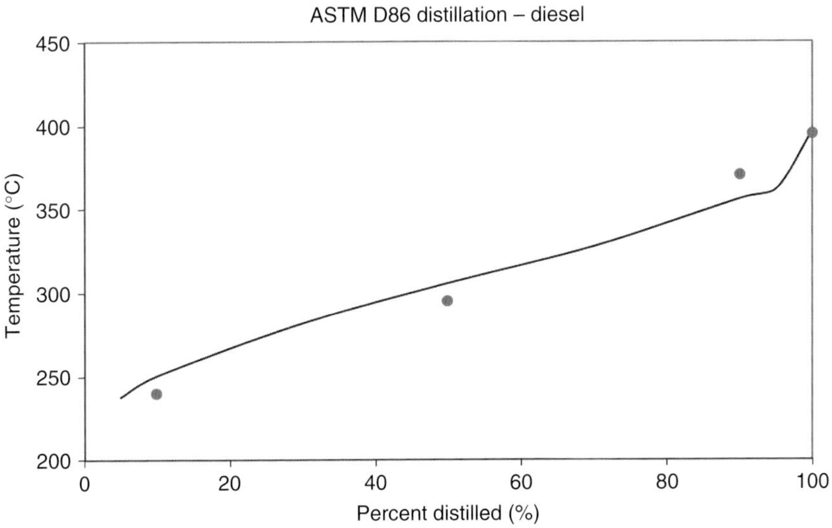

Figure 4.17 ASTM D86 distillation for the product diesel from the main fractionator (VALID-1).

properties. Using the results from the fractionator model, we can calculate the distillation curves of the liquid products. Figures 4.17 and 4.18 show the distillation curves for one of the validation cases. In general, the model predicts key points from the D86 curve (5%, 95%) within plant tolerance. Further refinement of this prediction requires accurate measurements of the pumparound rates and

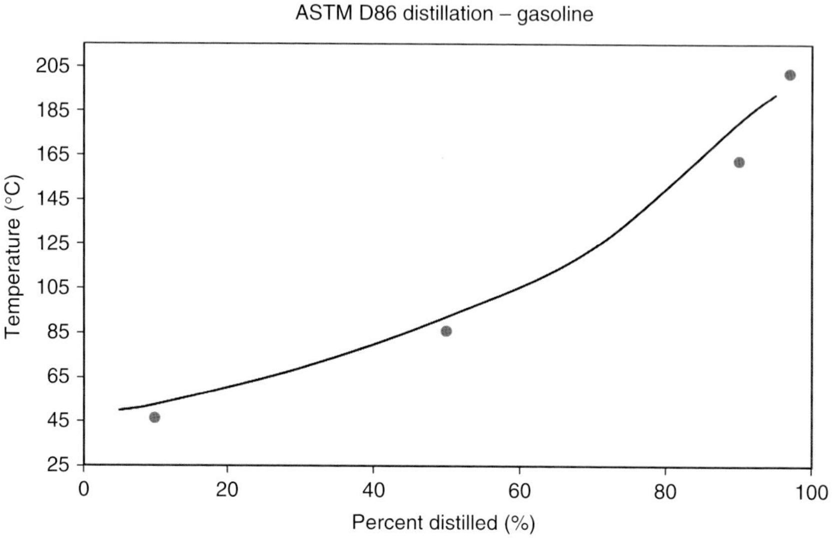

Figure 4.18 ASTM D86 distillation for the product gasoline from debutanizer column (VALID-1).

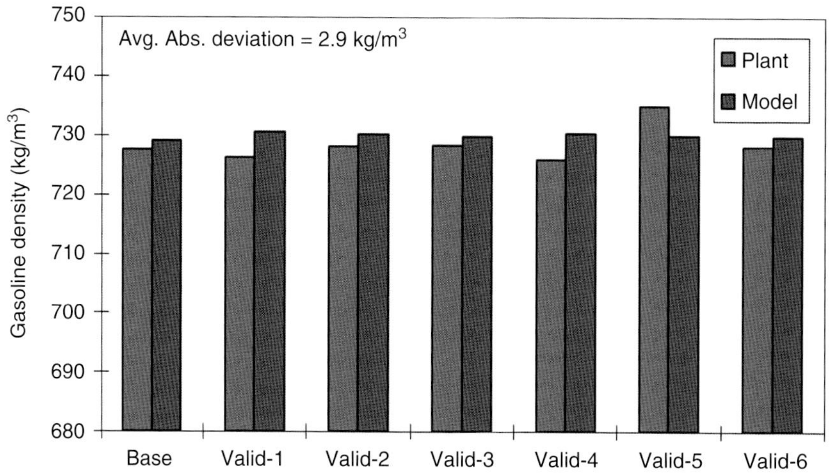

Figure 4.19 Gasoline density comparison.

the heat duty for each pumparound in the main fractionator. These data are not routinely measured.

We can use the predicted D86 curves to calculate several other properties of interest. There are several methods to calculate the flash point and other volatility properties in using the distillation curve and density. Figures 4.19 and 4.20 show the prediction of the densities for gasoline and diesel. We also see good agreement between the measured and predicted results for density. In Figure 4.21,

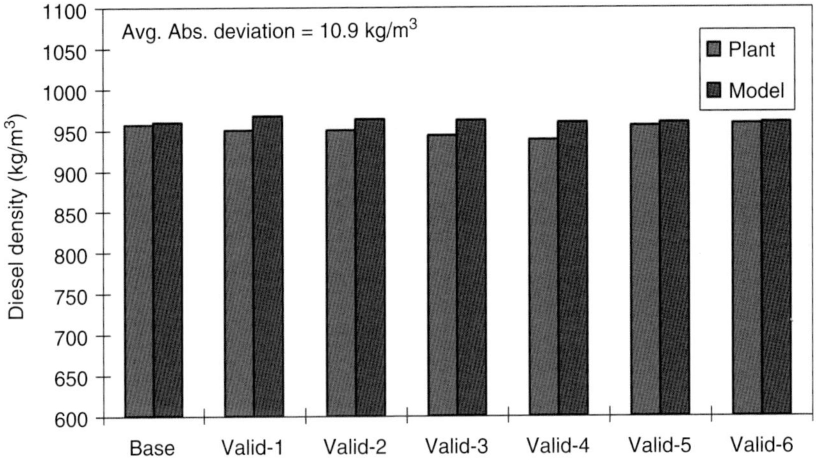

Figure 4.20 Diesel density comparison.

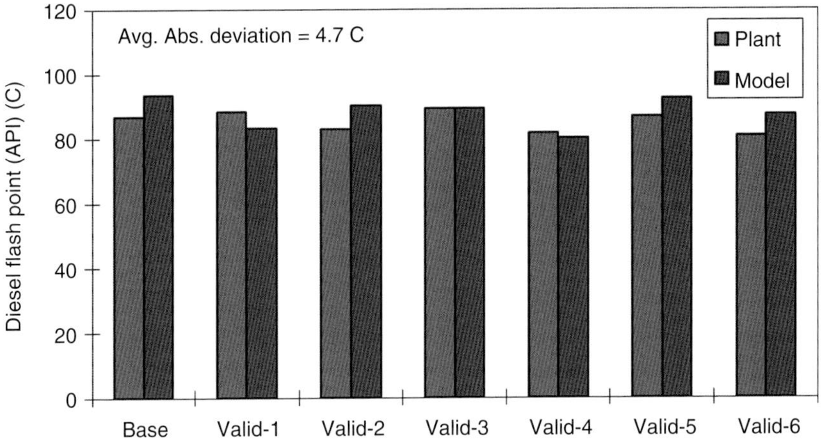

Figure 4.21 Diesel flash point comparison.

we compare our predictions using the API flash point correlations [54] to the measured data. We note good agreement for the flash point.

Roughly, 20–25% of the product in this FCC is LPG, which primarily consists of propane, propylene, butanes, and butenes. The presence of significant amounts (greater than 0.5%) of C5+ products in LPG indicates that the fractionation process is not operating well. Therefore, the prediction of the composition of all the gas and LPG products is essential to validate the model. Tables 4.14 and 4.15 compare the operating data and model predictions for LPG and dry gas. The AAD for the predictions of mole compositions in LPG and dry gas are 1.2% and 1.8%, respectively. We note that there is often more significant error in the prediction of hydrogen and nitrogen.

Table 4.14 Comparison of LPG composition, AAD = 1.2%.

LPG	VALID-1		VALID-2		VALID-3	
Mol%	*Model*	*Plant*	*Model*	*Plant*	*Model*	*Plant*
C3	13.9	15.5	13.9	14.9	14.7	13.3
C3=	36.6	38.3	35.1	35.9	38.3	38.4
NC4	4.5	5.3	4.1	5.6	4.0	5.6
IC4	17.5	17.1	16.9	18.8	16.1	18.0
IC4=	12.8	13.1	12.1	12.8	11.5	13.4
T-2-C4=	6.0	6.0	5.5	6.1	5.3	6.1
C-2-C4=	4.4	4.7	4.0	5.0	3.9	4.7
LPG	**VALID-4**		**VALID-5**		**VALID-6**	
Mol%	*Model*	*Plant*	*Model*	*Plant*	*Model*	*Plant*
C3	14.2	13.2	15.6	12.2	15.5	13.0
C3=	34.5	39.0	35.9	41.7	37.0	39.4
NC4	4.3	4.9	4.5	3.4	4.5	4.5
IC4	16.6	18.4	18.2	18.0	17.5	18.6
IC4=	12.3	13.1	13.1	13.1	12.7	13.2
T-2-C4=	5.7	6.1	6.0	5.7	6.0	6.3
C-2-C4=	4.1	4.8	4.5	4.8	4.5	4.6

Table 4.15 Comparison of dry gas composition, AAD = 1.8%.

Dry gas	VALID-1		VALID-2		VALID-3	
Mol%	*Model*	*Plant*	*Model*	*Plant*	*Model*	*Plant*
H_2	24.3	29.9	23.1	31.8	24.7	29.3
N_2	21.0	20.1	19.5	16.7	19.7	19.1
CO	1.6	1.6	1.5	2.0	1.6	1.8
CO_2	1.8	1.8	2.2	1.6	1.1	1.8
C1	24.8	23.0	24.5	24.8	25.6	23.1
C2	10.9	10.2	12.1	9.9	11.2	10.3
C2=	11.7	10.5	12.3	10.5	13.0	11.8

Table 4.15 (Continued)

Dry gas	VALID-4		VALID-5		VALID-6	
Mol%	*Model*	*Plant*	*Model*	*Plant*	*Model*	*Plant*
H_2	20.5	28.2	21.6	27.5	20.8	28.1
N_2	19.7	22.5	19.7	20.3	18.9	19.8
CO	1.6	1.7	1.6	1.7	1.5	1.4
CO_2	1.7	2.0	1.8	2.0	3.6	1.6
C1	27.7	21.4	26.6	23.1	24.5	23.6
C2	10.6	10.5	11.7	10.1	11.7	10.3
C2=	13.8	11.6	12.9	11.2	11.9	11.2

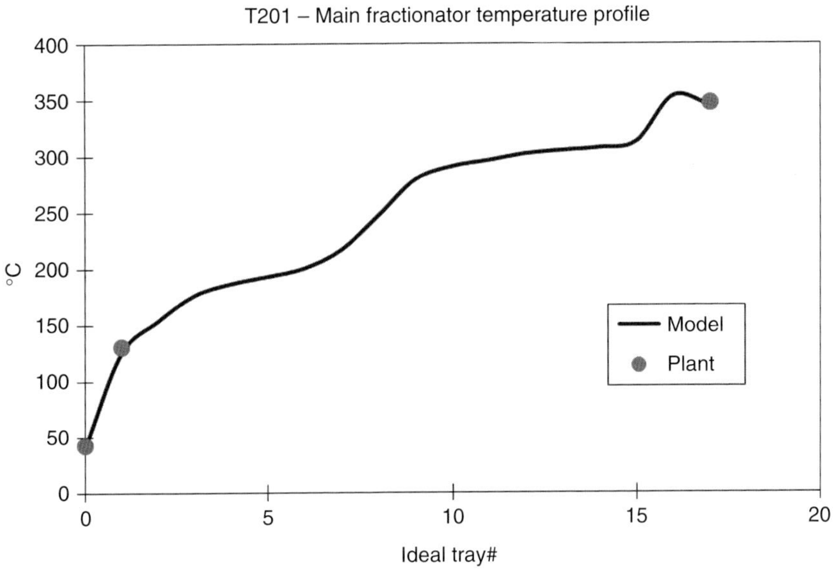

Figure 4.22 Main fractionator temperature profile.

We also apply the model to predict all temperature profiles of columns for each validation case and compare the results with plant operation. We find good agreement between plant measurements for all columns with the exception of the debutanizer column (T302) (see Figure 4.24). This column is very sensitive to the LPG composition in the model. We recall that the base calibration case shows some error in matching the LPG yield from the plant. It is possible to improve this prediction by including catalyst-specific parameters in the kinetic model to match the plant performance. However, we avoid this procedure at this time; thus, we can provide a more broadly useful model. Figures 4.22–4.26 compare model and plant values for temperature profiles for a single validation case (VALID-4).

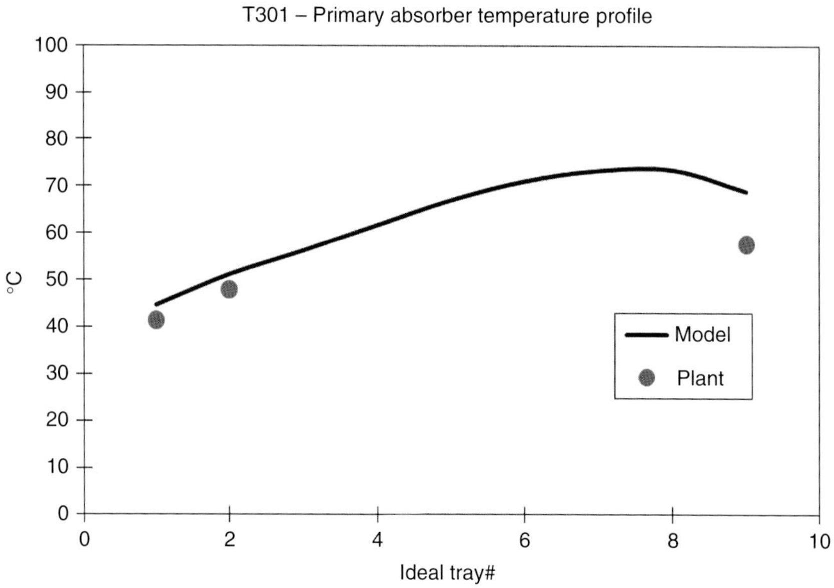

Figure 4.23 Primary absorber temperature profile.

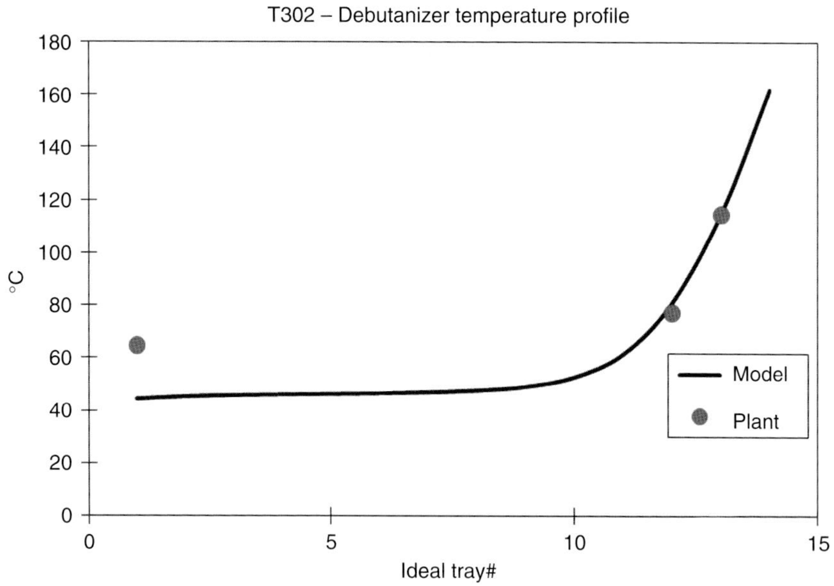

Figure 4.24 Debutanizer temperature profile.

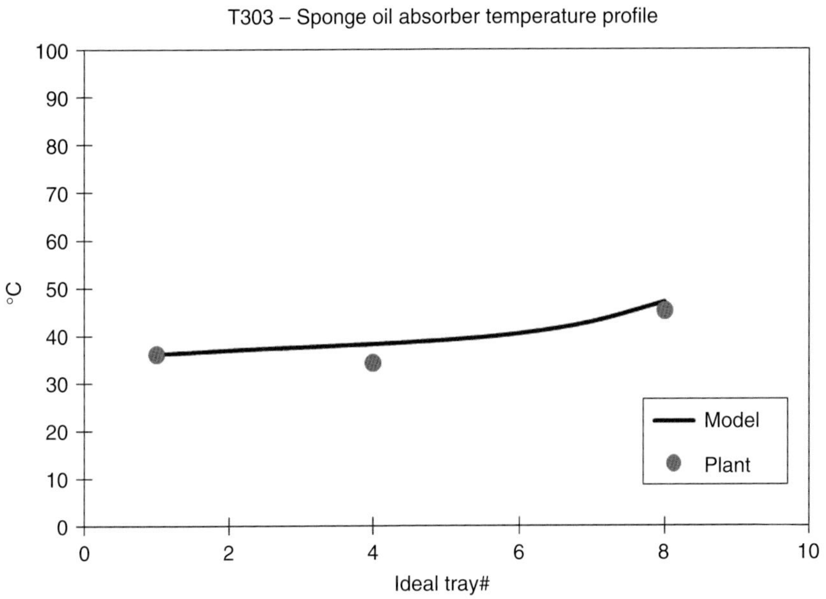

Figure 4.25 Sponge oil absorber temperature profile.

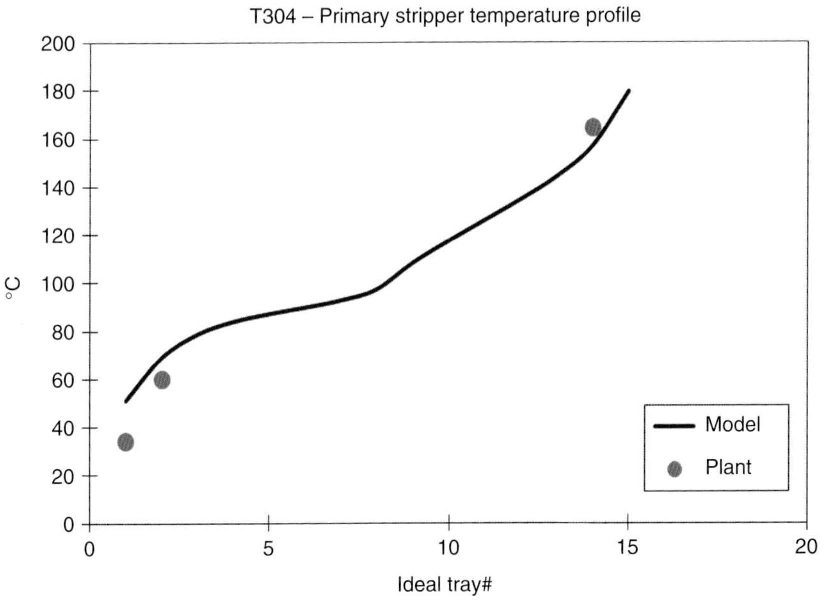

Figure 4.26 Primary stripper temperature profile.

4.11 Applications

Refiners are very interested in obtaining optimal operating conditions that maximize the yield of a profitable product slate. However, unlike traditional chemical plants, the FCC unit generates several products that have different profit margins. Further complicating matters is that these profit margins may change depending on refinery constraints, market conditions, and government regulations. Therefore, it is critical to understand how to manage the FCC unit under different operating scenarios. We consider two common scenarios in FCC operation: improving gasoline yield and increasing the throughput of the unit.

4.11.1 Improving Gasoline Yield

Gasoline yield is a typical complex function of temperature, pressure, feed quality, and catalyst-to-oil ratio [8]. We consider the case where the feed quality is fixed. An easily manipulated operating variable is the riser outlet temperature (ROT). Allowing the ROT to increase improves gasoline yield by promoting cracking and aromatic chain scission reactions that increase the yield of C5+ components. We compute the gasoline yield at various temperatures and Figure 4.27 shows the results. The current ROT is 510 °C and is marked with a yellow square. The ROT that leads to the highest yield of gasoline is roughly 530 °C. Does this mean that we should allow the ROT to increase to 530 °C? To answer this question, we plot the yields of other valuable products from the FCC in Figure 4.28.

Figure 4.28 shows that while gasoline yield reaches the maximum at an ROT of 530 °C, the yields of other valuable products (e.g., diesel) drop significantly. In addition, the yield of fuel/dry gas (light gases) rises quickly. This indicates that we are "overcracking" the feed. The high temperature accelerates the production of

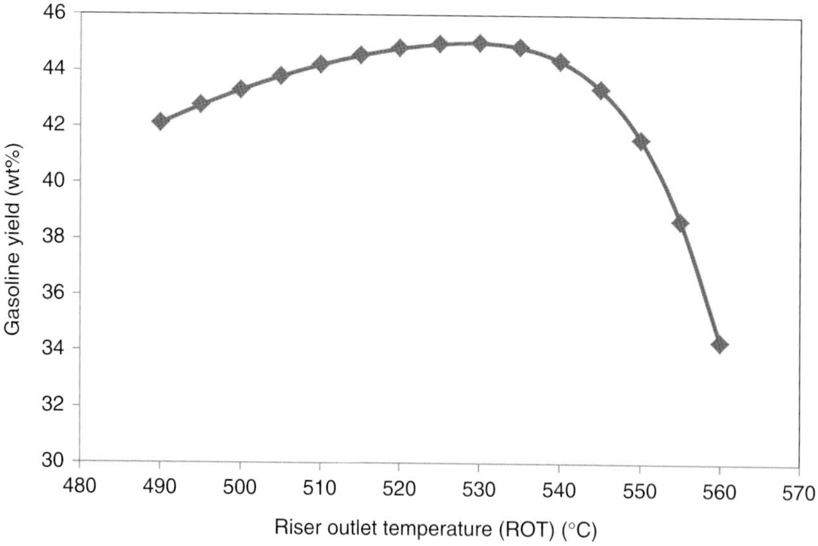

Figure 4.27 Gasoline yield profile as a function of ROT.

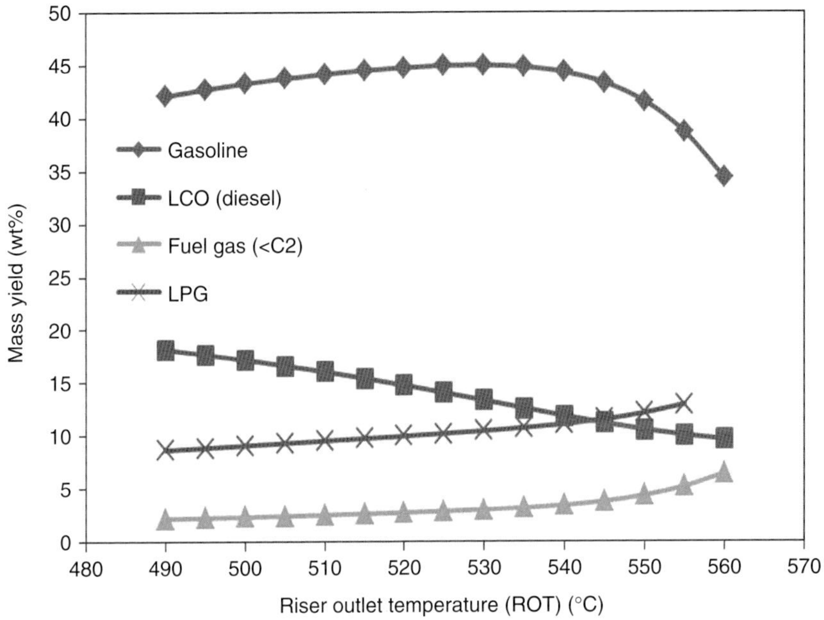

Figure 4.28 Yields of key products as functions of ROT.

C1–C2 components (i.e., fuel/dry gas) through the catalytic and thermal cracking pathway. This is clearly an undesired result. Dry gas is not of significant value and can easily overload the overhead wet gas compressor. In addition, Figure 4.29 shows the coke yield on the catalyst as a function of ROT. The amount of coke present on the catalyst leaving the riser is a strong function of ROT. Regenerating catalyst with higher coke deposits increases the utilities required to regenerate the coke to the same level. These side effects shrink the acceptable range of values for the ROT.

We can combine the results from these graphs and consider scenarios where a refiner needs to maximize different products. For example, refiner may need to maximize the production of gasoline and diesel or maximize the production of gasoline and LPG, depending on external constraints. We can easily use the model to generate a case study, as shown in Figure 4.30. This figure shows that there are different optimum ROT values for different scenarios. The maximum gasoline and diesel production occurs in the range of 505–510 °C (confirming the refiner's assertion where these data are obtained), whereas the maximum for gasoline and LPG production occurs in the range of 530–540 °C.

This example shows the importance of a model that accounts for all products, including light gases as a distinct lump. In addition, the integrated heat balance between the riser and regenerator allows us to provide useful estimates for the coke yield. We have not included the effect of these process changes on the downstream fractionation unit in this study. However, we note that there are often significant equipment and process constraints (a prime example is the wet gas compressor) that restrict the acceptable range for the ROT.

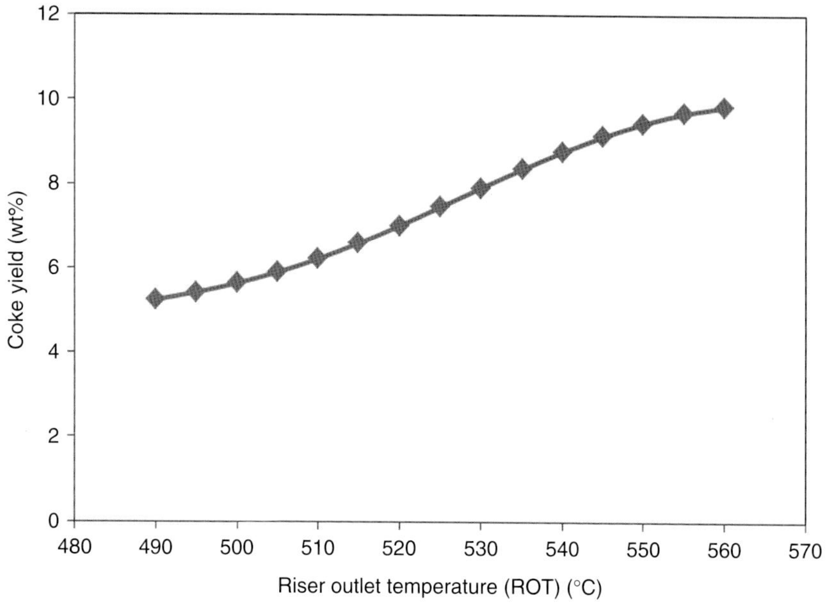

Figure 4.29 Coke yield as a function of ROT.

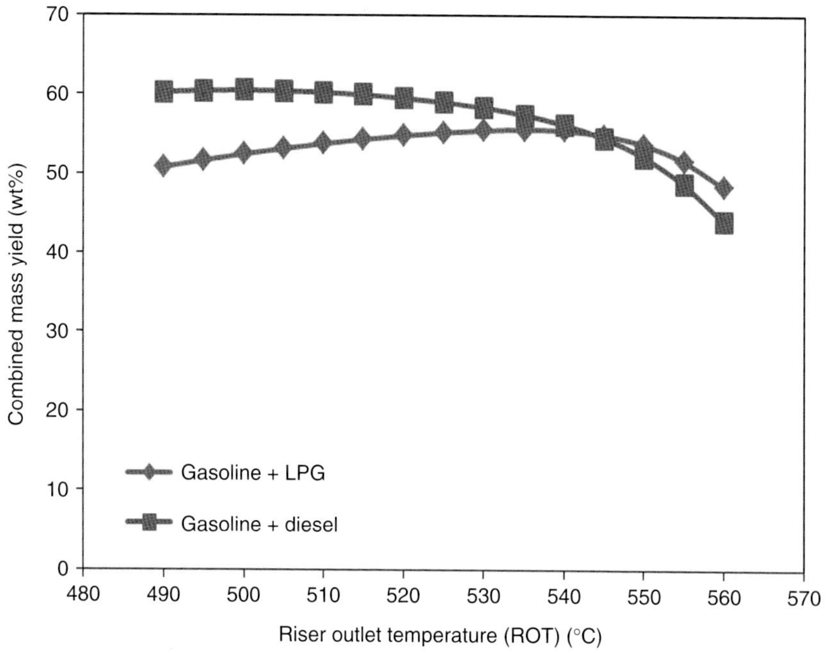

Figure 4.30 Maximizing production of key products as a function of ROT.

4.11.2 Increasing Unit Throughput

Let us consider another scenario where we need to increase the throughput of the unit. The refiner typically needs to process the largest volume of feedstock possible. Ideally, we would like the FCC to maintain a similar mass yield of the most valuable product (i.e., gasoline). Figure 4.31 shows the mass yield of gasoline as a function of feed rate to the unit. The mass yield decreases almost linearly with increasing feed rate. How can we explain this phenomenon? Figure 4.31 also shows the catalyst-to-oil ratio as a function of increasing feed rate. We note that the cat-to-oil ratio also decreases linearly.

The decreased cat-to-oil ratio determines that there is less contact time between the catalyst and the feed oil. Lower contact time will result in fewer species cracking and subsequently reduce the gasoline yield. However, we must not confuse this effect with "overcracking" described in the previous case study. Figure 4.31 also illustrates the difference between "overcracking" and a reduced cat-to-oil ratio. We note that yield of light products (dry gas and LPG) does not increase. This indicates that high-temperature thermal or catalytic cracking is not taking place.

Let us now consider the scenario where we need to increase or maintain gasoline yield that corresponds to the base unit throughput. We will allow the ROT to increase, while also increasing the feed rate to the unit. Figure 4.32 shows the effect of the increasing feed rate and ROT. We note that the gasoline yield increases with rising ROT. However, once we reach the ROT of 540 °C, the gasoline yield drops quickly. This occurs because we have passed the "overcracking" peak for this particular feed.

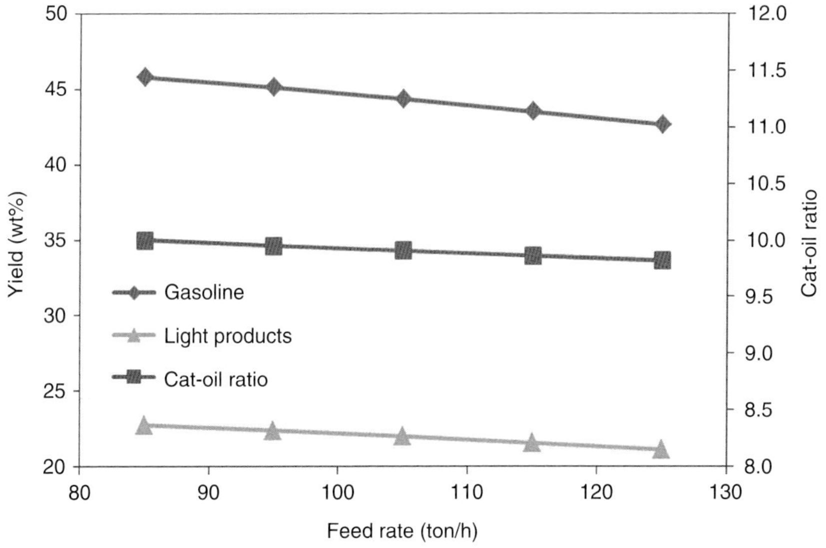

Figure 4.31 Mass yield and cat-to-oil ratio as a function of feed rate.

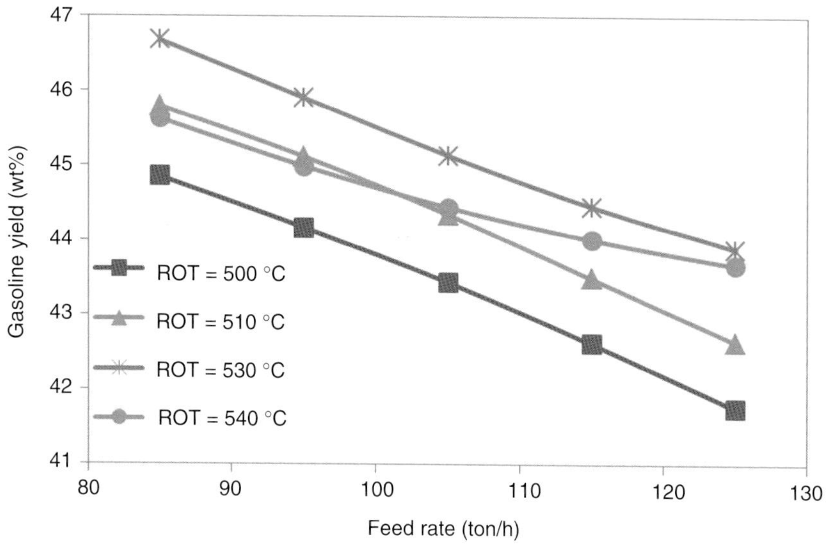

Figure 4.32 Gasoline yield as a function of feed rate.

4.11.3 Sulfur Content in Gasoline

Sulfur content in gasoline is an important regulatory constraint for refiners. Many schemes are in use to reduce the sulfur content in refinery products. In the case of the FCC unit, a significant portion of sulfur in the feed leaves the process as a dry gas. However, the remaining sulfur leaves through the key liquid products.

Sadeghbeigi [1] and Gary *et al.* [7] indicated that hydrotreating the feed significantly reduces the sulfur content in the nonslurry products. However, there may be an economic disadvantage in hydrotreating the feed to the FCC unit. In addition, low sulfur constraints may result in an excess of low-value resid feeds in the refinery. Often, the refiner looks for ways to blend this high-sulfur resid feeds into processing units that can tolerate higher sulfur content. In both cases, we need to understand how the changes in feed sulfur affect the sulfur distribution in the products.

Let us consider the situation where a cheaper feedstock, vacuum residue (VR), is available. The refiner may need to maximize the profitability of the unit by blending the VR with the existing VGO feed. Currently, 5.7 wt% of the feed to the FCC unit is the VR-type feed. We would like to know how much VR we could blend into the VGO feed while meeting the constraint of stabilized gasoline.

To study this question, we must also consider that sulfur content in the VGO feed is changing as well. We vary both the sulfur content in the VGO feed and the amount of VR that is blended. Figure 4.33 shows the outline of the case study process.

We vary the feed ratio of VR from 0% to 11.3% and the associated sulfur content in the VGO. The corresponding sulfur limit for FCC gasoline in this refinery is 800 ppm wt. We use the model to predict the sulfur content in different cases of feed ratio and sulfur in VGO. We note that for the base case of 0.71 wt% sulfur in feed VGO, we could blend more than 10% VR while still meeting the sulfur

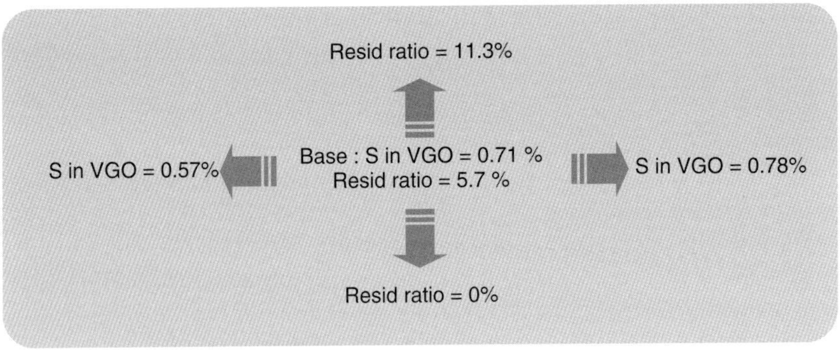

Figure 4.33 Scenario of feed sulfur change.

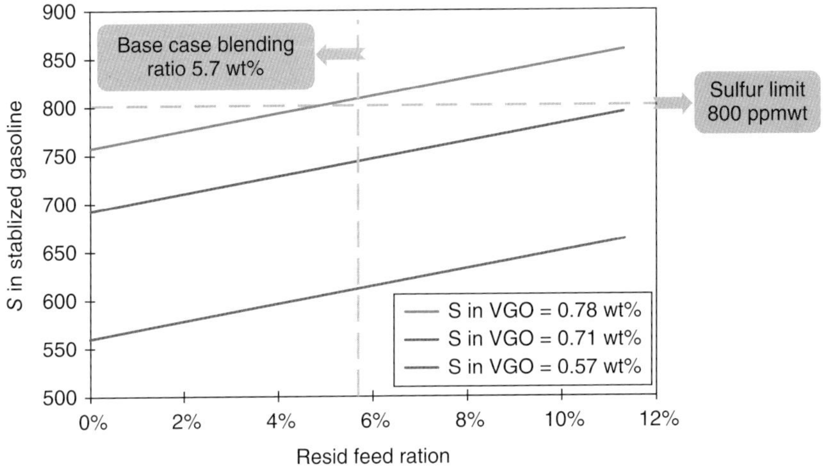

Figure 4.34 Blending varying amounts of residue feed.

constraint. However, if the sulfur content in the VGO increases to 0.78 wt%, we cannot blend more than 4.5 wt% of VR if we need to meet the sulfur constraint (Figure 4.34).

We note that all the above case studies and scenarios are limited to the FCC unit and the associated fractionation system. Modern refineries are highly integrated and changes that appear beneficial in one plant may not benefit another plant in the refinery. One way to apply these models in a larger context (in an existing refinery process) is through the LP for refinery production planning.

4.12 Refinery Planning

We briefly alluded to the complex nature of managing an FCC unit in the previous section. The typical refinery has many units in addition to the FCC (such as catalytic reforming and hydroprocessing) that have their own product distribution and associated profit margins. It is difficult to produce high profit margins dealing with each unit individually when the actual refinery process is highly integrated. The refiner needs methods to optimize feeds to each unit and related products on a refinery-wide scale.

Refiners have typically solved this problem by using LP methods, which have been used extensively in refineries since 1950. Gary *et al.* [7] stated that "*A site-wide model of the refinery is, therefore, usually required in order to properly determine refinery economics.*"

LP involves the maximization of a linear objective function of many variables subject to linear constraints on each variable [57]. In the context of a refinery, the objective function can refer to overall profit generated from processing a particular set of crudes. The variables that affect this objective function are typically the amounts of different crudes purchased. The goal is to determine an optimal set of crudes that maximize the profit margin of the refinery. This scenario is an example of crude oil evaluation. Refiners typically use LP methods in other scenarios as well. Prominent examples are product blending (where two or more products from different units are mixed to form a single product) and production planning (determining the most profitable distribution of products while meeting site constraints).

A key issue in using LP methods is that the relationships between variables must be linear. In other words, all the equations used in the model must be linear with respect to the variables involved. At first, this requirement appears very confining. In fact, the FCC and gas plant models developed in previous sections of this work are highly nonlinear. However, it is important to note that many units in the refinery have a small window of operating conditions during regular operation of the refinery. This allows us to linearize highly nonlinear processes around the regular operating window of the refinery.

That being said, modern LP software such as Aspen PIMS includes many tools to deal with nonlinear relationships. Aspen PIMS uses techniques such as "recursion" (a form of successive LP where the linear model runs many times with different coefficients to approximate nonlinear behavior) and nonlinear programming (NLP) techniques. These techniques can alleviate many problems that frequently arise, especially in product blending and property estimation, with linearized models. The focus of our application study is to improve an existing LP model for the FCC unit alone. Therefore, we do not consider more sophisticated techniques to deal with nonlinear behavior.

Figure 4.35 represents a highly simplified view of a FCC unit. We can consider the FCC unit as a black box that converts different types of feed into products with varying profit margins. The LP model expects that the profits or values of the products are readily available. If we consider that only straight-run VGO enters the unit at fixed operating conditions (riser temperature, catalyst-to-oil ratio, etc.), we can represent the yield of the unit as

$$1.0\,(\text{Normalized feed rate}) = \sum_{i=1}^{N} \text{Yield}_i \tag{4.12}$$

where we know all terms on the right-hand side to be fixed constants. The yield coefficients, Yield_i, correspond to each measured product of the FCC.

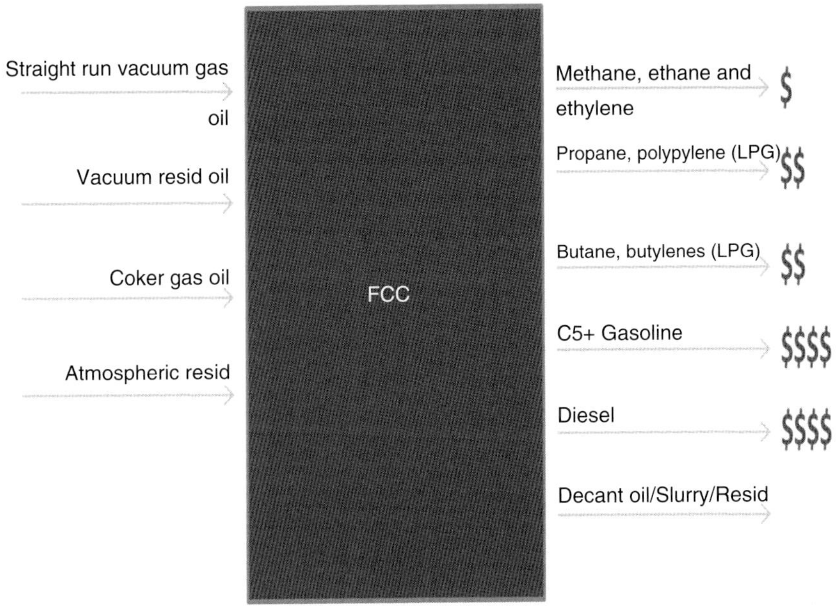

Figure 4.35 Simplified view of FCC unit for a LP application.

Table 4.16 Sample base vector with typical yields for a gasoline-maximizing FCC unit.

Row	Product	Base
1	Feed	1.00
2	Dry gas	−0.04
3	LPG	−0.18
4	Gasoline	−0.40
5	Diesel	−0.30
6	Loss and coke	−0.08

We consider the above equation to represent the base yield of the unit. In Aspen PIMS and other similar LP software, we call the base yield as *the base vector*. We typically encode the base vector in a form shown in Table 4.16. The negative signs arise from moving all the terms from the right-hand side of the equation to the left-hand side.

This base vector is sufficient to model a FCC unit that processes a single type of feed at fixed operating conditions. However, most FCC units do not operate this manner. They accept multiple feeds with varying composition and may operate at different conditions. To account for variations in feed composition, the concept of the delta vector is useful. Every attribute (specific gravity concarbon, sulfur content, etc.) of the feed that can affect the yield of the unit has its own delta vector.

The delta vector can be thought of a slope that modifies the base yield of each product. If we consider the specific gravity (SPG) of the feed as an attribute that can change the product yields, we can now rewrite the yield equation as

$$1.0 = \sum_{i=1}^{N} \text{yield}_i + \sum_{i=1}^{N} (\text{yield modifier or delta})_i * \text{SPG} \tag{4.13}$$

where the SPG of the feed is a known quantity, yield and delta coefficients are known for each product *i*. The products typically are dry gas, LPG, gasoline, diesel, and resid/coke/loss. Note the value of the delta coefficients correspond to the units of measurement of the particular feed attribute (in this case SPG). Table 4.17 gives sample base and delta vectors for a typical gasoline-maximizing FCC unit.

Refiners can typically obtain the base yield of the FCC unit by averaging the measured yields over some period. The delta vectors often come from estimations, refiner's internal correlations, or published correlations [7, 58, 59]. Previous work by Li *et al.* [60] uses correlations from Gary *et al.* [7] to generate FCC delta-base vectors. These vectors are then combined with a blending model and a crude distillation unit (CDU) model. This process results in two significant problems. The first problem is that the true yield of the FCC unit is not available to LP (only averaged yields). This leads to situations where the LP model can optimize the product distribution based on poor yield information. The second problem is that the delta vectors are fixed to particular correlations or estimates. These correlations may not correctly predict changes in yield accurately when the composition of the feed changes.

We overcome these problems by using the detailed FCC model developed in this work. We have shown that the FCC model can predict yields accurately for varying process conditions. To apply the FCC model into the refinery LP, we must first convert the large nonlinear model in to a linear yield model. We can then use coefficients from this generated linear yield model directly in the LP for the refinery. We show the process for generating the linear yield coefficients in Figure 4.36. We have found that 4–5% is a reasonable value for CHANGE% (variable perturbation) for most of the important feed attributes in the FCC process. For example, to generate the delta vector for sulfur content (SUL), we first run the model at the base conditions and record these yields as the base vector. Next, we perturb the

Table 4.17 Base and delta vectors with typical yields for a gasoline-maximizing FCC unit.

Row	Product	Base	SPG
1	Feed	1.00	–
2	Dry gas	−0.04	−0.01
3	LPG	−0.18	0.02
4	Gasoline	−0.40	0.01
5	Diesel	−0.30	−0.01
6	Loss and coke	−0.08	−0.02

1. Identify base operating conditions
2. Run model and record base yields
3. Pick attributes that influence product yields
4. For each attribute
   - Modify attribute value by CHANGE%
   - Run the model with modified attribute
   - Record yield of each product
   - Generate delta vector coefficient for each product by dividing the difference between the base yields and current yields by the change in the attribute
   - Record yield of each product
5. Export delta vectors to LP/PIMS software

Figure 4.36 Process to generate delta-base vectors.

SUL variable by 5% and record the perturbed product yields. We divide the difference in base yields and perturbed yields by the change in the perturbed value to obtain the delta vector corresponding to the SUL variable.

It is important to note that the process in Figure 4.36 essentially generates an approximation to the Jacobian of the nonlinear FCC unit model. If we consider that vector y represents the model outputs, then the $\overline{y}$ vector represents the base case in our planning scenario and the Δx vector represents the change in model inputs from the base case. We then have a matrix of $\Delta y/\Delta x$, which represents the change from the base condition as a function of the selected feed attributes (or possibly process conditions). Equation (4.14) illustrates the connection between the Jacobian and the delta-base vectors:

$$\begin{bmatrix} y_1 \\ y_2 \\ \vdots \\ y_m \end{bmatrix} \text{(Predicion)} = \begin{bmatrix} \overline{y_1} \\ \overline{y_2} \\ \vdots \\ \overline{y_m} \end{bmatrix} \text{(Base)} + \begin{bmatrix} \frac{\Delta y_1}{\Delta x_1} & \cdots & \frac{\Delta y_1}{\Delta x_n} \\ \vdots & \cdots & \vdots \\ \frac{\Delta y_m}{\Delta x_1} & \cdots & \frac{\Delta y_m}{\Delta x_n} \end{bmatrix}$$

$$\text{(Delta-base)} \cdot \begin{bmatrix} \Delta x_1 \\ \vdots \\ \Delta x_n \end{bmatrix} \text{(Delta)} \tag{4.14}$$

Table 4.18 shows the existing base and delta vectors for the FCC unit. The base vectors come from averaged yields of the FCC unit during the previous quarter (ending December 08). The delta vectors come from refiner's internal correlations. The delta vectors refer to the specific gravity of the feed (SPG), Conradson carbon (concarbon) in the feed (CON), and sulfur in the feed (SUL). We note that this particular set of base and delta vectors do not accurately reflect the operation of the unit. As shown earlier in this work, the actual gasoline yield of the FCC unit ranges from 42% to 46%. The LP model underestimates the gasoline yield.

Table 4.18 Existing delta-base vectors for FCC unit (normalized to a feed rate of 1.0).

Row	Feed/product	Base	SPG	CON	SUL
1	Feed	1.00	–	–	–
2	Sour gas	−0.0065	−0.0003	−0.0004	−0.0082
3	Dry gas	−0.0394	−0.0011	−0.0014	0.0000
3	LPG	−0.1740	0.0025	0.0041	0.0000
4	Gasoline	−0.3929	0.0098	0.0081	0.0000
5	Diesel	−0.2899	−0.0057	−0.0033	0.0000
6	Slurry	−0.0381	−0.0032	−0.0038	0.0082
7	Coke	−0.0544	−0.0020	−0.0034	0.0000
8	Loss	−0.0048	0.0000	0.0000	0.0000

Table 4.19 Delta-base vectors generated using rigorous model.

Row	Feed/product	Base	SPG	CON	SUL
1	Feed	1.00	–	–	–
2	Sour gas	−0.00439	0.00068	0.0001	−0.0057
3	Dry gas	−0.02527	0.00069	0.00033	0.00025
4	LPG	−0.19386	0.02213	0.00271	0.00164
5	Gasoline	−0.4421	0.09480	0.00621	0.00330
6	Coke	−0.06218	−0.05913	−0.00453	0.00038

In addition, as the FCC unit is the most significant producer of gasoline in the refinery, using the LP in crude selection context can lead to nonoptimal crude selection.

Table 4.19 shows the delta-base vectors we generated using the procedure in Figure 4.36. The new base vector accurately reflects the current base gasoline and LPG yields of the FCC unit. In addition, as a consistency check, we note that SUL coefficient for the sour gas (row 1) has a negative coefficient. This indicates that sour gas increases as the sulfur in the feed increases. A similar consistency test with CON coefficient and coke (row 5) shows the same result. We can use the LP model optimally, knowing that LP model does not underestimate key product yields.

The advantage of this method is that LP now reflects the actual capabilities of the unit and not the perceived capabilities based on historical data or correlations. In addition, if the rigorous simulation is updated alongside with plant retrofits, we can modify the LP model quickly to track these retrofits. The workflow we described in Figure 4.36 is easy to integrate into existing process simulation and LP software. Aspen HYSYS Petroleum Refining includes tools to automate the workflow and export the updated delta-base vectors to Aspen PIMS (LP software) directly. This automation allows quick updates of the LP model to accurately reflect the unit performance.

4.13 Workshop 4.1 – Guide for Modeling FCC Units in Aspen HYSYS Petroleum Refining

4.13.1 Introduction

In Sections 4.13–4.17, we demonstrate how to organize data and build and calibrate a model for a FCC unit using Aspen HYSYS Petroleum Refining. We discuss some key issues in model development and how to estimate missing data required by Aspen HYSYS Petroleum Refining. We divide this section into five workshops:

a) Workshop 4.1: building a basic FCC model;
b) Workshop 4.2: calibrating the basic FCC model;
c) Workshop 4.3: build a model for the main fractionator and gas plant system;
d) Workshop 4.4: perform case study to identify different gasoline production scenarios;
e) Workshop 4.5: generate delta-base vectors for LP-based planning.

4.13.2 Process Overview

Figures 4.37–4.39 show PFD for the FCC unit and downstream fractionation units that we use to build the model in question. We extensively discussed the features and operating issues associated with this type of unit early in the chapter. Figure 4.40 shows the FCC and fractionation simulation flowsheet that we are to develop in Workshops 4.1–4.3. In Figure 4.40, we showed the four intercolumn stream paths that are labeled in Figures 4.38 and 4.39, including (1) the unstabilized gas oil labeled A from T201_MainFractionator to T301_Absorber; (2) the rich sponge oil labeled B from T303_ReAbsorber to T201_MainFractionator; (3) the LCO product labeled C from T201_MainFractionator to T303_ReAbsorber; and (4) the wet gas labeled D from T201_MainFractionator to WetGas Compressors and then T302_Stripper.

4.13.3 Process Data

Tables 4.20–4.23 give detailed feeds, products, and operation data for a typical UOP FCC process. Values that have been estimated are marked with an *. Operating conditions for the fractionation section largely depend on the FCC unit effluent and are relatively static, so they are not given here.

4.13.4 Aspen HYSYS and Initial Component and Thermodynamics Setup

We start by opening Aspen HYSYS. The typical path to Aspen HYSYS is to enter the Start → Programs → AspenTech → Aspen Engineering Suite → Aspen HYSYS. We dismiss the "Tip" dialog and select File → New → Case. We wish to include fractionation, so we do not choose "FCC" alone. Save the resulting file as *FCC Components and Properties.hsc.*

The first step in creating the model is the selection of a standard set of components and a thermodynamic basis to model the physical properties of these components. When we create a new simulation, we must choose the

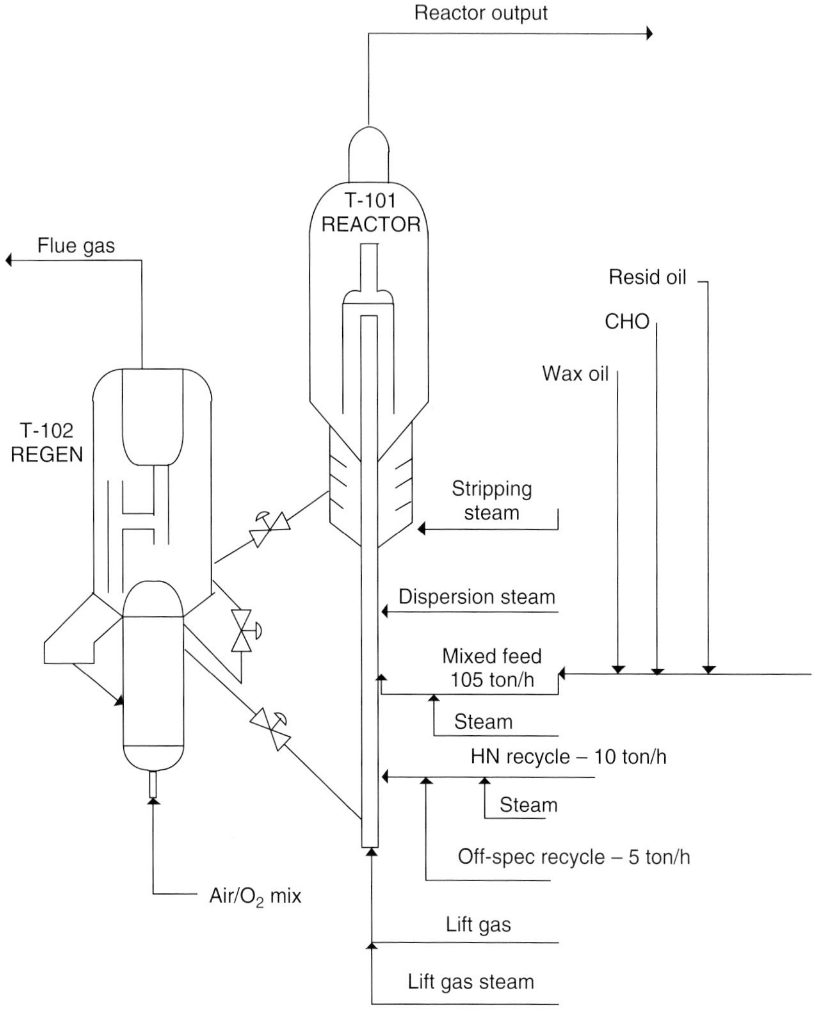

Figure 4.37 Reaction section of FCC unit.

components and thermodynamic model appropriate for the process using the simulation basis manager. The simulation basis manager allows us to define components and associated thermodynamic model in Aspen HYSYS. We may add components through the import button in Figure 4.40. However, we have a predetermined set of the components for the FCC model (Figure 4.41).

To import these components, we click "Import" and navigate to the directory location, "C.\Program Files\AspenTech\Aspen HYSYS V9.0\Paks" and select the "FCC Components Celsius.cml" as the component list (Figure 4.42). The path shown in this figure reflects a standard installation of Aspen HYSYS Petroleum Refining software.

Once we import a component list, HYSYS will create a new component list called "Component List-1." We can view the elements of this component list

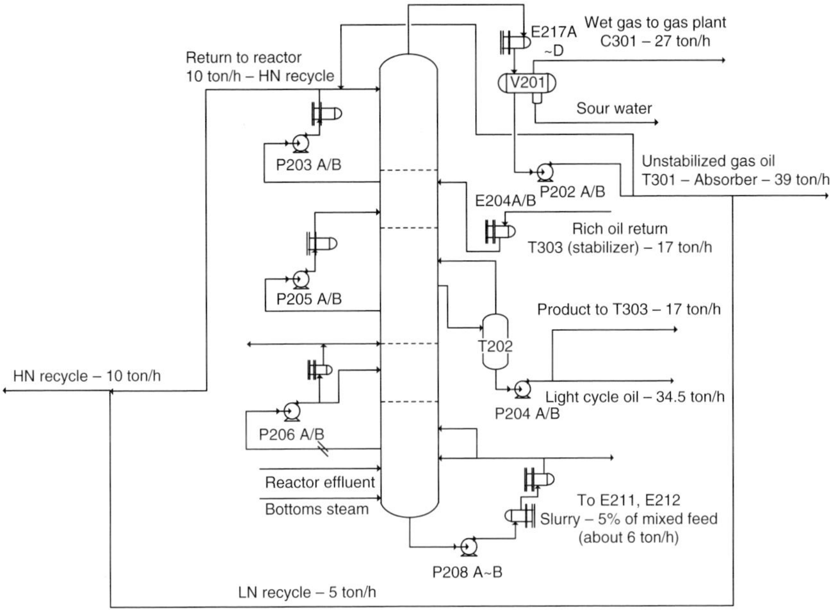

Figure 4.38 Main fractionator associated with FCC.

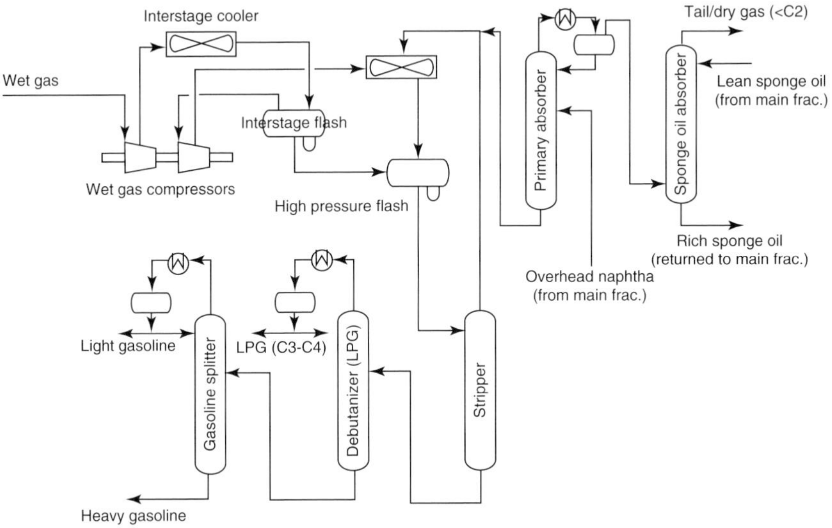

Figure 4.39 Gas plant associated with FCC unit.

by selecting "Component List-1" and clicking on "View" in the simulation basis manager (Figure 4.42). We can add additional components or modify the order of the elements in the component list. We note that the standard FCC component list is quite complete and model most refining processes. The rigorous FCC model does not predict components that are not part of the "FCC Components Celsius.cml" list. However, these additional components may be used in production

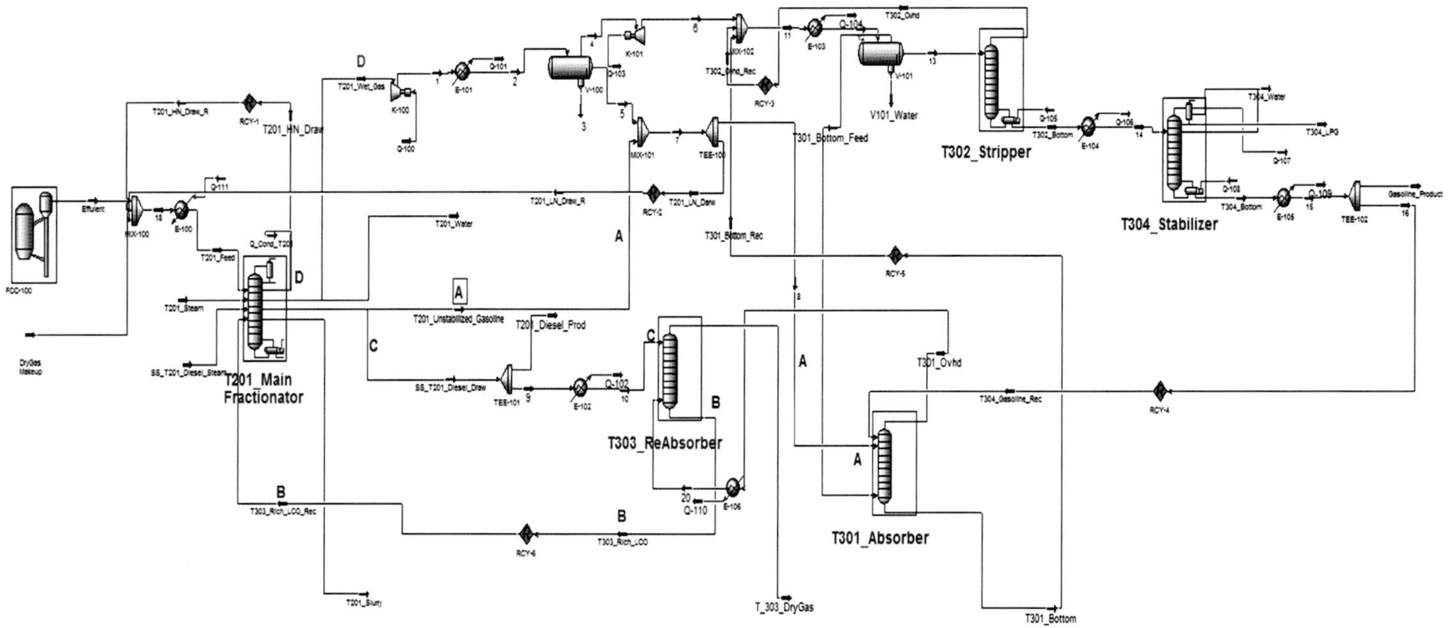

Figure 4.40 A simulation flowsheet of the FCC and fractionation system (see *Workshop 4.3-done.hsc*).

Table 4.20 Summary of liquid feeds and products.

Feed/products	Feed	Naphtha	Light cycle oil	Bottoms
Flow rate (kg/h)	108 208	46 583	24 333	4125
Specific gravity	0.9	0.7	1.0	1.0
Distillation type	D1160	D86	D86	TBP
Initial point (°C)	269.0	35.7	217.9	221
5%	358.6	40.8	235.9	314
10%	376.4	45.6	246.6	343.3
30%	419.0	64.7	275.7	382.2
50%	452.3	86.4	300.3	426.7
70%	488.0	115.0	326.9	468.3
90%	541.8	165.4	365.4	496.1
95%	567.9	191.4	382.5	545.1
End point	665.8	255.4	418.9	649
Nitrogen (ppm wt)	2409.0	9.0	127.8	324.3
Sulfur (wt%)	0.56	0.06	0.91	1.96
CCR (wt%)	1.86	0.01	0.11	0.38
Vanadium (ppm wt)	0.3	–	–	–
Nickel (ppm wt)	3	–	–	–
Sodium (ppm wt)	0.3	–	–	–
Iron (ppm wt)	2.1	–	–	–
Copper (ppm wt)	0.1	–	–	–
RON/MON	–	92/82	–	–
Paraffins (liquid vol%)	28.5	–	–	–
Naphthenes (liquid vol%)	8.529	–	–	–
Aromatics (liquid vol%)	23.6	–	–	–
Cloud point (°C)	–	–	−10	–

of the fractionation models associated with the FCC model. For the purposes of this simulation, we will add *benzene* (Figure 4.43).

The next step is the selection of a "Fluid Package" for this model. The "Fluid Package" refers to the thermodynamic model associated with the chosen list of components. We move to the "Fluid Pkgs" tab in the simulation basis manager and click "Add" (Figure 4.44). Aspen HYSYS will automatically choose the component list and present options for a "Property Package" for these components. The FCC system is mostly of pseudocomponents and light hydrocarbons. Consequently, the Peng–Robinson equation of state is sufficient. We discuss the implications of the process thermodynamics in Chapter 2. In the case of the FCC model, equation of state or hydrocarbon correlation methods (Grayson–Streed, etc.) can sufficiently model the process.

Table 4.21 Summary of gas flow rates and compositions.

	Dry gas	Sour gas	LPG	Regenerator flue gas
Flow rate (kg/h)	4833	667	19 542	–
Composition	mol%	mol%	vol%	mol%
N_2	22.5	0.6	–	NA
CO	1.7	–	–	NA
CO_2	1.8	30.5	–	NA
O_2	–	–	–	2.8
H_2S	0.0	68.5	–	NA
H_2	25.5	–	–	NA
C1	23.3	0.2	–	NA
C2	11.2	0.2	–	NA
C2=	11.3	–	–	–
C3	0.3	–	13.5	–
C3=	1.0	–	41.5	–
*n*C4	0.2	–	4.7	–
*i*C4	0.4	–	18.0	–
*i*C4=	0.4	–	12.5	–
1-C4=	–	–	–	–
*c*2–C4=	–	–	4.0	–
*t*2–C4=	–	–	5.7	–
*c*2–C5=	0.2	–	–	–
*t*2–C5=	0.2	–	–	–

Table 4.22 Riser and regenerator operating conditions.

	Flow rate (kg/h)	Temperature (°C)	Pressure (kPa)
Riser feed preheat temperature	–	175	–
Riser inlet steam	5000	200	1301
Riser outlet temperature	–	518	–
Stripping steam	5000	200	1301
Regenerator dense bed temperature	–	680	–
Regenerator pressure	–	–	

Table 4.23 Equilibrium catalyst properties.

Metal contents (V/Ni/Na/Fe/Cu) (ppm wt)	5000/4044/3103/5553/57
Equilibrium activity (%)	66
Inventory (kg)	150 000

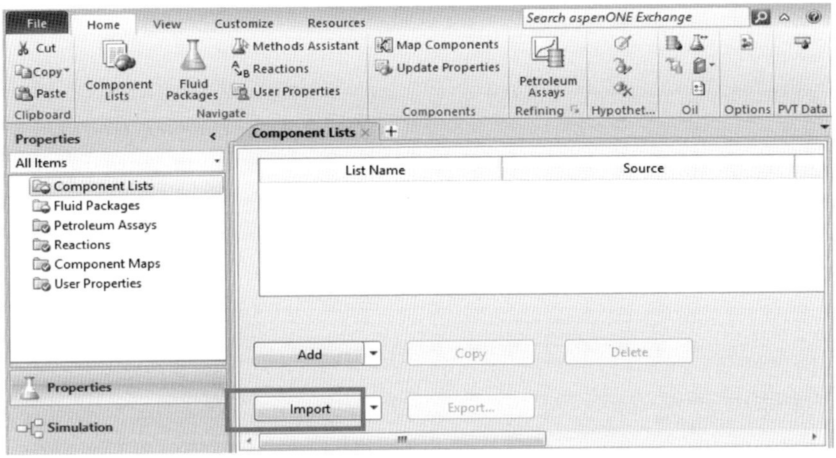

Figure 4.41 Adding a component list.

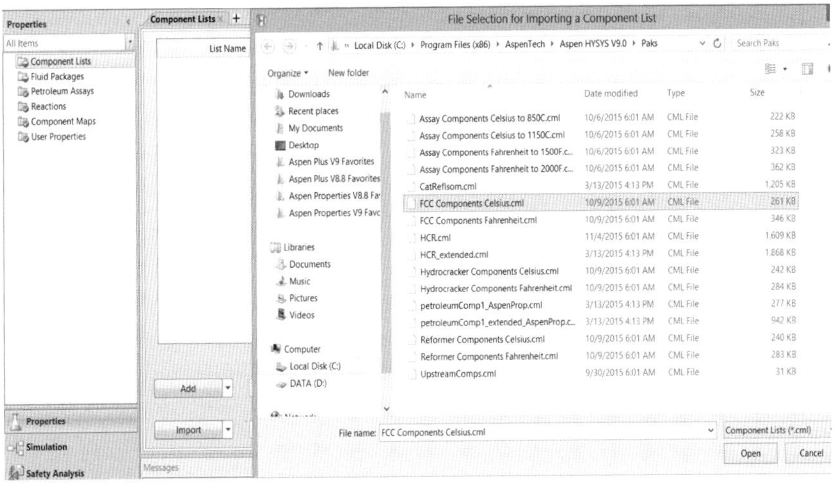

Figure 4.42 Adding FCC component list.

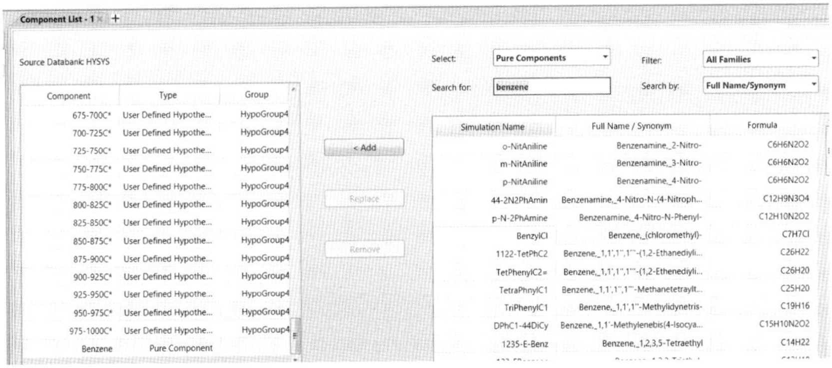

Figure 4.43 Adding additional components to FCC component list.

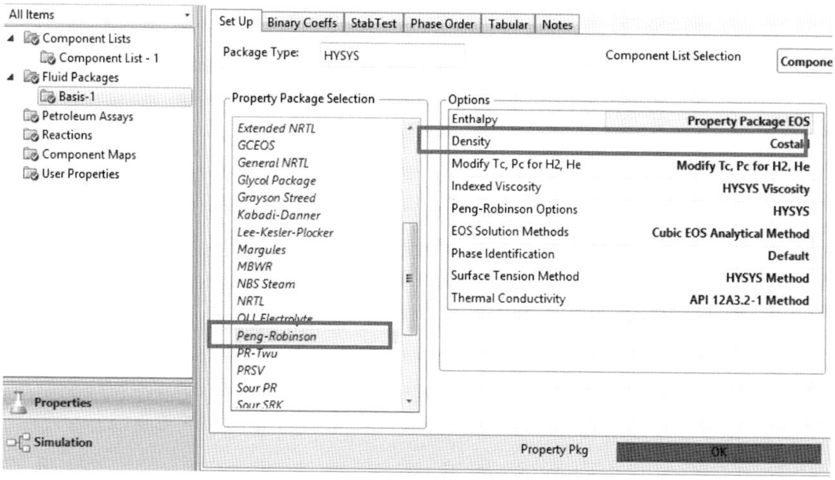

Figure 4.44 Select thermodynamics for fluid package.

It is important to note that even when we choose an equation-of-state approach, Aspen HYSYS does not calculate all physical properties from the equation of state. For hydrocarbons, equations of state do not generally predict the equilibrium properties of very light components such as hydrogen. In addition, density predictions (especially in the heavy hydrocarbon range) can be quite poor. We usually modify the equation of state to account for these deficiencies. For the FCC process, we choose the COSTALD method to predict the liquid density (Figure 4.44).

The last step before building the FCC flowsheet is to verify the interaction parameters (Figure 4.45). If we choose a correlation-based approach (Grayson–Streed, etc.), we do not have to examine the interaction parameters. As we choose an equation-of-state approach, we must ensure that the binary interaction parameters for the equation of state are meaningful. In Aspen HYSYS, the interaction parameters for defined components (such as methane, ethane, etc.) come from an internal data bank based on experimental data. For petroleum pseudocomponents, we can either set the interaction parameters to zero or estimate these values based on correlations. Note that there is little difference in practice whether or not the interactions are set to zero or estimated for lumped components. Especially for the FCC process, both methods yield nearly identical results. Once we have chosen an option for the interaction parameters, we can return to the simulation basis manager and click on "Enter Simulation Environment" to begin building the process model.

4.13.5 Basic FCC Model

We continue with the file, *FCC Components and Properties.hsc*, and save it in a new name, *Workshop 4.1-1.hsc*.The initial flowsheet presents a blank interface where we can place different objects from the Object palette shown in Figure 4.46.

We select the FCC icon from the Refining Reactors palette, click on the FCC icon, and place the icon in the flowsheet. Placing the icon invokes the several

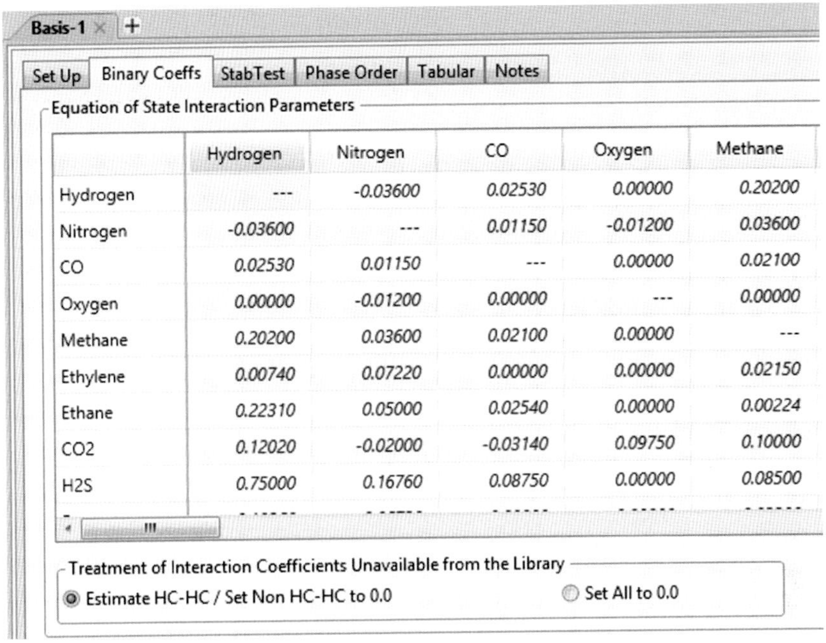

	Hydrogen	Nitrogen	CO	Oxygen	Methane
Hydrogen	---	-0.03600	0.02530	0.00000	0.20200
Nitrogen	-0.03600	---	0.01150	-0.01200	0.03600
CO	0.02530	0.01150	---	0.00000	0.02100
Oxygen	0.00000	-0.01200	0.00000	---	0.00000
Methane	0.20200	0.03600	0.02100	0.00000	---
Ethylene	0.00740	0.07220	0.00000	0.00000	0.02150
Ethane	0.22310	0.05000	0.02540	0.00000	0.00224
CO2	0.12020	-0.02000	-0.03140	0.09750	0.10000
H2S	0.75000	0.16760	0.08750	0.00000	0.08500

Figure 4.45 Binary interaction parameters for fluid package.

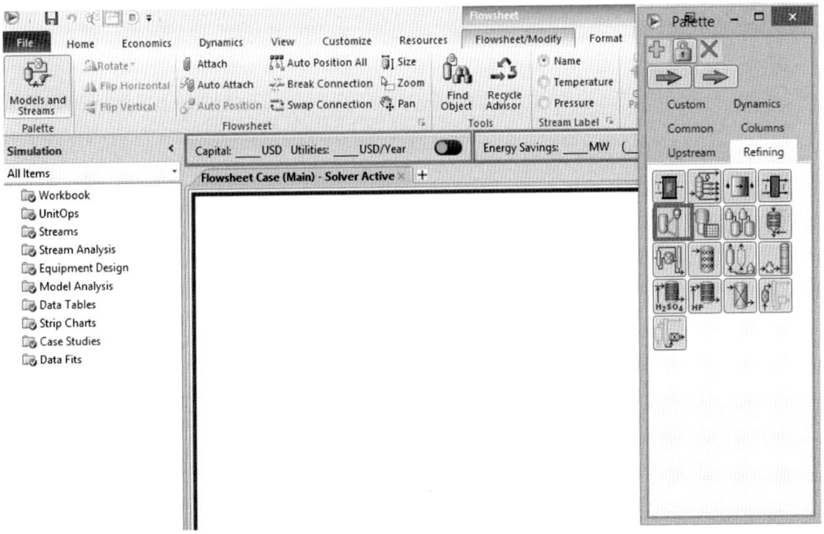

Figure 4.46 Initial Aspen HYSYS Flowsheet.

submodels that prepare the flowsheet for additional objects and creates a large depiction of the FCC object in the flowsheet.

The first step is to choose whether to use a FCC template or configure a new unit. Aspen HYSYS has several FCC templates that reflect the popular types of industrial FCC configurations. Figure 4.47 shows the initial window when we

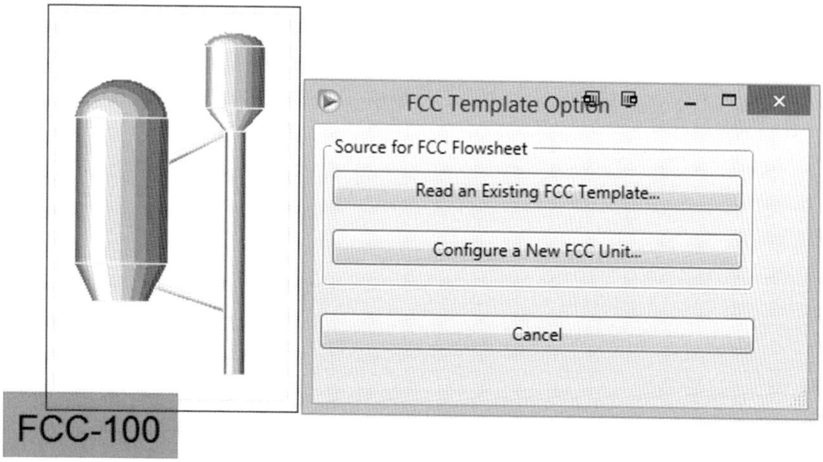

Figure 4.47 Adding the FCC unit.

place a FCC object in the flowsheet. If we choose a template, we do not have to assign the reactor dimensions and select catalyst configuration. However, in this workshop, we will build a FCC unit from scratch, so we choose "Configure a New FCC Unit."

The FCC configuration requires choosing the riser configuration, number and type of regenerators, and catalyst configuration. We may also specify additional downstream fractionation in the form of a simplified main fractionator for the FCC effluent. However, we note that a simplified model for fractionators may not be appropriate for a detailed and integrated process flowsheet. We recommend building a rigorous flowsheet based on standard Aspen HYSYS fractionation objects. In subsequent sections, we will build a complete fractionation section using rigorous stage-by-stage models. In Figure 4.48, we selected a FCC unit with one riser, one-stage regenerator and no fractionation model and click "Next>." We may also use the "Allow Midpoint Injection" to allow for a FCC riser that has multiple injection points.

In the next window, we must specify the key dimensions for the FCC unit. The values in Figure 4.49 reflect typical values for a one-riser, one-regenerator FCC unit. While all measurements are required, the key measurements are the length and diameter of the riser and the height and diameter of the dense and dilute phase in the regenerator. We can estimate all other values (i.e., use values in Figure 4.49) without significantly affecting model results. We click "Next>" after entering all measurements.

Aspen HYSYS now requests to enter to the heat loss for each section of the FCC unit, as shown in Figure 4.50. In general, these values are not available and we recommend using the default values of 0 for all heat losses. These heat losses can account for changes due to external cooling or heating surrounding the unit. Generally, these values are not significant and we may safely ignore them. We click "Next>" to complete the initial unit configuration.

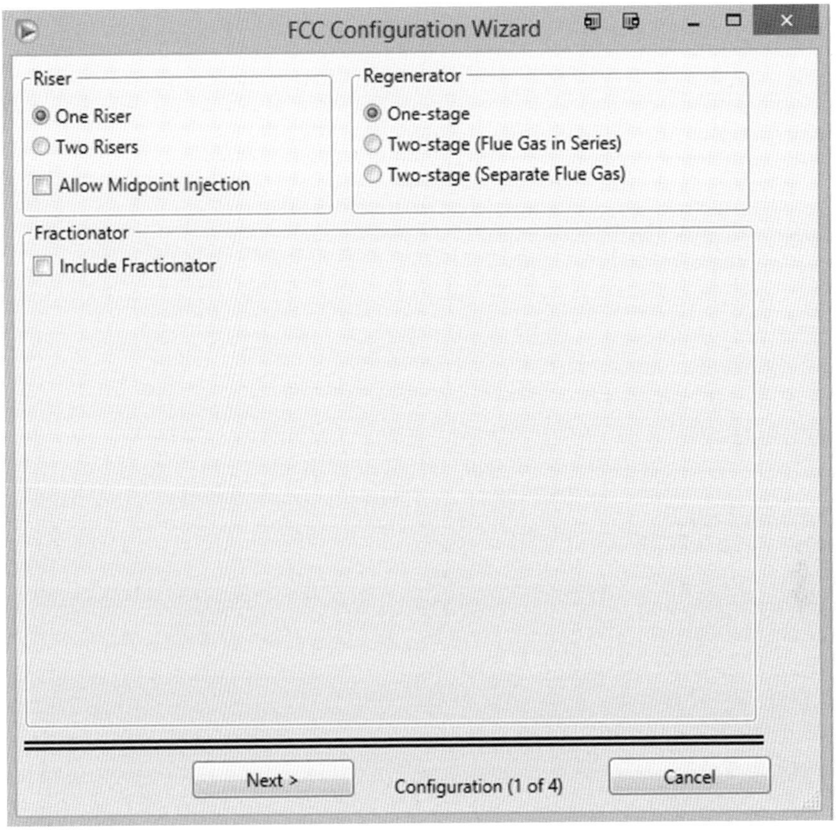

Figure 4.48 Selecting FCC configuration.

The last step is the calibration factors for this particular unit (Figure 4.51). The calibration factors refer to the tuning factors for a specific unit. These tuning factors allow us to match model results with current plant performance. As we need to adjust or calibrate these tuning factors, we choose the "Default" factors. It is possible to have several different sets of tuning factors corresponding to a variety of process, feedstock, and catalyst configurations. However, we recommend that each file should not have more than one set of tuning factors in addition to the "Default" calibration factors.

4.13.6 FCC Feed Configuration

After we complete the basic FCC configuration, we must specify the feed details. We double-click on the FCC icon on the flowsheet to bring up the FCC configuration window shown in Figure 4.52. We enter "Effluent" within Reactor Effluent. We delay entering "Feed-1" within Riser Feed-External, and "VGO" to a later step after we have imported the feed type from the Aspen HYSYS online FCC feed library and have defined the feed stream with its properties.

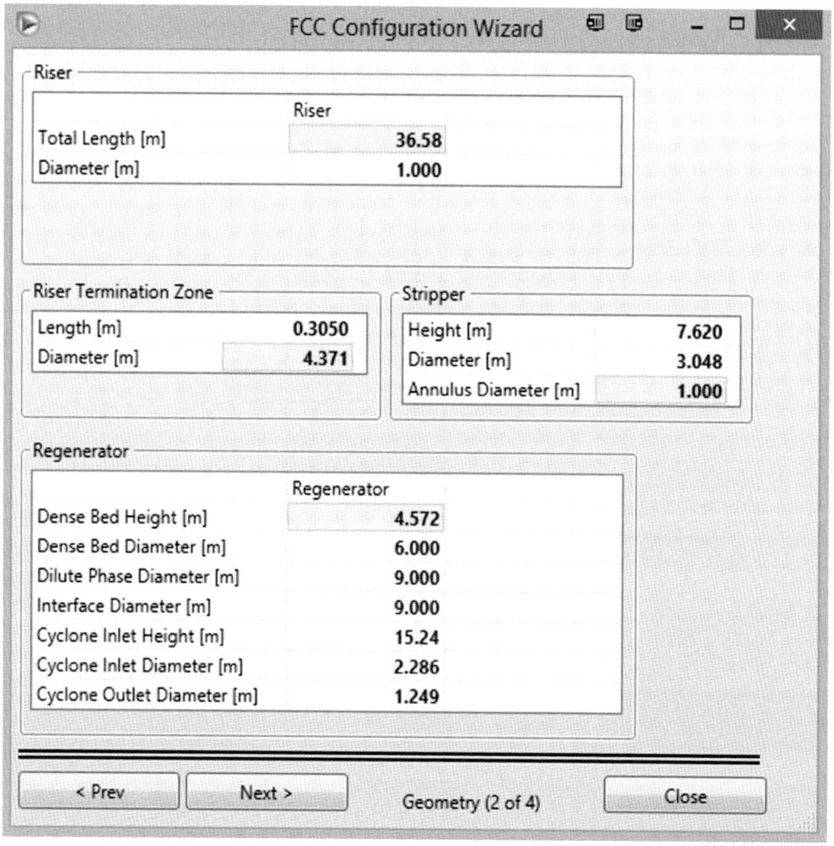

Figure 4.49 Specify the dimensions of the FCC unit.

We then click on "Feed Type Library" on the lower right corner of the screen in Figure 4.52 in order to assign a feed type for this model. A feed type refers to how Aspen HYSYS will translate the bulk property information into kinetic lumps. Aspen HYSYS supplies a variety of feed type templates for FCC feeds from a various sources such as VGO, hydrotreated vacuum gas oil (HTVGO), and so on. We click "Import" to import feed types from the feed library. The location of the feed library is shown in Figure 4.53. After choosing the VGO feed type, we delete the "Default" feed type. We also replace "Default" by "VGO" under Feed Type in the Connections screen shown in Figure 4.52.

For this model, we will only choose "fccfeed_vgo.csv." We note that it is possible to include multiple feed types in the same model. In most cases, the VGO feed type is appropriate for most FCC configurations. Even if the FCC feed is a mixture of gas oil from various sources, we recommend using the VGO feed type. If the FCC feed is a largely residue-type feed, then we recommend using the "fccfeed_resid.csv" feed type.

When we import the feed type, Aspen HYSYS shows the details of the feed type, as shown in Figure 4.54. The "Kinetic Lump Weight Percents" indicate the starting composition of the kinetic lumps and the "Methyls and Biases" indicate

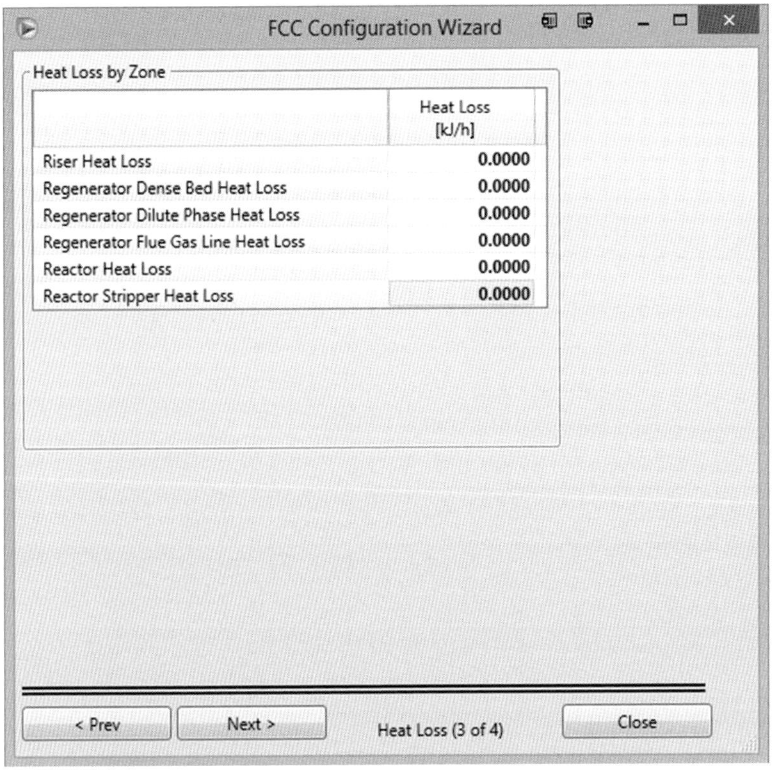

Figure 4.50 Specifying heat loss from different locations of the FCC unit.

how various bulk properties affect the final lump composition. Aspen HYSYS uses the biases to calculate actual kinetic lumps with the bias vectors. The bias vectors essentially correct the kinetic lump composition for the measured bulk properties (which we will enter) from the reference bulk properties in the feed type. We do not modify any information in this window and simply close it to continue the feed configuration process.

We go to the "Feed Data" tab and select the "Properties" section. We click on "Add" to insert "Feed-1" and begin entering the bulk properties of the feed based on the feed information given in Table 4.20 (Figure 4.55). The minimum required feed data are bulk properties (specific gravity, basic or total nitrogen, sulfur content, CCR, and metal contents) and the distillation curve of the feed. We expect that these properties are part of the routine analysis of the feed to the FCC unit. If both total and basic nitrogen are not available, we typically use a value of 3.0 for the total to basic nitrogen ratio. In addition, we typically use 0.5–0.6 for the fraction of feed sulfur processed. Residue-type feeds typically have lower amounts of the fraction of feed sulfur processed. While these values are not exact, they will suffice for our initial model. We also provide some guidelines for related feed information estimates in Table 4.24. However, it is important to provide reasonably accurate values for the metal contents of the feed. The metal contents significantly contribute to the coke production in the unit. As the riser and

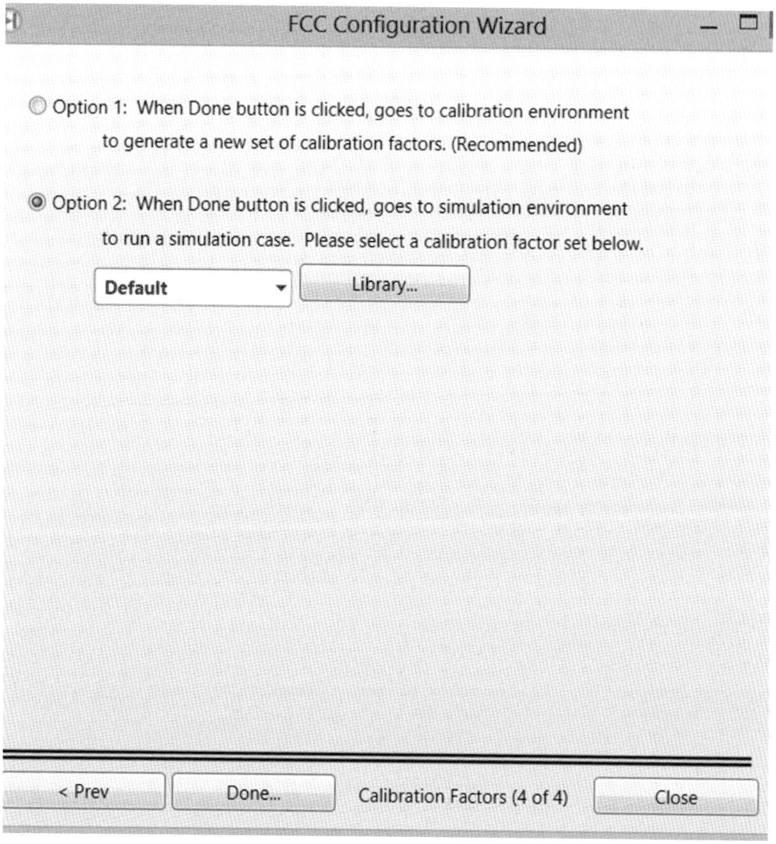

Figure 4.51 Choose option 2: develop an initial simulation model with default calibration factors.

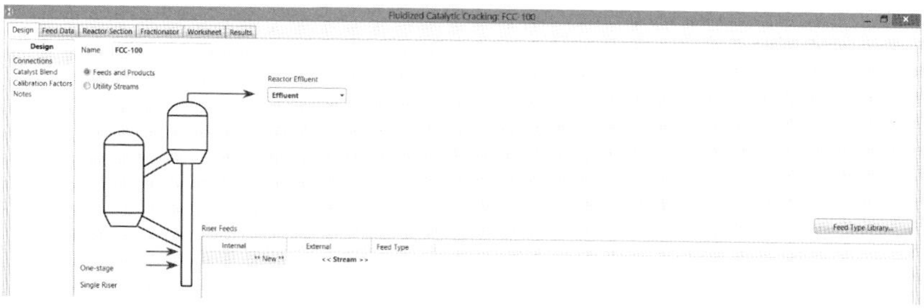

Figure 4.52 Enter the effluent stream, effluent, and delay entering feed stream and feed type to a later step.

regenerator are heat-integrated in the FCC unit, this can affect the overall yield prediction from the unit.

We now return to FCC-100 → Design → Connections to enter the Feed Type, VGO. As we are using the default calibration factors for our initial simulation before our model calibration with plant data, we ignore the recommendation

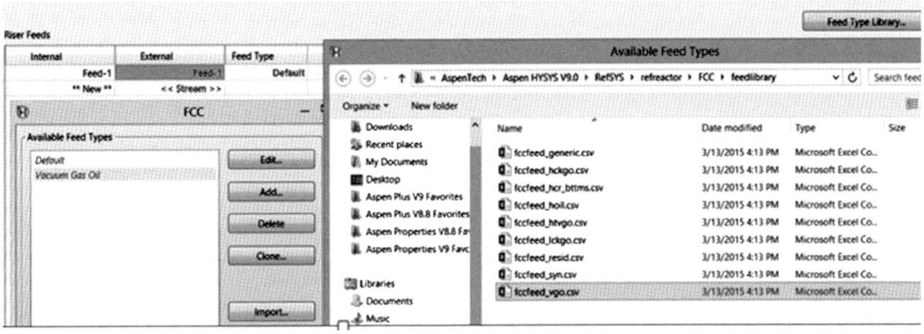

Figure 4.53 Import from the FCC feed library. C. → Program Files (×86) →AspenTech → Aspen HYSYS → RefSYS → refractor → FCC → feedlibrary → fccfeed_vgo.csv.

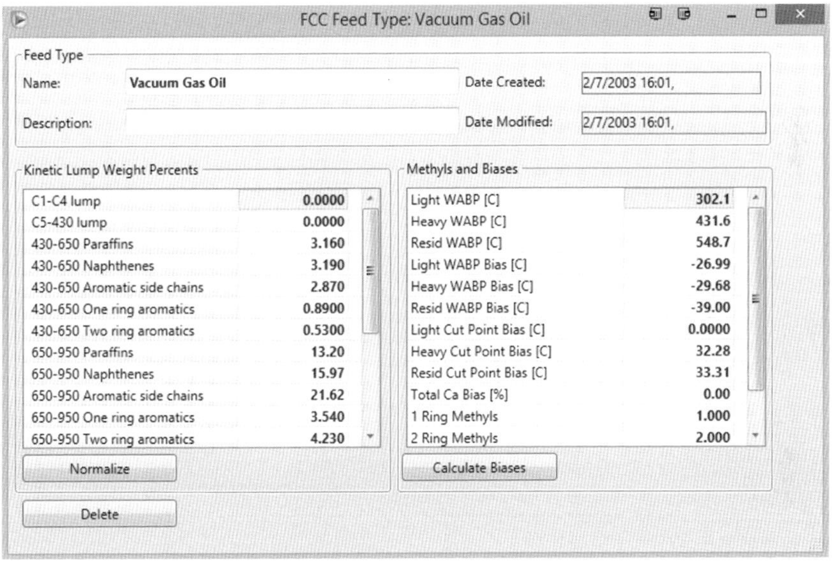

Figure 4.54 Feed type template.

of Aspen HYSYS to enter the feed stream under "New" internal steam (see Figure 4.56).

We do not enter specific utility streams in Figure 4.56 under "Utility Streams," but will enter the required stream temperature, pressure, and flow rate of essential utilities in subsequent steps.

Table 4.24 gives typical values for straight-run VGO and can serve a reality check for data collected during analysis. The nitrogen and sulfur contents can increase the rate of catalyst deactivation significantly, while the high metal contents can promote excessive production of hydrogen and light gas. We must be aware of these factors when developing the FCC model. This completes the feed configuration of the FCC unit. We may add additional feeds to the unit at this point (with the same feed type). For this simulation, we only use one feed.

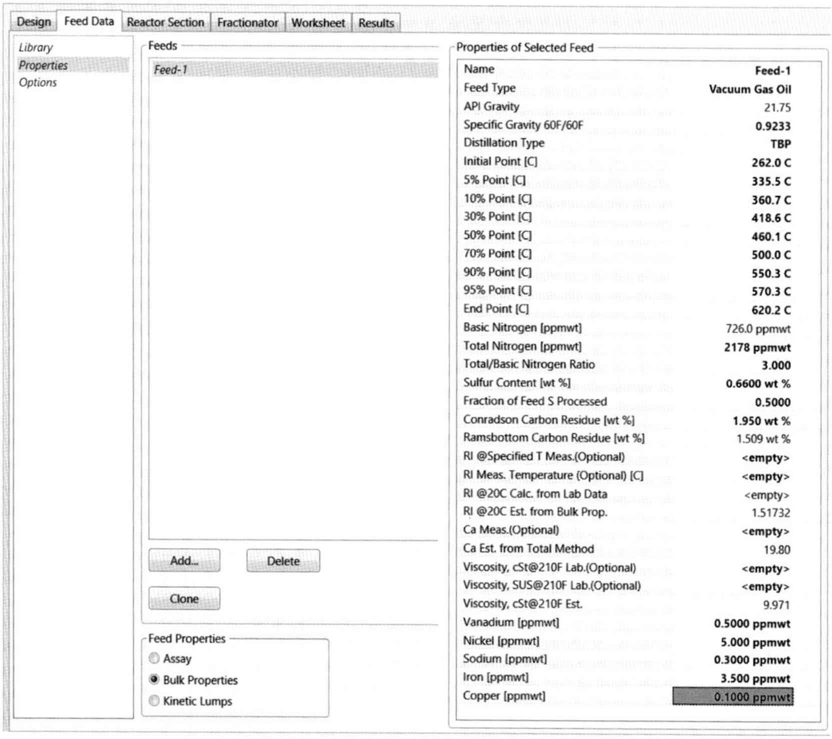

Figure 4.55 Completed feed bulk property information window: Feed Data → Properties → Add → Feed-1 → Feed Properties → Bulk Properties → Properties of Selected Feed.

Table 4.24 Typical range of properties for FCC feed.

Bulk property	Typical range or guideline
Specific gravity	0.8–1.2
Concarbon (wt%)	1–3
Basic nitrogen (ppm wt)	500–1000
Total/basic nitrogen ratio	3.0
Sulfur content (ppm wt)	<2
Fraction of feed sulfur processed	0.5–0.6
Total aromatic content (wt%)	20–30 (for straight-run VGO)
Nickel and iron content (ppm wt)	10–100x (vanadium + sodium + copper)

4.13.7 FCC Catalyst Configuration

The next step in building the model is selecting the catalyst blend in the unit. We select the "Catalyst Blend" tab in the Design Section Window, as shown in Figure 4.57. The process for importing a catalyst blend is similar to that for importing feed types. We click on the "Catalyst Library" button to bring up the import window for the catalysts.

Figure 4.58 shows the location of the catalyst library and lists the available catalyst types. A catalyst type essentially contains the tuning or calibration

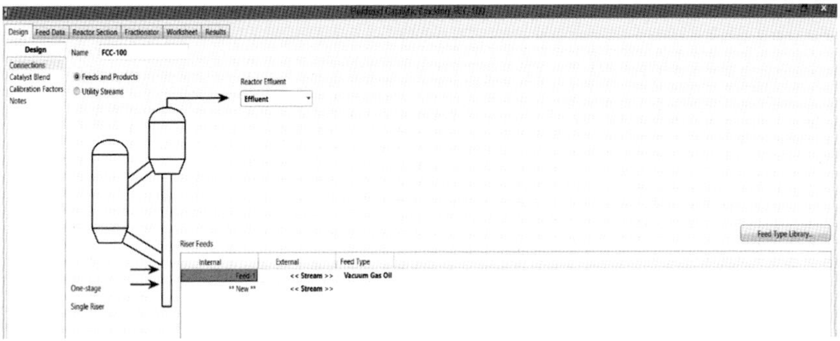

Figure 4.56 Specification of the feed type, Vacuum Gas Oil. We previously introduced the internal riser feed, Feed-1, in Figure 4.55, and do not enter the external riser feed.

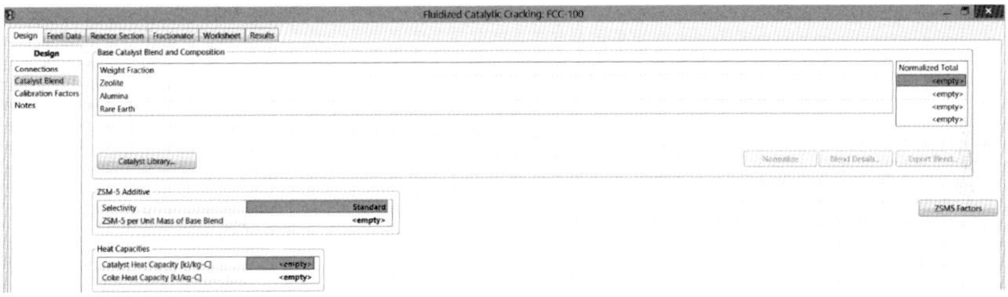

Figure 4.57 Initial catalyst blend window.

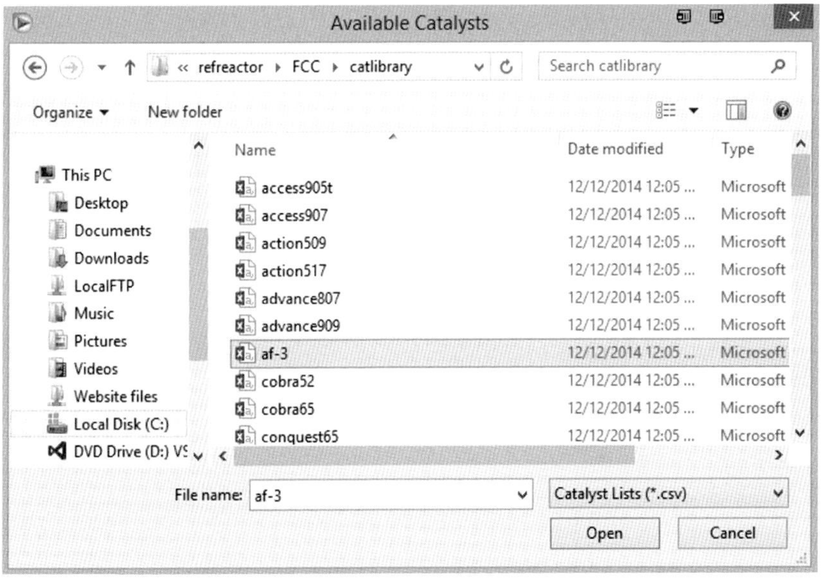

Figure 4.58 Catalyst library. C. → Program Files →AspenTech → Aspen HYSYS → RefSYS → refractor → FCC → catlibrary → af-3.csv.

factors responsible for light gas distribution, small adjustments to product bulk properties (RON, MON, etc.), and distribution of coke produced by the metal function of the catalyst. The catalyst library contains catalyst data from a variety of manufacturers and sources. If the exact catalyst is not available, we recommend using a similar match. It is possible to tune away variations in the tuning factors due to catalyst type, but this may produce an overcalibrated model with unrealistic yield predictions. For this model, we use the Akzo A/F-3 catalyst and choose "af-3.csv" from the catalyst library.

Once we choose a catalyst, Aspen HYSYS will display a summary of the key features of the catalyst (see Figure 4.59). We can use this list to compare with the true product specifications from the catalyst manufacturer. If the catalyst is not acceptable, we can click "Delete" to remove the catalyst and try another entry from the catalyst library. As we mentioned in the previous paragraph, it is not critical to find an exact match. Once we have added all catalysts we require, we can close the catalyst information window and return to the "FCC Reactor Section."

Next, we must specify the catalyst blend. The catalyst blend refers to two or more different kinds of catalysts from the catalyst library. We can assign individual weight fractions for each of the catalysts in the blend. In our model, we are using only one type of catalyst, so we set the weight fraction to 1.0, as shown in Figure 4.60. We use the default values for the heat capacities of the catalyst and coke. These values are generally not measured; however, we expect only small deviations from the default value in the actual FCC unit.

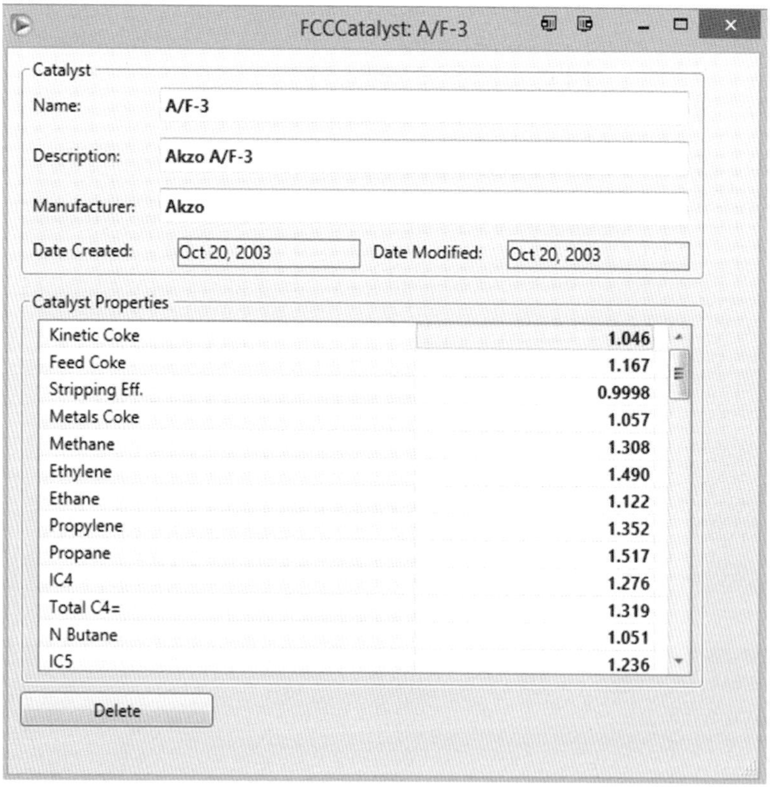

Figure 4.59 Catalyst parameters.

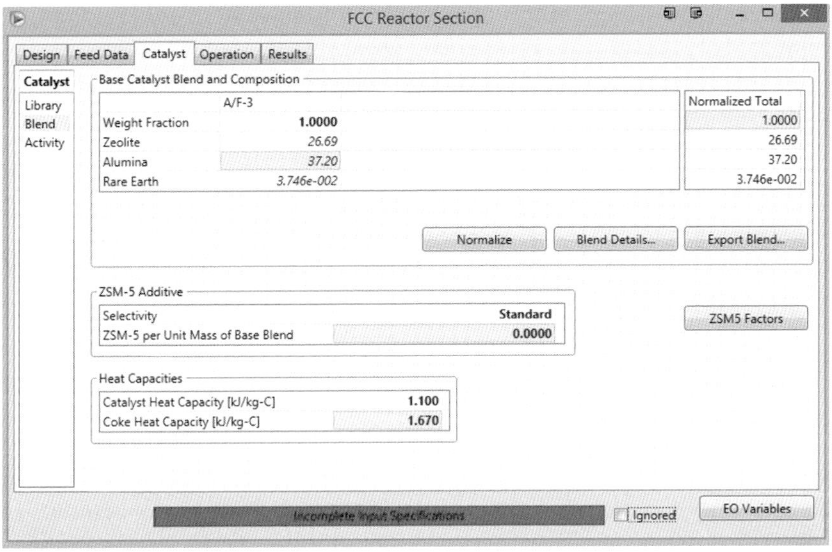

Figure 4.60 Catalyst parameters.

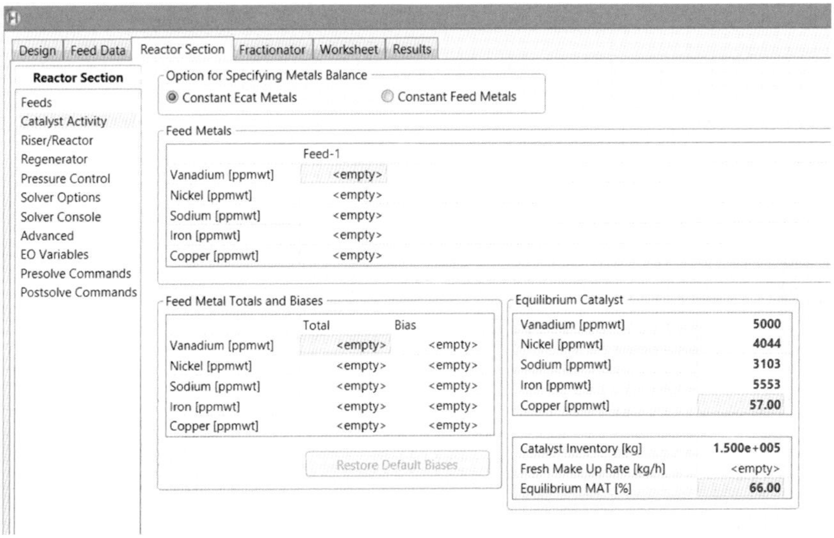

Figure 4.61 Catalyst activity factor and equilibrium metal contents.

We must also specify if any ZSM-5 additive is present in the catalyst. The "ZSM-5 per Unit Mass" variable acts as another tuning factor to adjust model yields of the unit. We may use an average value or set the ZSM-5 content to 0 if the information is not available. As we will tune the unit to an actual product distribution, it is not essential that this value is exactly the same as the actual unit.

The last step in catalyst configuration is to specify the "Activity" section of the "Catalyst" tab in the FCC Reaction Section Window, as shown in Figure 4.61. The activity of the catalyst essentially refers to the effect of metals on catalyst deactivation. We can either maintain a constant level of metals on the catalyst or keep adjusting the feed metal contents to match makeup rates, and equilibrium catalyst activity (Ecat). We recommend using "Ecat Metals" option, as the information required is available from routine equilibrium catalyst analysis of the FCC catalyst.

We will specify the metal contents of the equilibrium catalyst and equilibrium microactivity test (MAT) value. When we use this option, Aspen HYSYS will automatically calculate the makeup of catalyst required to maintain the equilibrium MAT and keep the metal contents on the catalyst fixed. The total catalyst inventory refers to the total amount of catalyst available to the FCC unit. We can now specify the operating variables for the FCC unit model.

4.13.8 FCC Operating Variable Configuration

Before we specify the operating variables of the FCC unit, we use the main application toolbar to hold the solver. Holding the solver ensures that the solver will not immediately solve, once we specify all variables for the FCC unit. It is generally *a good idea to hold the solver before changing many operating variables* as we do in the following sections. We hold the solver by clicking on the red stop sign in the main application toolbar. We can release the solver by clicking on the green go sign in the toolbar.

We specify the feed rate, temperature, and pressure into the preheater before the feed enters the riser, as shown in Figure 4.62. If we have multiple injection points, we can specify the feed into the injection points as well. To specify the actual temperature of the feed entering the riser, we must either set the preheat duty or a preheat temperature. As we have a single feed, we set the preheat outlet temperature to plant value. We must also specify the steam flow and conditions

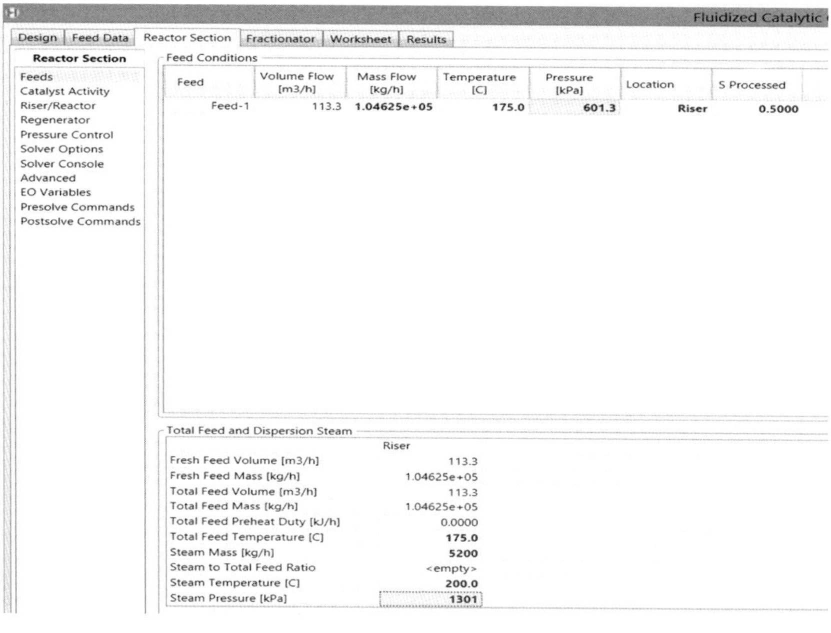

Figure 4.62 Specify feed conditions.

associated with the feed into the riser inlet. Typical values for dispersion steam are 1–5 wt% of the fresh feed rate.

The next step is to specify the operating variables for the riser and reactor, as shown in Figure 4.63. In most FCC units, control strategies generally fix the riser outlet temperature (ROT) as a set point, so the ROT is a natural specification for the riser. It is also possible to specify the Cat/Oil ratio or circulation rate, but these specifications make the model quite difficult to converge. We recommend using the ROT as an initial specification and then shifting to other possible specifications.

We also specify the flow rate and conditions of the Lift Gas and Reactor Stripping Zone shown in Figure 4.63. The lift gas is typically an inert in the cracking process and the steam for the reactor stripping zone minimizes thermal cracking due to high temperatures. We must at least supply the stripping steam rate to ensure that the model converges to a reasonable solution. The stripping steam rate is roughly about 1–5 wt% of the fresh feed. The next step is to specify the regenerator operating variables.

In Figure 4.64, we specify regenerator operating variables. The key variables are the dense bed temperature, flue gas oxygen (O_2) composition, and catalyst inventory. The flue gas composition and dense bed temperature fix the airflow rate and coke combustion rate for the regenerator. Some FCC units include side coolers and enriched oxygen streams to completely combust the coke on the catalyst. We may specify these as well; however, they are not common with mostly straight-run VGO type feeds. We enter nominal values for the ambient air conditions and blower discharge temperature. In the typical range for these variables, there is little effect on process performance.

We show the last step in configuring operating variables in Figure 4.65. All refiners continuously measure the reactor and regenerator pressure to ensure

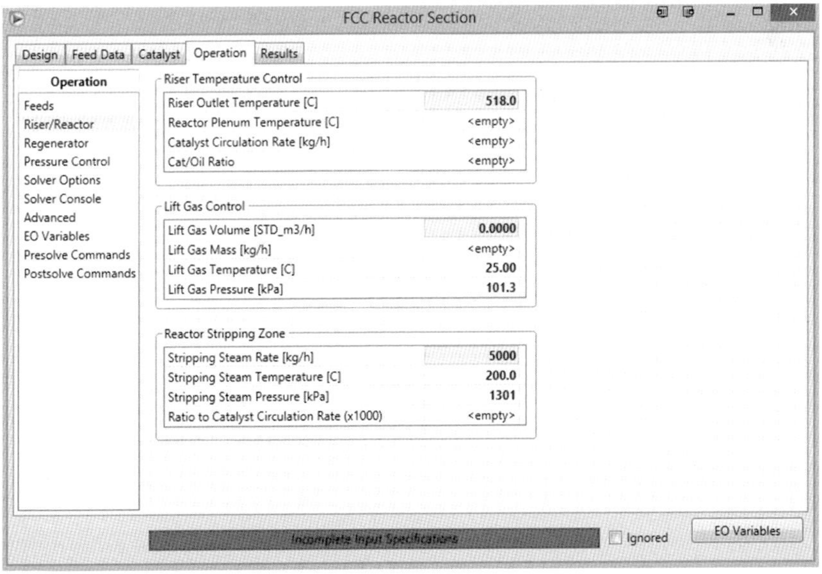

Figure 4.63 Riser conditions and steam input.

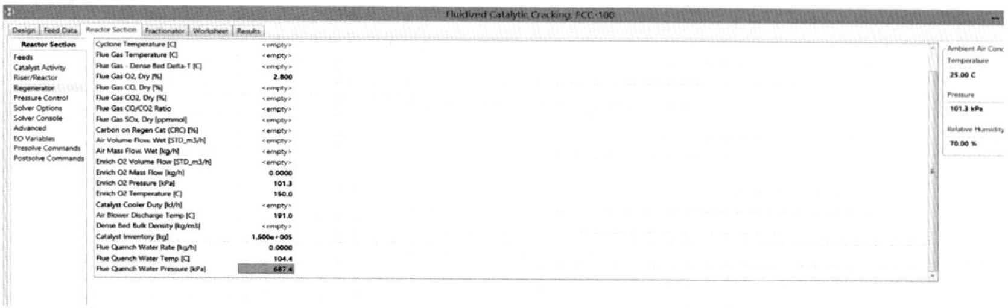

Figure 4.64 Regenerator operating parameters. Note that a required dense bed temperature (°C) of 608 is not visible in this image.

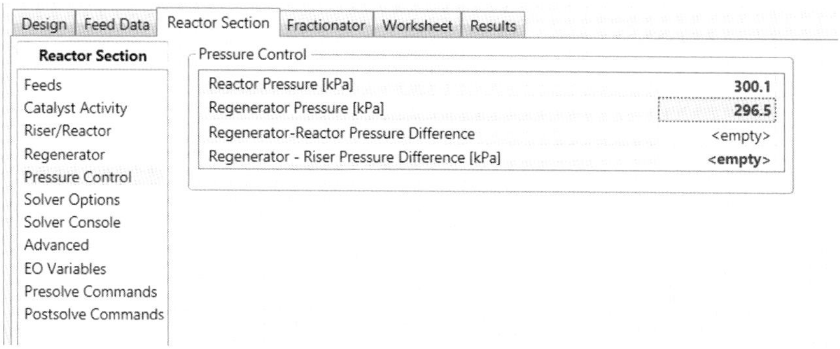

Figure 4.65 Pressure control (reactor pressure should be greater than regenerator pressure).

that catalyst is flowing through the unit. Accurate values here will aid in better predictions of catalyst circulation rate through the riser and the catalyst-to-oil ratio. We also note that once we enter the pressure measurements given in Figure 4.65, Aspen HYSYS will indicate that we are ready to solve the model.

4.13.9 Initial Model Solution

Before solving the model, we must ensure that the solver parameters will lead to robust convergence. We bring up the solver options by selecting the "Solver Options" section in "Operation" tab. Figure 4.66 shows the recommend values for the solver options. We have chosen these values based on our experience with running with model.

In general, we do not recommend modifying the constraints for the residual, Hessian parameters, and line search parameters. When running the model for the first time, we increase the number of creep iterations and maximum iterations. Creep iterations refer to initial small changes in the process variables when the starting guesses are very poor (the Jacobian cannot indicate a direction that will decrease the residual). The maximum iterations refer to how many times the solver will iterate through the model before exiting. Depending on process parameters, the initial solution may take up to 30–40 iterations.

We activate the solver by clicking on the green go button in the main application bar. The solver output appears in the lower right-hand corner of the PFD window. The solver output for the configured model is shown in Table 4.25. Column 1 of

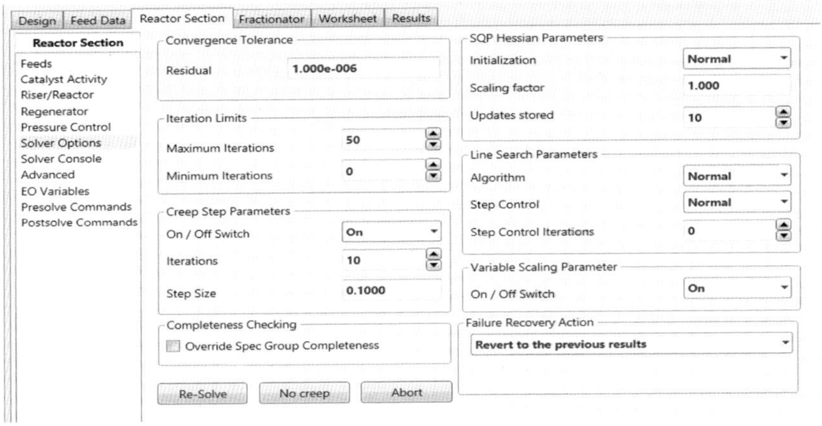

Figure 4.66 Solver convergence options.

the table indicates the number of iterations performed since starting the solver. The residual convergence function indicates how far we are from satisfying the process model equations. When we run the model for the first time, residuals on the order of 1E7 are expected. As we approach the solution, the residual drops to closer and closer to zero. Columns 3 and 4 refer to the residual of the objective function. We use the objective function only during calibration; therefore, it is zero for this model run. The solver used by Aspen HYSYS converges very quickly to solution once the changes in the process equations start appearing to be linear. This is the case when we are in the vicinity of the solution. The solver indicates the vicinity of the solution through columns 5 and 6. The Worst model column indicates which part of the reformer model is furthest from the solution. This is useful for tracking down issues when the model fails to converge. The last lines of the output show several running statistics for the solver.

In general, the FCC model should converge with 20 s on recent computer hardware. If solution requires more than 20 s, it is likely that some conflicts exist in the specifications.

4.13.10 Viewing Model Results

Figure 4.67 shows the converged FCC unit operation window after Aspen HYSYS has successfully solved the model. We connect an effluent stream by bringing up the "Connections" section of the Design tab and typing in "Effluent" for the Reactor Effluent stream. A stream titled "Effluent" will appear on the PFD and we can use this stream to build further downstream fractionation units.

The "Results" tab in Figure 4.68 summarizes various model results in different categories. The Feed Blend tab in Figure 4.69 shows the bulk property information and kinetic lumping for each feed entering the riser. An important check is the sum of the adjusted aromatic core compositions. In Figure 4.69, the sum of the highlighted one-ring, two-ring, and three+ ring aromatic cores is 22.60 wt%. This value should be close to the value of 16.93% of "Ca. Est. from Total Method"

Table 4.25 Initial solver output.

Iteration	Residual Convergence Function	Objective Convergence Function	Objective Function Value	Overall Nonlinearity Ratio	Model Nonlinearity Ratio	Worst Model
0	1.641D+02	0.000D+00	0.000D+00	9.991D–01	9.973D–01	
	<Line Search Creep Mode ACTIVE>		==>	Step taken 1.00D–01		
1	1.314D+02	0.000D+00	0.000D+00	9.761D–01	–9.852D+00	
	<Line Search Creep Mode ACTIVE>		==>	Step taken 1.00D–01		
2	1.059D+02	0.000D+00	0.000D+00	9.788D–01	–4.397D+00	
	<Line Search Creep Mode ACTIVE>		==>	Step taken 1.00D–01		
3	8.563D+01	0.000D+00	0.000D+00	9.811D–01	–2.454D+00	
	<Line Search Creep Mode ACTIVE>		==>	Step taken 1.00D–01		
4	6.950D+01	0.000D+00	0.000D+00	9.831D–01	–1.460D+00	
	<Line Search Creep Mode ACTIVE>		==>	Step taken 1.00D–01		
5	5.654D+01	0.000D+00	0.000D+00	9.849D–01	–8.596D–01	
	<Line Search Creep Mode ACTIVE>		==>	Step taken 1.00D–01		
6	4.608D+01	0.000D+00	0.000D+00	9.865D–01	–4.611D–01	
	<Line Search Creep Mode ACTIVE>		==>	Step taken 1.00D–01		
7	3.760D+01	0.000D+00	0.000D+00	9.879D–01	–1.778D–01	
	<Line Search Creep Mode ACTIVE>		==>	Step taken 1.00D–01		
8	3.070D+01	0.000D+00	0.000D+00	9.891D–01	3.205D–02	
	<Line Search Creep Mode ACTIVE>		==>	Step taken 1.00D–01		
9	2.508D+01	0.000D+00	0.000D+00	9.902D–01	1.928D–01	
10	2.049D+01	0.000D+00	0.000D+00	8.772D–01	–1.091D+01	
11	1.151D–01	0.000D+00	0.000D+00	9.974D–01	9.853D–01	
12	2.523D–06	0.000D+00	0.000D+00	1.000D+00	1.000D+0	
13	1.325D–14	0.000D+00	0.000D+00			
Successful solution.						
Optimization Timing Statistics			Time	Percent		
MODEL computations			0.88 secs	49.67 %		
DMO computations			0.66 secs	37.31 %		
Miscellaneous			0.23 secs	13.02 %		
Total Optimization Time			1.77 secs	100.00 %		
Problem converged						

heighted in Figure 4.68 that represents the aromatic content of feed. If these values differ significantly (>10 wt%), we may have chosen a feed type that does not represent the actual feed to the unit accurately.

We can view the overall product yields in "Product Yields" section. The yields shown in Figure 4.70 are the so-called *standard cut grouped yields*, or *square cut yields*. These yields refer to product yields with fixed end points; typical cuts include *C1-C4 lump*, *C5-430 lump* (C5-430°F or 221°C), *430–650 lump* (430–650 °F, or 221–343 °C), *650–950 lump* (650–950 °F, or 343–510 °C), and

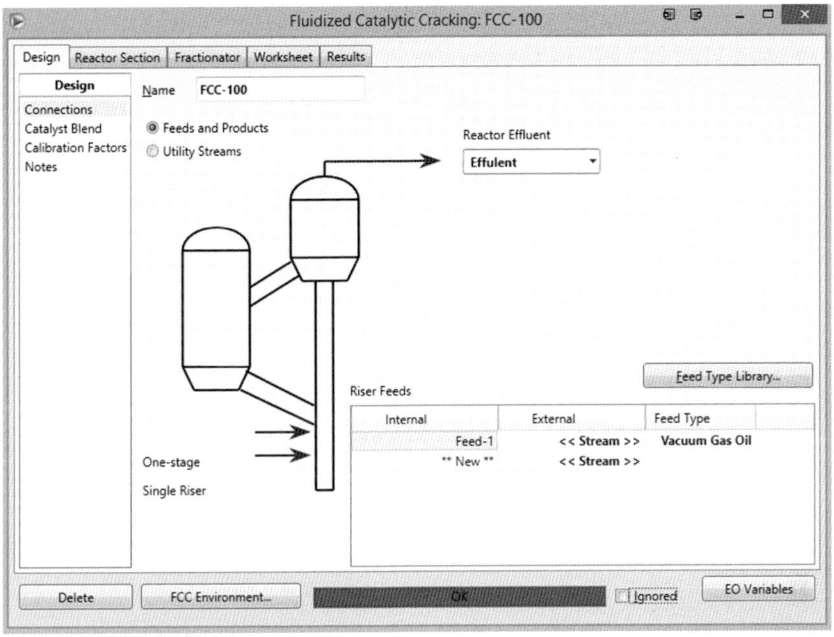

Figure 4.67 Add effluent stream to PFD.

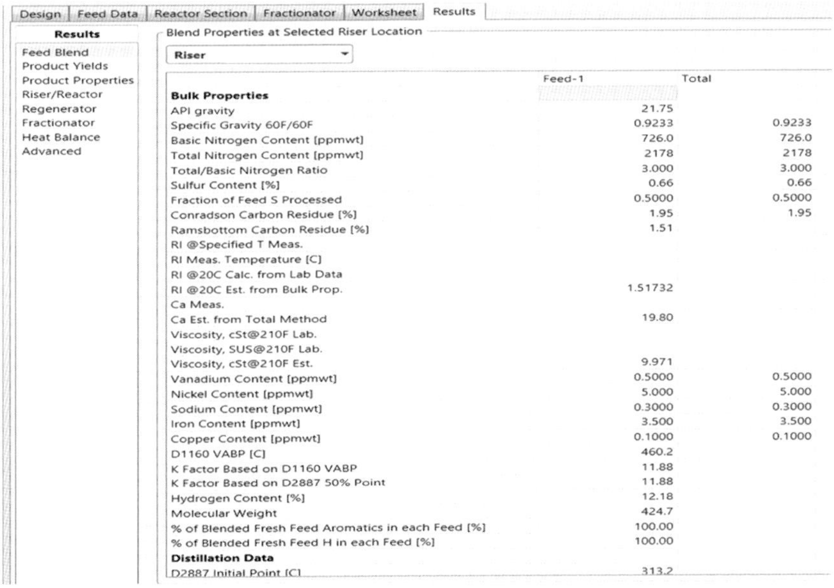

Figure 4.68 Navigate FCC results.

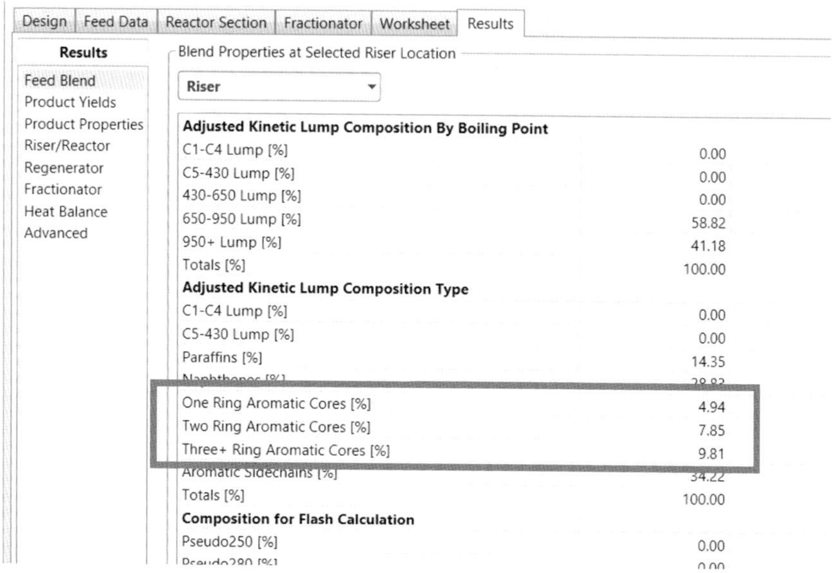

Figure 4.69 Adjusted kinetic lumps. The sum of highlighted one-ring, two-ring, and three+ ring aromatic cores (not side chains) is 22.60%.

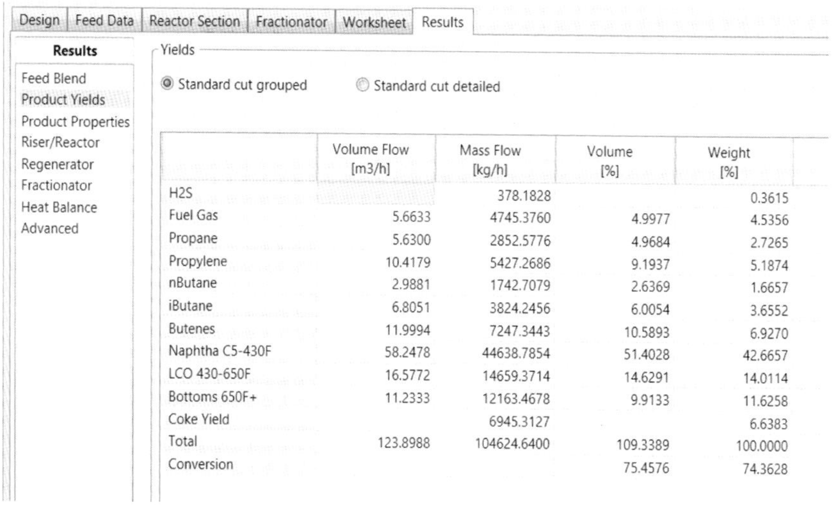

Figure 4.70 "Standard cut grouped" product yields.

950+ lump (>950 °F or 510 °C). These cut points may be different from those being used in the plant. The end point of the plant naphtha cut is generally lower; therefore, the square cut yield is often much higher than the plant yield. We will produce a true plant cut using rigorous fractionation in Workshop 4.3, Section 4.15.

Figure 4.71 shows the "Product Properties" of each square cut from the model. As the square cut yields do not reflect the plant yields directly, model results for

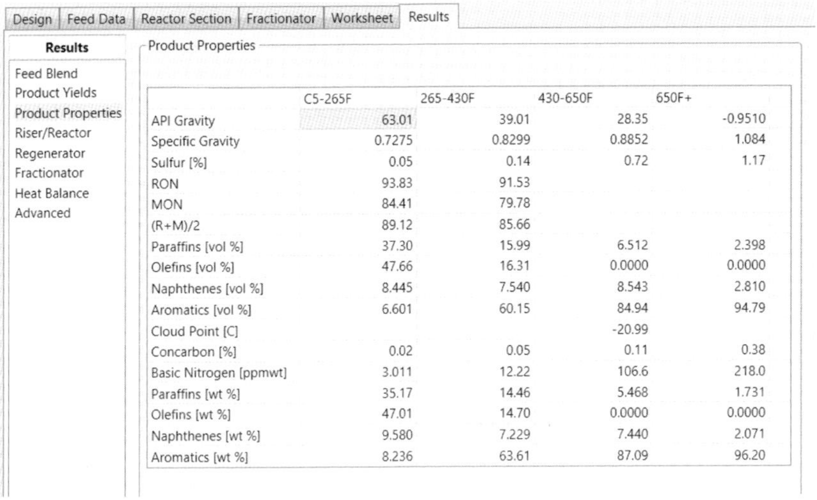

	C5-265F	265-430F	430-650F	650F+
API Gravity	63.01	39.01	28.35	-0.9510
Specific Gravity	0.7275	0.8299	0.8852	1.084
Sulfur [%]	0.05	0.14	0.72	1.17
RON	93.83	91.53		
MON	84.41	79.78		
(R+M)/2	89.12	85.66		
Paraffins [vol %]	37.30	15.99	6.512	2.398
Olefins [vol %]	47.66	16.31	0.0000	0.0000
Naphthenes [vol %]	8.445	7.540	8.543	2.810
Aromatics [vol %]	6.601	60.15	84.94	94.79
Cloud Point [C]			-20.99	
Concarbon [%]	0.02	0.05	0.11	0.38
Basic Nitrogen [ppmwt]	3.011	12.22	106.6	218.0
Paraffins [wt %]	35.17	14.46	5.468	1.731
Olefins [wt %]	47.01	14.70	0.0000	0.0000
Naphthenes [wt %]	9.580	7.229	7.440	2.071
Aromatics [wt %]	8.236	63.61	87.09	96.20

Figure 4.71 Properties of square cut products.

each property may not exactly match plant values. We need rigorous fractionation to compare model results with plant measurements. In addition, we will likely improve the agreement of product properties when we calibrate the model in the next workshop.

The last set of significant results is the "Heat Balance" section in Figure 4.72. The heat balance shows the overall coke yield and delta coke for the process. *Delta coke* is simply the difference between the coke on spent catalyst (CSC) at the stripper outlet and the coke on regenerated catalyst (CRC), expressed as a weight percent of catalyst. According to Figure 4.72, the difference between 0.65% and 0.04% is the delta coke of 0.61%.

In addition, the model calculates the catalyst-to-oil ratio (C/O) and catalyst circulation rate. A "coke balance" around the regenerator results in a useful expression for the delta coke [61]. The coke yield must be equal to the difference in the coke entering and leaving the regenerator. Therefore, we write

$$\text{Coke yield} = (\text{C/O})[\text{CSC} - \text{CRC}] \tag{4.15}$$

or

$$\text{CSC} - \text{CRC} = (\text{coke yield})/(\text{C/O}) = \text{delta coke} \tag{4.16}$$

In the equations, "coke yield" is wt% of feed; C/O is catalyst circulation, lb of catalyst, per lb of feed; CSC is the coke on spent catalyst, wt% of catalyst; CRC is the coke on regenerated catalyst, wt% of catalyst. Note that the coke yield and C/O must be expressed in the same feed basis, fresh or fresh plus recycle, to result in a meaningful number [61].

Aspen HYSYS uses the delta coke, catalyst circulation rate, and kinetic lumps to calculate an Apparent Heat of Cracking. This value represents the combined heat release from all the cracking reactions. In addition, we can calculate a theoretical heat of cracking with overall mass and heat balance constraints alone.

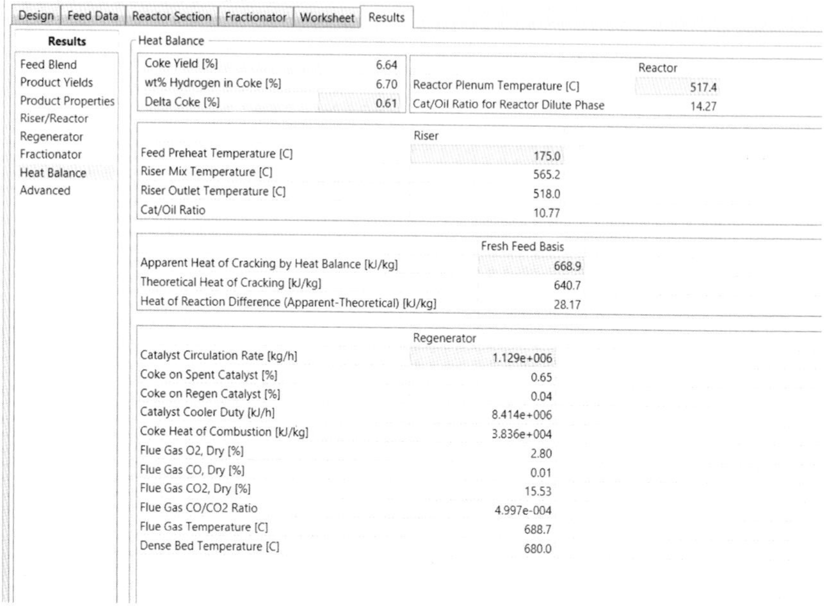

Figure 4.72 Overall heat balance between riser and regenerator.

In most cases, the apparent and theoretical heats of cracking should be quite similar (<15% relative error). In Figure 4.72, the relative error is less than 3%. The 4% error between the theoretical (640 kJ/kg) and apparent (668.9 kJ/kg) heats of cracking indicates that the kinetic model satisfies the thermodynamic constraints within the error tolerance of the plant measurements. Additionally, based on Figure 4.72, we find that the simulation results are within 5% error of plant data when applying Eq. (4.16). We save the completed simulation file as *Workshop 4.1-done.hsc*.

Once we verify that the model is making reasonable initial predictions, we can proceed to the calibration phase. In the calibration phase, we will adjust the tuning factors that come from the choice of feed and catalyst types.

4.14 Workshop 4.2 – Calibrating Basic FCC Model

In this section, we calibrate the model based on known product yields and reactor performance. Calibration involves four distinct steps:

1) Pulling data from current simulation.
2) Enter measured process yields and performance based on that current simulation.
3) Update the activity factors to match these plant yields and performance results.
4) Push the calibration data back to the simulation.

We begin the first step of model calibration procedure using our converged initial model, *Workshop 4.1-done.hsc*, and save it under a new name, *Workshop*

4.2-1.hsc. The converged initial model provides the initial guesses for the activity factors, which greatly simplify the model calibration procedure. We enter the model calibration environment by first entering the FCC sub-flowsheet and then selecting the "FCC > Calibration" menu option from the application menu bar (as shown in Figure 4.73). Figure 4.74 shows the FCC calibration environment.

The first step is to "Pull data" from the simulation. When Aspen HYSYS pulls data, current operating conditions, feed stock information, and process

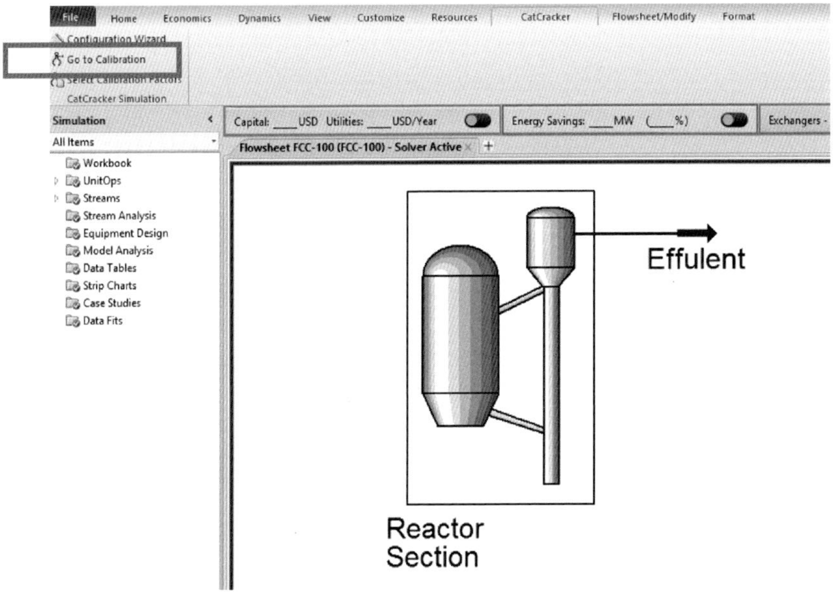

Figure 4.73 Entering FCC calibration environment.

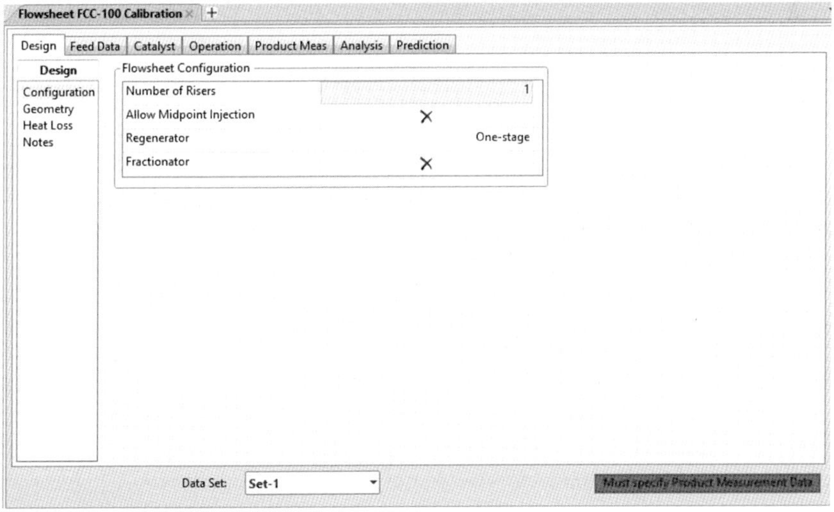

Figure 4.74 FCC calibration window.

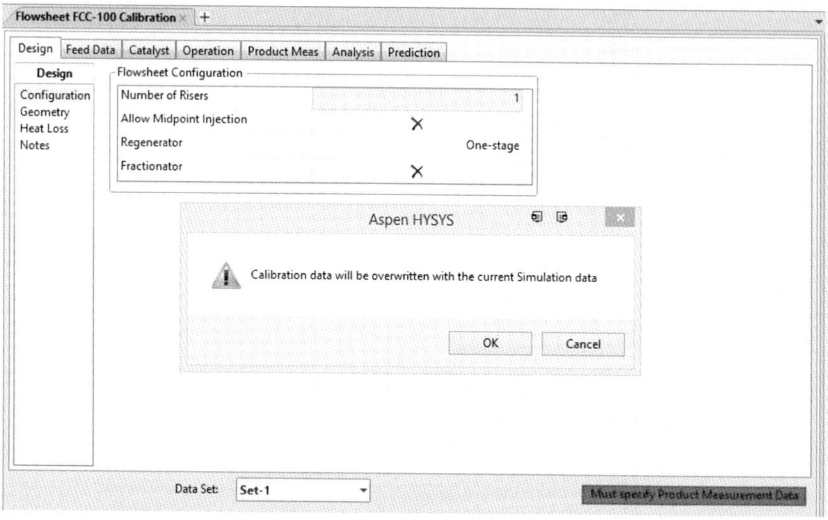

Figure 4.75 Pull current simulation data into calibration environment.

conditions entered in the FCC environment. A calibration refers to the set of the activity factors that produce a given set of product yields and reactor performance (which we provide to the calibration environment) based on current model state. We pull data by clicking on the "Pull Data from Simulation" button (Figure 4.75).

When we pull data from the simulation, Aspen HYSYS will warn us that current calibration data will be overwritten by the current model results, as shown in Figure 4.75. We can use the "Manage Data Sets" feature to allow multiple calibration data sets. This may be useful if the industrial FCC unit runs under very different operating scenarios. However, for this workshop, we use only one calibration data set. Aspen HYSYS will pull all the feedstock information and process operating conditions after we confirm the calibration data overwrite. The status bar now indicates that we must specify product measurements to begin the calibration process. If necessary, we can modify the operating variables (such as ROT, etc.) of the FCC unit, in addition to the measured values. However, we recommend *creating a new model file* if the operating scenarios are very different.

The second step in model calibration is specifying the measured yields and process performance. Click on the "Prod Meas." tab to bring up the Cuts interface (see Figure 4.76). In the Cuts interface, we can specify how many plant cuts of light gases, LPG, naphtha, LCO or diesel, and bottoms that this particular FCC unit has. FCC units typically have two light gas cuts: the dry gas (C1–C2) and the output from the desulfurization unit (H_2S). The LPG (C3–C4) stream typically leaves from the gasoline stabilizer. The remaining liquid cuts leave from the main fractionator unit. Depending on the type of the FCC unit, there may be two naphtha cuts (light and heavy) and two cycle oil cuts (LCO and HCO).

Once we select the number of cuts, we must enter the data from the light ends and the heavy liquids, as shown in Figure 4.77. If the plant draws multiple light gas streams, we recommend using the same number of streams. Aspen HYSYS will

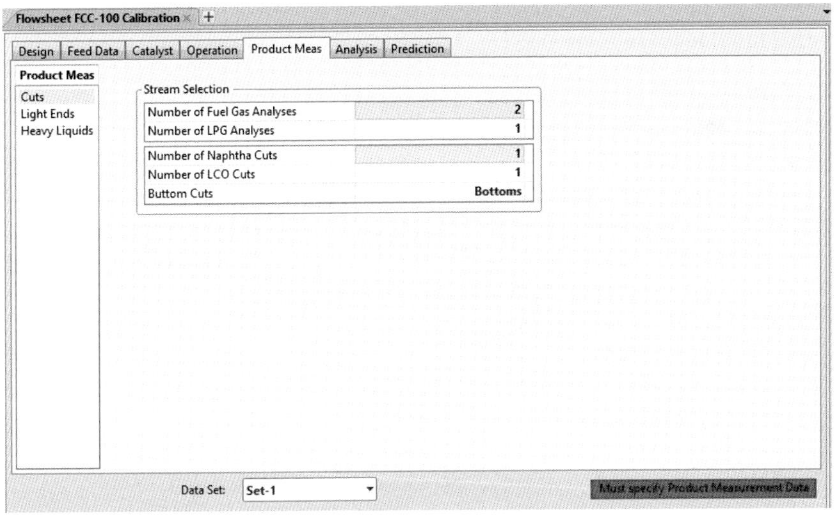

Figure 4.76 Specify cuts for plant measurement data.

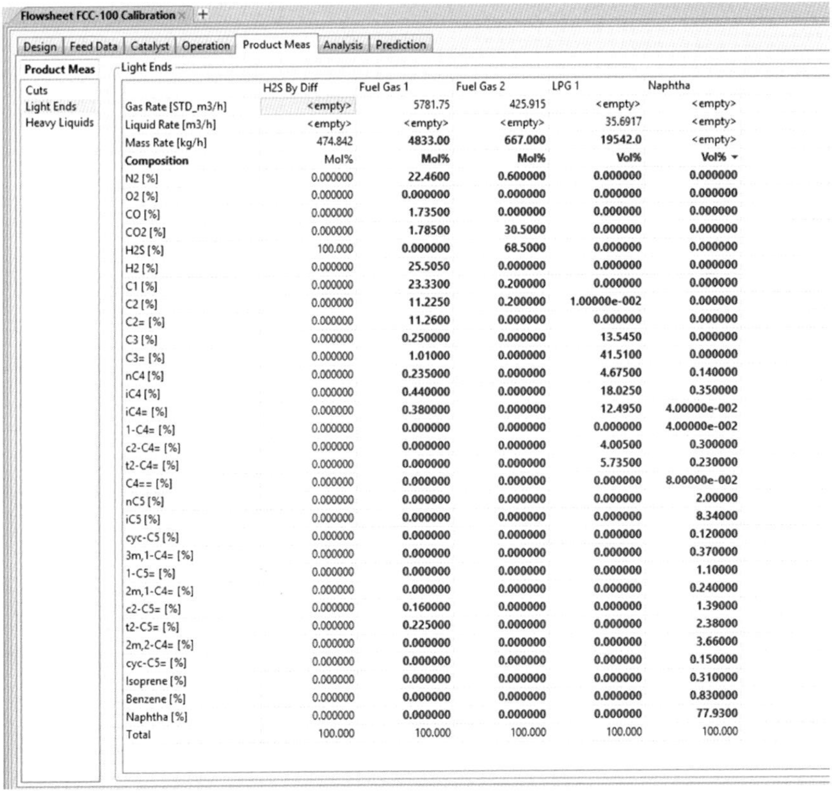

	H2S By Diff	Fuel Gas 1	Fuel Gas 2	LPG 1	Naphtha
Gas Rate [STD_m3/h]	<empty>	5781.75	425.915	<empty>	<empty>
Liquid Rate [m3/h]	<empty>	<empty>	<empty>	35.6917	<empty>
Mass Rate [kg/h]	474.842	4833.00	667.000	19542.0	<empty>
Composition	Mol%	Mol%	Mol%	Vol%	Vol%
N2 [%]	0.000000	22.4600	0.600000	0.000000	0.000000
O2 [%]	0.000000	0.000000	0.000000	0.000000	0.000000
CO [%]	0.000000	1.73500	0.000000	0.000000	0.000000
CO2 [%]	0.000000	1.78500	30.5000	0.000000	0.000000
H2S [%]	100.000	0.000000	68.5000	0.000000	0.000000
H2 [%]	0.000000	25.5050	0.000000	0.000000	0.000000
C1 [%]	0.000000	23.3300	0.200000	0.000000	0.000000
C2 [%]	0.000000	11.2250	0.200000	1.00000e-002	0.000000
C2= [%]	0.000000	11.2600	0.000000	0.000000	0.000000
C3 [%]	0.000000	0.250000	0.000000	13.5450	0.000000
C3= [%]	0.000000	1.01000	0.000000	41.5100	0.000000
nC4 [%]	0.000000	0.235000	0.000000	4.67500	0.140000
iC4 [%]	0.000000	0.440000	0.000000	18.0250	0.350000
iC4= [%]	0.000000	0.380000	0.000000	12.4950	4.00000e-002
1-C4= [%]	0.000000	0.000000	0.000000	0.000000	4.00000e-002
c2-C4= [%]	0.000000	0.000000	0.000000	4.00500	0.300000
t2-C4= [%]	0.000000	0.000000	0.000000	5.73500	0.230000
C4== [%]	0.000000	0.000000	0.000000	0.000000	8.00000e-002
nC5 [%]	0.000000	0.000000	0.000000	0.000000	2.00000
iC5 [%]	0.000000	0.000000	0.000000	0.000000	8.34000
cyc-C5 [%]	0.000000	0.000000	0.000000	0.000000	0.120000
3m,1-C4= [%]	0.000000	0.000000	0.000000	0.000000	0.370000
1-C5= [%]	0.000000	0.000000	0.000000	0.000000	1.10000
2m,1-C4= [%]	0.000000	0.000000	0.000000	0.000000	0.240000
c2-C5= [%]	0.000000	0.160000	0.000000	0.000000	1.39000
t2-C5= [%]	0.000000	0.225000	0.000000	0.000000	2.38000
2m,2-C4= [%]	0.000000	0.000000	0.000000	0.000000	3.66000
cyc-C5= [%]	0.000000	0.000000	0.000000	0.000000	0.150000
Isoprene [%]	0.000000	0.000000	0.000000	0.000000	0.310000
Benzene [%]	0.000000	0.000000	0.000000	0.000000	0.830000
Naphtha [%]	0.000000	0.000000	0.000000	0.000000	77.9300
Total	100.000	100.000	100.000	100.000	100.000

Figure 4.77 Measured light gas yields and compositions based on the product measurements in Table 4.21.

automatically combine the light end analyses to reconstruct the reactor effluent. Following the plant data in Table 4.21, Section 4.13.3, we enter the data for fuel gas 1 (dry gas in Table 4.21), fuel gas 2 (sour gas in Table 4.21), LPG 1 (gasoline stabilizer overhead; LPG in Table 4.21), and naphtha (light end analysis). Often, the light end analyses for the naphtha cuts may be missing, as in the case of Table 4.21. We recommend using the nominal values given in Figure 4.77. In addition, we can use a simple material balance around the gasoline stabilizer to estimate the C4 composition of the naphtha cut. However, we note that if we use any estimation method for the C4 content during calibration, the model will likely produce poor predictions for gasoline RVP (Reid vapor pressure) and overhead temperatures for the gasoline stabilizer column.

Figure 4.78 shows the entry window for the Heavy Liquid section of the Prod Meas. tab. The measurements required for the naphtha and LCO cuts are routine measurement data. The distillation curve, density, concarbon, sulfur content, and nitrogen content are required for all the heavy liquid cuts. In addition, the olefins, naphthenes, and aromatics contents are required for at least one of the cuts. In addition, we must enter cloud point for all LCO-type cuts. In most cases, we

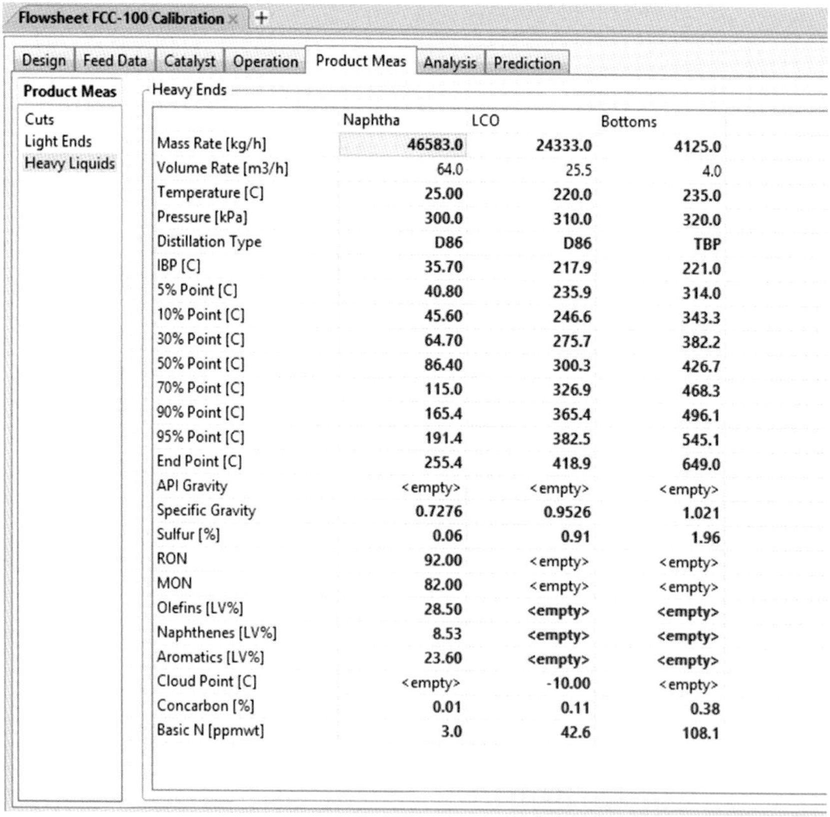

	Naphtha	LCO	Bottoms
Mass Rate [kg/h]	46583.0	24333.0	4125.0
Volume Rate [m3/h]	64.0	25.5	4.0
Temperature [C]	25.00	220.0	235.0
Pressure [kPa]	300.0	310.0	320.0
Distillation Type	D86	D86	TBP
IBP [C]	35.70	217.9	221.0
5% Point [C]	40.80	235.9	314.0
10% Point [C]	45.60	246.6	343.3
30% Point [C]	64.70	275.7	382.2
50% Point [C]	86.40	300.3	426.7
70% Point [C]	115.0	326.9	468.3
90% Point [C]	165.4	365.4	496.1
95% Point [C]	191.4	382.5	545.1
End Point [C]	255.4	418.9	649.0
API Gravity	<empty>	<empty>	<empty>
Specific Gravity	0.7276	0.9526	1.021
Sulfur [%]	0.06	0.91	1.96
RON	92.00	<empty>	<empty>
MON	82.00	<empty>	<empty>
Olefins [LV%]	28.50	<empty>	<empty>
Naphthenes [LV%]	8.53	<empty>	<empty>
Aromatics [LV%]	23.60	<empty>	<empty>
Cloud Point [C]	<empty>	-10.00	<empty>
Concarbon [%]	0.01	0.11	0.38
Basic N [ppmwt]	3.0	42.6	108.1

Figure 4.78 Measure liquid product yields and properties based on the product measurements in Table 4.20.

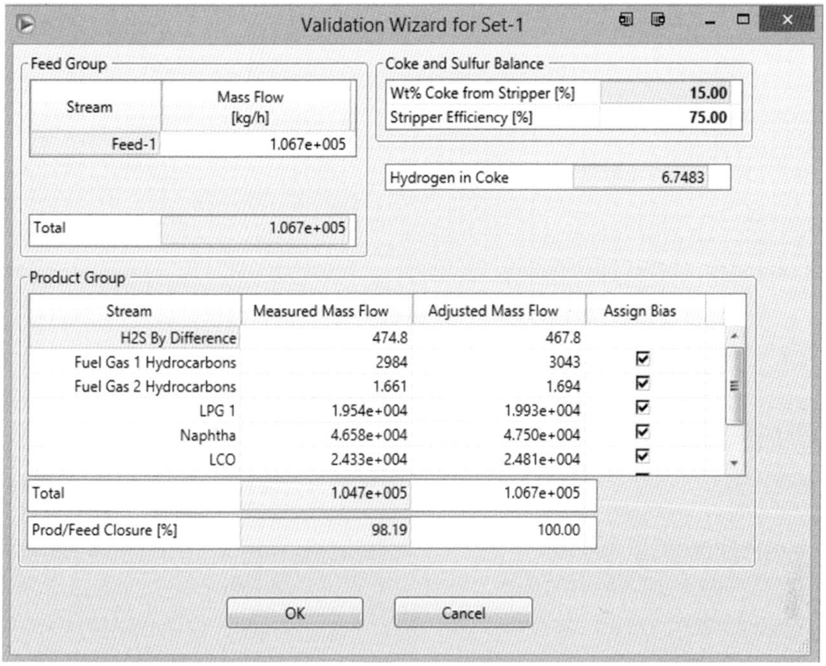

Figure 4.79 Mass balance validation wizard.

cannot obtain the distillation curve of the bottoms cut (routinely not measured or only partial measurement available). Kaes [51] gave a simple correlation to estimate the TBP curve of a bottoms cut as a function of density only. In general, we do not require accurate values for the TBP curve of the bottoms, as it is typically not a significant product.

Once we finish entering the heavy liquid product measurements shown in Figure 4.78, the status button of the calibration will turn yellow and indicate that the model is "Not Solved." At this point, we begin step 3 of the calibration process.

We click "Run Calibration" to bring up the Validation Wizard, as shown in Figure 4.79. The Validation Wizard allows us to assign biases to each measured flow rate, as the sum of the all flow measurements typically does not completely match the feed flow rate. The bias allows us to slightly adjust the measured flow rates to ensure an overall material balance. If the adjustments due the biases are small, we do not recommend removing biases from any product measurements. However, if the adjustments are significant, we should go back and check if all product flow rates and measurements are accurate. Lastly, we note that mass flow rates for the Fuel Gas cuts are much smaller than the values we entered in Light Ends section (see Figure 4.77). This is because inorganic compounds (H_2, N_2, O_2, CO_2, H_2S, etc.) are not included in the overall material balance. We begin the calibration by clicking "OK" in the Validation Wizards. Table 4.26 shows the progress of the solver during the calibration run.

Table 4.26 Solver output during calibration run.

Iteration	Residual Convergence Function	Objective Convergence Function	Objective Function Value	Overall Nonlinearity Ratio	Model Nonlinearity Ratio	Worst Model
0	5.039D+03	0.000D+00	0.000D+00	9.989D−01	−1.592D+01	REGEN
	<Line Search Creep Mode ACTIVE>		==>	Step taken 1.00D−01		
1	2.891D+03	0.000D+00	0.000D+00	9.536D−01	−8.333D+00	RISCOKE
	<Line Search Creep Mode ACTIVE>		==>	Step taken 1.00D−01		
2	1.600D+03	0.000D+00	0.000D+00	9.575D−01	−3.353D+00	RISCOKE
	<Line Search Creep Mode ACTIVE>		==>	Step taken 1.00D−01		
3	8.297D+02	0.000D+00	0.000D+00	9.605D−01	−1.579D+00	RISCOKE
	<Line Search Creep Mode ACTIVE>		==>	Step taken 1.00D−01		
4	3.764D+02	0.000D+00	0.000D+00	9.615D−01	−6.095D−01	RISCOKE
	<Line Search Creep Mode ACTIVE>		==>	Step taken 1.00D−01		
5	2.446D+02	0.000D+00	0.000D+00	9.619D−01	3.906D−02	RISCOKE
	<Line Search Creep Mode ACTIVE>		==>	Step taken 1.00D−01		
6	2.675D+02	0.000D+00	0.000D+00	9.631D−01	5.303D−01	RISCOKE
	<Line Search Creep Mode ACTIVE>		==>	Step taken 1.00D−01		
7	2.899D+02	0.000D+00	0.000D+00	9.693D−01	5.861D−01	RISER
	<Line Search Creep Mode ACTIVE>		==>	Step taken 1.00D−01		
8	3.088D+02	0.000D+00	0.000D+00	9.737D−01	6.337D−01	RISER
	<Line Search Creep Mode ACTIVE>		==>	Step taken 1.00D−01		
9	3.215D+02	0.000D+00	0.000D+00	9.760D−01	1.659D+00	RISCOKE
	<Line Search Creep Mode ACTIVE>		==>	Step taken 1.00D−01		
10	3.265D+02	0.000D+00	0.000D+00	6.975D−01	−9.358D+00	RISCOKE
11	1.105D+02	0.000D+00	0.000D+00	9.326D−01	−5.526D−01	REGEN
12	8.204D−02	0.000D+00	0.000D+00	9.994D−01	9.955D−01	PRTCALC
13	1.124D−08	0.000D+00	0.000D+00			

Successful solution.

Optimization Timing Statistics	Time	Percent
MODEL computations	0.91 secs	53.42 %
DMO computations	0.63 secs	36.99 %
Miscellaneous	0.16 secs	9.59 %
Total Optimization Time	1.71 secs	100.00 %

Problem converged

The calibration process for the FCC is "square." This implies that *there are no user adjustable tuning factors* unlike the Aspen HYSYS Reformer or Hydrocracking models. In other words, the number of tuning parameters equals the number of available measurements, and the calibration is a much simpler, root-finding exercise. In general, the calibration process is quick and converges within 20 iterations. If there is difficulty during calibration, it is mostly likely due to inconsistent product measurements.

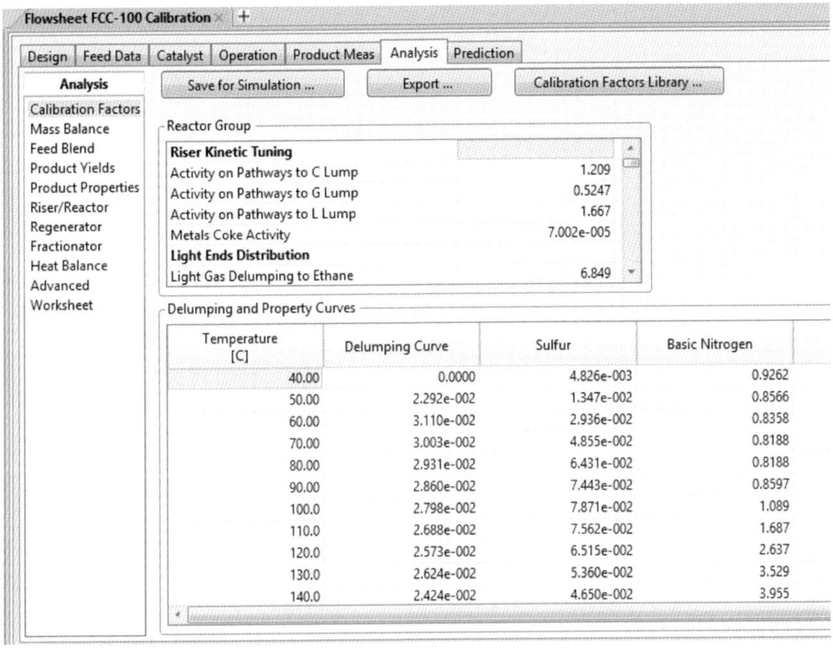

Figure 4.80 Calibrated activity factors.

Figure 4.80 shows the key results of the calibration procedure. The reactor group tuning parameters control the activity of each group of kinetic pathways and the light ends distribution. The delumping curves convert the kinetic lumps into fractionation lumps appropriate for a petroleum refining component slate. An important check of the calibration is shown in Figure 4.81. The theoretical and apparent heats of cracking should not be significantly different (<5% relative error), which we have previously discussed with reference to Figure 4.72 in Section 4.13.10. If we meet this error threshold, we conclude that the calibration procedure is successful.

The last step in the calibration procedure is to export the calibration factors back into the main flowsheet. To do this, we select Calibration factors section Analysis tab. Then, we click the "Save for Simulation ..." button to save current calibration factors as Set-1, as shown in Figure 4.82.

To return to the FCC unit PFD environment, we click on "Push Data to Simulation" to return the calibration factors back to the main environment, as shown in Figure 4.83. Aspen HYSYS may prompt to hold solver when returning to the main environment. As the FCC unit solves very quickly, we can choose "No" and force the solver to run when we return the main environment.

This completes the calibration workshop for the FCC unit. We save the converged simulation file as *Workshop 4.2-done.hsc.* At this point, we can perform case studies and build additional downstream fractionation units. In the next workshop, we go through some of the issues involved in building a complete downstream fractionation process for this FCC Unit.

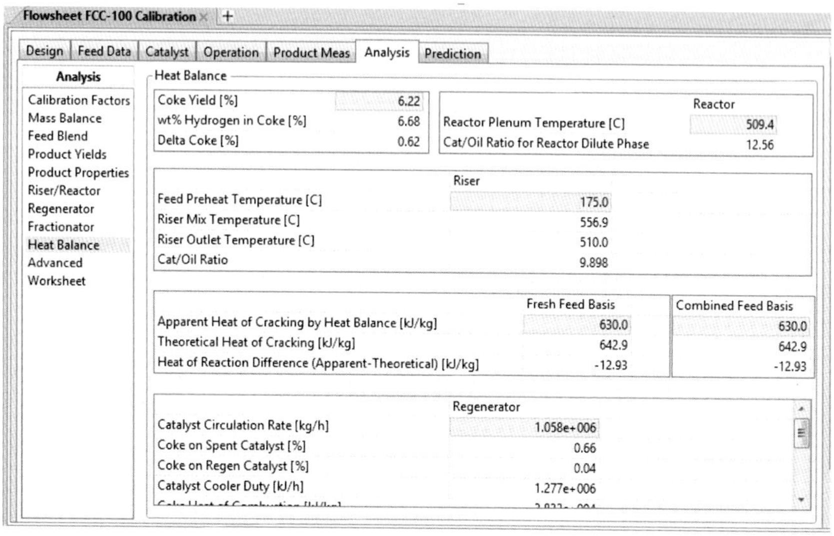

Figure 4.81 Calibrated heat balance between riser and regenerator.

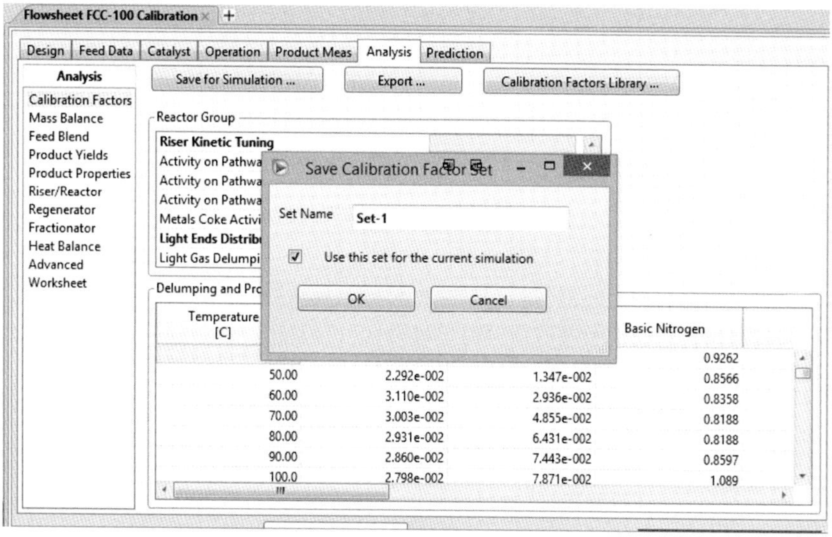

Figure 4.82 Save calibration factors for current calibration.

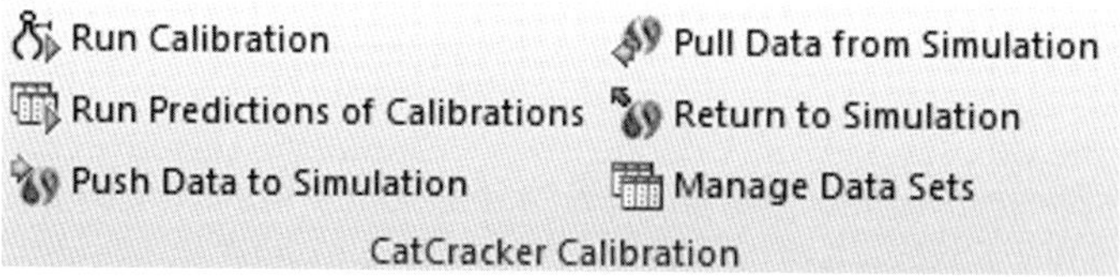

Figure 4.83 Return calibration factors to main FCC environment.

4.15 Workshop 4.3 – Build the Model for Main Fractionator and Gas Plant System

The effluent from the FCC unit is a broad mixture of light gases and liquid products that will be recovered as LPG, gasoline, and diesel (light and heavy cycle oil). The downstream fractionation units separate the reactor effluent into the product cuts through a series of distillation and absorption columns. As illustrated previously in Figures 4.38–4.40, Section 4.13.2, the main components of the downstream fractionation are as follows:

- *Main fractionator column (T201_MainFractionator)* – Recovers most naphtha, cycle oil, and bottoms product.
- *Overhead wet gas system* – Recompresses main fractionator overhead gas product to recover additional naphtha.
- *Primary absorber column (T301_Absorber)* – Returns light naphtha to the gasoline stream.
- *Primary stripper column (T302_Stripper)* – Removes heavy components from naphtha and returns these components to the diesel or LCO section of the main fractionator.
- *Secondary absorber or reabsorption column, or sponge oil absorber column (T303_ReAbsorber)* – Uses an LCO draw to remove very light components (<C2) from the primary absorber overhead vapor.
- Debutanizer or gasoline stabilization column (T304_Stabilizer) – Separates LPG (C3–C4) from product gasoline stream.

Workshop 4.3 is significant, as it is the first detailed, step-by-step demonstration of how to build a simulation model for the FCC main fractionator and gas plant system that has ever appeared in a textbook or in the literature. Refinery engineers can apply the same procedure to build the simulation models for the fractionation systems for reformers, hydrocrackers, delayed cokers, and so on. We demonstrate these applications in Chapters 5–7.

4.15.1 T201_MainFractionator

To build the main fractionator, we follow the same procedure as that of a CDU described in Chapter 2. We begin by opening a converged FCC reactor section, *Workshop 4.3-1.hsc* (see Figure 4.84).

Step 1. Create a stream for dry gas makeup. Connect this stream and the effluent from the FCC unit to a mixer (MIX-100) to create the main feed stream, T201_Feed, to the fractionator. Refer to *Workshop 4.3-Additional Specifications.xlsx* and *Workshop 4.3-Dry Gas Makeup Specifcations.xlsx* for required stream specifications. Create a heater block to preheat the feed stream up to 510 °C (see Figure 4.85). Save the file as *Workshop 4.3-2.hsc.*

Step 2. Table 4.7, Section 4.7, recommends the use of 17 equilibrium stages for a FCC main fractionator. We create a refluxed absorber of 17 equilibrium stages as *T201_MainFractionator* (Figures 4.86–4.89).

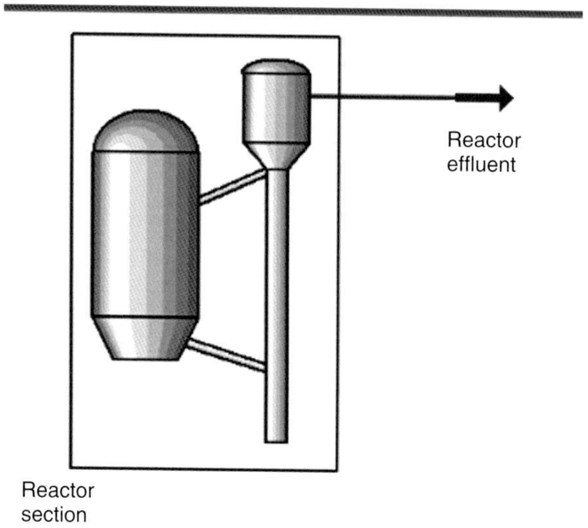

Figure 4.84 The FCC reactor section.

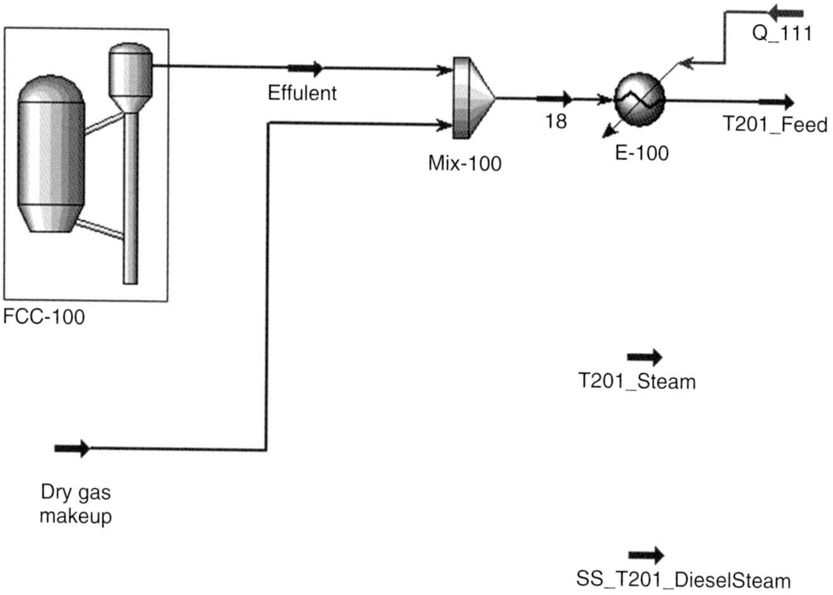

Figure 4.85 Flowsheet of step 1.

Step 3. Create a side draw for the heavy naphtha stream (T201_HN_Draw) from stage 9 and specify its draw rate of 1E4 kg/h. Run the simulation to update the temperature profile and save the converged simulation as *Workshop 4.3-3.hsc* (see Figures 4.90–4.92).

Step 4. Refer to Figures 4.93 and 4.94. Add a three-stage diesel side stripper (T201_SSDiesel) along with the side stripper steam flow (T201_SS_DieselSteam at 240 °C, 1351 kPa and 370 kg/h of H_2O) and specify the draw rate of SS_T201_DieselProd of 4.4E4 kg/h. Send T201_HN_Draw to a recycle

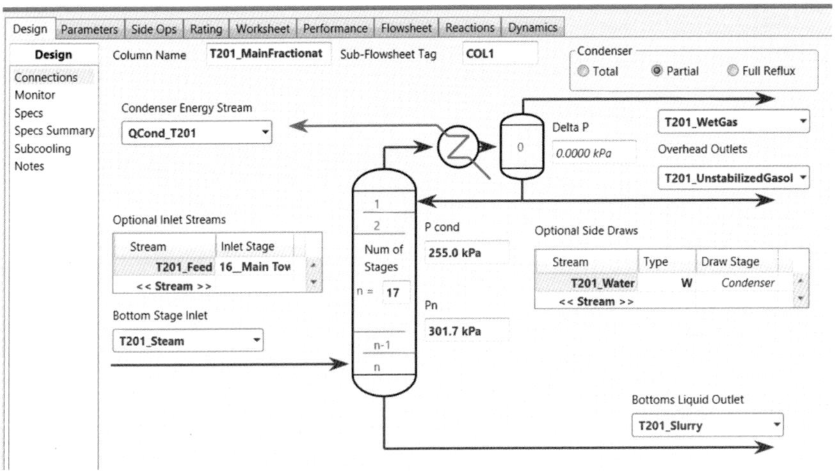

Figure 4.86 Specifications of T201_MainFractionator (1).

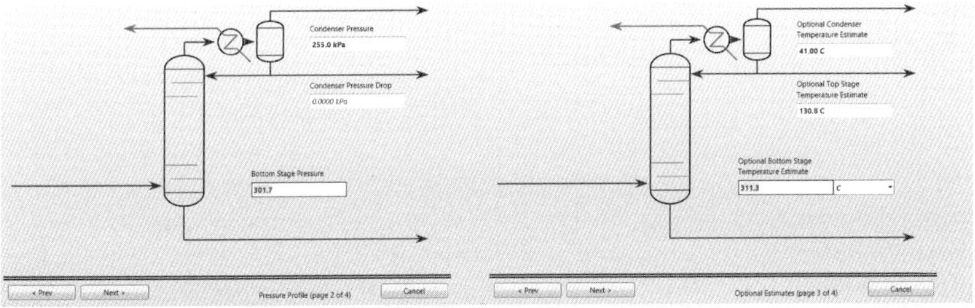

Figure 4.87 Specifications of T201_MianFractionator (2).

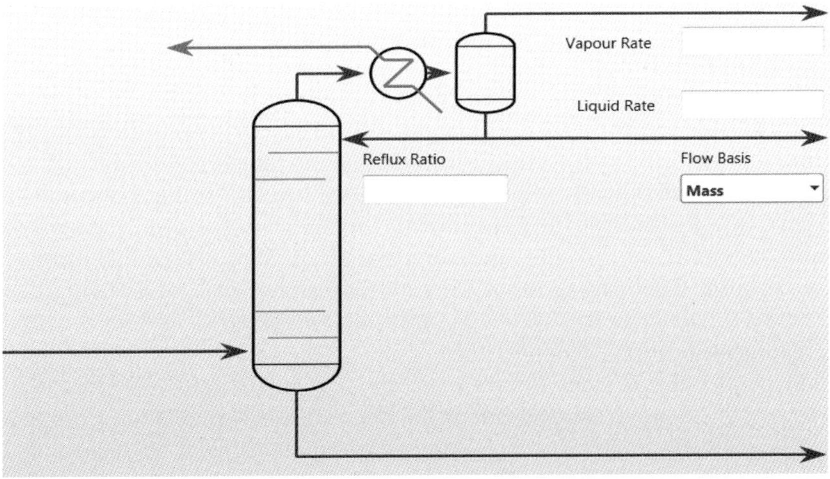

Figure 4.88 Specifications of T201_MainFractionator (3).

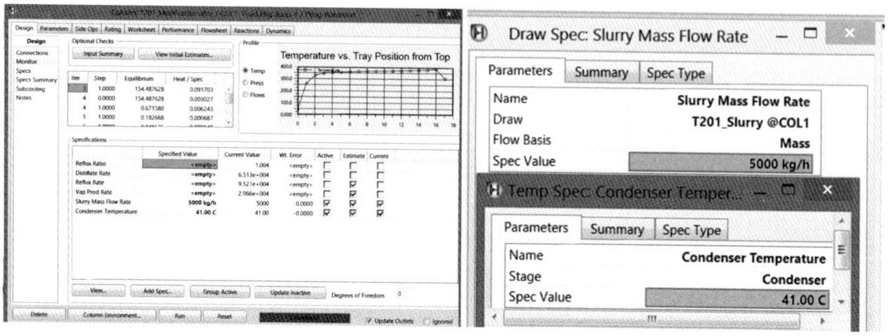

Figure 4.89 Specifying a slurry mass flow rate of 5000 kg/h and a condenser temperature of 41 °C enables column convergence.

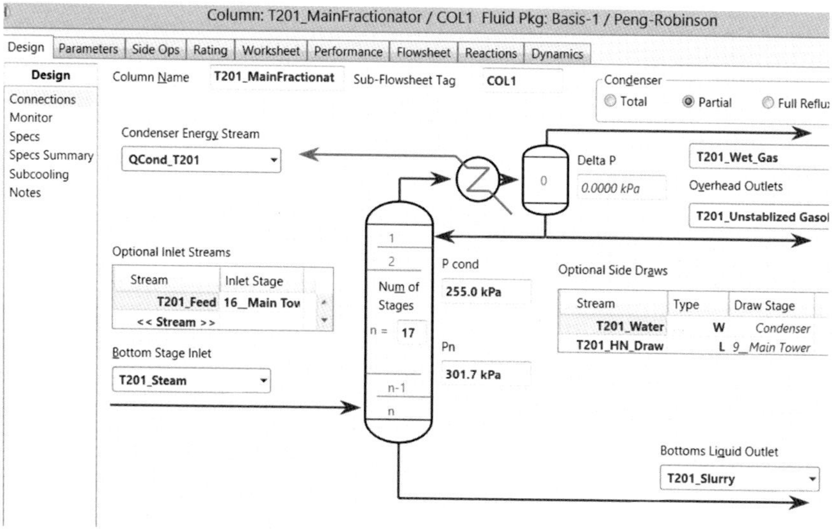

Figure 4.90 Add a liquid side draw, T201_HN_Draw from stage 9.

block RCY-1 and then send its output T201_HN-HNRecycle to MIX-100. Note that we have added the specification of the temperature estimates for stages 1 and 17. Create a new stream T201_RichLCO to represent return from T303_ReAbsorber to T201_MainFractionator. Set this new stream to same temperature, pressure, and composition as SS_T201_DieselProd and fix the mass flow rate at 5% of SS_T201_DieselProd (0.05 × 4.4E4 kg/h = 2200 kg//h). Refer to *Workshop 4.3-Additional Specifications.xlsx* for required stream specifications. Connect this stream to the return stage of the diesel side stripper, that is, stage 7 of the main tower.

Figure 4.95 shows the specifications for the converged simulation. Note that we have added the specification of the temperature estimates for stages 1 and 17. Save the converged simulation as *Workshop 4.3-4.*

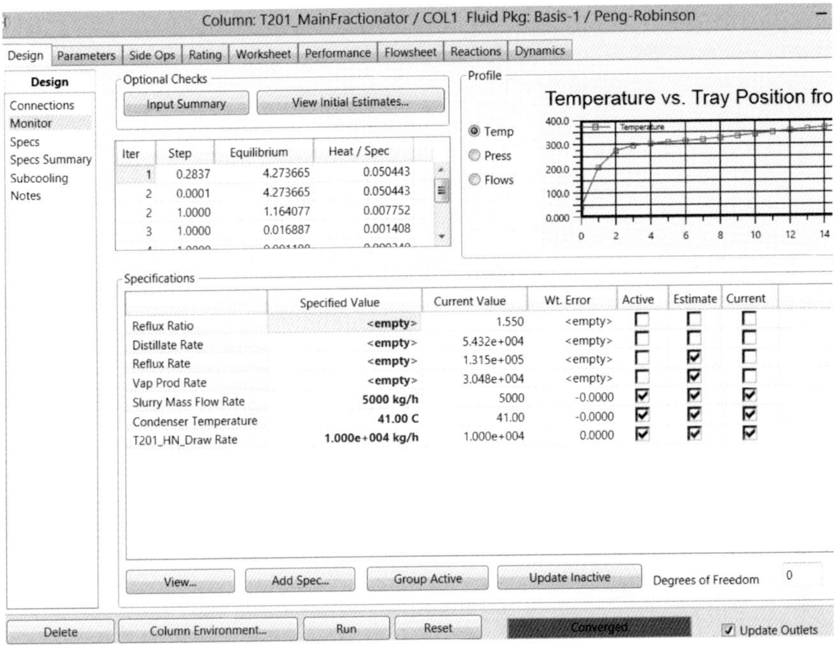

Figure 4.91 Specifying a mass flow rate of 1E4 kg/h for T201_HN_Draw leads to simulation convergence.

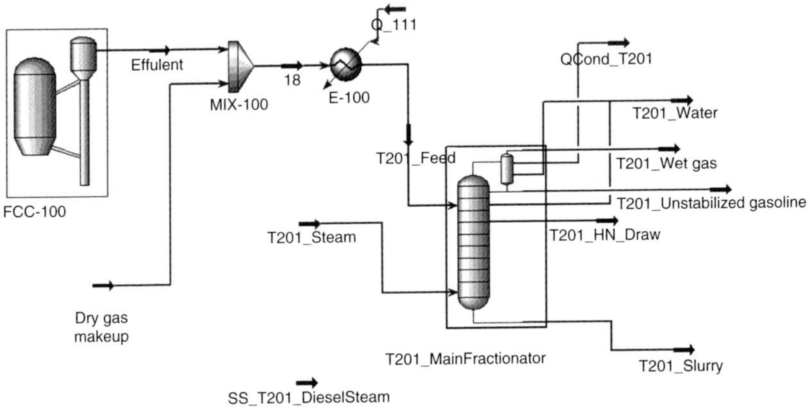

Figure 4.92 Flowsheet of step 3.

Step 5. Following the procedure illustrated in Figures 2.53–2.55 and in Table 2.15, we implement the five pumparounds specified in Table 4.27. Figures 4.96–4.98 show the column details with the pumparounds and the overall flowsheet for T201_MainFractionator, and the specifications for simulation convergence. We save the converged simulation as *Workshop 4.3-5*.

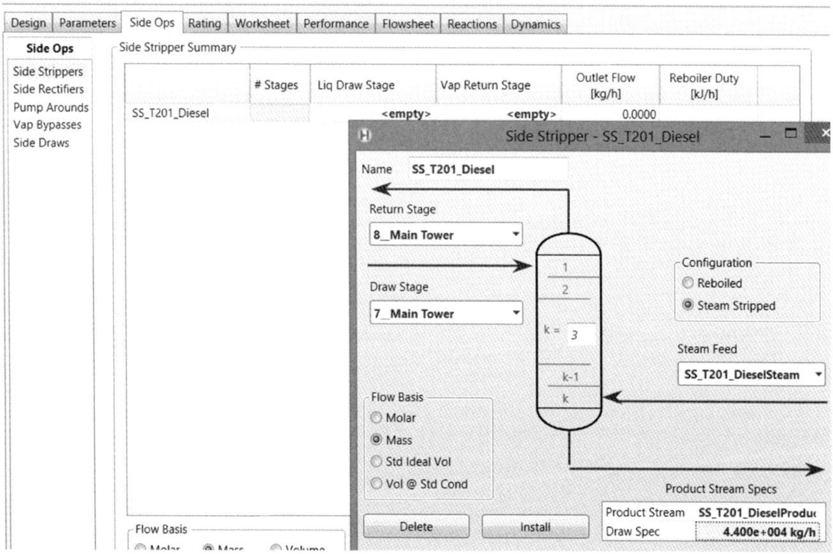

Figure 4.93 Add the diesel side stripper.

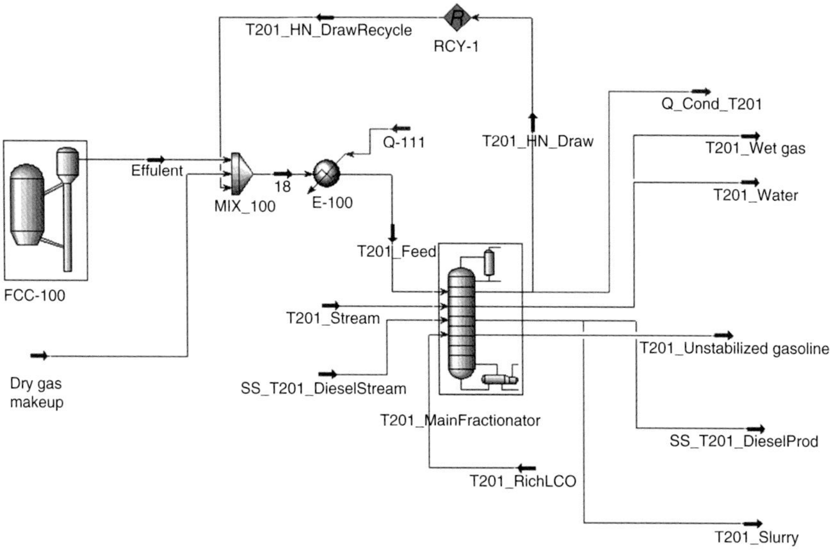

Figure 4.94 Flowsheet of step 4.

The standard inside-out algorithm (Section 2.4.4) can solve the main fractionator with ease when we follow the procedure mentioned above. However, flat distillation cuts or very tight specifications may not allow the standard method to converge robustly. We suggest the following changes to improve the convergence behavior in Aspen HYSYS.

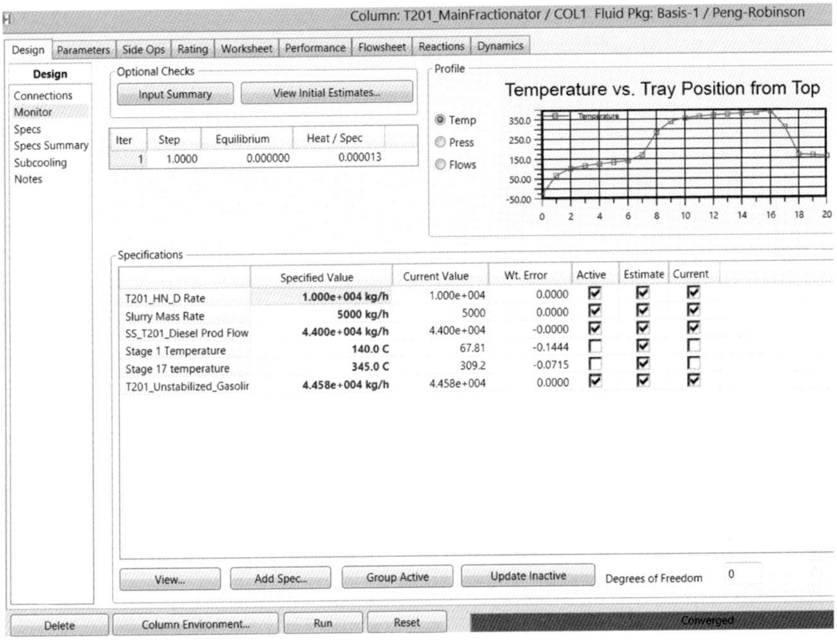

Figure 4.95 Specifications of step 4 for simulation convergence.

Table 4.27 Specifications of pumparounds for T201_MainFractionator.

Pumparound	PA_HN	PA_LCO	PA_HCO	PA_Quench	PA_Subcooling
Draw stage	2	8	14	17	17
Return stage	1	7	11	15	17
PA rate (kg/h)	1E4	1E5	4.6E4	2.4E5	8330
Return temperature (°C)	82	182	272	257	257

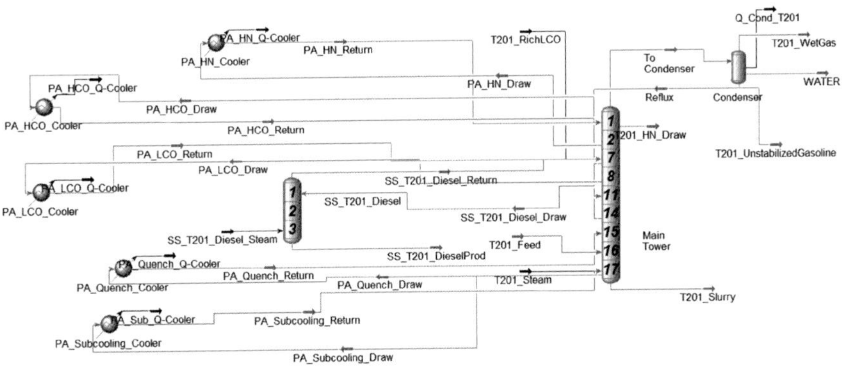

Figure 4.96 Column details with pumparounds for T201_MainFractionator.

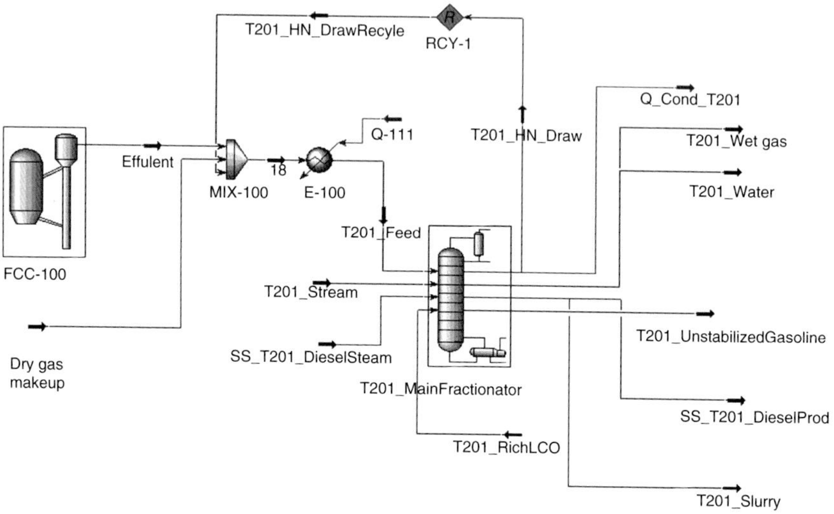

Figure 4.97 Flowsheet of step 5.

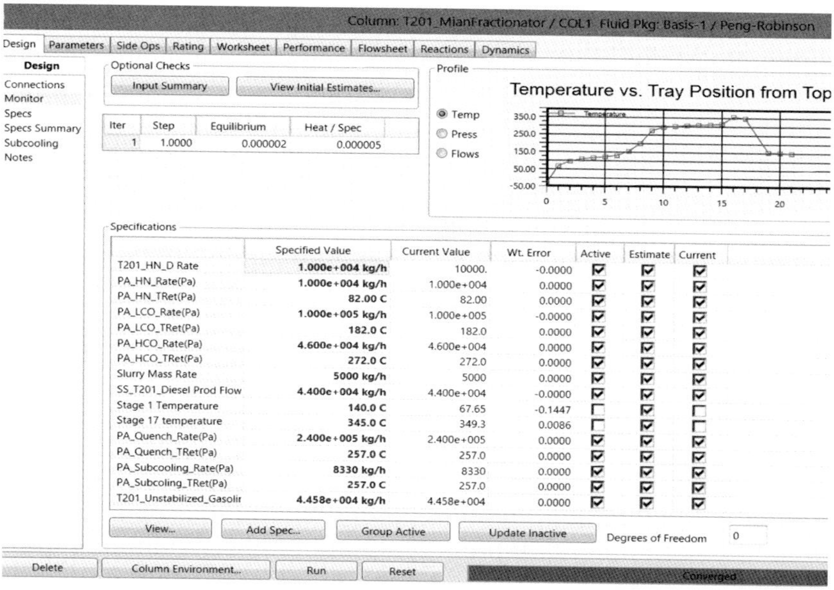

Figure 4.98 Specifications with T201_MainFractionator with side stripper and pumparounds.

1) Use the modified inside-out method with adaptive damping (Section 2.4.4, Figures 2.8 and 2.9; see also Figure 4.99). The modified method deals much better with tight product specifications.
2) Decrease the tolerance for Heat/Spec error from 1E-5 (dfault) to 5E-4 (see Figure 4.99). This method can significantly improve convergence when reconciling the recycle loops in the overall fraction model.

Once we solve the column using the above procedure, we use alternative specifications to allow more flexibility in the column model. This is especially relevant

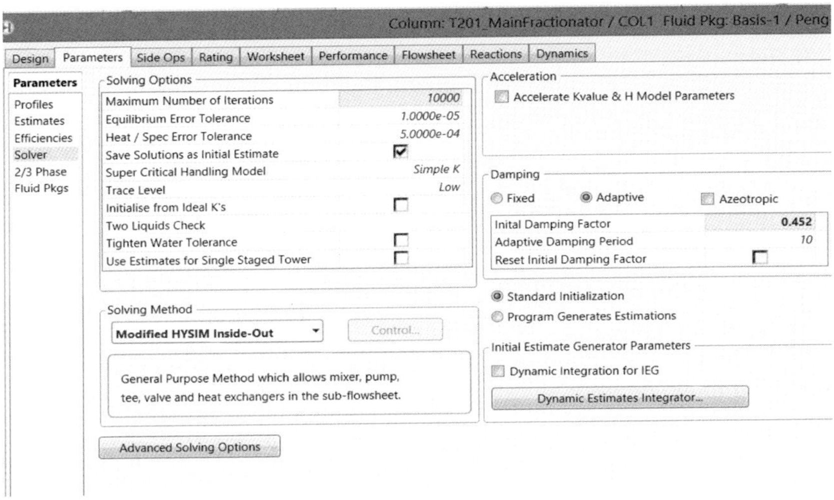

Figure 4.99 Convergence parameters for the inside-out method in Aspen HYSYS.

Table 4.28 Valid specifications for main fractionator.

Original specification	Flexible specification
Overhead liquid draw rate	Condenser temperature
Heavy naphtha draw rate	Heavy naphtha 95% D86 cut point
Pumparound temperature change	Pumparound duty (loose specification)
	Pumparound return temperature (tight specification)

when the flow rate to the column changes significantly. Table 4.28 lists possible replacements for the original specifications.

4.15.2 Overhead Wet Gas System, Primary Stripper Column T302_Stripper, and Debutanizer or Gasoline Stabilization Column T304_Stabilizer

Step 6. We build the model for the overhead wet gas system and the feed sections to T302_Stripper and to T304_Stabilizer. Refer to the flowsheet given in Figure 4.100.

1) Create a compressor (K-100) to compress the wet gas stream from the main fractionator to 748 kPa.
2) Create a cooler block to cool down the compressed gas stream to 35 °C (stream 2).
3) Connect stream 2 to a three-phase separator (V-100) with the exit streams – vapor stream 4, light liquid 5, and heavy liquid 3.
4) Connect the light liquid stream (stream 5) and the unstabilized gasoline stream from the main fractionator (T201_UnstablilizedGasoline) to a mixer MIX-101 to create a mixed stream (stream 7).

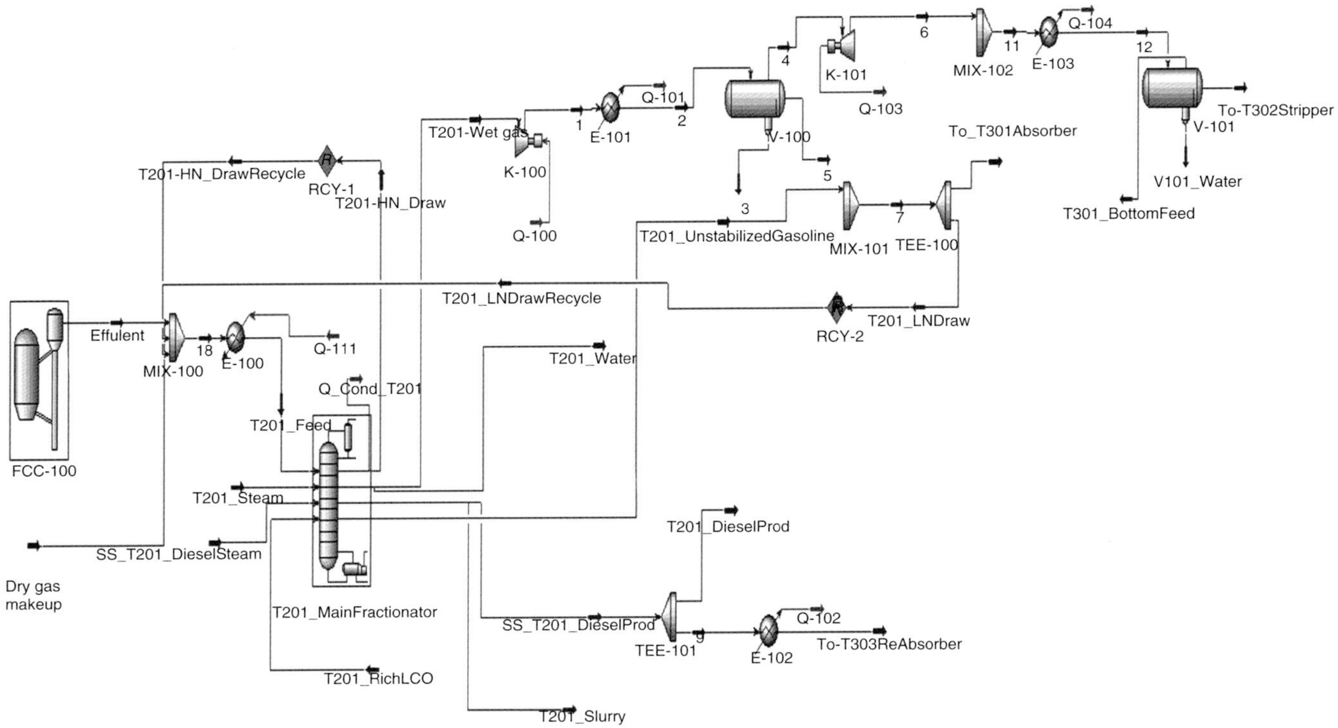

Figure 4.100 Flowsheet of the overhead wet gas system and feed sections to T301_Absorber, to T302_Stripper and to T303_ReAbsorber.

5) Connect the vapor stream 4 at 748.3 kPa from V-100 to a compressor K101 raising its pressure to 1802 kPa for stream 6. Then, send stream 6 to a mixer, MIX102 (to mix later in the following step with a recycle stream from T302_Strpper yet to be specified). The exit stream from the mixer MIX-102, stream 11, is cooled through a cooler E-103 to 40 °C for stream 12. Send stream 12 to a three-phase separator V-101, generating three exit streams_ vapor steam as T301_BottomFeed, light liquid stream C101_Water, and heavy liquid stream, To_T302Stripper.
6) Connect stream 7 to a splitter (TEE-100) with two exit streams, To_ T301Absorber and T201_LNDraw. Specify 90.76% of stream 7 mass flow rate going to stream To_T301Absorber.
7) Send SS_T201_DieselProd to a splitter (TEE-101), with 56.16% of its mass flow rate going to steamT201_DieeselProd and the remaining going to stream 9. Add a cooler (E-102) to cool stream 9 to 35 °C for stream To_T303ReAbsorber.
8) Send T201_LNDraw to a recycle block, RCY-2, and then send its exit stream, T201_LNDrawRecycle, to MIX-100 to feed into column T201_ MainFractionator. Refer to the specifications given in Figure 4.101. Save the converged simulation as *Workshop 4.3-6.hsc*.

Step 7. Referring to the flowsheet of Figure 4.102, we complete T302_Stripper and T304_Stabilizer as follows.

1) Create the primary stripper, column T302_Stripper, with 13 equilibrium stages suggested by Table 4.7 and feed the light liquid stream from V-101, To_T302Stripper, to the top stage. See specifications in Figures 4.103–4.105.

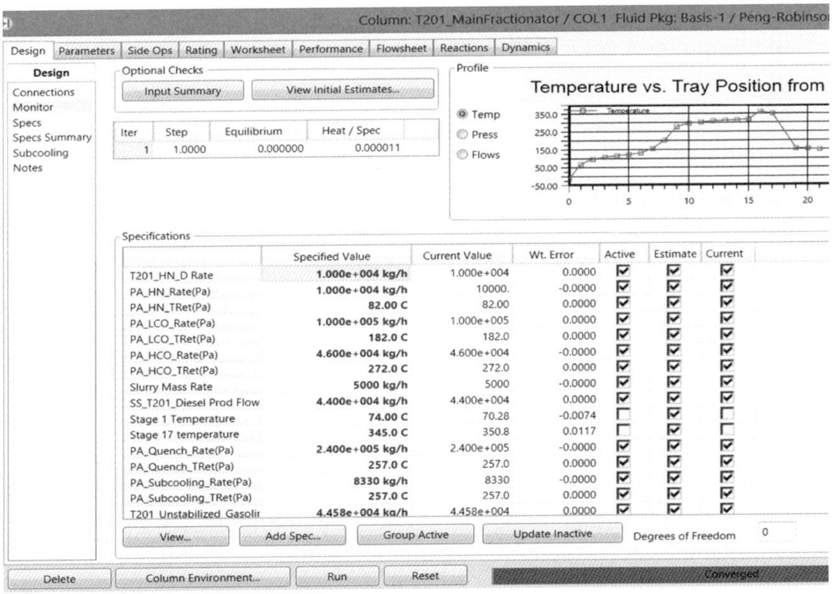

Figure 4.101 Specifications of step 6 for simulation convergence.

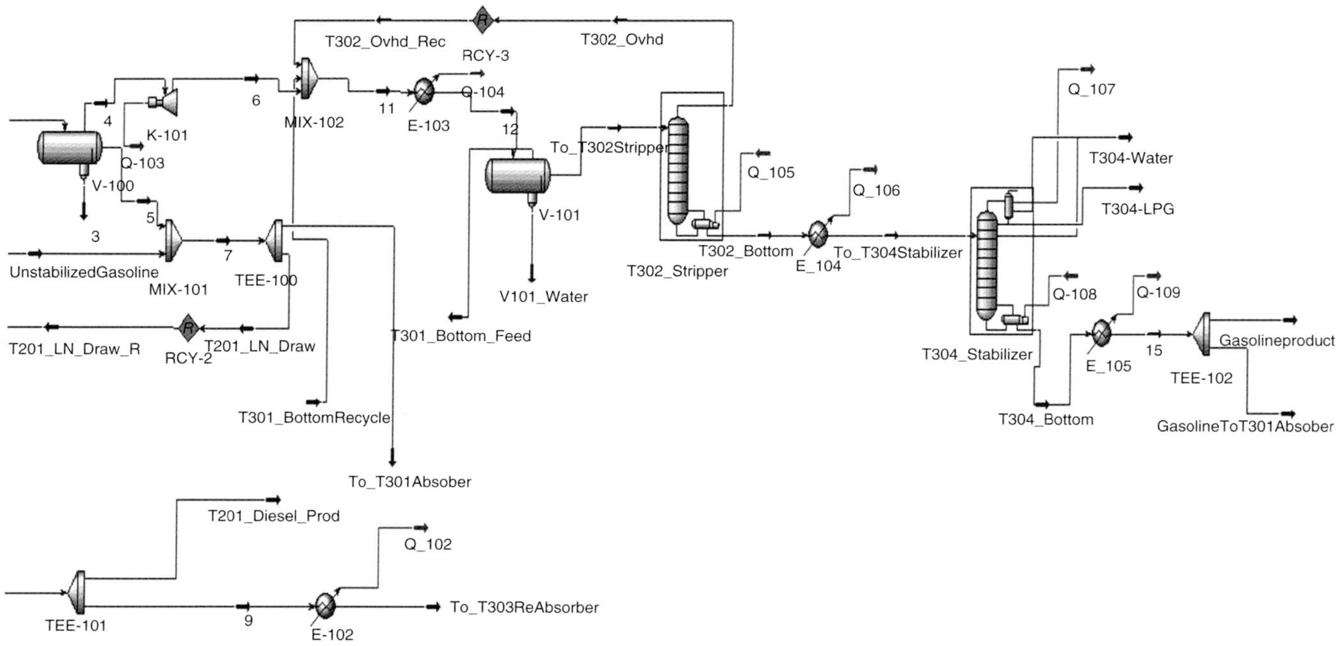

Figure 4.102 Add T302Stripper and T304Stabilizer.

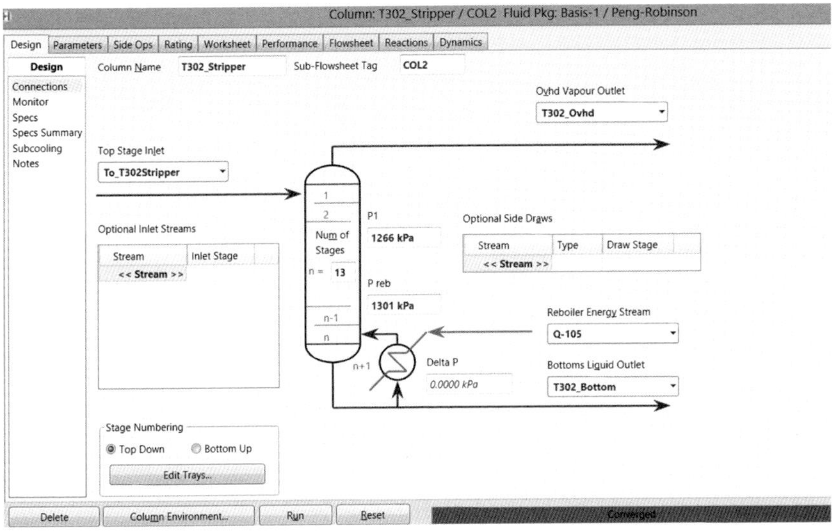

Figure 4.103 Specification of T302_Stripper (1).

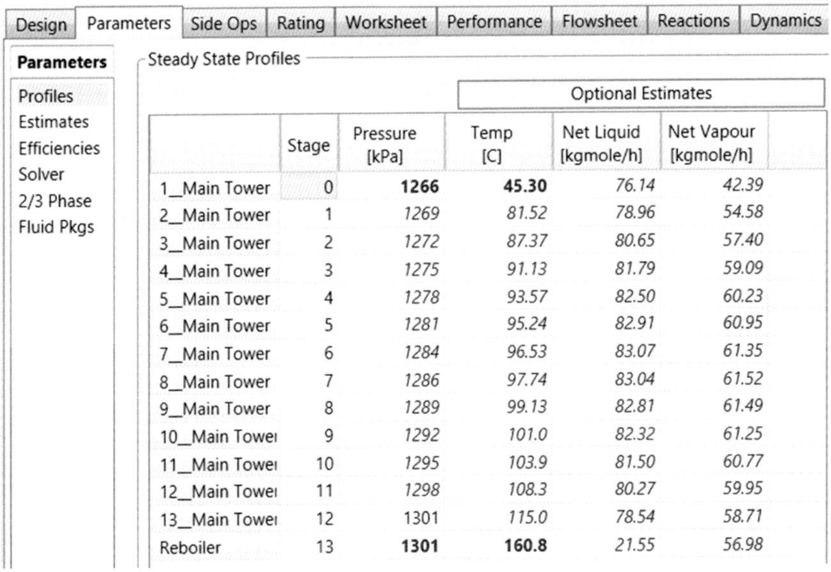

Steady State Profiles

	Stage	Pressure [kPa]	Temp [C]	Net Liquid [kgmole/h]	Net Vapour [kgmole/h]
1_Main Tower	0	1266	45.30	76.14	42.39
2_Main Tower	1	1269	81.52	78.96	54.58
3_Main Tower	2	1272	87.37	80.65	57.40
4_Main Tower	3	1275	91.13	81.79	59.09
5_Main Tower	4	1278	93.57	82.50	60.23
6_Main Tower	5	1281	95.24	82.91	60.95
7_Main Tower	6	1284	96.53	83.07	61.35
8_Main Tower	7	1286	97.74	83.04	61.52
9_Main Tower	8	1289	99.13	82.81	61.49
10_Main Towei	9	1292	101.0	82.32	61.25
11_Main Towei	10	1295	103.9	81.50	60.77
12_Main Towei	11	1298	108.3	80.27	59.95
13_Main Towei	12	1301	115.0	78.54	58.71
Reboiler	13	1301	160.8	21.55	56.98

Figure 4.104 Specifications of T302_Stripper (2): Pressure and temperature of stage 1 and reboiler.

2) Following the flowsheet of 4.102, recycle the overhead stream from T302_Stripper, (T302_Ovhd) back to MIX-102. Add a cooler (E-104) to cool the bottom outlet stream from T-302 (T302_Bottom) to stream 14 at 40 °C. Run the simulation of T302_Stripper again; save the converged simulation as *Workshop 4.3-7a*.

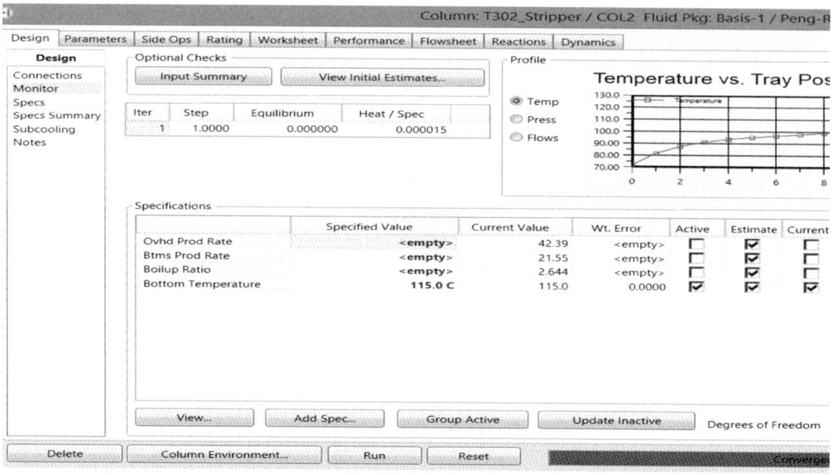

Figure 4.105 Specification of bottom (stage 13) temperature enables simulation convergence.

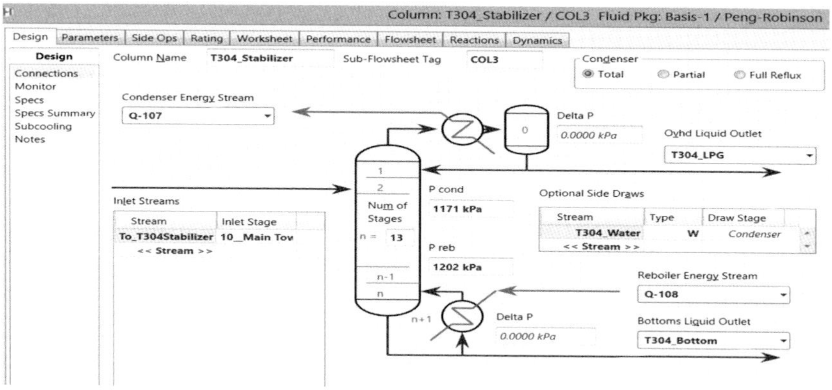

Figure 4.106 Specifications of T304_Stabilizer with a total condenser and a partial reboiler.

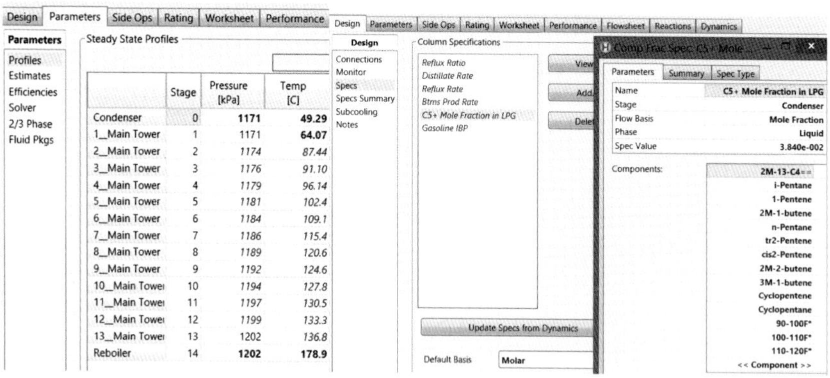

Figure 4.107 Specifications of temperatures of condenser, and stages 1 and 17, and C5+ component mole fraction in LPG.

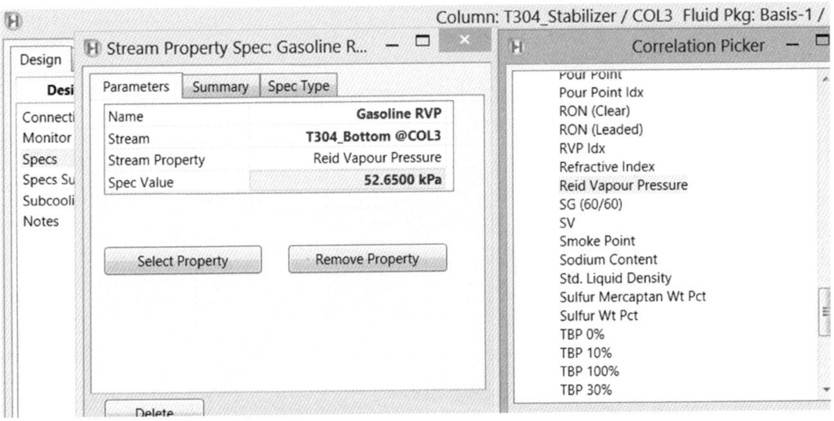

Figure 4.108 Specification of Gasoline RVP: Design → Specs → Add → Column Stream Property Spec → Name: Gasoline RVP; Stream-T304_Bottom@COL3; Stream Property → Select Property → Correlation Picker – Petroleum – Reid Vapor Pressure → Select; Spec Value → 52.65 kPa. Repeat the same procedure to set up the petroleum spec, Gasoline IBP.

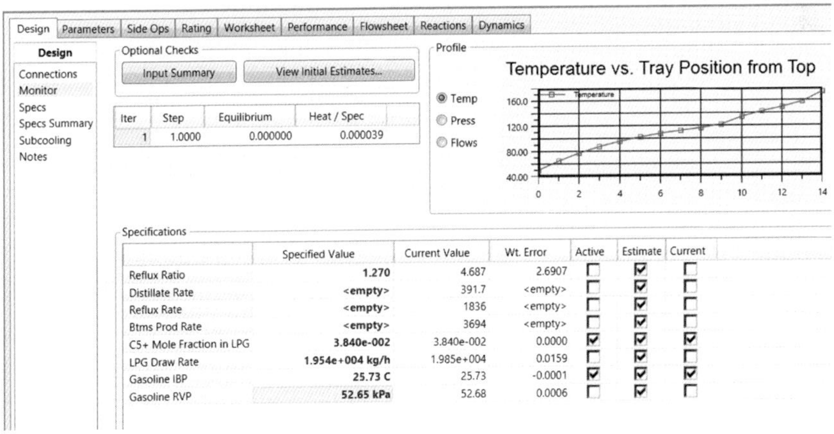

Figure 4.109 Specifications of T304_Absorber for simulation convergence.

3) Create T304_Stabilizer, the debutanizer or gasoline stabilization column, with specifications according to Figures 4.106–4.109. Save the converged simulation as *Workshop 4.3-7b*.
4) Refer to the flowsheet given in Figure 4.110. Create a cooler block (E-105) to cool down the bottom stream from T-304 to 35 °C for stream 15.
5) Create a tee (TEE-105) to split stream 15 into two streams, gasoline (Gasoline_Product) with a mass flow rate of 43630 kg/h and To_T301Absorber. Save the converged simulation as *Workshop 4.3-7c*.

4.15.3 T301_Absorber, Primary Absorber and T303_ReAbsorber, Sponge Oil Absorber, or Reabsorption Column

We continue by saving *Workshop 4.3-7c* to a new file, *Workshop 4.3-8*. Refer to the flowsheet given in Figure 4.111.

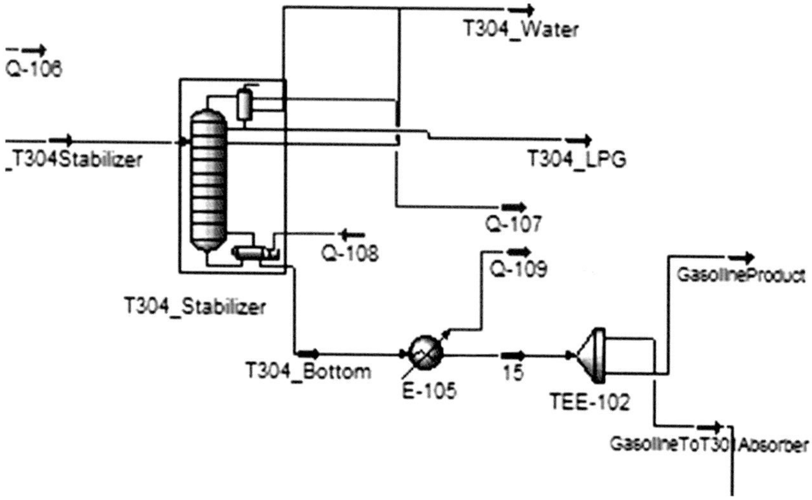

Figure 4.110 Flowsheet of the T304_Stabilizer product streams.

1) Create T301_Absorber, the primary absorber, with nine equilibrium stages according to Table 4.7, with three feeds: (1) To_T301Absorber from TEE-100 to stage 1; (2) vapor stream T301_BottomFeed from three-phase separator V-101 to stage 9; and (3) T304_GasolineRecycle, exiting a new recycle block RCY-4 that recycles the stream GasolineToT301Absorber from TEE-102, to stage 1.
2) Complete the column specifications following Figure 4.112.
3) Connect the overhead stream from T-301 (T301_Ovhd) to an exchanger block (E-106), setting the vapor fraction of the exit stream, stream 20 to 1.0.
4) Create a tee (TEE-105) to split the diesel side draw from the main fractionator (SS_T201_DieselProd) into two streams, T201_DieselProd with a mass flow rate of 24710 kg/h and stream 9. Create a cooler block (E-102) to cool down stream 9–50 °C.
5) To resolve possible convergence issues with the appearance of two liquid phases (as we demonstrated previously in Figures 2.88–2.91), we create water side draws from stage 1 to stage 9 of T301_Absorber (see Figures 4.113 and 4.114).
6) Create T303_ReAbsorber, the sponge oil absorber, or the reabsorption column with nine equilibrium stages (see Table 4.7) and two feeds: (1) To_T303ReAbsorber to stage 1, which is a part of the SS_T201_DieselProd through splitter TEE-101 and exchanger E-102 and (2) stream 20 to stage 8, which is T301_Ovhd from T301_Absorber going through exchanger E-100. See the specifications in Figure 4.115. As with T301_Absorber, we do not need additional specifications for T303_ReAbsorber to achieve simulation convergence. We can repeat the same procedure of adding water side draws to stage 1 to stage 9 of T303_ReAbsorber to resolve the converged simulation as *Workshop 4.3-8.hsc*. This concludes Workshop 4.3. Refer to Figure 4.111 for the final flowsheet and save the converged simulation as *Workshop 4.3-8*.

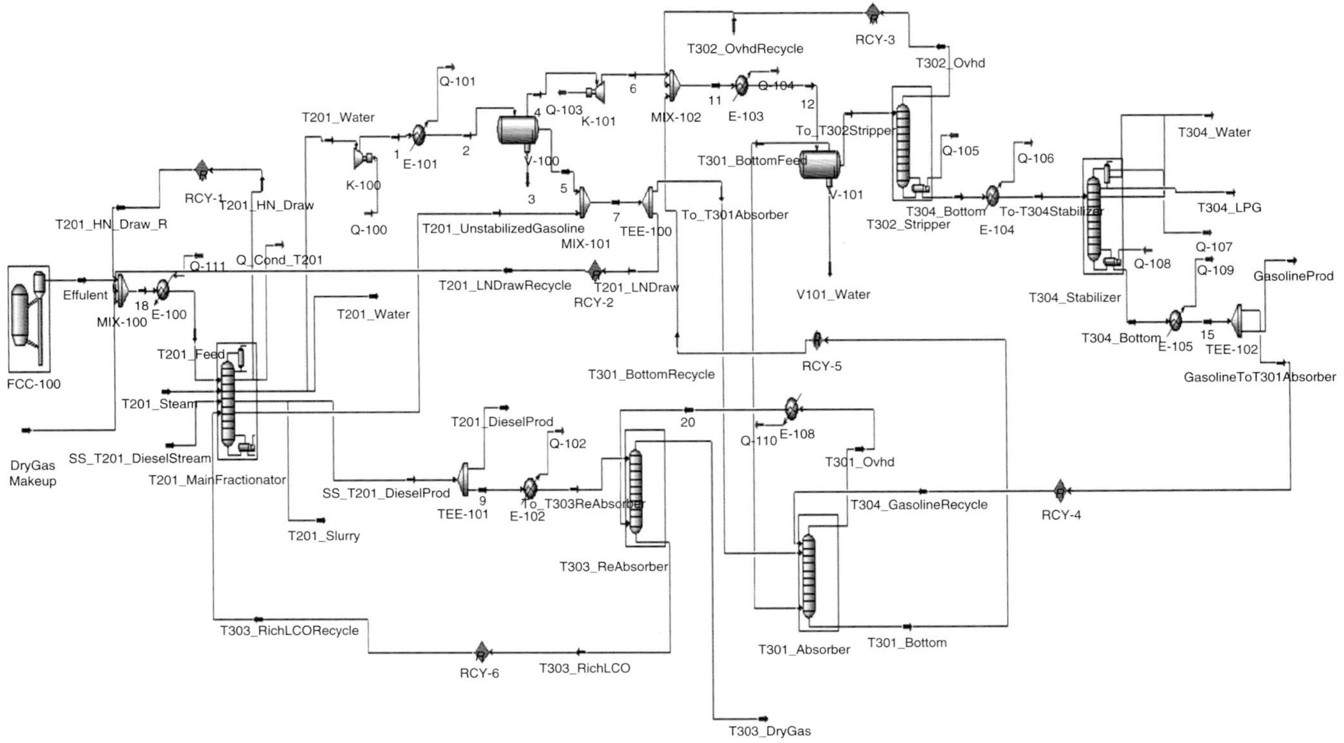

Figure 4.111 Overall Aspen HYSYS model of FCC unit and associated gas plant.

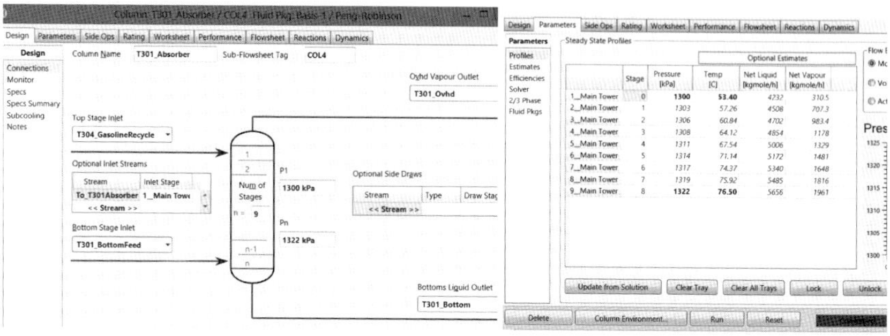

Figure 4.112 Specifications of T301_Absorber, which requires only feed stage locations and product streams, and pressure and temperature estimates to achieve simulation convergence.

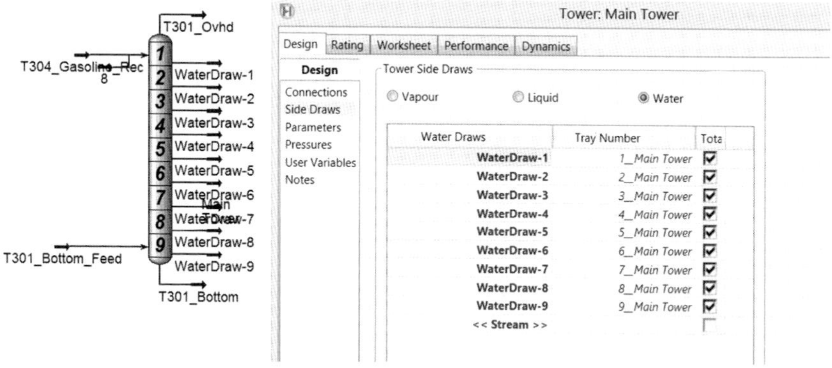

Figure 4.113 Create water side draws from stages 1 to 9 of T301_Absorber.

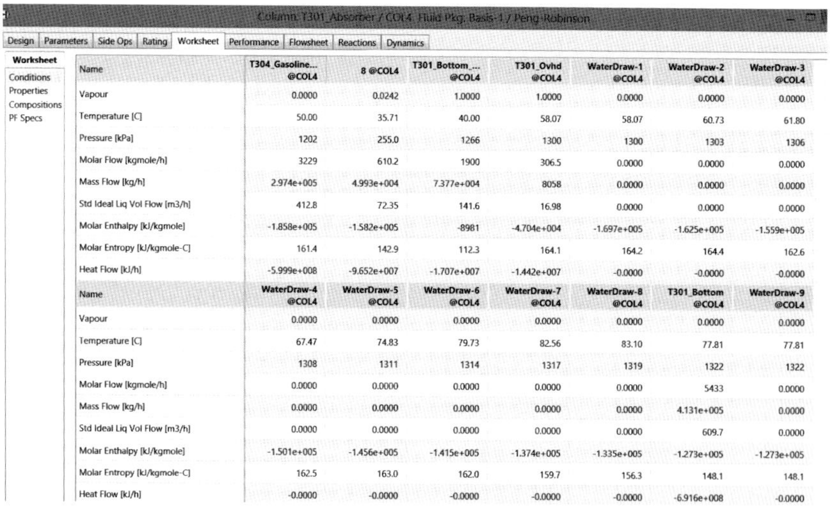

Name	T304_Gasoline... @COL4	8 @COL4	T301_Bottom_... @COL4	T301_Ovhd @COL4	WaterDraw-1 @COL4	WaterDraw-2 @COL4	WaterDraw-3 @COL4
Vapour	0.0000	0.0242	1.0000	1.0000	0.0000	0.0000	0.0000
Temperature [C]	50.00	35.71	40.00	58.07	58.07	60.73	61.80
Pressure [kPa]	1202	255.0	1266	1300	1300	1303	1306
Molar Flow [kgmole/h]	3229	610.2	1900	306.5	0.0000	0.0000	0.0000
Mass Flow [kg/h]	2.974e+005	4.993e+004	7.377e+004	8058	0.0000	0.0000	0.0000
Std Ideal Liq Vol Flow [m3/h]	412.8	72.35	141.6	16.98	0.0000	0.0000	0.0000
Molar Enthalpy [kJ/kgmole]	-1.858e+005	-1.582e+005	-8981	-4.704e+004	-1.697e+005	-1.625e+005	-1.559e+005
Molar Entropy [kJ/kgmole-C]	161.4	142.9	112.3	164.1	164.2	164.4	162.6
Heat Flow [kJ/h]	-5.999e+008	-9.652e+007	-1.707e+007	-1.442e+007	-0.0000	-0.0000	-0.0000

Name	WaterDraw-4 @COL4	WaterDraw-5 @COL4	WaterDraw-6 @COL4	WaterDraw-7 @COL4	WaterDraw-8 @COL4	T301_Bottom @COL4	WaterDraw-9 @COL4
Vapour	0.0000	0.0000	0.0000	0.0000	0.0000	0.0000	0.0000
Temperature [C]	67.47	74.83	79.73	82.56	83.10	77.81	77.81
Pressure [kPa]	1308	1311	1314	1317	1319	1322	1322
Molar Flow [kgmole/h]	0.0000	0.0000	0.0000	0.0000	0.0000	5433	0.0000
Mass Flow [kg/h]	0.0000	0.0000	0.0000	0.0000	0.0000	4.131e+005	0.0000
Std Ideal Liq Vol Flow [m3/h]	0.0000	0.0000	0.0000	0.0000	0.0000	609.7	0.0000
Molar Enthalpy [kJ/kgmole]	-1.501e+005	-1.456e+005	-1.415e+005	-1.374e+005	-1.335e+005	-1.273e+005	-1.273e+005
Molar Entropy [kJ/kgmole-C]	162.5	163.0	162.0	159.7	156.3	148.1	148.1
Heat Flow [kJ/h]	-0.0000	-0.0000	-0.0000	-0.0000	-0.0000	-6.916e+008	-0.0000

Figure 4.114 Converged results of T301_Absorber showing zero water side draw flow rates from stages 1 to 9.

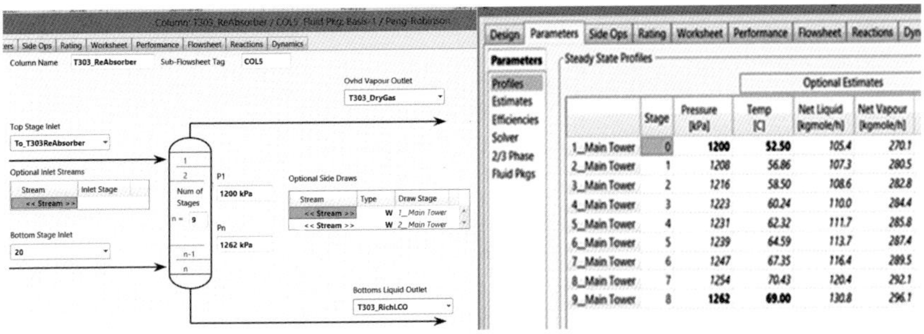

Figure 4.115 Specifications of T303_ReAbsorber, which requires only feed stage locations and product streams, and pressure and temperature estimates to achieve simulation convergence.

4.16 Workshop 4.4 – Perform Case Studies to Quantify Effects of Key FCC Operating Variables

In this workshop, we focus on methods to perform different case studies using a calibrated model. We generally do not need the rigorous fractionation model for many types of yield-related case studies. An important consideration during FCC operation is to improve the yield of a particular key product. As the FCC unit is a large producer of gasoline, we generally need to maximize the throughput and conversion of feed to gasoline. In Section 4.11 regarding FCC modeling and kinetics, we extensively discussed how changes in feed rate and operating temperatures can affect the yield of the unit. We perform two case studies below using Aspen HYSYS that illustrate the effects of feed rate and ROT in practice.

Open simulation file of the calibrated model, *Workshop 4.2-done.hsc*, and save it as *Workshop 4.4-1*. This workshop follows the procedure of Workshop 2.1, Figures 2.69–2.75, Section 2.1. We begin a new case study according to Figure 4.116.

We define the independent variables in Figures 4.117 and 4.118.

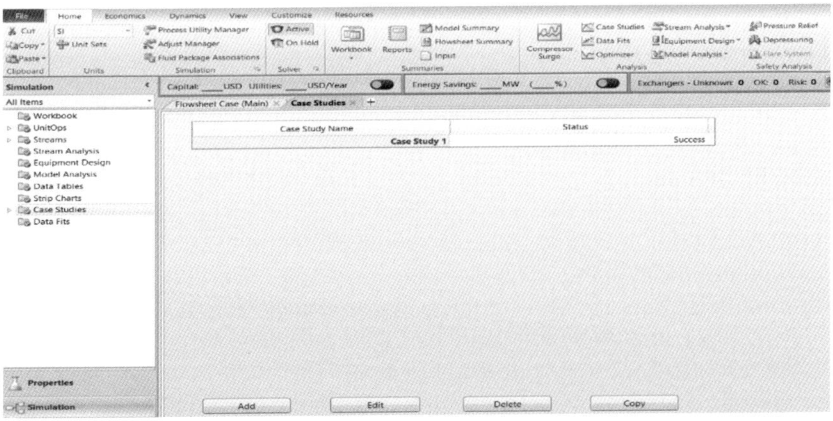

Figure 4.116 Initialize a new case study: Case Studies → Add → Case Study 1 → Edit.

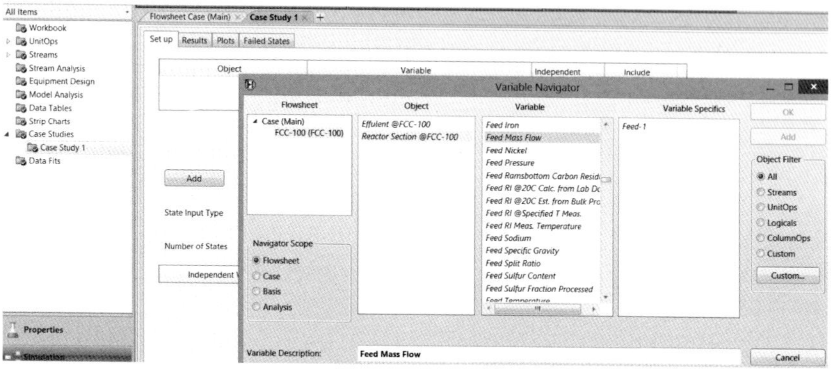

Figure 4.117 Define independent variables – feed mass flow (Feed-1) and riser outlet temperature (Riser).

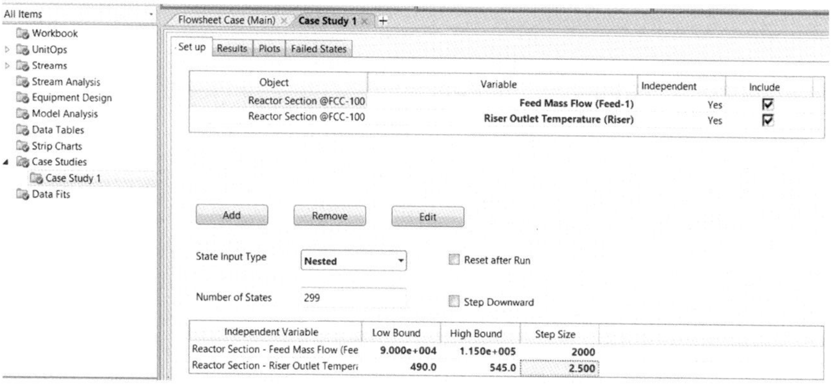

Figure 4.118 Specify the lower and upper bounds of independent variables and their step sizes.

As we focus only on the product yields, we use the "standard cuts grouped" from the model directly. Previously, we have defined the standard cuts or square cuts" in Section 4.13.10 and in Figure 4.71. Specifically, *standard cut grouped yields or square cut yields* refer to product yields with fixed end points; typical cuts include C1–C4 lump, C5-430 lump (C5-430 °F or 221 °C), 430–650 lump (430–650 °F, or 221–343 °C), 650–950 lump (650–950 °F, or 343–510 °C), and 950+ lump (>950 °F or 510 °C). Figure 4.119 shows our chosen cut yields for dependent variables.

It is possible to perform the same case study based on plant cuts. In that case, we would add a simple component splitter to separate the reactor effluent based on the initial and end points of the cuts (Figures 4.120 and 4.121).

We summarize the case study of the feed rate change (at a ROT of 510 °C) in Figure 4.122. We save the simulation file as *Workshop 4.4-1.hsc*.

As the feed rate increases to the unit, we note that there is a significant loss in the standard naphtha cut yield. In addition, both the LCO and Bottoms yields

FCC-100	Feed Mass Flow (Feed-1)
FCC-100	Yield, Std Cut Grouped (H2S-Weight %)
FCC-100	Yield, Std Cut Grouped (Naphtha C5-430F-Weight %)
FCC-100	Yield, Std Cut Grouped (Fuel Gas-Weight %)
FCC-100	Yield, Std Cut Grouped (Propane-Weight %)
FCC-100	Yield, Std Cut Grouped (Propylene-Weight %)
FCC-100	Yield, Std Cut Grouped (nButane-Weight %)
FCC-100	Yield, Std Cut Grouped (iButane-Weight %)
FCC-100	Yield, Std Cut Grouped (Butenes-Weight %)
FCC-100	Yield, Std Cut Grouped (Naphtha C5-430F-Weight %)
FCC-100	Yield, Std Cut Grouped (LCO 430-650F-Weight %)
FCC-100	Yield, Std Cut Grouped (Bottoms 650F+-Weight %)
FCC-100	Yield, Std Cut Grouped (Coke Yield-Weight %)
FCC-100	Yield, Std Cut Grouped (Conversion-Weight %)

Figure 4.119 The standard cut grouped yields as our dependent variables.

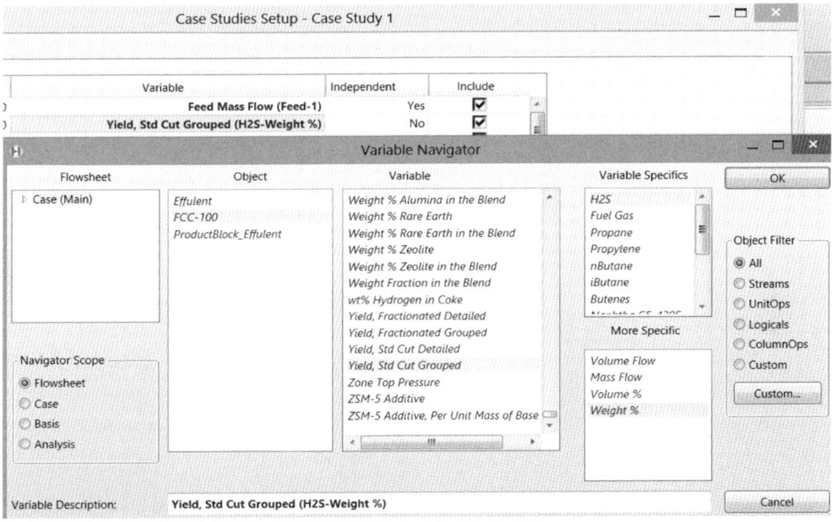

Figure 4.120 An illustration for the selection of dependent variable.

increase significantly. We discuss the reason for loss of naphtha yield extensively in the previous chapter. The loss is essentially a result of low residence time in the riser, which prevents catalytic cracking of the feed. In fact, most of the bottoms product can likely be recovered as LCO at a lower feed rate. *Therefore, if we try to increase the feed rate of the unit, we must also increase the cracking temperature to account for the lowered residence time.* We investigate an increase in cracking temperature in the next case study.

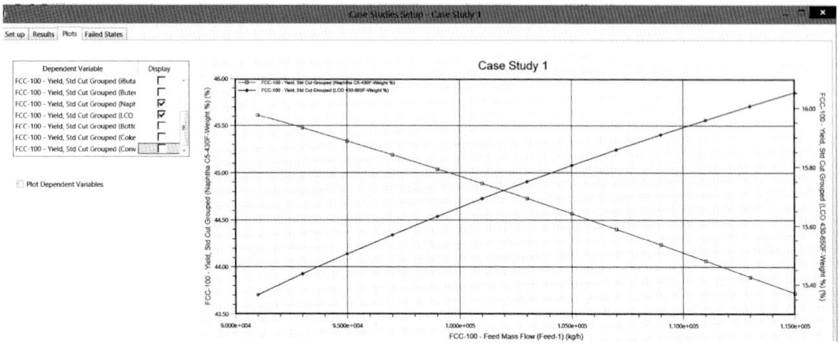

Figure 4.121 Effect of the FCC feed mass flow rate on the standard naphtha and LCO cut yields.

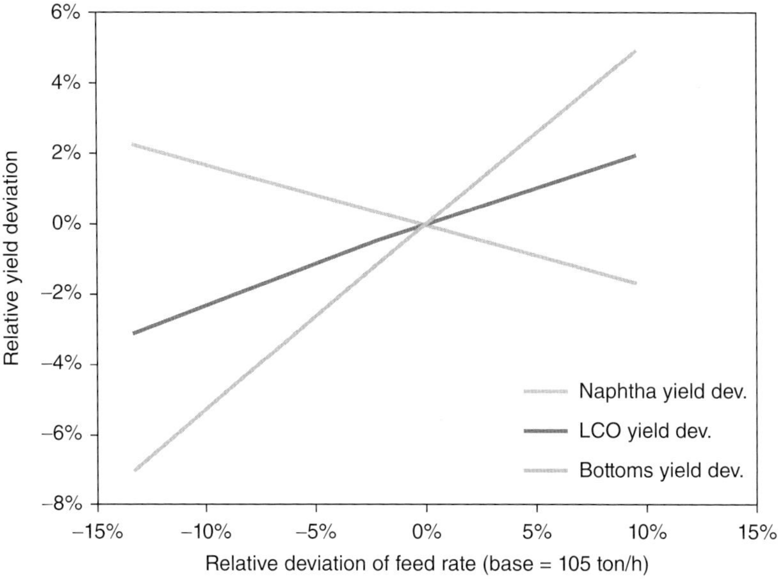

Figure 4.122 Effect of feed rate change on product yield change.

To study the effect of riser temperature at higher unit throughput, we create a case study where we vary the ROT. First, we increase the feed flow rate to the unit Reactor Section of the FCC unit operation window. For this example, we set the feed flow rate to 115 ton/h, as shown in Figure 4.123 and solve the model. If the model does not converge, we can increase the number of creep and total iterations in the Solve Options Section (Figure 4.124).

The results of the case study 2 appear in Figures 4.125 and 4.126. We summarize the results of the case study in Figure 4.127. We save the simulation file as *Workshop 4.4-2.hsc*.

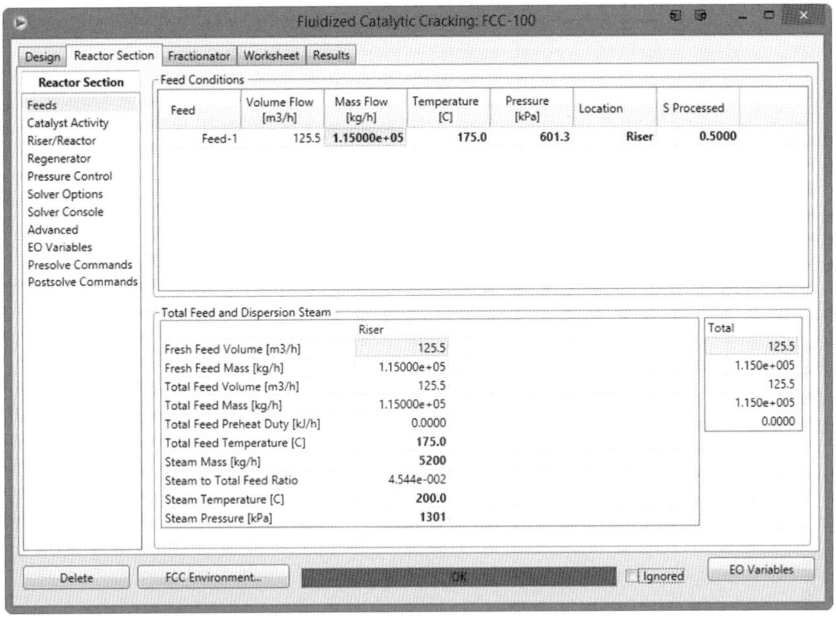

Figure 4.123 Increase feed flow rate for riser outlet temperature case study.

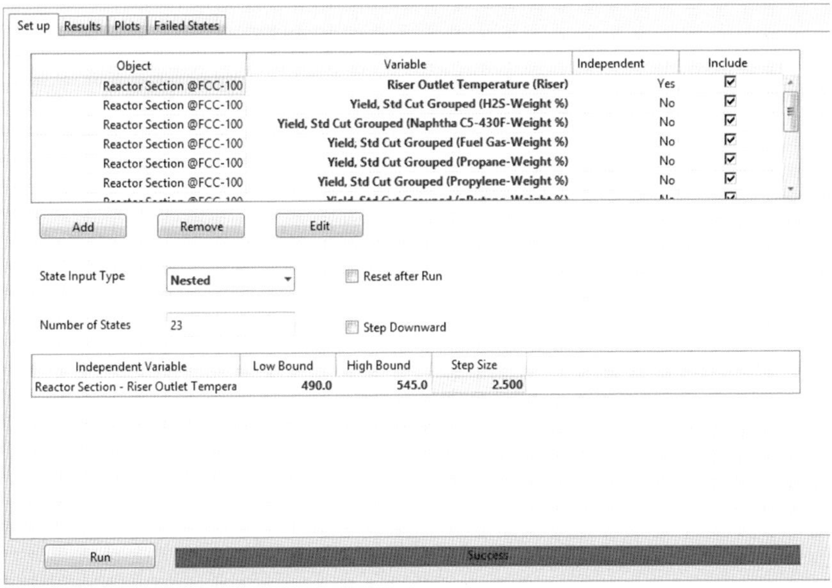

Figure 4.124 Case study setup for riser outlet temperature.

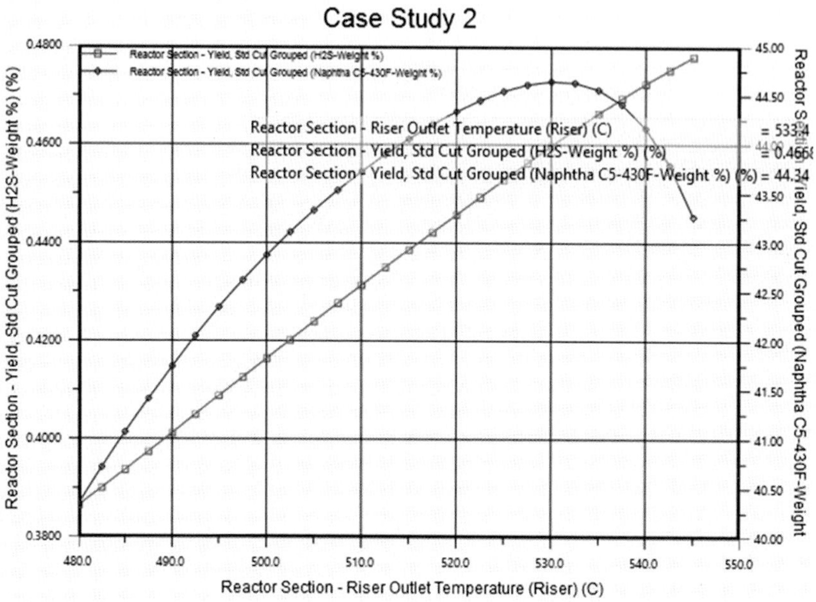

Figure 4.125 Graphical results from the case study.

State	Reactor Section - Riser Outlet Temperature (Riser) [C]	Reactor Section - Yield, Std Cut Grouped (H2S-Weight %) [%]	Reactor Section - Yield, Std Cut Grouped (Naphtha C5-430F-Weight %) [%]	Reactor Section - Yield, Std Cut Grouped (Fuel Gas-Weight %) [%]	R Gr
State 1	480.0	0.39	40.34	2.01	
State 2	482.5	0.39	40.71	2.04	
State 3	485.0	0.39	41.07	2.07	
State 4	487.5	0.40	41.41	2.11	
State 5	490.0	0.40	41.74	2.14	
State 6	492.5	0.40	42.05	2.18	
State 7	495.0	0.41	42.34	2.22	
State 8	497.5	0.41	42.62	2.26	
State 9	500.0	0.42	42.87	2.30	
State 10	502.5	0.42	43.11	2.35	
State 11	505.0	0.42	43.33	2.39	
State 12	507.5	0.43	43.53	2.44	

Figure 4.126 Tabular results from the case study.

Figure 4.127 shows that as we increase the ROT, the yield of naphtha also increases until we reach about 532 °C. At this point, the naphtha yield drops and we have a dramatic increase in the production of light gases and coke. In addition, there is also a significant decrease in the LCO yield. All of these trends are a result of the naphtha "overcracking" curve. We discussed this

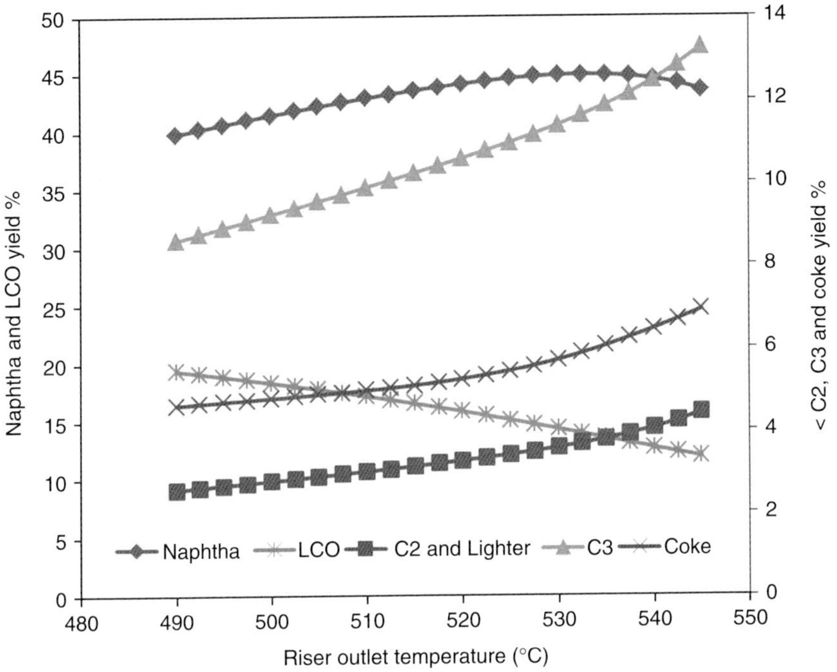

Figure 4.127 Product yield as a function of riser outlet temperature.

phenomenon extensively in the preceding chapter. Gasoline "overcracking" is a result of excessive thermal cracking and catalyst activation. Thermal cracking tends to produce many light compounds (C1–C4). This explains the increase in C2 and C3 yields. In addition, Coke yield increases because of increased coke deposits in the riser and subsequent catalyst deactivation. The lack of catalytic cracking activity explains the loss in LCO yield (as most of the feed that could have been cracked to LCO is now cracked directly in light gases). Figure 4.127 in conjunction with case study can help identify operating scenarios (flow rate and temperatures) to increase yield or shift product distribution slate from the FCC Unit.

4.17 Workshop 4.5 – Generate Delta-Base Vectors for Linear Programming (LP)-Based Planning

An important application of the calibrated model is the generation of LP delta-base vectors for refinery planning. The delta-base vectors essentially represent a linearized model of FCC unit as a function of several key variables. We have extensively discussed linear models in a previous chapter. In this workshop, we will demonstrate how to generate LP delta-base vectors for the calibrated FCC for use with a specific planning software, Aspen PIMS.

We open our calibrated model, *Workshop 4.2-done*, and save it as a new file, *Workshop 4.5.hsc*.

We can attempt to linearize the model by identifying key operating parameters and manually running the model for each chosen operating parameter. However, Aspen HYSYS provides a utility to automate this process. We can access the utility by going to the Analysis → Model Analysis → PIMS Support in home application menu, as shown in Figure 4.128.

Figure 4.129 shows the delta-base utility configuration window. We must first identify the scope of the delta-base utility. The scope refers to flowsheet objects we will modify during the course of the study. We choose the entire FCC unit as the scope of the utility, as shown in Figure 4.130.

To use the delta-base utility, we must first choose independent and dependent variables. The independent variables refer to model drivers or key operating parameters that control the yield of the unit. In the case of the FCC unit, the key operations parameters are feed-specific gravity, concarbon, and sulfur content.

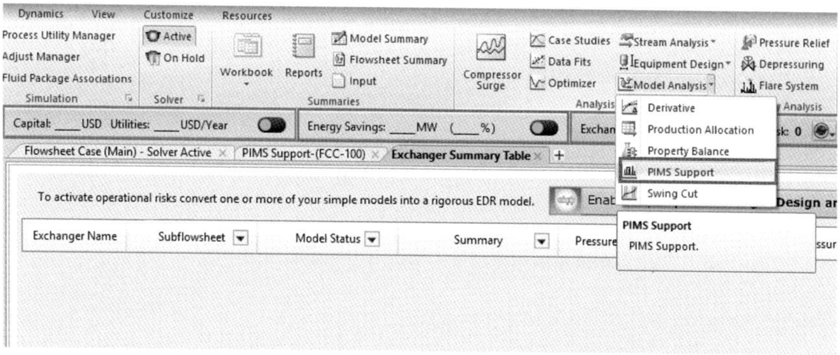

Figure 4.128 Creating the delta-base utility from main application menu bar.

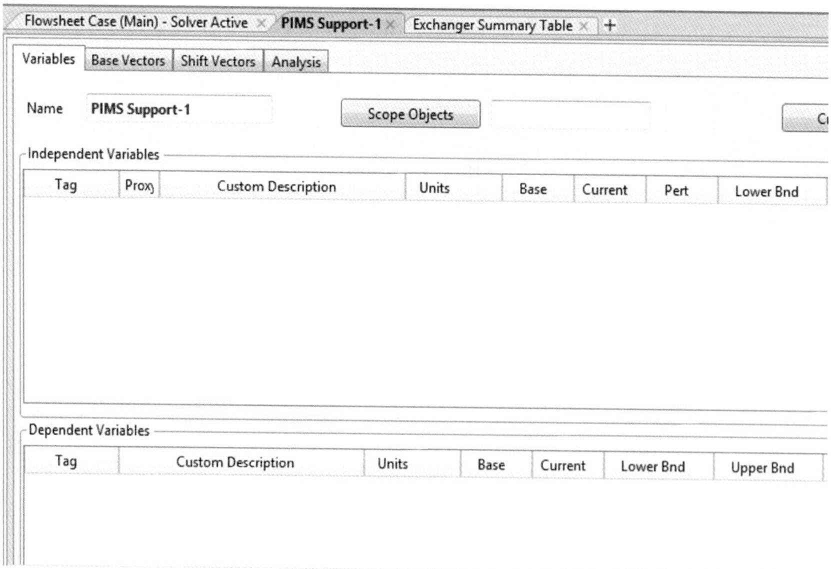

Figure 4.129 Delta-base utility configuration window.

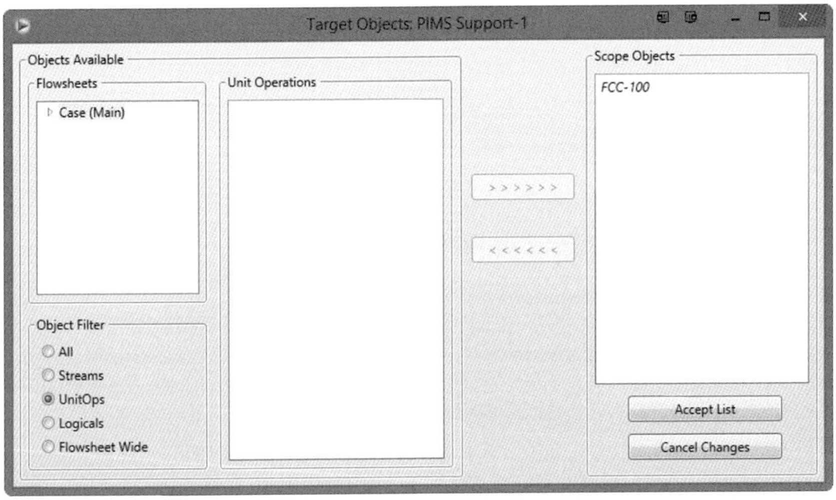

Figure 4.130 Scope of delta-base utility.

We add independent variables by clicking on the "Add Independent variables" button on the configuration window. The Variable Navigator (used in earlier workshops) appears and we select the following variables.

- FCC – 100 > Reactor Section > Feed Specific Gravity > Feed – 1
- FCC – 100 > Reactor Section > Feed Conradson Carbon > Feed – 1
- FCC – 100 > Reactor Section > Feed Sulfur Content > Feed – 1

Figure 4.131 shows how we can add the specific gravity to the independent variables. We repeat this process for the other independent variables. A description of each variable added appears in the "Desc." section.

After adding all the independent variables (Figure 4.132), we must add the dependent variables. The dependent variables in the case of refinery planning

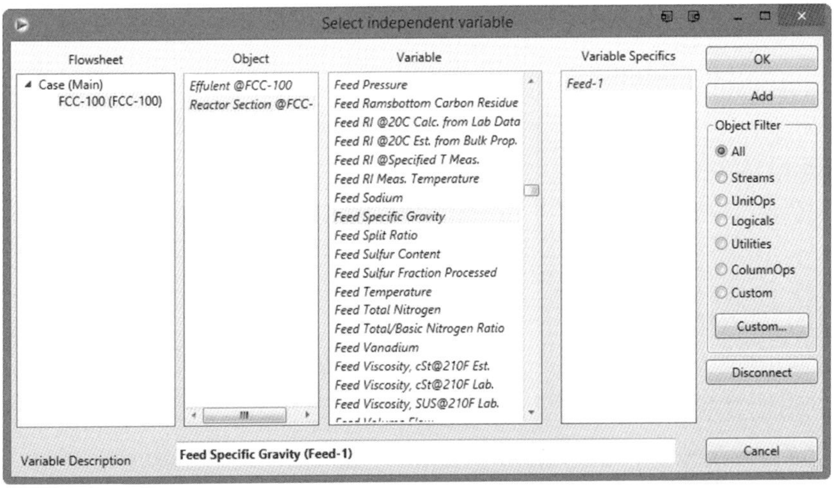

Figure 4.131 Adding specific gravity as an independent variable.

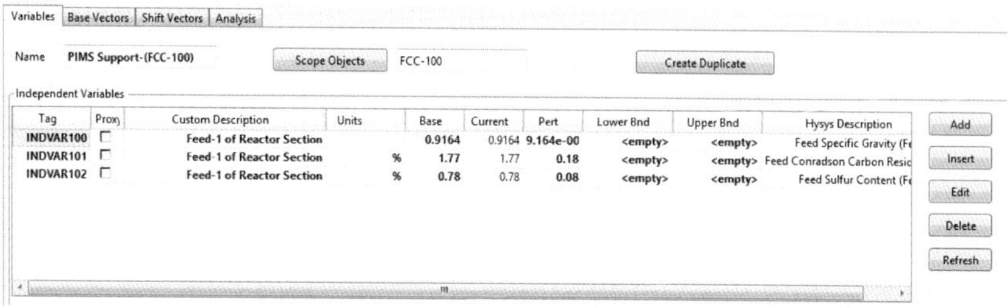

Tag	Proxy	Custom Description	Units	Base	Current	Pert	Lower Bnd	Upper Bnd	Hysys Description
INDVAR100		Feed-1 of Reactor Section		0.9164	0.9164	9.164e-00	<empty>	<empty>	Feed Specific Gravity (Fe
INDVAR101		Feed-1 of Reactor Section	%	1.77	1.77	0.18	<empty>	<empty>	Feed Conradson Carbon Resid
INDVAR102		Feed-1 of Reactor Section	%	0.78	0.78	0.08	<empty>	<empty>	Feed Sulfur Content (Fe

Figure 4.132 All independent variables added to delta-base utility.

almost always refer to the yields of the key products from the FCC unit. In this workshop, we use the square cut yields of the products. However, if we wish to use plant cut yields, we can use a simple component splitter to remap the product effluent from the FCC unit plant cuts based on a TBP cut points.

We add dependent variables by click "Add dependent variables" button. The variable navigator appears and we choose the group yields of all products as dependent variables. We show an example of adding H_2S yield to the dependent variable list in Figure 4.133.

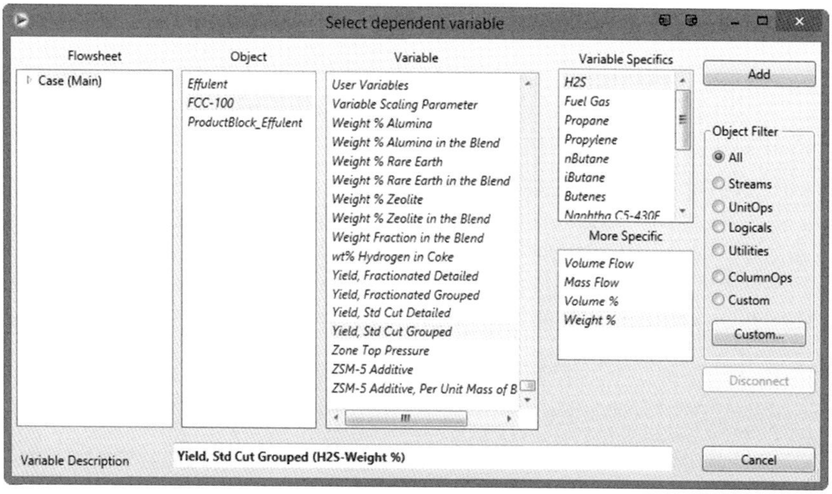

Figure 4.133 Adding H_2S yield as a dependent variable.

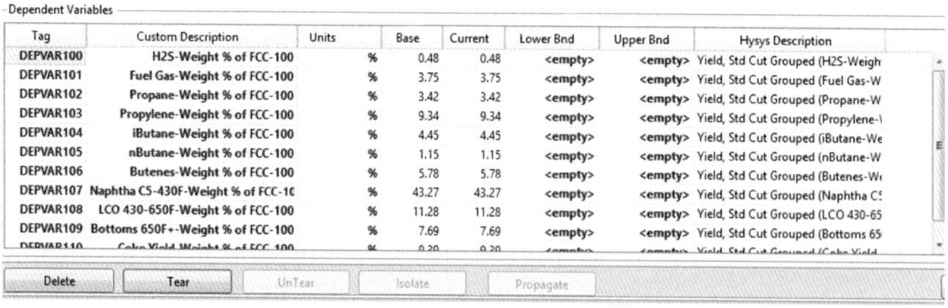

Tag	Custom Description	Units	Base	Current	Lower Bnd	Upper Bnd	Hysys Description
DEPVAR100	H2S-Weight % of FCC-100	%	0.48	0.48	<empty>	<empty>	Yield, Std Cut Grouped (H2S-Weigh
DEPVAR101	Fuel Gas-Weight % of FCC-100	%	3.75	3.75	<empty>	<empty>	Yield, Std Cut Grouped (Fuel Gas-W
DEPVAR102	Propane-Weight % of FCC-100	%	3.42	3.42	<empty>	<empty>	Yield, Std Cut Grouped (Propane-W
DEPVAR103	Propylene-Weight % of FCC-100	%	9.34	9.34	<empty>	<empty>	Yield, Std Cut Grouped (Propylene-\
DEPVAR104	iButane-Weight % of FCC-100	%	4.45	4.45	<empty>	<empty>	Yield, Std Cut Grouped (iButane-We
DEPVAR105	nButane-Weight % of FCC-100	%	1.15	1.15	<empty>	<empty>	Yield, Std Cut Grouped (nButane-W
DEPVAR106	Butenes-Weight % of FCC-100	%	5.78	5.78	<empty>	<empty>	Yield, Std Cut Grouped (Butenes-W
DEPVAR107	Naphtha C5-430F-Weight % of FCC-1C	%	43.27	43.27	<empty>	<empty>	Yield, Std Cut Grouped (Naphtha C!
DEPVAR108	LCO 430-650F-Weight % of FCC-100	%	11.28	11.28	<empty>	<empty>	Yield, Std Cut Grouped (LCO 430-65
DEPVAR109	Bottoms 650F+-Weight % of FCC-100	%	7.69	7.69	<empty>	<empty>	Yield, Std Cut Grouped (Bottoms 65
DEPVAR110	Coke Yield Weight % of FCC-100	%	0.20	0.20	<empty>	<empty>	Yield, Std Cut Grouped (Coke Yield

Figure 4.134 All dependent variables added to delta-base utility.

We use the variable navigator to add the following variables as dependent variables (Figure 4.134).

- Case > FCC-100 > Yield, Std. Cut. Grouped > H_2S
- Case > FCC-100 > Yield, Std. Cut. Grouped > Fuel Gas
- Case > FCC-100 > Yield, Std. Cut. Grouped > Propane
- Case > FCC-100 > Yield, Std. Cut. Grouped > Propylene
- Case > FCC-100 > Yield, Std. Cut. Grouped > *n*Butane
- Case > FCC-100 > Yield, Std. Cut. Grouped > *i*Butane
- Case > FCC-100 > Yield, Std. Cut. Grouped > Butenes
- Case > FCC-100 > Yield, Std. Cut. Grouped > Naphtha C5-430 °F
- Case > FCC-100 > Yield, Std. Cut. Grouped > LCO 430F – 650 °F
- Case > FCC-100 > Yield, Std. Cut. Grouped > Bottoms 650+ °F
- Case > FCC-100 > Yield, Std. Cut. Grouped > Coke

The next step is to choose a perturbation amount for each variable. As the delta-base utility generates a linearized model of the FCC unit, we must choose the range over which we need to linearize the model. For this workshop, we will perturb each independent variable by 10% of its original base value, as shown in Figure 4.135. We can click "Generate Derivatives" to begin running the model.

Once we click the "Generate Derivatives" button, the model runs several times at the base and perturbed values of the independent variables. The delta-base values appear in the table shown in Figure 4.136. These values may be directly copied into an Excel spreadsheet for Aspen PIMS or exported for further study. We can export the table to a PIMS style interface by clicking the "Export Data." The exported data are shown in Figure 4.137.

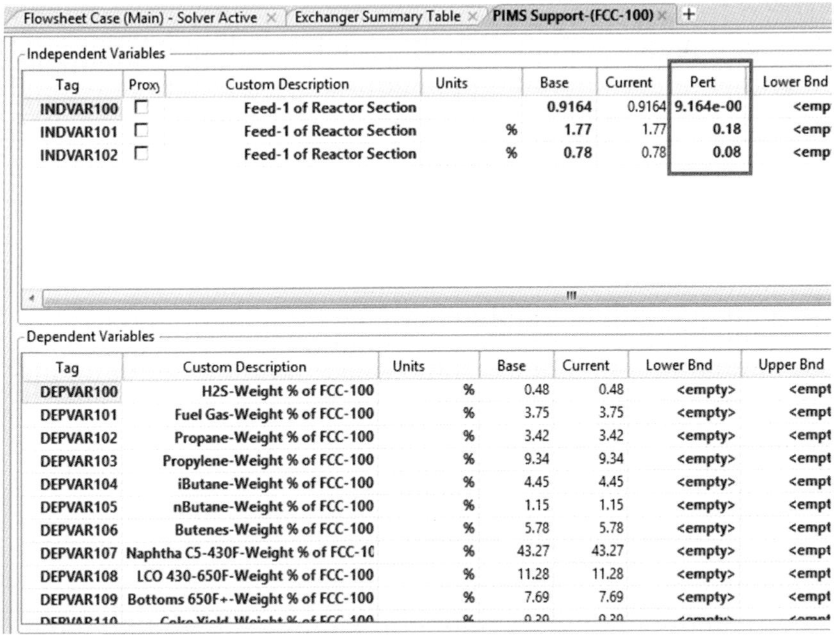

Tag	Proxy	Custom Description	Units	Base	Current	Pert	Lower Bnd
INDVAR100	☐	Feed-1 of Reactor Section		0.9164	0.9164	9.164e-00	<emp
INDVAR101	☐	Feed-1 of Reactor Section	%	1.77	1.77	0.18	<emp
INDVAR102	☐	Feed-1 of Reactor Section	%	0.78	0.78	0.08	<emp

Tag	Custom Description	Units	Base	Current	Lower Bnd	Upper Bnd
DEPVAR100	H2S-Weight % of FCC-100	%	0.48	0.48	<empty>	<empt
DEPVAR101	Fuel Gas-Weight % of FCC-100	%	3.75	3.75	<empty>	<empt
DEPVAR102	Propane-Weight % of FCC-100	%	3.42	3.42	<empty>	<empt
DEPVAR103	Propylene-Weight % of FCC-100	%	9.34	9.34	<empty>	<empt
DEPVAR104	iButane-Weight % of FCC-100	%	4.45	4.45	<empty>	<empt
DEPVAR105	nButane-Weight % of FCC-100	%	1.15	1.15	<empty>	<empt
DEPVAR106	Butenes-Weight % of FCC-100	%	5.78	5.78	<empty>	<empt
DEPVAR107	Naphtha C5-430F-Weight % of FCC-1C	%	43.27	43.27	<empty>	<empt
DEPVAR108	LCO 430-650F-Weight % of FCC-100	%	11.28	11.28	<empty>	<empt
DEPVAR109	Bottoms 650F+-Weight % of FCC-100	%	7.69	7.69	<empty>	<empt
DEPVAR110	Coke Yield-Weight % of FCC-100	%	0.20	0.20	<empty>	<empt

Figure 4.135 Perturb independent variables.

Variables | Base Vectors | Shift Vectors | Analysis

Name PIMS Support-(FCC-100)

Parameters (Optional)

Independent Variables

Tag	INDVAR100	INDVAR101	INDVAR102
Units		%	%
Pert	9.164e-002	0.18	0.08
Use proxy	☐	☐	☐
Desc	Feed-1 of Reacto	Feed-1 of Reacto	Feed-1 of Reacto
Base Value	0.9164	1.77	0.78
Skip Deriv Calc	☐	☐	☐

Dependent Variables / Derivatives

Tag	Desc	Units	Base	Allow Edit	Derivatives		
DEPVAR100	H2S-Weight 9	%	0.48	☐	-3.745e-002	-4.369e-003	0.6173
DEPVAR101	Fuel Gas-Weig	%	3.75	☐	-0.1422	-3.874e-002	-4.300e-002
DEPVAR102	Propane-Weig	%	3.42	☐	-0.3659	-3.512e-002	-3.782e-002
DEPVAR103	Propylene-We	%	9.34	☐	-0.4879	-8.365e-002	-8.858e-002
DEPVAR104	iButane-Weig	%	4.45	☐	-0.4936	-3.905e-002	-4.189e-002
DEPVAR105	nButane-Weig	%	1.15	☐	-0.1305	-9.192e-003	-9.837e-003
DEPVAR106	Butenes-Weig	%	5.78	☐	-0.2164	-4.883e-002	-5.136e-002
DEPVAR107	Naphtha C5-4	%	43.27	☐	-3.271	-0.2808	-0.2933
DEPVAR108	LCO 430-650F	%	11.28	☐	6.767	4.028e-002	-1.715e-002
DEPVAR109	Bottoms 650F	%	7.69	☐	4.897	1.006e-002	-3.094e-003
DEPVAR110	Coke Yield-W	%	9.39	☐	-6.519	0.4894	-3.120e-002

Figure 4.136 Results from delta-base utility.

A	B	C	D	E	F	G
	TEXT	BAS	IndVar100	IndVar101	IndVar102	***
*	Units			%	%	
*	Base Perturbation		0.09164	0.177	0.078	
*						
DepVar10	H2S-Weight % of FCC-100 %	-0.47791	0.03745	0.00437	-0.61727	
DepVar10	Fuel Gas-Weight % of FCC-100 %	-3.74881	0.14215	0.03874	0.043	
DepVar10	Propane-Weight % of FCC-100 %	-3.42432	0.36591	0.03512	0.03782	
DepVar10	Propylene-Weight % of FCC-100 %	-9.339	0.48788	0.08365	0.08858	
DepVar10	iButane-Weight % of FCC-100 %	-4.45333	0.49363	0.03905	0.04189	
DepVar10	nButane-Weight % of FCC-100 %	-1.14871	0.13053	0.00919	0.00984	
DepVar10	Butenes-Weight % of FCC-100 %	-5.78458	0.21641	0.04883	0.05136	
DepVar10	Naphtha C5-430F-Weight % of FCC-100 %	-43.2683	3.271	0.28077	0.29333	
DepVar10	LCO 430-650F-Weight % of FCC-100 %	-11.2755	-6.76671	-0.04028	0.01715	
DepVar10	Bottoms 650F+-Weight % of FCC-100 %	-7.68896	-4.89723	-0.01006	0.00309	
DepVar11	Coke Yield-Weight % of FCC-100 %	-9.39057	6.51897	-0.48937	0.0312	
*						
	Feed-1 of Reactor Section	0.9164	1			
	Feed-1 of Reactor Section	1.77		1		
	Feed-1 of Reactor Section	0.7761			1	

Figure 4.137 PIMS style output for delta-base vectors.

If necessary, we can also rename all the variables to be consistent with PIMS delta-base vectors. To rename variables, we enter new names for each entry in the corresponding "Tag" box, as shown in Figure 4.138. When we reexport the delta-base table, we replace all variables with the new tags, as shown in Figure 4.139.

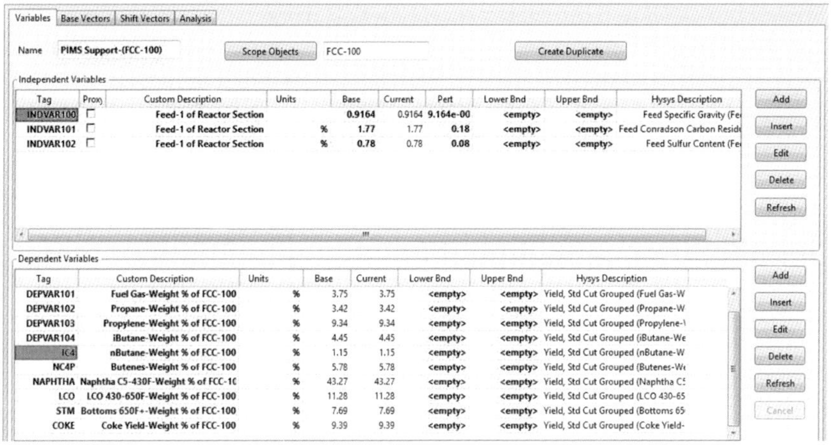

Figure 4.138 Renaming variables in delta-base utility.

	TEXT	BAS	IndVar100	IndVar101	IndVar102	***
*	Units			%	%	
*	Base Perturbation		0.09164	0.177	0.078	
*						
DepVar10	H2S-Weight % of FCC-100 %	-0.47791	0.03745	0.00437	-0.61727	
SOURGAS	Fuel Gas-Weight % of FCC-100 %	-3.74881	0.14215	0.03874	0.043	
DRYGAS	Propane-Weight % of FCC-100 %	-3.42432	0.36591	0.03512	0.03782	
C3	Propylene-Weight % of FCC-100 %	-9.339	0.48788	0.08365	0.08858	
C3P	iButane-Weight % of FCC-100 %	-4.45333	0.49363	0.03905	0.04189	
IC4	nButane-Weight % of FCC-100 %	-1.14871	0.13053	0.00919	0.00984	
NC4P	Butenes-Weight % of FCC-100 %	-5.78458	0.21641	0.04883	0.05136	
NAPHTHA	Naphtha C5-430F-Weight % of FCC-100 %	-43.2683	3.271	0.28077	0.29333	
LCO	LCO 430-650F-Weight % of FCC-100 %	-11.2755	-6.76671	-0.04028	0.01715	
STM	Bottoms 650F+-Weight % of FCC-100 %	-7.68896	-4.89723	-0.01006	0.00309	
COKE	Coke Yield-Weight % of FCC-100 %	-9.39057	6.51897	-0.48937	0.0312	
*						
	Feed-1 of Reactor Section	0.9164	1			
	Feed-1 of Reactor Section	1.77		1		
	Feed-1 of Reactor Section	0.7761			1	
*						

Figure 4.139 Renamed variables and tags in PIMS interface.

4.18 Conclusions

In this work, we have developed a model for a FCC unit that includes a significant implementation of the associated gas plant using Aspen HYSYS. The key highlights of this work are as follows:

1) Brief summary of existing literature for modeling a typical FCC unit.
2) Description of the Aspen HYSYS FCC model and 21-lump kinetics.
3) Technique to fill out partial distillation curves using statistical functions.
4) Regression of parameters for a new PNA correlation for petroleum fractions.
5) Technique to infer molecular composition of FCC feedstock from routine analysis.

6) Strategy to develop reasonable process models using industrial plant data.
7) Application of the model to a large-scale refinery process showing less than 2.0% AAD for key product yields and satisfactory predictions of product composition and product quality (composition/distillation data, density, and flash point).
8) Case studies that use the model to investigate industrially useful changes in operation.
9) Strategy to transfer results from this model into LP-based refinery planning tool.

Earlier work in this area has focused mostly on isolated parts (kinetic model, riser/regenerator, and gas plant) of the FCC process. In this work, we show how to use routinely collected plant data with well-known commercial software tools to present an integrated process model that includes both reaction and fractionation systems. An integrated model allows users to identify opportunities to improve yield, to increase profitability, and to monitor the unit for predictable operation. This approach is critical for modern refineries that have increasingly complex process flows and require engineers to examine the performance of refinery units holistically.

Nomenclature

VGO	Vacuum gas oil
CGO	Coker gas oil
LCO	Light cycle oil
HCO	Heavy cycle oil
TBP	True boiling point
C1	Methane
C2	Ethane
C3	Propane and propylene
C4	Butanes and butenes
C5	Pentanes and pentenes
PNA	Paraffin, naphthene, and aromatics
φ	Slip factor, unitless
ε	Voidage factor, unitless
D	Riser diameter, m
G	Acceleration due to gravity, $m/s^2 = 9.81\ m/s^2$
u_o	Superficial gas velocity, m/s
u_t	Terminal catalyst particle settling velocity, m/s
Fr	Froude number, unitless
Fr_t	Particle Froude number, unitless
ϕ_{COKE}	Total coke deactivation function, unitless
ϕ_{KCOKE}	Deactivation function due to kinetic coke, unitless
ϕ_{MCOKE}	Deactivation function due to metal coke, unitless
C_{KCOKE}	Kinetic coke on catalyst, kg kinetic coke/kg catalyst
C_{MCOKE}	Metal coke on catalyst, kg metal coke/kg catalyst

C_{METALS}	Metal composition on catalyst ppm metals/kg catalyst
a_{KCOKE}	Activity factor due to kinetic coke, unitless
a_{MCOKE}	Activity factor due to metal coke, unitless
E	Murphree stage efficiency factor
x_n	Mole fraction of liquid leaving stage n
y_n	Mole fraction of vapor leaving stage n
X	Normalized liquid recovery, unitless
x_{exp}	Normalized experimental liquid recovery, unitless
RSS	Sum of least squares
AAD	Average absolute deviation
A, B, α, β	Fitting parameters for cumulative beta distribution
θ	Normalized temperature
T_0	Lower reference temperature, °C
T_1	Upper reference temperature, °C
$\%X_P$	Mole composition of paraffins, unitless
$\%X_N$	Mole composition of naphthenes, unitless
$\%X_A$	Mole composition of aromatics, unitless
R_i	Refractive index, unitless
VGC	Viscosity gravity constant, unitless
VGF	Viscosity gravity factor, unitless
a, b, c, d	Fitting parameters for PNA correlation
SG, SPG	Specific gravity
K_W	Watson K factor, unitless
MeABP	Mean average boiling point temperature, R
RON	Research octane number
MON	Motor octane number
CCR, CON	Conradson carbon residue, wt%
Yield_i	Yield coefficients for LP model, unitless
SUL	Sulfur content, wt%

Bibliography

1 Sadeghbeigi, R. (2000) *Fluid Catalytic Cracking Handbook. Design, Operation and Troubleshooting of FCC Facilities*, Gulf Publishing Company, Houston, TX.
2 Arbel, A., Huang, Z., Rinard, I.H., Shinnar, R., and Sapre, A.V. (1995) *Industrial and Engineering Chemistry Research*, **34**, 1228–1243.
3 McFarlane, R.C., Reineman, R.C., Bartee, J.F., and Georgakis, C. (1993) *Computers and Chemical Engineering*, **3**, 275–300.
4 Chitnis, U.K. and Corripio, A.B. (1998) *ISA Transactions*, **37**, 215–226.
5 Khandalekar, P.D. and Riggs, J.B. (1995) *Computers and Chemical Engineering*, **19**, 1153–1168.
6 Hsu, C.S. and Robinson, P.R. (2006) *Practical Advances in Petroleum Processing. Volume 1 &2*, Springer, New York.
7 Gary, J.H. and Handwerk, G.E. (2001) *Petroleum Refining Technology and Economics*, 4th edn, Marcel-Dekker, New York.

8 Raseev, S.D. (2003) *Thermal and Catalytic Processing in Petroleum Refining*, CRC Press, Boca Raton, FL.
9 Takatsuka, T., Sato, S., Morimoto, Y., and Hashimoto, H. (1987) *International Chemical Engineering*, **27**, 107–116.
10 Lee, E. and Groves, F.R. Jr. (1985) *Transactions of the Society for Computer Simulation International*, **2**, 219–236.
11 Blanding, F.H. (1953) *Industrial and Engineering Chemistry*, **45**, 1193–1197.
12 Gupta, R.K., Kumar, V., and Srivastava, V.K. (2007) *Chemical Engineering Science*, **62**, 4510–4528.
13 Jacob, S.M., Gross, B., Voltz, S.E., and Weekman, V.W. (1976) *AIChE Journal*, **22**, 701–713.
14 Oliviera, L.L. and Biscasia, E.C. Jr. (1989) *Industrial and Engineering Chemistry Research*, **28**, 264–271.
15 Pitault, I., Nevicato, D., Forissier, M., and Bernard, J.R. (1994) *Chemical Engineering Science*, **49**, 4249–4262.
16 Van Landeghem, F., Nevicato, D., Pitault, I., Forissier, M., Turlier, P., Derouin, C., and Bernard, J.R. (1996) *Applied Catalysis A*, **138**, 381–405.
17 Aspen RefSYS Option Guide (2006) AspenTech, Cambridge, MA.
18 Aspen Plus FCC User's Guide (2006) AspenTech, Cambridge, MA.
19 Froment, G.F. (2005) *Catalysis Reviews – Science and Engineering*, **47**, 83.
20 Quann, R.J. and Jaffe, S.B. (1992) *Industrial and Engineering Chemistry Research*, **31**, 2483.
21 Quann, R.J. and Jaffe, S.B. (1996) *Chemical Engineering Science*, **51**, 1615.
22 Quann, R. (1998) *Environmental Health Perspectives Supplements*, **106**, 1501.
23 Christensen, G., Apelian, M.R., Hickey, K.J., and Jaffe, S.B. (1999) *Chemical Engineering Science*, **54**, 2753–2764.
24 Klein, M.T. (2006) *Molecular Modeling in Heavy Hydrocarbon Conversions*, CRC Press, Boca Raton, FL.
25 Kumar, S., Chadha, A., Gupta, R., and Sharma, R. (1995) *Industrial and Engineering Chemistry Research*, **34**, 3737–3748.
26 Ellis, R.C., Li, X., and Riggs, J.B. (1998) *AIChE Journal*, **44**, 2068–2079.
27 Secchi, A.R., Santos, M.G., Neumann, G.A., and Trierwiler, J.O. (2001) *Computers and Chemical Engineering*, **25**, 851–858.
28 Mo, W., Hadjigeorge, G., Khouw, F.H.H., van der Werf, R.P., and Muller, F. (October 2002) *Hydrocarbon Asia*, 30–42.
29 Elnashaie, S.S.E.H., Mohamed, N.F., and Kamal, M. (2004) *Chemical Engineering Communications*, **191**, 813–831.
30 Rao, R.M., Rengaswamy, R., Suresh, A.K., and Balaraman, K.S. (2004) *Trans IChemE: Part A*, **82**, 527–552.
31 Araujo-Monroy, C. and Lopez-Isunza, F. (2006) *Industrial and Engineering Chemistry Research*, **45**, 120–128.
32 Bollas, G.M., Vasalos, I.A., Lappas, A.A., Iatridis, D.K., Voutetakis, S.S., and Papadopoulou, S.A. (2007) *Chemical Engineering Science*, **62**, 1887–1904.
33 Fernandes, J.L., Pinheiro, C.I.C., Oliveira, N.M.C., Inverno, J., and Ribeiro, F.R. (2008) *Industrial & Engineering Chemistry Research*, **47**, 850–866.
34 Shaikh, A.A., Al-Mutairi, E.M., and Ino, T. (2008) *Industrial & Engineering Chemistry Research*, **47**, 9018–9024.

35 Fernandes, J.L., Pinheiro, C.I.C., Oliveira, N.M.C., Neto, A.I., and F. Ramôa, R. (2007) *Chemical Engineering Science*, **62**, 6308–6322.
36 Chang, S.L. and Zhou, C.Q. (2003) *Computational Mechanics*, **31**, 519–532.
37 Arandes, J.M., Azkoti, M.J., Bilbao, J., and de Lasa, H.I. (2000) *The Canadian Journal of Chemical Engineering*, **78**, 111–123.
38 Han, I.S., Riggs, J.B., and Chung, C.B. (2004) *Chemical Engineering and Processing*, **43**, 1063–1084.
39 Paraskos, J.A., Shah, Y.T., McKinney, J.D., and Carr, N.L. (1976) *Industrial and Engineering Chemistry Process Design and Development*, **15**, 165–169.
40 Shah, Y.T., Huling, G.P., Paraskos, J.A., and McKinney, J.D. (1977) *Industrial and Engineering Chemistry Process Design and Development*, **16**, 89–94.
41 Arandes, J.M., Abajo, I., Fernandez, I., Lopez, D., and Bilbao, J. (1999) *Industrial and Engineering Chemistry Research*, **38**, 3255–3260.
42 De Lasa, H.I. and Grace, J.R. (1979) *The Canadian Journal of Chemical Engineering*, **25**, 984–990.
43 Rice, N.M. and Wojciechowski, B.W. (1991) *The Canadian Journal of Chemical Engineering*, **69**, 1100–1105.
44 Harriot, P. (2003) *Chemical Reactor Design*, Marcel Dekker, New York, NY.
45 Malay, P., Milne, B.J., and Rohani, S. (1999) *The Canadian Journal of Chemical Engineering*, **77**, 169–179.
46 Corella, J. and Frances, E. (1991) *Fluid Catalytic Cracking-II. Concepts in Catalyst Design*, ACS Symposium Series, vol. **452**, American Chemical Society, Washington, DC, pp. 165–182.
47 Bolkan-Kenny, Y.G., Pugsley, T.S., and Berutti, F. (1994) *Industrial and Engineering Chemistry Research*, **33**, 3043–3052.
48 Han, I.S. and Chung, C.B. (2001) *Chemical Engineering Science*, **56**, 1951–1971.
49 Froment, G.F., Bischoff, K.B., and Wilde, J.D. (2010) *Chemical Reaction Analysis and Design*, 3rd edn, Wiley.
50 Kister, H.Z. (1992) *Distillation Design*, McGraw-Hill, Inc., New York, NY.
51 Kaes, G.L. (2000) *Refinery Process Modeling A Practical Guide to Steady State Modeling of Petroleum Processes*, The Athens Printing Company, Athens, GA.
52 Bollas, G.M., Vasalos, I.A., Lappas, A.A., Iatridis, D.K., and Tsioni, G.K. (2004) *Industrial and Engineering Chemistry Research*, **43**, 370–3281.
53 Sanchez, S., Ancheyta, J., and McCaffrey, W.C. (2007) *Energy & Fuels*, **21**, 2955–2963.
54 Daubert, T.E. and Danner, R.P. (1997) *API Technical Data Book – Petroleum Refining*, 6th edn, American Petroleum Institute, Washington DC.
55 Riazi, M.R. (2005) *Characterization and Properties of Petroleum Fractions*, 1st edn, American Society for Testing and Materials, West Conshohocken, PA.
56 Goosens, A.G. (1997) *Industrial and Engineering Chemistry Research*, **36**, 2500.
57 Bazaraa, M.S., Jarvis, J.J., and Sherali, H.D. (2009) *Linear Programming and Network Flows*, John Wiley and Sons, Hoboken, NJ.
58 Xu, C., Gao, J., Zhao, S., and Lin, S. (2005) *Fuel*, **84**, 669–674.
59 Ancheyta-Juarez, J. and Murillo-Hernandez, J.A. (2000) *Energy & Fuels*, **14**, 373–379.

60 Li, W., Chi-Wai, H., and An-Xue, L. (2005) *Computers and Chemical Engineering*, **29**, 2010–2028.
61 Davision, G. (1993) *Guide to Fluid Catalytic Cracking*, W.R. Grace & Co., Columbia, MD, pp. 65–66.
62 Saleh, K., Ibrahim, H., Jayyousi, M. and Diabat, A. (2013) A Novel Optimization Formulation of Fluid Catalytic Cracking Unit. *5th International Conference on Industrial Engineering and Systems Management (IESM)*, Rabat, Morocco, October.
63 Gao, H., Wang, G., Li, R., Xu, C., and Gao, J. (2012) *Energy and Fuels*, **26**, 1880–1891.
64 Xu, J., Chen, Z., Fan, Y., Shi, G., and Bao, X. (2015) *Fuel Processing Technology*, **130**, 117–126.
65 Zhang, J., Chang, J., Chen, H., Yang, Y., Meng, F., and Wang, L. (2012) *Chemical Engineering Science*, **78**, 128–143.
66 Pashikanti, K. and Liu, Y.A. (2011) Predictive modeling of large-scale integrated refinery reaction and fractionation systems from plant data: 2. Fluid catalytic cracking (FCC) process. *Energy and Fuels*, **25**, 5298–5319.
67 Lancu, M. and Agachi, P.S. (2010) Optimal process control and operation of an industrial heat integrated fluid catalytic cracking plant using model predictive control (MPC). *Computer Aided Chemical Engineering*, **28**, 505–510.
68 Lancu, M., Criestea, M.V., and Agachi, P.S. (2013) Retrofit design of heat exchanger network of a fluid catalytic cracking plant and control based on MPC. *Computers and Chemical Engineering*, **49**, 205–216.
69 Radu, S.; Ciuparu, D., Modeling and simulation of an industrial fluid catalytic cracking unit, *Revista de Chimie*, 2014, **65**, 113–119. http://www.revistadechimie.ro/pdf/RADU%20S.pdf%201%2014.pdf.
70 Kumar, S., Lange, J.-P., and Rossum, G.V. (2015) Liquefaction of lignocellulose in fluid catalytic cracker feed: A process concept study. *ChemSusChem*, **8**, 4086–4094.
71 Khandeparker, A. (2012) Study of Different Operating Parameters of FCC Unit with Aspen-HYSYS, National Institute of Technology, Rourkela, http://ethesis.nitrkl.ac.in/3436/.
72 Yusuf, R.O., El-Nafaty, V.A., and Jibril, M. (2012) Effects of operating variables on fluid catalytic cracking unit (FCCU) using HYSYS. *International Journal of Computer Applications*, **3** (2), 1–9.
73 Azubuike, L. C.; Okonkwo, E.; Egbujuo, W.; Chilke-Onyegbula, C. Optimization of propylene production process from fluid catalytic cracking unit, European Journal of Advances in Engineering and Technology, 2016, **3**, No. 9, 81–87. http://www.ejaet.com/PDF/3-9/EJAET-3-9-81-87.pdf.
74 Rajeev, N., Prasad, R.K., and Ragula, U.B.R. (2015) Process simulation and modeling of fluidized catalytic cracker performance in crude refinery. *Petroleum Science and Technology*, **33**, 110–117.
75 Takeda, K. (2011) Refinery Margin Improvement, Taiyo Oil Company Ltd., AspenTech Global Conference: OPTIMIZE 2011, Washington, DC, May 2011.

5

Predictive Modeling of Continuous Catalyst Regeneration (CCR) Reforming Process

This chapter presents the methodology for developing a predictive model for the rating and optimization of an integrated catalytic reforming process with a continuous catalyst regeneration (CCR) using Aspen HYSYS Petroleum Refining. The model relies on routinely monitored data such as ASTM distillation curves, paraffin–naphthene–aromatic (PNA) analysis, and operating conditions. We use a lumped kinetic network with 64 species over a broad C1–C14 range. This network can represent the key dehydrogenation, dehydrocyclization, isomerization, and hydrocracking reactions that typically occur with petroleum feedstock. The lumped kinetic scheme allows us to make accurate predictions of benzene, toluene, ethylbenzene, and xylenes (BTEX). Also, this work accounts for the coke deposited on the catalyst and the associated catalyst regeneration. We implement the hydrogen recycle and product recontacting sections as separate unit operations connected to the CCR reformer model. Also, we include rigorous tray-by-tray simulation models for primary product recovery.

We validate this model using 6 months of plant data from a commercial CCR reforming process handling a feed capacity of 1.4 million tons per year in the Asia Pacific. The validated model predicts key process yields and aromatic yields to within an average absolute deviation (AAD) of 1%. In addition, the model predicts liquid petroleum gas (LPG) composition to within 2.0% AAD. We also present several industrially useful case studies that display common interactions among process variables such as feed composition, reaction temperature, space velocity, and hydrogen-to-hydrocarbon ratio (H_2HC). These case studies accurately quantify the effects of key process variables on process performance and demonstrate the model applications for improving energy efficiency and for optimizing the reformer performance for chemical feedstock production.

This chapter differentiates itself from the reported studies in the literature through the following contributions: (1) detailed kinetic model that accounts for coke generation and catalyst deactivation; (2) complete implementation of a recontactor and primary product fractionation; (3) feed lumping from limited feed information; (4) detailed procedure for kinetic model calibration; (5) industrially relevant case studies that highlight the effects of changes in key process variables; and (6) application of the model to refinery-wide production planning.

The contents of this chapter are as follows. Section 5.2 gives the motivation for our model development and applications. Section 5.3 describes a typical

Petroleum Refinery Process Modeling: Integrated Optimization Tools and Applications,
First Edition. Y. A. Liu, Ai-Fu Chang, and Kiran Pashikanti.

catalytic reforming unit with CCR. Section 5.4 discusses the chemistry of the catalytic reforming process. Section 5.4.1 presents a literature review relevant to the predictive modeling of catalytic reforming processes, covering lumped kinetic models and unit-level models. Section 5.5 describes the features of the Aspen HYSYS Petroleum Refining CatReform model. Section 5.6 discusses the thermophysical properties required for our model development and the suggested methods for estimating them. Section 5.7 discusses the modeling of the downstream fractionation units. Section 5.8 presents the important aspects of feed characterization for model development. Section 5.9 outlines the overall strategy for the model implementation, covering data consistency, feed characterization, and model calibration with plant data. Section 5.10 describes the overall modeling strategy and Section 5.11 compares the model predictions with plant data. Section 5.12 presents case studies on the effects of reactor temperature, feed rate, and feed quality on process yields; on the optimization of process operations for chemical feedstock production; and on energy utilization and process performance. Section 5.13 demonstrates the model applications to refinery production planning. Sections 5.14–5.17 present four hands-on workshops of development and validation of catalytic reforming reaction and fractionation systems from plant data, together with model applications to process optimization and production planning. Section 5.18 gives the conclusion, followed by nomenclature and bibliography.

5.1 Introduction

Catalytic reforming has long been a significant source of high-octane gasoline and aromatic feedstocks for chemical processes. Recently, there has been renewed interest in processing nonconventional feedstock, synthetic crude, bio-oil, and so on. Even with those technologies, which generally produce mostly paraffin-like feedstocks, the refinery needs reforming to convert these paraffins into high-octane components. With all these factors in play, it becomes critical to quantitatively understand the reforming process on an industrial scale. This understanding must not be limited to the catalyst behavior itself but also include the associated reforming technology and fractionation equipment.

It is in this context that we present the current work regarding the integrated modeling of the CCR process. There is significant previous work in the area, particularly those by Ancheyta-Juarez *et al.* [1–3] and Taskar *et al.* [4, 5]. Although previous authors have provided significant details on reaction kinetics, there is not much information concerning the associated fractionation system and industrially useful case studies using a rigorous kinetic model. This work fills the gap between the development of a rigorous kinetic model and industrial application in a large-scale refinery.

5.2 Process Overview

The catalytic reforming unit exists primarily to upgrade the octane for gasoline-producing refineries or a rich source of aromatics for petrochemical complexes.

The modern catalytic reforming process was first introduced by UOP in 1940 [6]. Since then, there have been many different types of reforming processes developed. In general, current processes are of three distinct types.

1) Semiregenerative
2) Cyclic
3) Moving bed or CCR.

Semiregenerative processes generally involve a single reactor that processes feed. As the reactor processes feed, the catalyst begins to lose activity. At some point, typically around the middle of the catalyst life cycle, the reactor is taken offline and the catalyst is regenerated. The advantages of this process are low capital investment and simple process configuration. However, depending on the type of the feed that the refiner processes, the regeneration cycle may be too long to maintain desired levels of production.

Cyclic processes involve a series of beds that operate on a rotating basis. There is a set of five to six reactors; however, only three to four may be active at any given time. When the catalyst activity for a given reactor falls below a certain value, that reactor is taken offline and the feed flow is shunted to a reactor with recently regenerated catalyst [6].

Moving bed or CCR involves the continuous regeneration of the catalyst. This is possible through the construction of a special reactor that allows the continuous withdrawal of catalyst while the reactor is on-stream. The withdrawn catalyst enters a regeneration section [6]. Figure 5.1 shows representative reactors from each of these processes.

The UOP CCR process is by far the most popular reforming process. Over 50% of current reforming capacity originates from this process. This process relies on

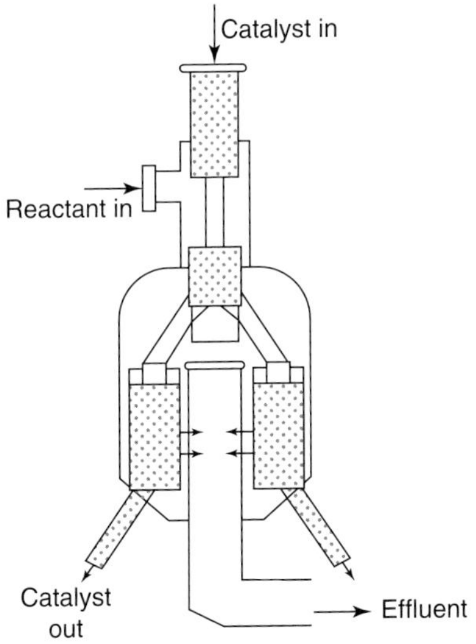

Figure 5.1 Different types of reactors used in reforming processes. (a) Continuous catalyst regeneration (CCR) reactor. (b) Fixed-bed axial flow reactor. (c) Fixed-bed radial flow reactor. (Adapted from Ref. [6].)

the continuous regeneration of the catalyst. This type of unit is the focus of our work and we document the process flow in the following section.

Figure 5.2 shows the process flow diagram of a commercial CCR reforming process in the Asia Pacific. This unit typically converts 1.4 million tons per year (28,100 BPD) of straight-run naphtha into high-octane gasoline and aromatic feedstocks for use in subsequent chemical processes. The CCR unit is organized as a series of reaction sections, each having a different loading (weight) of catalyst. Typically, the first unit has the least amount of catalyst and the last unit has the most. This distribution of catalyst loadings is common to all reformers and reflects the fact that during the initial stages of the reaction, highly endothermic reactions dominate the process. This effect slows down the reaction rate; therefore, the interstage heaters reheat the reactor effluent from each section.

Reactor effluent heats the heavy naphtha (from Unit #200 in Figure 5.2) entering the process through a cross exchanger. The hot feed enters the first interstage heater where the temperature rises to the reaction temperature. The feed contacts the moving bed of the catalyst. The components in the feed undergo several reactions, such as dehydrogenation, dehydrocyclization, isomerization, and hydrocracking. However, for a typical feed, the endothermic reactions (namely, dehydrogenation) dominate and the temperature drops significantly as the reactants flow radially through the catalyst bed. The effluent leaves this reactor bed and enters the second interstage heater. A key process variable is the temperature of the feed entering each reaction section. Heaters typically operate to return the reactor effluent at a fixed temperature. The effluent from the first reactor enters the second interstage heater and leaves again at a set reaction temperature. This follows because most of the desirable reactions in reforming are endothermic. This process of heating and reaction continues until the effluent leaves the last reactor and heats up the feed into the reforming reactors. The effluent then enters the recontacting and hydrogen separation section of the process.

At the same time, small amounts of catalyst typically flow through the basket and enter the next reactive section. This is possible because of the special gravity-assisted reactant flow shown in Figure 5.3. The CCR process is unique in which only relatively small amounts of catalyst leave the system for regeneration. As the unit continuously regenerates the catalyst, the unit is designed to operate at much lower pressure than other reforming processes. Low-pressure operation not only encourages high severity but also increases the coke generation rate.

We show the process flow of a typical regeneration cycle in Figure 5.4. The spent catalyst leaves the last reactor and enters the regeneration unit. Several activities occur as the catalyst travels down the regeneration tower. Little [6] indicated five operations that must take place during the catalyst regeneration process: burn the coke, oxidize the active metal promoters on the catalyst, adjust the chloride balance, dry the catalyst to remove unwanted moisture, and finally reduce the metal promoters [7]. These processes occur in a stepwise, semiregenerative manner and can operate independently of the reforming process. In addition, the regeneration process operates at a much different timescale. It typically takes 5–7 days for the spent catalyst to return back to the reforming reactors [7, 8]. This is in stark contrast with the fluid catalytic cracking (FCC) process, where the reaction unit and regeneration unit are highly coupled. A key modeling implication of this

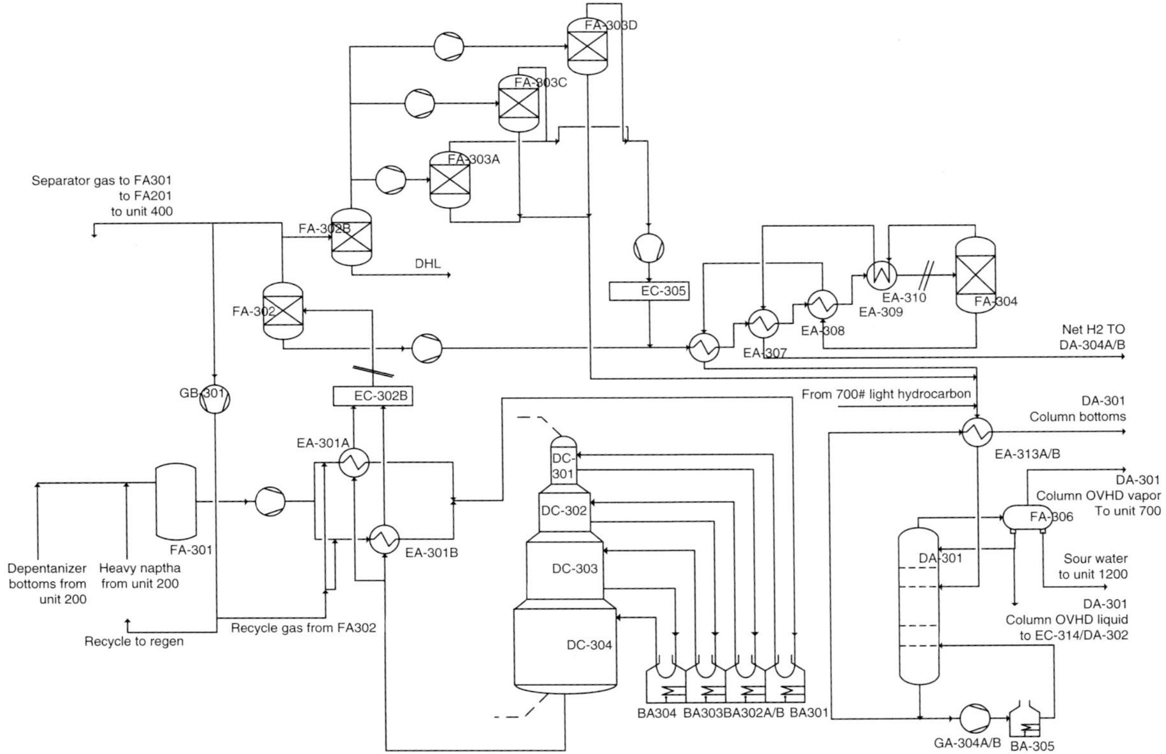

Figure 5.2 Process flow diagram for CCR reforming process.

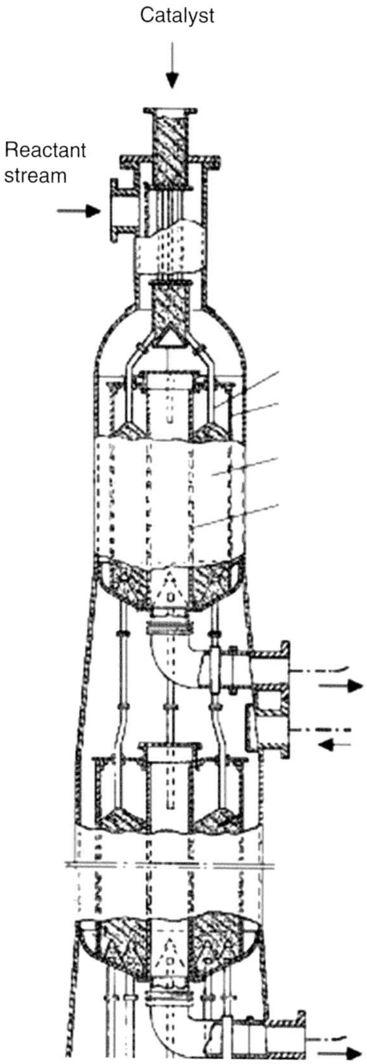

Figure 5.3 Cutaway of gravity-assisted reactor [9].

regeneration time scale and process flow is that we do not need a rigorous model of the regeneration cycle to simulate the reforming process effectively.

The cooled reactor effluent enters a series of separators (shown in Figure 5.2 as FA302 through FA304) that operate at increasing pressure. This process accounts for the fact that the CCR generally operates a much lower pressures than other reforming units. The objective is to improve the recovery of light LPG components (C3–C4) and some C5 components. The liquid product from each of the separators is subsequently cooled in several cross exchangers to recover significant amounts of heat and to condense additional light components in the liquid product. The combined liquid product enters a final separator where significant pressure change occurs and a H2-rich (94–95 mol%) stream leaves as the vapor. This H2-rich stream can typically supply hydrotreating and

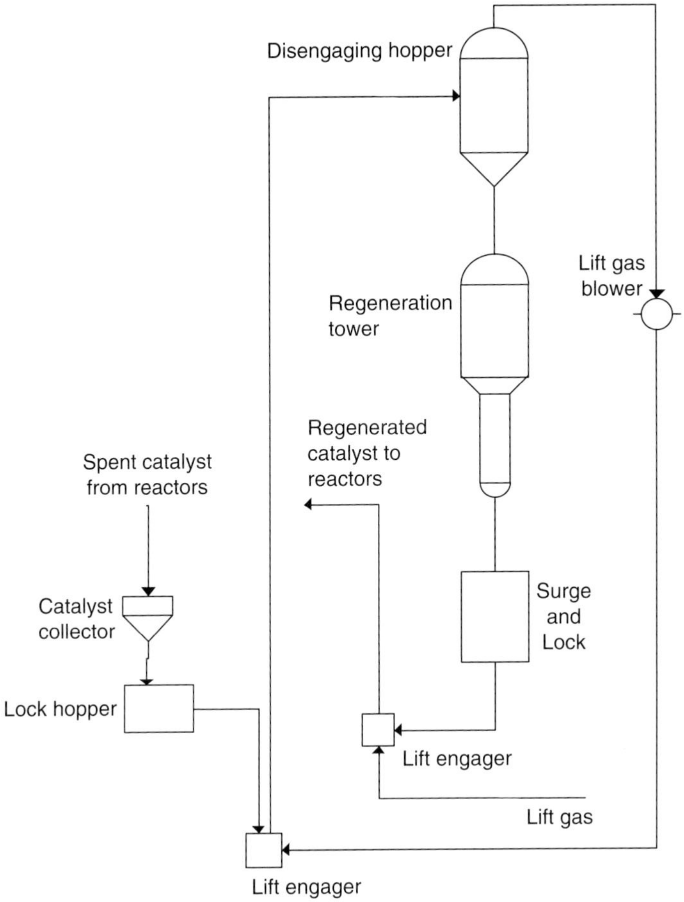

Figure 5.4 Schematic of catalyst regeneration process [6].

hydrocracking process in the refinery. The liquid product combined with other products (containing a significant quantity of aromatics) enters the fractionation section of the process.

Depending on the end use of the reforming product (often called *the reformate*), there are two possible paths for production fractionation. If the purpose of the unit is *gasoline production*, the reformate enters a stabilization fractionator. This fractionator typically only separates the LPG-like portion of the reformate as the overhead product and the bottom product leaves as high-octane gasoline destined for the refinery blending pool. However, if the purpose of the unit is *aromatics production* to support a petrochemical complex, the stabilizer operates differently as a depentanizer (shown as DA-301 in Figure 5.2). The depentanizer separates all the C5 and lighter components as the overhead product. The bottom product largely contains all the aromatics, remaining paraffin, and naphthenic content greater than C6, and it then enters the BTX (benzene–toluene–xylene) separation plant, which may be located in a different area of the refinery altogether.

The separation of product aromatics into discrete aromatic species depends on the refinery configuration. This process can be quite large and complex, especially in the case of petrochemical refineries where aromatics can be recovered from many sources. Typically, a special solvent (e.g., sulfolane or polyglycols) separates out the benzene and toluene components from the feed to BTX separation plant. The separation of xylenes requires additional processing.

Fractionation towers can separate *ortho*-xylene and ethylbenzene isomers. However, the *meta*-xylene and *para*-xylene isomers typically require a crystallization or adsorption on molecular sieves (e.g., IFP process and UOP Parex process) [8]. Due to the complexity of the BTX separation plant, we do not include BTX fractionation in this work. However, our recent textbook covers the design, simulation and optimization of simulated moving beds for xylene purification in the UOP Parex process [56].

The feed to the reforming unit is an important process consideration. The feedstock to a reformer is typically a straight-run naphtha cut or hydrotreated gasoline cut from an FCC unit. In general, a feed that has an end boiling point (EBP) of 205–210 °C is not included. This feed encourages hydrocracking reactions and excessive coke generation. The feed is usually hydrotreated because sulfur, nitrogen, and other trace components can deactivate the catalyst significantly. In fact, many processes may also include several "guard reactors" to prevent sulfur entering the reforming unit. Table 5.1 shows a typical distillation curve and basic compositional analysis of reformer feedstock.

Refiners often consider the total naphthene (N) and aromatics (A) content of the feed as an indicator of how high an octane rating a feedstock can produce. This is referred to as N + A or N +2A indicator for the feed. Many correlations for reformer yield exist based on these indicators. However, Little [6] indicated that these correlations often have strong built-in assumptions such as catalyst type and operating conditions. Although it may serve for simple feedstock selection, it is not the only significant indicator of unit performance.

The catalyst in the unit is the most important consideration for optimal operation. Little [6] identified three key characteristics of reforming catalysts, namely, *activity*, *selectivity*, and *stability*. The activity is a measure of how efficiently

Table 5.1 Typical reforming feedstock.

ASTM D86 (vol%)	(°C)	Group	Paraffin (wt%)	Naphthene (wt%)	Aromatics (wt%)
IBP	76	C5	1.00	0.47	–
5%	90	C6	6.85	6.66	0.88
10%	94	C7	11.25	13.17	2.31
30%	104	C8	9.42	14.02	3.02
50%	116	C9	7.35	10.79	3.04
70%	131	C10	4.45	5.31	0.00
90%	152	Total	40.32	50.42	9.25
95%	160	Specific gravity (SG)			0.745
EBP	170	Sulfur/nitrogen/halide content (ppm)			0.5/0.5/NA

the catalyst can help convert the reactants into products. In general, current reforming catalysts can operate at higher temperatures and maintain high reaction conversion when the reactant flow rate increases. The selectivity refers to the catalyst ability to produce more of the high-value products (aromatics) than low-value products. The stability refers the ability of the catalyst to maintain high activity and selectivity over long periods. The catalyst in modern reforming units is only changed once in every 1–2 years [7].

Modern reforming catalysts consist of an alumina base that supports platinum and rhenium particles to catalyze the desired reactions. Current consensus indicates that the platinum sites promote the dehydrogenation reactions; the alumina, acting as an acid site, promotes cyclization, isomerization, and hydrocyclization [7, 10–12]. These types of catalysts are known as *bimetallic* (and sometimes *bifunctional* catalysts). As the catalyst spends more time on stream, coke deposits and lack of acid sites prevent additional reaction. The rate of coke deposition is a function of olefin-like precursors that lead to the formation of a multiaromatic ring [13]. At this point, the catalyst is taken off-stream and regenerated through several processes to restore its function. The reaction chemistry that occurs on these catalysts can be quite complex, and published experimental studies often do not reflect the conditions that a catalyst operates under in an industrial process. In the following section, we briefly survey some of the key process chemistry and operating parameters.

5.3 Process Chemistry

Table 5.2 lists the major reactions observed in the reforming process. This is by no means an exhaustive list. In general, the desired reactions take the following paths: (1) paraffins in the feed convert to isoparaffins or are cyclized into the naphthenes; (2) the naphthenes present convert to aromatic groups; and (3) olefins convert to paraffins through hydrogenation [14].

A detailed study of many of the reactions is out of the scope of this work. We refer readers to Froment *et al.* [10–12] for detailed experimental and mechanistic studies. These studies are very useful in the course of detailed catalyst design and kinetic network generation [15–18]. However, neither of these topics is the subject of the current work. We present these reactions in the context of an integrated process model. As mentioned earlier in this work, the typical reactions in

Table 5.2 Examples of reactions from key reaction classes.

Dehydrogenation of alkylcycloalkanes to aromatics	$MCH \rightarrow TOL + H_2$
Dehydroisomerization of alkylcyclopentanes	$MCP \rightarrow MCH$
Dehydrocyclization of paraffins to aromatics	$NP_7 \rightarrow TOL + H_2$
Isomerization of normal paraffins to isoparaffins	$NP \rightarrow IP$
Isomerization of alkylcylcopentanes to cyclohexanes	$MCP \rightarrow MCH$
Hydrocracking reactions	$P_X \rightarrow P_Y + P_Z$
Hydrogenolysis	$P_7 + 6H_2 \rightarrow 7P_1$

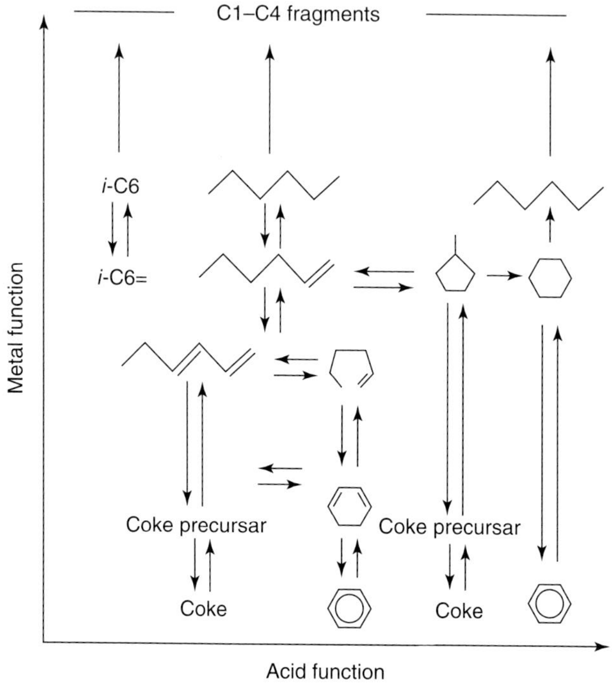

Figure 5.5 Relationship between catalyst features and reaction classes [13, 14].

the reforming process are dehydrogenation, dehydrocyclization, isomerization, and hydrocracking. Table 5.2 shows examples of these reaction classes.

Figure 5.5 shows the relationship between the acid and metal functions of the catalyst and particular classes of reactions. The acidic function of the catalyst promotes the isomerization reactions, namely, reactions that convert paraffins into naphthenes and isoparaffins. Isoparaffins are important contributors to high-octane number. The metal function promotes the dehydrogenation reactions, where the naphthenes are dehydrogenated into aromatics. The metal function is also a significant source of coke (or polyaromatic compound) that adsorbs to the catalyst surface. In addition, the olefins are hydrogenated producing paraffins for further reaction.

The degree to which each reaction propagates is a function of temperature and pressure. High temperature and pressure tend to promote hydrocracking and the undesirable hydrogenolysis. The effect of pressure is quite significant on hydrogenolysis and modern reformers tend to operate at much lower pressures than their predecessors. Table 5.3 summarizes the effect of key operating variables on yields. In all cases, increase in reactor temperature increases the reaction rate.

In addition to the operating variables of reactor, the feed composition also plays an important role in determining the distribution of products. Industrial experience and experimental studies of the chemistry of reforming reactions indicate several key trends [7.19].

Table 5.3 Behavior summary key reaction classes.

Reaction	Rate	Heat	Pressure	Hydrogen
Dehydrogenation (naphthene)	Very fast	Endothermic	Negative	Produces
Isomerization (naphthene)	Fast	Exothermic (mild)	None	None
Isomerization (paraffin)	Fast	Exothermic (mild)	None	None
Cyclization	Slow	Exothermic (mild)	Negative	Produces
Hydrocracking	Slowest	Exothermic	Positive	Consumes
Hydrogenolysis	Slowest	Exothermic (high)	Positive	Consumes

Source: Adapted from Little [6], Antos [7], and Gary [8].

- The primary source of benzene in the reactor products is methylcyclopentane (MCP).
- Dimethylcyclopentane and cycloheptane form a key pathway to produce additional toluene.
- Dimethylcyclohexane and methylcyclohexane produce additional xylene in the product.

In industrial operations, it is difficult to control many process variables to drive reactions to optimal product distributions. There are four primary control variables for reformers, namely, *reactor inlet temperatures, reactor pressures, hydrogen content*, and *feed rate*. There are other variables such as feedstock properties and catalyst type, but these variables are generally fixed for a given period of time.

Refiners generally control the inlet temperature to each reactor bed or section. The inlet temperatures are typically averaged (weighted by the ratio of the catalyst in the given bed to the total catalyst) and presented as *the weight-averaged inlet temperature (WAIT)*. The pressure in sections of the reactor is typically fixed by design and does not vary significantly during operation. This is especially the case in CCR units where the pressure balance drives the catalyst flow. Another important variable is the amount of hydrogen that is recycled back to the unit along with fresh feed. Current reformers typically operate at high conversions and a significant quantity of hydrogen is required to prevent coke formation. During normal operation, the H_2HC ratio (ratio of hydrogen to hydrocarbons) ranges from 3 to 4. The final control variable is typically the feed to the unit. High feed rates typically indicate the low contact time between the catalyst and the feed.

5.4 Literature Review

There is a significant body of literature on the topic of modeling catalytic reformers. They consist of two types of models, kinetic models and unit-level models. Kinetic analysis refers to detailed studies of the reaction mechanism and catalyst behavior. This work is necessarily experimental and based on laboratory studies of various feed compounds. Model development work uses the insights from the kinetic analysis to develop a kinetic network with associated rate constants and reaction orders. This work typically results in rate expressions that are verified using bench-scale reactors. The unit-level models focus on models that integrate the kinetic model in the context of pilot-scale or commercial reactors. This work often includes models for multiple reactor beds and associated process equipment (interstage heaters, etc.). We provide a brief survey of the current state of knowledge in each of these areas.

5.4.1 Kinetic Models and Networks

Mechanistic and experimental studies generally result in the creation of a kinetic network that quantitatively describes the path that a particular reactant takes. Given the complexity of the reforming reactions and the number of species involved, many researchers have taken a "lumped" approach toward describing the kinetics. In a lumped approach, many different molecules are placed into a single group or lump. The reaction kinetics then assumes that all species in a lump behave identically. Recently, some researchers have presented models that involve hundreds of reaction species and thousands of reactions [16, 18]. However, there is little published information about these complex kinetic models validated against industrial operation.

The earliest kinetic model for reforming is that of Smith [20], which assumes that the feed is a combination of three lumps: paraffins (P), naphthenes (N), and aromatics (A). We show a basic schematic of the network in Figure 5.6a. The kinetic network accounts for dehydrocyclization (P → N), dehydrogenation (N → A), and hydrocracking (A → P). The hydrocracking reactions in this model result in an equilibrium distribution of paraffins. This model does not include the effect of reaction parameters such as pressure and excess hydrogen present. In addition, there is no deactivation factor due to the presence of coke or heavy adsorbed hydrocarbons. Krane *et al.* [21] further refined this model by splitting up each P, N, and A lump into groups corresponding to the number of carbons. This model has 20 lumps and 53 reactions. Equation (5.1) shows the basic form for each rate expression.

$$\frac{dN_i}{d\left(\frac{A_c}{W}\right)} = -k_i N_i \tag{5.1}$$

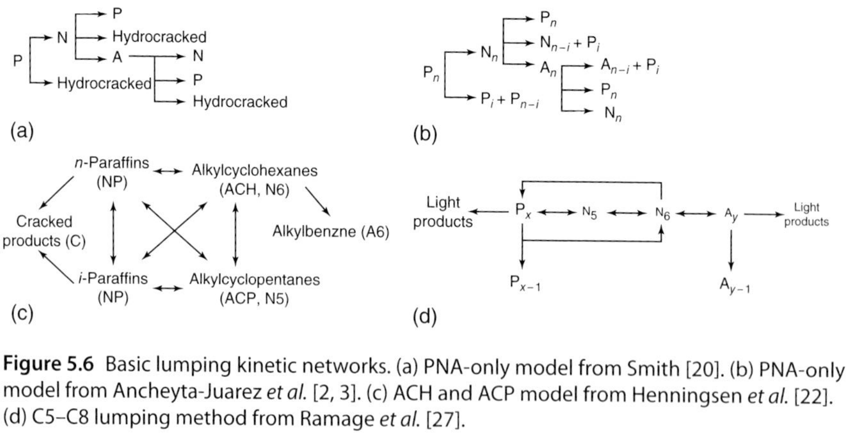

Figure 5.6 Basic lumping kinetic networks. (a) PNA-only model from Smith [20]. (b) PNA-only model from Ancheyta-Juarez *et al.* [2, 3]. (c) ACH and ACP model from Henningsen *et al.* [22]. (d) C5–C8 lumping method from Ramage *et al.* [27].

A significant oversight in Krane's model is the lack of the effect of catalyst activity and pressure. Henningsen *et al.* [22] introduced a network that considers the different rates of reactivity between C5 and C6 naphthenes and an activity factor for catalyst deactivation. Jenkins *et al.* [23] included empirical correction factors for acid and pressure in the rate expression. Ancheyta-Juarez [2, 3] also introduced a similar pressure correction term to account for pressures other than 300 psig specified in the Krane *et al.* model. Later work by Ancheyta and coworkers included additional pathways to consider MCH as a primary precursor to benzene [19] in the product pool and to deal with nonisothermal operation. Models derived from Krane *et al.* [21] and Ancheyta *et al.* [19] have been used to model a variety of reforming processes, ranging from pilot plants to commercial operations. Hu *et al.* [24] used a similar approach to generate a kinetic network. Ancheyta's modifications to Krane's original model still remain in use and work published recently shows good agreement with measured data and model predictions [19, 25, 26].

$$\frac{\mathrm{d}N_i}{\mathrm{d}\left(\frac{A_c}{W}\right)} = -k_i e^{(E_i/R)\left(\frac{1}{T_0}-\frac{1}{T}\right)} \left(\frac{P}{P_0}\right)^{\alpha} P_i \tag{5.2}$$

Krane's original model and modifications by Ancheyta do not treat kinetic network as a catalytic process occurring heterogeneously and do not consider the difference in reactivities of cyclopentanes and cyclohexanes. Figure 5.6c shows the kinetic network from Henningsen *et al.* [22] that includes separate pathways for cyclopentanes and cyclohexanes. Henningsen *et al.* apply this model in conjunction with a heat balance to account for the nonisothermal operation of the reactor. These works have generally shown excellent agreement with commercial and pilot plant data.

$$\frac{\mathrm{d}C_i}{\mathrm{d}t} = \sum k_i e^{(E_i/RT)} P_i \tag{5.3}$$

A key limitation of the models derived from Krane *et al.* and Henningsen *et al.* is that the reaction network is not treated as a catalytic process. A catalytic reaction kinetics network must include terms to allow for inhibition and decrease in activity due to variety of factors. Raseev *et al.* [14] present the earliest model treating the reaction network as a catalytic system. However, this study is limited due to the lack of experimental data. Figure 5.6d shows the kinetic network from an extensive study by Ramage *et al.* [27] where independent pathways for cyclohexanes and cyclopentanes exist in addition to adsorption and pressure effects. However, this model is limited by the lumping into only C5− and C5+. Kmak [28] presented a similar model that extends the lumping to include C7 components.

$$\frac{\mathrm{d}w_i}{\mathrm{d}v} = \frac{\left(\frac{PV}{FRT}\right) k_\phi}{1 + K_\mathrm{H} P_\mathrm{H} + (PF_\mathrm{c}/F)\sum K_{w_i} w_i} \sum k_i w_i \tag{5.4}$$

Key work by Froment and coworkers [7, 10–12] has produced a nearly complete lumping-based reaction network for C5–C9 (and C1–C5 for paraffins) components of reforming feed. This model includes several insights from experimental studies. They consider that the metal sites on the catalyst promote only

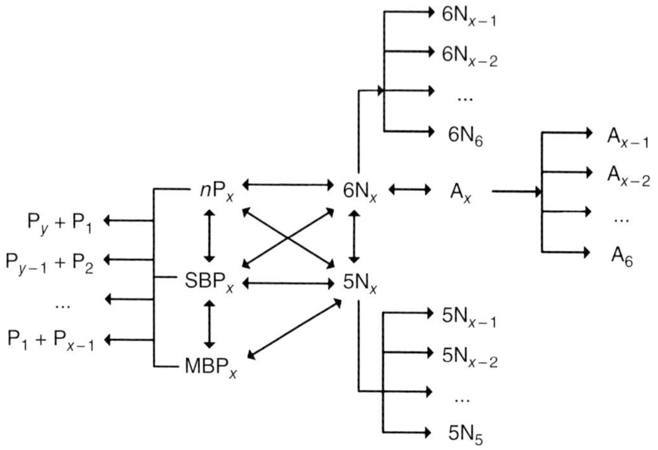

Figure 5.7 Lumped kinetic network from Froment where $5 < x < 9$ [12].

the dehydrogenation reactions, whereas the acid site promotes the cyclization, isomerization, and hydrocracking reactions. We show the network in Figure 5.7.

The kinetic network in Figure 5.7 includes separate pathways for N5 and N6 components and accounts explicitly for light component production (C1–C5). This is critical to maintaining a good prediction of light gas components from industrial models. In addition, the adsorption factors include terms to account for hydrogen content, total pressure, and adsorbed hydrocarbons. Additional work by Taskar *et al.* [4, 5] modifies this network to include the effects of catalyst deactivation. Table 5.4 shows the key rate equations for each class and the deactivation factor due to Taskar *et al.*

Recent advances in computational power and theoretical insight have led to the creation of mechanistic reaction pathways that can involve thousands of reactions and hundreds of species. The approach of Froment [15–17] is called *the single-event approach.* In this approach, an algorithm generates a reaction network based on fundamental mechanisms such as hydride shifts and beta scission. The use of structural relationships such as Evans–Polanyi reduces the number of

Table 5.4 Key rate equations from Taskar *et al.* [4, 5].

Isomerization of paraffins	$\phi \cdot A_0 e^{-E/RT}(P_A - P_B/K_{AB})/\Gamma$	(5.5)
Hydrocracking of paraffins	$\phi \cdot A_0 e^{-E/RT}(P_A P_B)/\Gamma$	(5.6)
Ring closure of paraffins	$\phi \cdot A_0 e^{-E/RT}(P_A - P_B P_H/K_{AB})/\Gamma$	(5.7)
Ring expansion (C5–C6)	$\phi \cdot A_0 e^{-E/RT}(P_A - P_B)/\Gamma$	(5.8)
Dehydrogenation	$\phi \cdot A_0 e^{-E/RT}(P_A - P_B P_H{}^3/K_{AB})/(P_H\theta)^2$	(5.9)
Adsorption due to acid function	$\Gamma = (P_H + K_{C6-}P_{C6-} + K_{P7}P_{P7} + K_{N7}P_{N7} + K_{TOL}P_{TOL})$	(5.10)
Adsorption due to metal function	$\theta = 1 + K_{MCH1}P_{MCH} + K_{MCH2}(P_{MCH}/P_H{}^2)$	(5.11)
Deactivation term	$\phi = e^{-\alpha C_c}$	(5.12)

Source: Adapted from Taskar 1996 [4] and Taskar 1997 [5].

parameters required for modeling significantly. Experimental data may be used to fit the remaining parameters (roughly 30–50). This approach has been successfully used for a variety of processes including methanol-to-olefins (MTO) and FCC that exhibit similar features as the catalytic reforming process. Due to limitations of feedstock analysis, this technique makes several assumptions to lump together components in the feedstock and presents rate equation that is the summation of many rate equations drawn from fundamental chemistry.

Another approach is the molecular modeling work by Klein and coworkers [18]. In their work, they propose technique of pathway modeling where a series of chemical reaction paths are applied to many hundreds (if not thousands) of feed species. They then construct a reaction path that only contains the allowable reaction chemistry. Klein *et al.* also simplified the process of estimating kinetic parameters through the linear free energy relationships (LFER). The final network for naphtha reforming involves 116 species and 546 reactions. Several works report the success of this model through several pilot plant studies. A key issue is the feedstock characterization. Klein *et al.* [29] used a stochastic approach where they pick combinations of thousands of species and attempt to match the calculated bulk properties (specific gravity, molecular weight, sulfur content, etc.) of a particular combination to measured bulk properties.

In the course of applying a model to a commercial plant, it is best to rely on kinetic models that only require *the minimal amount of feedstock information and calibration*. Feed to reformers may change quickly, and without laboratory analysis, there is often no choice but to lump components together. In addition, it may not be possible to incorporate large complex models into existing highly integrated flowsheet models. These factors generally drive model developers to choose lumped kinetic networks.

5.4.2 Unit-Level Models

After choosing a representative kinetic model, we must decide how to represent the remaining units for a truly integrated model. Researchers have applied many of the kinetic networks described in the previous section in integrated process models. Figure 5.8 is an overview of the key features of an integrated process model for a three-section reformer. This overview applies to both semiregenerative fixed bed and CCR reformers.

First, the model must be able to take bulk property measurements and convert them into appropriate lumps for kinetic network. This step may be quite simple if the kinetic model chosen only includes total PNA content for the total fraction. However, if the kinetic lumping requires detailed composition information, we must provide some way of estimating these lumps from limited composition information. Taskar *et al.* [4, 5] discussed a possible method based on the measurements of certain bulk properties such as gravity and distillation curve. We discuss the approach used in this work in Sections 5.9 and 5.9.3.

The second consideration is the model for the interstage heaters, product separators, and compressors. In order to model these units meaningfully, we must have reasonable estimates for the key thermophysical properties of the lumps. In the case of the reformer, we must make reasonable prediction of reactant

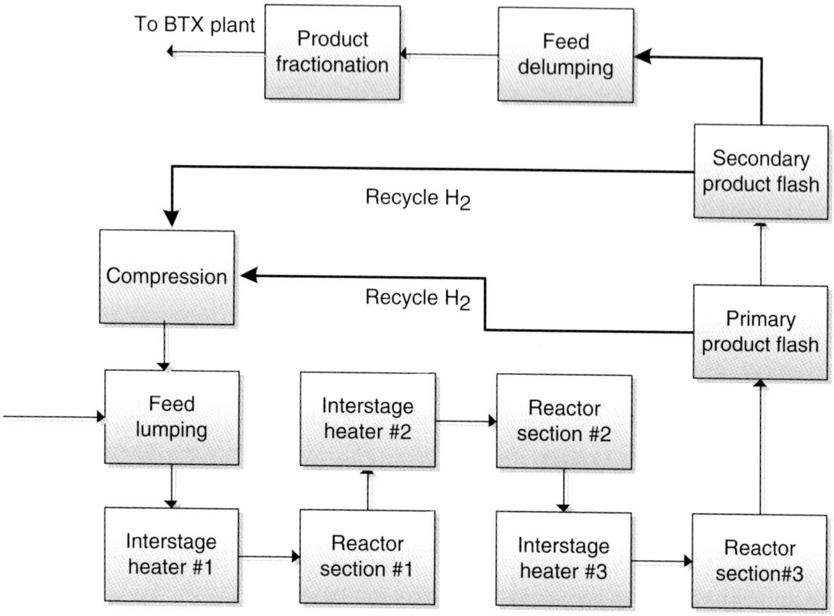

Figure 5.8 Basic process flow for an integrated reformer model.

concentration (at system pressure), K-values (for the product separator), and heat capacity (to model the reactor temperature drop and product temperatures correctly). The reforming process generally operates at temperatures and pressures where the ideal gas law applies for hydrocarbon species in the reactor section. Ancheyta-Juarez *et al.* [1, 2] used the ideal gas assumption to calculate the concentration of reactant species. In addition, they use the polynomial heat capacity correlations for pure components to approximate the heat capacity of the mixture. Work by Bommannan *et al.* [30] and Padmavathi *et al.* [31] uses a fixed value for the heat capacity and K-value correlation to predict compositions in the primary product separator.

Most authors model the reactor section as a plug-flow reactor (PFR) of fixed length. This length is typically the size of the packing bed for a fixed-bed semiregenerative unit. This assumption works well with all the kinetic networks mentioned above. Modeling the flow through the CCR unit is slightly different in that reactants travel through a moving bed of catalyst particles. Hou *et al.* [32] described how to modify the standard PFR to account for a radial flow unit. Szczygiel [33] studied mass transfer and diffusional resistance in reforming reactors. However, these types of studies are difficult to apply in the context of commercial plants and many authors of integrated models have ignored these effects.

The final step in an integrated model is the delumping of kinetic lumps back to bulk properties and lumps suitable for fractionation models. Many authors do not consider this delumping process as they do not include a rigorous fractionation section. Typically, many studies report only properties such as RON and MON. If the kinetic lumping method used spans to a significant range, then fractionation models can work directly with the kinetic lumps. Works by Hou *et al.* [32] and Li *et al.* [34] use the kinetic lumps directly.

Table 5.5 Summary of unit-level models reported in the literature.

Reference	Application	Kinetics	Feed lumping	Calibration	Planning (LP)
Ramage *et al.* [27]	Semiregenerative	C5–C8(P, N5, N6, A) lumps	None	Yes	Yes
Bommannan *et al.* [30]	Semiregenerative	Simple lumps (P, N, A)	None	None	None
Ancheyta *et al.* [1, 2]	Semiregenerative	C5–C10 (P, N, A)	None	None	None
Taskar [45]	Semiregenerative	C5–C10 (P, N5, N6, A) lumps	Yes	Yes	None
Lee *et al.* [35]	CCR	Simple lumps (P, N, A)	None	None	None
Padmavathi *et al.* [31]	Semiregenerative	C6–C9 (P, N5, N6, A) lumps	None	Yes	None
Ancheyta-Juarez *et al.* [19]	Pilot plant	C5–C11 (P, MCP, N6, A) lumps	None	Yes (kinetic regression)	None
Hu *et al.* [36]	CCR	C6–C9 (P, N, A) lumps	None	Yes	None
Li *et al.* [34]	Semiregenerative	C1–C9 (P, N5, N5, A) lumps	None	Yes	None
Hou *et al.* [32]	CCR	C1–C9 (P, N, A) lumps	None	Yes	None
Stijepovic *et al.* [25, 37]	Semiregenerative	C6–C9 (P, N, A) lumps	No	No	None
This work	CCR	C1–C14 (P, N5, N6, A) lumps	Yes	Yes	Yes

Table 5.5 summarizes the key features in reported unit-level models (using lumped kinetics) applied to reforming processes. We have only included studies where the authors compare their results to pilot plant or industrial data. In addition, we include those studies where the authors use the model for case studies and plant optimization.

5.5 Aspen HYSYS Petroleum Refining Catalytic Reformer Model

This section discusses the key features of the Aspen HYSYS Petroleum Refining model we use throughout this work. Although the features we discuss are specific to Aspen HYSYS Petroleum Refining, there are other simulation programs that have similar functionality. The goal of this section is to discuss the key features of the simulator that are relevant to developing an integrated reaction and fractionation model.

Figure 5.9 shows a basic outline of the key submodels in Aspen HYSYS Petroleum Refining. This model contains all the key submodels identified in

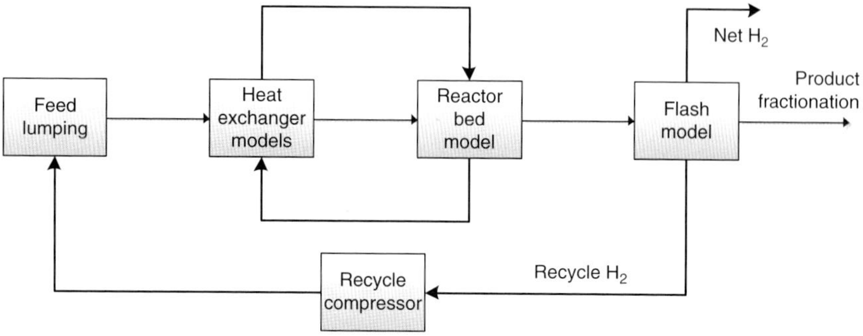

Figure 5.9 Organization of Aspen HYSYS Petroleum Refining CatReform model.

the previous section. The model presented in this work includes the additional fractionation units to model the separation of LPG (<C4) and the reformate into gasoline and high-octane compounds for blending and chemical purposes.

The feed-lumping technique in the Aspen HYSYS Petroleum Refining model relies on a base of compositions and a method to correct those measured compositions based on changes in measured bulk properties. The feed is broken into many (4–14) lumps for each chemical group. Typically, these measured properties are the distillation curve and total PNA content. In our work, we had access to detailed feed composition information, so we did not use this technique. However, we have developed an alternate technique of feed lumping based on minimal base composition data and bulk property requirements. We discuss this technique in Section 5.9.

The reaction network in the reactor model is similar to the network presented by Froment *et al.* [12] and Taskar [4]. However, the reaction network supports higher aromatics up to C14. Although these typically are not expected in reformer feeds, the kinetic model can handle them as well. In addition, the reactor model includes paths for the undesired hydrogenolysis reactions. These highly exothermic reactions do not occur in any significant degree in stable reforming units. However, older reactors may display this behavior, so it is important to model them as well (Table 5.6).

Equations (5.13)–(5.17) show the general form of the kinetic rate expression. The important thing to note is that there are two activity correction factors associated with each rate expression. The first correction factor, a_{class}, is fixed for a

Table 5.6 Key reaction classes in Aspen HYSYS Petroleum Refining catalytic reformer model.

Isomerization of paraffins	$a_{\text{class}} a_{\text{reaction}} A_0 e^{-E/RT} (P_A - P_B/K_{AB})/\Gamma$	(5.13)
Hydrocracking of paraffins	$a_{\text{class}} a_{\text{reaction}} A_0 e^{-E/RT} (P_A P_B)/\Gamma$	(5.14)
Ring closure of paraffins	$a_{\text{class}} a_{\text{reaction}} A_0 e^{-E/RT} (P_A - P_B P_H/K_{AB})/\Gamma$	(5.15)
Ring expansion (C5–C6)	$a_{\text{class}} a_{\text{reaction}} A_0 e^{-E/RT} (P_A - P_B)/\Gamma$	(5.16)
Dehydrogenation	$a_{\text{class}} a_{\text{reaction}} A_0 e^{-E/RT} (P_A - P_B {P_H}^3/K_{AB})/(P_H \theta)^2$	(5.17)

given class of reactions. For example, all the isomerization reactions may have a rate constant of 1.0. The second correction factor, a_{reaction}, refers to correction for an individual pathway. For example, the activity factor for the isomerization of C6 paraffins may have a correction factor of 0.5. The product of these two factors presents the overall activity correction for that reaction. The individual rate constant and activation energy remain fixed. These factors have been derived from experimental data over a variety of catalysts. In practice, however, even significant changes in unit operations do not require significant changes in values of these reaction activity factors.

Another significant feature is that the coke generation is rigorously modeled and included in the deactivation and adsorption factor, Γ, for each reaction. The deactivation factor is a function of reactor pressure, adsorbed hydrocarbons, coke on catalyst, and acid/metal function of the catalyst. This feature allows us to calibrate the model to a variety of operating conditions and catalyst behavior. In this work, we model a CCR with a hydrotreated feed; therefore, we do not include any significant changes in catalyst activity due to changes in acid component of the catalyst.

The reactor model is based on a modified PFR for a moving bed that accounts for catalyst flow in the CCR system. A key consideration in the reactor is the phenomenon of "pinning" [38, 39] in CCR reformers. "*Pinning*" refers to the catalyst that is held immobile against the wall due to cross-flow of reactants. It is important to model this effect, as pinning imposes a maximum flow rate on reactants. The reactor also correctly models the temperature drop due to heat of reaction in the exothermic and endothermic reactions. The other key variables are the WAIT and weight-averaged bed temperature (WABT), both defined in Section 5.4, and the weighted hourly space velocity (WHSV), that is, the weight of feed per hour per unit weight of catalyst loaded in the reactor.

As mentioned in a previous section, integrated model for CCR must also include rigorous models for interstage heaters to predict energy consumption of the unit correctly. We may model the unit as rigorous fired heaters or basic heat exchangers. We include a model to recompress the vapor from the primary product flash. Our work also includes the complete model for the product recontacting section. We must model this section correctly in order to predict the composition of the recycle stream entering the reformer. All of these units require thermophysical properties and methods to predict equilibrium. We use the Peng–Robinson (PR) equation of state modified for hydrogen-containing systems. We describe how to obtain the relevant thermophysical properties for each lump in Section 5.7.

The final step in the integrated model before fractionation is the delumping of products and prediction of bulk properties. As our lumping system is quite broad, we can just calculate key properties of the reformer effluent as combination of the individual properties of the lumps.

$$\text{RON}_{\text{MIX}} = \sum w_i \text{RON}_i \tag{5.18}$$

$$\text{MON}_{\text{MIX}} = \sum w_i \text{MON}_i \tag{5.19}$$

where RON_{MIX} and MON_{MIX} refer to the research and motor octane number of product measured in bulk, w_i refers to the weight fraction of each lump, and RON_i and MON_i refer to the research and motor octane number of each lump.

As we wish to use this model to simulate BTX production as well, we need to predict the composition of all the relevant isomers of A8 (ethylbenzene, *ortho*-xylene, *para*-xylene, and *meta*-xylene). In our model, we assume that these isomers take on fixed equilibrium ratios as a function of temperature. Figure 5.10 shows the equilibrium distribution of these isomers at various temperatures [40, 41]. The distributions correspond to expected temperatures in the reforming process. Figure 5.11 shows the observed A8 isomer distribution measured at the plant. We note that it is remarkably stable over a lengthy operating period (6 months) and a variety of feed conditions.

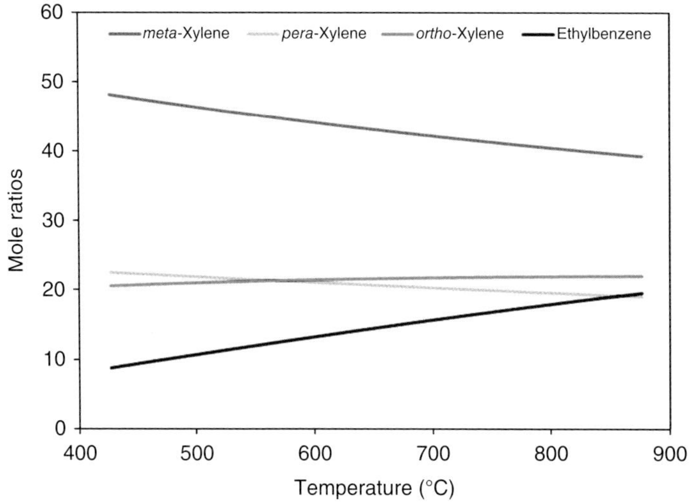

Figure 5.10 Equilibrium composition of A8 isomers (assuming ideal gas conditions).

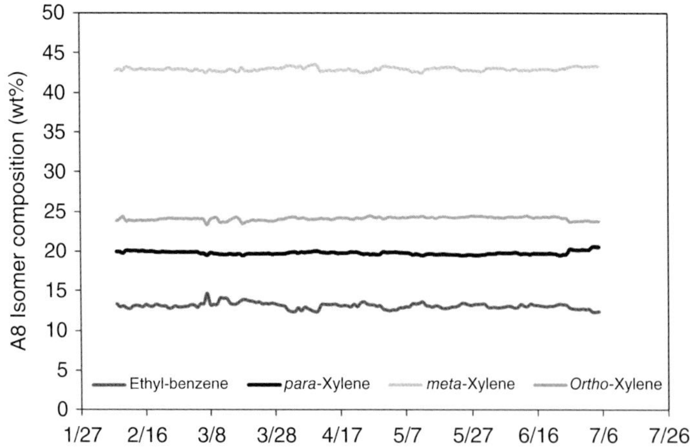

Figure 5.11 Composition of A8 isomers over the study period.

This completes our description of the Aspen HYSYS Petroleum Refining model. In subsequent sections, we discuss issues of thermophysical properties, fractionation, and feed lumping. These issues are not specific to a simulation program and apply generally to any model of a reforming process.

5.6 Thermophysical Properties

The requirements for thermophysical properties depend on the kinetic lumping chosen for the process. Typically, the reactor model requires only the heat capacity and molecular weight. The fractionation section may require a correlation to predict K-values or critical parameters when an equation of state is used. One approach is to use one set of lumps for the reactor model and another set for the fractionation. However, this approach may cause problems when recycling material is back into the reactor and makes producing an integrated model difficult. If possible, we suggest the use of uniform lumps across the reactor and fractionation models.

If the reactor lumps resemble real measured products (e.g., A8), then it is sufficient to use the known properties of one of the compounds comprising the lump as the properties of the lump. The kinetic lumps in this work resemble real lumps, so we use known compound properties. If this information is not available, we can use Riazi's correlations [42] to estimate the relevant critical properties for different classes of compounds (paraffins, naphthenes, and aromatics) given the molecular weight of a particular lump.

$$\theta = a(\mathrm{MW})^b(\mathrm{CH})^c \tag{5.20}$$

where θ represents critical temperature (T_c), critical pressure (P_c), critical volume (V_c), specific gravity (SG), or refractive index (I); CH denotes the carbon-to-hydrogen weight ratio. Riazi [42] provided values for a, b, and c for different classes of compounds.

5.7 Fractionation System

We use the standard inside-out method [43] discussed in Section 2.4.4. This work only includes the primary product debutanizer and deheptanizer. These columns prepare the reactor effluent for further aromatic extraction in the BTX plant.

We refer the reader to review our discussion in Sections 2.4.2 and 2.4.3 regarding the concepts of individual stage efficiency and the overall stage efficiency. In particular, we agree with the recommendation of Kister [43] and Kaes [44] who strongly advised against the use of individual stage efficiencies such as the Murphree vapor stage efficiency, defined in Section 2.4.2.

We recommend the use of overall stage efficiency, which is the ratio of the number of theoretical stages and the number of actual physical trays. This is a single value that can range from 30% to 90%. If we consider the case of a distillation column having 20 physical trays and overall efficiency of 0.5, we would model it as a column with 10 theoretical stages. With this approach, every tray remains

Table 5.7 Summary of overall column efficiencies for product fractionation in CCR.

Fractionator	Theoretical trays	Overall efficiency (%)
Reformate splitter (debutanizer)	27	60–70
Deheptanizer	36	60–70

Table 5.8 Key specifications in fractionation section.

Fractionator	Initial specifications	Final specifications
Reformate splitter (debutanizer)	1) Reflux ratio 2) Overhead (or bottom) draw rate 3) Control stage temperature	1) Reflux ratio 2) Mole purity of C5 in the overhead 3) Control stage temperature
Deheptanizer	1) Reflux ratio 2) Overhead draw rate	1) Reflux ratio 2) Control stage temperature

in thermodynamic equilibrium and predictions away from the base operating scenario are reasonable. In this chapter, we model the DA301, a reformate splitter, and DA302, a deheptanizer. Table 5.7 shows the relevant overall efficiencies for these columns [44].

An important consideration is the selection of specifications to converge columns. Modern simulation software makes it quite easy to choose a wide range of specifications. However, software generally does not provide a guide to choosing reasonable specifications. In our work, we use a two-stage process. We first choose specifications that we know to converge easily for a given feed rate to the column. For a simple distillation column, these are typically the reflux ratio and overhead draw rate. In addition, we also provide temperature estimates. Once we obtain an initial solution, we introduce more difficult specifications such as temperature, mole recovery, and control temperatures. Table 5.8 gives the specifications for relevant columns in the CCR fractionation process.

Another significant consideration is that when modeling an existing plant, model developers should be aware of what the key control variables in the column are. The final specifications in the column must reflect actual plant control variables. For example, we should not fix the temperature of a condenser in the model when the plant actually controls the column based on an overhead draw rate.

5.8 Feed Characterization

The most important consideration for a reactor model is an accurate measure of the feed composition. This is particularly troublesome when modeling refinery reaction processes. Feed to units may change quickly and unpredictably. Although refinery techniques for online measurements of feed composition have improved, many still do not perform detailed molecular based analysis required for complex kinetic models. Without an accurate and up-to-date feed composition, kinetic models fail to make reasonable predictions of product yield and process performance.

There are several methods to alleviate this issue. One method is to work from a standard set of preanalyzed feeds and generate a set of base compositions. In addition, a large database of standard preanalyzed feeds can provide a process to generate the composition shift vectors. This is very similar to the process of generating delta-base vectors for refinery planning discussed in Section 4.12. We attempt to quantify the effects of changes in easily and routinely measured bulk properties such as TBP curves, specific gravity, molecular weight, and viscosity on the changes in the feed composition. Aspen HYSYS Petroleum Refining provides a method based on the presence of several feed types. The feed types refer to the origin of the feedstock entering the reforming unit. Depending on the size of the database used to generate these shift vectors, this method can be very accurate in practice.

Another method is to try and estimate the composition of the reactors based only on bulk property information. This bulk property information typically refers to routinely measured properties such as density and distillation curves. Klein and coworkers [29] have used a much more sophisticated version of this approach to probabilistically sample candidate molecules and generate a very large list of molecules whose combined properties match the measured bulk properties. Hu *et al.* [24] used a probability distribution method to estimate the PNA compositions for their approach toward refinery reactor modeling. The approach we describe is similar, but much simpler to use, as it is targeted only for reformer feeds.

A key assumption in this method is that each class of molecules (i.e., paraffins, naphthenes, and aromatics) is statistically distributed around a certain mean value. For the case of reformer feed, we know that significant portion (80+ wt%) lies between the C6–C9 range. With this information, we assume that each class centers around the C6–C9 range following a statistical distribution. Sanchez *et al.* [45] applied various statistical distributions to fit a variety of distillation data. They recommend the use of the beta statistical function to accurately represent distillation data.

A key criterion is that the statistical function can be normalized and distributed, but nonsymmetrical, as a certain class of compounds may exist in very narrow ranges. In addition, we would like a function that is easily accessible in popular software tools (e.g., Microsoft Excel) and has as few parameters as possible. Based on the observations by Sanchez *et al.* [45] and our criterion, we find that a two-parameter normalized beta statistical distribution for each class of molecules is sufficient for characterizing a reformer feed. We discussed this statistical beta function in Section 1.2, Eq. (1.7), and in Section 1.4, Workshop 1.2. We renumber this equation as Eq. (5.21):

$$f(x,\ \alpha,\beta,A,B) = \int_{A}^{x\leq B} \left(\frac{1}{B-A}\right) \frac{\Gamma(\alpha+\beta)}{\Gamma(\alpha)\Gamma(\beta)} \left(\frac{x-A}{B-A}\right)^{\alpha-1} \left(\frac{B-x}{B-A}\right)^{\beta-1} \tag{5.21}$$

where α, β, A, and B refer to the positive-valued parameters that control the shape of the distribution, Γ refers to the standard gamma function, and x identifies a given lump.

We apply the method in the following steps:

1) Choose the lumping range. In our work, we choose the PNA lumps in C5–C11 range.
2) Precompute the individual properties of each of the lumps (i.e., associate each lump with normal boiling point, standard liquid density, and molecular weight). It is possible to compute each of the properties using correlations from Riazi [42].
3) Obtain as much bulk data about the feed as possible. The minimum requirements are specific gravity and true boiling point (TBP) curve.
4) If a TBP curve is not available, use API correlations to convert a D86 distillation curve to a TBP curve (see Section1.3).
5) This method requires the total PNA content expressed in either weight %, volume %, or mole %. If this information is not available, the API correlation [42] (requiring viscosity) can provide these values.
6) Guess values for the mean and standard deviation for each distribution to compute the fraction of each component in the C5–C11 (a total of six parameters). As we know the total PNA (from step 5), we can normalize each distribution to make sure that the sum of fractions of each class of lumps matches the total PNA.
7) Compute the bulk property information using the candidate lump compositions.
8) Arrange all the candidate lumps in the order of increasing boiling point to generate candidate TBP curve.
9) Compute a residual between the measured or known bulk properties and calculated bulk properties in step 7.
10) Return to step 6 unless the residual is minimized to some small value.

In our experience, the last end points of a typical 5-point TBP curve (the end point or EBP, 90% vaporization point, and 70% vaporization point), the molecular weight (measured or estimated from API correlation) and specific gravity are good candidate bulk properties to minimize against. This is a basic optimization problem. We have used the SOLVER add-in in Microsoft Excel with considerable success. We note that once an optimized solution has been reached for a base feed, it is often very simple (even manually) to adjust the parameters of the statistical distribution to fit a new feed type. We report the optimal values for the fitting parameters in Table 5.9.

We apply this method to the feed specified in Table 5.1 using the ASTM D86 distillation, specific gravity, and individual PNA composition. We convert the ASTM D86 distillation curve to a TBP curve and estimate the molecular weight (using standard API correlations). We then optimize the parameters to match the EP, 90%, and 70% of the TBP curve, molecular weight, and specific gravity. We compare the calculated and measured values in Figure 5.12 and Table 5.10. For details of our calculations, please refer to Excel file *Alternate_Feed_Lump.xlsm* in the supplement to this text.

Table 5.9 Optimized parameters for PNA beta distribution functions.

Group	α	β
P	3.9145	6.6190
N	1.2454	4.5050
A	3.0402	6.9700

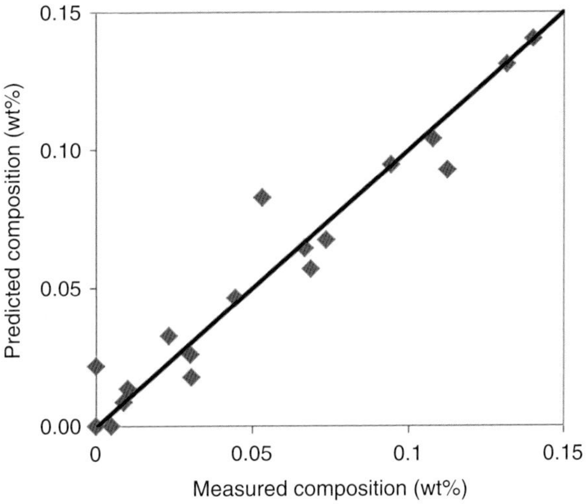

Figure 5.12 Correlation between prediction and measured composition.

Table 5.10 Predicted PNA composition from parameter estimation process.

	Predicted			Measured		
	P	N	A	P	N	A
C5	1.36%	0.00%	–	1.00%	0.47%	–
C6	5.70%	6.43%	0.85%	6.85%	6.66%	0.88%
C7	9.29%	13.09%	3.26%	11.25%	13.17%	2.31%
C8	9.46%	14.01%	2.57%	9.42%	14.02%	3.02%
C9	6.74%	10.38%	1.78%	7.35%	10.79%	3.04%
C10	4.64%	8.27%	2.17%	4.45%	5.31%	0.00%

There is good agreement between the measured TBP (converted from ASTM D86 data) and calculated TBP curve. Note that we have not included all the TBP points in the optimization routine, but the optimized solution makes good predictions for the lower TBP points as well (Figure 5.13).

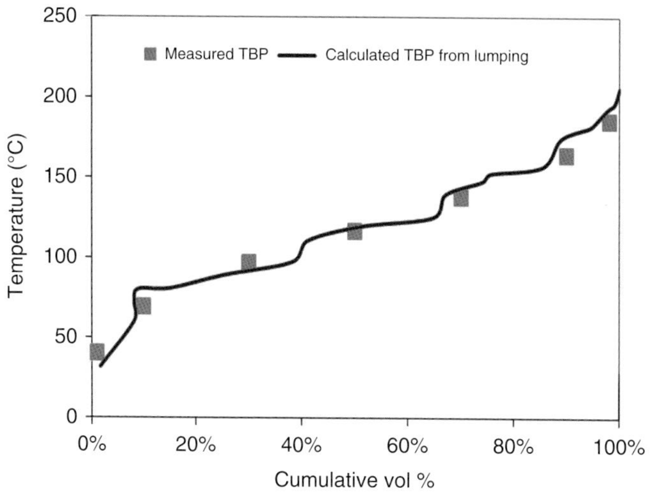

Figure 5.13 Comparison between measured and calculated TBP based on PNA lumping.

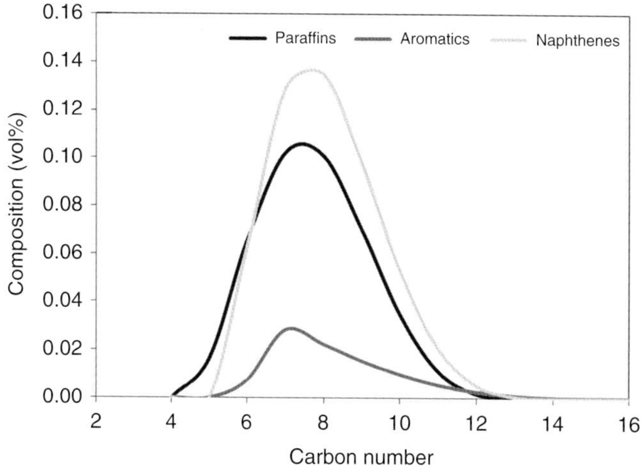

Figure 5.14 Optimized distribution of paraffin, naphthene, and aromatic for a given feed type.

Figure 5.14 shows the optimized distribution of PNA for this feed. As the distribution function predicts an A5 lump (a physically impossible solution), we ignore this component while calculating the lump composition. We note that each of the distributions has a different shape that reflects the different nature of a specific component class. If we use a simple normal distribution function, it is unlikely that we would be able to represent many different types of feed.

5.9 Model Implementation

There are three important considerations while building any reaction model based on plant data:

- Ensure the data consistency through accurate mass balance.
- Characterize the feed based on limited information.
- Calibrate the reactor model to a reasonable level of accuracy.

In the following sections, we discuss several steps and tools to help with the implementation of the model. We provide access to the tools mentioned in this section in the supplement to this text. Finally, we discuss an overall modeling strategy to model an existing reforming unit.

5.9.1 Data Consistency

An important task during data collection and model calibration is the overall mass and hydrogen balance across the reformer unit. The overall mass balance is simply a difference between the sum of all the feeds entering the unit and sum of all products leaving the unit. Although this concept is fundamentally simple, it can be difficult to realize in a real production plant. Many reformer units include feeds from other units that only enter plant through the fractionation section. This is typically the case when the refiner maximizes aromatics recovery produced by other units in the refinery.

We provide a spreadsheet, *Hydrogen_Balance.xls* (Figure 5.15), in the supplement to this text to account for feeds to the reforming plant that enter the reactor section and fractionation section. We can either subtract the feeds entering the unit or make sure they are accounted in the overall balance. We have successfully closed the mass balance to under 0.2–0.3% by making sure to account for all products. The advantages of a closed mass balance are not limited to the kinetic modeling process itself, as other refinery-wide modeling (such as production planning) often relies on accurate mass balances.

A secondary issue relates to the calibration and predictions from the rigorous reformer model. It is critical to ensure that the hydrogen balance is satisfactorily closed before beginning model development. We define the hydrogen balance as follows:

$$\text{Mass flow rate of hydrogen in the feed} \\ = \text{Mass flow rate of hydrogen leaving the unit} \tag{5.22}$$

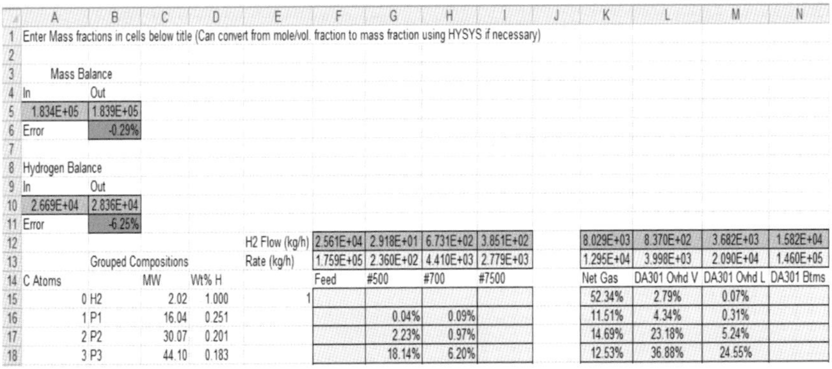

Figure 5.15 Microsoft Excel-based spreadsheet tool for mass and hydrogen balance calculations (see *Hydrogen_Balance.xls* in the supplement to this text).

Turpin [46] provided a simple formula for calculating the hydrogen content. We use a similar equation to verify the balance of the unit.

$$H_{\mathrm{FACTOR}_{ij}}(\text{for } \mathrm{C}_i\mathrm{H}_j) = \frac{j \cdot 1.01}{i \cdot 12.01 + j \cdot 1.01} \tag{5.23}$$

$$\text{Hydrogen flow of } \mathrm{C}_i\mathrm{H}_j = H_{\mathrm{FACTOR}_{ij}} \cdot \text{Mass flow of } \mathrm{C}_i\mathrm{H}_j \tag{5.24}$$

Turpin [46] recommended that hydrogen mass balances should be closed to less than 0.5% error. This can be difficult without detailed verification of measured flow rates. We recommend that calibration proceeds even if the hydrogen balance cannot be closed. However, it may not be possible to perform a finely tuned calibration as a result.

5.9.2 Feed Characterization

Section 5.8 discusses a method to obtain estimates for the composition when only limited feed characterization data (distillation curves and density) are available. Although the method produces good estimates of the feed composition, it may fail to predict the correct amount of N5 and N6 in the feed. Good estimates of N5 and N6 are critical for a meaningful calibration as these components are the primary pathways to obtain benzene in the reformate.

We recommend that analysis can be performed to determine the N5 and N6 composition before calibrating a detailed model of the reformer. Once feed analysis establishes the baseline N5 and N6 contents, we can expect the calibration to reflect reactor operation more accurately. Figure 5.16 shows the variation in N5 and N6 contents of the hydrotreated reformer feed over the course of our work. There is significant variation in the N6 content, which justifies a detailed feed analysis before the model calibration.

5.9.3 Calibration

Because of the number of unit-level and kinetic models available in the literature and commercially, it is impossible to prescribe a single calibration method that will work for all models and methods. However, there are significant common

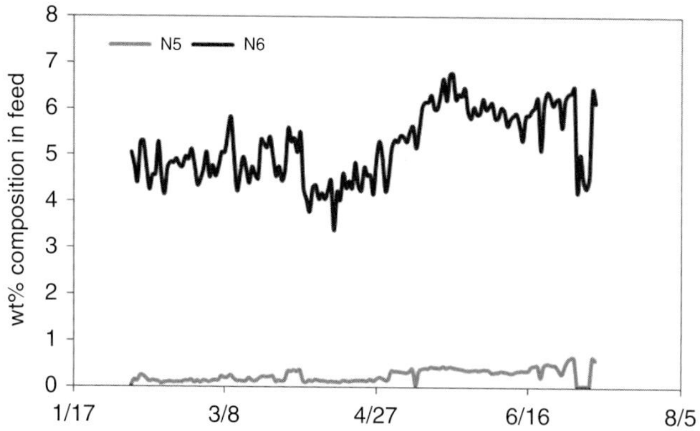

Figure 5.16 Variation in N5 and N6 contents in feed.

features in all of the models to allow for general recommendations. These recommendations form a simple workflow to manage the large number of parameters that may occur in many models.

Modern calibration methods in current software allows users to change many (if not all) parameters in a particular model with ease. Although this is a simple procedure, it is easy to "overcalibrate" the model and generate calibration values that ignore process chemistry and other phenomena. We believe that it is better to follow a step-by-step process where we only change a few parameters (of the same class and with bounds) at a time.

We perform the calibration in two passes. The first pass is *the coarse calibration* of the model, whereas the second pass performs *the fine calibration*. The quality of the model calibration relies on consistent and reliable data. If we cannot find these data, it may be difficult to justify performing the fine calibration of the model. In fact, performing the fine calibration with poor-quality data may result in an "overcalibrated" model. With that in mind, we propose a step-by-step process for calibration.

The key steps in calibration in applying Aspen HYSYS Petroleum Refining are as follows:

1) Verify that the material and hydrogen balances are closed.
   a) If the material balance has an error exceeding 1–2%, this data set should not be used for calibration.
   b) If the total hydrogen balance has an error exceeding 2–3%, it is unlikely that fine-tuning of the reactor model will be successful.
2) Obtain the feed composition.
   a) Use detailed PNA information if possible.
   b) If detailed PNA information is not available, use total PNA information and feed characterization method described earlier.
   c) If total PNA information is not available, it is possible to use bulk measurements such as viscosity, density, and distillation data to estimate the PNA composition required for (b). These correlations are available from our previous work and Riazi [42]. In this case, fine-tuning of the reactor model can become difficult.
3) Select the objective function criteria.
   a) Define the objective function to minimize as $\sum w_i(\text{Measured}_i - \text{Predicted}_i)^2$.
   b) Table 5.11 suggests the terms and associated weightings for both coarse and fine calibrations. A zero entry in the weighting factor indicates that the term should not be part of the objective function.
   c) If a detailed analysis of the reactor effluent is available, do not include every component in the objective function.
4) Coarse tuning
   a) Select the overall reactor selectivity only.
   b) Use Table 5.11 to select the terms for coarse tuning form of the objective function.
5) Second pass
   a) Select the overall reactor selectivity and the overall reaction activity.

b) Use Table 5.12 to select terms for fine-tuning form of the objective function.
c) Calibrate the model.
d) Adjust the selectivity for light ends (P1–P3) as the last step in the calibration.

It is important to not include the yield of every significant component. Including every possible measurement for optimization often results in a poor calibration. A poor calibration means that the model is essentially fixed to a single data point, and it will result in a model that responds widely even to small

Table 5.11 Major terms and their recommended weighting factors in the reformer model objective function for calibration.

Term	Coarse calibration	Fine calibration
Reactor delta temperature (s)	1.0	1.5–2.0
Total aromatics (wt%)	5.0	10.0
Benzene (wt%)	5.0	10.0
Toluene (wt%)	5.0	10.0
Xylenes (wt%)	1.0	10.0
A9+ (wt%)	0.0 (ignore)	5.0
Paraffins (P1–P3)	0.0 (ignore)	0.5 (last)
Paraffins (P4+)	0.0 (ignore)	1.0
Paraffins (P8+)	0.0 (ignore)	1.0
Naphthenes (N5, N6)	1.0	10.0
Ratio of isomer to normal paraffins	0.0 (ignore)	0.5 (may not be predicted)
Net gas flow	1.0	1.0
Total heavy (wt%)	0.0 (ignore)	1.0

Table 5.12 Typical adjustment factors to calibrate reformer model.

Parameter	Range of deviation from the base
Overall reactor activities	0.1–10
Reaction class	
Dehydrogenation	0.1–1.1
Hydrocracking	0.1–1.1
Isomerization	0.1–1.1
Ring closure	0.1–1.1
Ring expansion	0.1–1.1
Light ends tuning	
C1/C2/C3	0.1–5.0

changes in the input variables. It is better to avoid poor calibration even at the expense of not agreeing to plant measurements. When this situation happens, it means that there is likely mass imbalance or hydrogen imbalance in the feed and product measurements. It is best to recheck model inputs before attempting any further calibration.

We use the ranges for adjustment factors and weightings for the error residual to generate constraints for an optimization procedure. As the model is developed in an equation-oriented (EO) format, it is not difficult to apply an optimization procedure to generate optimal values for the adjustment factors. An objective value of less than 250 (using coarse weightings) is sufficient for coarse adjustment when significant feed information (such as composition) is missing or estimated. For fine adjustment, which is required in the case of accurate prediction for aromatic component composition, an objective value of less than 200 (using fine weightings) is required. Obtaining a reasonable calibration using fine-tuning requires accurate composition, feed rate, hydrogen yield, and reactor operating parameter (temperature and pressure) measurements.

The adjustment factors in Table 5.12 are sufficient to represent a wide variety of operating behavior. Models may allow users to individually tune each reaction in the kinetic network. Reaction-specific tuning may result in very good agreement with plant data, but the model may lose predictive ability. The reaction-specific tuning essentially fixes it to one operating point. In addition, models may include adjustment factors for the primary product separation. We do not adjust these values routinely as part of the calibration.

We note that it may not be possible to fine-tune the model to the prescribed limits earlier. Plant mass balance error, poor measurements, and unexpected process variation may limit how well the model agrees with the plant data. However, by following the above calibration procedure, we can ensure that we do not "over-calibrate" the model and subsequently produce poor predictions.

5.10 Overall Modeling Strategy

Figure 5.17 outlines the overall modeling strategy used in this work. We implement and calibrate the model while it is in regular operation in the refinery. Many factors such as abrupt changes in feed quality, operating parameters, and poor and inconsistent measurements impede this process. Work by Fernandes *et al.* [47] documented the same difficulties while modeling an FCC unit. In our work, data set refers to a collection of measurements that reflect plant operation for a short period (less than 1 day). We propose the following steps to ensure that calibration results in a model that is predictive and not fixed to a single operating scenario:

- Record data on a continuous basis from the plant.
  - Reconcile data from multiple sources (DCS, inventory, etc.).
  - Check consistency of the data set by performing material and hydrogen balance. Use the criteria in previous section to accept or reject certain data.

  - Accept a data set (or conditionally accept acknowledging that there may be significant error in calibration and prediction).
  - Track variation in the data set to ensure that we verify the model against significant changes in feed and operating parameters. We show the significant changes in feed quality in our work in Figure 5.18.
- Develop fractionation models by backblending the measured reactor products and verify that the models agree with plant measurements.
  - We provide guidelines for developing the fractionation system in Section 5.17, Workshop 5.3.
- Calibrate reactor model.
  - Use calibration procedure to produce a coarsely and finely calibrated model.
  - The product yields from the finely calibrated model should be within 1% of actual plant yield. If this is not the case, it is likely that the material balance and hydrogen was not closed sufficiently.
  - The outlet temperatures from the finely calibrated model should be within 3–5°C of measured plant values.
- Validation
  - Use of accepted data sets track the performance of the reformer and fractionation sections with the model.
  - If possible, examine the yield of the reactor effluent directly with measure products. We can identify if errors arise from the reactor model or the fractionation section and isolate the section for further validation or calibration.
  - It is typically possible to predict yields of key products (BTX) on a feed normalized mass basis with AAD of less than 2–3%.
- Recalibration
  - We suggest recalibration when significant changes occur in the catalyst or regeneration section. The model can generally account for significant changes in feedstock and operating parameters.

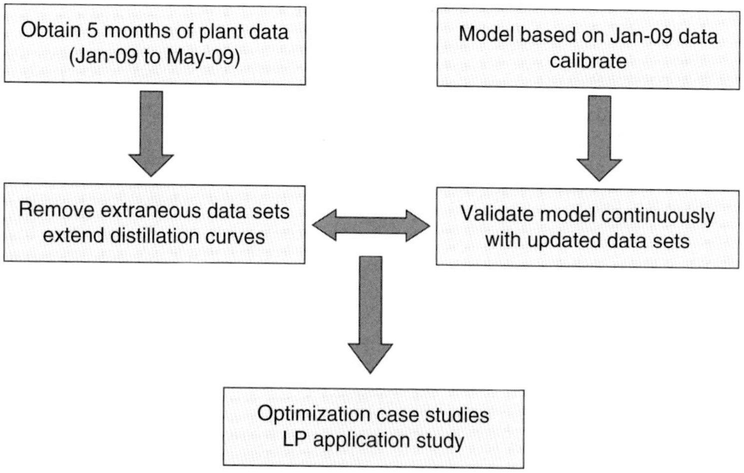

Figure 5.17 Overall modeling strategy.

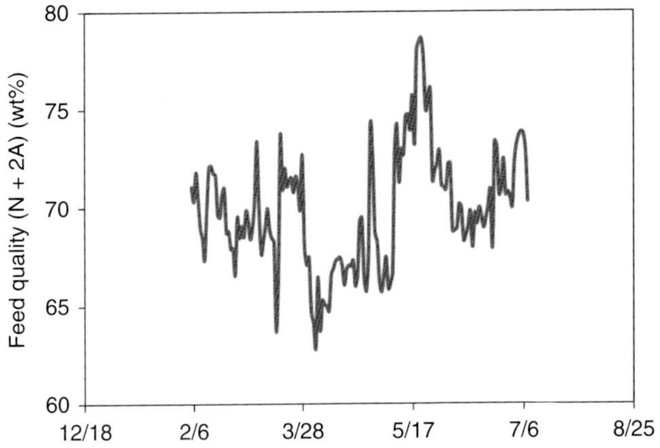

Figure 5.18 Variation in feed quality over the study period.

5.11 Results

Figures 5.19–5.21 show the completed HYSYS/Refining simulation models for the CCR reformer studied in this work. See simulation file *Workshop 5.3.hsc*. We evaluate the model using over 6 months of operating data from a refinery in the Asia Pacific processing hydrotreated naphtha. Key factors for the evaluation of

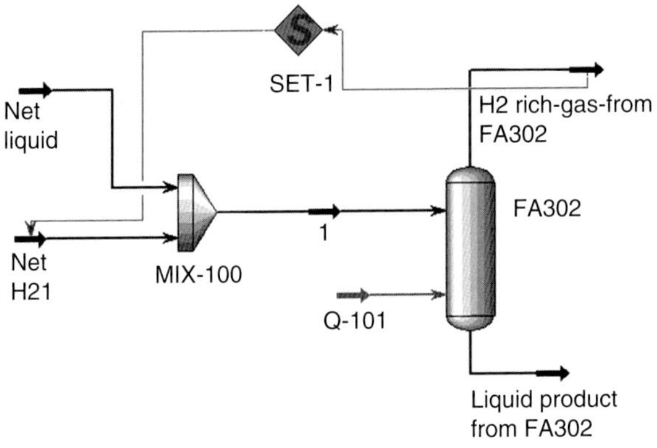

Figure 5.19 Remixing section.

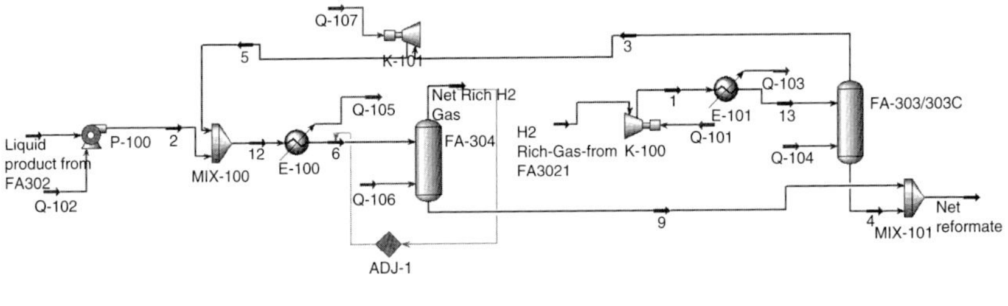

Figure 5.20 Recontacting section.

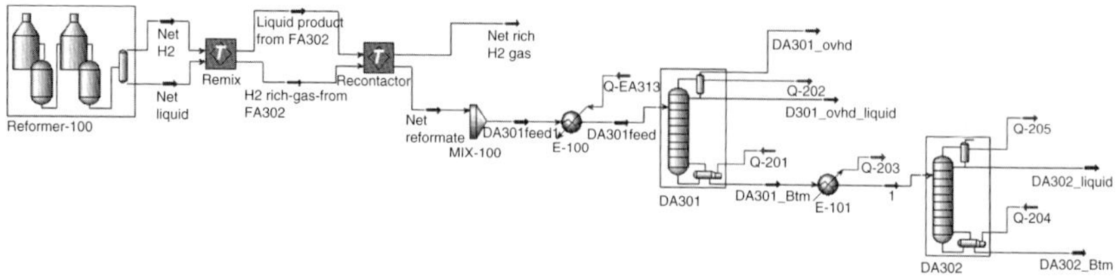

Figure 5.21 Combined reformer and primary fractionation.

the model are comparisons of overall product yields and operating profiles for key equipment in the gas plant. In general, the model accurately predicts the product yields, compositions, and operating profiles over a wide range of feed conditions.

The fractionation section of model uses the PR equation of state and the kinetic lumps directly as the fractionation lumps.

The remixing section is a simple way to reconstruct the plant effluent as the reactor model produces separate streams for the hydrogen product, Net H21, and liquid product, Net Liquid. Remixing the streams allows us to model the recontacting sections to predict compositions reported in the actual hydrogen product and liquid product streams. In Figure 5.19, the Net Liquid is the reformer liquid effluent. Essentially, all of the vapor product from the flash unit FA302, stream "H2 Rich-Gas-from FA302," is recycled back and remixed with the Net Liquid stream entering MIX-100. The liquid product from flash unit FA302, stream "Liq Product from FA302," goes to the recontacting section in Figure 5.20 as stream "H2 Rich-Gas-from FA3021," entering compressor K100.

In Figure 5.20, FA303/303C is a *low-pressure* flash unit at 21.62 bar, and FA304 is a *high-pressure* flash unit at 56.81 bar. The goal of the recontacting section is *to condense more liquid reformate* from the stream "H2 Rich-Gas-from FA3021" as stream 4 going to the mixer MIX-101 has to be collected as Net Reformate; at the same time, we wish *to recycle more pure hydrogen*, stream 3, from the low-pressure flash unit, FA303/303C, through mixer MIX-100 and exchanger E-100, to the high-pressure flash unit, FA304, *as a high-purity hydrogen stream*, Net Rich H2 Gas. The ADJ-1 block adjusts stream 6 temperature to ensure that the mole fraction of hydrogen in stream Net Rich H2 Gas is 0.9406. The liquid product from the high-pressure flash unit, FA304, stream 9, joins stream 4, from the low-pressure flash unit, FA303/303C, through the mixer, MIX-101, as stream NetReformate, going to the subsequent fractionation units.

The recontacting section in Figure 5.20 is different from the process flow shown in the plant PFD in Figure 5.2. We find that we do not require as many flash stages as the real process to obtain results similar to the plant. This is expected as each of the separators of in-plant PFD are likely not operating at equilibrium conditions. This is similar to the concept of using overall efficiency in our tray-by-tray fractionation models. We acknowledge that the simplified model of the recontacting section does not report the energy consumption (especially by the secondary compressors) correctly; in practice, the total energy consumption reported by model and the plant is similar.

A well-calibrated model produces significant and repeatable predictions over a wide range of operating conditions. Tables 5.13–5.16 summarize the predictions from our model developed and calibrated according to the previous sections. Each validation case represents roughly 1-month interval of the reformer.

Table 5.13 Comparison of overall reactor model and plant yields, AAD = 0.85%.

Yield	VALID-1		VALID-2		VALID-3	
mass%	**Model**	**Plant**	**Model**	**Plant**	**Model**	**Plant**
Rich H_2	6.1	6.9	6.4	7.2	6.5	7.0
DA301 Ovhd. vapor	1.2	1.8	1.9	1.9	1.5	1.5
DA301 Ovhd. liquid	13.0	12.0	14.2	12.4	12.6	12.4
DA301 Bttm. liquid	79.6	79.3	77.5	78.6	79.4	79.1
DA301 Ovhd. liquid	43.4	45.1	43.4	44.1	42.6	44.6
DA301 Bttm. liquid	56.6	54.9	56.6	55.9	57.4	55.4
Yield	VALID-4		VALID-5		VALID-6	
mass%	**Model**	**Plant**	**Model**	**Plant**	**Model**	**Plant**
Rich H_2	6.3	7.0	6.5	6.7	6.6	6.8
DA301 Ovhd. vapor	1.7	1.7	1.8	1.8	1.8	1.7
DA301 Ovhd. liquid	13.6	12.2	11.2	12.0	11.1	12.3
DA301 Bttm. liquid	78.3	79.1	80.5	79.5	80.5	79.2
DA302 Ovhd. liquid	45.9	45.3	45.4	46.2	45.4	43.2
DA302 Bttm. liquid	54.1	54.7	54.6	53.8	54.6	56.3

Table 5.14 Comparison of key reactor temperature drop in model and plant values, AAD (Total) = 1.7 °C.

Reactor temperature drop	VALID-1		VALID-2		VALID-3	
(°C)	**Model**	**Plant**	**Model**	**Plant**	**Model**	**Plant**
Reactor #1	108.2	109.9	107.3	106.0	114.1	111.5
Reactor #2	61.6	63.1	60.6	59.9	67.8	64.9
Reactor #3	33.7	35.2	32.1	33.9	38.0	37.0
Reactor #4	20.5	23.3	18.7	22.3	22.7	25.5
Reactor temperature drop	VALID-4		VALID-5		VALID-6	
(°C)	**Model**	**Plant**	**Model**	**Plant**	**Model**	**Plant**
Reactor #1	107.4	107.6	113.9	112.8	113.3	111.7
Reactor #2	60.7	61.9	66.7	67.0	66.1	66.2
Reactor #3	32.8	34.9	37.0	37.1	36.4	37.0
Reactor #4	19.6	23.3	22.1	24.2	21.7	24.6

Table 5.15 Comparison of key model and plant yields in the reformate, AAD (total) = 1.05; AAD (aromatics) = 0.85.

Reformate yields	VALID-1		VALID-2		VALID-3	
(wt%)	**Model**	**Plant**	**Model**	**Plant**	**Model**	**Plant**
Benzene (B)	7.5	7.9	7.7	7.1	7.0	6.4
Toluene (T)	21.3	20.7	22.0	21.1	20.9	19.9
Ethylbenzene (EB)	3.6	3.5	3.6	3.4	3.5	3.4
para-Xylene (PX)	5.5	5.1	5.6	5.3	5.5	5.1
meta-Xylene (MX)	11.9	11.1	12.1	11.7	11.8	11.2
ortho-Xylene (OX)	6.7	6.3	6.8	6.5	6.6	6.3
Higher aromatics (A9+)	40.5	38.1	39.2	41.6	41.5	43.3
Paraffins (P)	1.4	2.0	1.2	1.0	1.3	1.1
Naphthenes (N)	12.5	14.5	11.9	14.0	12.7	14.5
Reformate yields	VALID-4		VALID-5		VALID-6	
(wt%)	**Model**	**Plant**	**Model**	**Plant**	**Model**	**Plant**
Benzene (B)	8.4	7.7	8.0	8.1	8.0	8.0
Toluene (T)	22.7	21.5	23.2	20.8	23.2	20.5
Ethylbenzene (EB)	3.6	3.3	3.6	3.4	3.6	3.4
para-Xylene (PX)	5.5	5.3	5.6	5.0	5.6	4.9
ortho-Xylene (OX)	11.9	11.4	12.1	11.0	12.1	10.7
meta-Xylene (MX)	6.7	6.4	6.8	6.3	6.8	6.1
Higher aromatics (A9+)	35.8	38.0	34.5	41.2	34.5	40.1
Paraffins (P)	1.4	1.2	1.5	1.4	1.5	1.4
Naphthenes (N)	12.6	14.6	12.1	13.7	12.1	14.7

The most important predictions from the reactor model are the overall yields of all the key products from the unit. In case of the reformer, they are the net gas production, LPG (DA301 Ovhd. liquid), and reformate (DA301 Bttm. liquid). The yields in the above table are from the rigorous tray-by-tray fractionation section. Therefore, the effect of downstream fractionation is also included in these predictions. We note good agreement with the plant values. The AAD (counting all products) is less than 1.0%.

The reactor performance is also a key indicator of model's calibration and prediction. We note that reactor model tracks reactors #1 to #3 with roughly the same accuracy. We observe the larger error in reactor #4 because we do not allow significant changes in individual tuning of the reactions. In the final reactor, more exothermic reactions start to dominate and push the reactor into a region where paraffin cracking becomes significant. However, even this higher deviation of outlet temperature is within the expected deviation at the plant.

As this reformer is part of a petrochemical complex, the predictions of individual molecules in the reformate are quite significant. An accurate prediction of the compositions of benzene, toluene, ethylbenzene, and xylenes

Table 5.16 Comparison of LPG composition from model and plant, AAD = 2.0 mol%.

DA301 Ovhd. liquid	VALID-1		VALID-2		VALID-3	
(mol%)	**Model**	**Plant**	**Model**	**Plant**	**Model**	**Plant**
Ethane (C2)	8.7	8.3	8.3	8.1	7.5	9.5
Propane (C3)	25.4	28.3	24.9	26.8	23.6	28.0
Isobutane (*i*C4)	23.4	20.3	23.9	19.3	23.6	19.2
n-Butane (*n*C4)	19.6	18.0	19.4	18.4	20.1	17.5
Isopentane (*i*C5)	14.1	16.0	14.6	17.7	16.0	15.9
n-Pentane (*n*C5)	6.2	7.6	6.2	8.5	6.7	7.8
DA301 Ovhd. liquid	VALID-4		VALID-5		VALID-6	
(mol%)	**Model**	**Plant**	**Model**	**Plant**	**Model**	**Plant**
Ethane (C2)	8.4	9.4	7.1	7.4	7.0	8.0
Propane (C3)	26.0	29.5	25.1	28.5	25.0	26.9
Isobutane (*i*C4)	23.5	20.6	23.6	20.7	23.6	19.6
n-Butane (*n*C4)	19.2	17.1	19.5	18.6	19.6	18.1
Isopentane (*i*C5)	13.9	15.3	15.1	16.4	15.1	17.3
n-Pentane(*n*C5)	5.9	6.4	6.4	7.1	6.5	8.5

(collectively referred to as BTEX) validates our model and provides feed values for the downstream model for the BTX separation plant. Table 5.15 compares the predicted values and plant data. The AAD for all the components is 1.05 wt%, whereas the aromatics show a deviation of only 0.85 wt%.

A key part of this work in the development of fractionation sections for the reformate and A6 splitter. We compare the model predictions of the temperature profiles of the LPG column DA301 and reformate separator DA302. We note good agreement with plant measurements (Figures 5.22 and 5.23).

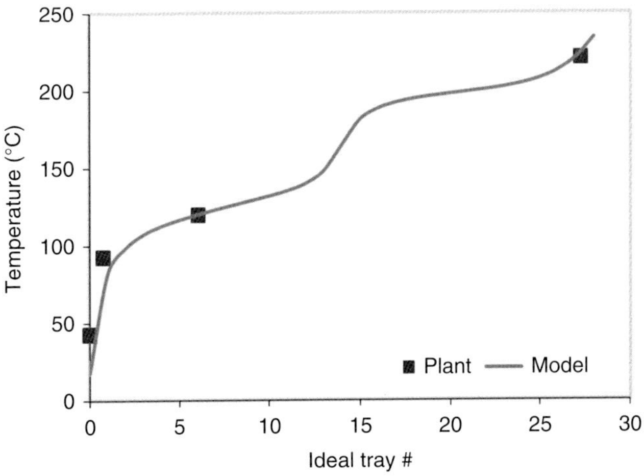

Figure 5.22 Temperature profile of column DA301.

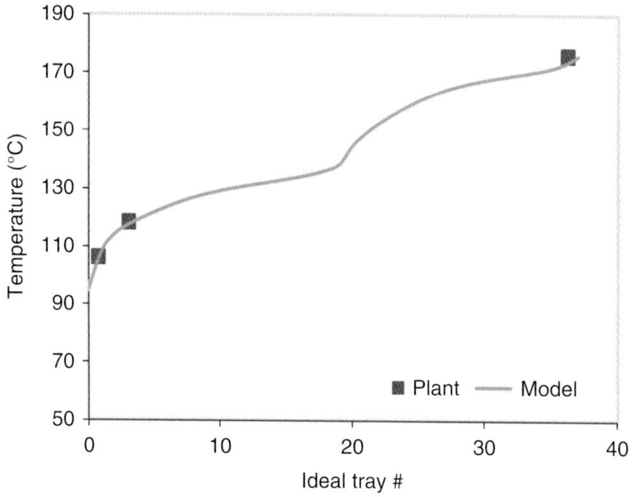

Figure 5.23 Temperature profile of column DA302.

5.12 Applications

Refiners typically face two types of operating scenarios with reformers. The first type of scenario is *the "what-if" scenario.* In this scenario, we want to predict the process performance given a change in a key process variable. For CCR reformers, the typical operating variables for a given feedstock are *the reactor temperature, feed rate (or space velocity), reactor pressure, hydrogen to hydrocarbon ratio (H_2HC),* and *activity of the catalyst.* By making changes in the process variables, refiners can make significant shifts in the product distribution.

The second type of scenario is *the "how-to" scenario.* Modern reforming units may consume a variety of different feedstocks while facing changing product demands. Due to the highly integrated nature of refineries, it is important to consider the effects of the upstream and downstream units on the reformer's performance. There are several typical questions that form the "how-to" scenario. How can we reduce benzene in the reformer outlet? How can we use (or blend in) an additional feedstock? How can we account for changes in the reformer process on an economic basis?

Refiners often rely on performance charts, empirical correlations, and historical data to study these types of scenarios. Gary *et al.* [8] and Little [6] gave examples of several types of these correlations. These methods can be unreliable because they assume a fixed feedstock and a fixed set of operating conditions. In addition, these methods often ignore the interaction between process variables and can mask optimal operating conditions. It is in this context that we consider the use of rigorous models to study various operating scenarios. Rigorous models can account for complex changes in process variables and provide detailed predictions of reactor performance.

5.12.1 Effect of Reactor Temperature on Process Yields

A typical operating scenario is the increase of reactor temperature to promote higher severity operation to produce high-octane reformate and aromatics. Figures 5.24–5.31 indicate key changes in the reformer performance as a function of WAIT in the process. In addition, we must also consider the effect of hydrogen partial pressure in the reactors. We study this effect by changing the WAIT and various values for the H_2HC ratio.

Increasing the reactor temperature through WAIT generally increases the yield of the aromatic components and the octane number. However, for a given H_2HC ratio, there is a maximum aromatic yield and octane number. This results from the increased relative extent of hydrocracking versus dehydrogenation due to the

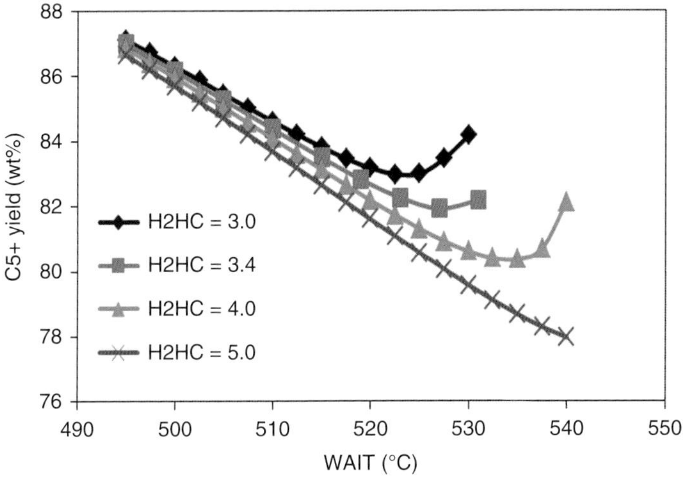

Figure 5.24 Change in C5+ yield (wt%) as a function of WAIT and H_2HC ratio (WHSV = 1.37).

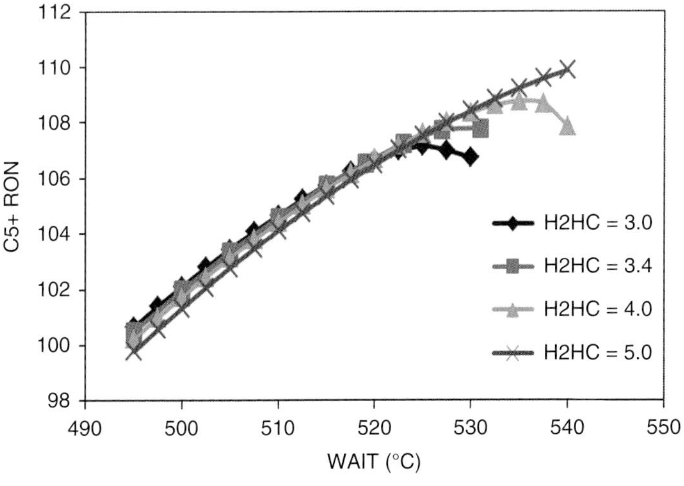

Figure 5.25 Change in C5+ RON as a function of WAIT and H_2HC ratio (WHSV = 1.37).

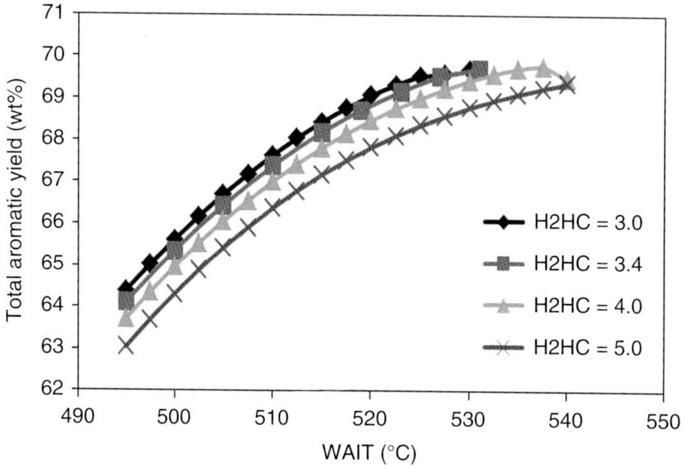

Figure 5.26 Change in total aromatic yield (wt%) as a function of WAIT and H_2HC ratio (WHSV = 1.37).

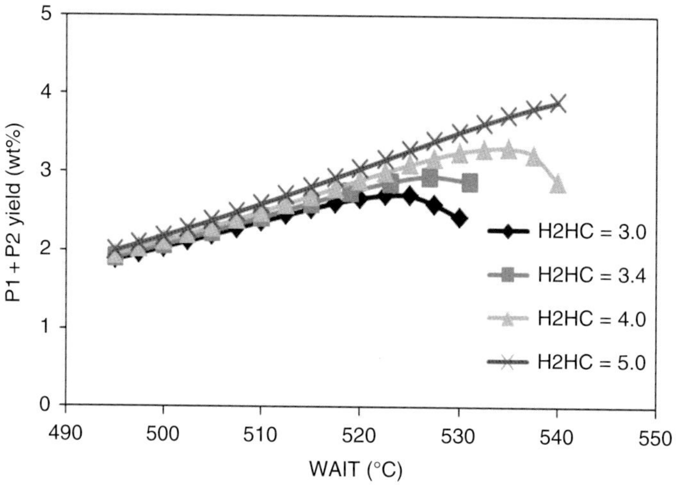

Figure 5.27 Change in light gas yield as a function of WAIT and H_2HC ratio (WHSV = 1.37).

temperature increase. Correspondingly, the C5+ yield (sum of all components greater than C4) decreases with increasing octane number.

To consider operating at high WAIT conditions, it is possible to run the reactor at much higher H_2HC ratios. Figures 5.24–5.26 show that we can reach a much higher octane number at high WAIT values. However, when the WAIT is low (compared to the octane peak), Figures 5.25 and 5.26 show correspondingly lower aromatic yield. Therefore, there must be a balance between the H_2HC ratio and WAIT to produce optimal octane number and aromatic yield. Another important consideration in increasing WAIT is production of undesirable side products and excessive coke generation. Figures 5.27 and 5.28 show the effect of WAIT on the production of dry gas (methane and ethane) and the coke laydown rate.

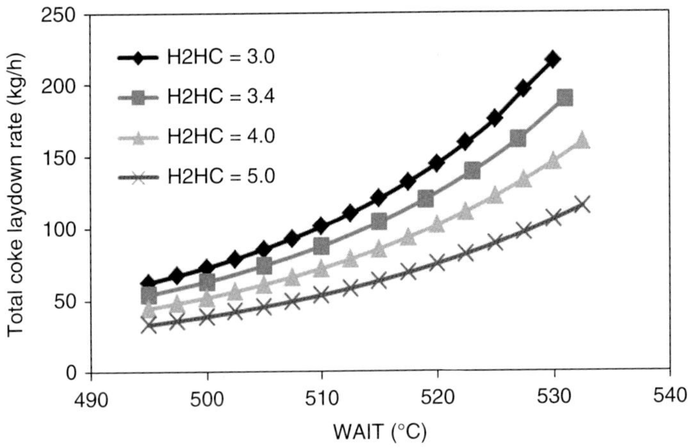

Figure 5.28 Coke laydown rate (kg/h) as a function of WAIT and H_2HC ratio (WHSV = 1.37).

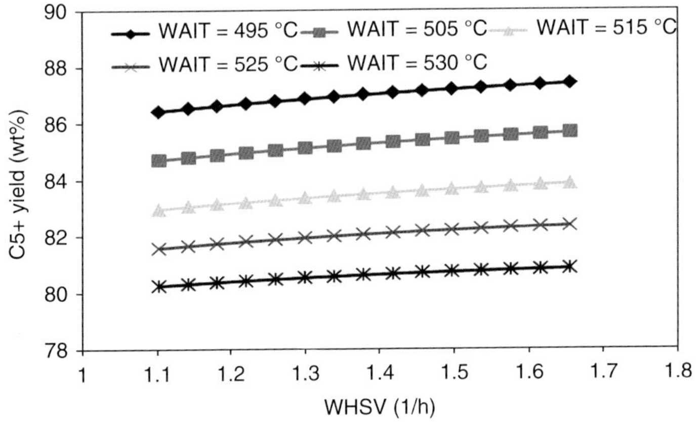

Figure 5.29 Change in C5+ yield (wt%) as a function of WHSV and WAIT.

Before we approach the octane-maximizing peak in Figure 5.25, increasing the reactor temperature increases the yield of light gas and the coke generation rate according to Figures 5.27 and 5.28. The increase in the light gas yield can be problematic. The light gas typically has little economic value and causes bottlenecks in the recycle compressors in the product separation section. Increasing the H_2HC ratio typically does not help in reducing the light gas yield (Figure 5.27), as high partial pressure of hydrogen in the reactor promotes hydrocracking and subsequently increases the light gas yield. In addition, the coke laydown rate increases exponentially with increased temperature (Figure 5.28) and can put significant pressure on the regenerator section of the CCR. Operating at high temperatures may require a significant addition of fresh catalyst to maintain the same level of catalyst activity.

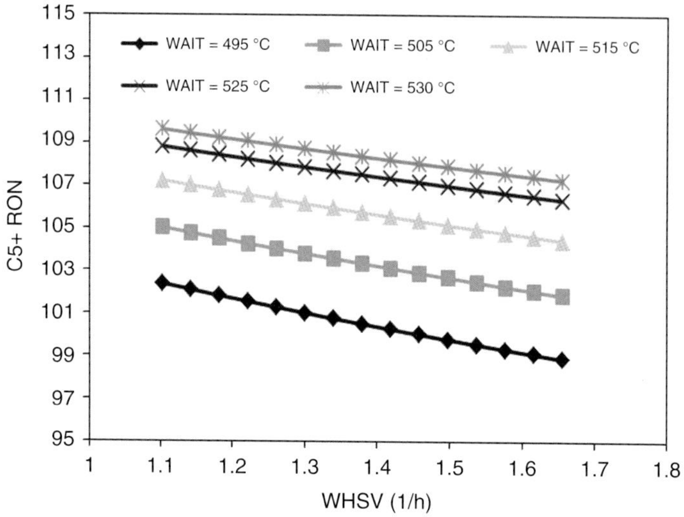

Figure 5.30 Change in C5+ RON as a function of WHSV and WAIT.

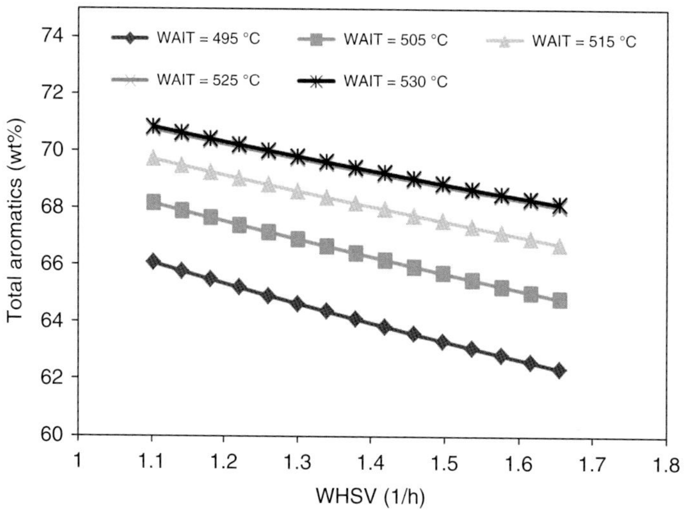

Figure 5.31 Change in total aromatics yield as functions of WHSV and WAIT.

5.12.2 Effect of Feed Rate on Process Yields

The reactor temperature is a primary method of shifting the reactor yield to produce more valuable product distributions. Another process variable is the feed rate to the unit. The feed rate cannot take on drastically different values due to the demands of other units in the refinery. However, small changes in feed rate can influence the product distribution. This occurs because of the change in contact time with the catalyst. Higher contact times increase the conversion of feed to products.

Figures 5.29–5.31 show the change in key reactor yields as functions of weight hourly space velocity (WHSV) and the WAIT. The figures show that as WHSV increases (feed rate increases), the conversion to aromatics decreases and the corresponding octane number decreases. This is consistent with our expectation of lower contact time. In general, the impact of changing feed rate is less than changing the reactor temperature. For significant changes in the RON and total aromatic yield, the reactor temperature is still the primary driver. In Figures 5.30 and 5.31, the lines for high WAIT approach a minimum slope, as we are approaching the octane peak for the baseline H_2HC ratio.

5.12.3 Combined Effects on Process Yields

Therefore, changes in octane number and total aromatic yield reflect the coupled effects of feed rate and reactor temperature. We can use the model to provide reactor temperatures that correspond to a fixed RON and varying feed rate. Figure 5.32 shows the relevant WAIT and WSHV to obtain a given C5+ RON. We note that at high C5+ RON operation and high WHSV, the required reactor temperature increases significantly. As shown in Figures 5.27 and 5.28, this increases the unwanted dry gas yield and produces an excessive amount of coke. By using Figure 5.32, we can determine how to change process variables to achieve desired C5+ RON.

We show a related study (Figure 5.33), indicating how the C5+ yield changes with increasing values for C5+ RON. Figure 5.33 helps the refiner identify the range of values that H_2HC ratio may take to obtain the same C5+ RON. Combined with Figure 5.32, we can identify possible operating windows for WAIT, H_2HC, and WHSV for a given feedstock composition.

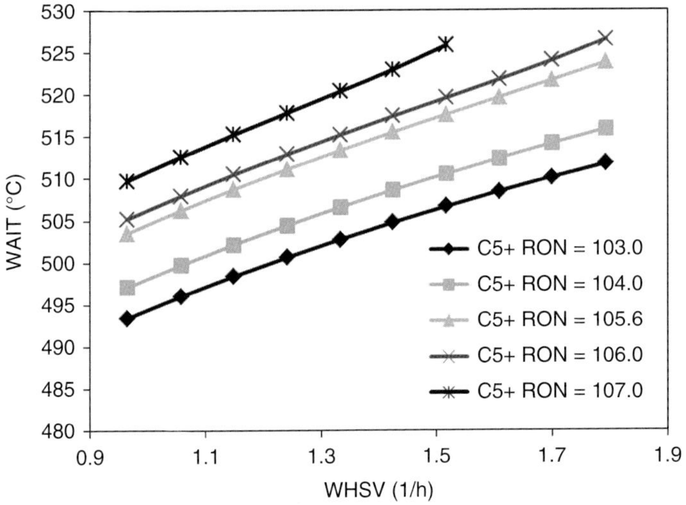

Figure 5.32 Corresponding WAIT and WHSV to obtain various C5+ RON in reactor products.

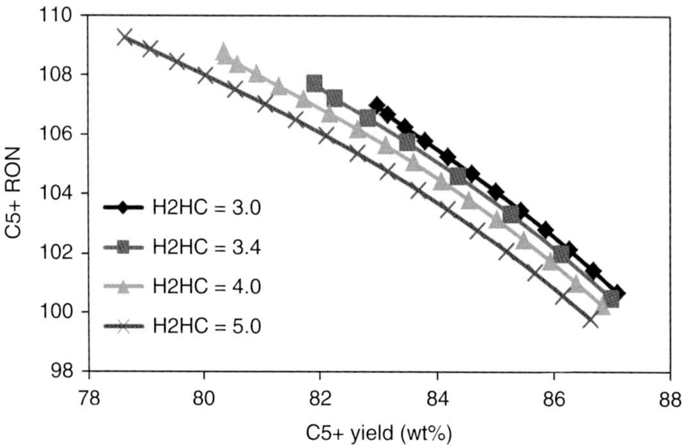

Figure 5.33 Effect of C5+ yield (wt%) on C5+ RON.

5.12.4 Effect of Feedstock Quality on Process Yields

All the previous studies involve a uniform feedstock composition. In practice, however, feed composition can change significantly over the course of regular refinery operation (see Figure 5.18). Therefore, it is important to study changes in product distribution when the feed composition varies. The benzene content of reformate is of particular interest to refiners. Recent regulations have imposed strict limits on the amount of benzene present in the gasoline pool. As the reformer is the primary source of benzene, we look for ways to reduce the benzene in reformate.

The primary contributors to benzene and toluene are methylcyclopentane (MCP) and methylcyclohexane (MCH). Various authors have commented on the significance of this pathway to produce aromatics [48, 49]. We study the effect of the MCP composition in the feed on the yields of benzene, toluene, and xylenes in Figure 5.34. We use the standard operating parameters as with other

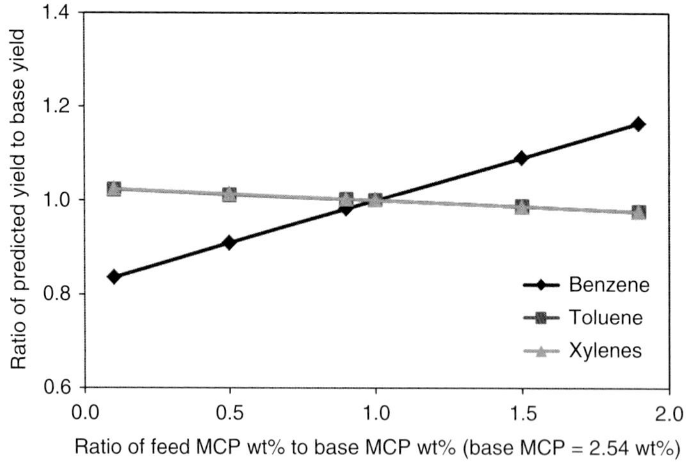

Figure 5.34 Effect of changing feed MCP composition on aromatic yields.

case studies in this work. Figure 5.34 shows that increasing MCP concentration has a strong effect on the benzene yield in the reformate. In addition, MCP composition has little effect on the composition of the higher aromatics.

In practice, a refiner does not directly control the feed composition of MCP to the unit. Typically, we blend in additional feed that has an IBP greater than 95–100 °C. Feeds with IBPs greater than 95–100 °C contain little amounts of MCP and this ratio can be used to control the benzene yield of the unit. By contrast, a refiner who wants to increase the production of benzene (to supply a chemical process) may need to increase the feed of MCP instead of operating the reformer at increasing severity and converting reactor products to benzene. Using a rigorous model can help us find and understand these types of trade-offs.

5.12.5 Chemical Feedstock Production

Many reformers are now part of integrated petrochemical complexes and produce aromatics (benzene, toluene, and xylenes or BTX) to feed into chemical processes for polystyrene, polyesters, and other commodity chemicals. As such, it is important to consider how models can help in optimizing the BTX operation. Model developers and users must also be aware that complete BTX operation may not be the most profitable reformer operation scenario. Economic analyses are required to justify changes from a gasoline-producing scenario to a BTX-producing scenario.

In general, many of the case studies show in earlier sections (relating to higher-octane operation) apply to the BTX scenario as a well. Figures 5.25 and 5.26 show the relationship between octane number and aromatic yield. We repeat some of the case studies shown in previous sections, showing the effect of process variables on BTX yields. In Figures 5.35 and 5.36, we take the yields

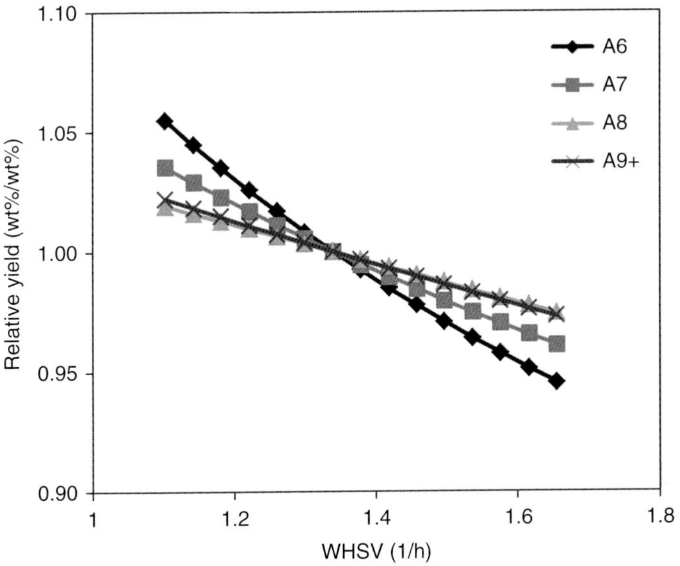

Figure 5.35 Relative yields of aromatic components (where A6 refers to benzene, A7 refers to toluene, and A8 refers to xylenes) as a function of WHSV and WAIT = 495 °C.

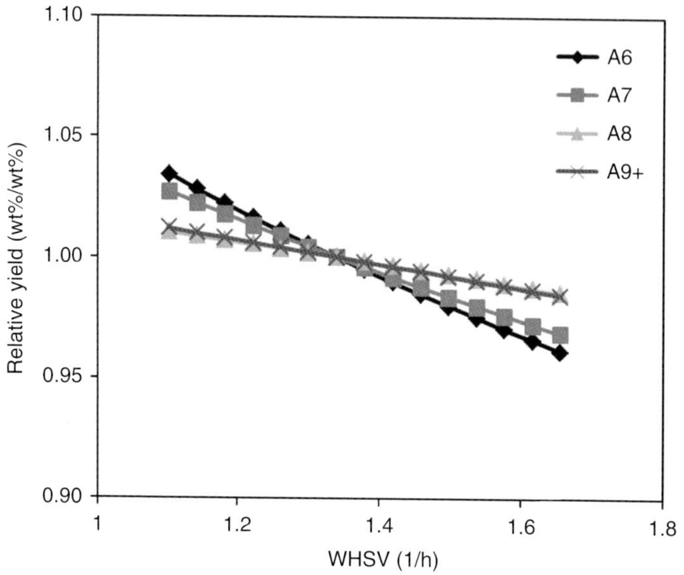

Figure 5.36 Relative yields of aromatic components (where A6 refers to benzene, A7 refers to toluene, and A8 refers to xylenes) as a function of WHSV and WAIT = 525 °C.

of aromatic yields at WHSV of 1.34 and H_2HC ratio of 3.41 as base yields. Table 5.17 shows the base yields at several temperatures.

As the reaction temperature (WAIT) increases, the yields of aromatic components increase significantly. However, at higher temperatures (greater than 520 °C), the H_2HC ratio is not sufficient to prevent undesired hydrocracking reactions. These reactions will decrease the yields of higher aromatics and favor light gas production. Table 5.17 shows that the rate of increase in the production A7 and higher components decreases rapidly. In the case of A9+ yield, we actually show a decrease in yield even though the reactor is operating at higher temperatures. In this case, the refiner may choose to increase H_2HC ratio to continue producing high yields of aromatic components at the expense of increased recycle compressor duties and increased severity during catalyst regeneration. If the recycle compressor is already operating close to the design limit, then an extensive (and costly) revamp may be required to produce additional aromatics. In such a situation, the use of a model to predict alternative scenarios can be very cost-effective.

Table 5.17 Base yields of aromatic components at various WAIT and H_2HC ratio of 3.4.

WAIT	A6 yield (wt%)	A7 yield (wt%)	A8 yield (wt%)	A9+ yield (wt%)
495 °C	4.15	15.90	21.70	22.63
515 °C	6.09	17.13	22.16	23.01
525 °C	6.88	17.56	22.17	22.94

Another important issue is the effect of feed rate (WHSV) on the yield of key aromatics. We note that at lower reaction temperature (WAIT), the effect of WHSV is more pronounced. High feed rates and low reaction temperatures tend to make the process more selective toward toluene (see Figure 5.35). At higher temperatures and high feed rates, as Figure 5.36 shows, there is little difference between the yields of benzene and toluene. It is possible to take advantage of these differences in selectivity to help favor one aromatic component over another. In addition, changes in aromatic precursors (such as MCP) can also significantly shift the aromatic production profile. Figure 5.34 shows the effect of feed composition in the previous section in the context of reducing benzene content in gasoline.

5.12.6 Energy Utilization and Process Performance

The modern refinery is not only concerned with meeting product specifications and demands but also the energy and utility (cooling water, power) consumption of various units. Table 5.18 lists some of the utility consumption data based on various catalytic reforming processes.

In the reforming process, significant energy-consuming steps are interstage heating and recycle compression. About 65–80% of the energy input into the reformer drives the fired heaters responsible for interstage heating. Modest changes in the operation of these fired heaters can provide significant energy savings. Improving the operation of the fired heaters directly can be a significant undertaking [50] and is outside the scope of this work. However, we can study the effect of changing the reactor inlet temperatures (fired heater outlet temperatures) on the product yield and required heater duty.

We consider the scenario in Table 5.19, where the reactor inlet temperature for each reactor bed is fixed to certain values. The values in parenthesis indicate change from the base case. We change the reactor inlet temperatures by values given in the table for four subsequent model runs. We choose these values

Table 5.18 Utility consumption data for catalytic reformers [6, 14].

Fuel (BTU/barrel of feed)	200e3–350e3
Power (kW-h/barrel of feed)	0.6–6
Cooling water (gal/barrel of feed)	40–200

Table 5.19 Reactor inlet temperature deviations.

Scenario	Bed #1(°C)	Bed #2(°C)	Bed #3(°C)	Bed #4(°C)	WAIT (°C)
Base	515.9	513.6	513.6	515.0	514.5
Case-1	510.9 (−5.0)	513.6 (0.0)	513.6 (0.0)	515.0 (0.0)	514.0
Case-2	510.9 (−5.0)	513.6 (0.0)	513.6 (0.0)	510.0 (−5.0)	511.6
Case-3	515.9 (0.0)	508.6 (−5.0)	508.6 (−5.0)	515.0 (0.0)	512.5
Case-4	515.9 (0.0)	513.6 (0.0)	513.6 (0.0)	515.0 (−5.0)	512.2

Table 5.20 Key model yields for fired duty case study.

Scenario	Total fired duty (kJ/kg)	Aromatic yield (wt%)	C5+ RON	C5+ yield (wt%)	Fired duty deviation (%)
Base	1001.4	66.26	101.1	91.52	0.00
Case-1	996.0	66.08	100.9	91.59	−0.54
Case-2	987.0	65.76	100.4	91.74	−1.92
Case-3	987.8	65.82	100.5	91.74	−1.35
Case-4	987.5	65.94	100.7	91.67	−1.39

to highlight the effect of reactor inlet temperatures on the initial, final, and intermediate beds independently. The results of this case study appear in Table 5.20.

Although initially the fired duty reductions appear quite small (0.5–1.4%), this may lead to significant energy savings in fuel costs for the fired heater. Vinayagam [51] stated that even a 1% reduction in fuel consumption can give significant cost savings. We note that these energy savings appear because of small octane loss and aromatic yield loss. If the reformer is already operating at high severity, this type of energy analysis may allow for some flexibility in the operating costs of the unit. In addition, this type of analysis serves as a starting point for a larger heat integration analysis to understand how to reduce energy consumption of the overall unit.

5.13 Refinery Production Planning

Production planning is an important activity in modern refineries. The modern refinery is a combination of many complex units such as catalytic reforming, FCC, and hydroprocessing. Although it is possible to tune each unit to an optimal yield, the optimum yield of a particular unit may not reflect a true optimum because of the demands and prices for the wide range of product that the refiner produces. Therefore, it is important to consider each unit in the context of the whole refinery. *Production planning* refers to the activity of choosing feedstock to refinery (and its constituent units) that produces optimal economic benefit while meeting equipment, business, and regulatory constraints.

The refinery production planning problem has been traditionally solved using linear programming (LP) techniques. LP is a mathematical technique that maximizes a linear objective function of many variables with respect to linear constraints on these variables. Bazaraa *et al.* [51] have described the theory and applications of LP techniques extensively. It is well known that LP techniques have several deficiencies, which include linearization of inherently nonlinear process behavior. This often results in finding local optimum instead of a global optimum. Many authors have worked on several different techniques to use nonlinear programming in refinery production planning. However, LP techniques are still popular because they are easy to use and incorporate into existing refineries.

A refinery LP and linear unit model represents a set of linear correlations that predict yield given an average yield value and changes in the certain operating variables. In this section, we discuss how to apply the rigorous reforming model in the context of a linear unit model. The key information for a linear model of a nonlinear process is the delta-base vector.

$$\begin{bmatrix} y_1 \\ y_2 \\ \vdots \\ y_m \end{bmatrix} (\text{PREDICTION}) = \begin{bmatrix} \overline{y_1} \\ \overline{y_2} \\ \vdots \\ \overline{y_m} \end{bmatrix} (\text{BASE}) + \begin{bmatrix} \frac{\Delta y_1}{\Delta x_1} & \cdots & \frac{\Delta y_1}{\Delta x_n} \\ \vdots & \cdots & \vdots \\ \frac{\Delta y_m}{\Delta x_1} & \cdots & \frac{\Delta y_m}{\Delta x_n} \end{bmatrix} \times (\text{DELTA-BASE}) \cdot \begin{bmatrix} \Delta x_1 \\ \vdots \\ \Delta x_n \end{bmatrix} (\text{DELTA}) \tag{5.25}$$

The delta-base relates the prediction of a new reactor yield (PREDICTION, y_m) given an average starting prediction value (BASE, y_m) and the change in operating variables (DELTA, Δx_n). We note that this delta-base matrix ($\Delta y_n / \Delta x_n$) is essentially the Jacobian matrix for our nonlinear process model centered on a given operating point.

Refiners often develop the linear yield correlations for the LP in a simple fashion. The average value of historical unit yields over a significant period of time (e.g., one operating quarter) form the base yield of the unit. We may calculate the delta-base from published or internal refiner correlations for the given unit. Alternatively, we may generate the delta-base vectors from the change in yields recorded while operating conditions of the unit change. In either approach, the base yield and delta-base matrix represent average values (fixed to certain operating conditions) and may not correctly reflect the true operation of unit. In this work, we use the rigorous nonlinear model to supply the base and delta-base values. Figure 5.37 (previously shown as Figure 4.36) illustrates the process to generate the delta-base vectors.

1. Identify base operating conditions
2. Run model and record base yields
3. Pick attributes that influence product yields
4. For each attribute
   - Modify attribute value by CHANGE%
   - Run the model with modified attribute
   - Record yield of each product
   - Generate delta vector coefficient for each product by dividing the difference between the base yields and current yields by the change in the attribute
   - Record yield of each product
5. Export delta vectors to LP/PIMS software

Figure 5.37 Process to generate delta-base vectors from rigorous model.

Another important consideration is the choice of operating variables to manipulate in the delta vector. It is not useful to map the entire nonlinear model with all of its variables into the LP. We must choose key operating variables that we can track throughout the whole LP. Typically, each unit model only includes the feedstock characteristics. For catalytic reforming, the choice of operating variables depends on how the refiner deals with the reformer products. If the reformer is primarily a generator of high-octane gasoline for the gasoline pool, it is sufficient to include only a few feed-quality parameters such as N + 2A and feed IBP (initial boiling point). However, if the reformer is a source of aromatics destined for a chemical complex, there may be a cause to include additional feed-quality descriptions such as feed content of cyclopentane (CP) and methylcyclopentane (MCP).

In this work, we restrict ourselves to the gasoline-producing reformer. We choose the feed N(naphthenes content) + 2A (aromatics content) as the single input variable and the output variables as yields of hydrogen, dry gas yield, and yield of the reformate. We also generate the base and delta-base vectors for several cases of varying C5+ reformate RON. Table 5.21 shows the relevant yields of the reactor model. The feed composition for the given N + 2A corresponds to measured plant data. We fix the RON of the C5+ reformate and calculate the required WAIT during model execution.

We use the yield information from the rigorous model in Table 5.21 to construct the LP yield vectors. The base vector is the average of the yields in each RON case. We choose the average value of N + 2A (64) to compute the Δx_n. We then use one of the N + 2A data points to compute the delta-base vector. We show the steps and results of this calculation for RON 102 case in Table 5.22. We compare the results of the linear yield vector predictions and model predictions for an intermediate N + 2A value of 66.6 in Table 5.23. Table 5.24 shows the delta-base calculated for all the RON cases.

Table 5.21 Reformer yields at various N + 2A and C5+ reformate RON from rigorous model.

WAIT (°C)	501.1	500.8	508.5	508.1	517.2	516.5
N + 2A	64	72	64	72	64	72
Product			Yield (wt%)			
Hydrogen	2.96	3.13	3.03	3.23	3.10	3.31
Methane	0.59	0.47	0.66	0.53	0.75	0.61
Ethane	1.76	1.41	1.98	1.59	2.25	1.82
Propane	3.38	2.86	3.87	3.27	4.46	3.77
Isobutane	3.36	2.63	3.81	2.99	4.35	3.43
n-Butane	3.10	2.46	3.24	2.58	3.36	2.70
C5+ 102 RON reformate	84.82	87.00	–	–	–	–
C5+ 104 RON reformate	–	–	83.37	85.78	–	–
C5+ 106 RON reformate	–	–	–	–	81.69	84.34
Other	0.03	0.02	0.03	0.03	0.04	0.03

Table 5.22 Calculating the delta-base vectors for the C5+ RON = 102 case.

		Dev. to N+2A = 72	Dev. to N+2A = 64
Avg. N+2A	68	4	−4
	(wt%)	Delta-base	Prediction
Hydrogen	3.05	0.022	2.96
Methane	0.53	−0.014	0.59
Ethane	1.59	−0.043	1.76
Propane	3.12	−0.066	3.38
Isobutane	3.00	−0.091	3.36
n-Butane	2.78	−0.079	3.10
Reformate	85.91	0.273	84.82

Table 5.23 Comparison of yield predictions from rigorous model and LP yield model.

	Rigorous model prediction	LP vector prediction	AAD
N + 2A	66.6	66.6	
	(wt%)	(wt%)	
Hydrogen	3.18	3.17	0.01
Methane	0.73	0.71	0.02
Ethane	2.17	2.11	0.06
Propane	4.45	4.24	0.21
Isobutane	4.14	4.05	0.09
n-Butane	3.16	3.14	0.02
Reformate	82.13	82.55	0.41

Table 5.24 Delta-base vectors for different RON cases.

	RON = 102 Base	N + 2A = 68 Delta-base	RON = 104 Base	N + 2A = 68 Delta-base	RON = 106 Base	N + 2A = 68 Delta-base
Hydrogen	3.05	0.022	3.13	0.024	3.20	0.027
Methane	0.53	−0.014	0.60	−0.016	0.68	−0.018
Ethane	1.59	−0.043	1.79	−0.049	2.04	−0.055
Propane	3.12	−0.066	3.57	−0.075	4.12	−0.086
Isobutane	3.00	−0.091	3.40	−0.103	3.89	−0.116
n-Butane	2.78	−0.079	2.91	−0.081	3.03	−0.082
Reformate	85.91	0.273	84.57	0.301	83.01	0.331

We can repeat the process outlined in Figure 5.37 and Table 5.24 for any number of feed composition variables. In general, for typical process changes in feed quality (10–15%), the LP yield vectors can provide reasonable predictions for the process yield. A potential problem is that LP yield prediction can be poor when operating close to process minima or maxima (such as octane number at fixed H_2HC ratio). In addition, N + 2A may not be a detailed descriptor for feedstock changes. If these problems occur in practice, the LP may require more frequent updates to reflect the true unit operation.

5.14 Workshop 5.1 – Guide for Modeling CCR Units in Aspen HYSYS Petroleum Refining

5.14.1 Introduction

In Sections 5.14.1–5.17, we show how to organize data, build and calibrate a model for a catalytic reformer using Aspen HYSYS Petroleum Refining. We discuss some key issues in model development, particularly on how to estimate missing data required by Aspen HYSYS Petroleum Refining. We divide this task into four workshops:

a) Workshop 5.1 – Building a basic catalytic reformer model
b) Workshop 5.2. – Calibrating the basic catalytic reformer model
c) Workshop 5.3. – Building a downstream fractionation system
d) Workshop 5.4. – Performing case study to identify different RON scenarios.

5.14.2 Process Overview and Relevant Data

Figure 5.2 in Section 5.3 shows a typical CCR unit that we use to build the model in question. We also build models for the remixing and hydrogen recontactor section of this flowsheet. Tables 5.25–5.29 list some typical operating data for this unit.

Table 5.25 Feed properties.

ASTM D86		(wt%)	P	N	A
IBP	78	C2		–	–
5%	90	C3		–	–
10%	96	C4	0	–	–
30%	108	C5	0.78	0.18	–
50%	119	C6	5.4	5.01	0.91
70%	133	C7	10.72	12.05	2.56
90%	152	C8	9.62	13.68	0.93/0.67/1.74/0.71
95%	160	C9	8.13	11.14	2.61
EBP	170	C10+	6.42	6.74	–
S. G.	0.745	Sum	41.07	48.8	10.13

Table 5.26 Product composition profile.

Comp. (vol%)	Recycle H_2	Rich H_2	DA301 top vapor	DA301 top liquid
H_2	86.72	94.06	36.89	0.66
CH_4	2.61	2.40	5.64	0.44
C_2H_6	2.86	1.78	18.50	8.29
C_3H_8	3.33	1.10	22.04	28.32
C_3H_6	0.01	0.00	0.00	0.12
iC_4H_{10}	1.56	0.31	7.82	20.32
nC_4H_{10}	1.24	0.19	5.53	18.02
iC_5H_{12}	1.08	0.11	2.56	15.95
nC_5H_{12}	0.59	0.05	0.95	7.62
C_4=			0.07	0.26

Table 5.27 DA301 liquid product composition.

ASTM D86		(wt%)	P	N	A
IBP	74	C2		–	–
5%	85	C3		–	–
10%	94	C4	0	–	–
30%	112	C5	0	0.27	–
50%	128	C6	0.2	0.53	7.925
70%	145	C7	7.22	0.65	20.72
90%	165	C8	5.87	0.54	3.4/5.11/11.1/6.3
95%	173	C9	1.17	–	20.62
EBP	208	C10+	–	–	8.75
SG	0.83	Sum	14.46	1.99	83.55

Table 5.28 Overall product flow rate and yield.

Stream	Rate (tons/h)
Feed	175.9
Net rich H_2	12.4
DA301 Ovhd. vapor	3.3
DA301 Ovhd. liquid	21.7
DA301 Bttm. liquid	138.5

Table 5.29 Reactor configuration.

Reactor bed	Length (m)	Loading (kg)	Inlet temperature (°C)	ΔT (°C)
#1	0.54	1.275e4	516.0	110.4
#2	0.69	1.913e4	513.6	64.2
#3	0.96	3.188e4	513.1	36.4
#4	1.41	6.375e4	515.0	23.1

5.14.3 Aspen HYSYS and Initial Component and Thermodynamic Setup

We start by opening Aspen HYSYS. The typical path to Aspen HYSYS is to enter the Start → Programs → AspenTech → Aspen Process Modeling v9.0 → Aspen HYSYS (shown in Figure 5.38). We save the simulation file as *Workshop 5.1.hsc.*

The first step in creating the model is the selection of a standard set of components and a thermodynamic basis to model the physical properties of these components. When we create a new simulation, we must choose the components and the appropriate thermodynamic model. Components may be added manually through the Add button shown in Figure 5.39. However, we have a predetermined set of components for the reformer model.

To import these components, we click "Import" and navigate to the directory location, "C.\Program Files\AspenTech\Aspen HYSYS V9.0\Paks" and select the "CatRefIsom.cml" as the component list (Figure 5.40). The path shown reflects a standard installation of Aspen HYSYS Petroleum Refining software.

Once we import a component list, HYSYS will create a new component list called "Component List-1". We can view the elements of this component lists

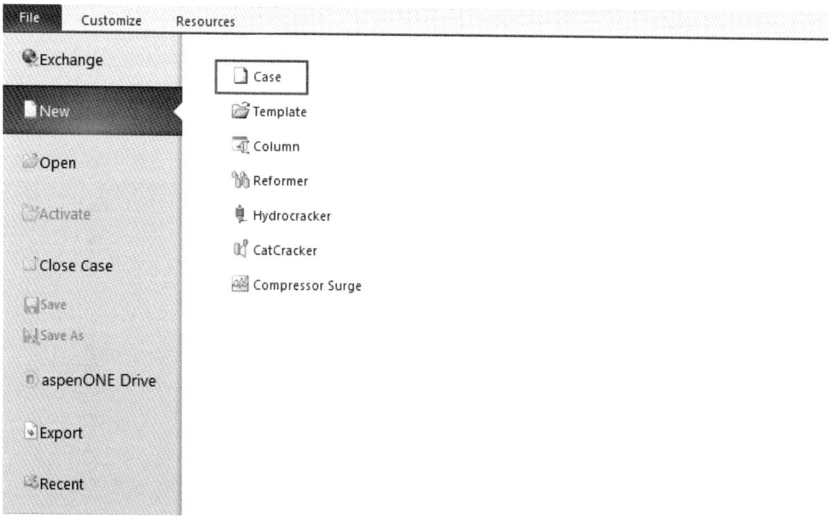

Figure 5.38 Initial start-up of Aspen HYSYS.

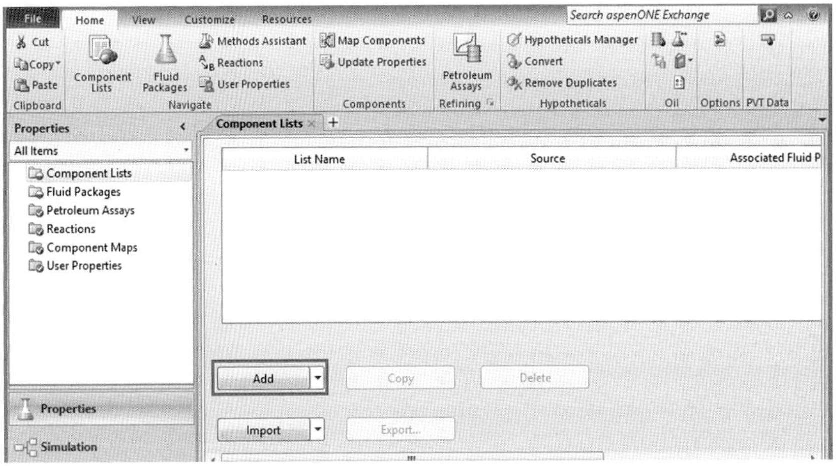

Figure 5.39 Adding a component list.

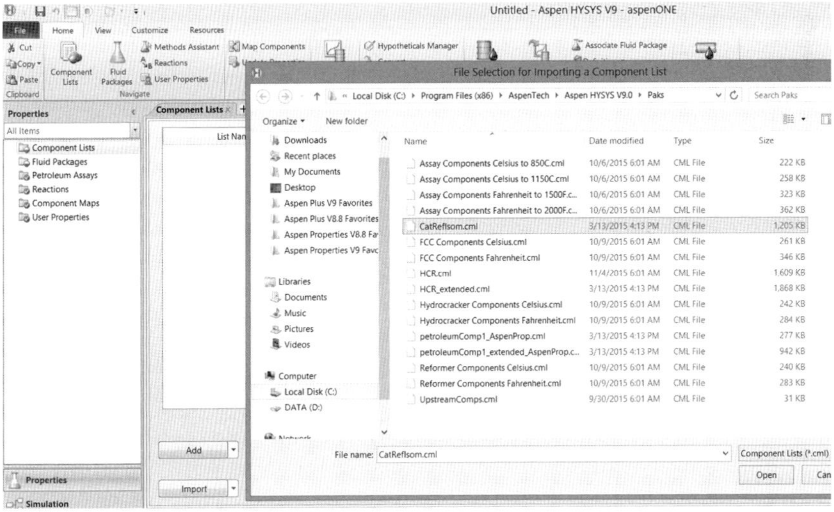

Figure 5.40 Importing reformer component list.

by selecting "Component List-1" and clicking on "View" in the Simulation Basis Manager (Figure 5.41). We can add more components or modify the order of the elements in the component list. We note that the standard reforming component list is quite complete and model most refining processes. The rigorous reforming model does not predict components that are not part of the "CatReform.cml" list. However, these additional components may be used in production fractionation models associated with the reformer model.

The next step is the settings of a "Fluid Package" for this model. The "Fluid Package" refers to the thermodynamic system associated with the chosen list of components. After we import the component list, Apsen HYSYS will automatically set up a fluid package, which is named "REFSRK" (Figure 5.42). The reformer

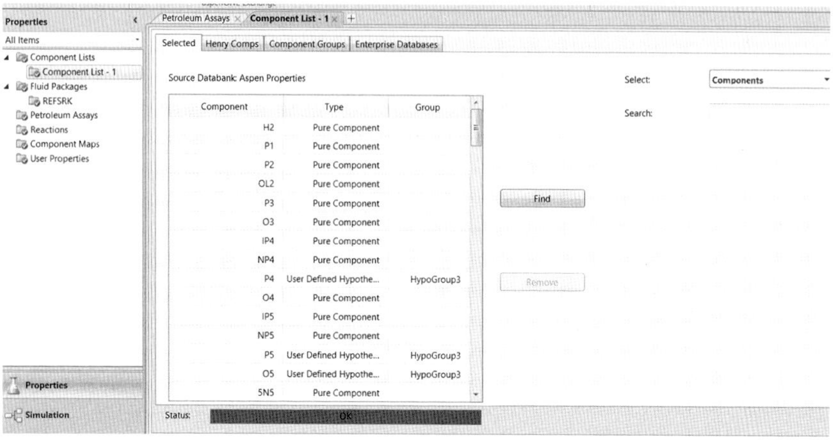

Figure 5.41 Initial component list for reforming process.

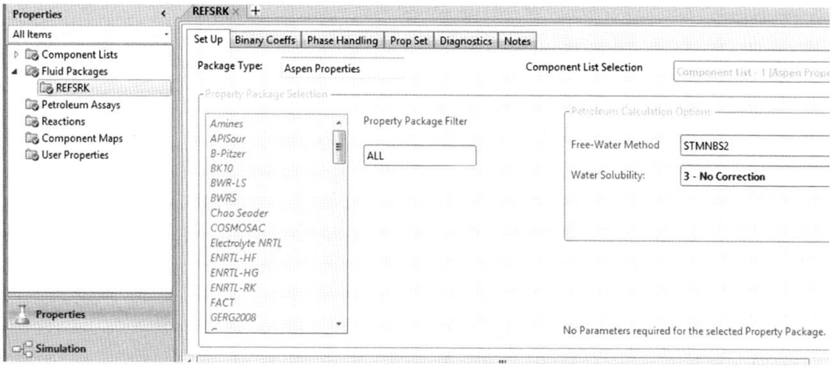

Figure 5.42 Select thermodynamics for fluid package.

system is mostly hydrocarbons and consequently the Soave–Redich–Kwong equation of state is sufficient. We discussed the implications of the process thermodynamics in Section 1.9. For the reformer model, equation of state or hydrocarbon correlation methods (Grayson–Streed, etc.) can sufficiently model the process.

5.14.4 Basic Reformer Configuration

The initial flowsheet presents a blank interface where we can place different objects from the Object palette shown in Figure 5.43. The initial tool palette only shows typical unit operations and does not show the advanced Aspen HYSYS Petroleum Refining objects. We use both toolbars to build the complete reformer model. We can bring up the advanced palette by pressing F6.

We select the Reformer icon from the Refining Reactors palette, click on the Reformer icon, and place the icon the flowsheet. Placing the icon invokes the several submodels that prepare the flowsheet for additional objects and creates a large depiction of the reformer object on the flowsheet (Figure 5.44).

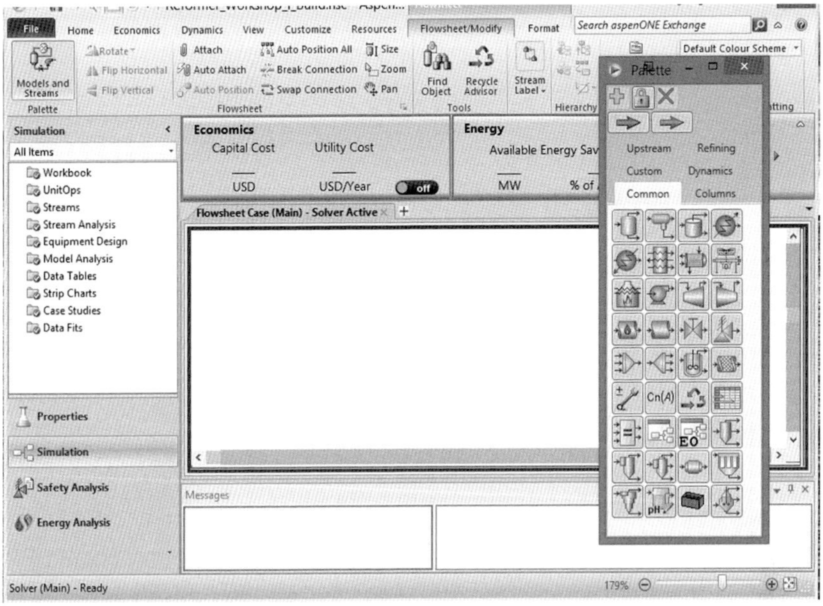

Figure 5.43 Refining reactor palette.

Figure 5.44 Reformer icon in refining reactors palette.

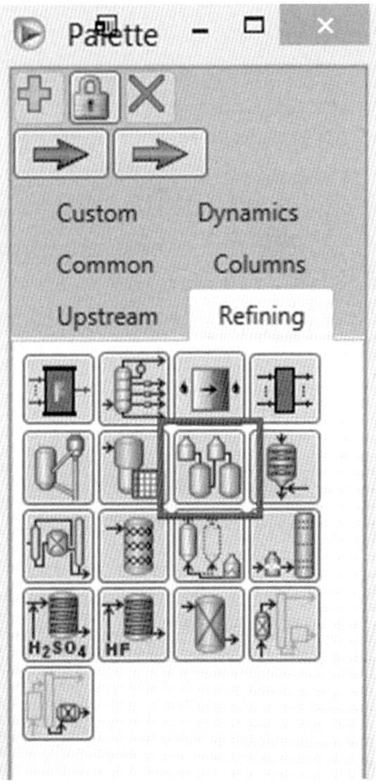

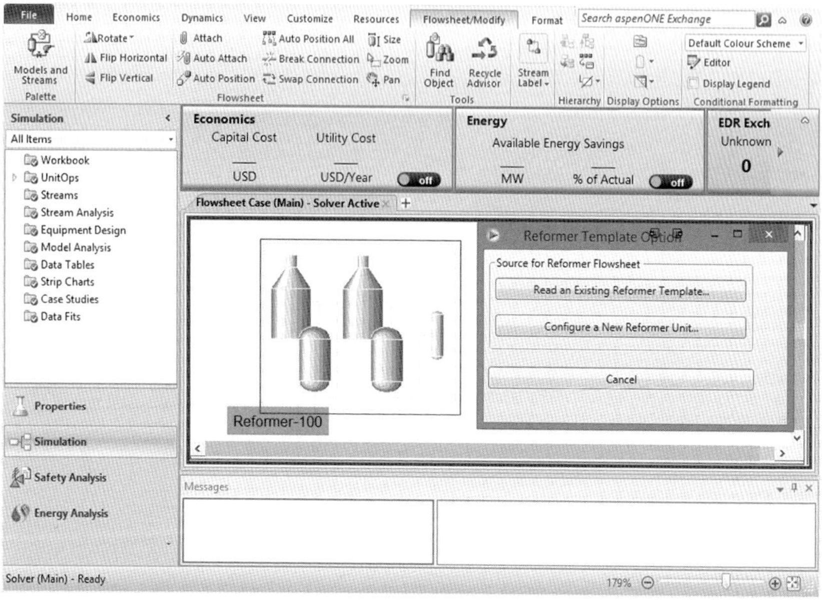

Figure 5.45 Initial reformer window.

The first step is to choose whether to use a reformer template or configure a new unit. Aspen HYSYS has several reformer templates that reflect the popular types of industrial reformer configurations. Figure 5.45 shows the initial window when placed a Reformer object on the Flowsheet. If we choose a template, we do not have to assign the reactor dimensions and catalyst loadings. However, in this workshop, we will build a reformer from scratch, so we choose "Configure a New Reformer Unit."

The reformer configuration requires choosing the type of reformer, number of reactors, and their dimensions and catalyst loadings for each reactor. In addition, we may also specify additional downstream fractionation equipment such as hydrogen recontactor and stabilizer tower. However, we note that the option of including the stabilizer tower actually corresponds to a *simplified* model for fractionation that may not be appropriate for *a detailed and integrated process flowsheet*. We recommend building a rigorous flowsheet based on standard Aspen HYSYS fractionation objects. In Figure 5.46, we select a CCR reformer with four reactor beds and click "Next>."

The primary catalyst configuration is the dimension of the catalyst bed and associated catalyst loading. Here, the catalyst loading refers to the amount of catalyst exposed to feed in each reactor bed. The length refers to the distance the feed travels radially through the catalyst bed. The most important parameters are the catalyst loadings for all reactor beds and it is important to obtain accurate values from industrial data. An important operating variable of a reformer system is *the WAIT* of each reactor bed or section. We find WAIT by summing up the inlet temperature of each bed that is multiplied (weighted) by the ratio of the catalyst in the given bed (see Figure 5.47) to the total catalyst. Likewise, we find *the WABT*

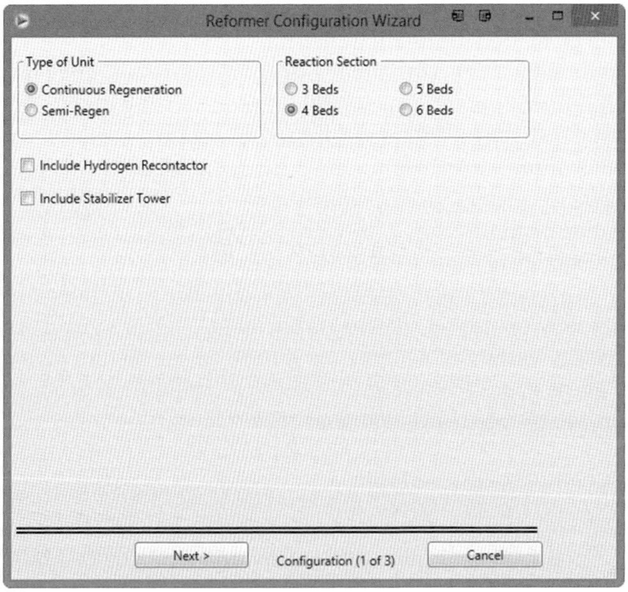

Figure 5.46 Basic reformer configuration.

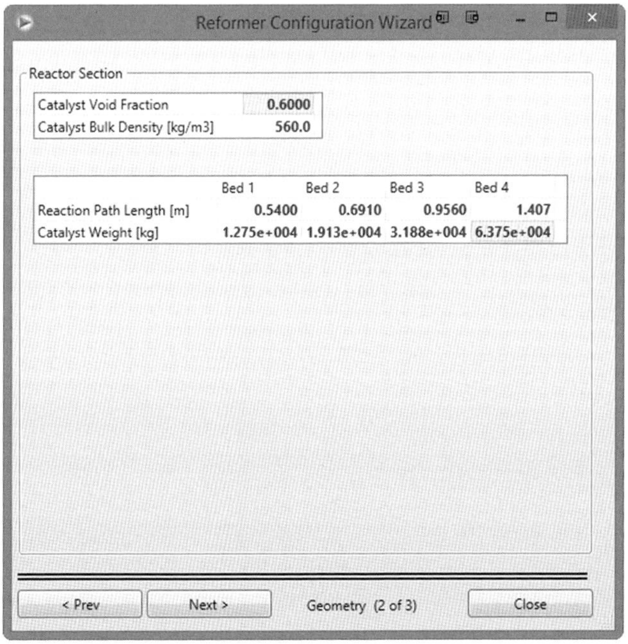

Figure 5.47 Reactor dimensions and catalyst loadings.

by summing up the temperature of each bed that is multiplied (weighted) by the ratio of the catalyst in the given bed to the total catalyst.

We use the data given in Table 5.29. The values shown in Figure 5.47 may not be applicable to all CCR reformer plants but provide a good starting point. The void fraction and catalyst density are not that significant for product predictions, but they affect predictions of pressure drop across the reactor beds. The default values given are acceptable for many types of reformers.

The last step in reformer configuration is to choose option 2 for calibration factors for the model as shown in Figure 5.48. The calibration factors refer to the various reaction and process parameters that we will calibrate to match plant performance and predict new operating scenarios. The default values given are based on calibrations from a variety of sources. In general, these factors give a reasonable set of initial guess that we can refine through the calibration process. For the initial model run, we choose the default and click "Done."

5.14.5 Input Feedstock and Process Variables

Figure 5.49 shows the primary control window for the reformer model. Through this window, we can enter feed and process information and view model results. To manipulate the feedstock information, we must drill down to the Reformer submodel. We enter the Reformer submodel by clicking on "Reformer" environment.

Figure 5.50 shows the reformer submodel. We note that the Net Hydrogen and Net Liquids streams are already attached to the reformer model. The reformer model depiction appears red because there is not enough information to solve

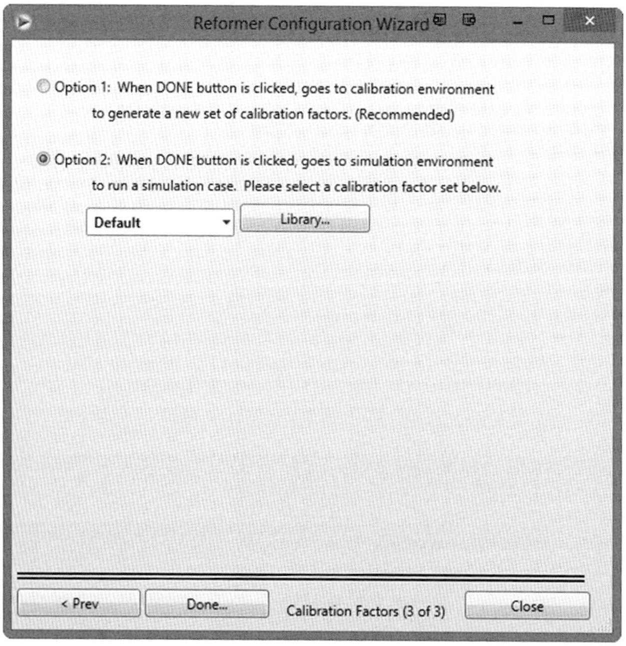

Figure 5.48 Choose calibration factors.

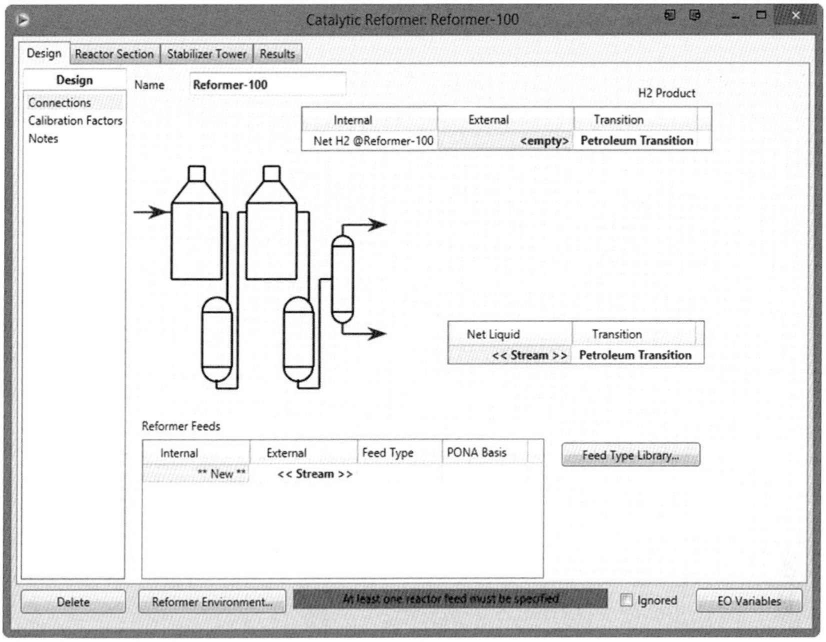

Figure 5.49 Primary control window for reformer.

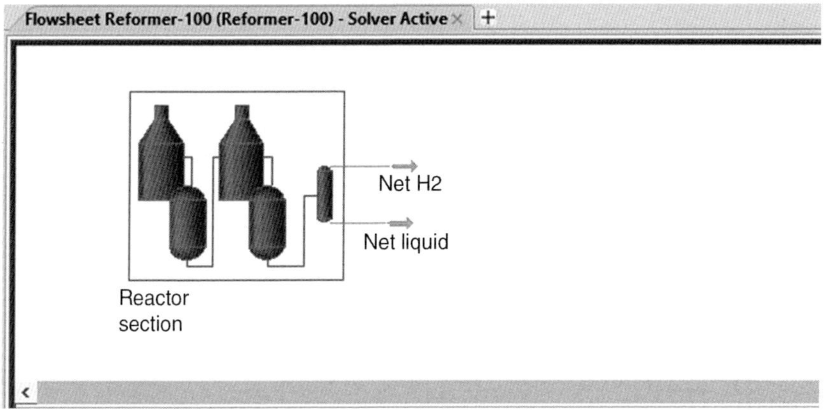

Figure 5.50 Reformer submodel flowsheet.

the model. When enough information is available, the depiction turns yellow and we can proceed to solve. We manipulate the feedstock information by double-clicking on the reactor submodel icon to bring up the reactor submodel window.

Figure 5.51 shows the Feed Data tab from the Reformer submodel. The Feed Type is a basic set of relationships and initial values for all the kinetic lumps in the reactor model. Aspen HYSYS uses bulk property information such as density, distillation curves, and total PNA content in conjunction with the feed type to predict the composition of feed lumps to the model. The "Default" type is sufficient

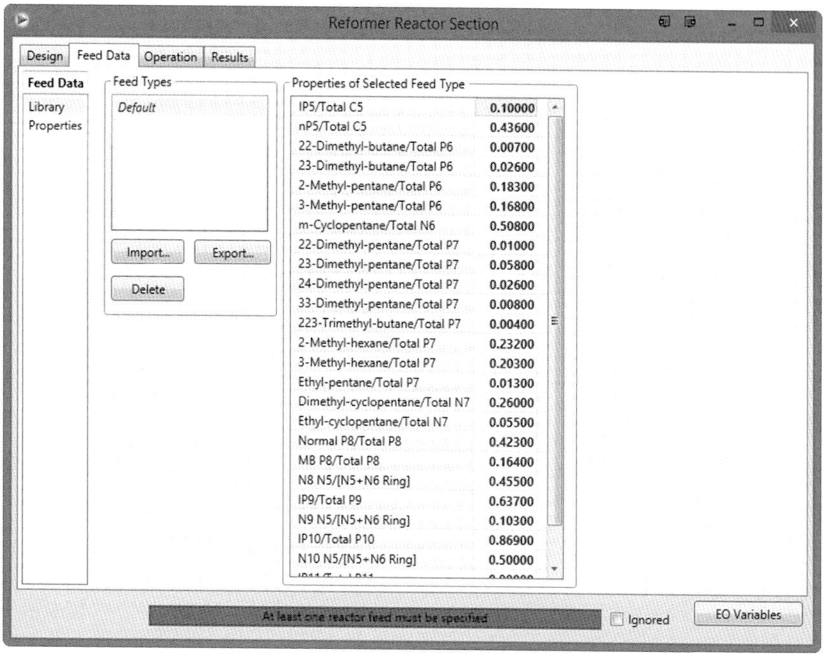

Figure 5.51 Feed data tab.

for light-to-heavy naphtha. However, there is no guarantee that a particular feed type represents the actual feed accurately. Aspen HYSYS will attempt to manipulate the feed composition to satisfy bulk property measurements given. In general, we advise users to develop a few sets of compositional analysis to verify the kinetic lumps calculated by Aspen HYSYS. We discuss a process to verify these lumps in Section 5.15.

We enter the measured bulk property information in the “Properties” section of the Feed Data tab as shown in Figure 5.52. These data come from sample process data given in Table 5.25. Once we enter the bulk feed information, it is important to “Hold” the solver. By design, Aspen HYSYS will attempt to recalculate the model the instant we make a change. This can be inconvenient and may cause convergence problems when we change many variables. To “Hold” the solver, simply select the Red Stop sign in the top toolbar of the flowsheet window (Figure 5.53).

We now input other operation details by navigating to the “Operation” tab and “Feeds” section of the reformer submodel (see Figure 5.54). The flow rates and process parameters should reflect an operating schedule where the actual reformer is running smoothly. It is difficult to use a model based on upset data for future predictions of stable operating scenarios. We discussed some techniques and approaches previously in Section 5.11 to ensure that the data collected for the model reflects stable operation.

After we enter the feedstock information, we must define operating temperatures and associated process variables. We enter the “Reactor Control” section and define the operating temperature of each bed. There are two ways to specify

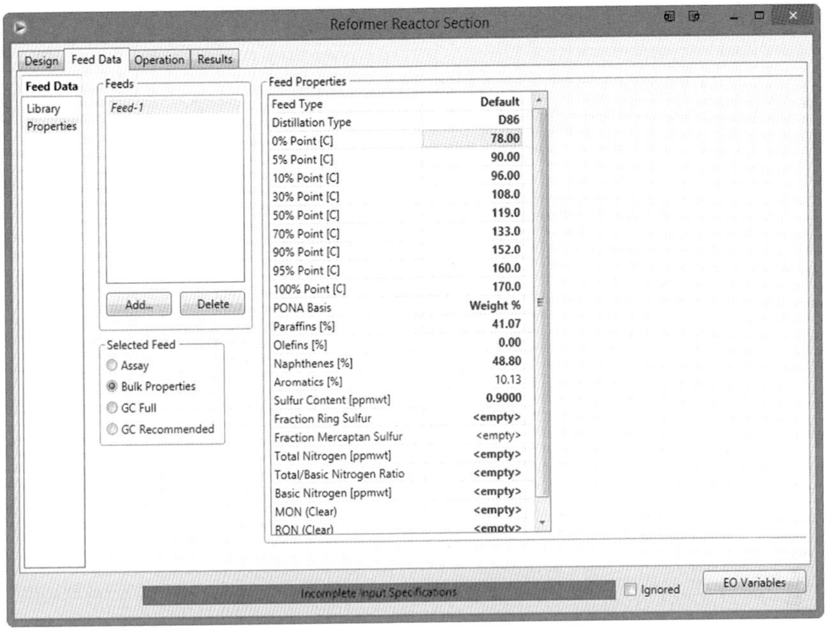

Figure 5.52 Bulk property information.

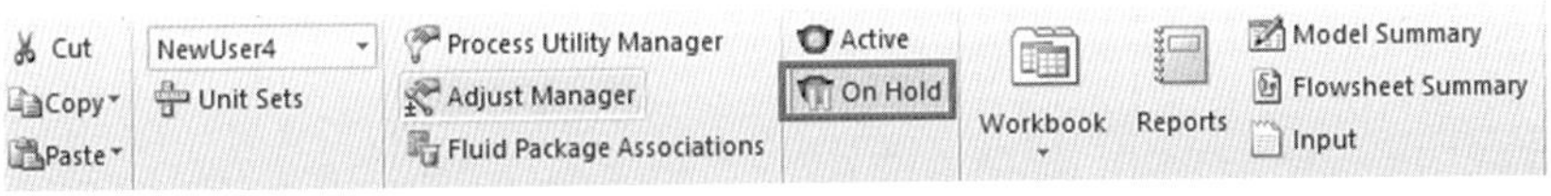

Figure 5.53 Hold Aspen HYSYS solver.

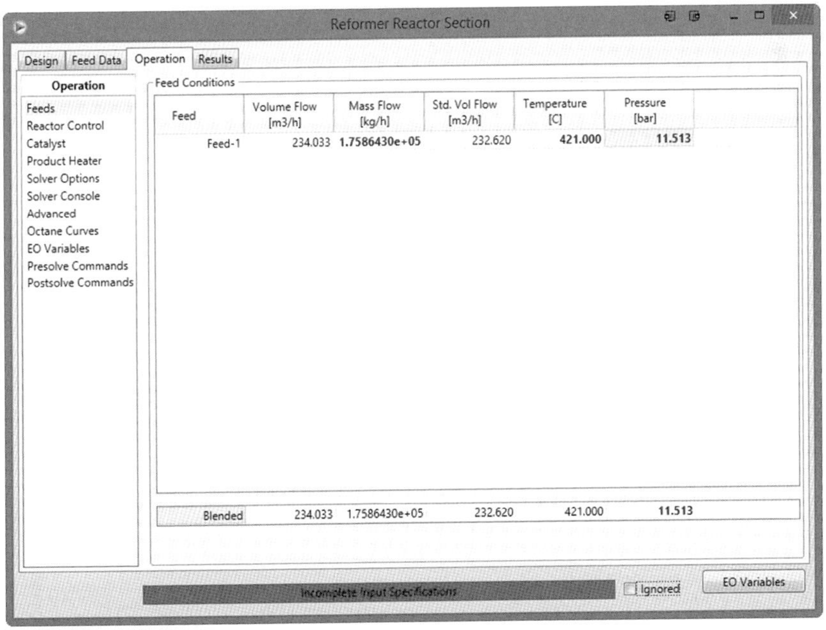

Figure 5.54 Feed flow rate specifications.

reactor inlet temperature. In the first method, we enter *the WAIT* for all the reactors and specify a bias for each reactor. In the second method, we enter a reactor reference temperature and specify a bias for each reactor. We use the second method to fix the inlet temperature of each bed accurately. We recommend this method when running the model for the first time. This ensures that inlet temperatures are accurate for the purposes of calibration. We show how to input the reactors temperatures in Figure 5.55.

The full EO nature of the reformer model technically allows us to enter the octane number of the product and back-calculate the required inlet temperatures to achieve the specified octane number. However, it is very unlikely that an uncalibrated model will converge with those specifications. We recommend entering reactor temperatures directly.

In addition, we must also enter the hydrogen-to-hydrocarbon ratio for the recycle process in the reformer model. The typical range of this value for CCR reforming units is 3–4. Reforming plants routinely measure this value and we expect to enter accurate values. The product separator refers to the conditions of the first separator after leaving the last reforming reactor. This value should be accurate if we do not plan to build a downstream fractionation model.

In Figure 5.56, we enter the "Catalyst" section of "Operation" tab. We must enter an estimate for the catalyst circulate rate as we are modeling a CCR unit. Users will note that it is possible to enter other specifications in the Catalyst Section; however, only the circulation rate ensures robust convergence.

The last process operation parameters are the product heater specifications. As we are building a rigorous fractionation section in this example, we only enter estimated values. If there is no fractionation model planned, we can enter

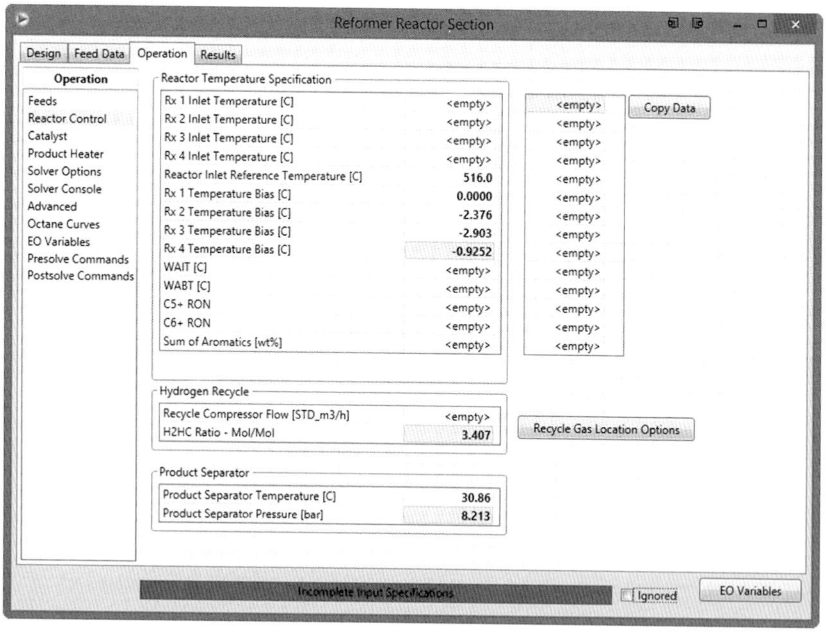

Figure 5.55 Reactor temperature specifications.

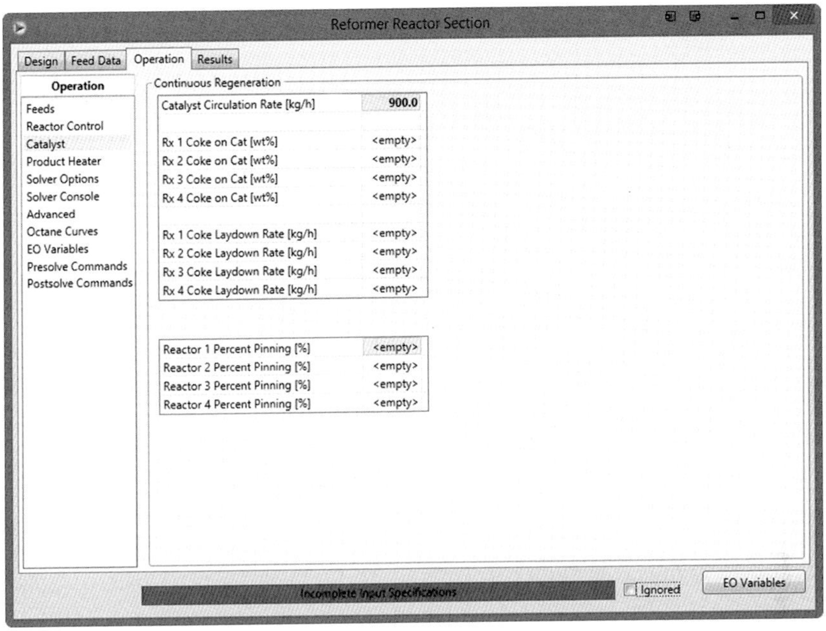

Figure 5.56 Catalyst specifications.

measured values for the heater immediately preceding the gasoline stabilization tower. In Figure 5.57, once we enter the product heater specifications, we notice a yellow bar indicating that we are ready to solve the model. In the following section, we will discuss how to solve the model and ensure robust convergence.

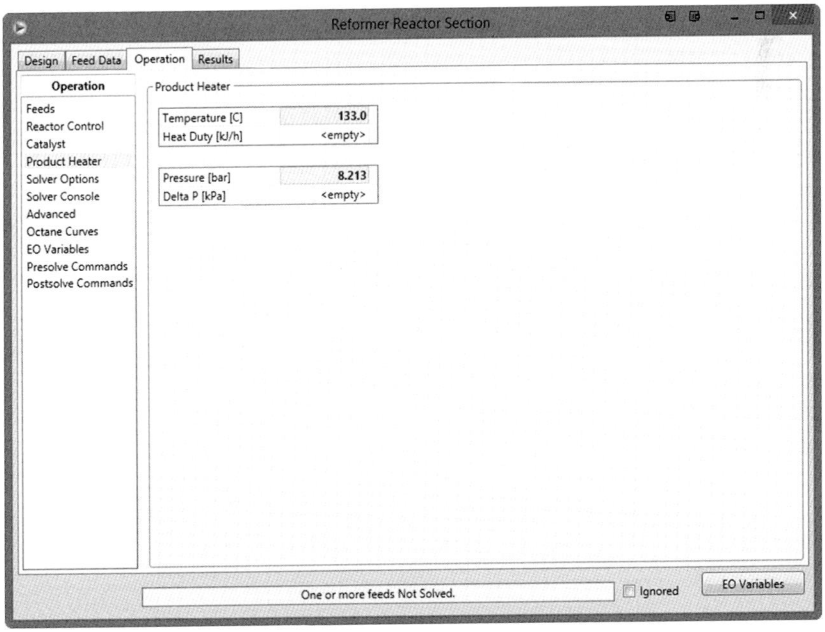

Figure 5.57 Product heater specifications.

5.14.6 Solver Parameters and Running the Initial Model

Before solving the model, we must ensure that the solver parameters will lead to robust convergence. We bring up the Solver options by selecting the "Solver Options" section in "Operation" tab. Figure 5.58 shows the recommended values for the solver options. We have chosen these values based on our experience with running refinery models.

In general, we do not recommend modifying the constraints for the Residual, Hessian parameters, and Line search parameters. When running the model for the first time, we increase the number creep iterations and total maximum iterations. Creep iterations refer to initial small changes in the process variables when the starting guesses are very poor. The maximum iterations refer to how many times the solver will iterate through the model before exiting. Depending on process parameters, the initial solution may take up to 30–40 iterations.

To begin solving the model, we select the green start icon in the flowsheet toolbar as shown in Figure 5.59. Several initialization steps will appear in the lower right corner window of the application. The solution process may take several minutes and the software appears disabled shortly, whereas solver status messages appear in the lower right corner window.

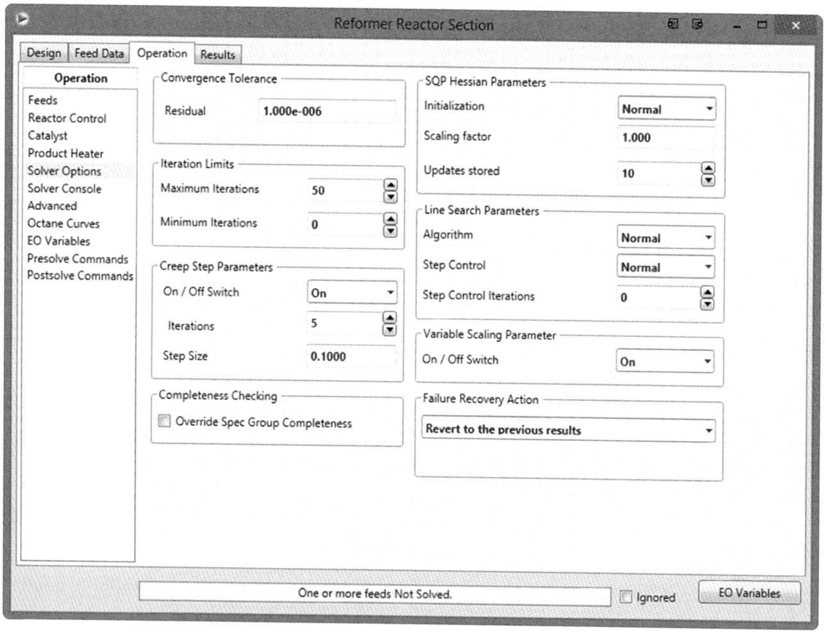

Figure 5.58 Solver parameters.

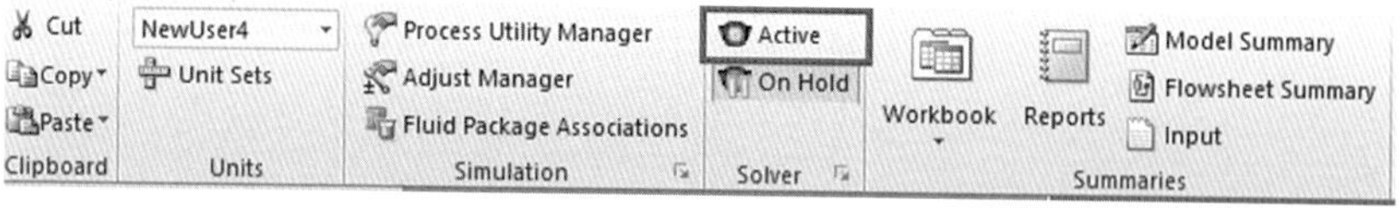

Figure 5.59 Main application toolbar.

Table 5.30 Initial solver output.

Iteration	Residual Convergence Function	Objective Convergence Function	Objective Function Value	Overall Nonlinearity Ratio	Model Nonlinearity Ratio	Worst Model
0	7.223D+08	0.000D+00	0.000D+00	1.000D+00	−4.404D+00	PRODHTR
	<Line Search Creep Mode ACTIVE> ==> Step taken 1.00D−01					
1	5.835D+08	0.000D+00	0.000D+00	9.895D−01	−8.761D+00	RXR3.RXHTR
	<Line Search Creep Mode ACTIVE> ==> Step taken 1.00D−01					
2	4.712D+08	0.000D+00	0.000D+00	9.903D−01	−3.505D+00	CCRDMO
	<Line Search Creep Mode ACTIVE> ==> Step taken 1.00D−01					
3	3.806D+08	0.000D+00	0.000D+00	9.898D−01	−1.947D+00	RXR4.RXACT
	<Line Search Creep Mode ACTIVE> ==> Step taken 1.00D−01					
4	3.076D+08	0.000D+00	0.000D+00	9.907D−01	−4.639D+00	RXR2.RXHTR
	<Line Search Creep Mode ACTIVE> ==> Step taken 1.00D−01					
5	2.487D+08	0.000D+00	0.000D+00	9.022D−01	−3.586D+01	RXR2.RXHTR
6	5.236D+04	0.000D+00	0.000D+00	9.640D−01	9.310D−01	RXR2.RXR
7	5.204D+02	0.000D+00	0.000D+00	9.901D−01	−9.009D−01	RXR2.RXHTR
8	1.165D−02	0.000D+00	0.000D+00	1.000D+00	9.999D−01	RXR4.RXACT
9	1.066D−10	0.000D+00	0.000D+00			

Successful solution.

Optimization Timing Statistics	Time	Percent
MODEL computations	1.56 secs	19.36 %
DMO computations	6.05 secs	75.11 %
Miscellaneous	0.45 secs	5.53 %
Total Optimization Time	8.05 secs	100.00 %

Problem converged<invoke postsolve.ebs>

We show the solver output for the configured model in Table 5.30. Column 1 indicates the number of iterations performed since starting the solver. The residual convergence function indicates how far we are from satisfying the process model equations. When we run the model for the first time, residuals on the order of 1E9 and 1E10 are expected. As we approach the solution, the residual becomes closer to zero. Column 3 and Column 4 refer to the residual of the objective function. We use the objective function *only during calibration*; therefore, it is zero for this model run. The solver used by Aspen HYSYS converges very quickly to the solution, once the changes in the process equations begin approaching linear. This is the case when we are near the solution. The solver indicates the vicinity of the solution through columns 5 and 6. The Worst model column indicates which part of the reformer model is farthest from the solution. This is useful for tracking down issues when the model fails to converge. The last lines of the output show several running statistics for the solver.

5.14.7 Viewing Model Results

After we complete the initial model solution, we can view the model results by navigating to “Results” tab and clicking the “Summary” section. The Summary section shows the yields of the many products relevant to the reforming process. Figure 5.60 shows that the results from the initial model run. We note that the results are mostly close to the plant measurements. This indicates that we will not have to do significant amounts of calibration to match model predictions with plant performance and yields. We can view the detailed yield results for each lump by going to the “Product Yields” section and select Grouped or Detailed yields as shown in Figure 5.61.

We can also view the reactor temperature and flow profile by selecting the “Reactors” section in the Results tab, as shown in Figure 5.62. Again, we note that the predicted temperature drop for each reactor bed compares well with the measured temperature drop. Most of the temperature change is due to the naphthene dehydrogenation reactions. As the model gives reasonable predictions of the aromatic content, we expect the reactor temperatures to agree as well.

This completes in the initial model solution based on bulk property information. We save the converged simulation as *Workshop 5.1-1.hsc*. We can return to the parent flowsheet by clicking the green up arrow on the flowsheet toolbar (shown in Figure 5.63). Once we return to the main flowsheet, we can attach true product streams by entering names for the Net H2 and Net Liquid Streams and selecting the Basic Transition (see Figure 5.64).

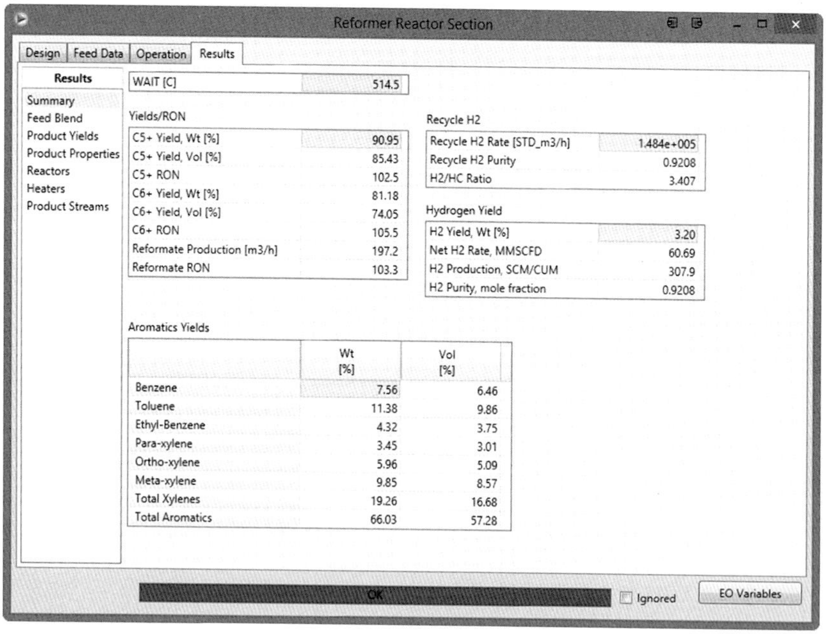

Figure 5.60 Reformer result summary.

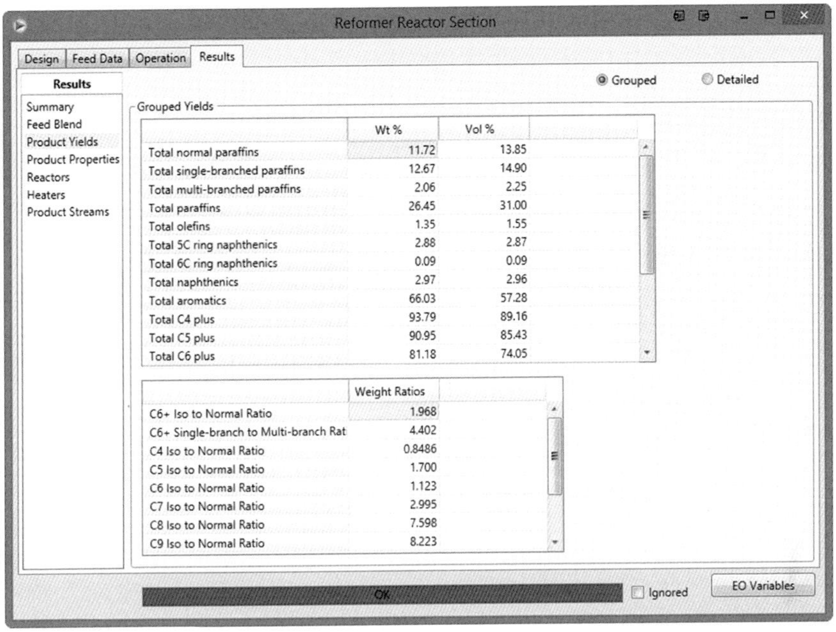

Figure 5.61 Reformer yield results.

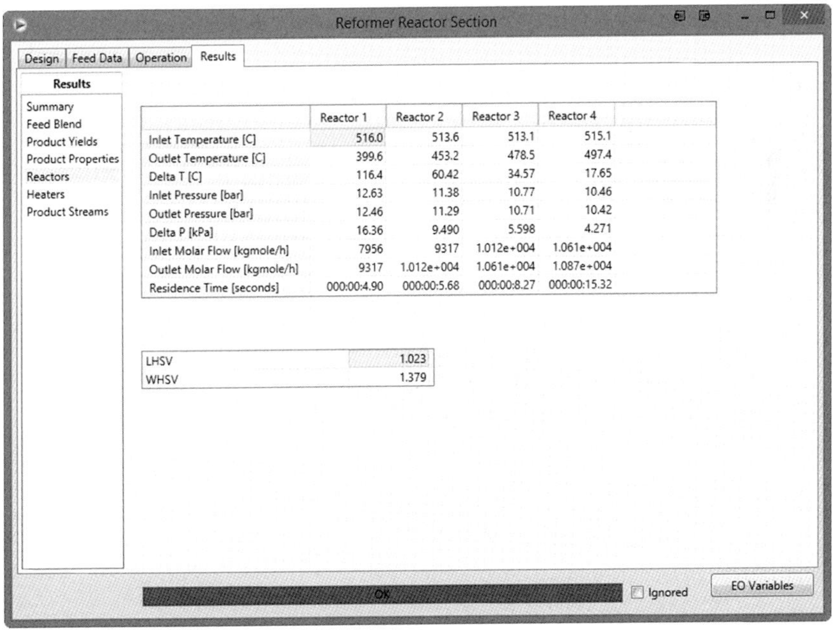

Figure 5.62 Reactor performance results.

Figure 5.63 Returning to the main flowsheet.

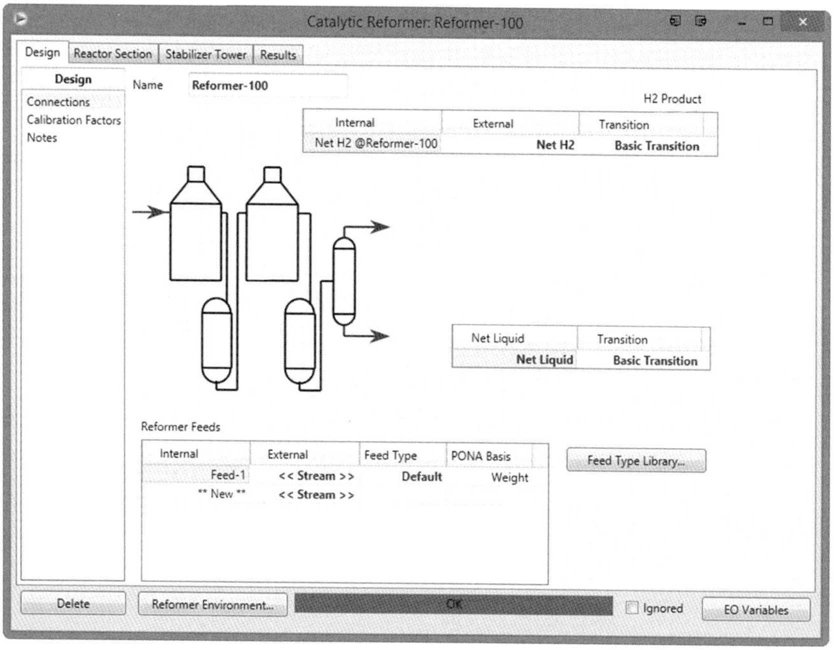

Figure 5.64 Connect external streams to reformer flowsheet.

5.14.8 Updating Results with Molecular Composition Information

In the previous section, we built and solved the reformer model using bulk property and total PNA information only. This approach works reasonably, when the actual feedstock is quite similar to the "Default" or selected feed type. In actual refinery operation, the feed type may change quickly or may not have been analyzed for feed type information. In this section, we discuss an approach to integrate measured molecular composition analysis with the feed type to improve modeling results. This method has shown significant improvement in model predictions, especially in the petrochemical reformers where accurate predictions of aromatic content are significant.

Once we solve the model using the bulk property information, we can obtain the feed lump composition from the "Feed Blend" section of the "Results" tab as shown in Figure 5.65. The composition in mole fraction represents Aspen HYSYS's best estimate of the composition from the bulk information and chosen feed type. In our example, we also have the detailed compositional analysis by PNA and carbon number. We show these measured compositions

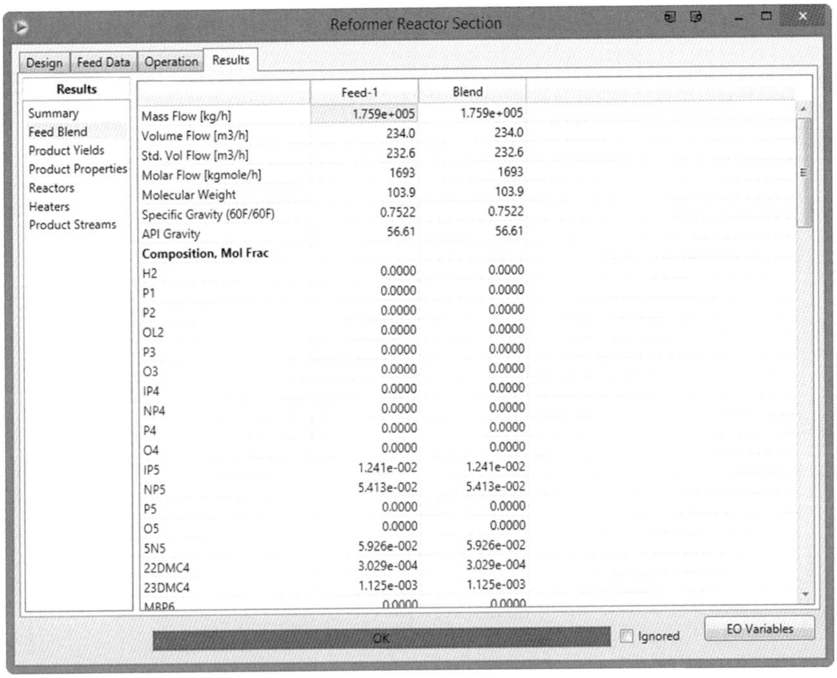

Figure 5.65 Feed blending results.

in the Sample Data section of this chapter. We resave the converged simulation, *Workshop 5.1-1.hsc* (using bulk property information), as *Workshop 5.1-2.hsc* (using measured molecular composition).

Figure 5.66 shows a spreadsheet in Microsoft Excel, *Feed_AspenHYSYS_Transform.xlxs*, available in the supplement to this text, that accepts the measured molecular information and Aspen HYSYS's best estimate of the composition. Using both sets of data, we can rescale Aspen HYSYS's estimate to match the measure molecular composition. Essentially, we rescale the estimates to match plant data for each compositional and carbon group number, while keeping isomer ratios constant.

We perform this rescaling by copying the results of the "Feed Blend" (Figure 5.65) from Aspen HYSYS into Column I of the spreadsheet (Figure 5.66). We also enter the measured compositional information in Column C. Cells C5–C21 and cells C26–C42 represent plant data for PNA (paraffins–naphthenes–aromatics) given in Table 5.25, Section 5.14.3. The results of the rescaling appear in Column U. We must now enter the rescaled feed information back into the reformer model. We must reenter the Reformer subflowsheet and enter the Feed Data tab.

Figure 5.67 shows the Feed Data tab. We select *GC Full (Kinetic Lump)* instead of *Bulk Properties*.Aspen HYSYS now prompts to indicate that we are discarding the bulk property information. We confirm this change and edit the GC Full directly. We copy the results from Column U of the spreadsheet into the Edit Lumps dialog as shown in Figure 5.68. We enter the new feed lump composition

	A	B	C	D	E	F	G	H	I	J	K	L	M	N	O	P	Q	R	S	T	U	V
1	Feed transform for HYSYS model																					
2																						
3																						
4		Actual Plant Data							Old mol% Input												New wt% Input	
5	P	C5	0.73%	41.61%					Feed-1	MW		Wt %							Lump	wt%	wt PERCENT	
6		C6	5.46%			H2		Hydrogen	0	2.016	0	0.00%							H	0.00%	0.00	
7		C7	11.80%			P1	P01-1	Methane	0	16.04	0	0.00%	P01-1	6	A06-1	0.73%	0.95%		P01-1	0.00%	0.00	
8		C8	10.58%			P2	P02-1	Ethane	0	30.07	0	0.00%	P02-1	7	A07-1	3.37%	2.14%		P02-1	0.00%	0.00	
9		C9	8.57%			OL2	O02-1	Ethylene	0	28.05	0	0.00%	O02-1	8	A08-1	0.56%	0.74%		O02-1	0.00%	0.00	
10		C10	4.47%			P3	P03-1	Propane	0	44.1	0	0.00%	P03-1	8	A08-2	0.56%	0.74%		P03-1	0.00%	0.00	
11	N	C5	0.15%	50.20%		O3	O03-1	1-Propene	0	42.08	0	0.00%	O03-1	8	A08-3	0.56%	0.74%		O03-1	0.00%	0.00	
12		C6	5.13%			IP4	P04-1	i-Butane	0	58.12	0	0.00%	P04-1	8	A08-4	0.56%	0.74%		P04-1	0.00%	0.00	
13		C7	13.05%			NP4	P04-2	n-Butane	0	58.12	0	0.00%	P04-2	8	A08-5	0.00%	0.00%		P04-2	0.00%	0.00	
14		C8	15.28%			P4	P04-3	Hypo	0	58.12	0	0.00%	P04-3	9	A09-1	2.66%	2.15%		P04-3	0.00%	0.00	
15		C9	12.15%			O4	O04-1	1-Butene	0.00E+00	56.11	0	0.00%	O04-1	10	A10-1	1.11%	0.00%		O04-1	0.00%	0.00	
16		C10	4.44%			IP5	P05-1	i-Pentane	1.24E-02	72.15	0.895673559	0.86%	P05-1	11	A11-1	0.00%	0.00%		P05-1	0.14%	0.14	0.1
17	A	C6	0.95%	8.19%		NP5	P05-2	1-Pentane	5.41E-02	72.15	3.905136715	3.76%	P05-2	12	A12-1	0.00%	0.00%		P05-2	0.59%	0.59	0.5
18		C7	2.14%			P5	P05-3	Hypo	0.00E+00	72.15	0	0.00%	P05-3	13	A13-1	0.00%	0.00%		P05-3	0.00%	0.00	
19		C8	2.95%			O5	O05-1	Hypo	0.00E+00	70.13	0	0.00%	O05-1	14	A14-1	0.00%	0.00%		O05-1	0.00%	0.00	
20		C9	2.15%			5N5	N05-1	Cyclopentane	5.93E-02	70.13	4.15568071	4.00%	N05-1	5	N05-1	4.00%	0.15%		N05-1	0.15%	0.15	
21		C10	0.00%			22DMC4	P06-1	22-Mbutane	3.03E-04	86.18	0.026105726	0.03%	P06-1	6	N06-1	5.47%	2.61%		P06-1	0.04%	0.04	0.0
22						23DMC4	P06-2	23-Mbutane	1.13E-03	86.18	0.096964126	0.09%	P06-2	6	N06-2	5.30%	2.52%		P06-2	0.14%	0.14	0.
23						MBP6	P06-3	Hypo	0E+00	86.18	0	0.00%	P06-3	7	N07-1	1.85%	3.39%		P06-3	0.00%	0.00	
24						2MC5	P06-4	2-methyl-pentan	7.92E-03	86.18	0.68247827	0.66%	P06-4	7	N07-2	0.39%	0.72%		P06-4	1.00%	1.00	0.
25		Distribution calculated by HYSYS				3MC5	P06-5	3-methyl-pentan	7.27E-03	86.18	0.626537428	0.60%	P06-5	7	N07-3	0.00%	0.00%		P06-5	0.92%	0.92	0.
26	P	C5	4.62%	37.97%		SBP6	P06-6	Hypo	0.00E+00	86.18	0	0.00%	P06-6	7	N07-4	4.88%	8.94%		P06-6	0.00%	0.00	
27		C6	3.59%			NP6	P06-7	n-Hexane	2.67E-02	86.18	2.297303904	2.21%	P06-7	8	N08-1	8.89%	6.95%		P06-7	3.36%	3.36	3.
28		C7	9.14%			O6	O06-1	Hexenes	0.00E+00	84.16	0	0.00%	O06-1	8	N08-2	10.65%	8.33%		O06-1	0.00%	0.00	
29		C8	6.53%			5N6	N06-1	Mcyclopentan	6.75E-02	84.16	5.682045216	5.47%	N06-1	9	N09-1	0.44%	1.25%		N06-1	2.61%	2.61	2.
30		C9	5.55%			A6	A06-1	Benzene	9.77E-03	78.11	0.76291256	0.73%	A06-1	9	N09-2	3.84%	10.90%		A06-1	0.95%	0.95	
31		C10	8.52%			6N6	N06-2	Cyclohexane	6.54E-02	84.16	5.503083162	5.30%	N06-2	10	N10-1	1.54%	2.22%		N06-2	2.52%	2.52	2.

Figure 5.66 Feed rescaling spreadsheet (*Feed_AspenHYSYS_Transform.xlxs*).

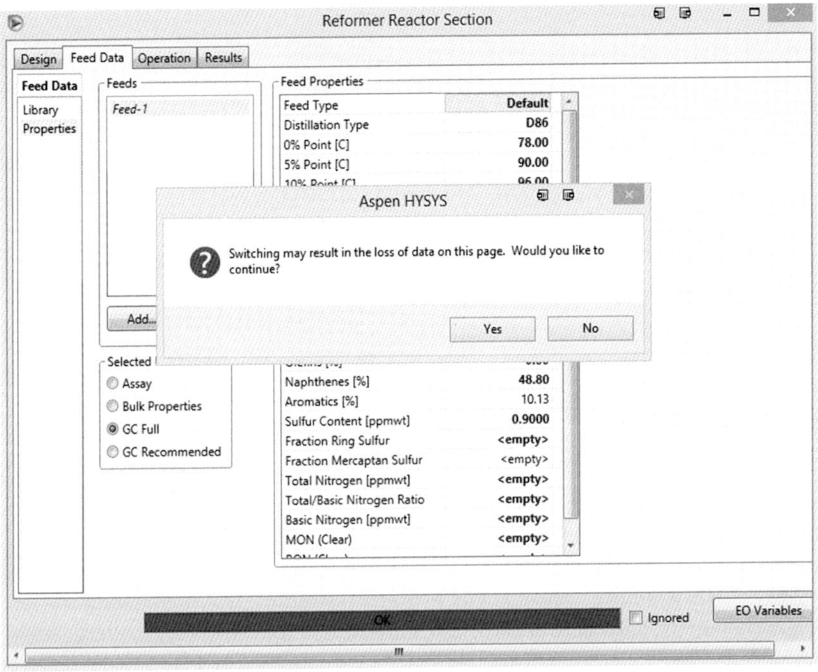

Figure 5.67 Changing from the bulk property data to GC Full (kinetic lumps).

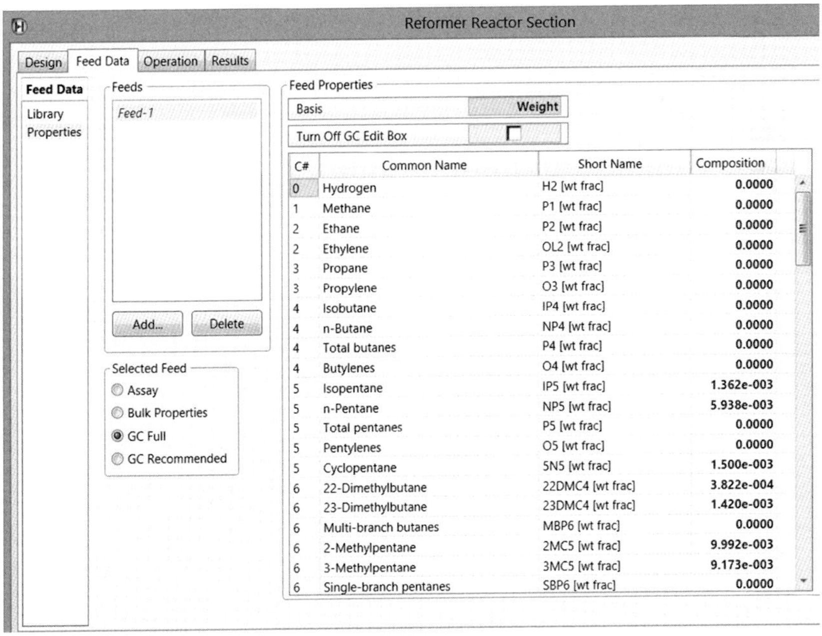

Figure 5.68 Kinetic lump composition entry window.

by weight and normalize to make sure the sum of all the lump compositions is 1. The solver will automatically resolve the model using the new feed lump composition. In general, the initial residual should be on the order of 1E3–1E4, which indicates that the only changes to the model are the feed lump compositions.

5.15 Workshop 5.2. – Model Calibration

In this section, we calibrate the model based on known product yields and reactor performance. Calibration involves four distinct steps:

1) Pull data from the current simulation.
2) Enter measured process yields and performance based on that current simulation.
3) Update the activity factors to match the plant yields and performance.
4) Push calibration data back to the simulation.

We begin the first step of model calibration using a converged initial model using measured molecular information, *Workshop 5.1-2.hsc*. We resave the simulation file as *Workshop 5.2.hsc*. The converged initial model will provide initial guesses for the activity factors, which greatly simplify the model calibration. We enter the model calibration environment by first entering the reformer subflowsheet and then selecting the "Reformer > Go to Calibration" menu option from the application menu bar (as shown in Figure 5.69). Figure 5.70 shows the reformer calibration environment.

The first step is to "Pull data" from the simulation. When Aspen HYSYS pulls data, all of the current operating conditions, feed stock information, and process parameters enter the reforming environment. A calibration refers to the set of the activity factors that produce given product yields and reactor performance (which we provide to the calibration environment) based on current model state. We pull data by clicking on the "Pull Data from Simulation" button (Figure 5.71).

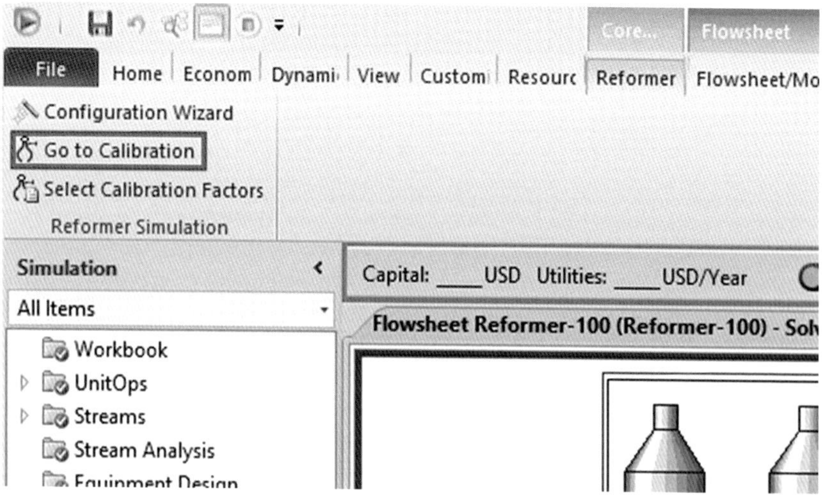

Figure 5.69 Starting the reformer calibration environment.

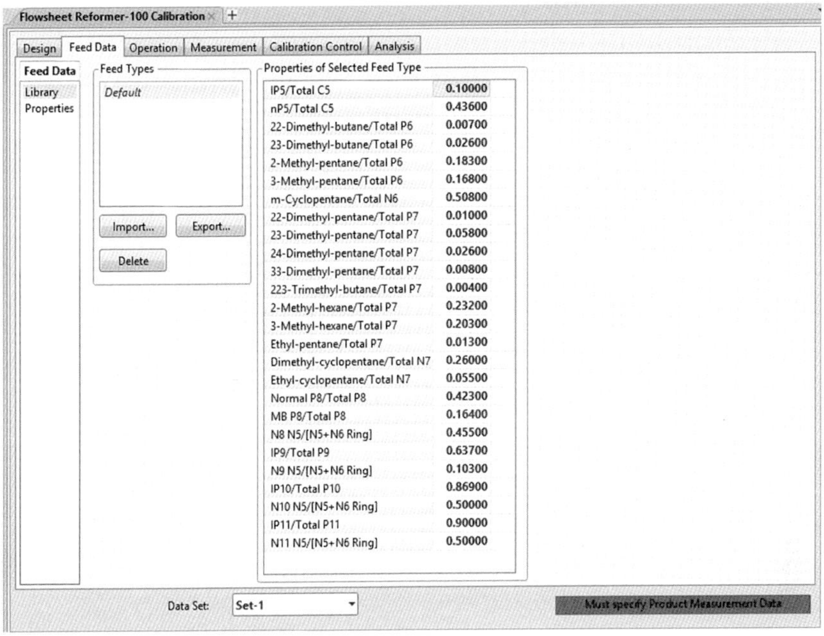

Figure 5.70 Reformer calibration environment.

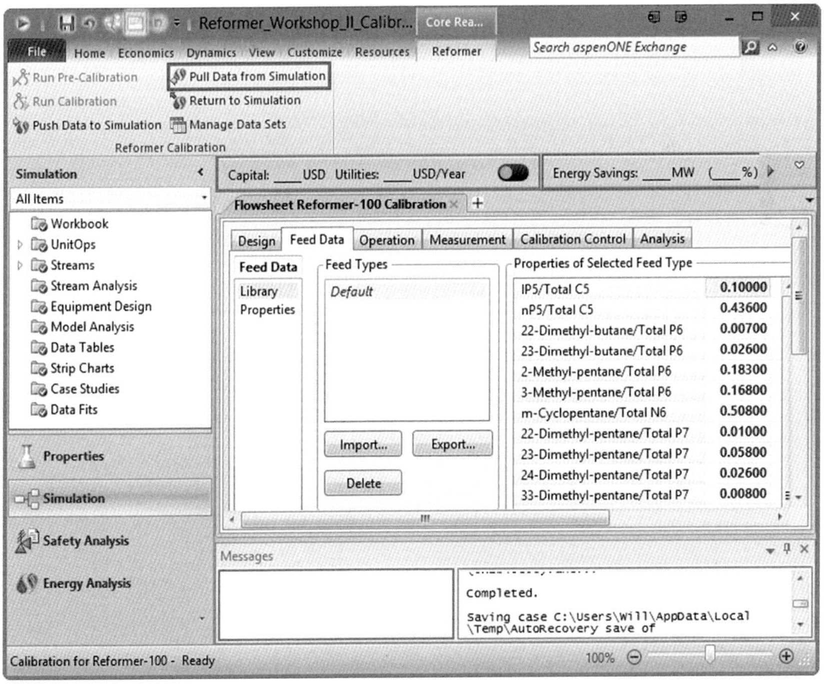

Figure 5.71 Pull data from model results.

When we pull data from the simulation, Aspen HYSYS will warn us that current calibration data will be overwritten by the current model results as shown in Figure 5.72. We can use the Data Set feature (in Figure 5.73) to allow multiple calibration data sets. This may be useful if the industrial reformer runs under very different operating scenarios. However, for the purposes of this workshop, we use only one calibration data set.

Aspen HYSYS will pull all the feedstock information and process operating conditions after we confirm the calibration data overwrite. The status bar now

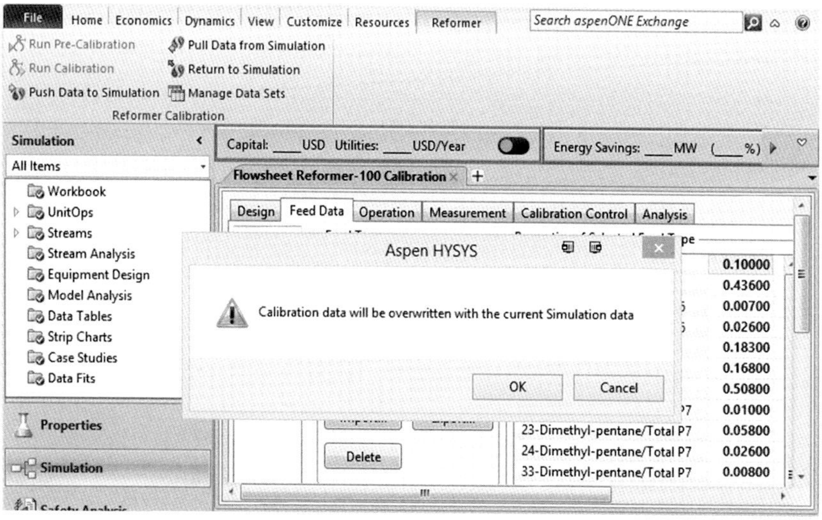

Figure 5.72 Importing initial model solution.

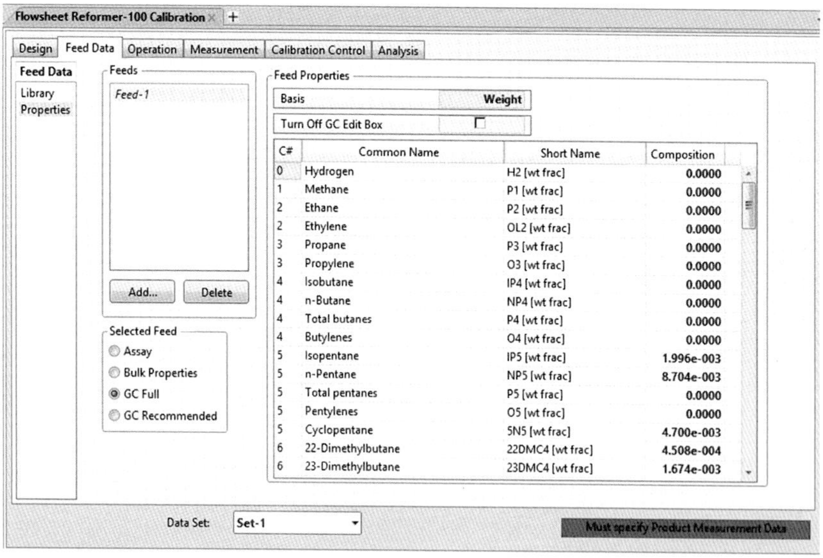

Figure 5.73 Feed composition on weight basis.

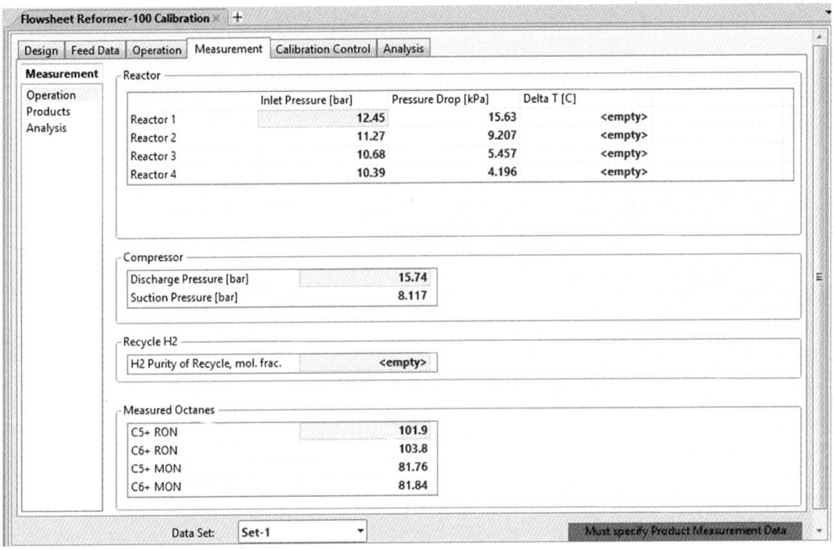

Figure 5.74 Reactor performance tab.

indicates that we must specify product measurements to begin the calibration process. If necessary, we can modify the operating variables (such as WAIT) of the reformer in addition to the measured values. However, we recommend creating a new model file if the operating scenarios are very different.

The second step in model calibration is specifying the measured yields and process performance. Click on the Measurement tab to bring up the Operation interface (see Figure 5.74). In the Operation section, we must enter values for reactor temperature drop and recycle hydrogen purity. We can enter in the pressure drops and measure octane values of the product. The default values come from the current model results. Entering new pressure drops allows us to account for unexpected flow behavior in the reforming reactors. Figure 5.75 shows the complete input window for the Operation section.

Next, we specify the flow rates, yields, and composition of all the key streams from the reformer (Figure 5.76). A compositional analysis is necessary to make sure that we model key reaction paths accurately. We recommend that users enter all compositional information for gas streams in mol% and all compositional information for liquid streams in vol% or wt%. Given the data available, we can enter the flow rates of each steam on a gas flow or mass flow basis. We note that internally, Aspen HYSYS converts all measurements into a mol% to achieve an overall material balance in the model results.

We suggest a few guidelines when entering the compositional data:

- If analysis for H_2 to fuel stream is not available, we can enter 85–87 mol% H_2 as the composition for the stream.
- Measurements for the stabilizer overhead liquid can often be confusing. Often there is little difference in the model results if we choose mol% or vol% for the original data. The molar volumes of these light components are roughly similar, so errors due to mistaken mol% or vol% are often quite small.

- If we do not have all isomers of a given kinetic lump (such as P8, SBP8, and MBP8), then it is possible to distribute the total measured lump over the three components. However, we must make sure *not* to include the isomer ratio as a calibration activity factor. This comment does *not* apply to xylenes. We must have the isomer ratio of xylenes to proceed with the calibration.
- We can group the aromatics higher than A9 into a single lump as A10. This is acceptable as we do not calibrate on aromatics higher than A9 and allow the model to calculate aromatic composition higher than A9 freely.

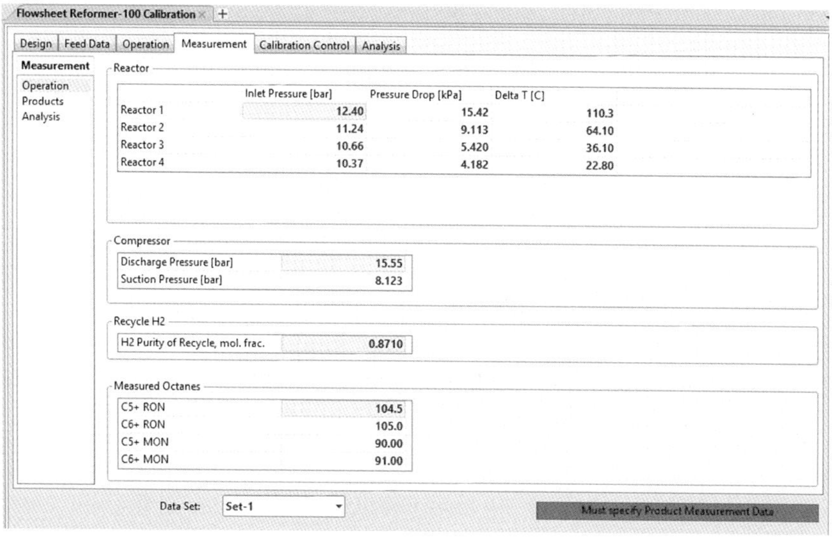

Figure 5.75 Completed reactor performance tab.

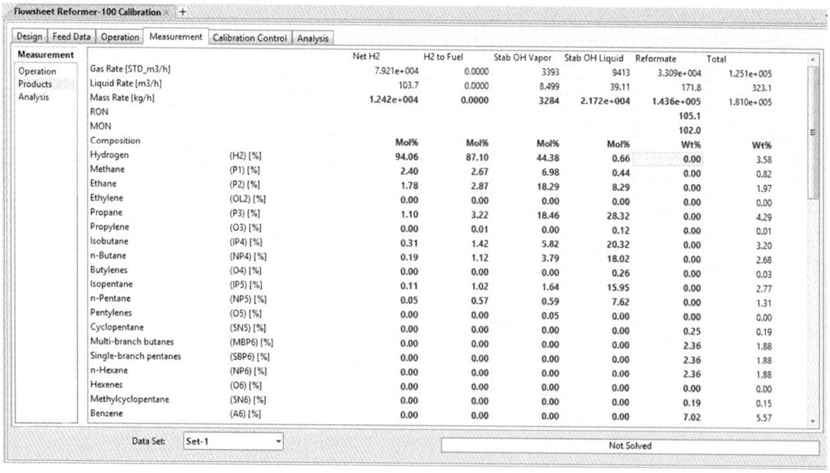

Figure 5.76 Product measurement tab.

Once we enter the composition information correctly, the status bar will turn yellow (see Figure 5.76), indicating that we are ready to begin varying activity factors.

In step 3 of the calibration, we use Aspen HYSYS to vary several activity factors in order to minimize the objective function. We define the objective function as the weighted sum of the absolute deviations from the model predictions and measure data. We can select terms in the objective function by going to the "Objective" section of the Calibration Control tab. We show this interface in Figure 5.77.

The initial objective function is quite strict and requires significantly detailed analysis for calibration purposes. We suggest an alternative objective function that works well when the compositional analysis is limited. In addition, less strict objective function helps make sure that the model does not become fixed or overcalibrated to a single data set.

Terms that do not appear in Table 5.31 are not part of the initial calibration. Low weightings indicate that agreement with a given term is more significant than other terms. We generally do not include isomer ratios as part of the initial calibration. Once we have completed an initial calibration, we use another data set to further calibrate the model using the original strict objective function. For the purposes of this workshop, we perform the calibration only once.

Once we select the objective function, it is a good practice to run *a model precalibration.* The model precalibration ensures that we are starting the model in a feasible location and indicates if the calibration process will succeed. We run the precalibration by clicking the "Pre-Calib" button in the calibration environment (Figure 5.78).

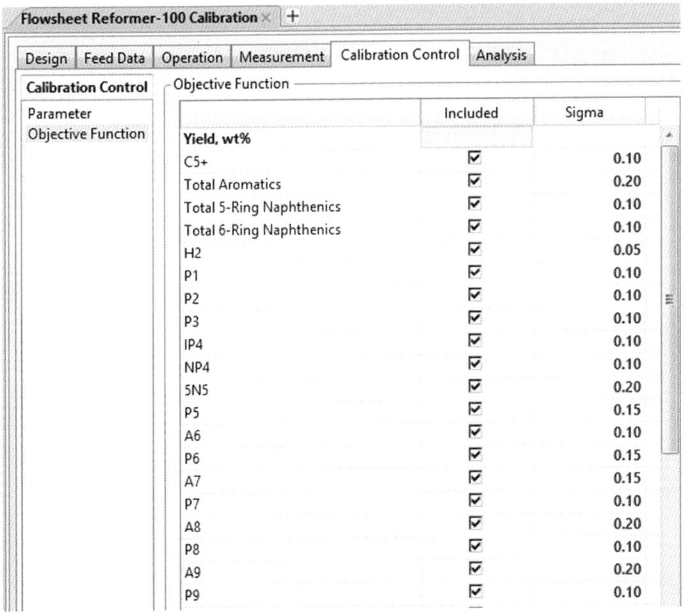

Figure 5.77 Initial objective function.

Table 5.31 Weighting factors for a less strict objective function.

Model prediction	Weight
C5+ yield	0.10
Total aromatic yield	0.20
H_2 yield	0.05
P1 yield	0.10
P2 yield	0.10
P3 yield	0.10
IP4 yield	0.10
NP4 yield	0.10
5N5 yield	0.20
P5 yield	0.15
A6 yield	0.10
P6 yield	0.15
A7 yield	0.15
P7 yield	0.10
A8 yield	0.20
P8 yield	0.10
A9 yield	0.20
A10 yield	0.20
P10 yield	0.10
Recycle gas purity	0.01
Reactor 1 ΔT	0.75
Reactor 2 ΔT	0.75
Reactor 3 ΔT	0.75
Reactor 4 ΔT	0.75

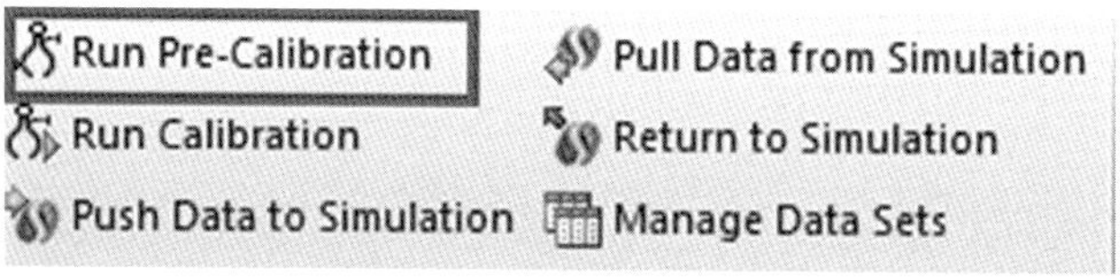

Figure 5.78 Precalibration in the Reformer Calibration.

When we run the precalibration of the model, Aspen HYSYS presents the Validation Wizard for this data set. The key results in this wizard are mass and hydrogen closure of this data set. Figure 5.79 shows the initial state of the wizard. We note that there is a significant mass and hydrogen imbalance. We can attempt to correct the error by changing the bias for each stream. The biases refer to how the stream flow will be adjusted to ensure that mass and hydrogen balance is closed. Figure 5.80 shows that we can improve the imbalance by unselecting the bias for the reformate.

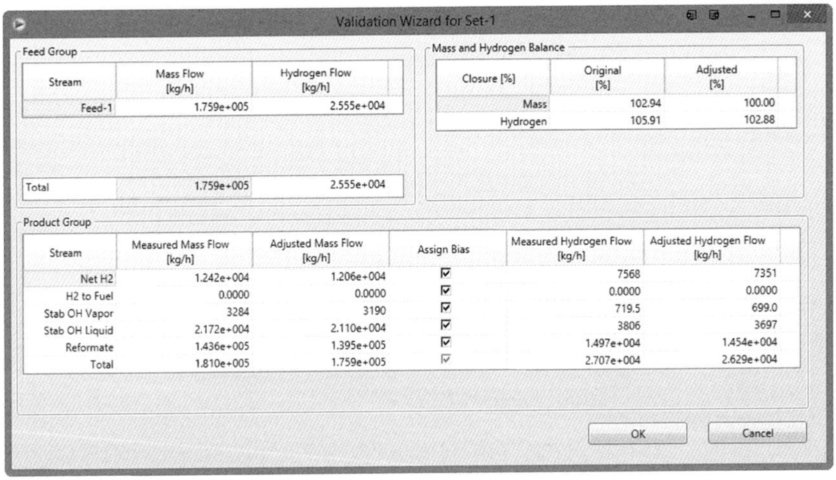

Figure 5.79 Assign bias.

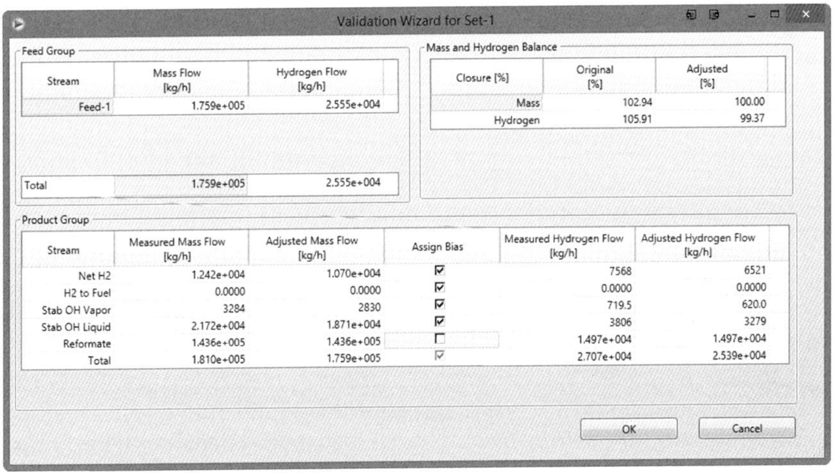

Figure 5.80 Assign bias – select reformate.

Changing the Assign Bias may not improve the calibration. Significant mass and hydrogen imbalance indicates that the data set may be inconsistent. The first step is to verify the measurement data and obtain updated measurements if necessary. If we cannot close the mass balance, we can proceed with calibration. However, we must realize that a close calibration may not be possible and we must view model prediction with extra caution.

The next step is to choose model activity factors to vary during the calibration run. We select activity factors by navigating to the Parameter section of the Calibration Control tab (Figure 5.81). To include a factor in the calibration, we must check the "Included box" for that factor and specify an upper and lower bound for that factor as shown in Figure 5.82. The bounds for the upper and lower factor must be reasonable to avoid overcalibrating the model. We discuss upper and

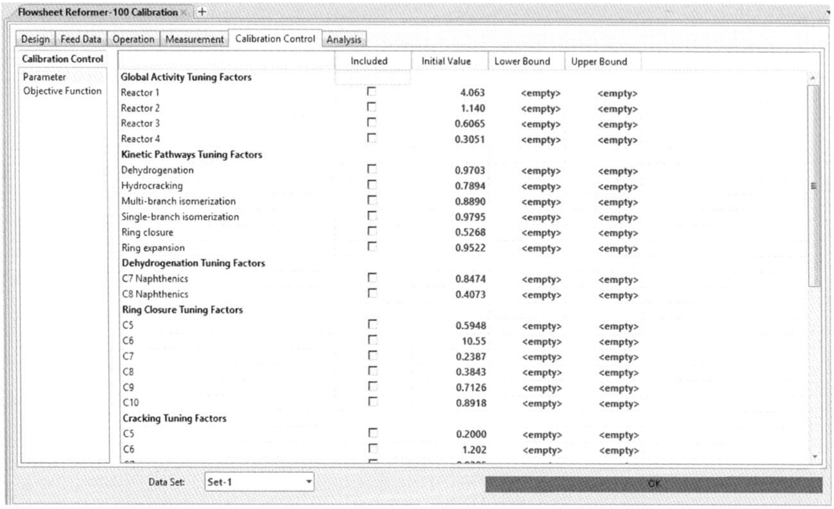

Figure 5.81 Calibration parameters.

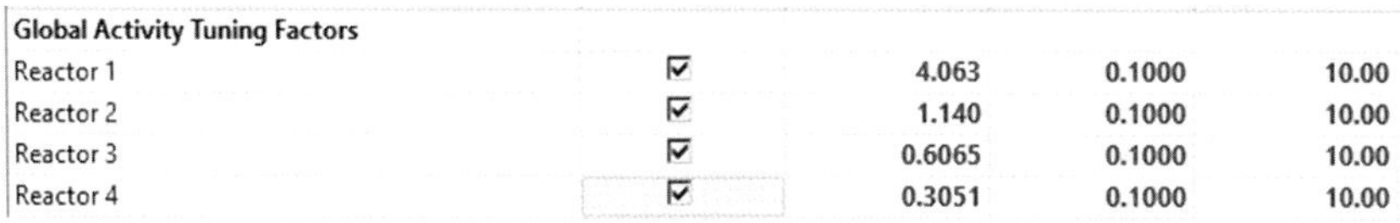

Figure 5.82 Set upper and lower bounds for global activity tuning factors.

lower bounds for adjustment factors previously in Table 5.12. Table 5.32 also presents some reasonable upper and lower bounds for the most common activity factors.

We calibrate the model by selecting each group of factors in Table 5.32 once at a time and subsequently run for each group selection. For example, when we calibrate the model for the first time, we should select the Global Activity Tuning Factors and enter the appropriate bounds from Table 5.32 (Figure 5.81). Then, we click on Run Calib to start the optimization process. We run the process at least five times, selecting a different group to calibrate each time.

The output from the solver appears in Table 5.33. Our goal is to reduce the final value of column 4, "Objective Function Value," to some small value. For an

Table 5.32 Suggested activity factors for calibration.

Group #	Terms	Range
1	Global activity tuning factors	1–20
2	Dehydrogenation and hydrocracking tuning factors	0.1–1
3	Isomerization, ring closure, and expansion tuning factors	0.1–1
4	Individual tuning factors for C7 and C8	0.1–1
5	Light gas yield (C1 and C2 only)	0.1–10

Table 5.33 Solver output during calibration.

Iteration	Residual Convergence Function	Objective Convergence Function	Objective Function Value	Overall Nonlinearity Ratio	Model Nonlinearity Ratio	Worst Model
0	1.332D−02	9.878D−03	1.247D+05	7.076D−01	8.211D−01	RXR2.RXR
	<Line Search Creep Mode ACTIVE> ==> Step taken 3.00D−01					
1	9.110D−03	7.029D−03	1.250D+05	9.739D−01	−6.060D−01	NETCALV
	<Line Search Creep Mode ACTIVE> ==> Step taken 3.00D−01					
2	6.273D−03	4.953D−03	1.253D+05	9.813D−01	3.392D−01	NETCALV
	<Line Search Creep Mode ACTIVE> ==> Step taken 3.00D−01					
3	4.340D−03	3.478D−03	1.255D+05	9.866D−01	6.410D−01	NETCALV
	<Line Search Creep Mode ACTIVE> ==> Step taken 3.00D−01					
4	3.014D−03	2.438D−03	1.256D+05	9.904D−01	7.827D−01	NETCALV
	<Line Search Creep Mode ACTIVE> ==> Step taken 3.00D−01					
5	2.098D−03	1.707D−03	1.257D+05	9.652D−01	5.367D−01	NETCALV
6	1.191D−05	1.186D−05	1.259D+05	9.999D−01	9.997D−01	ISOMP4
7	2.669D−09	1.338D−09	1.259D+05			

Successful solution.

Optimization Timing Statistics	Time	Percent
MODEL computations	1.60 secs	24.17 %
DMO computations	4.57 secs	69.15 %
Miscellaneous	0.44 secs	6.68 %
Total Optimization Time	6.61 secs	100.00 %

Problem converged

accurate calibration, the objective function should be lower than 250–300 using the weightings given in Table 5.31.

Each time we successfully run a calibration, we can verify how far model predictions are from measured input values given to Aspen HYSYS. We go to the Calibration Factors section (see Figure 5.83) in the Analysis tab of the Calibration Environment. The "Delta" column indicates the difference between the measured and model values for a given term of the objective function. Contribution indicates the given term's contribution to the objective function (Delta/Weighting). Using the steps in Table 5.32, we can reduce the objective function value to 180. This is below our 250–300 criterion for a reasonable model.

Once we finish calibrating the model to some small residual (<250–300), we should export the results back to the main reformer flowsheet. This is step 4, the last step, of the model calibration that we discussed in the beginning of this section.

We save the model calibration by clicking "Save for Simulation …" in the Analysis tab of the Reformer Calibration Environment. Aspen HYSYS will prompt us (see Figure 5.84) to save this calibration as "Set-1." We can have multiple calibrations for the same reformer and use different calibration sets for different operating scenarios. We recommend only having only calibration set per reformer model file.

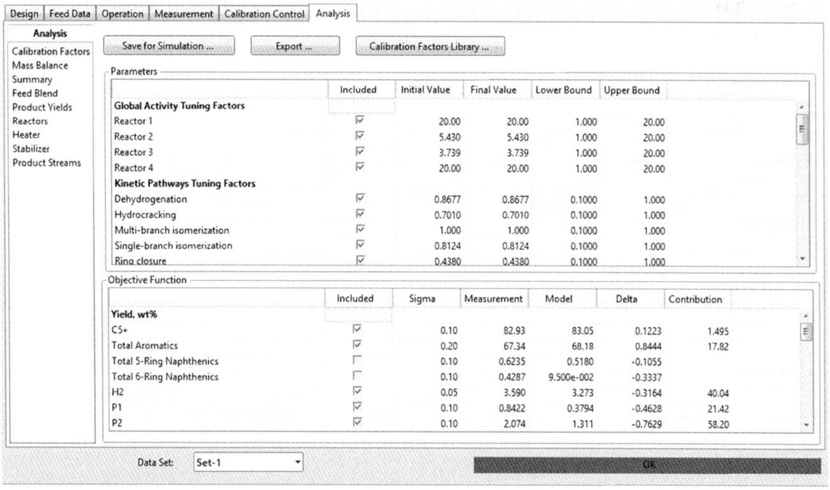

Figure 5.83 Calibration factors analysis.

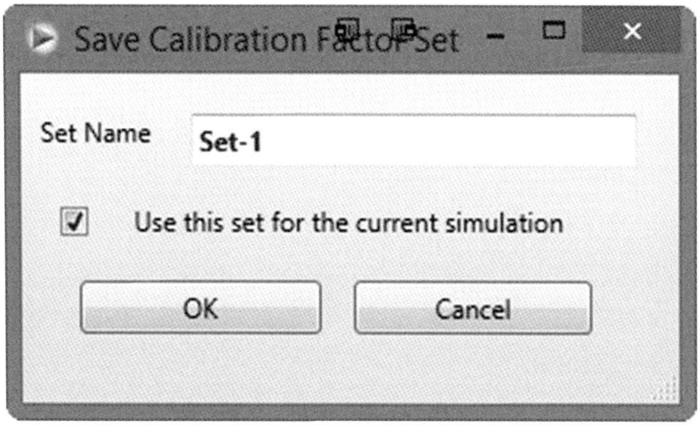

Figure 5.84 Save calibration factor set.

After saving the Calibration, we should put the solver in holding mode to make sure that Aspen HYSYS exported the calibration factors properly (Figure 5.85). We return to the Reformer Subflowsheet environment. We recommend that users go through each one of the tabs in Reformer Subflowsheet environment to make sure the input data have not changed. It is also important to make sure

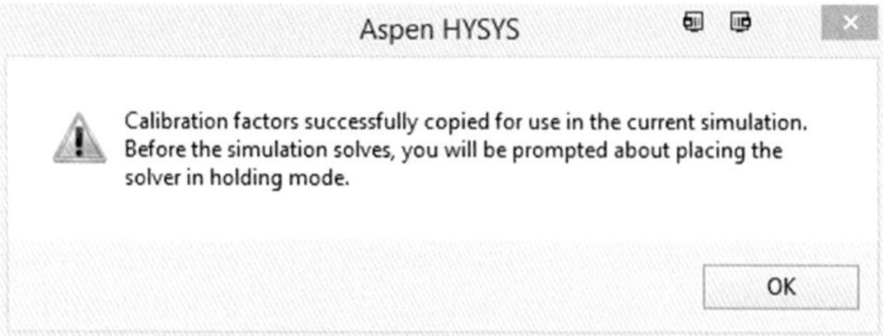

Figure 5.85 Prompt to hold Aspen HYSYS solver.

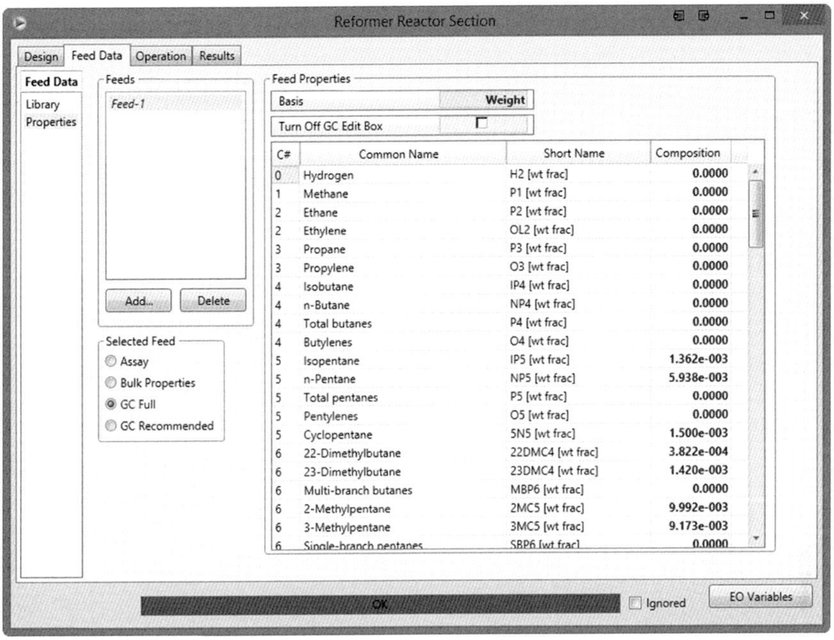

Figure 5.86 Verify feed basis for Feed Data.

the feed basis for the kinetic lumps is same as what was chosen initially (in this work, we always use wt%, see Figure 5.86). We can release the solver to allow Aspen HYSYS to solve the model as shown in Figure 5.86.

We return to the main flowsheet to complete the calibration process for the Reformer Model.

5.16 Workshop 5.3 – Build a Downstream Fractionation System

The next step is to build the downstream fractionation system. The downstream fractionation system for this CCR reformer has three distinct parts:

1) Product remixer.
2) Hydrogen recontactor.
3) Primary gasoline/LPG stabilizer and aromatics recovery.

We have previously explained these three parts in conjunction with Figures 5.19–5.21 in Section 5.12. We open the simulation file, *Workshop 5.2.hsc*, and save it as *Workshop 5.3.hsc*, the starting file for the current workshop.

We build a subflowsheet environment for the product remixer by returning to the main flowsheet and creating a subflowsheet. We create a subflowsheet using the "Sub-Flowsheet" icon in the Aspen HYSYS toolbar palette shown in Figure 5.87. The new subflowsheet appears on the main flowsheet as large icon

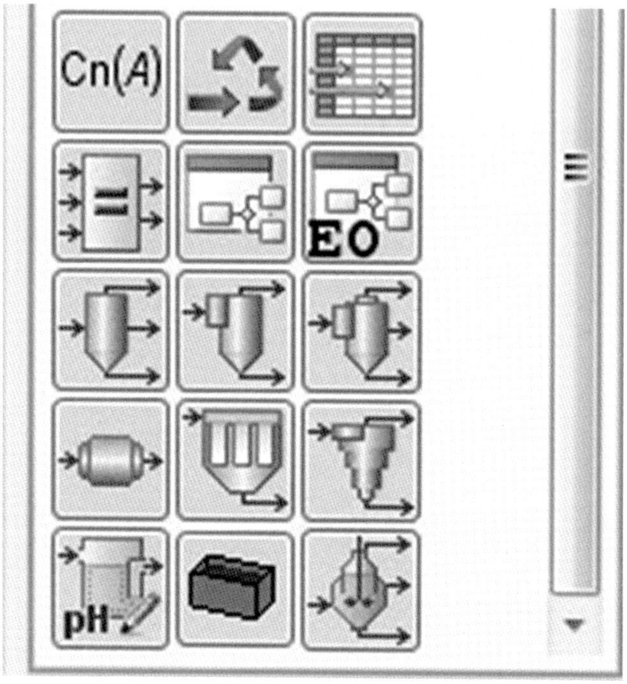

Figure 5.87 Flowsheet unit operation in Aspen HYSYS palette.

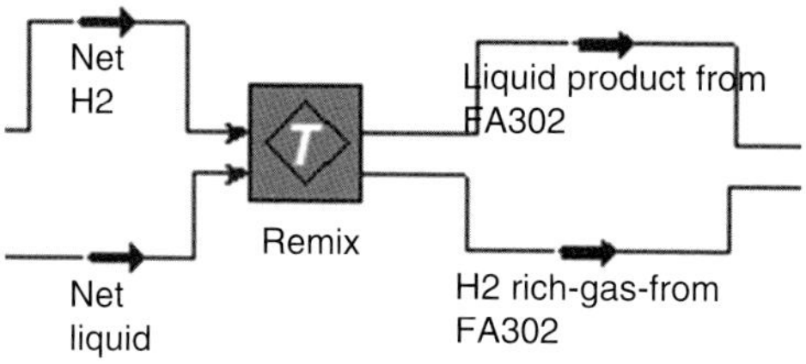

Figure 5.88 Remixer subflowsheet configuration.

with "T" marker (Figure 5.88). We can double-click the icon to bring up the subflowsheet connections window (Figure 5.89).

We attach the inlet connections to the subflowsheet and begin building the internal structure of the subflowsheet (Figure 5.90). We attach the outlet connections once we have completed building the flowsheet (Figure 5.89).

Using the standard Aspen HYSYS objects, we build a simple mixer and separator to remix the product streams and flash the mixed product at the temperature and pressure of the primary product separator. The outlet gas from FA302 represents the initial release of net gas. We use a Set object to ensure that the temperature of the flash is the same as the Net H2 product from the reformer model. Once we finish building this subflowsheet, we can connect outlet feeds as shown in Figure 5.91.

We now proceed to build the hydrogen recontacting section of the fractionation system. Using the same procedure as before, we create a subflowsheet for the Recontactor. The goal of the recontacting section is to improve separation of the light ends from the net gas stream and recover aromatics lost in the initial

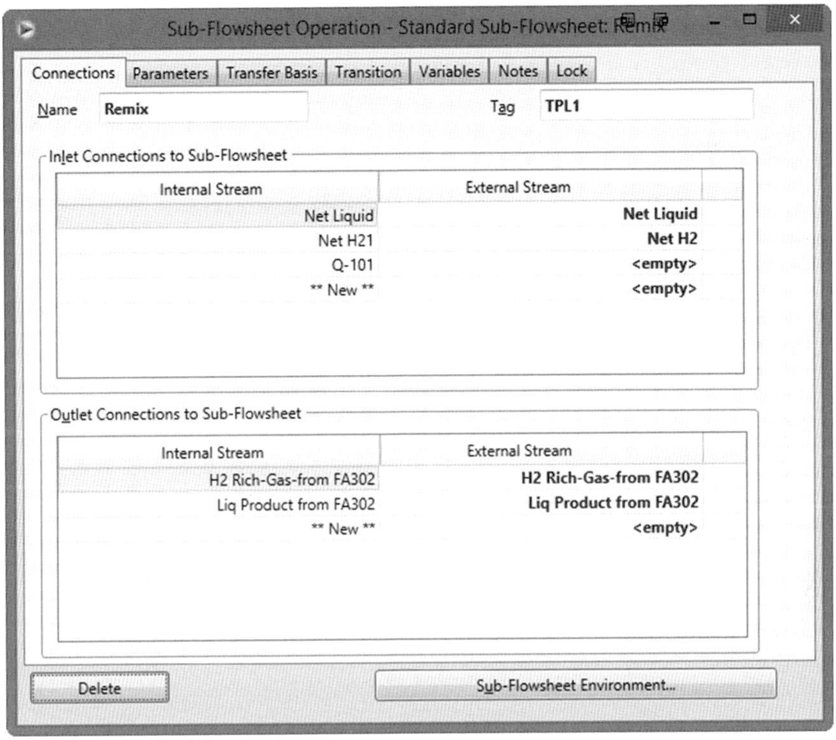

Figure 5.89 Inlet–outlet connections for remixer subflowsheet.

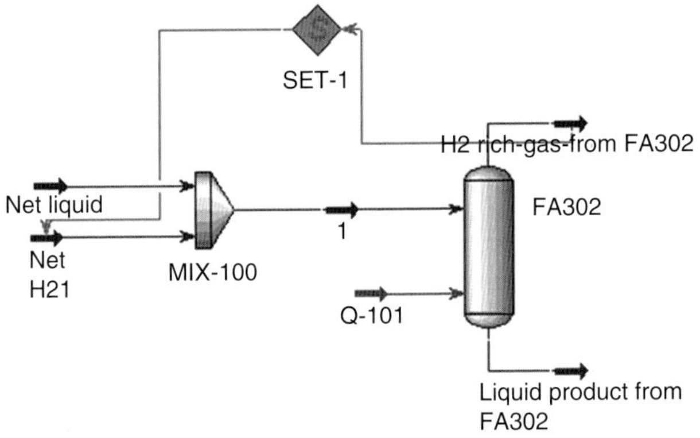

Figure 5.90 Subflowsheet for remixer.

Figure 5.91 Subflowsheet for recontactor.

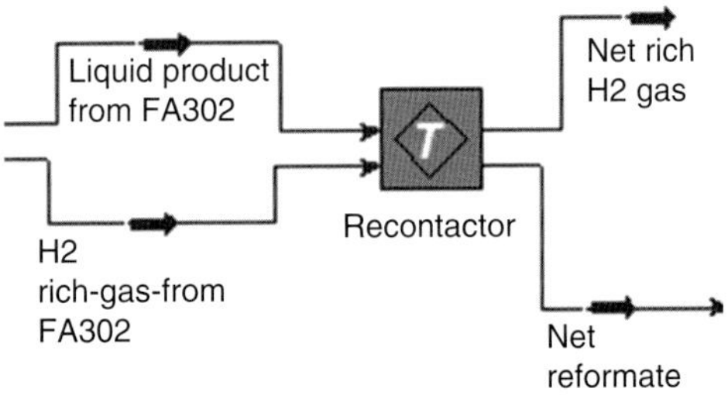

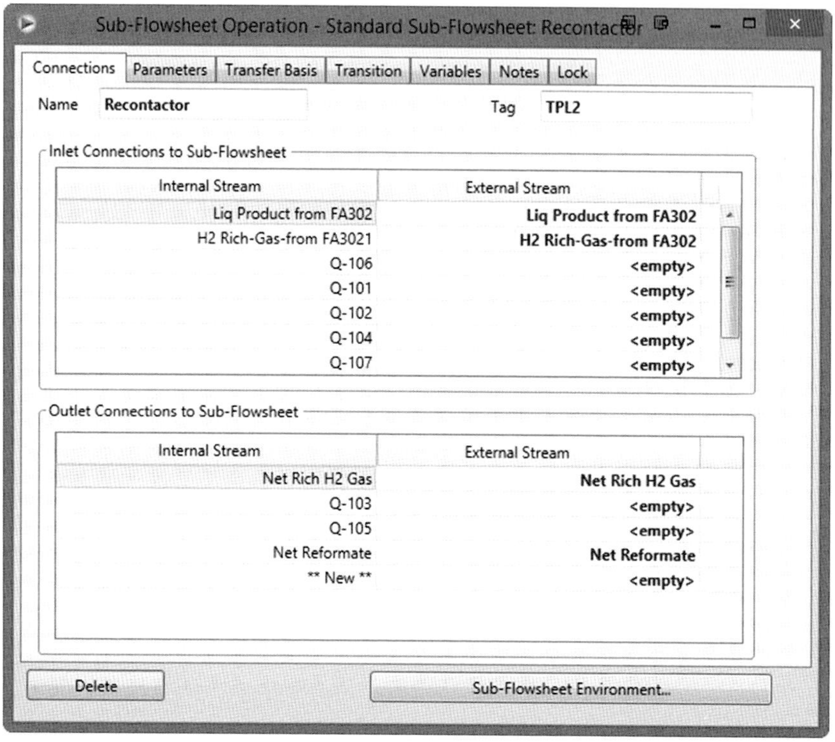

Figure 5.92 Subflowsheet for recontactor.

net gas stream. Figure 5.92 shows the relevant inlet and outlet stream names and variables for the subflowsheet.

We use standard Aspen HYSYS objects to recreate the recontacting section (Figure 5.93). Typically, a real recontacting section may have several stages to improve product separation. In general, two ideal separators can model multiple real separators as real separators do not typically operate at thermodynamic equilibrium. We also include an Adjust block to ensure that the temperature of the Net H2 Rich Gas leaving matches the plant value. This is often the only calibration required to model plant performance accurately. Table 5.34 shows the specifications we enter for each of the streams in the subflowsheet. We note that these values are not exact but approximated from various sources. When developing a

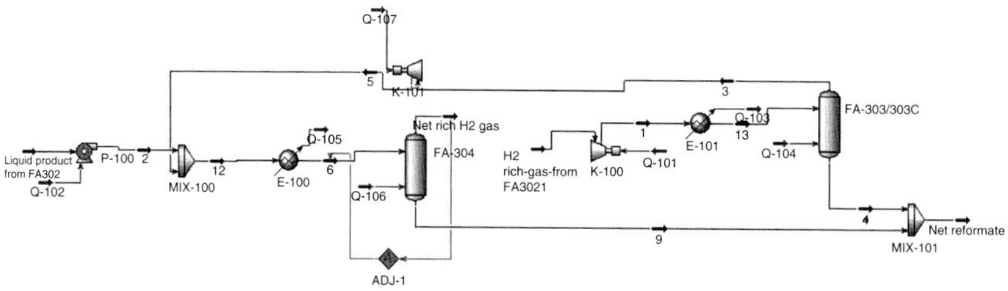

Figure 5.93 Flowsheet for recontacting section.

Table 5.34 Stream specifications for recontactor.

Stream	Temperature (°C)	Pressure (kPa)
1	–	2612
2	–	5681
5	–	5681
6	10.11	–
13	30.00	–
Net rich H_2 gas	–	5681

model for industrial use, we must make sure to use actual plant values. Table 5.34 indicates the specifications for each stream. Values given by "–" in Table 5.34 indicate that this value should not be specified.

Before the Net Liquid enters the gasoline stabilizer, we must heat the product to a temperature suitable for fractionation. In the actual refinery process, the product heater is often integrated with the bottoms outlet of the gasoline splitter or other columns. However, for the purposes of this simulation, we use a simple heat exchanger instead. For more detailed simulations, we advise the use of cross exchangers to accurately model the duty required for the fractionation (Figure 5.94).

Figure 5.95 shows the stream configuration for the primary gasoline stabilizer. The overhead gas contains mostly light C1–C2 components that did not leave the Net H2 stream. The overhead liquid draw is mostly C3–C4 components, which

Figure 5.94 DA301 preheater.

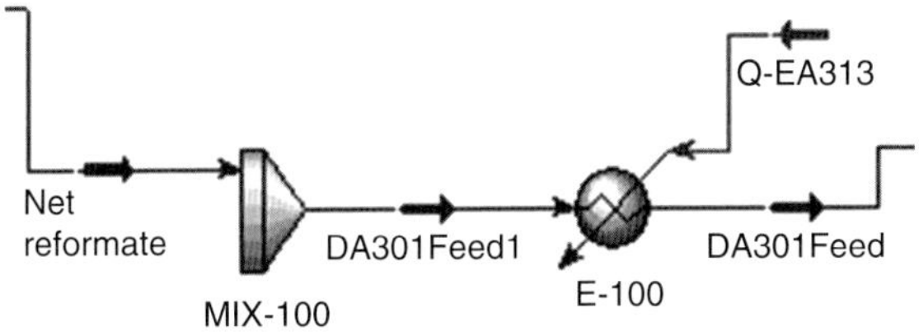

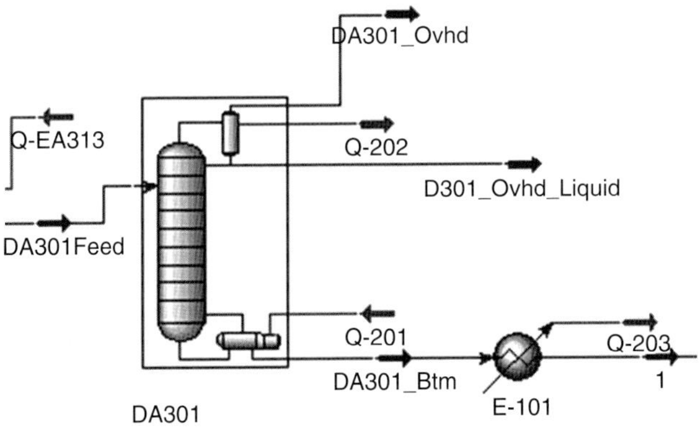

Figure 5.95 DA301 flowsheet.

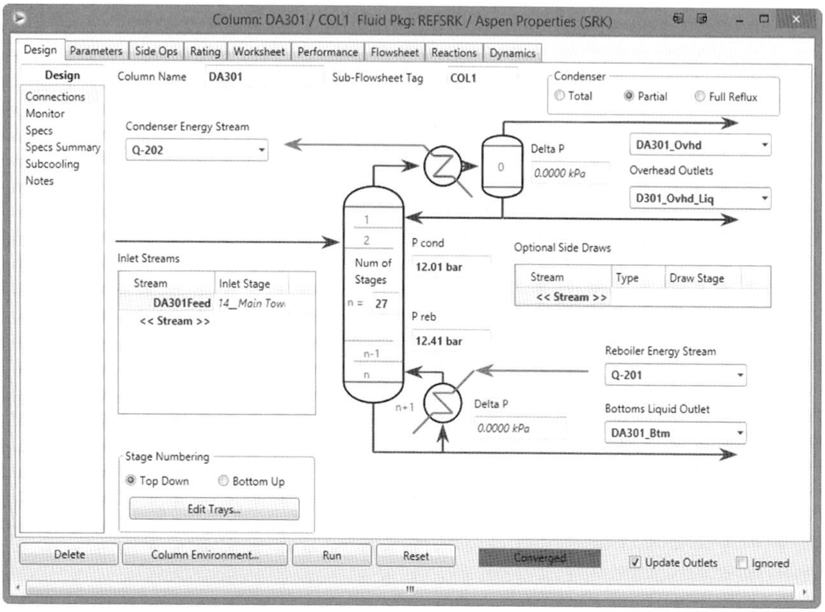

Figure 5.96 DA301 configuration.

form an LPG-like stream. The bottoms stream represents the stabilized gasoline or aromatic enriched liquid product from the reformer.

We show the pressure profile and number of stages required for the gasoline splitter in Figure 5.96. We use 27 theoretical stages to represent the stabilizer in our reformer system according to Table 5.7 in Section 5.8. We use this approach to approximate the column's overall efficiency at 60–70%. We discussed the importance of using the overall stage efficiency over stage-by-stage efficiencies in Section 2.4.3. In general, using the overall stage efficiency approach leads to more robust and predictable column model operation.

As we have three draw streams on DA301, we will require three independent specifications for the column to converge robustly as shown in Figure 5.97. Typically, we use the reflux ratio, temperature of a particular stage, and mole purity (either C4 or C5 in the overhead liquid or vapor) as specifications for the column (see Figure 5.98). If the column is operating as a gasoline splitter, we may want to use the Reid vapor pressure (RVP) of the bottoms as a performance specification. If the column does not converge, we can use the alternate specifications of overhead draw rate, reflux ratio, and bottoms draw to ensure that the column converges to a solution. Once we have a solution, it is quite easy to converge on a performance specification.

As this reformer is part of a petrochemical complex, the product from the gasoline splitter enters an aromatics fractionation column. Column DA302 (see Figure 5.99) has 36 equilibrium stages according to Table 5.7, Section 5.8. It separates toluene and lighter components from xylenes and heavier components. The bottoms product of DA301 enters a heat exchanger to bring down the temperature of the gasoline product to a suitable fractionation temperature (Figure 5.100).

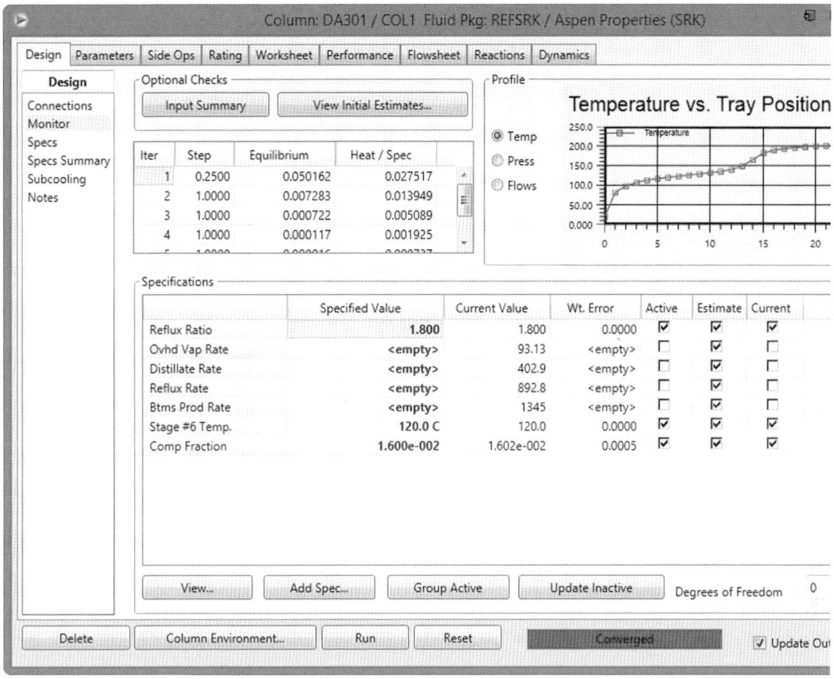

Figure 5.97 DA301 specifications.

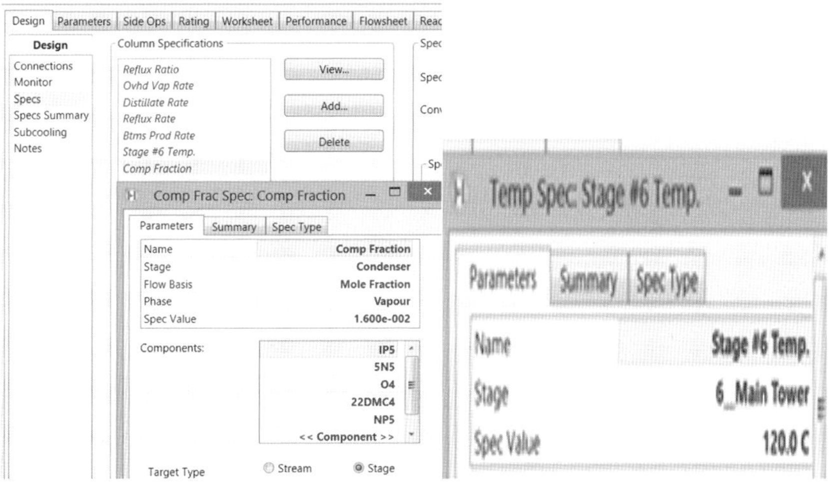

Figure 5.98 Two specifications for DA301 simulation convergence: Components – IP5(*i*-pentane), 5N5(cyclopentane), O4 (1-butene), 22DMC4(22-methyl-butane), NP5 (1-pentane).

Figure 5.101 shows the pressure profile and number of stages required for the aromatics column. Again, we use the same principle of overall stage efficiency (60–70%) to calculate the number of theoretical stages required for the column model. We note that the industrial columns may include a small vent stream in the

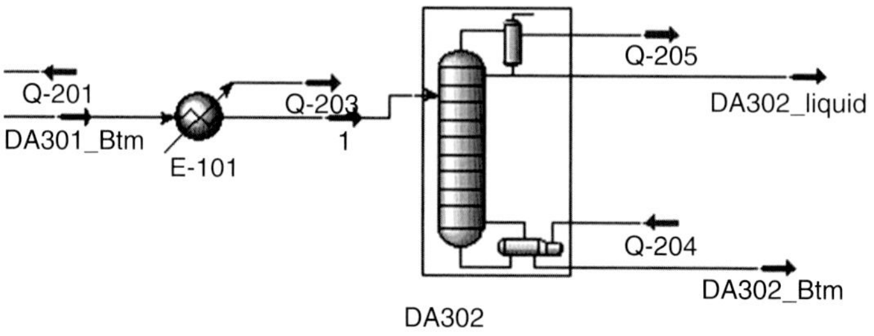

Figure 5.99 DA302 flowsheet.

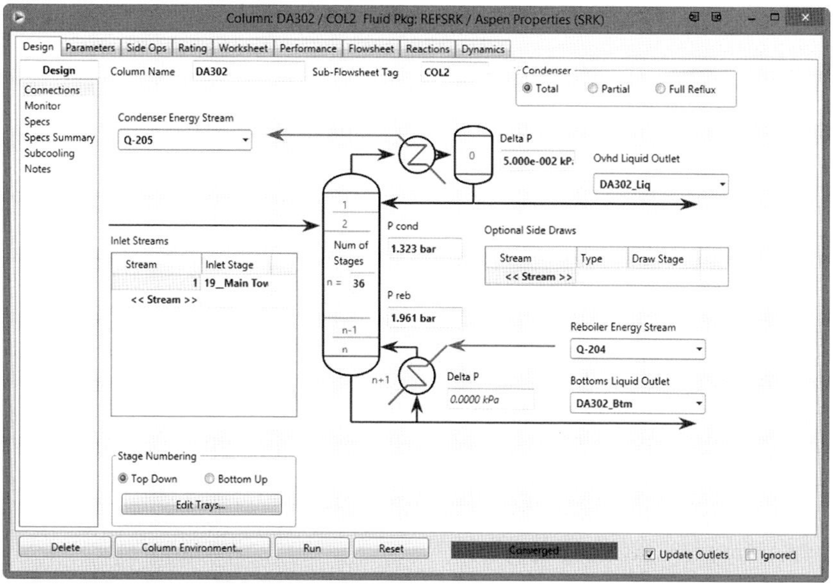

Figure 5.100 DA302 column configuration.

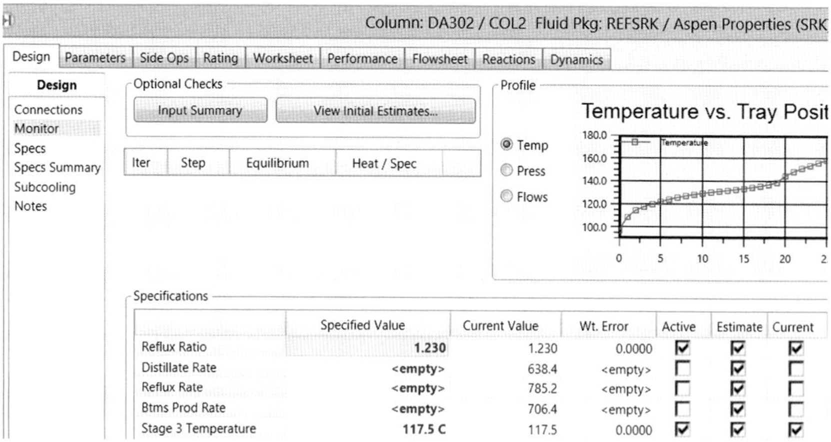

	Specified Value	Current Value	Wt. Error	Active	Estimate	Current
Reflux Ratio	1.230	1.230	0.0000	☑	☑	☑
Distillate Rate	<empty>	638.4	<empty>	☐	☑	☐
Reflux Rate	<empty>	785.2	<empty>	☐	☑	☐
Btms Prod Rate	<empty>	706.4	<empty>	☐	☑	☐
Stage 3 Temperature	117.5 C	117.5	0.0000	☑	☑	☑

Figure 5.101 DA302 specifications for aromatic splitter.

condenser for this column. However, depending on the thermodynamic model chosen, the feed to DA302 may not contain any light components. If we create a vent stream, it is likely that the column will have difficulty in converging, as we expect the vent stream to be very small.

As DA302 has two draws in the column, we require two specifications. We typically run the column with the overhead draw rate and reflux ratio as the initial set of specifications (see Figure 5.101). Once have a converged solution, we can use stage temperature as a performance specification to match plant operation. We save the converged simulation as *Workshop 5.3.hsc*.

5.17 Workshop 5.4. – Case Study to Vary RON and Product Distribution Profile

We begin with our calibrated simulation file, *Workshop 5.2.hsc*, and resave it as *Workshop 5.4.hsc*.

In this workshop, we use the calibrated model to perform a case study to determine the operating conditions to produce a desired product yield. The composition of the feed to the reformer may change quickly, and the composition of lighter naphthenes (N5, N6) can change dramatically with the changes to the IBP of the feed. In Section 5.12.1, we discussed several situations that change the product yields with changes in operating conditions and feedstock composition. The most basic, yet useful, case study is to vary the reactor temperature and H_2HC ratio and the effect on product RON and aromatic yields.

We developed the initial model using Reactor Inlet Temperature and associated temperature biases for each reactor. This is useful for a specific reformer plant; however, this method can mask the effect of reactor temperature on the process. We will instead use the WAIT to control the reactor temperature (Figure 5.102).

We change the reactor to the WAIT basis by first holding the solver and prevent it from running while we change the reactor temperature. We note the calculated WAIT from the current solution and copy the value. We paste the value back into the WAIT textbox and release the solver. The solution process should be quite quick with the initial residual on the order of 1E-3 or lower. Higher residuals may indicate that the model was overcalibrated or the model is very sensitive to the operation conditions. In both cases, we will likely have to recalibrate the model with more recent data.

We follow the procedure for case studies demonstrated in Figures 2.69–2.73, Section 2.10.3. Our goal is to the observe product yields as functions of the WAIT and H_2HC ratio. It is possible to manually change each WAIT and H_2HC ratio and rerun the model each time. However, given the typical run time for the reformer solver, this quickly becomes a tedious process. It is better to use the case study features of Aspen HYSYS to automate this process. In addition, as the case study feature will run the model at a variety of conditions and if we successfully solved a model, we can make sure that the model is not overcalibrated.

We add variables from the Reformer object, and we select the Reformer object in the Flowsheet List. The Variable List will show all variables that belong to the Reformer object. We can scroll through this list and click "Add" to add a particular

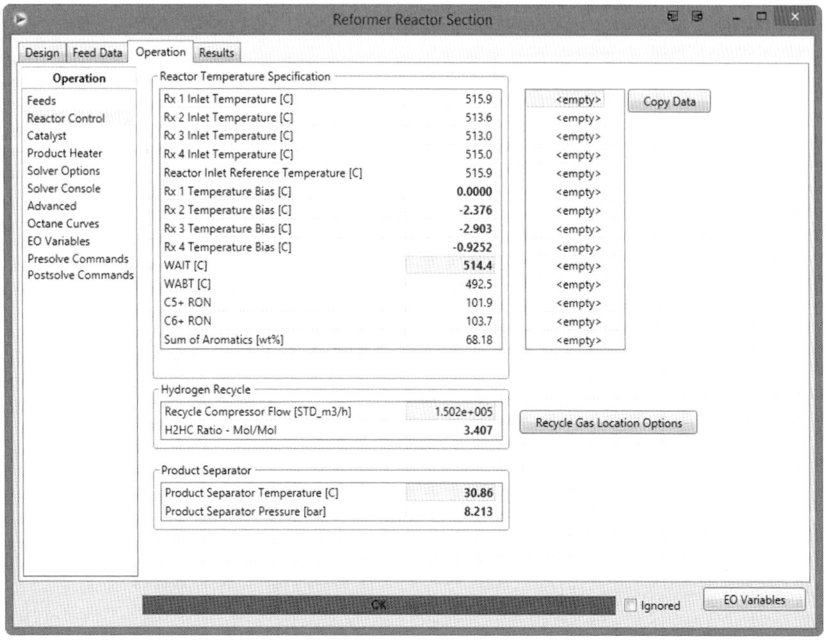

Figure 5.102 Change reactor temperature to WAIT basis.

Table 5.35 Variables for RON case study.

Variable	Type
WAIT, 495–525°C (an increment of 5°C)	Independent
C5+ RON and C6+ RON	Dependent
H_2HC Ratio, 3–4 (an increment of 0.25)	Independent
Detailed yields (total aromatics, total C8 aromatics)	Dependent
Detailed yields (A6, A7)	Dependent
Detailed yields (H2, P1, P2, and P3)	Dependent

variable to the case study. Table 5.35 shows the variables we need for this case study. Figure 5.103 shows the independent and dependent variables for our case study.

We click View to set the upper and lower bounds for the case study. We change the WAIT from 495 to 525 °C inclusively in 5 °C increments. We also change the H_2HC ratio from 3.0 to 4.0 with an increment of 0.25. The number of states indicates how many times the reformer model will run with various inputs. We generally advise against running more than 40–50 states at a time since the total run time for more than 50 states can be quite significant. In most cases, the reformer operating temperatures are not more than 10 °C or so during normal operation. We click on Start to begin running the case studies. We observe the solver running in the lower right corner of the flowsheet.

Figure 5.104 shows the results of the case study. We can view this graph by clicking the "Results ..." button. The default is to show the graph with the results

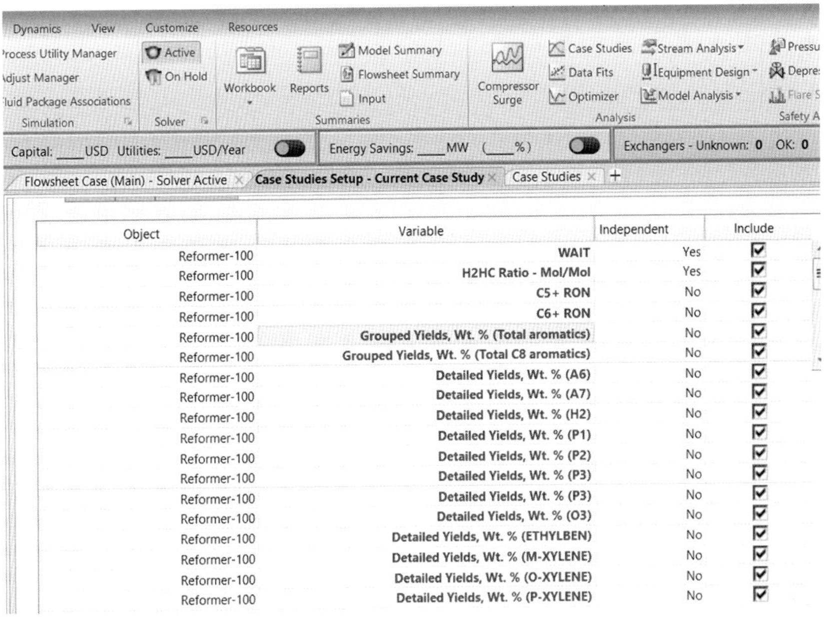

Figure 5.103 Independent and dependent variables of the case study.

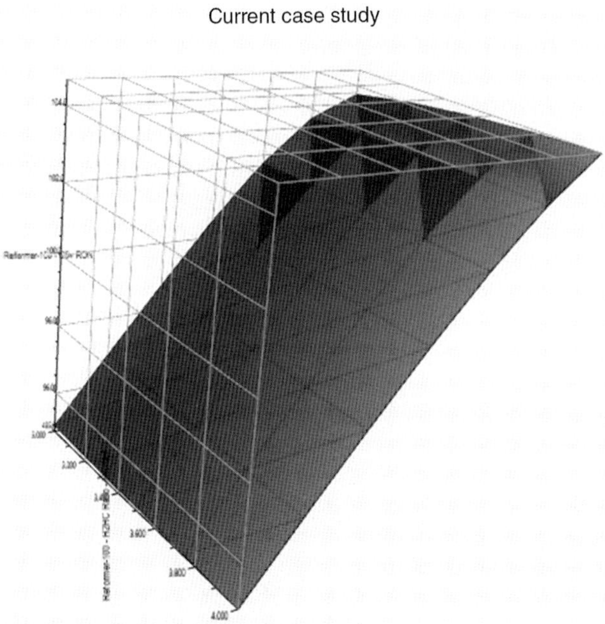

Figure 5.104 Graphical results of case study.

of the case study. In general, we can see the general trend at the high reactor temperatures and low H_2HC ratios increase the RON of the product. We can view the numerical results of the case study by selecting the "Transpose Table" option. The results appear in the order of increasing WAIT and H_2HC ratio. Figure 5.105 shows the results table for this case study. We copy these results into Microsoft Excel and create the graphs in Figures 5.105–5.109.

When we graph the results using Microsoft Excel, we find several interesting trends in the data that are not readily apparent from the initial result graph and numerical results. The case study shows that as temperature increases, the RON and yield of aromatic products increase as well (Figures 5.106 and 5.107). However, at around 520 °C for a H_2HC ratio of 3.0, we find that the yield begins to

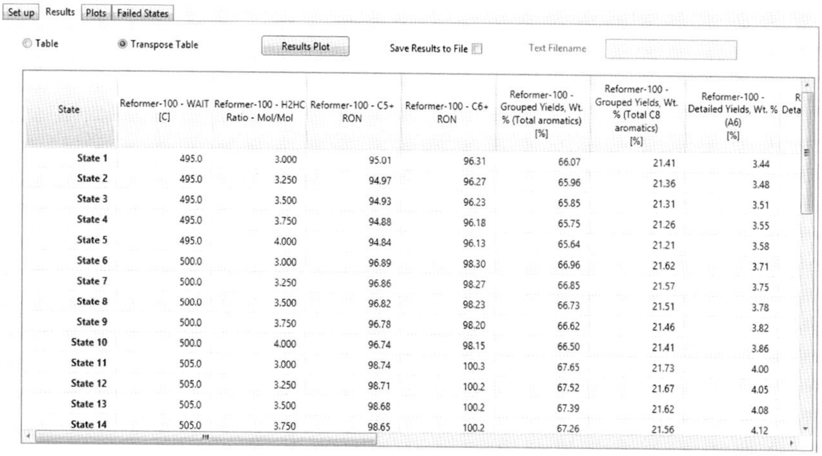

State	Reformer-100 - WAIT [C]	Reformer-100 - H2HC Ratio - Mol/Mol	Reformer-100 - C5+ RON	Reformer-100 - C6+ RON	Reformer-100 - Grouped Yields, Wt. % (Total aromatics) [%]	Reformer-100 - Grouped Yields, Wt. % (Total C8 aromatics) [%]	Reformer-100 - Detailed Yields, Wt. % (A6) [%]	R... Deta...
State 1	495.0	3.000	95.01	96.31	66.07	21.41	3.44	
State 2	495.0	3.250	94.97	96.27	65.96	21.36	3.48	
State 3	495.0	3.500	94.93	96.23	65.85	21.31	3.51	
State 4	495.0	3.750	94.88	96.18	65.75	21.26	3.55	
State 5	495.0	4.000	94.84	96.13	65.64	21.21	3.58	
State 6	500.0	3.000	96.89	98.30	66.96	21.62	3.71	
State 7	500.0	3.250	96.86	98.27	66.85	21.57	3.75	
State 8	500.0	3.500	96.82	98.23	66.73	21.51	3.78	
State 9	500.0	3.750	96.78	98.20	66.62	21.46	3.82	
State 10	500.0	4.000	96.74	98.15	66.50	21.41	3.86	
State 11	505.0	3.000	98.74	100.3	67.65	21.73	4.00	
State 12	505.0	3.250	98.71	100.2	67.52	21.67	4.05	
State 13	505.0	3.500	98.68	100.2	67.39	21.62	4.08	
State 14	505.0	3.750	98.65	100.2	67.26	21.56	4.12	

Figure 5.105 Numerical results for case study.

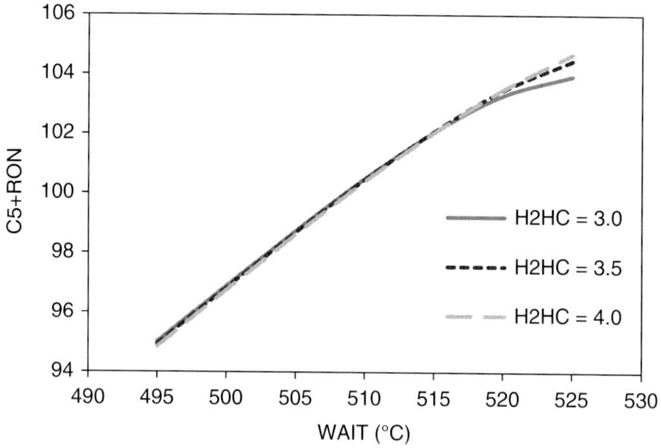

Figure 5.106 RON as a function of WAIT and H_2HC ratio.

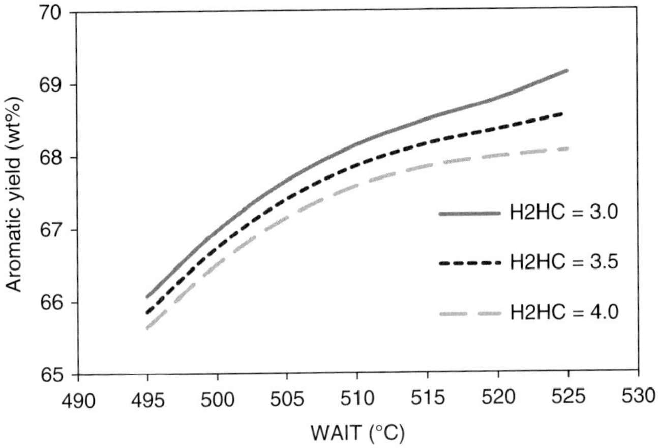

Figure 5.107 Aromatic yield as a function of WAIT and H_2HC ratio.

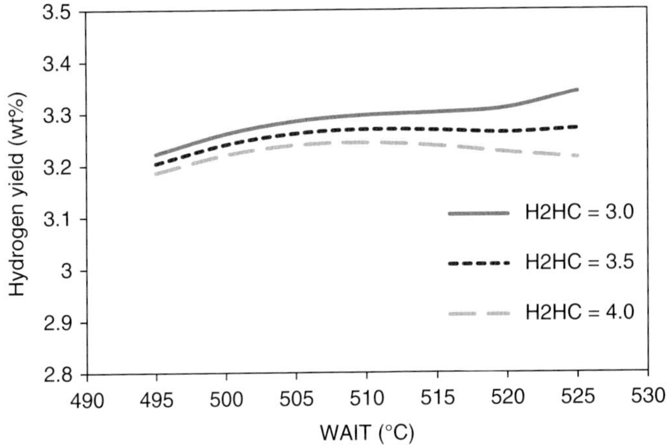

Figure 5.108 H_2 yield as a function of WAIT and H_2HC ratio.

drop. This is due to the increased deactivation of the catalyst at high temperature and low H_2HC ratio. We observe that we can alleviate this situation by increasing the H_2HC ratio.

An interesting side effect of increasing the H_2HC ratio is that around 520 °C, we start to see marked increases in the production of light gases and hydrogen yield (Figures 5.108 and 5.109). Although initially these increases appear small, they can have a significant effect on downstream fractionation. Excessive amounts of light gas can overload recycle compressors and increase the condenser duty requirements for stabilizer columns.

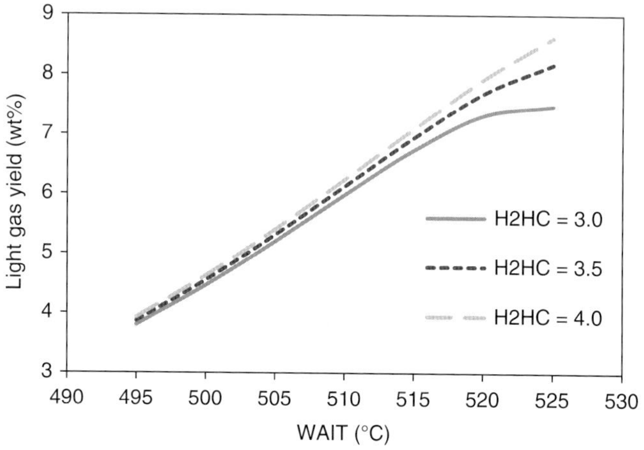

Figure 5.109 Gas yield as a function of WAIT and H_2HC ratio.

5.18 Conclusion

In this work, we have developed an integrated model for an UOP CCR unit in Aspen HYSYS Petroleum Refining. We used detailed feed composition (PNA content) and the routinely collected data such as operating profiles of reactor, product yields, and fractionator temperature profiles. The key highlights of this work are as follows:

- Detailed process description and overview of process chemistry relevant to modeling the reactor.
- Brief survey of existing kinetic and unit-level models for reforming processes.
- Discussion of kinetic and reactor model in Aspen HYSYS Petroleum Refining.
- Guidelines for dealing with the physical properties of the kinetic lumps in the context of the radial flow reactors and fractionator.
- Detailed process to infer molecular composition of feed when little plant information is available.
- Identified key issues relevant to calibration and how to prevent overcalibration of reactor model.
- Used industrial plant data to obtain workflow that produces a reasonable model.
- Applied model to industrial plant data and showed good agreement with plant measurements in yield and composition of key products.
- Investigation of the effects of various process parameters on product yield and composition.
- Transitioned the results from rigorous nonlinear model to the LP model for the refinery.

Nomenclature

α	Beta distribution shape parameter
α_j	Pressure effect exponent for reaction class j

A_c	Catalyst activity factor
A_i	Aromatic lump containing i carbon atoms ($i \geq 6$)
A_o	Preexponential factor in rate constant (1/s)
a_x	Activity factor for reaction group x
β	Beta distribution shape parameter
BEN	Benzene
CH	Carbon-to-hydrogen weight ratio
C_i	Concentration of component i
CP	Cyclopentane
E	Tray efficiency factor
EBP	End boiling point (°C)
E_i	Activation energy associated with reaction I (J/kmol)
ϕ	Catalyst deactivation due to coke on catalyst
F	Total molar flow rate (kmol/h)
F_i	Molar flow rate of component i (kmol/h)
Γ	Combined adsorption factor due to acid function
H_2HC	Hydrogen-to-hydrocarbon mole ratio
$H_{\mathrm{FACTOR}ij}$	Hydrogen-to-carbon weight ratio for component C_iH_j
I_{BP}	Initial boiling point (°C)
IP_x	Iso (or branched) paraffin containing x carbon atoms
k_i	Rate constant associated with reaction or component i (kmol/kg cat s)
K_i	Adsorption factor for component i (1/kPa)
MBP_x	Multiple-branched paraffin containing x carbon atoms
MCH	Methylcyclohexane
MCP	Methylcyclopentane
MON	Motor octane number
MON_i	Motor octane number of component or lump i
MW	Molecular weight
N_i	Naphthene lump containing i carbon atoms ($i \geq 5$)
N_i	Weight or mole faction of given lump i
$5N_i$	Five-membered naphthene lump containing i carbon atoms ($i \geq 5$)
$6N_i$	Six-membered naphthene lump containing i carbon atoms ($i \geq 6$)
NP_x	Normal paraffin containing x carbon atoms
P	Pressure (kPa)
P_i	Partial pressure of component i (kPa)
P_o	Reference pressure (kPa)
P_x	General paraffin containing x carbon atoms ($x \geq 1$)
θ	Combined adsorption factor due to metal function
R	Universal gas constant (J/kmol K)
RON	Research octane number
RON_i	Research octane number of component or lump i
SBP_x	Single-branched paraffin containing x carbon atoms
T	Temperature (K)
TBP	True boiling point curve (°C)
T_o	Reference temperature (K)
TOL	Toluene
W	Space velocity (1/h)

WAIT	Weight-averaged inlet temperature (°C)
WHSV	Weight hourly space velocity (1/h)
w_i	Weight fraction of component i
x_n	Molar composition of liquid leaving a given tray
y_n	Molar composition of vapor leaving a given tray

Bibliography

1 Ancheyta-Juarez, J. and Villafuerte-Macias, E. (2000) *Energy & Fuels*, **14**, 1032–1037.
2 Aguilar-Rodriguez, E. and Ancheyta-Juarez, J. (1994) *Oil & Gas Journal*, **92**, 80–83.
3 Ancheyta-Juarez, J. and Aguilar-Rodriguez, E. (1994) *Oil & Gas Journal*, **92**, 93–95.
4 Taskar, U.M. (1996) Ph.D. Dissertation. Texas Tech University, Lubbock, TX.
5 Taskar, U.M. and Riggs, J.B. (1997) *AIChE Journal*, **43**, 740–753.
6 Little, D. (1985) *Catalytic Reforming*, Penwell Books, Tulsa, OK.
7 Antos, G.J. and Aitaini, A.M. (2004) *Catalytic Naphtha Reforming*, 2nd edn, Marcel Dekker, New York, NY.
8 Gary, J.H., Handwerk, G.E., and Kaiser, M.J. (2007) *Petroleum Refining. Technology and Economics*, 5th edn, CRC Press, Boca Raton, FL.
9 UOP. (1971) US Patent 3,706,536 (A. R. Greenwood *et al.*).
10 Hosten, L.H. and Froment, G.F. (1971) *Industrial & Engineering Chemistry Process Design and Development*, **10**, 280–287.
11 Selman, D.M. and Voorhies, A. (1975) *Industrial & Engineering Chemistry Process Design and Development*, **14**, 118–123.
12 Froment, G.F. (1987) *Chemical Engineering Science*, **42**, 1073–1087.
13 Menon, P.G. and Paal, Z. (1997) *Industrial & Engineering Chemistry Research*, **36**, 3282–3291.
14 Raseev, S. (2003) *Thermal and Catalytic Processes in Petroleum Refining*, Marcel Dekker, New York, NY.
15 Svoboda, G.D., Vynckier, E., Debrabandere, B., and Froment, G.F. (1995) *Industrial & Engineering Chemistry Research*, **34**, 3793–3800.
16 Froment, G.F. (2005) *Catalysis Reviews*, **47**, 83–124.
17 Sotelo-Boyas, R. and Froment, G.F. (2009) *Industrial & Engineering Chemistry Research*, **48**, 1107–1119.
18 Wei, W., Bennet, C.A., Tanaka, R., Hou, G., and Klein, M.T. (2008) *Fuel Process Technology*, **89**, 344–349.
19 Ancheyta-Juarez, J., Macias-Villafuerte, E., Schachat, P., Aguilar-Rodriguez, E., and Gonzales-Arredondo, E. (2002) *Chemical Engineering & Technology*, **25**, 541–546.
20 Smith, R.B. (1959) *Chemical Engineering Progress*, **55**, 76–80.
21 Krane, H.G., Groth, B.A., Schulman, L.B., and Sinfelt, H.J. (1959) *Fifth World Petroleum Congress Section III*, World Petroleum Council, London, New York, p. 39.

22 Henningsen, J. and Bundgaard-Nielson, M. (1970) *British Chemical Engineering*, **15**, 1433–1436.
23 Jenkins, H. and Stephens, T.W. (1980) *Hydrocarbon processing* November, 163 - 167.
24 Hu, S. and Zhu, X.X. (2004) *Chemical Engineering Communications*, **191**, 500–512.
25 Stijepovic, M.Z., Vojvodic-Ostojic, A., Milenkovic, I., and Linke, P. (2009) *Energy & Fuels*, **23**, 979–983.
26 Tailleur, R.G. and Davila, Y. (2008) *Energy & Fuels*, 2892–2901.
27 Ramage, M.P., Graziani, K.R., Schipper, P.H., Krambeck, F.J., and Choi, B.C. (1987) *Advances in Chemical Engineering*, **13**, 193–266.
28 Kmak, W.S. (1972) *AIChE National Meeting*, Houston, TX.
29 Klein, M.T. (2006) *Molecular Modeling in Heavy Hydrocarbon Conversions*, CRC Press, Boca Raton, FL.
30 Bommannan, D., Srivastava, R.D., and Saraf, D.N. (1989) *The Canadian Journal of Chemical Engineering*, **67**, 405–411.
31 Padmavathi, G. and Chaudhuri, K.K. (1997) *The Canadian Journal of Chemical Engineering*, **75**, 930–937.
32 Hou, W., Su, H., Hu, Y., and Chu, J. (2006) *Chinese Journal of Chemical Engineering*, **14**, 584–591.
33 Szczygiel, J. (2005) *Energy & Fuels*, **19**, 7–21.
34 Li, J., Tan, Y., and Liao, L. (2005) *IEEE Conference on Control Applications*, **2005**, 867–872.
35 Lee, J.W., Ko, Y.C., Lee, K.S., and Yoon, E.S. (1997) *Computers & Chemical Engineering*, **21**, S1105–S1110.
36 Hu, Y., Su, H. and Chu, J. (December 2003) Proceedings of the 42nd IEEE, pp. 6206-6211.
37 Stijepovic, M.Z., Linke, P., and Kijevcannin, M. (2010) *Energy & Fuels*, **24**, 1908–1916.
38 Ginestra, J.C. and Jackson, R. (1985) *Industrial and Engineering Chemistry Fundamentals*, **24**, 121–128.
39 Doyle, F.J. III, Jackson, R., and Ginestra, J.C. (1986) *Chemical Engineering Science*, **41**, 1485–1495.
40 Bhatia, S., Chandra, S., and Das, T. (1989) *Industrial & Engineering Chemistry Research*, **28**, 1185–1190.
41 Chirico, R.D. and Steele, W.V. (1997) *Journal of Chemical & Engineering Data*, **42**, 784–790.
42 Riazi, M.R. (2005) *Characterization and Properties of Petroleum Fractions*, 1st edn, American Society for Testing and Materials, West Conshohocken, PA.
43 Kister, H.Z. (1992) *Distillation Design*, McGraw-Hill, Inc., New York, NY.
44 Kaes, G.L. (2000) *Refinery Process Modeling. A Practical Guide to Steady State Modeling of Petroleum Processes*, The Athens Printing Company, Athens, GA.
45 Sanchez, S., Ancheyta, J., and McCaffrey, W.C. (2007) *Energy & Fuels*, **21**, 2955–2963.
46 Aitani, G.M. and Parera, J.M. (1995) *Catalytic Naphtha Reforming (Science and Technology)*, 1st edn, Marcel Dekker, New York.

47 Fernandes, J.L., Pinheiro, C.I.C., Oliviera, N.M.C., Inverno, J., and Ribiero, F.R. (2008) *Ind. Eng. Chem. Res.*, **47**, 850–866.
48 Van Trimpont, P.A., Marin, G.B., and Froment, G.F. (1986) *Industrial and Engineering Chemistry Fundamentals*, **25**, 544–553.
49 Van Trimpont, P.A., Marin, G.B., and Froment, G.F. (1988) *Industrial & Engineering Chemistry Research*, **27**, 51–57.
50 Garg, A. (June 1997) *Hydrocarbon processing*, pp. 97 – 105.
51 Vinayagam, K. (October 2007) *Hydrocarbon processing*, 95–104.
52 Bazaraa, M.S., Jarvis, J.J., and Sherali, H.D. (2009) *Linear Programming and Network Flows*, John Wiley and Sons, Hoboken, NJ.
53 Pashikanti, K. and Liu, Y.A. (2011) Predictive modeling of large-scale integrated refinery reaction and fractionation systems from plant data: 3. Continuous catalyst regeneration (CCR) reforming process. *Energy and Fuels*, **25**, 5320–5344.
54 Ayodele, B. and Cheng, C.K. (2015) Process modelling, thermodynamic analysis and optimization of dry reforming, partial oxidation and auto-thermal methane reforming for hydrogen and syngas production. *Chemical Product and Process Modeling*, **10**, 211–220.
55 Taghavi, B. and Fatemi, S. (2014) Modeling and application of response surface methodology in optimization of a commercial continuous catalytic reforming process. *Chemical Engineering Communications*, **201**, 171–190.
56 Wood, K.R., Liu, Y.A., and Yu, Y. (2018) *Design, Simulation and Optimization of Adsorptive and Chromatographic Separations: A Hands-on Approach*, Wiley-VCH, Weinheim, Germany.

6

Predictive Modeling of the Hydroprocessing Units

This chapter presents a workflow to develop, validate, and apply predictive models for rating and optimization of large-scale integrated hydrocracking (HCR) reaction and fractionation systems from plant data. In practice, a HCR process includes hydrotreating (HT) reactors for hydrodesulfurization (HDS) and hydrodenitrogenation (HDN) of the feedstock to HCR reactors. Thus, this chapter actually covers two important types of hydroprocessing operations: *hydrotreating* and *hydrocracking*.

Section 6.1 illustrates the typical HCR process and summarizes the previous work in the development of kinetic, reactor, and process models for HCR operations. Section 6.2 presents the features of the Aspen HYSYS Petroleum Refining HCR modeling tool and discusses in detail the kinetic lumping and reaction networks involved in the HCR model. Section 6.3 describes two commercial processes: a medium-pressure hydrocracking (MP HCR) unit with a feed capacity of 1 million tons per year and a high-pressure hydrocracking (HP HCR) unit with a feed capacity of 2 million tons per year in the Asia Pacific. The units include reactors, fractionators, and hydrogen recycle system. With catalyst and hydrogen, the process converts heavy feedstocks, such as vacuum gas oil (VGO), into valuable low-boiling products, such as gasoline and diesel.

Section 6.4 presents in detail the workflow of developing predictive models of integrated HCR reaction and fractionation systems. We present the detailed procedure for data acquisition to ensure accurate mass balances and for implementing the workflow using Excel spreadsheets and a commercial software tool, Aspen Petroleum Refining. Our procedure is equally applicable to other commercial software tools. The workflow includes special tools to facilitate an accurate transition from lumped kinetic components used in reactor modeling to the pseudocomponents based on boiling point ranges required in the rigorous stage-by-stage simulation of fractionators.

In Sections 6.5 and 6.6, we validate the MP HCR and HP HCR models with 2–3 months of plant data, and the resulting models accurately predict unit performance, product yields, and fuel properties from the corresponding operating conditions.

Section 6.7 illustrates applications of the validated plantwide model to quantify the effect of H_2-to-oil ratio on product distribution and catalyst life and the effect of HCR reactor temperature and feed flow rate on product distribution.

Petroleum Refinery Process Modeling: Integrated Optimization Tools and Applications,
First Edition. Y. A. Liu, Ai-Fu Chang, and Kiran Pashikanti.

The results agree well with experimental observations reported in the literature. Our resulting models only require typical operating conditions and routine analysis of feedstock and products and appear to be the only reported integrated HCR models that can quantitatively simulate all key aspects of reactor operation, fractionator performance, hydrogen consumption, product yield, and fuel properties.

Section 6.8 applies the developed model to generate the delta-base vector for production planning. Sections 6.9–6.12 present four hands-on workshops, including the development of a preliminary model for HCR reactor, calibration of the reactor model to match plant data, application of the calibrated model to case studies of the effects of key operating variables, and the fractionation system for a HCR reactor. This chapter concludes with the Nomenclature and Bibliography.

6.1 Introduction

Hydroprocessing units upgrade oil fractions with excess hydrogen and severe process conditions. HCR is one of the most significant hydroprocessing units in modern refinery. It is widely used to upgrade the heavy petroleum fraction, such as VGO. With catalyst and excess hydrogen, HCR converts heavy oil fractions, such as VGO, from crude distillation unit (CDU) into a broad range of valuable low-boiling products, such as gasoline and diesel. Figure 6.1 represents a typical process flow diagram of a single-stage HCR process with two reactors. The first

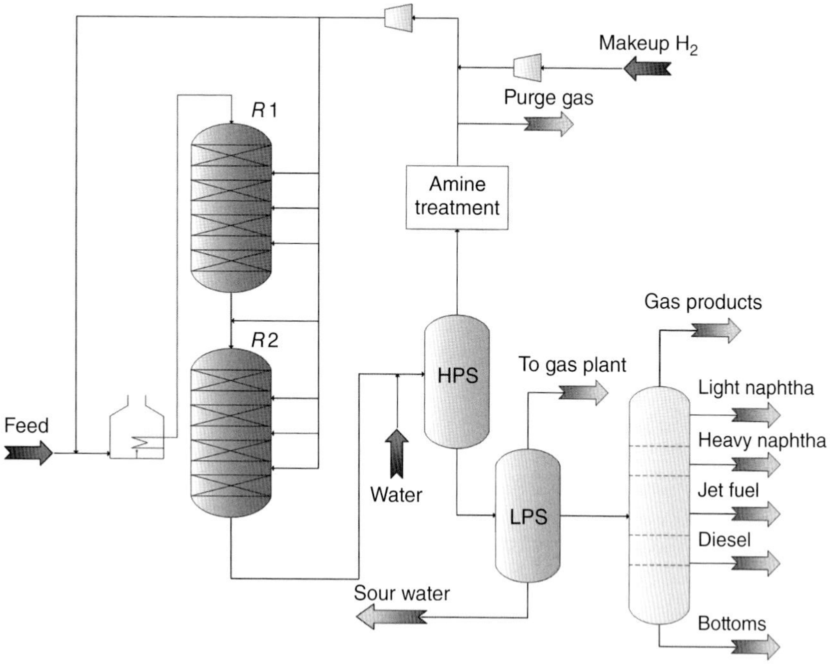

Figure 6.1 Flow diagram of a typical single-stage HCR process.

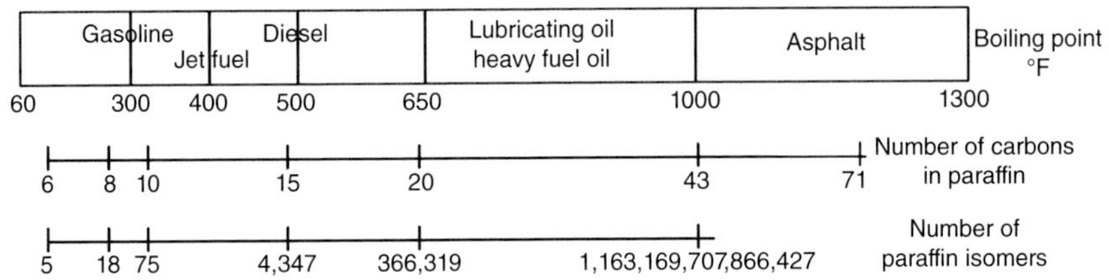

Figure 6.2 Complexity of petroleum oil. (Adapted from Aye and Zhang [1].)

reactor is usually loaded with HT catalyst to remove most of the nitrogen and sulfur compounds from feedstock. In addition, small extent of HCR also takes place in the first reactor. The effluent from the first reactor passes through the HCR catalyst loaded in the second reactor where most of the HCR is reached.

The petroleum fraction is a complex mixture that contains an enormous number of hydrocarbons. Figure 6.2 illustrates the compositional complexity of petroleum oil, displaying the number of paraffin isomers rapidly increasing with the boiling point (BP) and carbon number [1]. Therefore, it is difficult to identify the molecules involved in petroleum oil and study reaction kinetics of the HCR process based on the "real compositions" of the feed oil. To overcome this difficulty, refiners apply lumping technique to partition the hydrocarbons into multiple lumps (or model compounds) based on the molecular structure or/and BP and assume that the hydrocarbons of each lump have an identical reactivity to build the reaction kinetics of HCR. Since Qader and Hill [2] presented first kinetic model of HCR process by using two lumps, a large number of kinetic lumping models of HCR have appeared in the literature.

Figure 6.3 illustrates the scope of published HCR models classified according to a three-layer onion. The core of the onion is the kinetic model, focusing on the microkinetic analysis of reaction mechanisms. It allows for the study of the catalyst selection, feedstock effect, and influence of reaction conditions. The reactor model quantifies the reactor performance (e.g., product yields and fuel properties) under different operating conditions, such as flow rate, temperature profile, and hydrogen pressure. It helps the refiner determine the optimal unit operations. A process model aids in the optimization of plantwide operating conditions to maximize the profit, minimize the cost, and enhance the safety. However, there is little attention paid on developing a plantwide HCR process model in the modeling literature. On the other hand, lumping techniques of kinetic model, as the core of HCR modeling work, have been widely reported in the literature. Most of the modeling literature is concerned about developing detail kinetic lumping model to identify the reaction of the chemistry of HCR process. There are two major classes of lumping techniques: (1) lumping based on nonmolecular composition, and (2) lumping based on molecular composition.

Lumping based on molecular composition defines the kinetic lumps according to structural and reactive characterizations of hydrocarbon species and tracks interactions among a large number of kinetic lumps and reactions. It selects lumped components to characterize the feed oil, build the reaction network, and represent the product composition. In contrast, lumping based on nonmolecular composition considers molecules of different homologous families. For example,

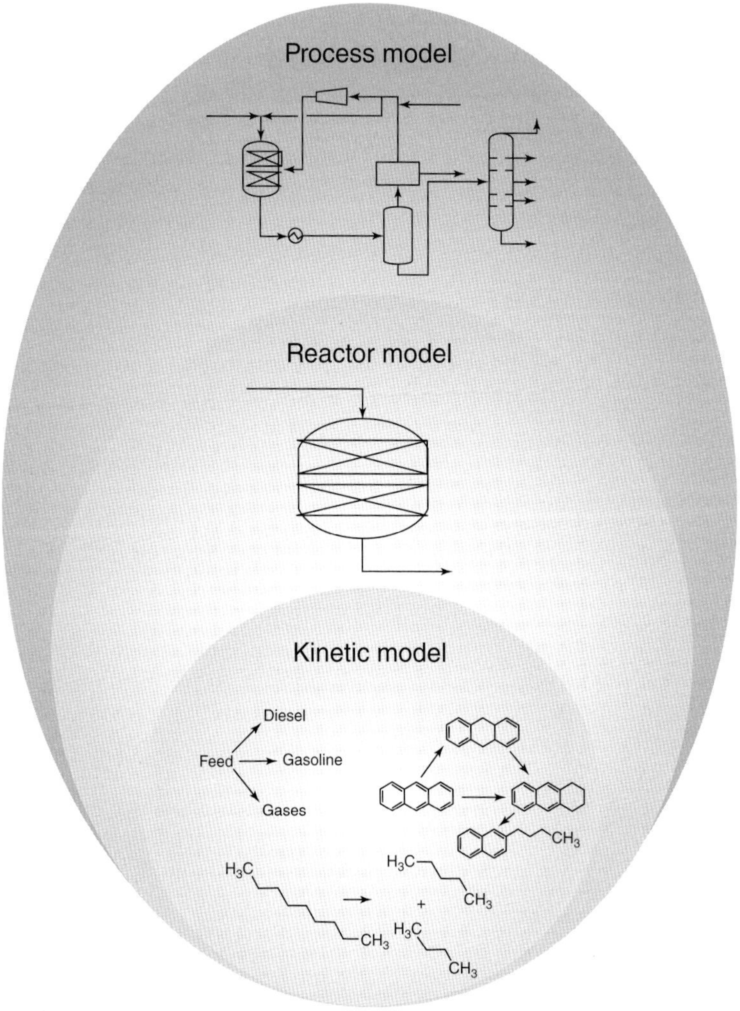

Figure 6.3 Three-layer onion for modeling scope.

a kinetic lump of the BP cut assumes the hydrocarbons within a certain boiling point range to have the same reactivity and cannot differentiate between the different hydrocarbon types in the same boiling point range. When applying a lumping scheme based on molecular composition, the feed oil composition has small or no effect on the resulting kinetic scheme and it allows for predictions of fuel qualities from molecular composition. The most well-known lumping techniques based on molecular composition are the structure-oriented lumping (SOL) [3–5] and the single-event model [6]. The SOL technique has been applied to plantwide process models such as HDS [7] and fluid catalytic cracking (FCC) units [8]. In addition, there is report of the single-event model of HCR kinetics of the oil fraction that includes as many as 1266 kinetic lumps [9]. The lumping based on molecular composition usually requires more computation time

Table 6.1 Key features of published HCR models built by lumping based on nonmolecular composition.

	Nature of the model					Model capability			
	Modeling scope	Lumping technique	Data source	Data requirement (feed)	Data requirement[b)] (product)	Reactor operation	Product yield	Column simulation	Fuel quality estimation
Qader and Hill [2]	Kinetic model	2 Lumps	Laboratory	None	Yield	N/A	Yes	N/A	N/A
Valavarasu *et al.* [12]	Kinetic model	4 Lumps	Laboratory	None	Yield	N/A	Yes	N/A	N/A
Sánchez *et al.* [13]	Kinetic model	5 Lumps	Pilot	None	Yield	N/A	Yes	N/A	N/A
Verstraete *et al.* [14]	Kinetic model	37 Lumps	Laboratory	TBP curve/ SARA analysis/ elemental analysis – C, H, S, N, O, Ni, V	Yield/TBP curve/SARA analysis/ elemental analysis – C, H, S, N, O, Ni, V	N/A	Yes	N/A	N/A
Stangeland [15]	Kinetic model	Discrete lumps[a)]	Pilot/ commercial	TBP curve	Yield/TBP curve	N/A	Yes	N/A	TBP curve
Mohanty *et al.* [16]	Reactor model	Discrete lumps	Commercial	TBP curve/ density distribution	Yield/TBP curve	Temperature profile/ hydrogen consumption	Yes	N/A	N/A

(Continued)

Table 6.1 (Continued)

	Nature of the model					Model capability			
	Modeling scope	**Lumping technique**	**Data source**	**Data requirement (feed)**	**Data requirement[b)] (product)**	**Reactor operation**	**Product yield**	**Column simulation**	**Fuel quality estimation**
Pacheco and Dassori [17]	Reactor model	Discrete lumps	Commercial	TBP curve/ density distribution	Yield/TBP curve	Temperature profile/ hydrogen consumption	Yes	N/A	N/A
Bhutani *et al.* [18]	Reactor model	Discrete lumps	Commercial	TBP curve/ density distribution	Yield/TBP curve	Temperature profile/ hydrogen consumption	Yes	N/A	N/A
Laxminarasimhan *et al.* [19]	Kinetic model	Continuous lumping[a)]	Pilot	TBP curve	Yield/TBP curve	N/A	Yes	N/A	N/A
Basak *et al.* [20]	Reactor model	Continuous lumping	Commercial	TBP curve/PNA distribution along with TBP curve	TBP curve/PNA distribution along with TBP curve	Temperature profile/hydrogen consumption	Yes	N/A	PNA composition of product
Fukuyama and Terai [21]	Kinetic model	Seven lumps	Laboratory	SARA analysis	Yield/SARA analysis	N/A	Yes	N/A	N/A

a) Discrete lump and continuous lump are defined by boiling points.
b) TBP = true boiling point; SARA = saturates, aromatics, resins, and asphaltenes; PNA = paraffins, naphthalene, and aromatics.

and makes it difficult to incorporate equipment simulations, such as reactor hydrodynamics. It also requires more data than what the routine chemical analysis in a refinery can provide. This limits its application to kinetics and catalyst studies and can rarely apply to a plantwide process model. In addition to the SOL and single-event model, however, there are other noncomplex lumping techniques based on molecular composition, such as the approach of the Aspen HYSYS Petroleum Refining hydrocracker model that we will discuss in Section 6.2. Table 6.1 summarizes the key features of well-known published HCR models based on nonmolecular composition lumping. For a review and comparison on HCR reactor models, please see the study by Ancheyta *et al.* [10], and for a review of kinetic modeling of large-scale reaction systems through lumping, please refer to the study by Ho [11].

The objective of this chapter is to develop, validate, and apply a methodology for the predictive process model of large-scale integrated refinery reaction and fractionation systems from plant data. In particular, we model two commercial HCR units in the Asia Pacific. MP HCR unit processes 1 million tons feedstock per year with a reactor pressure of 11.5–12.5 MPa, whereas HP HCR unit processes 2 million tons of feedstock per year with a reactor pressure of 14.5–15.0 MPa.

6.2 Aspen HYSYS Petroleum Refining HCR Modeling Tool

In this chapter, we use Aspen HYSYS Petroleum Refining HCR to model the HCR reactors and Aspen HYSYS to develop the rigorous plantwide simulation, including fractionation units.

Figure 6.4 represents the built-in process flow diagram of Aspen HYSYS Petroleum Refining HCR for a single-stage HCR process. It can simulate the feed heater, reactor, high-pressure separator (HPS), hydrogen recycle system, amine treatment (optional), and distillation column (optional). To ensure that the simulation agrees with the real process, users have to configure *the process type (single- or two-stage), number of reactors, number of reactor beds for each reactor,* and *the operation of each unit.* The model of amine treatment is a shortcut component splitter that separates H_2S from the vapor product of the HPS, and the simulation of the distillation column is a shortcut petroleum distillation column discussed in Section 2.15 [22]. In addition, ammonia (NH_3) produced by HDN reactions is split from reactor effluent before its entry into the HPS that is modeled by rigorous thermodynamics.

The reactor model of Aspen HYSYS Petroleum Refining HCR utilizes 97-lump reaction kinetics. The selection of 97 model compounds is based on the carbon number and structural characteristics and is consistent with previous publications [14, 23–26]. The 97 model compounds belong to six groups – light gases, paraffin, naphthene, aromatics, sulfur compound, and nitrogen compound. Furthermore, the sulfur compounds are separated into eight groups of 13 components: thiophene, sulfide, benzothiophene, naphthobenzothiophene, dibenzothiophene, tetrahydrobenzothiophene, tetrahydrodibenzothiophene, and tetrahydronaphthobenzothiophene [22].

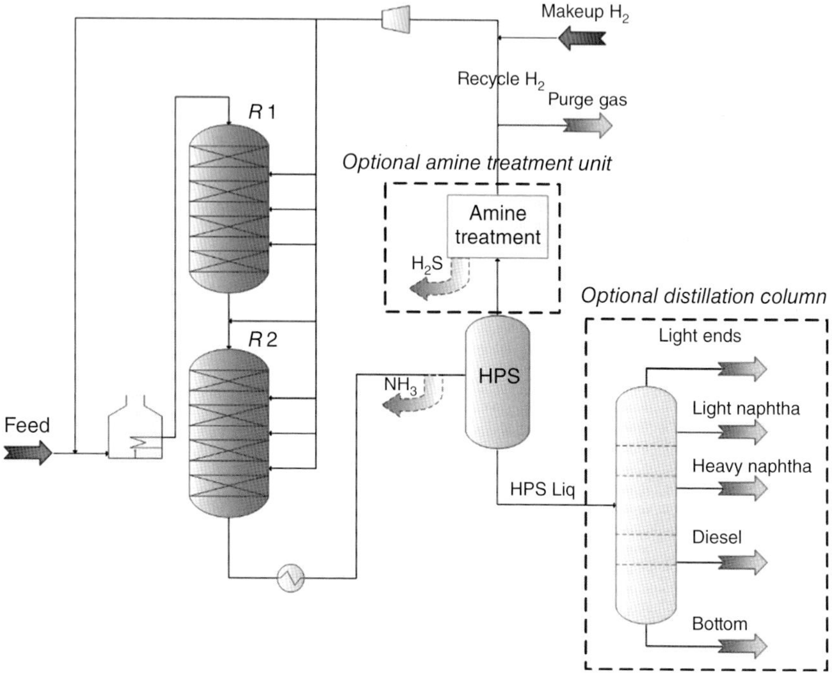

Figure 6.4 Built-in process flow diagram of Aspen HYSYS Petroleum Refining HCR.

In the literature, there are two approaches to develop the kinetic lumping compositions of the feedstock: *forward* and *backward approaches*. The forward approach requires detailed compositional and structural information by performing comprehensive analysis of the feedstock. However, refinery can seldom apply the forward approach because the routine analysis in the refinery does not include the required detailed structural analysis. This leads to the backward approach, which requires a reference library and only limited analytical data from routine measurements, such as density and sulfur content, to estimate kinetic lumping compositions. Brown *et al.* [27] reported a methodology estimating detailed compositional information for the SOL-based model, and Gomez-Prado *et al.* [28] developed molecular-type homologous series (MTHS) representation to characterize heavy petroleum fractions.

In Aspen HYSYS Petroleum Refining, *the forward approach* requires detailed compositional and structural information by performing comprehensive analysis of the feedstock, including API gravity, ASTM D-2887 distillation, refractive index, viscosity, bromine number, total sulfur, total and basic nitrogen, fluorescent indicator adsorption (FIA; total aromatics in vol%), nuclear magnetic resonance (NMR) (carbon in aromatic rings), ultraviolet (UV) method (wt% of mono-, di-, tri-, and tetra-aromatics), high-performance liquid chromatography (HPLC), and gas chromatography/mass spectrometry (GC/MS). With the detailed compositional and structural information, Aspen HYSYS Petroleum Refining quantifies the so-called "fingerprint" (molecular representation) of the feedstock based on 97 kinetic lumps [29]. On the other hand, *the backward approach* of Aspen HYSYS Petroleum Refining requires only the bulk properties (density, ASTM D-2887 distillation curve, and sulfur and nitrogen contents) of the feedstock. Aspen HYSYS Petroleum Refining contains a built-in fingerprint databank for various types of feedstock, such as light VGO, heavy VGO, FCC cycle oil. The backward approach assumes that the petroleum feedstock with the same fingerprint type maintains the same generic kinetic lump distribution as the initial composition. Aspen HYSYS Petroleum Refining uses a tool called "Feed Adjust" [29] to skew the kinetic lump distribution of the selected fingerprint type to minimize the difference between the measured and calculated bulk properties of the feedstock. We use the resulting kinetic lump distribution as the feed condition for the HCR model. If there is specific concern about compositional information, the user is able to customize the feed fingerprint to match the measurement. For example, the user can change sulfur lump distribution of selected feed fingerprint manually to ensure that the distribution of hindered and nonhindered sulfur compounds matches plant measurement.

The 97 lumps construct the reaction pathways of 177 reactions, including [30] (1) paraffin HCR; (2) ring opening; (3) dealkylation of aromatics, naphthenes, nitrogen lumps, and sulfur lumps; (4) saturation of aromatics, nonbasic nitrogen lumps, and hindered sulfur lumps; (5) HDS of unhindered sulfur lumps; and (6) HDN of nitrogen lumps. Figures 6.5–6.7 illustrate the reaction network.

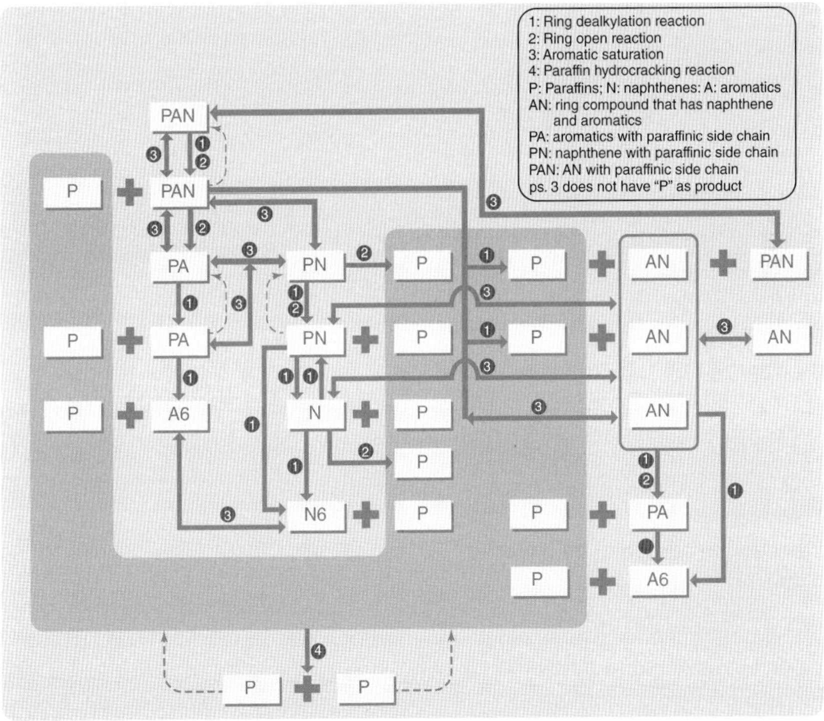

Figure 6.5 Reaction network of Aspen HYSYS Petroleum Refining HCR, paraffin HCR, ring open, ring dealkylation, and aromatic saturation.

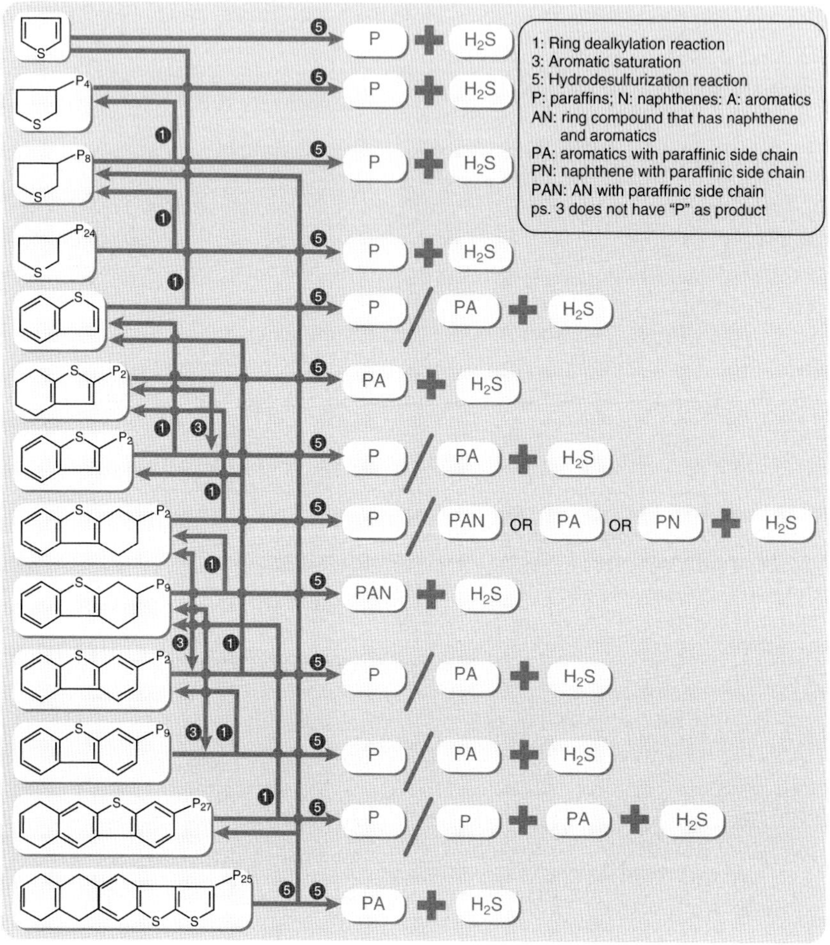

Figure 6.6 HDS reaction network of Aspen HYSYS Petroleum Refining HCR.

Rate equation of each reaction is based on Langmuir–Hinshelwood–Hougen–Watson (LHHW) mechanism with both reversible and irreversible reactions. The mechanism includes [30] the following:

- Adsorption of reactants to the catalyst surface
- Inhibition of adsorption
- Reaction of adsorbed molecules
- Desorption of products.

The kinetic scheme also includes the inhibition resulting from H_2S, NH_3, and organic nitrogen compounds [30]:

- Inhibition of HDS reactions by H_2S
- Inhibition of paraffin HCR, ring opening, and dealkylation reactions by NH_3 and organic nitrogen compounds.

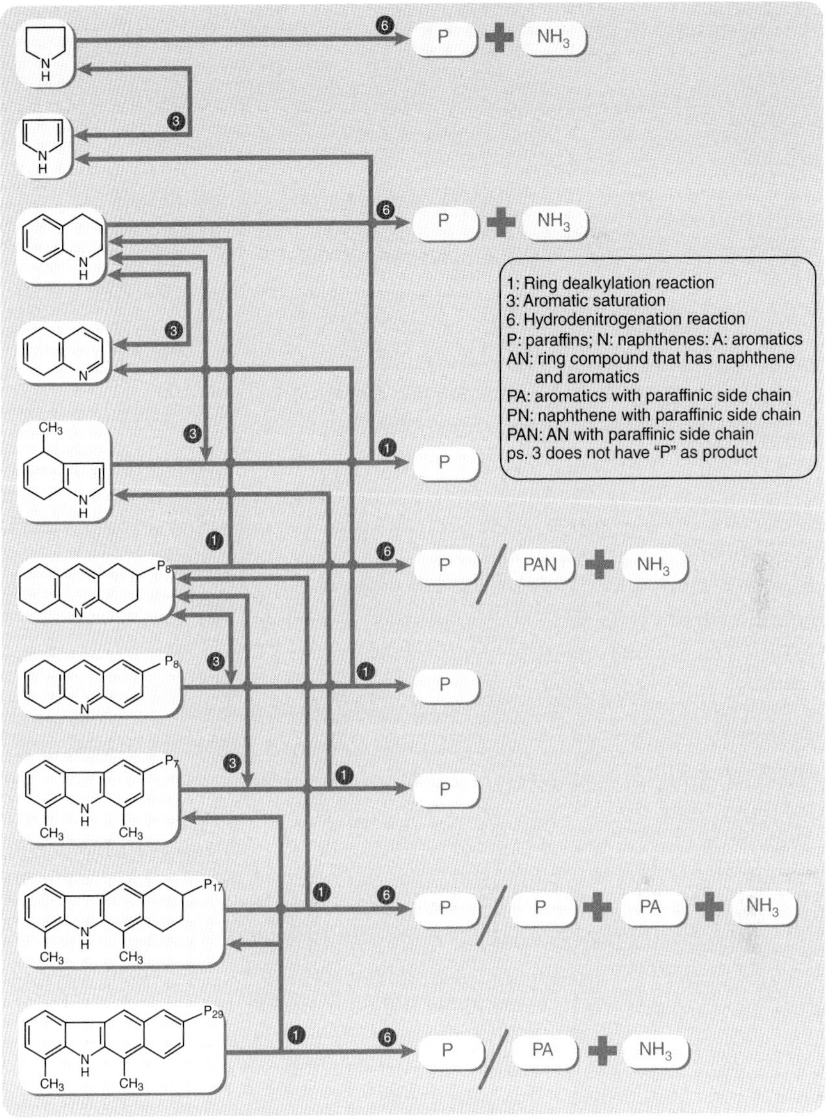

Figure 6.7 HDN reaction network of Aspen HYSYS Petroleum Refining HCR.

Equations (6.1) and (6.2) represent the LHHW-based rate equations for reversible and irreversible reactions, respectively [22].

$$\text{Rate} = K_{\text{total}} \times k \times \frac{((K_{\text{ADS},i}C_i \times K_{\text{ADS,H}_2}(P_{\text{H}_2})^x / K_{\text{eq}}) - K_{\text{ADS},j}C_j)}{\text{ADS}} \cdots \tag{6.1}$$

$$\text{Rate} = K_{\text{total}} \times k \times \frac{K_{\text{ADS},i}C_i \times K_{\text{ADS,H}_2}(P_{\text{H}_2})^x}{\text{ADS}} \tag{6.2}$$

where K_{total} is the overall activity and k is the intrinsic rate constant, which is assigned by fundamental research [22]. $K_{\text{ADS},i}$ and $K_{\text{ADS},j}$ are the adsorption

Table 6.2 Reaction types and the corresponding inhibitors.

Reaction type	Inhibitors
1. C–C scission (acid site reaction)[a]	NH_3, organic nitrogen compound and aromatic hydrocarbon
2. Aromatic saturation (metal site reaction)	Organic nitrogen compound, H_2S, and aromatic hydrocarbon
3. HDS (metal site reaction)	Organic nitrogen compound, H_2S, and aromatic hydrocarbon
4. HDN (metal site reaction)	Organic nitrogen compound, H_2S, and aromatic hydrocarbon

a) C–C scission includes HCR, ring open, and ring dealkylation reactions.

constants of hydrocarbons i and j, which are assigned by fundamental researches [22], C_i and C_j are the concentrations of hydrocarbons i and j, P_{H_2} is the partial pressure of hydrogen, and K_{eq} is the equilibrium constant of the reaction, which is assigned by fundamental researches [22]. ADS is the LHHW adsorption term, which represents competitive adsorption by different inhibitors including aromatic hydrocarbon, H_2S, NH_3, and organic nitrogen compound. Table 6.2 represents the inhibitors used for each reaction type in Aspen HYSYS Petroleum Refining.

In the rate expressions shown in Eqs. (6.1) and (6.2), K_{total} is the combination of a series of activity factors to represent apparent reaction rates of different reaction groups. For example, K_{total} of the hydrogenation reaction of a light aromatic hydrocarbon is the product of K_{global}, $K_{hdg,\ overall}$, and $K_{hdg,\ light}$. K_{global} is the global activity factor assigned to the each catalyst bed, $K_{hdg,\ overall}$ represents the group activity factor of all hydrogenation reactions, and $K_{hdg,\ light}$ indicates the activity factor of the hydrogenation reactions for the compounds belonging to light BP cut (below 430 °F). Section 6.4.4 includes more details about the idea of the reaction group and activity factors and gives the details of reaction activities in Aspen HYSYS Petroleum Refining. For reactor design and hydrodynamics, Aspen HYSYS Petroleum Refining HCR applies the design equations of ideal trickle bed and the hydrodynamics described by Satterfield [31], and each catalyst bed is modeled as a separate reactor.

6.3 Process Description

6.3.1 MP HCR Process

Figure 6.8 shows the process flow diagram of an MP HCR unit of a large-scale refinery in the Asia Pacific. The unit upgrades 1 million tons/year of VGO from the CDU into valuable naphtha, diesel, and bottom (the feedstock to ethylene plant) by HCR. The VGO feed from the CDU is mixed with a hydrogen-rich gas and preheated before entering the first reactor. The first reactor uses the HT catalyst to reduce nitrogen and sulfur contents. The second reactor uses the

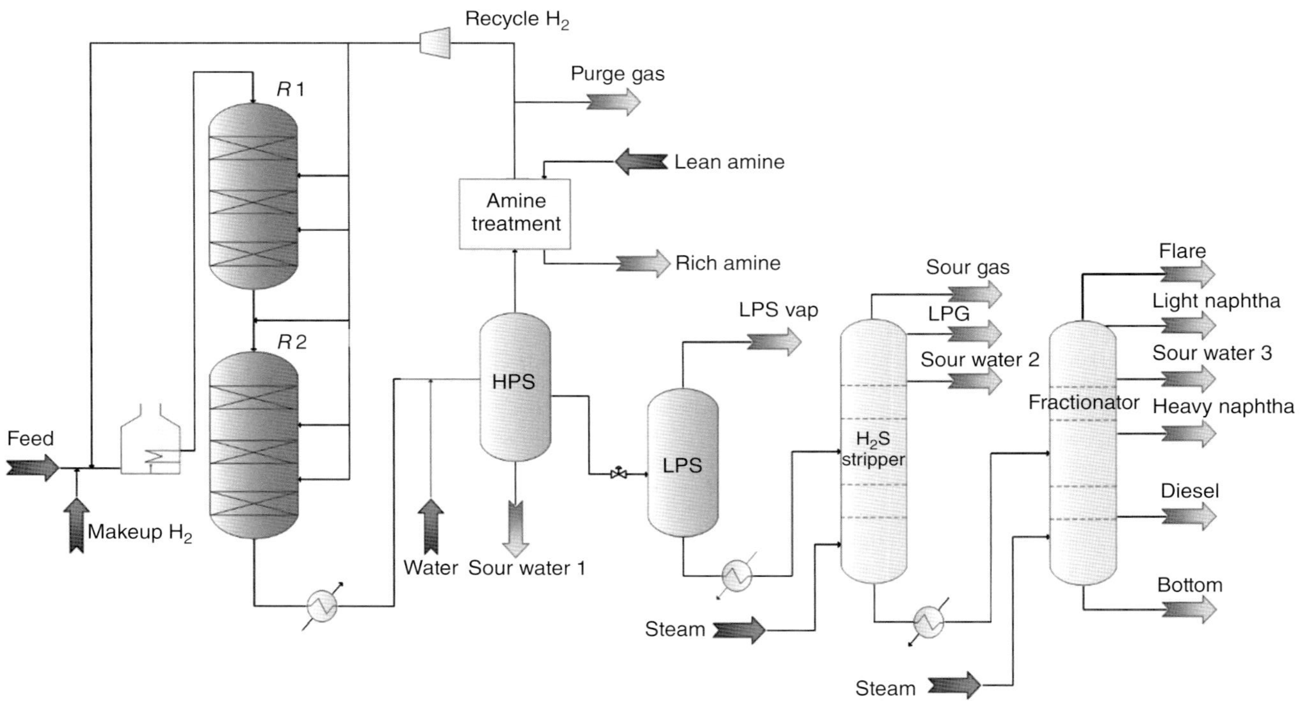

Figure 6.8 Simplified process flow diagram of the MP HCR unit.

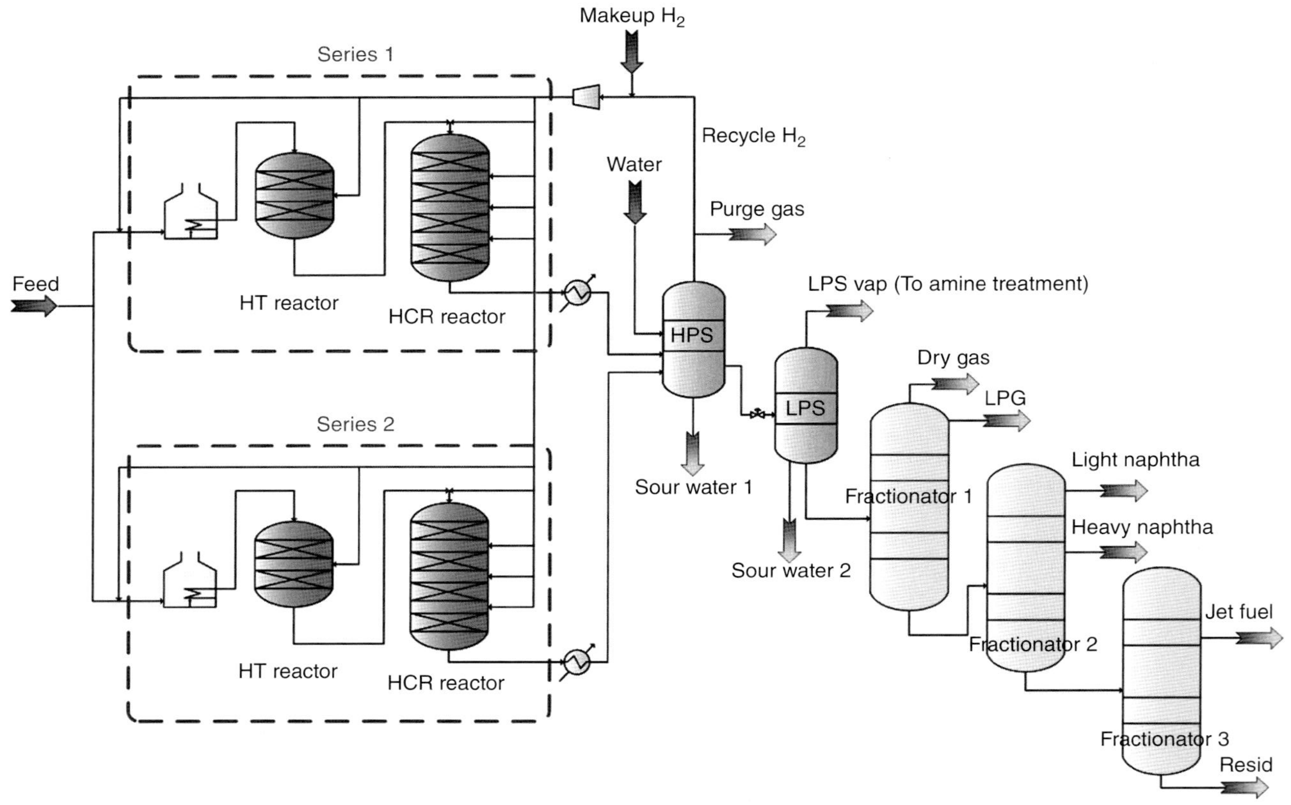

Figure 6.9 Simplified process flow diagram of the HP HPR unit.

HCR catalyst to crack heavy hydrocarbons into lighter oils: naphtha, diesel, and bottom. Following the two reactors, a HPS recovers unreacted hydrogen and a low-pressure separator (LPS) separates the light gases from the liquid outlet of HPS. An amine treatment scrubs sour gases from the vapor product of HPS to concentrate the hydrogen content of the hydrogen recycle stream. To balance the hydrogen in the system, a purge gas stream is removed from amine treatment. In the fractionation part, a H_2S stripper removes the dissolved H_2S from light hydrocarbons and a fractionator with two side strippers produces the major products: light naphtha, heavy naphtha, diesel, and bottom.

6.3.2 HP HCR Process

Figure 6.9 shows the process flow diagram of a HP HCR unit of a large-scale refinery in the Asia Pacific. The unit upgrades 2 million tons/year of VGO into valuable naphtha, jet fuel, and residue oil by HCR. Unlike typical HCR unit, this process includes two parallel reactor series and each series contains one HT reactor and HCR reactor. The VGO feed is mixed with a hydrogen-rich gas and preheated before being fed to the first reactors of both reactor series. The first reactors of both series are loaded with the HT catalyst to reduce nitrogen and sulfur contents. The second reactors of both series are loaded with HCR catalyst to crack heavy hydrocarbons into more valuable liquid products: liquefied petroleum gas (LPG), light naphtha, heavy naphtha, and jet fuel. Following the two reactor series, a HPS recovers unreacted hydrogen and an LPS separates the light gases from the liquid outlet of HPS. To balance the hydrogen in the system, a purge gas stream is removed from vapor product of HPS. In the fractionation part, the first fractionator separates light gases and LPG from light hydrocarbons, the second fractionator produces the most valuable products, namely, light naphtha and heavy naphtha, and the third fractionator further produces jet fuel and residue oil.

6.4 Model Development

6.4.1 Workflow of Developing an Integrated HCR Process Model

Figure 6.10 shows our workflow of developing an integrated HCR model using software tool Aspen HYSYS Petroleum Refining. We recommend that developing all HCR models should follow the same workflow, with only minor changes in the details of each block according to the selection of the kinetic model. For example, the different data requirement of feedstock analysis between wide distillation range lumping (distillation curve) and the SOL model [Fourier transform infrared (FTIR) spectroscopy, API gravity, distillation curve, viscosity, etc.] will make the procedure for data acquisition quite dissimilar. We discuss the details of each block when using Aspen HYSYS Petroleum Refining to build an integrated HCR process model.

The first step of model development is data acquisition, that is, to collect the required data for modeling and then to organize the gathered data and divide them into base and validation data sets. We use the base data set to develop the

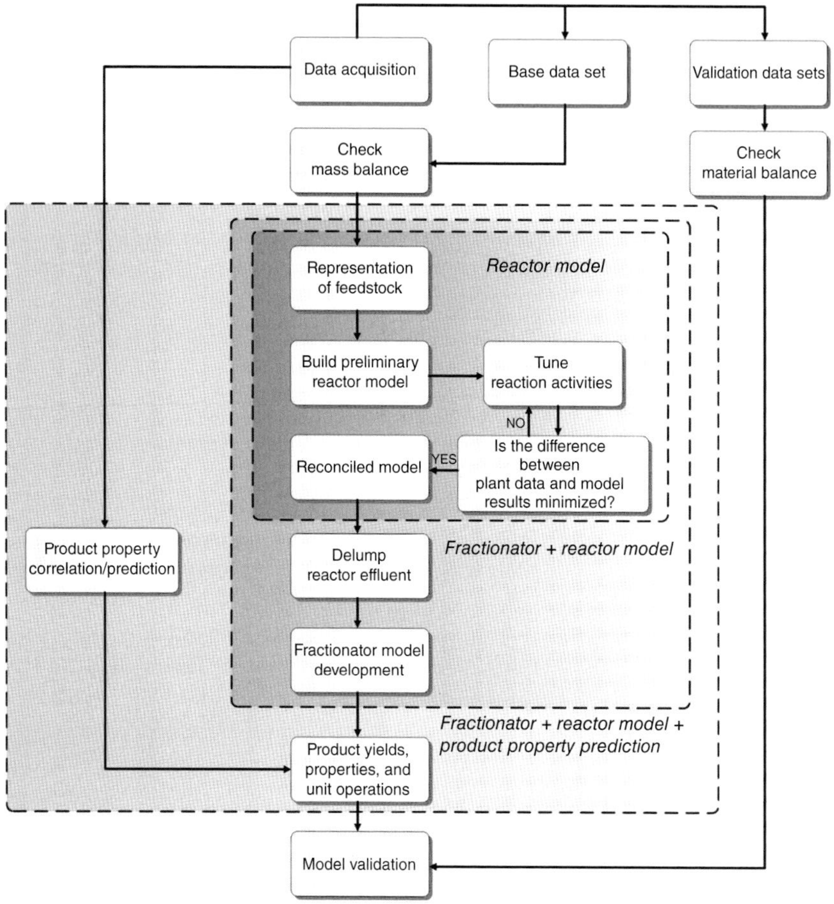

Figure 6.10 Workflow of building an integrated HCR process model.

process model and the validation data sets to test the prediction accuracy of the process model. Before developing the model, it is important to do an accurate mass balance, including the total fresh feed and product streams. If the total mass flow rates of inlets and outlets differ more than 2% or 3%, it is necessary to identify the cause of the imbalance [32]. Following the mass balance is the development of a reactor model. The steps to develop a reactor model also depend on the selection of the kinetic model. The procedures shown in Figure 6.10 correspond to the case using Aspen HYSYS Petroleum Refining. The development of a fractionator model in a HCR process is similar to a CDU. The only difference is the representation of the feed stream to the HCR fractionator. This follows because the HCR reactor effluent is characterized by kinetic lumps instead of the pseudocomponents based on the BP, which are widely used in a CDU model. Therefore, we use a step called *delumping* when the chosen kinetic lumps cannot appropriately characterize the feed stream to a HCR fractionator. Delumping is the most important step to build a plantwide model of the HCR process because it needs to capture the key properties of the reactor effluent for fractionator simulation during the component transition process. After completing the fractionator model, we incorporate the oil property correlations into the process model to calculate fuel properties, such as the flash point of diesel fuel. Lastly, we verify the model by comparing the predictions with multiple plant data sets.

6.4.2 Data Acquisition

Regardless of the selection of the kinetic model, data acquisition is always the first step of model development. We obtain 2 months of feedstock/product analysis, productions, and operation data from the plant and construct multiple data sets to build and validate the model. It is important to consult plant engineers about data consistency to ensure that each data set does not include the data in the period of operation upsets and significant operation changes. Moreover, it is always helpful to revisit the original data for the test run because we usually adjust the test run data to achieve reasonably accurate perfect mass and heat balances [32].

Data required for the modeling purpose are quite sensitive to the selection of the kinetic model and the modeling scope. This work only requires the operation and analysis data measured daily, and Table 6.3 lists the data requirement in this work. We collect the data from March 2009 to June 2009 and organize the data into eight complete data sets for the MP HCR process and 10 complete data sets for the HP HCR process. We only extract a small number of complete data sets from 4 months of plant data by considering the following: (1) each product stream has its own analysis period, and the analyses of all product streams performed on the same day are not available; (2) it is necessary to find out the date that includes most analysis data and fill up the missing data from adjacent days; and (3) some of the meters fail to record correct values during the period; and (4) some of the data sets fail in mass balance checking (see Section 6.4.3 for the procedure of mass balance calculation). Therefore, it is always useful to collect a long period (1–3 months) of data for modeling purposes, particularly for a commercial process. As it is common to have missing data or failed meters, we take the averages of data over a short period (1–3 days) (an industrial practice also recommended by Kaes [32]), or we make up the missing data by adjacent time periods to construct one complete data set for modeling.

6.4.3 Mass Balance

It is critical to review the collected information to ensure accurate model development, particularly mass balance. The calculation of mass balance should include all of the inlet streams (such as feed oil, makeup H_2, wash water, lean amine, and steam in the MP HCR process) and the outlet streams (such as LPS vapor, sour gas, LPG, flare, light naphtha, heavy naphtha, diesel, bottom, purge gas, sour water, and rich amine in the MP HCR process). However, the streams around amine treatment, wash water, and sour water streams are not routinely measured,

Table 6.3 Requirement of the HCR process model.

Reactor model

Flow rate

- Feed oil
- Makeup H_2
- Wash water
- All product streams, including purge gas and rich amine
- Recycle H_2 (before compressor)
- Hydrogen quench to each catalyst bed
- Lean amine

Pressure

- Feed oil
- Inlet and outlet of each catalyst bed
- Inlet and outlet of recycle H_2 compressor
- High-pressure separator
- Low-pressure separator

Temperature

- Feed oil
- Inlet and outlet of each catalyst bed
- Inlet and outlet of recycle H_2 compressor
- High-pressure separator
- Low-pressure separator

Laboratory analysis

- Feed oil (density, distillation curve, total sulfur, total nitrogen, and basic nitrogen)
- All gas products including purge gas (composition analysis)
- Composition analysis of light naphtha
- All liquid products from fractionator (density, distillation curve, and element analysis – C, H, S, N)
- Composition analysis of sour water
- Composition analysis of lean amine and rich amine
- Makeup H_2 (composition analysis)
- Recycle H_2 (composition analysis)
- Purge gas (composition analysis)
- Low-pressure separator gas (composition analysis)

Others

- Bed temperature at SOR (start of run) provided by catalyst vendor
- Bed temperature at EOR (end of run) provided by catalyst vendor

Fractionator model

Flow rate

- Steams
- All pumparound streams

Pressure

- Feed to the main column
- Steams
- Condenser of main column
- Top stage of main column
- Bottom stage of main column
- Feed stage of main column

Table 6.3 (Continued)

Temperature
- Feed to the main column
- Steams
- Inlet and outlet of pumparound
- Inlet and outlet of sides striper reboiler
- Condenser
- Top stage
- Bottom stage
- Feed stage
- Each stage with product draw
- Each stage with side draw
- Bottom stage of main column and side strippers

	A	B	C	D	E	F	G	H	I	J	K	L
2		**Feed Streams**										
3			Feed Oil	S in Feed Oil	N in Feed Oil	Make Up H_2						
4		kg/h	**92200**	**1844**	**83**	**4020**						
5		S wt%	2									
6		N wt%	0.09									
7			Total Feed	Total Sulfur Feec	Total Nitrogen Feed							
8		kg/h	**96220**	**1844**	**83**							
9												
10		**Product Streams**										
11			Purge Gas	LPS Vap	Sour Gas	LPG	Light Naphtha	Heavy Naphtha	Diesel	Bottom		
12		kg/h	**1660**	**1130**	**1740**	**3940**	**3480**	**20900**	**32670**	**29170**		
13		H_2S wt%	0	7.95	11.81	2.03						
14		H_2S kg/h	0	90	205	80						
15		S wt%					0	0.0002333	0.00198	0.0015		
16		S kg/h					**0**	**0.05**	**0.65**	**0.43**		
17		N wt%					0	0	0	0		
18		N kg/h					**0**	**0**	**0**	**0**		
19			Sweet PG	Sweet LV	Sweet SG	Sweet LPG					H_2S	NH_3
20		kg/h	**1660**	**1040**	**1535**	**3860**					**1958**	**101**
21			Sweet Gas products + Liquid Products	Total H_2S	Total NH_3							
22		kg/h	**94315**	**1958**	**101**							
23												
24		**Material Balance Deviation**										
25		**0.16%**										

Figure 6.11 Spreadsheet for the mass balance calculation of a HCR process, *Mass Balance.xls*.

and it is unlikely to include those streams in the calculation of material balance. As those streams only affect the mass balance of sulfur and nitrogen, we recommend doing a separate mass balance of sulfur and nitrogen by assuming that all of the removed sulfur and nitrogen atoms are reacted into H_2S and NH_3.

We calculate the mass balance as follows: (1) calculate the H_2S and NH_3 production by the severity of HDS and HDN reactions; (2) determine the production rates of "sweet" gas products and "sweet" LPG, which means subtracting any reported H_2S and NH_3 from all gas products and LPG; (3) sum up "sweet" gas products, "sweet" LPG, all liquid products, H_2S production, and NH_3 production to determine the total production rate of the reactor effluent; (4) sum up the flow rates of feed oil and makeup H_2 to obtain total feed rate to the reactor; and (5) calculate the ratio between the total production rate of the reactor effluent and the total feed rate.

Figure 6.11 illustrates an Excel spreadsheet (*Mass Balance.xls* available in the supplement) that we develop to do the mass balance calculations. Although we have developed the spreadsheet and the formulas for a specific HCR process, the reader can generalize the steps described earlier and apply the spreadsheet to do a mass balance of any HCR process with only minor changes.

6.4.4 Reactor Model Development

Reactor model development is the core of building a HCR process model. Although the procedure of building a reactor model depends on the selection of the kinetic model, we require the following tasks in developing a model for most commercial HCR processes: (1) do the feedstock analysis based on the selected kinetic model; (2) represent the feedstock as a mixture of kinetic lumps, which can be modeling compounds or pseudocomponents based on boiling point ranges; (3) build the reaction network, define rate equations, and estimate rate constants and heat of reaction; (4) apply the operation data (e.g., reactor temperature and feed rate) to solve rate equations and reactor design equations simultaneously; and (5) minimize objective functions (user-defined indices to represent the differences between model predictions and plant data) by tuning reaction activity parameters.

6.4.4.1 MP HCR Reactor Model

We describe in Section 6.2 the concept of the backward approach in representing the feedstock using the Aspen HYSYS Petroleum Refining. As the refinery does not conduct comprehensive analysis of HCR feedstock routinely, this work applies the backward approach to characterize the feedstock. We select the "LV-GO" fingerprint type for both HCR processes because the feeds to both processes are mainly VGO from CDU, and the selected fingerprint type should be as close to the real feeds as possible. This section demonstrates the last step of building the reactor model using Aspen HYSYS Petroleum Refining to minimize the difference between model predictions and plant data to make the model match plant operation.

Although Aspen HYSYS Petroleum Refining assigns the rate constants to the 177 reactions based on fundamental research, it is necessary to identify the activity factor to match plant operation because the reactor configuration, catalyst activity, and operating conditions vary for different refineries. The procedure of minimizing the difference between model predictions and plant data in Aspen HYSYS Petroleum Refining is called "*calibration*," which means to adjust the model parameters in order for the model predictions to agree with plant data.

Table 6.4 lists the 31 optional objective functions, and Table 6.5 shows the 48 reaction activity factors for selection. Aspen HYSYS Petroleum Refining combines the plant product distribution we enter to construct the reactor effluent. It also partitions the reactor effluent into C1, C2, C3, C4, C5, and four "square cuts," namely, naphtha (C6 to 430 °F cut), diesel (430–700 °F cut), bottom (700–1000 °F) cut, and residue (1000 °F+ cut), which are listed in Table 6.4. All of the objective functions listed in Table 6.4 are either the prediction errors of crucial operations or the important product yields for the HCR process. Aspen HYSYS Petroleum Refining allows us to select the desired objective functions during calibration. After selecting the objective functions, we choose appropriate activity factors to calibrate the reactor model. Figure 6.12 illustrates the relationships among the activity factor, catalyst bed, and reactor

Table 6.4 Objective functions in Aspen HYSYS Petroleum Refining.

	Note	Notation in this work
The predicting error of temperature rise of catalyst bed	One for each catalyst bed	OBJ_{TR_i} $i = 1–6$
The predicting error of hydrogen quench of catalyst bed	One for each catalyst bed	OBJ_{HQ_i} $i = 1–6$
The predicting error of flow rate of purge gas		OBJ_{PGF}
The predicting error of flow rate of makeup H_2		OBJ_{MHF}
The predicting error of chemical H_2 consumption		OBJ_{HC}
The predicting error of C6 to 430 °F cut (naphtha) volume flow		OBJ_{NVF}
The predicting error of 430–700 °F cut (diesel) volume flow		OBJ_{DVF}
The predicting error of 700–1000 °F cut (bottom) volume flow		OBJ_{BVF}
The predicting error of 1000 °F+ cut (resid) volume flow		OBJ_{RVF}
The predicting error of C6 to 430 °F cut (naphtha) mass flow		OBJ_{NMF}
The predicting error of 430–700 °F cut (diesel) mass flow		OBJ_{DMF}
The predicting error of 700–1000 °F cut (bottom) mass flow		OBJ_{BMF}
The predicting error of 1000 °F+ cut (resid) mass flow		OBJ_{RMF}
The predicting error of C1–C2 mass yield		OBJ_{C1C2}
The predicting error of C3 mass yield		OBJ_{C3}
The predicting error of C4 mass yield		OBJ_{C4}
The predicting error of sulfur content of 430–700 °F cut		OBJ_{SD}
The predicting error of sulfur content of 700–1000 °F cut		OBJ_{SB}
The predicting error of nitrogen content of 430–700 °F cut		OBJ_{ND}
The predicting error of nitrogen content of 700–1000 °F cut		OBJ_{NB}
The predicting error of nitrogen content in reactor 1 effluent		OBJ_{NR1}

type, and Table 6.5 shows the major effect of each activity factor on the model performance, such as global activity (K_{global}) on the bed temperature profile, to help the selection of activity factors.

The procedure of model calibration depends on the operational mode, product yields, and precision of plant data. For example, a hydrogen-insufficient refinery might pay more attention to hydrogen consumption and makeup hydrogen flow. In addition, it is necessary to have high precision of light end analysis (C1–C5) if we desire to have accurate predictions of light gas yields. For MPHCR process, the most important considerations to the plant management are the product yields, flow rate of makeup hydrogen, reactor temperature, and properties of liquid fuel products. We note that the reactor model cannot calculate some fuel properties, such as flash point and freezing point of diesel and jet fuel, because the square cuts defined by Aspen HYSYS Petroleum Refining have distillation ranges that are different from those of plant cuts. Therefore, we develop correlations to estimate such fuel properties (see Section 6.4.6).

Table 6.5 Reaction activity factors in Aspen HYSYS Petroleum Refining.

Notation in this work	Description	Major observation	Number of activity factors	Note
K_{global_i} $i = 1–6$	Global activity for each catalyst bed	Bed temperature profile	6[a)]	6 global activity factors for 6 catalyst beds
$K_{sul\_i\_j}$ i = HT, HCR j = O, L, M, H	HDS activity	Sulfur content	8	1 factor for overall HDS activity of hydrotreating beds 3 factors for 3 wide boiling point cuts[b)] of hydrotreating beds 1 factor for overall HDS activity of HCR beds 3 factors for 3 wide boiling point cuts of HCR beds
$K_{nit\_i\_j}$ i = HT, HCR j = O, L, H	HDN activity	Nitrogen content	6	1 factor for overall HDN activity of hydrotreating beds 2 factors for 2 wide boiling point cuts of hydrotreating beds 1 factor for overall HDN activity of HCR beds 2 factors for 2 wide boiling point cuts of HCR beds
$K_{crc\_i\_j}$ i = HT, HCR j = O, L, M, H	Activity of HCR and ring dealkylation	Product yield	8	1 factor for overall HCR activity of hydrotreating beds 3 factors for 3 wide boiling point cuts of hydrotreating beds 1 factor for overall HCR activity of HCR beds 3 factors for 3 wide boiling point cuts of HCR beds
$K_{hdg\_i\_j}$ i = HT, HCR j = O, L, M, H	Activity of hydrogenation (HDG, saturation of aromatic rings)	Hydrogen consumption/ reactor temperature	8	1 factor for overall HDG activity of hydrotreating beds 3 factors for 3 wide boiling point cuts of hydrotreating beds 1 factor for overall HDG activity of HCR beds 3 factors for 3 wide boiling point cuts of HCR beds

Table 6.5 (Continued)

Notation in this work	Description	Major observation	Number of activity factors	Note
$K_{ro\_i\_j}$ i = HT, HCR j = O, L, M, H	Activity of ring opening (RO)	Paraffin: naphthene ratio	8	1 factor for overall RO activity of hydrotreating beds 3 factors for 3 wide boiling point cuts of hydrotreating beds 1 factor for overall RO activity of HCR beds 3 factors for 3 wide boiling point cuts of HCR beds
K_{light_i} i = 1, 2, 3, 4	Light gas tuning factor	Distribute C1–C4	4	1 factor for each light gas (C1–C4)

a) Number of global activity factors depends on the number of catalyst beds.
b) The three wide boiling point cuts used to define activity factors are 430 °F– (L), 430–950 °F (M), and 950 °F+ (H).

Figure 6.13 illustrates the steps to identify activity factors in this work, which are divided into two phases. The first phase is applicable to any Aspen HYSYS Petroleum Refining HCR model and the second phase depends on the modeling priority of the refinery. As Aspen HYSYS Petroleum Refining assigns small values to K_{global} to ensure the initial convergence, all catalyst beds' performance is almost "dead" initially, which means that the reaction conversion is small. Thus, the first task is to tune the global activity factor of each catalyst bed to "activate" the reactors. After the reactors are activated, the reaction conversion must increase to some extent and we tune the cracking activity factors to minimize the difference between predicted and actual liquid product yields.

Due to heat effects of the reactions, the calculated reactor temperature profiles from previous steps would show deviations from actual plant data. We tune the global activity factors again to ensure that the deviations of reactor temperature predictions are within tolerance. We repeat the calibration of "reactor temperature profiles" and "mass yields of liquid products" several times until the errors of model predictions are within the acceptable tolerance. These back-and-forth procedures compose the first phase shown in Figure 6.13. This figure represents a generalized guideline of initial calibration for the Aspen HYSYS Petroleum Refining HCR model because reactor temperature profiles and major liquid product yields are always crucial considerations for any hydrocracker.

The second phase of Figure 6.13 shows the calibration procedure to reconcile the predictions of the reactor model to agree with the modeling priority of the refinery about process operations and productions. In this case, the flow rate of makeup hydrogen, volume yields of liquid products (crucial to density calculation), and light gas yields are important to the MP HCR process. Due to the lack

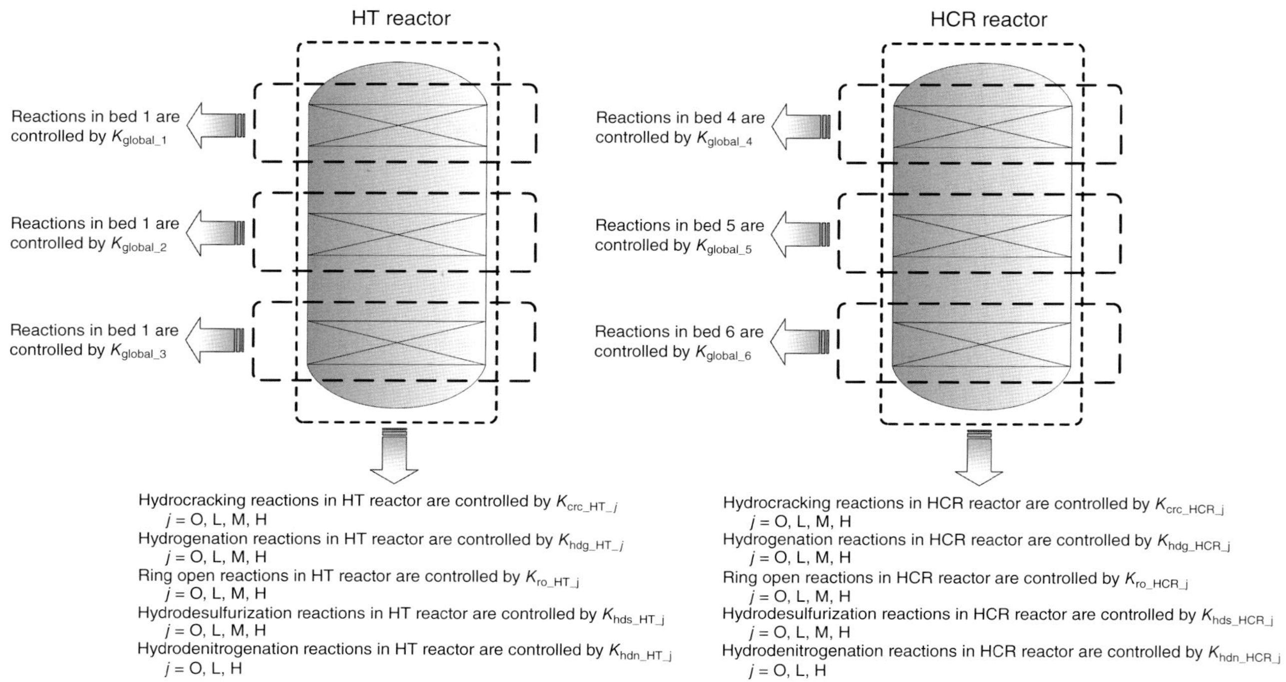

Figure 6.12 Relationships among activity factor, catalyst bed, and reactor type for HT and HCR.

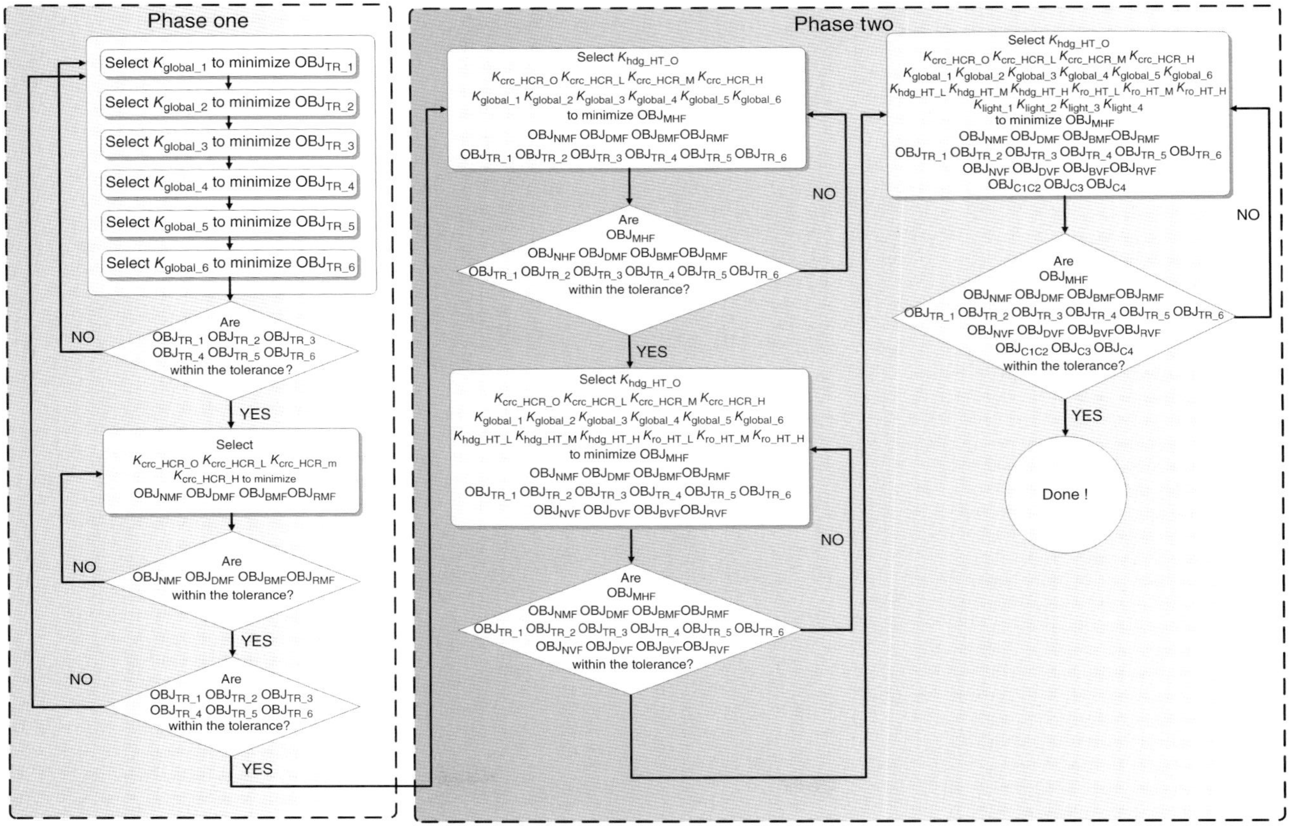

Figure 6.13 Procedure of model calibration.

of analysis data of nitrogen and sulfur contents of liquid product streams, the calibration procedure of this case (see Figure 6.13) does not include reconciliation of HDN and HDS activities.

Although the steps involved in second phase depend on the modeling priority of the refinery management, we can give some common guidelines. (1) Always check reactor temperature profiles and mass yields of liquid products. (2) By our experience, the overall model performance is most sensitive to K_{global} and least sensitive to K_{light}. The following list is in the order of decreasing sensitivity: K_{global}, K_{crc}, K_{hdg}, K_{hds}, K_{hdn}, K_{ro}, and K_{light}. (3) K_{global} has the most significant effect on all objective functions. (4) K_{crc} has a significant effect on the product yield, reactor temperature profile, hydrogen consumption, and flow rate of makeup hydrogen. (5) K_{hdg} affects the product yield, reactor temperature profile, hydrogen consumption, and flow rate of makeup hydrogen. (6) K_{hds} has a notable effect on the sulfur content, some effect on the hydrogen consumption and flow rate of makeup hydrogen, and a small effect on the product yield. (7) K_{hds} has a significant effect on the nitrogen content. (8) K_{light} only affects the distribution ratio between light gases. (9) Tuning K_{light} to distribute light gases (C1–C4) last because the total yields of light gases are determined by cracking reactions. K_{light} only redistributes the light gases and has little effect on the overall model performance.

The goal of model calibration is to seek an optimal solution for the reactor model to match real operation and there is no single optimum solution. It is important to assign reasonable tolerances to the objective functions and lose some of them when necessary.

6.4.4.2 HP HCR Reactor Model

We describe the generalized step-by-step instruction of the reactor model development for our MP HCR process in Section 6.4.4.1. However, the procedures are not applicable to the process with an unusual process flow diagram, such as the HP HCR process, which includes two parallel reactor series. The two parallel reactor series that share one fractionation unit make it unachievable to distinguish the production data from one series to the other. For example, there is no way to split the heavy naphtha product into two streams to represent the performance of each reactor series. In addition, it is difficult to start with building the model of two parallel reactor series because model reconciliation of two reactor series is a time-consuming and difficult task. Therefore, we develop the following procedures to build and reconcile HP HCR reactor model.

1) Construct an equivalent reactor to represent the two parallel reactor series.
2) Build and reconcile the equivalent reactor model.
3) Construct the preliminary models of the real process (two parallel reactor series).
4) Apply the reaction activities obtained from equivalent reactor model into the reactor model of two parallel reactor series.
5) Fine-tune the model of two parallel reactor series to match real operations and productions.

Equivalent Reactor This section demonstrates the concept of equivalent reactor. Considering a system with two parallel isothermal plug flow reactors (PFRs), where a first-order liquid-phase reaction takes place (see Figure 6.14), the relationship between conversion and residence time of each PFR is

$$\mathrm{CONV}_1 = 1 - \mathrm{Exp}(-k\tau_1) \tag{6.3}$$

$$\mathrm{CONV}_2 = 1 - \mathrm{Exp}(-k\tau_2) \tag{6.4}$$

where CONV is conversion, τ is the residence time, and k is the rate constant [33]. We define an equivalent reactor as a reactor that can convert the same amount of total feed flow into the same amount of total product. For the equivalent reactor,

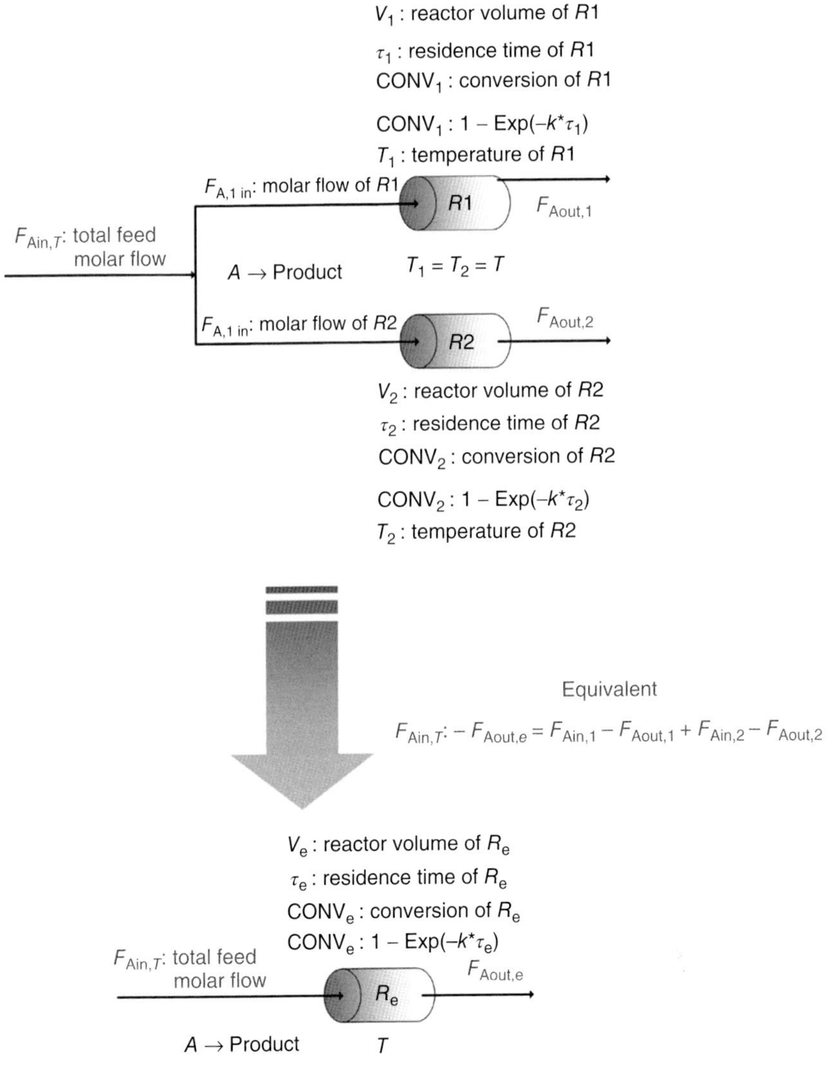

Figure 6.14 Concept of the equivalent reactor.

we represent the reaction conversion as

$$\mathrm{CONV}_e = 1 - \mathrm{Exp}(-k\tau_e) \tag{6.5}$$

As the equivalent reactor is defined by having an identical total production to the two parallel isothermal PFRs, we can obtain the following equation:

$$F_{\mathrm{Ain},T} - F_{\mathrm{Aout},e} = F_{\mathrm{Ain},1} - F_{\mathrm{Aout},1} + F_{\mathrm{Ain},2} - F_{\mathrm{Aout},2} \tag{6.6}$$

Substitute the relationship between the molar flow rate and conversion.

$$\mathrm{CONV}_e = \frac{F_{\mathrm{Ain},1} - F_{\mathrm{Aout},1} + F_{\mathrm{Ain},2} - F_{\mathrm{Aout},2}}{F_{\mathrm{Ain},T}} \tag{6.7}$$

Let $\theta_1 = F_{\mathrm{Ain},1}/F_{\mathrm{Ain},T}$ and $\theta_2 = F_{\mathrm{Ain},2}/F_{\mathrm{Ain},T}$ and we have

$$\mathrm{CONV}_e = \theta_1 \times \mathrm{CONV}_1 + \theta_2 \times \mathrm{CONV}_2 \tag{6.8}$$

Substituting Eqs. (6.3)–(6.5) into Eq. (6.8),

$$1 - \mathrm{Exp}(-k\tau_e) = \theta_1 \times [1 - \mathrm{Exp}(-k\tau_1)] + \theta_2 \times [1 - \mathrm{Exp}(-k\tau_2)] \tag{6.9}$$

After organizing,

$$\tau_e = \frac{-\ln(\theta_1 \times \mathrm{Exp}(-k\tau_1) + \theta_2 \times \mathrm{Exp}(-k\tau_2))}{k} \tag{6.10}$$

We can rewrite Eq. (6.10) into Eq. (6.11) in terms of space velocity (SV).

$$\mathrm{SV} = \frac{k}{-\ln(\theta_1 \times \mathrm{Exp}(-k\tau_1) + \theta_2 \times \mathrm{Exp}(-k\tau_2))} \tag{6.11}$$

With molar flow rate, conversion, and SV, we can calculate reactor volume to conduct reactor design. The idea of equivalent reactor provides us a convenient way to understand the performance of a complex reactor system, namely, two parallel PFRs.

Reconciliation of HP HCR Reactor Model As mentioned in the beginning of Section 6.4.4.2, there are five steps to build and reconcile the reactor model of the HP HCR process. We first build an equivalent reactor model to represent the two parallel reactor series. By doing so, we can obtain good initial values of reaction activities to further model the real process. However, the difficulty of building an equivalent reactor model is to assign the process variables by Eq. (6.11) because SV is a function of the rate constant. Qader and Hill [2] presented a two-lump kinetic model of the HCR process that characterizes the feedstock and product as a single lump (see Figure 6.15) and apply first-order kinetics to obtain rate constants under different operating conditions. Equation (6.12) represents the rate equation, and they apply an Arrhenius equation to correlate experimental data to obtain the preexponential term and activation energy. Equation (6.13) shows the temperature dependence of the rate constant.

$$-\frac{\mathrm{d[Gas\ oil]}}{\mathrm{d}t} = k_{\mathrm{GO}} \times [\mathrm{Gas\ oil}] \tag{6.12}$$

$$k_{\mathrm{GO}}(\mathrm{h}^{-1}) = 1 \times 10^7(\mathrm{h}^{-1}) \times \mathrm{Exp}\left[\frac{-21100(\mathrm{cal/mol})}{RT}\right] \tag{6.13}$$

In the equations, k_{GO} is rate constant of the gas oil HCR reaction. The experimental data were obtained at 10.34 MPa pressure, 400–500 °C, 0.5–3.0 h^{-1} SV, and a constant H_2:oil ratio of 500 standard (STD) m^3/m^3. As they conduct experiment under similar condition as the industrial reactor, it is practical to use kinetic data by Qader and Hill [2] to investigate the design of the equivalent reactor model. We apply feed flow rates, reactor volumes, and space velocities from the HP HCR process and calculate the reactor volume of the equivalent reactor under different rate constants.

Figure 6.15 Two-lump scheme developed by Qader and Hill. (Adapted from Qader and Hill [2].)

Figure 6.16 illustrates how the HCR rate constant affects equivalent reactor volume. The *y* axis represents the ratio of the equivalent reactor volume to the sum of reactor volumes of the two parallel HCR reactors ($V_e/(V_1 + V_2)$). As *k* approaches zero, the upper limit of 100% is also achieved. This reflects the physical limitation when no reaction takes place. On the other hand, the value of $V_e/(V_1 + V_2)$ drops while *k* increases. Under industrial operating conditions, the *k* value ranges from 0.5 to 3 h^{-1} (corresponding reactor temperatures are 360–430 °C) according to the kinetic data by Qader and Hill [2]. Therefore, typical values of $V_e/(V_1 + V_2)$ should always be greater than 90%.

As we build the equivalent reactor model merely for obtaining initial values of reaction activities, we will use the sum of the catalyst loading of the real process to construct the equivalent reactor. We also sum up all of the material streams, namely feed flows and hydrogen quenches, to ensure mass balance of the equivalent reactor. In addition, we apply the arithmetic averages of operating conditions, such as reactor temperatures, for the development of equivalent reactor (see Figure 6.17 for details). During the course of model reconciliation of the equivalent reactor model, we take the reactor temperature profile, flow rate of makeup hydrogen, mass and volume yields of liquid products, and light gas yields as objective functions because they are the major concerns of the HP HCR refiners. The objective functions of the HP HCR process are the same as the MP HCR

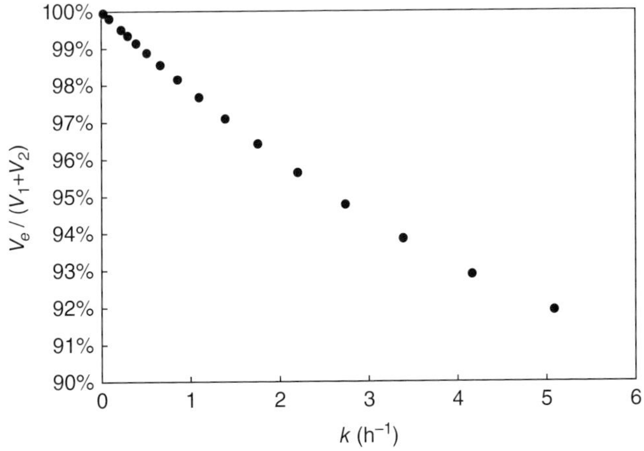

Figure 6.16 HCR rate constant versus equivalent reactor volume.

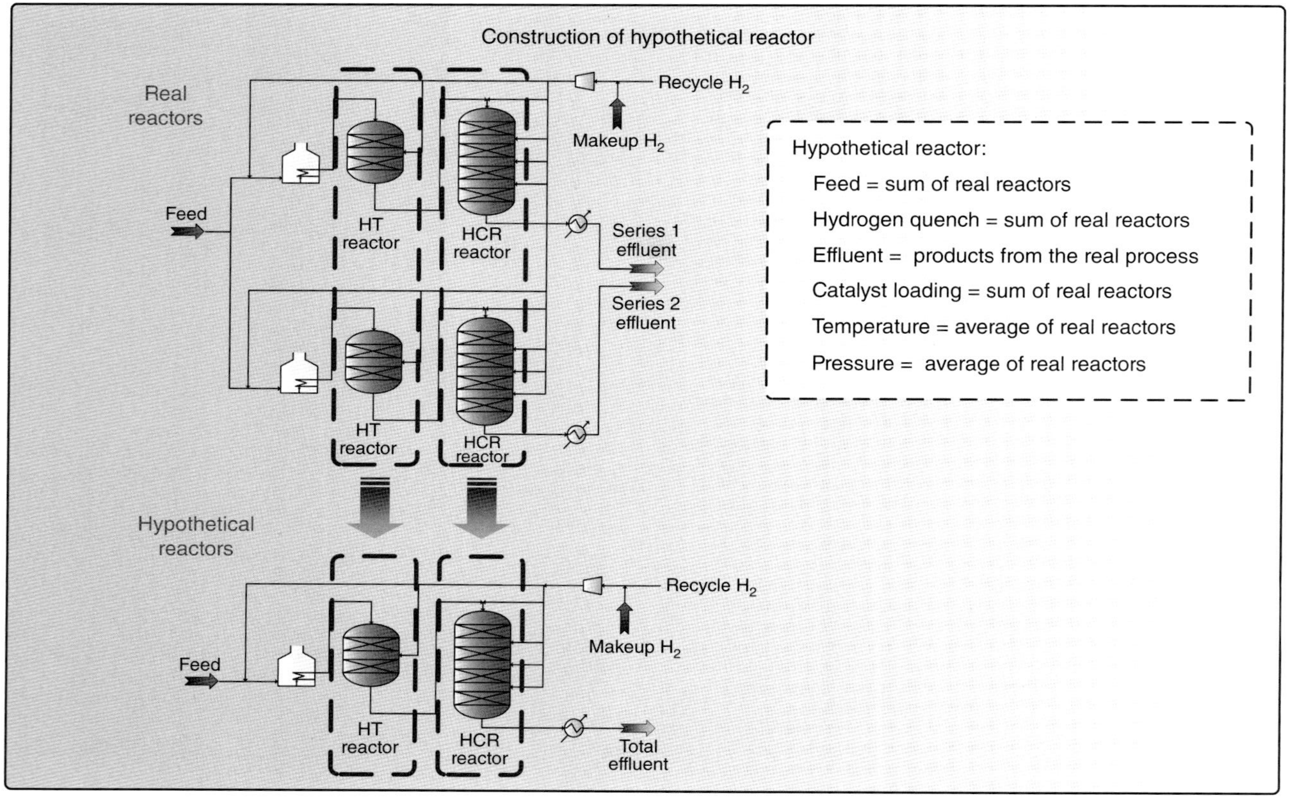

Figure 6.17 Construction of equivalent reactor.

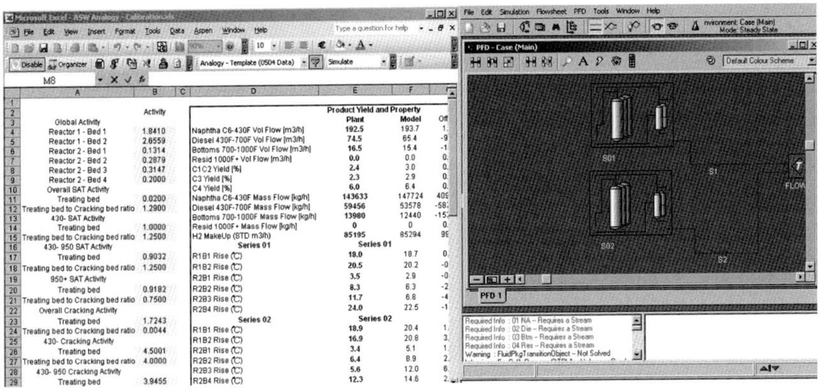

Figure 6.18 Model reconciliation by MS Excel.

process model; thus, we follow the procedures shown in Figure 6.13 to reconcile the equivalent reactor model.

Following reconciliation of the equivalent reactor model is using real operating data to build preliminary models for real HP HCR reactors. We apply the reaction activities from the equivalent reactor model into the preliminary reactor models. It is necessary to fine-tune the preliminary reactor models. From the Aspen Simulation Workbook, we create an MS Excel spreadsheet (Figure 6.18) to make it feasible to fine-tune reactor models of the two parallel series simultaneously. In the HP HCR model, we only fine-tune the HCR selectivity from 4.5 to 3.9 and the resulting model agrees well with real operation and production. The development of equivalent reactor model reduces time and makes it achievable to develop the HP HCR model of two parallel reactor series.

6.4.5 Delumping of the Reactor Model Effluent and Fractionator Model Development

Delumping the reactor model effluent is an essential step to integrate the reactor model with the fractionator model because kinetic lumps used in the reactor model are based on the structure and carbon number and cannot represent accurate thermodynamic behavior of the fractionator model. As BP (volatility) is the most important property for distillation operation, process modelers typically use pseudocomponents based on the true boiling point (TBP) curve to represent the feed oil to the HCR fractionators. We present five steps to develop pseudocomponents based on BP ranges to represent the petroleum fraction [32, 34].

1) Convert ASTM D86/ASTM D1160/simple distillation curve into the TBP curve if the curve is not available.
   - We develop a spreadsheet to enable the conversion from different ASTM distillation types to the TBP curve based on the correlations from Ref. [35] (see Figure 6.19 and Section 1.3, Workshop 1.1).
2) Cut the entire BP range into a number of cut point ranges to define the pseudocomponents based on boiling point ranges (see Figure 6.20).

	A	B	C	D	E	F	G	H	I	J
1	760 mmHg		760 mmHg	760 mmHg	760 mmHg		760 mmHg	760 mmHg	760 mmHg	760 mmHg
2	ASTM-D86 (C)	Vol. %	ASTM-D86 (F)	TBP (F)	TBP (C)		TBP (C)	TBP (F)	ASTM-D86 (F)	ASTM-D86 (C)
3	160.0	0%	320	259.1	126.2		126.2	259.1	320	160.0
4	176.7	10%	350	316.5	158.1		158.1	316.5	350	176.7
5	193.3	30%	380	372.6	189.2		189.2	372.6	380	193.3
6	206.7	50%	404	411.2	210.7		210.7	411.2	404	206.7
7	222.8	70%	433	451.2	232.9		232.9	451.2	433	222.8
8	242.8	90%	469	496.7	258.2		258.2	496.7	469	242.8
9	248.9	100%	480	503.0	261.7		261.7	503.0	480	248.9
10										
11										
12				760 mmHg	760 mmHg		760 mmHg	760 mmHg		
13	ASTM-D2887(C)	Wt%/Vol%	ASTM-D2887(F)	TBP (F)	TBP (C)		TBP (C)	TBP (F)	ASTM-D2887 (F)	ASTM-D2887(C)
14	145.0	5%	293	322.2	161.2		348.0	658.4	639.1711023	337.3
15	151.7	10%	305	327.7	164.3		369.0	696.2	685.3443333	363.0
16	162.2	30%	324	332.4	166.9		406.0	762.8	756.2204767	402.3
17	168.9	50%	336	336.0	168.9		433.0	811.4	811.4	433.0
18	173.3	70%	344	339.6	170.9		459.0	858.2	861.2301007	460.7
19	181.7	90%	359	350.1	176.7		495.0	923.0	922.5542047	494.8
20	187.2	95%	369	357.4	180.8		512.0	953.6	974.5478925	523.6
21	198.9	100%	390	366.2	185.7		556.0	1032.8	1038.378625	559.1
22										
23										
24				760 mmHg	760 mmHg		760 mmHg	760 mmHg		
25	ASTM-D2287 (C)	Wt%/Vol. %	ASTM-D2287 (F)	ASTM-D86 (F)	ASTM-D86 (C)		ASTM-D86 (C)	ASTM-D86 (F)	ASTM-D2887 (F)	ASTM-D2287 (C)
26	25.0	0%	77	121.3	49.6		298.8	569.9	446.4892018	230.3
27	33.9	10%	93	128.2	53.5		349.7	661.5	605.3731877	318.5
28	64.4	30%	148	154.8	68.2		392.0	737.5	715.3377437	379.6
29	101.7	60%	215	206.3	96.8		424.2	795.5	787.7262099	419.8
30	140.6	70%	285	270.6	132.5		459.0	858.2	856.5298061	468.1
31	182.2	90%	360	334.0	167.8		514.5	958.0	964.7774337	518.2
32	208.9	100%	408	367.5	186.4		577.9	1072.2	1273.441992	689.7
33										
34	760 mmHg		760 mmHg	760 mmHg	760 mmHg		760 mmHg	760 mmHg	760 mmHg	760 mmHg
35	ASTM-D1160 (C)	Vol%	ASTM-D1160 (F)	TBP (F)	TBP (C)		TBP (C)	TBP (F)	ASTM-D1160 (F)	ASTM-D1160 (C)
36	369.0	10%	696.2	686.2	363.4		143.1	289.5	300.1	149.0
37	406.0	30%	762.8	757.9	403.3		201.5	394.7	400.1	204.5
38	433.0	50%	811.4	811.4	433.0		246.1	475.0	475.0	246.1
39	459.0	70%	858.2	857.9	458.8		287.7	549.9	550.0	287.8
40	495.0	90%	923	922.5	494.7		343.3	650.0	650.0	343.4
41										
42										
43	Convert distillation curve at subatmospheric pressure to distillation curve at 1 atm.									
44	Pressure =	10	mmHg	2 =< P =< 760						
45	X	0.00195599			760 mmHg	760 mmHg	760 mmHg			
46	TBP/D1160 (C)	Vol%	TBP/D1160 (F)	TBP/D1160 (R)	TBP/D1160 (R)	TBP/D1160 (F)	TBP/D1160 (C)			
47	143.0637743	10%	289.5147938	749.2	997.0241391	537.3541391	280.7522995			
48	201.5221632	30%	394.7398938	854.4	1122.78319	663.1131895	350.6184386			
49	246.1111111	50%	475	934.7	1216.602752	756.9327522	402.7404179			
50	287.7201281	70%	549.8962306	1009.6	1302.560937	842.8909373	450.4949651			
51	343.3328559	90%	649.9991406	1109.7	1415.120783	955.4507826	513.0282125			

Figure 6.19 Interconversion between different ASTM distillation types (see Section 1.3).

3) Develop the density distribution of pseudocomponents if only the bulk density is available.
   - Assume that Watson K factor is constant throughout the entire boiling point range and calculate the mean average boiling point (MeABP). We develop a spreadsheet tool (see Section 1.2 and Section 1.5, Workshop 1.3) to perform the iteration of estimating MeABP based on the method presented by Bollas *et al.* [36].

$$K_{\text{avg}} = [\text{MeABP}]^{0.333} / \text{SG}_{\text{avg}} \tag{6.14}$$

   where K_{avg} is Watson K factor and SG_{avg} is the bulk specific gravity 60 °F/60 °F.
   - Calculate the density distribution of the entire boiling point range.

$$\text{SG}_i = [T_{i,b}]^{0.333} / K_{\text{avg}} \tag{6.15}$$

   where SG_i is the specific gravity 60 °F/60 °F of pseudocomponent i and $T_{i,b}$ is the TBP of pseudocomponent i.
4) Estimate molecular weight distribution of the entire boiling point range if it is not available.
   There are various correlations to estimate pseudocomponent molecular weight based on standard liquid density and TBP. Riazi [37] presented a comprehensive review and comparison of published correlations.
5) Estimate critical temperatures (T_c), critical pressures (P_c), critical volumes (V_c), and acentric factors (ω) of pseudocomponents. Refer to Riazi [37] for published correlations.

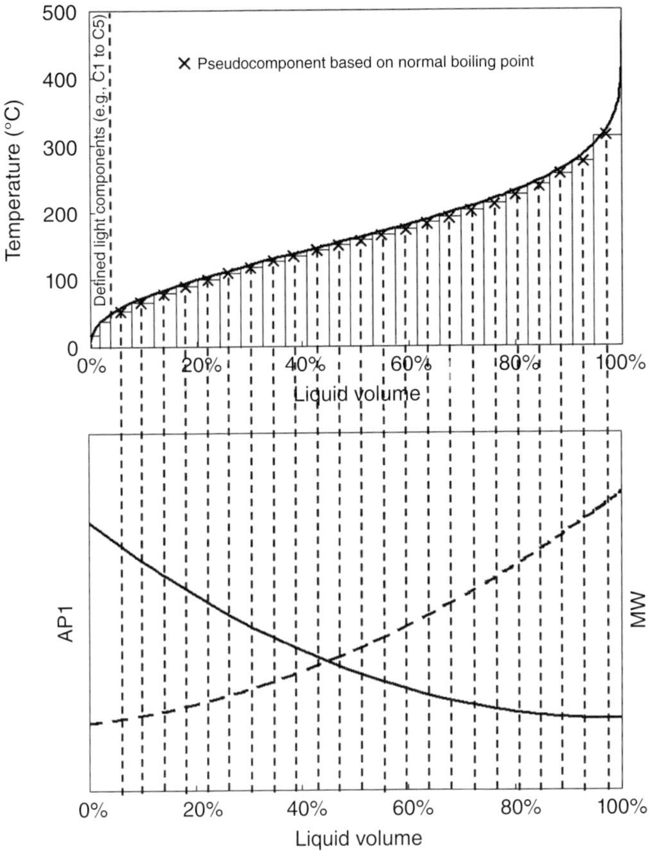

Figure 6.20 Relationship between pseudocomponent properties and the TBP curve (redrawn from ref. [32]).

As the reactor model provides the TBP curve, API gravity, and molecular weight distribution of the model effluent from kinetic lumps, the major issue of developing pseudocomponents in this work is to cut the TBP curve properly into a number of pseudocomponents based on boiling point ranges. However, the number and boiling point ranges of cut points are arbitrary, and there is no general rule to determine the cut point ranges. Having a large number of cuts does not always lead to a good representation, whereas having a small number of cuts may cause discontinuous prediction of column operation [32]. In addition, the discrete nature of kinetic lumps (see Figure 6.21) makes it difficult to cut the TBP curve of the reactor model effluent to define reasonable pseudocomponents based on boiling point ranges. In this work, we find that applying Gauss–Legendre quadrature to cut the reactor model effluent into 20 pseudocomponents based on boiling point ranges works well. The rest of this section will represent the delumping that we use to define these pseudocomponents, how to apply the stage efficiency model while building the fractionator model, and the sensitivity test of the fractionator model to verify the delumping method used in this work.

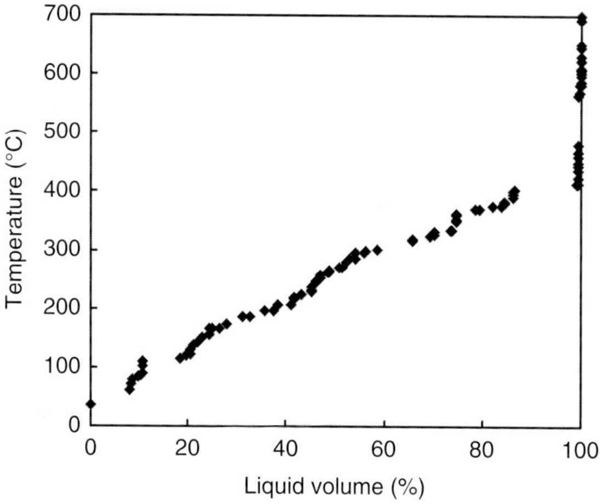

Figure 6.21 Discontinuity of the C6+ kinetic lump distribution of the reactor model effluent.

6.4.5.1 Applying the Gauss–Legendre Quadrature to Delump the Reactor Model Effluent

Haynes and Matthews [38] applied the Gauss–Legendre quadrature to predict the vapor–liquid equilibrium (VLE) of the hydrocarbon mixture derived from a continuous equation-of-state developed by Cotterman *et al.* [39]. Later, Mani *et al.* [40] extended the work of Haynes and Matthews [38] to partition the cut point ranges of the TBP curve of a petroleum fraction to define pseudocomponents based on boiling point ranges and the predicted VLE satisfactorily matches the experimental data. Hence, we extend the method represented by Mani *et al.* [40] to delump the reactor model effluent into pseudocomponents based on boiling point ranges.

In this work, we develop a method with six steps to delump the reactor model effluent into pseudocomponents based on boiling point ranges by the Gauss–Legendre quadrature.

1) Split the reactor model effluent into C6− and C6+ streams because the components below C6 are well-defined light components.
2) Obtain the TBP curve, API gravity, and molecular weight distribution of C6+ stream from reactor model.
3) Determine the number (n) of pseudocomponents to be used in delumping.
   - In this work, we delump the reactor model effluent into 20 pseudocomponents based on boiling point ranges.
4) We have included in the supplement to this text the Excel spreadsheet for the quadrature points and weight factors for the Gauss–Legendre integration, *GL_Quad Pt.xls*, which are used to partition the cut points over the TBP curve.
   - Use F_{vi} calculated from the equation below to partition the cut point (F_{vi}) over the TBP curve of C6+ stream.

$$F_{vi} = \frac{1}{2} \times [q_i + 1] \tag{6.16}$$

   Interpolate TBP curve to obtain the TBP associated with each cut point (F_{vi}). Figure 6.22 shows the case of $n = 6$.
   - Use the same interpolation procedure to obtain API gravities and molecular weights of the associated cut points (F_{vi}).

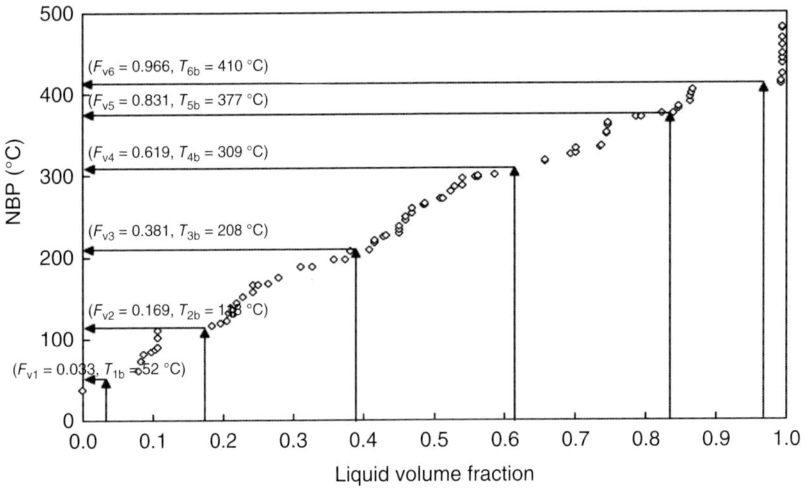

Figure 6.22 Demonstration of the allocating cut point over the TBP curve.

5) Estimate T_c, P_c, V_c, and ω of each pseudocomponent by using molecular weight and specific gravity 60 °F/60 °F, which can be converted from API gravity.
   For T_c and P_c, Haynes and Matthews [38] recommended to use the correlation developed by Riazi and Daubert [41].

$$T_c\ (\mathrm{K}) = 19.0627 \times T_b^{0.58848} \times \mathrm{SG}^{0.3596} \tag{6.17}$$

$$P_c\ (\mathrm{atm}) = 5.458 \times 10^7 \times T_b^{-2.3125} \times \mathrm{SG}^{2.3201} \tag{6.18}$$

   For the acentric factor ω, Haynes and Matthews [38] suggested to use the correlation developed by Lee and Kesler [42].

$$\omega = \frac{\left(-\ln(P_c, \mathrm{atm}) - 5.92714 + \dfrac{6.09648}{T_r} + 1.28862 \times T_r - 0.169347 \times T_r^{\,6}\right)}{\left(15.2518 - \dfrac{15.6875}{T_r} - 13.4721 \times \ln(T_r) + 0.43577 \times T_r^{\,6}\right)} \tag{6.19}$$

   - For V_c, to be consistent with the estimations of T_c and P_c, we also apply the correlation developed by Riazi and Daubert [41].

$$V_c(\mathrm{cm}^3/\mathrm{mol}) = 1.7842 \times 10^{-4} \times T_b^{\,2.3829} \times \mathrm{SG}^{-1.683} \tag{6.20}$$

6) The last step of delumping is to calculate mole fraction (x_i) of each pseudocomponent.
   - Use the equation below to calculate mole fraction of each pseudocomponent.

$$x_i = \frac{w_i \times \mathrm{SG}_i \times \mathrm{MW}_{\mathrm{avg}}}{2 \times \mathrm{SG}_{\mathrm{avg}} \times \mathrm{MW}_i} \tag{6.21}$$

   In the equation, w_i is the weight factor of Gauss–Legendre quadrature, SG_i and MW_i are the specific gravity and molecular weight of pseudocomponent I, which are calculated by interpolating the specific gravity and

Table 6.6 Pseudocomponents and their properties and compositions.

	x_i	q_i[a)]	w_i	TBP (°C)	MW	SG	T_c (°C)	P_c (kPa)	V_c (m³/kg mol)	ω
Pseudo 1	0.1559	−0.932470	0.171324	52	84.0	0.6694	223.6	3373.3	0.340	0.2326
Pseudo 2	0.2529	−0.661209	0.360762	118	128.8	0.7904	314.8	3233.3	0.400	0.2789
Pseudo 3	0.2550	−0.238619	0.467914	208	174.9	0.8346	403.3	2282.8	0.595	0.4286
Pseudo 4	0.1809	0.238619	0.467914	309	248.6	0.8411	486.3	1491.6	0.928	0.6792
Pseudo 5	0.1091	0.661209	0.360762	377	318.7	0.8438	538.3	1163.0	1.201	0.8968
Pseudo 6	0.0462	0.932470	0.171324	410	357.5	0.8438	562.3	1037.4	1.352	1.0252
$MW_{avg} = 175$										
$SG_{avg} = 0.8084$										

a) q_i are the zeros of the Legendre polynomial of order n and m_i are the associated weight factors.

molecular weight distributions of reactor model effluent, respectively, and SG_{avg} and MW_{avg} are the average specific gravity and molecular weight obtained from reactor model, respectively.

- Table 6.6 lists the resulting pseudocomponents and their properties and compositions for the case of $n = 6$.

6.4.5.2 Key Issue of the Building Fractionator Model – Overall Stage Efficiency Model

In building simulation models for fractionators, simulation software users often misunderstand the concept of "stage efficiency" [32]. The theoretical column model based on rigorous thermodynamics assumes that each stage is in perfect VLE. However, real distillation columns do not perform perfectly. The "overall stage efficiency," defined as the number of theoretical stages divided by the number of actual stages, indicates the difference of a real column to a theoretical column. We can apply the overall stage efficiency to the entire column or specific separation zones. For example, 20 theoretical stages are required to model an operating column with 40 actual stages and 50% overall stage efficiency. It is important to remember that all stages in this column still perform perfect VLE.

In Section 2.4.2, we discussed the concepts of the overall stage efficiency and the Murphree stage efficiency. In Section 2.4.3, we give recommendations on how to handle the efficiency issue correctly. In particular, in modeling refinery distillation columns, we recommend to use an overall stage efficiency to convert the actual number of stages to the equivalent number of theoretical stages. Refer to Table 2.3, Section 2.4.3, for typical values of overall stage efficiency for refinery distillation. In Section 6.13, we illustrate the development of the fractionation simulation model in HCR units in Workshop 6.4.

6.4.5.3 Verification of the Delumping Method – Gaussian–Legendre Quadrature

As we mentioned before, the number of cut point ranges is defined arbitrarily. Kaes [32] stated that it is necessary to perform a sensitivity test to study the relationship of the side draw rate to the side draw temperature and the associated distillation curve to ensure if the defined pseudocomponents based on boiling point ranges are able to provide reasonable results. If the relationship is stepwise rather than continuous, we need to redefine the number of pseudocomponents based on boiling point ranges. In this work, we cut the reactor model effluent into 20 TBP pseudocomponents to represent the feed to fractionators. To run the sensitivity test, we change draw rates of diesel fuel to investigate the relationship among draw rates, draw temperatures, and distillation curves of products.

To verify that the delumping method of Gauss–Legendre quadrature with 20 pseudocomponents based on boiling point ranges is sufficient for column models, we perform another sensitivity test as a contrast, which uses the even cut point range method to cut reactor model effluent into 46 pseudocomponents based on boiling point ranges. The even cut point range method is a built-in method available in Aspen HYSYS Petroleum Refining that converts the reactor model into pseudocomponents based on boiling point ranges with equal boiling point ranges.

Figures 6.23–6.26 represent the results of sensitivity tests for the even cut point range method and the Gauss–Legendre quadrature. The figures do not include initial points, end points, and 90% and 95% points because the modeled initial points and end points are usually not reliable [32], and the variations in 90% and 95% points are too flat to provide representative results (both are less than 1%). Apparently, both methods generate smooth and continuous relationships between the draw rates and draw temperatures (see Figures 6.23 and 6.24). However, Figures 6.25 and 6.26 illustrate that these two methods have different performances on predicting the relationships between the draw rate and the distillation curve. The Gauss–Legendre quadrature is able to predict smooth and continuous relationship between draw rate and distillation curve, whereas the even cut point range method is not. Using the Gauss–Legendre quadrature to delump the reactor model effluent, we are able to build well-behaved column models with a few pseudocomponents based on boiling point ranges.

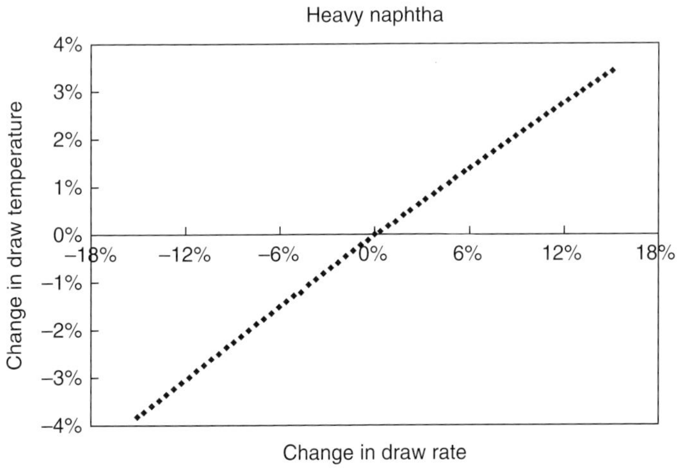

Figure 6.23 Relationship between the draw rate and the draw temperature of heavy naphtha (even cut point range method).

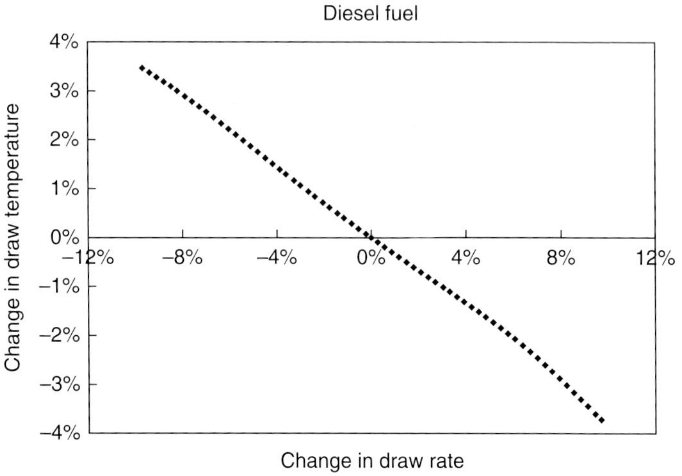

Figure 6.24 Relationship between the draw rate and the draw temperature of diesel fuel (Gauss–Legendre quadrature method).

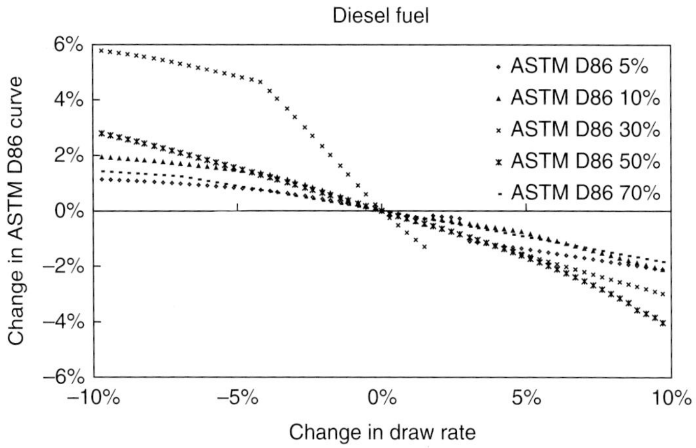

Figure 6.25 Relationship between the draw rate and the distillation curve of diesel fuel (even cut point range method).

6.4.6 Product Property Correlation

The last important issue of building an integrated HCR model is the fuel property estimation, particularly the flash point and freezing point of diesel fuel and specific gravities of liquid products. We can estimate the specific gravities of liquid products, once we have defined the pseudocomponents based on boiling point ranges and calibrated the model for product flow rates (mass and volume). The flash point is defined as the lowest temperature at which a flame or spark can ignite the mixture of air and the vapors arising from oils. The flash point indicates the highest temperature at which we can store and transport the oils safely. For a pure substance, the freezing point is the temperature at which liquid solidifies.

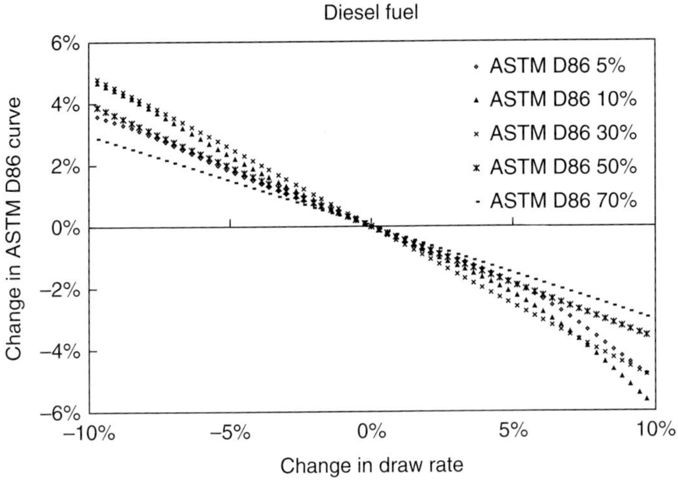

Figure 6.26 Relationship between the draw rate and the distillation curve of diesel fuel (Gauss–Legendre quadrature method).

For the petroleum fraction, which is the mixture of hydrocarbons, the freezing point is defined as the temperature at which solid crystals formed upon cooling disappear as the temperature is increased [35]. For both properties, we update the parameters used in API correlations [35], Eqs. (6.22) and (6.23).

$$\text{Flash point (fahrenheit)} = A \times 10\% \text{ of ASTMD86 (fahrenheit)} + B \quad (6.22)$$

$$\text{Freezing point } (R) = A + B \times \text{SG} + C \times \frac{\text{MeABP}^{1/3}}{\text{SG}} + D \times \text{MeABP} \quad (6.23)$$

For MP HCR process, we apply 130 and 115 data points collected from the plant to refit Eqs. (6.22) and (6.23), respectively. The average absolute deviations (AADs) of the new correlations for flash point and freezing point are 2.7 and 2.3 °C, respectively, and the resulting correlations are

$$\text{Flash point (fahrenheit)} = 0.677 \times 10\% \text{ of ASTM D86 (fahrenheit)} - 118.2 \quad (6.24)$$

$$\text{Freezing point}(R) = A + B \times \text{SG} + C \times \frac{\text{MeABP}^{1/3}}{\text{SG}} + D \times \text{MeABP} \quad (6.25)$$

For HP HCR process, we apply 142 and 63 data points collected from the plant to refit Eqs. (6.22) and (6.23), respectively. The AADs of the new correlations for flash point and freezing point are 1.2 and 1.6°C, respectively, and the resulting correlations are

$$\text{Flash point (fahrenheit)} = 0.51 \times 10\% \text{of ASTMD86 (fahrenheit)} - 57.7 \quad (6.26)$$

$$\text{Freezing point } (R) = -857.63 + 437.16 \times \text{SG} + 41.68 \times \frac{\text{MeABP}^{1/3}}{SG} - 0.483 \times \text{MeABP} \quad (6.27)$$

We apply Eqs. (6.24)–(6.27) to estimate the flash points and freezing points of diesel fuel in MP HCR process and jet fuel in HP HCR process by model predictions on distillation curve, specific gravity, and MeABP.

6.5 Modeling Results of MP HCR Process

6.5.1 Performance of the Reactor and Hydrogen Recycle System

Our MP HCR model includes three major parts of commercial HCR process, including reactors, fractionators, and hydrogen recycle system. Figures 6.27 and 6.28 show the model predictions of the weight average reactor temperatures (WARTs) of the HT reactor and HCR reactors. In the reactor model, we define the inlet temperature of each catalyst bed and the model will calculate the outlet temperature of each bed. The AAD of catalyst bed outlet temperatures of the HCR reactor is 1.9 °C. The model generates good predictions on temperature profile of the HCR reactor, which is important for estimating product yields. However, the predictions on temperature profile of the HT reactor are less accurate than those of the HCR reactor. As model calibration does not consider HDS and HDN reactions, the model is not able to estimate the reaction activity of the HT reactor well.

Figure 6.29 represents modeling result of makeup hydrogen flow rate, and the average relative deviation (ARD) is about 8%. The error results from two factors. First, the model is not good at predicting HDS and HDN activities, which affects the estimation of hydrogen consumption. Next, the allocation of the hydrogen recycle system of Aspen HYSYS Petroleum Refining (see Figure 6.4) is different from that of the MP HCR unit (see Figure 6.8). Aspen HYSYS Petroleum Refining considers makeup hydrogen mixes with recycle hydrogen before feeding into the hydrogen recycle system; however, in the MP HCR unit, the makeup hydrogen directly mixes with feed oil and does not influence the hydrogen recycle

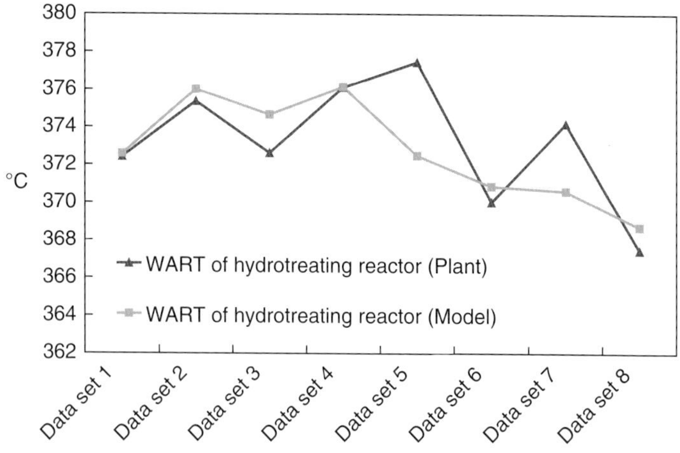

Figure 6.27 Predictions of WARTs of the HT reactor (MP HCR process).

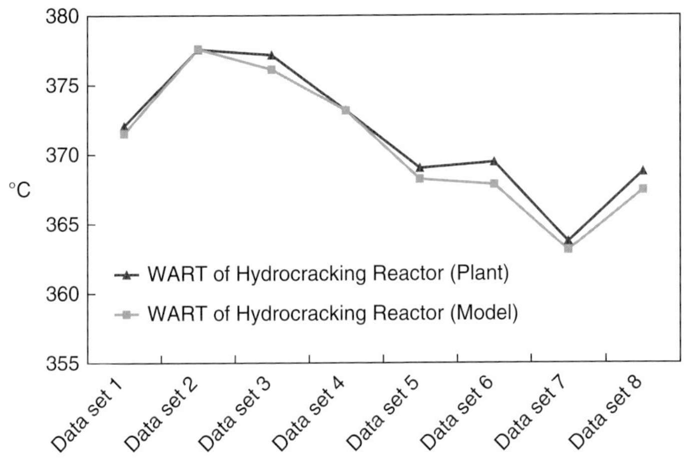

Figure 6.28 Predictions of WARTs of the HCR reactor (MP HCR process).

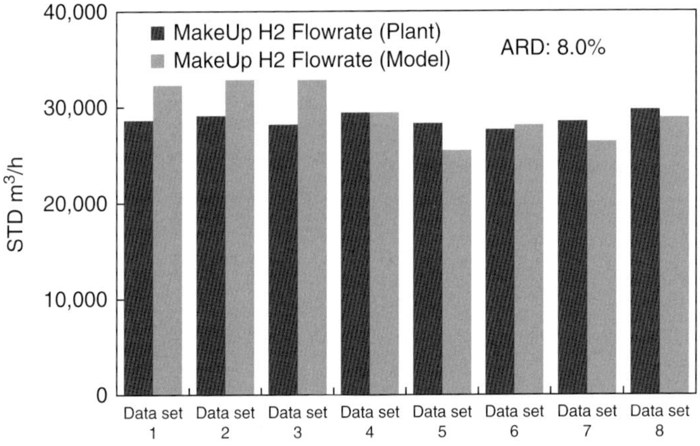

Figure 6.29 Predictions of the makeup hydrogen flow rate (MP HCR).

system. This will make the reactor model less accurate in calculating the hydrogen partial pressure of the reactors, causing deviation upon estimating hydrogen consumption.

6.5.2 Performance of Fractionators

The temperature profile of the distillation column is valuable for evaluating energy consumption and for helping plant operation of cut point and process optimization. Figures 6.30–6.33 illustrate selected modeling results on temperature profiles of distillation columns. Note that we apply the overall stage efficiency model to column simulations and the resulting stage number of the column model does not correspond to the stage number in the real column. Therefore, we use "top → bottom" in Figures 6.30–6.33 instead, showing the stage number to illustrate temperature distribution from the condenser to

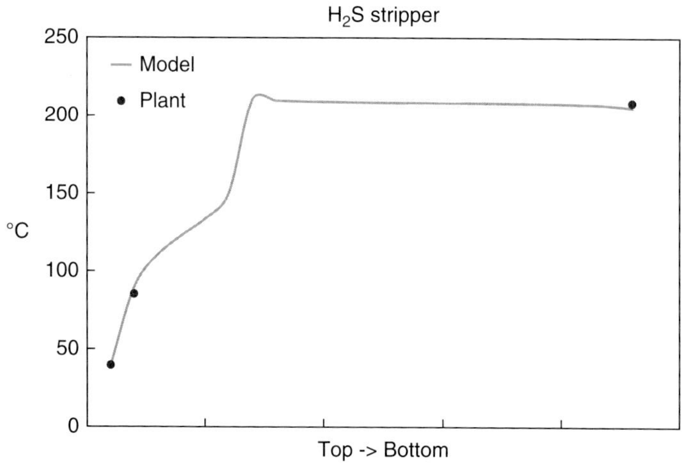

Figure 6.30 Prediction of the temperature profile of the H_2S stripper (data set 1 in MP HCR).

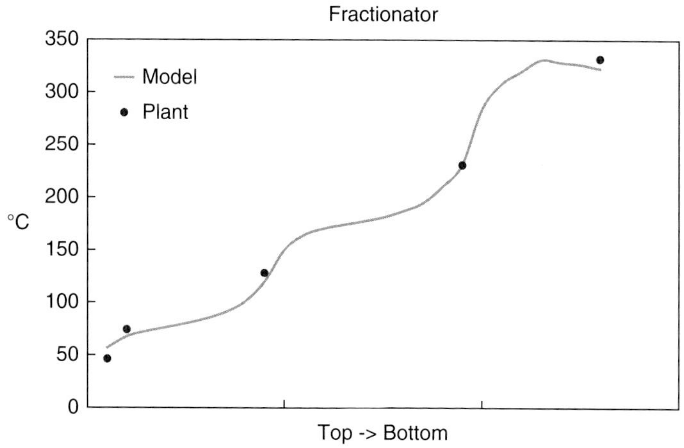

Figure 6.31 Prediction of the temperature profile of the fractionator (data set 1 in MP HCR).

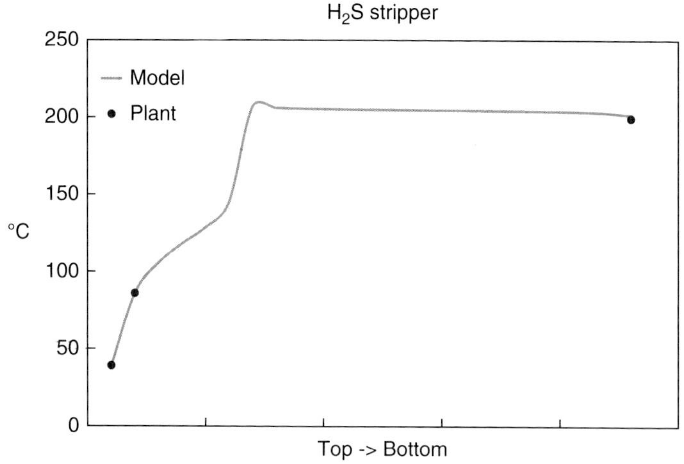

Figure 6.32 Prediction of the temperature profile of the H_2S stripper (data set 5 in MP HCR).

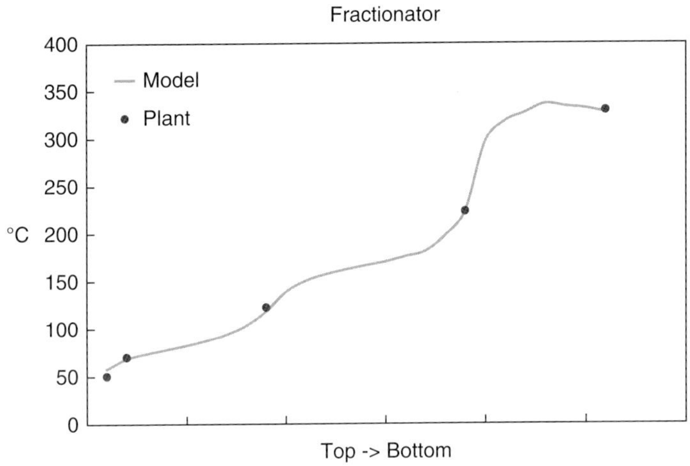

Figure 6.33 Prediction of the temperature profile of the fractionator (data set 5 in MP HCR).

bottom of the column. Obviously, the model is able to provide good predictions on column temperature profiles.

6.5.3 Product Yields

There are seven products in the MP HCR units, as depicted in Figure 6.8, namely, LPS vapor (LPS VAP), sour gas, LPG, light naphtha, heavy naphtha, diesel fuel, and bottom oil. Among these seven products, light naphtha, heavy naphtha, diesel fuel, and bottom oil are major products because they account for over 95 wt% of the overall production. Figures 6.34–6.37 illustrate the model predictions on light naphtha, heavy naphtha, diesel fuel, and bottom oil and the AADs are 0.3, 3.4, 2.4, and 2.4 wt%, respectively. We calculate AADs by averaging the absolute deviations (i.e., |predicted wt% − plant wt%|) of the eight

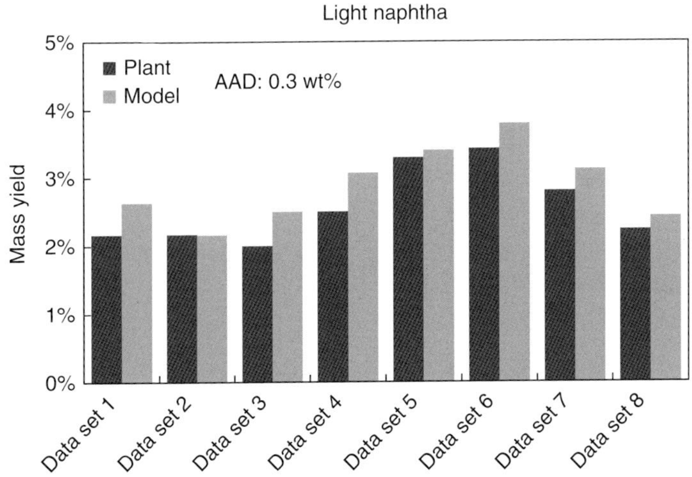

Figure 6.34 Predictions of the light naphtha yield (MP HCR).

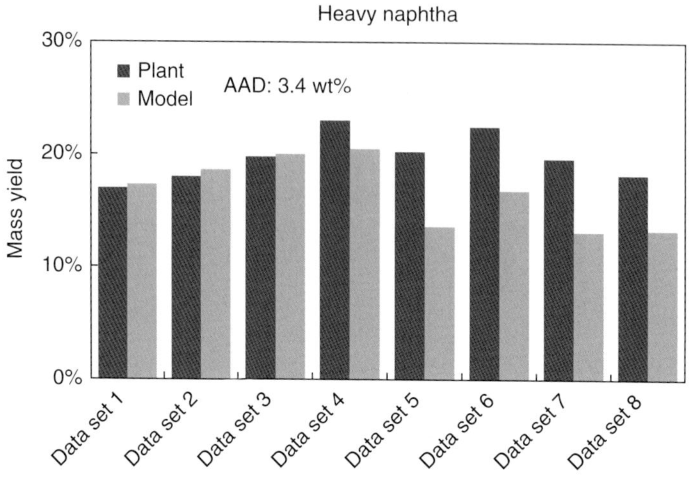

Figure 6.35 Predictions of the heavy naphtha yield (MP HCR).

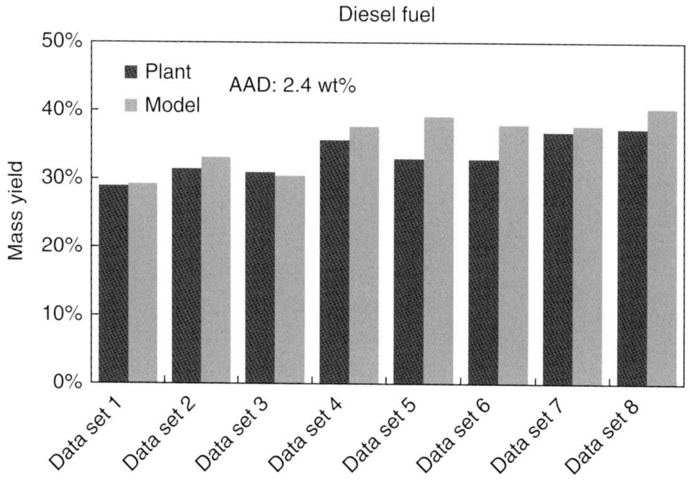

Figure 6.36 Predictions of the diesel fuel yield (MP HCR).

data sets. This follows because the relative deviation (i.e., |predicted wt% − plant wt%|/|plant wt%|) only represents the prediction of model on each product yield rather than the overall yield, which is the key profit concern to the refinery. On the other hand, absolute deviation indicates how the model affects the estimation of the profit of the refinery by considering the deviations in the same scale toward overall production. For example, the model shows 13% relative deviation on predicting the mass production of light naphtha but gives no clues about how the model affects the overall yield. Considering that the mass yield of light naphtha is about 2.6 wt%, the 13% relative deviation has only a very small effect (0.3 wt%) on the overall yield. The model gives good prediction results on product yields when considering the average values of product yields.

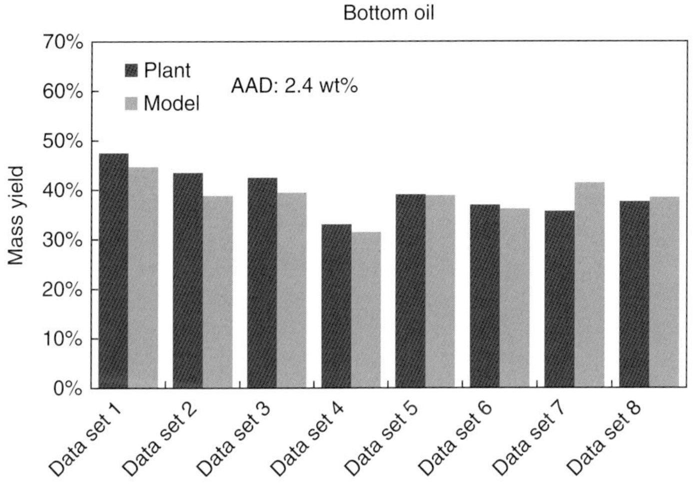

Figure 6.37 Predictions of the diesel fuel yield (MP HCR).

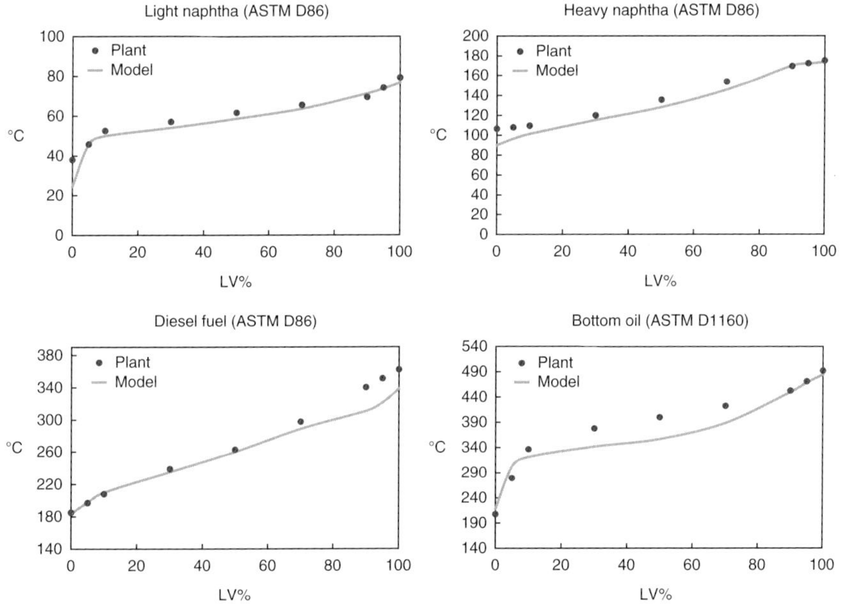

Figure 6.38 Predictions of distillation curves of liquid products (data set 1 in MP HCR).

6.5.4 Distillation Curves of Liquid Products

The distillation curve displays the vaporization temperature after having a certain amount of oil fraction vaporized. Figures 6.38 and 6.39 illustrate selected model predictions on distillation curves of light naphtha, heavy naphtha, diesel fuel, and bottom oil. The deviations of predicting distillation curves result from two factors. First, the fractionator simulation cannot provide reliable results of

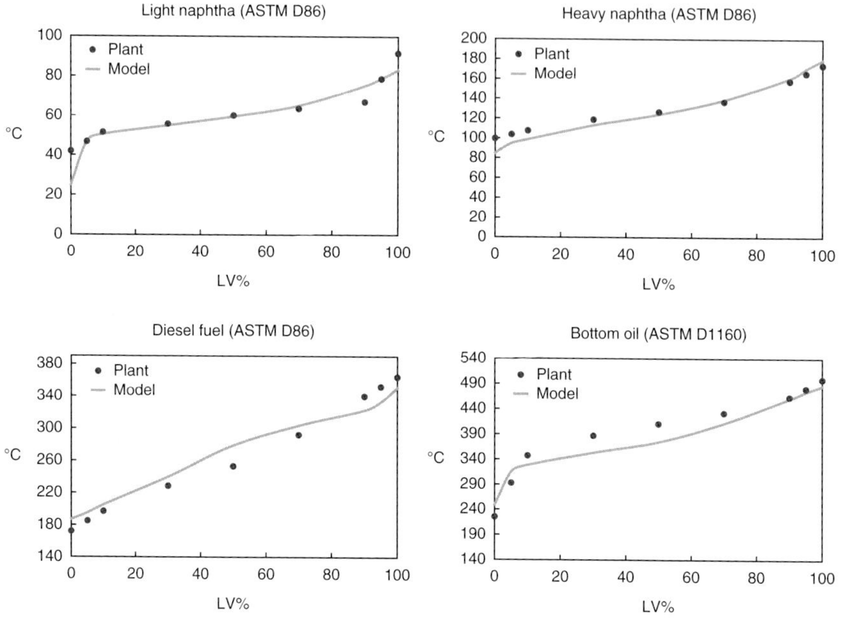

Figure 6.39 Predictions of distillation curves of liquid products (data set 5 in MP HCR).

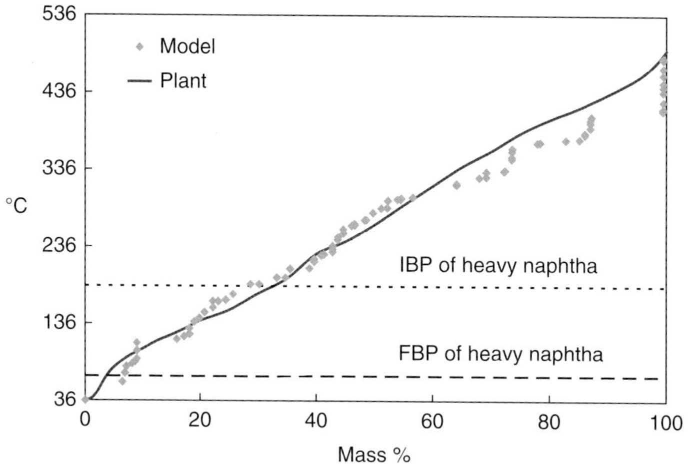

Figure 6.40 Comparison between the C5+ distribution of the plant reactor effluent and the model prediction within the boiling point range of heavy naphtha (data set 4 in MP HCR).

the initial and final BPs of liquid products [32]. Next, the reactor model cannot provide accurate predictions of the BP distribution of the reactor effluent. Although the model is able to predict product yield accurately after calibration, it does not predict the BP distribution (distillation curve) of the liquid product with an equal accuracy. This follows because of the nature of the discrete BP distribution of kinetic lumps. Figures 6.40–6.42 illustrate the differences between the C5+ distribution of the plant reactor effluent and the model prediction.

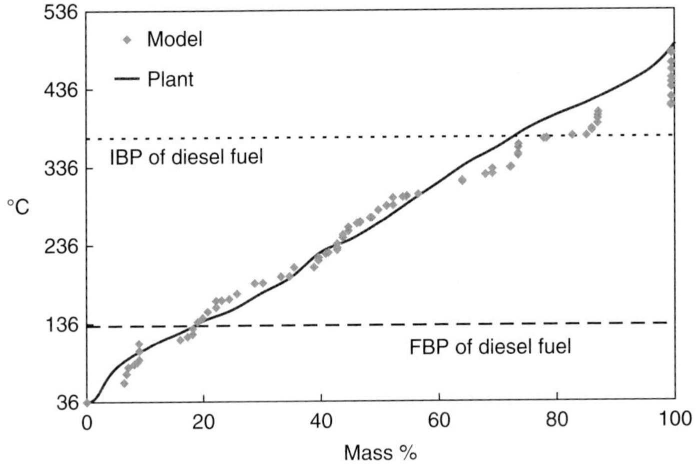

Figure 6.41 Comparison between the C5+ distribution of the plant reactor effluent and the model prediction within the boiling point range of diesel fuel (data set 4 in MP HCR).

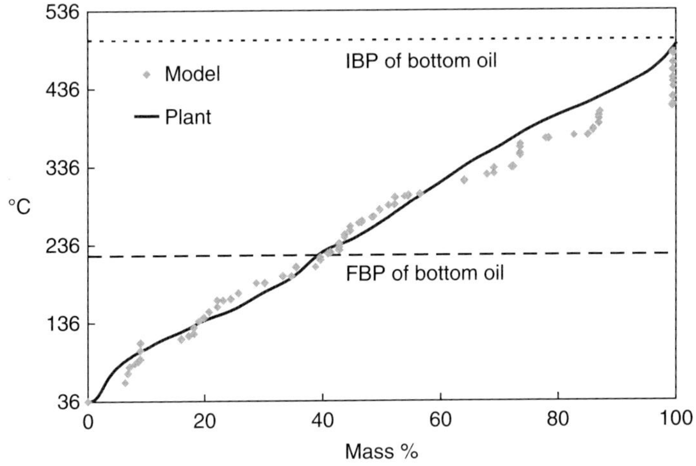

Figure 6.42 Comparison between the C5+ distribution of the plant reactor effluent and the model prediction within the boiling point range of bottom oil (data set 4 in MP HCR).

6.5.5 Product Property

Section 6.4.6 demonstrates the updated correlations for predicting the flash point and freezing point of diesel fuel. Figures 6.43 and 6.44 illustrate model predictions on flash point and freezing point of diesel fuel. The AADs are 3.6 and 4.1 °C, respectively, which are about the same values as that obtained from correlating plant data. Applying the updated correlations demonstrated in Section 6.4.6, we find satisfactory predictions on the flash point and freezing point of diesel fuel. Figures 6.45–6.48 illustrate the specific gravity predictions of liquid products, which are calculated by Aspen HYSYS Petroleum Refining. The accurate predictions reflect the accuracy of the model to predict specific gravity

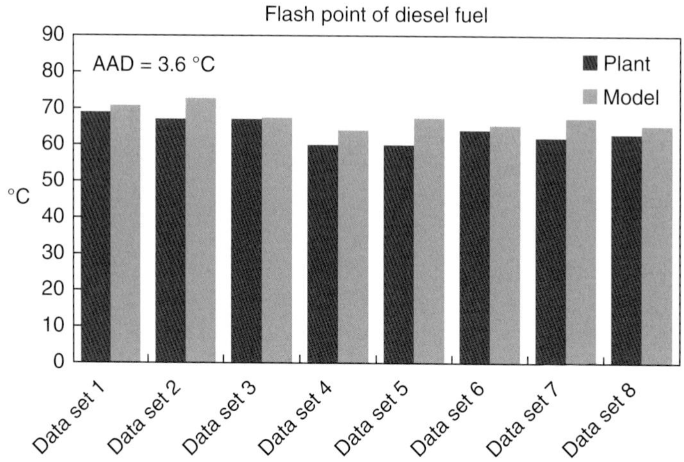

Figure 6.43 Predictions of the flash point of diesel fuel (MP HCR).

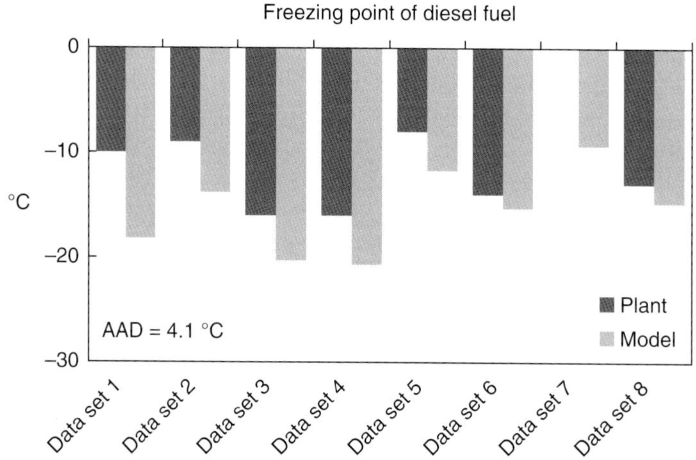

Figure 6.44 Predictions of the freezing point of diesel fuel (MP HCR).

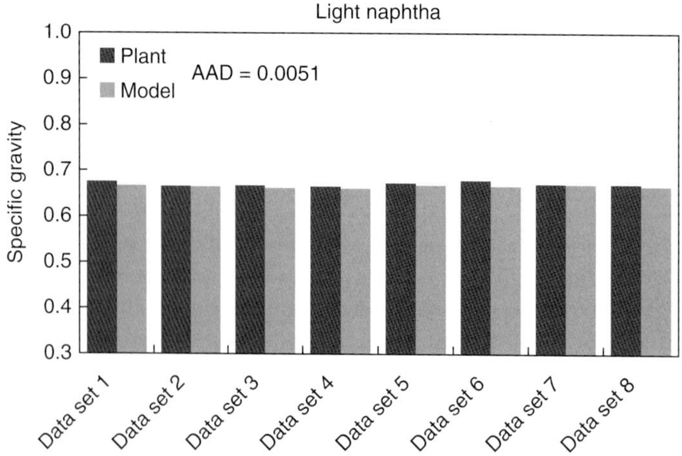

Figure 6.45 Predictions of the specific gravity of light naphtha (MP HCR).

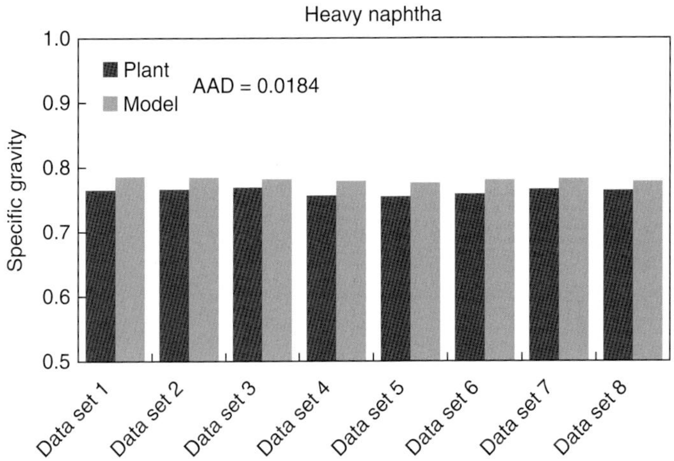

Figure 6.46 Predictions of the specific gravity of heavy naphtha (MP HCR).

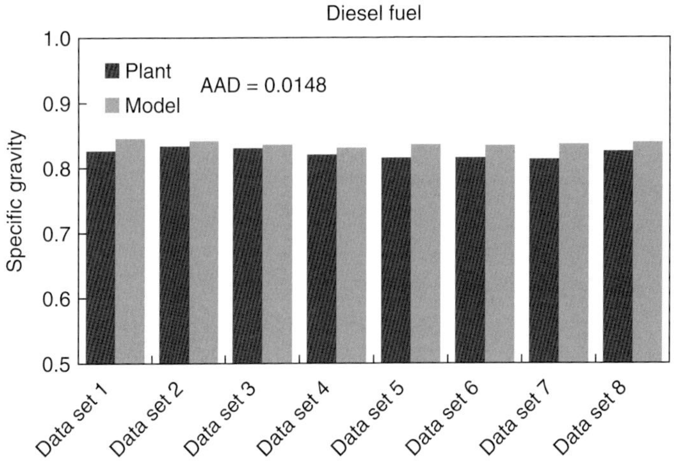

Figure 6.47 Predictions of the specific gravity of diesel fuel (MP HCR).

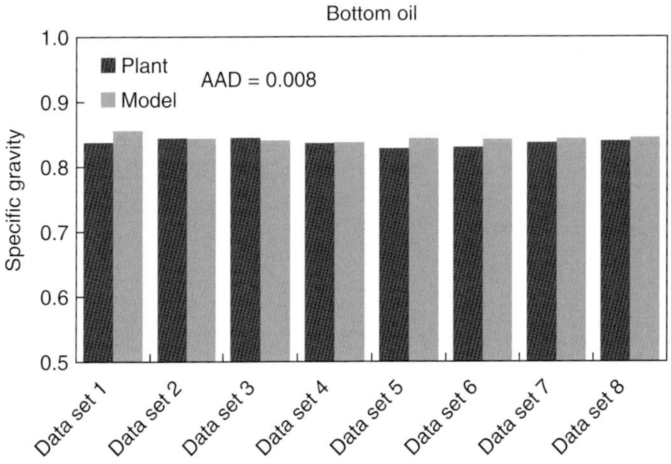

Figure 6.48 Predictions of the specific gravity of bottom oil (MP HCR).

of the liquid product and demonstrate that the delumping method described in Section 6.4.5 is able to carry over density distribution to pseudocomponents based on boiling point ranges.

6.6 Modeling Results of HP HCR Process

6.6.1 Performance of the Reactor and Hydrogen Recycle System

Our HP HCR model includes three major parts of the commercial HCR process, including reactors, fractionators, and hydrogen recycle system. In the reactor model, we define the inlet temperature of each catalyst bed, and the model will calculate the outlet temperature of each bed. The AADs of catalyst bed outlet temperatures of the two HCR reactors are 1.8 and 3.2 °C for series 1 and 2, respectively. Figures 6.49 and 6.50 show the model predictions of WARTs of HT and

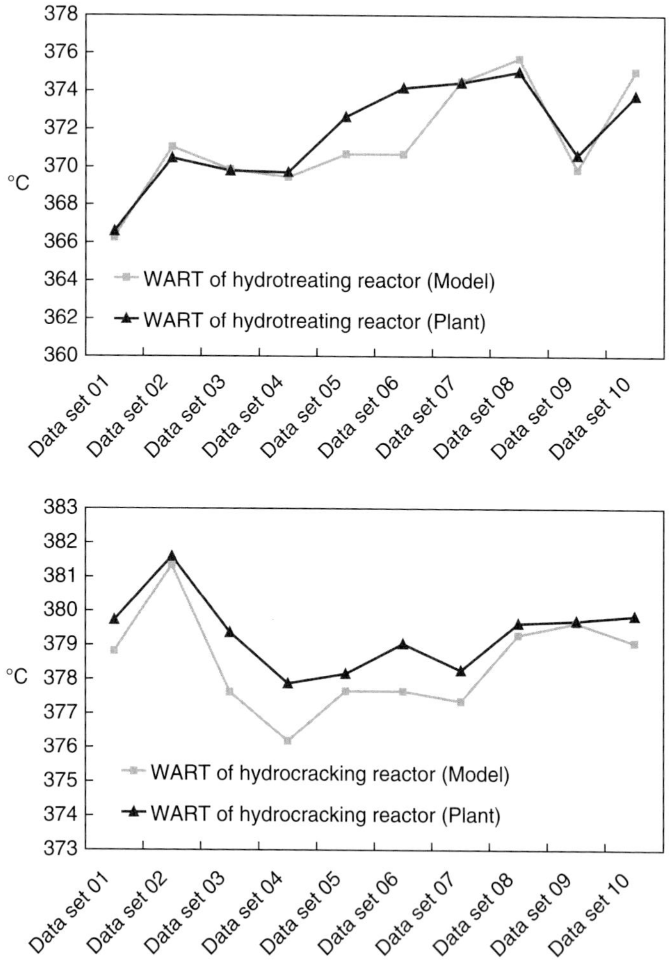

Figure 6.49 Predictions of WARTs of HT and HCR reactors (Series 1 in HP HCR).

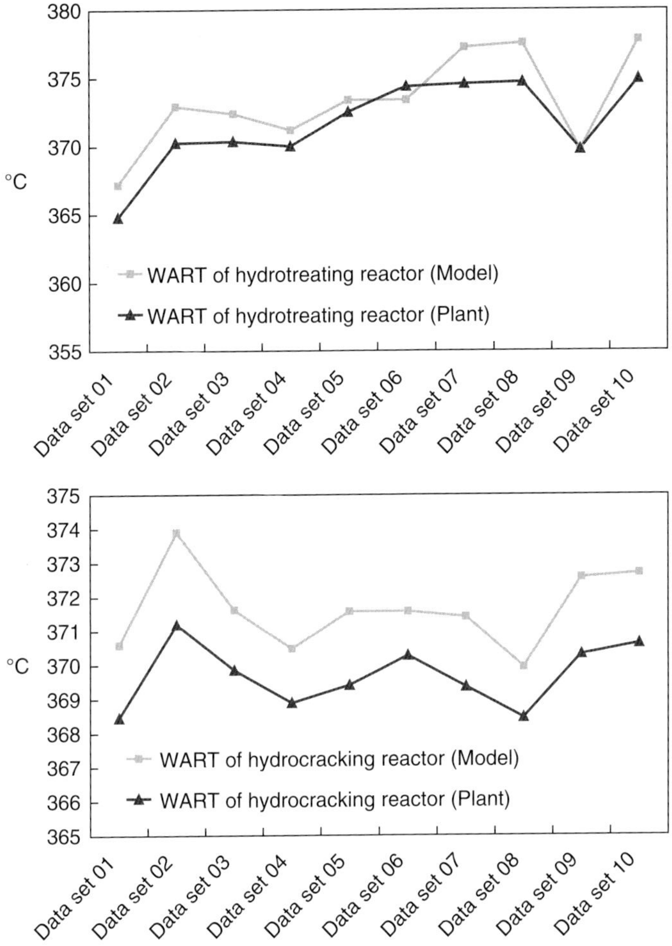

Figure 6.50 Predictions of WARTs of HT and HCR reactors (Series 2 in HP HCR).

HCR reactors. The model generates good predictions on the temperature profile of reactors. Figure 6.51 represents the modeling result of the makeup hydrogen flow rate, and the ARD is only 2%.

6.6.2 Performance of Fractionators

Figures 6.52 and 6.53 illustrate selected modeling results on temperature profiles of distillation columns. These figures are similar to Figures 6.30–6.33 for the MP HCR.

6.6.3 Product Yields

There are seven products in the HP HCR unit, namely, LPS VAP, dry gas, LPG, light naphtha, heavy naphtha, jet fuel, and residue oil. Among these seven products, LPG, light naphtha, heavy naphtha, jet fuel, and residue oil are

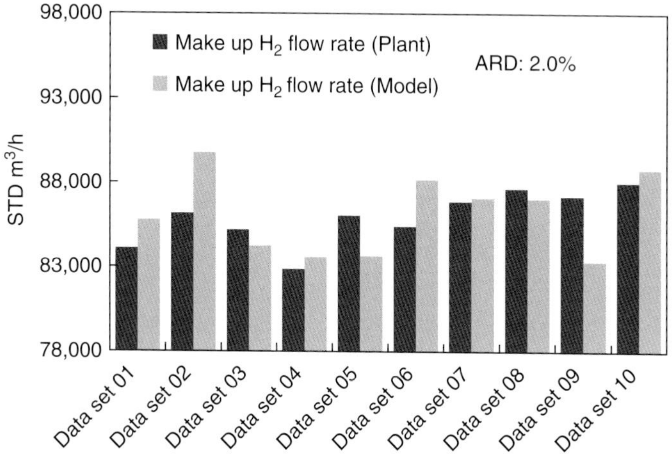

Figure 6.51 Predictions of the makeup hydrogen flow rate (HP HCR).

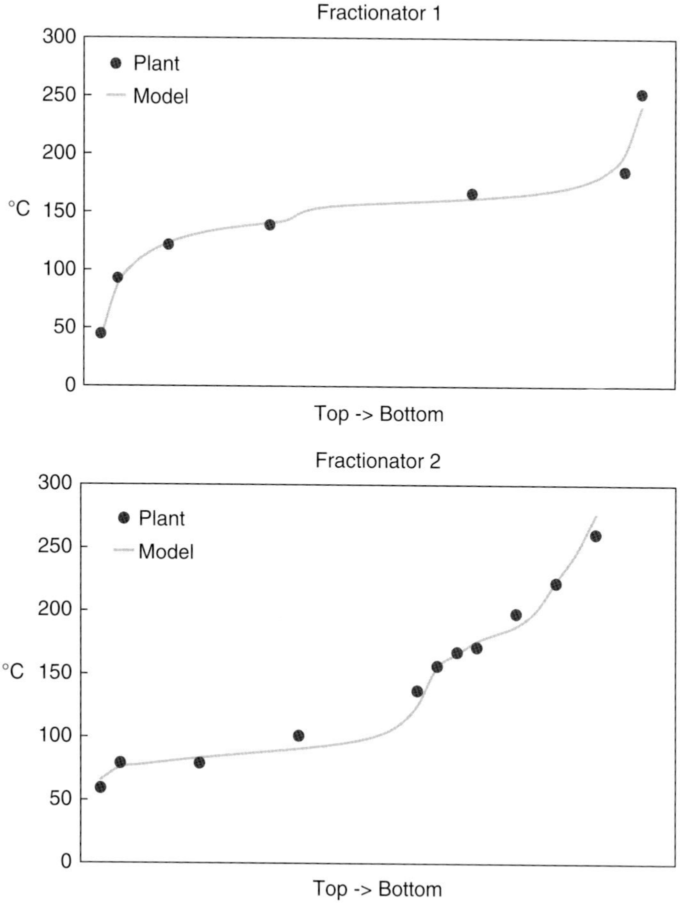

Figure 6.52 Prediction of the temperature profiles of fractionators (data set 1 in HP HCR).

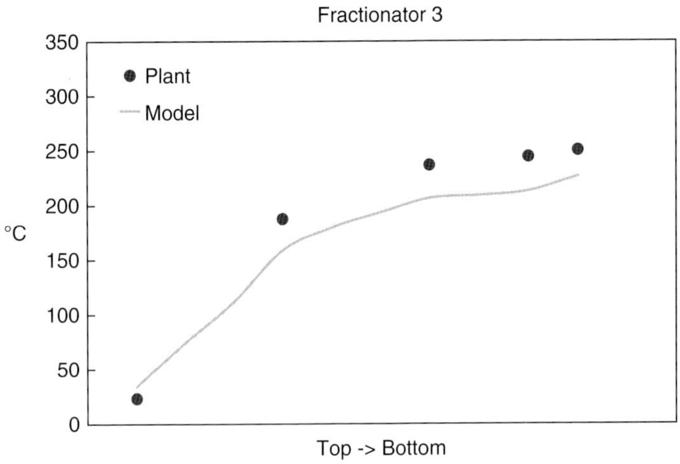

Figure 6.52 *(Continued)*

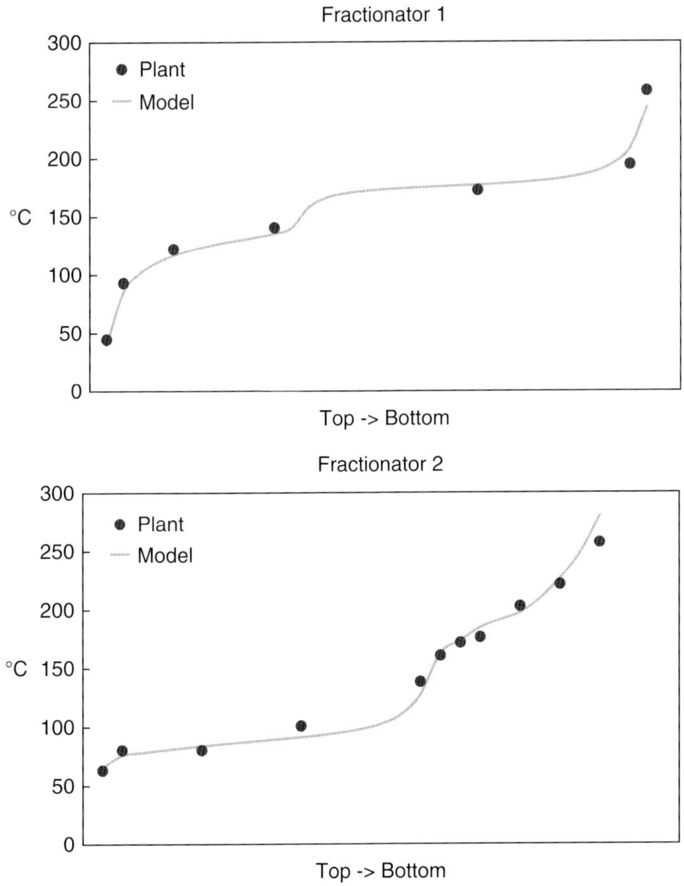

Figure 6.53 Prediction of the temperature profiles of fractionators (data set 7 in HP HCR).

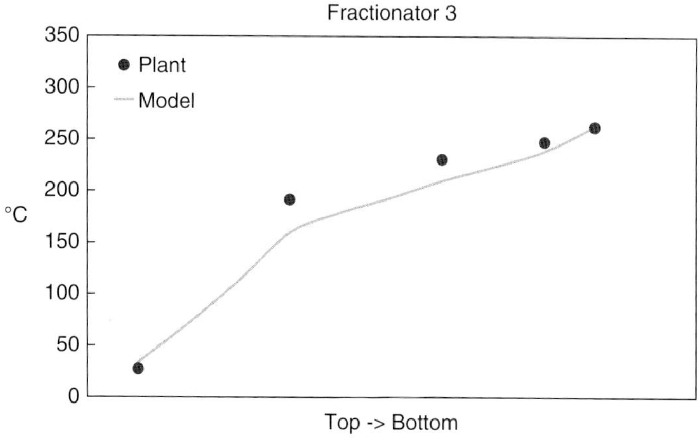

Figure 6.53 (*Continued*)

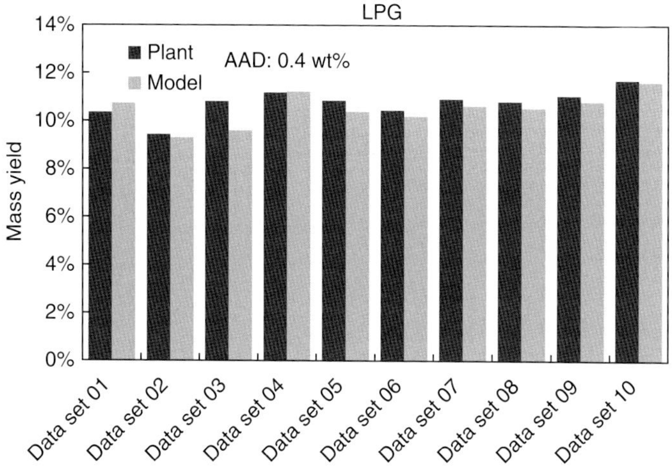

Figure 6.54 Predictions of the LPG yield (HP HCR).

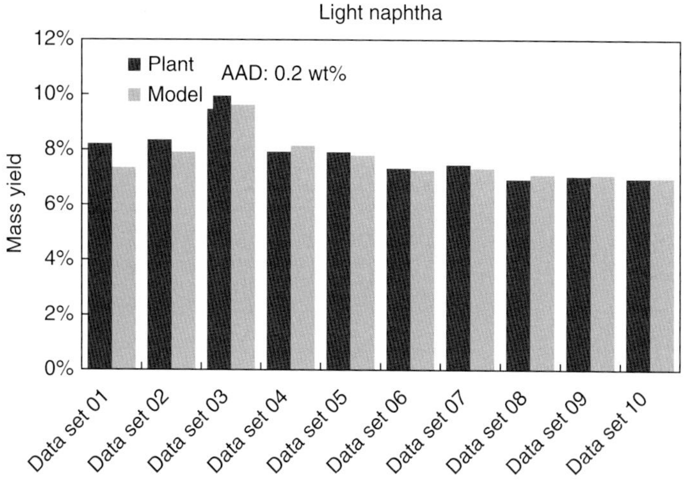

Figure 6.55 Predictions of the light naphtha yield (HP HCR).

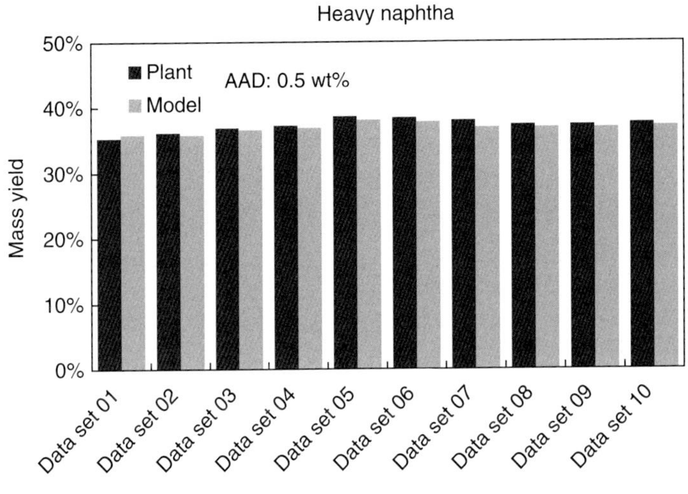

Figure 6.56 Predictions of the heavy naphtha yield (HP HCR).

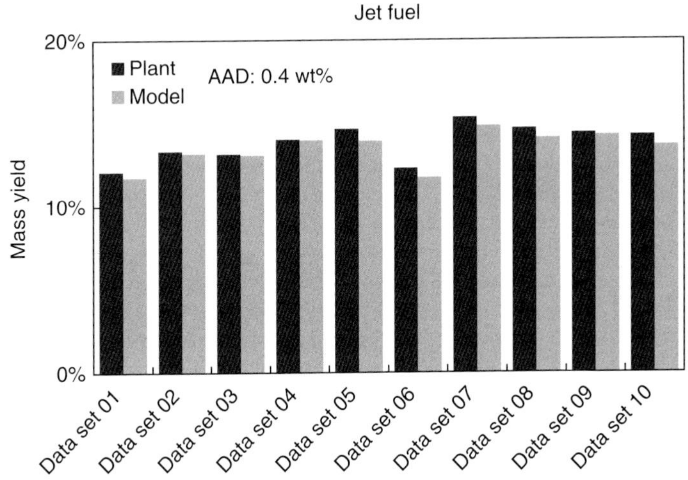

Figure 6.57 Predictions of the jet fuel yield (HP HCR).

major products because they account for over 95 wt% of the overall production. Figures 6.54–6.58 illustrate the model predictions on LPG, light naphtha, heavy naphtha, jet fuel, and residue oil and the AADs are 0.4, 0.2, 0.5, 0.4, and 1.7 wt%, respectively. The model provides good prediction results on product yields when considering overall production.

6.6.4 LPG Composition and Distillation Curves of Liquid Products

Composition, particularly C3 and C4, is the most important indicator to evaluate the quality of the LPG product. Figure 6.59 represents selected model predictions on LPG composition with AAD of each component. For the most important components, C3 and C4, the model shows only 0.021 and 0.058 AADs, respectively,

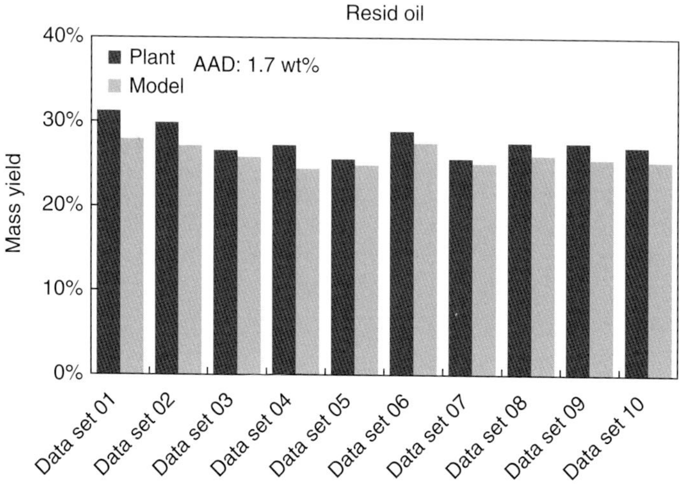

Figure 6.58 Predictions of the residue oil yield (HP HCR).

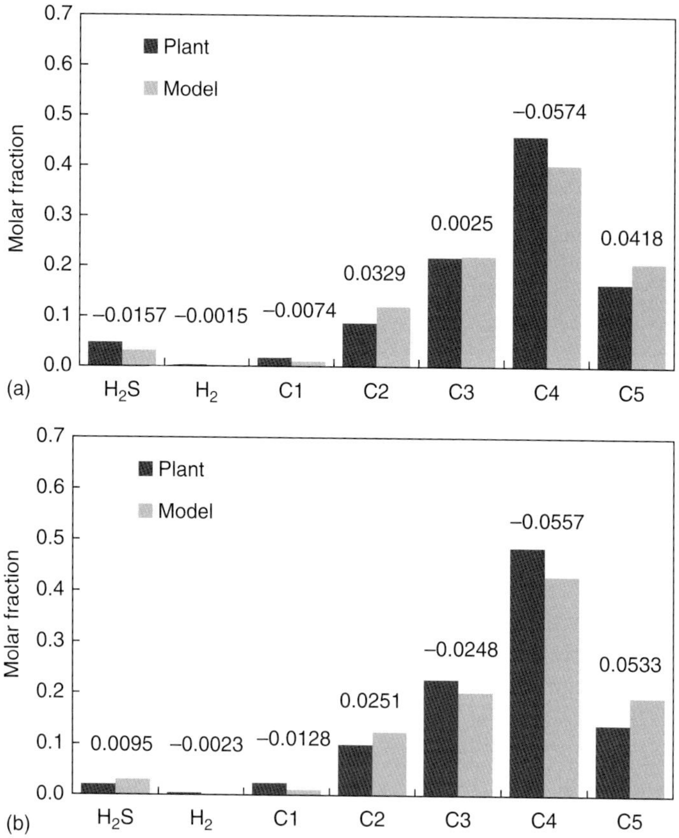

Figure 6.59 Predictions of LPG compositions (HP HCR).

in molar fraction predictions. For other liquid products, the distillation curve is the most popular analysis to indicate the vaporization temperature after having a certain amount of oil fraction vaporized. Figures 6.60 and 6.61 illustrate selected model predictions on distillation curves of light naphtha, heavy naphtha, jet fuel, and residue oil.

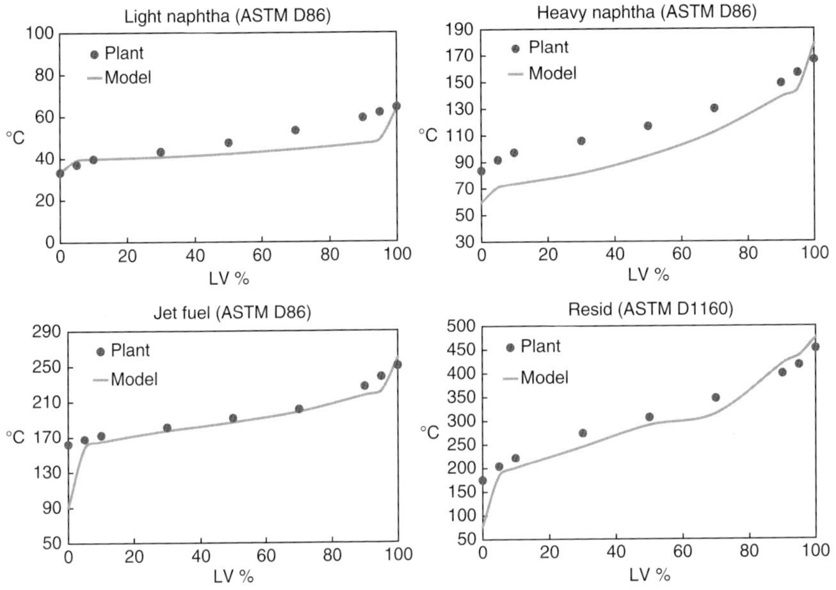

Figure 6.60 Predictions of distillation curves of liquid products (data set 1 in HP HCR).

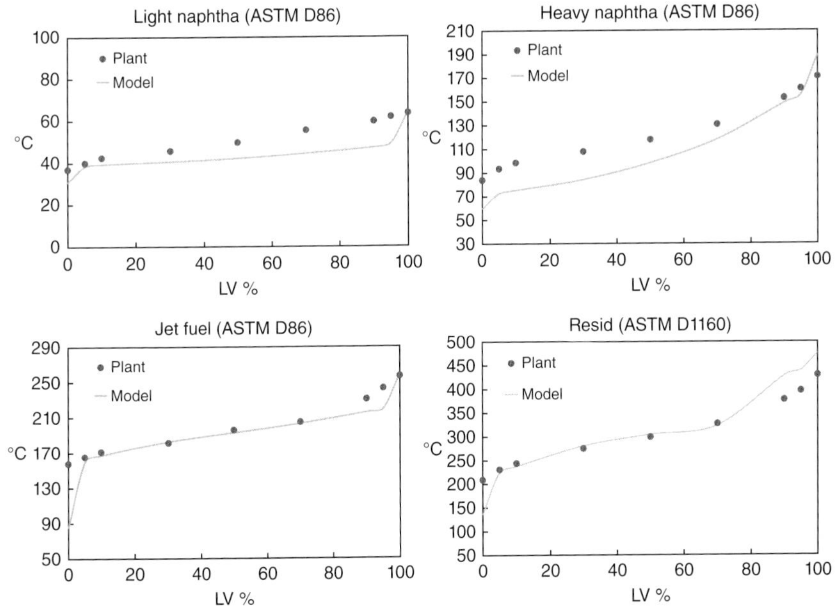

Figure 6.61 Predictions of the distillation curves of liquid products (data set 7 in HP HCR).

6.6.5 Product Property

We apply the updated correlations developed in Section 6.4.6 to estimate the flash point and freezing point of jet fuel. Figures 6.62 and 6.63 illustrate model predictions on the flash point and freezing point of jet fuel. The AADs are 1.6 and 2.3 °C, respectively, which are about the same values as that obtained from correlating plant data. The integrated model collaborated with updated correlations provides satisfactory predictions on the flash point and freezing point of jet fuel. Figures 6.64–6.67 illustrate the specific gravity predictions of liquid products, which are calculated by Aspen HYSYS Petroleum Refining. The AADs of the specific gravity predictions for light naphtha, heavy naphtha, jet fuel, and residue oil are 0.0049, 0.0062, 0.0134, and 0.0045, respectively.

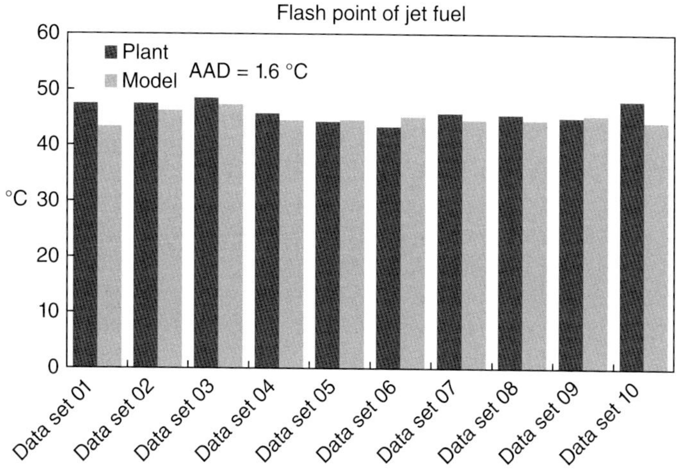

Figure 6.62 Predictions of the flash point of jet fuel (HP HCR).

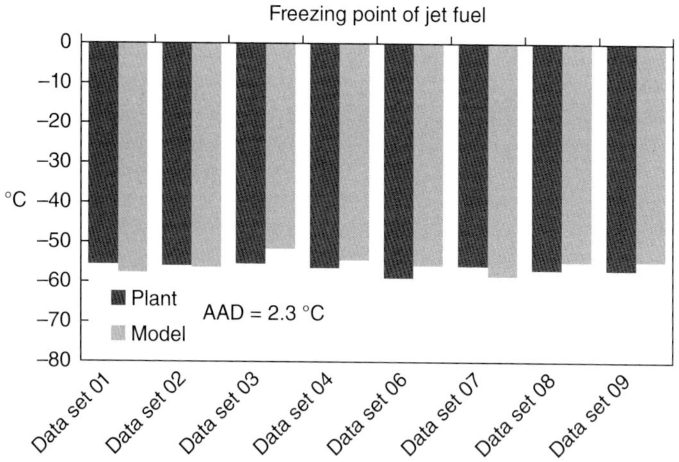

Figure 6.63 Predictions of the freezing point of jet fuel (HP HCR).

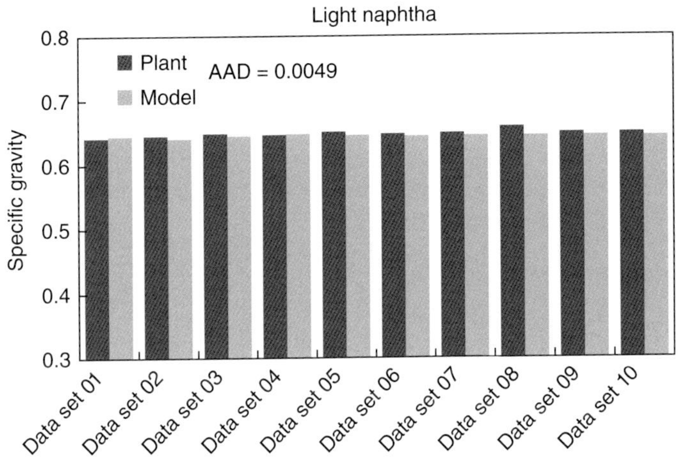

Figure 6.64 Predictions of the specific gravity of light naphtha (HP HCR).

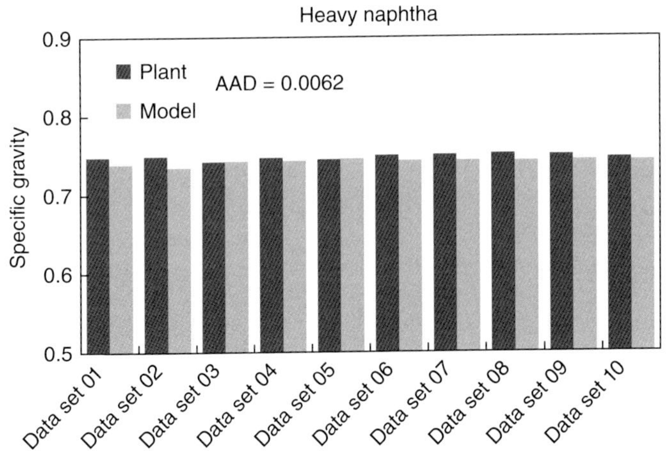

Figure 6.65 Predictions of the specific gravity of heavy naphtha (HP HCR).

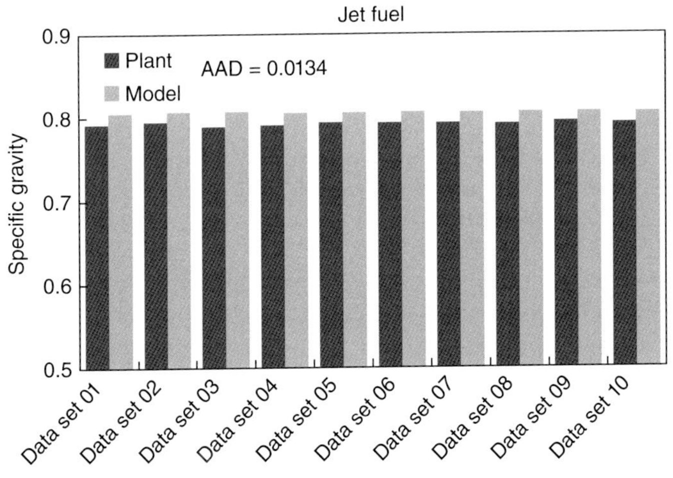

Figure 6.66 Predictions of the specific gravity of jet fuel (HP HCR).

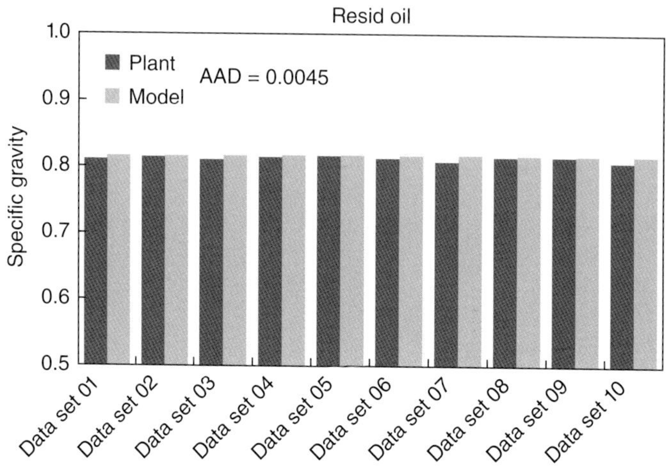

Figure 6.67 Predictions of the specific gravity of residue oil (HP HCR).

6.7 Model Applications

The major operating variables that affect the product distribution (yield) of the HCR process are the reactor temperature, hydrogen partial pressure, amount of ammonia present, and residence time. This section uses MP HCR model to illustrate how to quantify the effects of operating variables on process performance.

6.7.1 H_2-to-Oil Ratio versus Product Distribution, Remained Catalyst Life, and Hydrogen Consumption

Hydrogen partial pressure is a key operating variable for the HCR process. It has two opposite effects on product distribution and process profitability. A higher hydrogen partial pressure can enhance aromatic hydrogenation, increase the H/C ratio of products, and extend the catalyst life by reducing coke precursors (hydrogenation of multiring aromatics). Hydrogen also has a negative effect on paraffin HCR that is crucial for product distribution [44]. In addition, a higher hydrogen partial pressure leads to higher hydrogen consumption, which raises the processing cost.

In this section, we conduct a simulation experiment to study the relationship among the hydrogen partial pressure, product distribution, and remaining catalyst life. The catalyst deactivation model is built in Aspen HYSYS Petroleum Refining that estimates the remaining catalyst life by WART at the SOC (start of run), WART at the EOC (end of run, provided by the catalyst vendor), WART of the current operation, number of days in service, coke precursors (multiring aromatics) in the feed, and hydrogen partial pressure. As the industrial HCR process tunes hydrogen partial pressure through changing gas:oil ratio, we choose the gas:oil ratio as the operating variable rather than the hydrogen partial pressure. Figure 6.68 represents the selected H_2:oil ratios in our simulation experiment and the corresponding values of the hydrogen partial pressure.

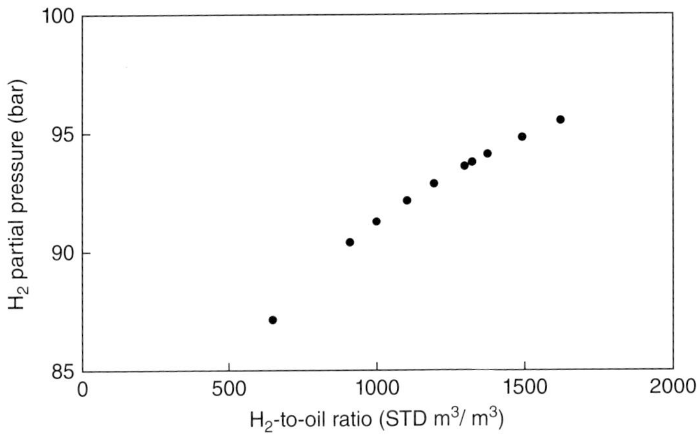

Figure 6.68 H_2:oil ratios and the corresponding values of H_2 partial pressure.

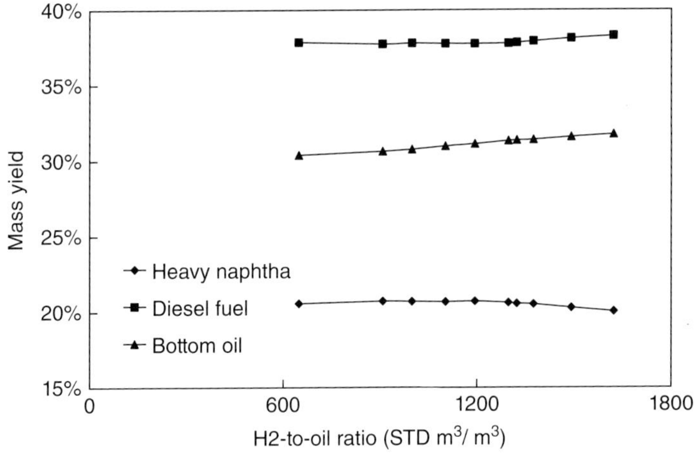

Figure 6.69 H_2:oil ratios and the corresponding values of H_2 partial pressure.

Figure 6.69 illustrates that the H_2:oil ratio (hydrogen partial pressure) has little effect on product distribution. The flat product distribution under various H_2:oil ratios (hydrogen partial pressures) is consistent with observations reported in the literature [45–47]. This implies that the current operation is around the maximum conversion and a further increase/decrease in the hydrogen partial pressure will not change the yields of valuable products, such as heavy naphtha and diesel. Even so, the H_2:oil ratio is still a double-edged knife for process profitability because it affects the hydrogen consumption and remaining catalyst life. Figure 6.70 represents how the H_2:oil ratio affects hydrogen consumption and remaining catalyst life. Obviously, the H_2:oil ratio has a positive effect on both variables. However, the two variables have opposite effects on process profitability, and we can use the model to study the optimal operating point.

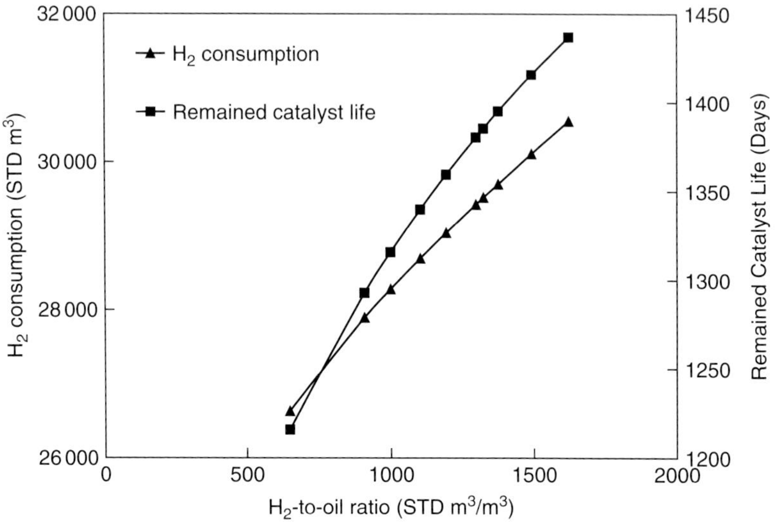

Figure 6.70 Effects of the H_2:oil ratio on H_2 consumption and catalyst life.

6.7.2 WART Versus Feed Flow Rate Versus Product Distribution

The most important operating variable for HCR process is the reactor temperature. Increasing the reactor temperature increases the reaction conversion and shifts the product distribution from heavier to lighter products. However, increasing the reactor temperature does not always benefit the refinery and may generate a process safety issue. This follows because a high reactor temperature will accelerate coke formation, and the secondary HCR of middle distillate oils (such as gasoline and diesel) will increase the product yield of gas products, which are less profitable. Thus, the refiners intend to raise the reactor temperature gradually to produce a desirable product distribution. For instance, the 2 months of operating data of the MP HCR unit show that the WART of the HCR reactor varies within ±8°C from the base data set.

Figures 6.71–6.73 illustrate the effects of the feed flow rate and WART (HCR reactor) on product distribution (yields). The heavy naphtha yield increases significantly, whereas WART increases and/or feed flow decreases. This follows because the rising HCR reactor temperature enhances HCR reactions, and the decreasing feed flow implies a longer residence time that also enhances HCR reactions. On the other side, the bottom oil yield presents the opposite trend to that of the heavy naphtha yield. This follows because bottom oil is the heaviest product and the higher severity of cracking reactions resulted from a rising WART and/or a falling feed flow rate, with the lower bottom oil yield. However, the most interesting observation comes from Figure 6.72 that represents the diesel fuel yield reaching a maximum value at some operating point. Both Tippett *et al.* [48] and Rossi *et al.* [49] reported that in a HCR process, the yield of the middle distillate fraction (diesel fuel in this case) would approach a maximum value with an increasing reactor temperature because of secondary

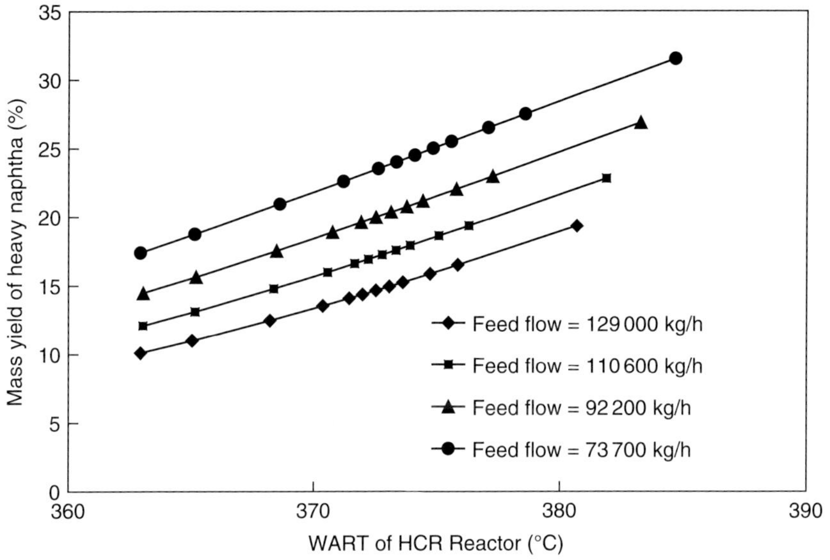

Figure 6.71 Effect of the feed flow rate and WART of the HCR reactor on the heavy naphtha yield.

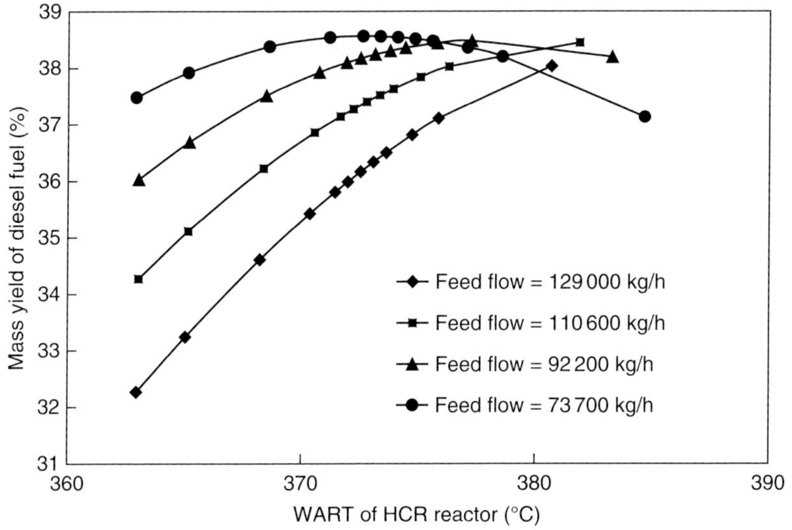

Figure 6.72 Effect of the feed flow rate and WART of the HCR reactor on the diesel fuel yield.

HCR reactions of middle distillate paraffins. We can conclude that with a lower feed flow rate, the diesel fuel yield tends to approach a maximum when increasing WART of the HCR reactor. When performing this simulation experiment, refiners can determine the optimal reactor temperature and feed flow rate to achieve maximum profits under various supply-and-demand situations.

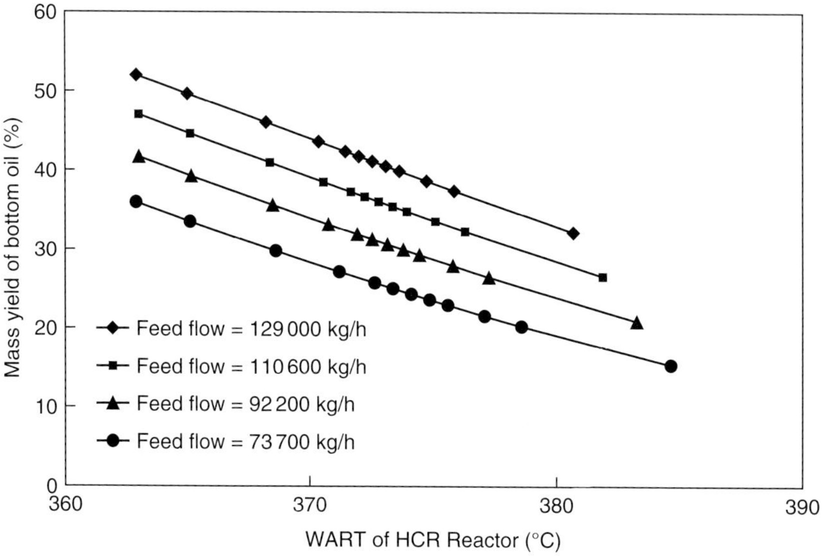

Figure 6.73 Effect of the feed flow rate and WART of the HCR reactor on the bottom oil yield.

6.8 Model Application – Delta-Base Vector Generation

Refining industry started to investigate the application of linear programming (LP)-based model since 1950s [50]. Nowadays, LP-based model is the most important optimized tool to schedule production, evaluate feedstock, study new process configuration, and adjust production plan after operational upsets. For a given refinery, the LP-based model is a combination of economic and technical databanks. The economic databank requires the availability and price of feedstocks, the demand and price of refining products, and operating cost of process units. The technical databank needs process product yields, product properties, product specifications, operating constraints, and use of utility.

Modern refiners gather and update most of the required information from market research, government regulation, design data, and operating history except for product yields. Instead of adopting historic data, refiners apply process model to generate required information of product yields for LP-based model. However, actual refining reaction processes are highly nonlinear and the responses of product yields to process variables, such as operating conditions and feed properties, are usually complex. Figure 6.74 illustrates the nonlinear relationship between the HCR reactor temperature and product distribution (redrawn from Ref. [51]). Yield of each product represents nonlinear variation along with the change of reactor temperature. To integrate the nonlinear relationship between product yields and process variables with LP-based model, refiners linearize product yields over a small range of process variables as illustrated in Figure 6.75a. The linear relationship between product yields and process variables is so-called "delta-base" technology in modern refinery production planning.

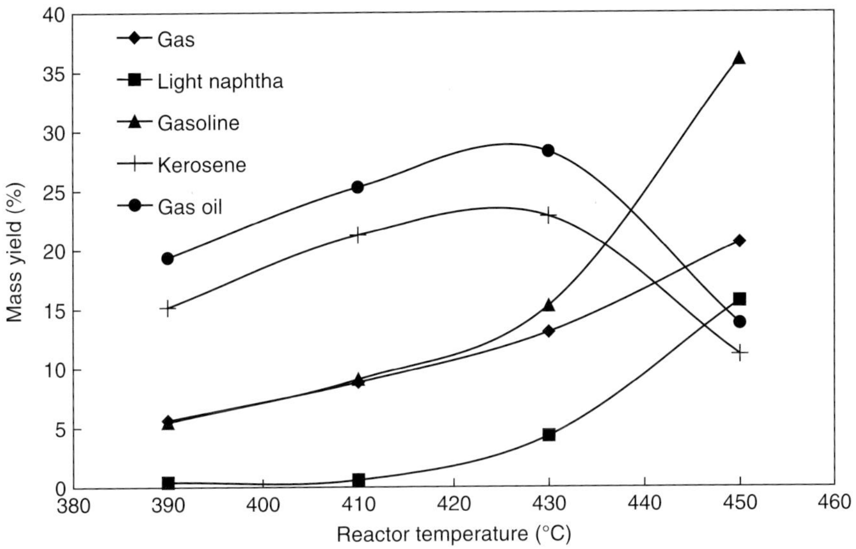

Figure 6.74 Nonlinear relationship between product distribution and reactor temperature: (a) linearization of process performance. (b) Scenarios for different operations.

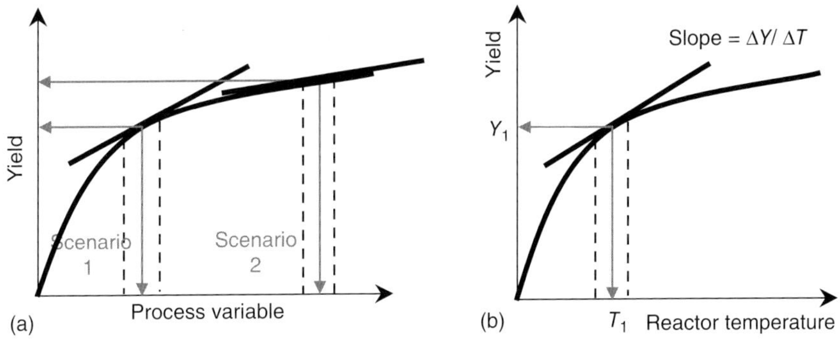

Figure 6.75 Linearization of production yield's response on process variable.

As shown in Eq. (6.28), refiners use delta-base technology to construct the linear representation of product yield's (Y) response to the change of process variable ($X - \overline{X} = \Delta X$), namely, delta vector. Base vector ($\overline{Y}$) represents the product yield under a selected operating condition and feedstock quality ($\overline{X}$). Delta-base vector ($\Delta Y / \Delta X$) indicates the departure of product yield (ΔY) from base vector ($\overline{Y}$) corresponding to a unit change of process variable (ΔX). The delta-base technology simplifies the nonlinearity of refining process and allows LP-based product planning to consider the product yield. However, the resulting LP-based model only provides good prediction of product yield over a small range of operating condition and feedstock qualities. To extend the application of LP-based model, refiners generate different sets of delta-base vectors to reflect various production scenarios as shown in Figure 6.75b. By doing this, the

	TEXT	R90	NA1	RF1	R94	NA2	RF2	R98	NA3	RF3	R02	NA4	RF4
*													
FREE	Free up Adjustors		1			1			1			1	
*													
WBALRFF	Reformer Feed			1.0			1.0			1.0			1.0
*													
WBALHYL	Low-Purity H2	-0.0173	-0.000		-0.0194	-0.000		-0.0214	-0.000		-0.0230	-0.0007	
WBALNC1	Methane	-0.0074	0.001		-0.0093	0.001		-0.0118	0.0013		-0.0150	0.0013	
WBALNC2	Ethane	-0.0136	0.002		-0.0171	0.002		-0.0216	0.0025		-0.0275	0.0025	
WBALNC3	Propane	-0.0192	0.003		-0.0241	0.003		-0.0304	0.0035		-0.0387	0.0035	
WBALIC4	Iso-Butane	-0.0087	0.001		-0.0109	0.001		-0.0137	0.0016		-0.0175	0.0016	
WBALNC4	N- Butane	-0.0129	0.002		-0.0163	0.002		-0.0206	0.0023		-0.0263	0.0023	
WBALR90	90 RONC Reformate	-0.9209	-0.010										
WBALR94	94 RONC Reformate				-0.9029	-0.010							
WBALR98	98 RONC Reformate							-0.8805	-0.010				
WBALR02	102 RONC Reformate										-0.8520	-0.0105	

Figure 6.76 Multiscenario delta-base vectors in a catalytic reforming process.

LP-based production planning can switch over delta-base vectors according to the production scenario. Figure 6.76 represents multiscenario delta-base vectors of a catalytic reforming process used in Aspen PIMS. The delta-base vector inside each red box indicates a scenario producing gasoline product with varying research octant number (RON).

$$
\begin{bmatrix} Y_1 \\ Y_2 \\ \vdots \\ \vdots \\ Y_m \end{bmatrix} = \begin{bmatrix} \overline{Y}_1 \\ \overline{Y}_2 \\ \vdots \\ \vdots \\ \overline{Y}_m \end{bmatrix} + \begin{bmatrix} \dfrac{\Delta Y_1}{\Delta X_1} & \dfrac{\Delta Y_1}{\Delta X_2} & \cdots & \dfrac{\Delta Y_1}{\Delta X_n} \\ \dfrac{\Delta Y_2}{\Delta X_1} & \dfrac{\Delta Y_2}{\Delta X_2} & \cdots & \dfrac{\Delta Y_2}{\Delta X_n} \\ \vdots & & & \vdots \\ \vdots & & & \vdots \\ \dfrac{\Delta Y_m}{\Delta X_1} & \dfrac{\Delta Y_m}{\Delta X_2} & \cdots & \dfrac{\Delta Y_m}{\Delta X_n} \end{bmatrix} \bullet \begin{bmatrix} X_1 - \overline{X}_1 \\ X_2 - \overline{X}_2 \\ \vdots \\ \vdots \\ X_n - \overline{X}_n \end{bmatrix} \tag{6.28}
$$

To generate delta-base vector, refiners produce case studies by running process model under varied feed and operating conditions by the following procedures:

1) Run the process model.
2) Choose the process variables to produce case studies.
   - In real practice, we choose feedstock qualities (such as specific gravity, Watson K, and PNA) rather than operating conditions.
3) Record base yields (base vector), $\overline{Y}$ in Eq. (6.28), and the values of the selected process variables, $\overline{X}$ in Eq. (6.28), in the process model.
4) Produce case studies by running the process model with changing selected process variable(s).
5) Record the changes of process variables, ΔX in Eq. (6.28), and the corresponding yields, Y in Eq. (6.28).
6) Apply Eq. (6.28) to run a linear regression to obtain delta-base vector.

In this section, we use HP HCR model to demonstrate how to generate the delta-base vector from computer simulation. We choose sulfur content, Watson K factor, and API gravity of feed oil as process variables to perform case study and generate delta-base vector. The product yields calculated from the base case of HP HCR model are defined as base vector, $\overline{Y}$ in Eq. (6.28). Then, we input different

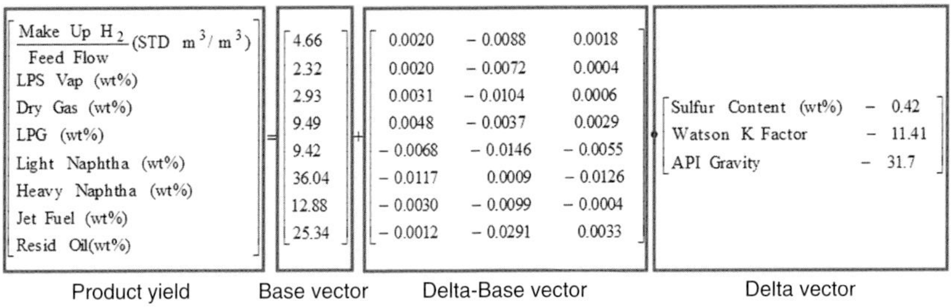

Figure 6.77 Delta-base vector of HP HCR process generated in this work.

feed analyses obtained from the refinery into the HP HCR model to produce case studies. We regress Eq. (6.28) with base vector, $\overline{Y}$ in Eq. (6.28). We record the product yields, Y in Eq. (6.28), and the corresponding change of process variable, namely, ΔX, $X - \overline{X}$ in Eq. (6.28), from 15 case studies to obtain delta-base vector. Figure 6.77 represents the resulting delta-base vector for HP HCR process. We generate one set of delta-base vector because the plant data collected for building the HP HCR model are based on the same production scenario.

The resulting delta-base vector shows that sulfur content of feed oil has a positive effect on the yields of light products and a negative effect on heavier liquid product because more H_2S is produced with increasing sulfur compounds of feed oil. Hu *et al.* [52] also reported that sulfur content of feed oil has opposite effects on light and heavy products. However, the trends of API gravity and Watson K factor on product yields are irregular. This follows because API factor and Watson factor are not independent, and the resulting delta-base vector represents the mutual effect of these two variables on product yields as well. It is worth noting that API gravity and Watson K factor are not good enough to generate a delta-base vector of HCR process because they provide little information of feed composition such as PNA content that is important to HCR modeling. The attributes relevant to feed composition should be included to obtain a more precise delta-base vector. Although we only use API gravity and Watson K factor to generate a delta-base vector due to the limitation of plant data, HP HCR model's good predictability among two and half months of plant data provides a good incentive for the application of delta-base vector.

6.9 Workshop 6.1 – Build a Preliminary Reactor Model of HCR Process

We start by opening a new case in Aspen HYSYS and go to Properties environment (see Figure 6.78). We save the simulation file as *Workshop 6.1.hsc.*

The first step in creating the model is the selection of a standard set of components and a thermodynamic basis to model the physical properties of these components. We choose to import a predetermined set of the components for the hydrocracker model (Figure 6.79).

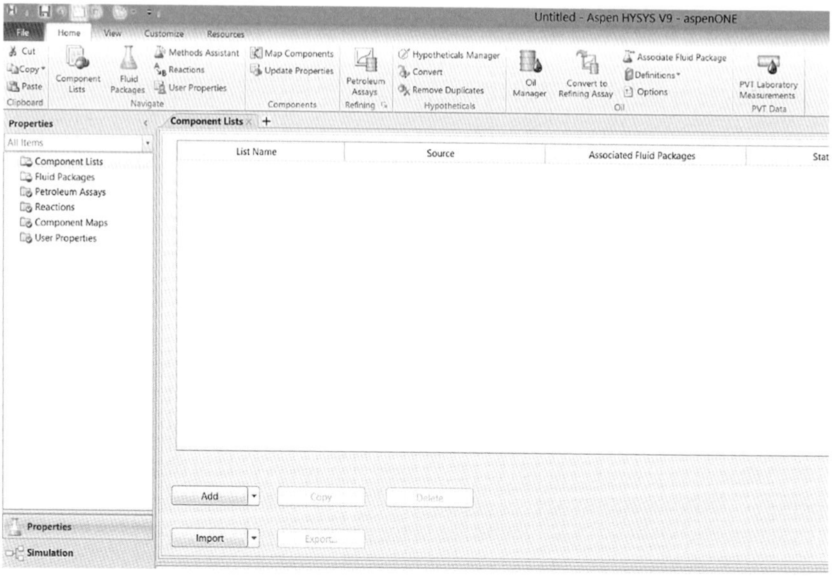

Figure 6.78 Initial start-up of Aspen HYSYS.

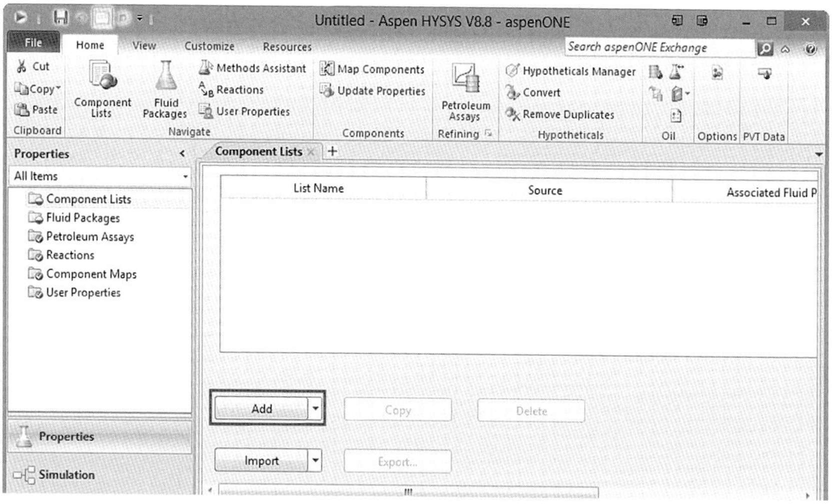

Figure 6.79 Adding a component list.

To import these components, we click "Import" and navigate to the directory location, "C.\Program Files\AspenTech\Aspen HYSYS V9.0\Paks" and select the "Hydrocracker Components Celsius.cml" as the component list (Figure 6.80). The path shown reflects a standard installation of Aspen HYSYS Petroleum Refining software.

Once we import a component list, HYSYS will create a new component list called "Component List-1." We can view the elements of this component lists

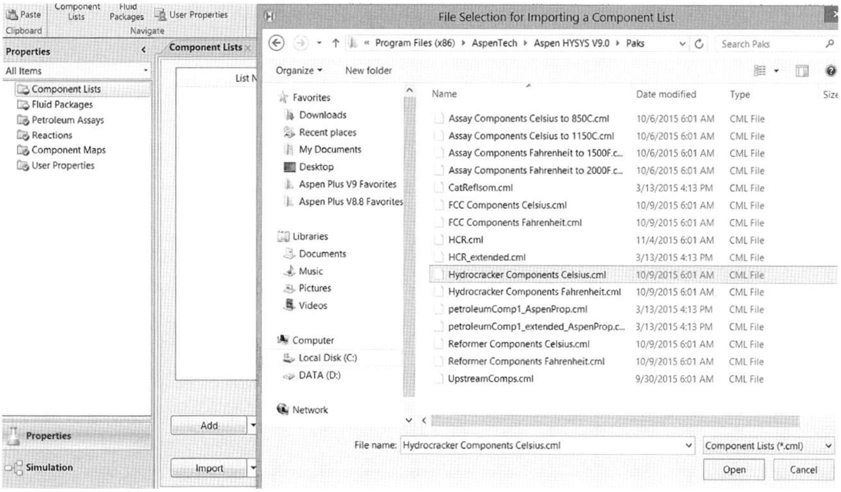

Figure 6.80 Importing hydrocracker component list.

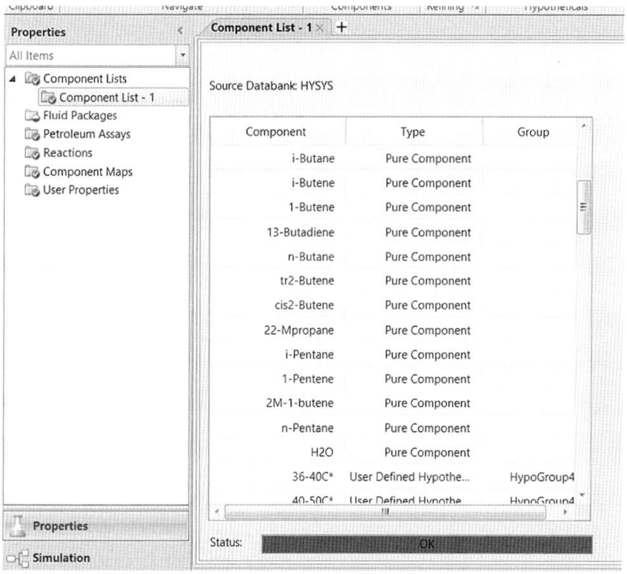

Figure 6.81 Initial component list for hydrocracking process.

by selecting "Component List-1" and clicking on "View" in the Simulation Basis Manager (Figure 6.81). We can add more components or modify the order of the elements in the component list. We note that the standard HCR component list is quite complete and model most refining processes.

The next step is the settings of a "Fluid Package" for this model. The "Fluid Package" refers to the thermodynamic system associated with the chosen list of components. We choose the SRK thermodynamic model for our fluid package

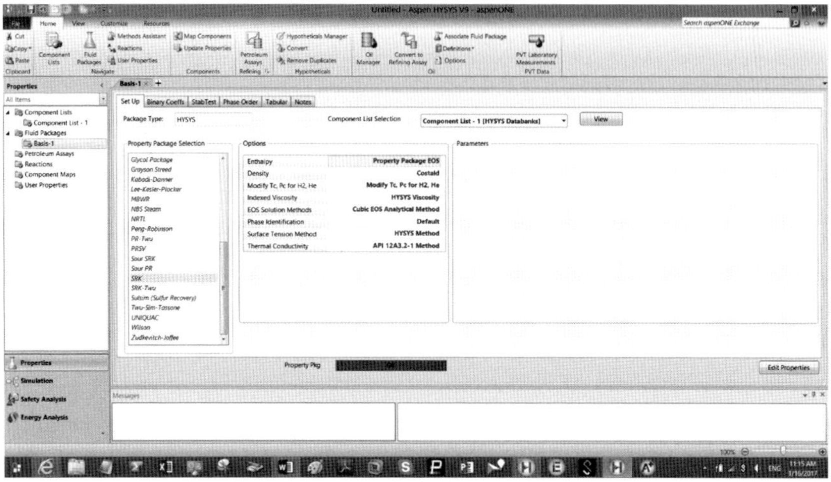

Figure 6.82 Select SRK thermodynamic model for fluid package.

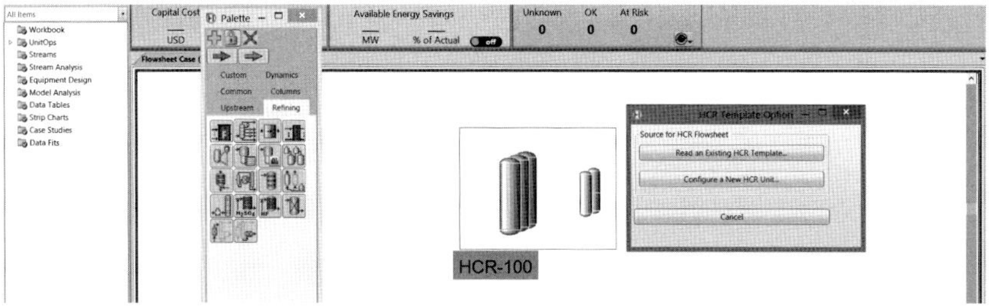

Figure 6.83 Hydrocracker reactor palette – F4 → Refining → Hydrocracker model → Configure a New HCR Unit → HCR configuration wizard.

(Figure 6.82). The HCR system is mostly hydrocarbons and consequently the Soave–Redlich–Kwong equation of state is sufficient. We discuss the implications of the process thermodynamics in Section 1.9.

The initial flowsheet presents a blank interface where we can place different objects from the Object palette shown in Figure 6.83.

We continue this workshop with the following step-by-step guidelines to demonstrate how to build a preliminary reactor model of HCR process:

Step 1. Define the process type (single-stage or two-stage), the number of reactors of each stage and the corresponding reaction beds, the number of HPS, and if the amine treatment is included in the model (Figure 6.84).

Step 2. Assign the dimensions and catalyst loading information of each reaction bed (Figure 6.85).

Step 3. Select data set of reaction activity. "Default" is suggested when building a preliminary model from scratch (Figure 6.86).

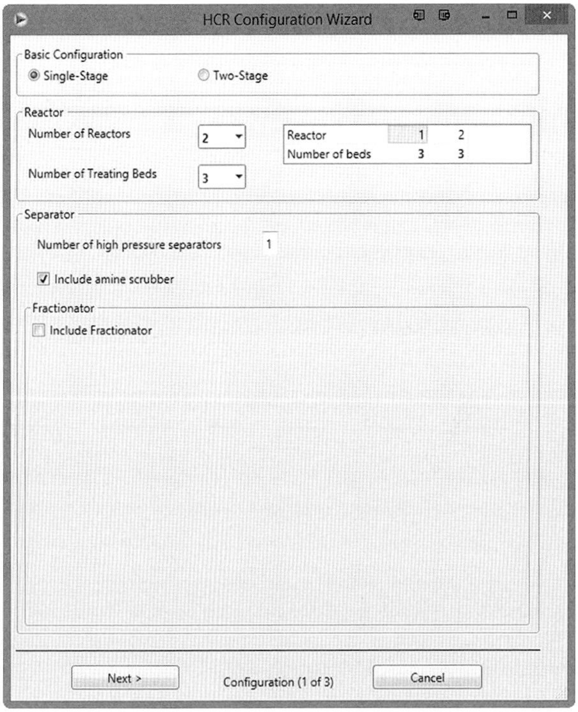

Figure 6.84 HCR configuration wizard – define reactors in HCR process.

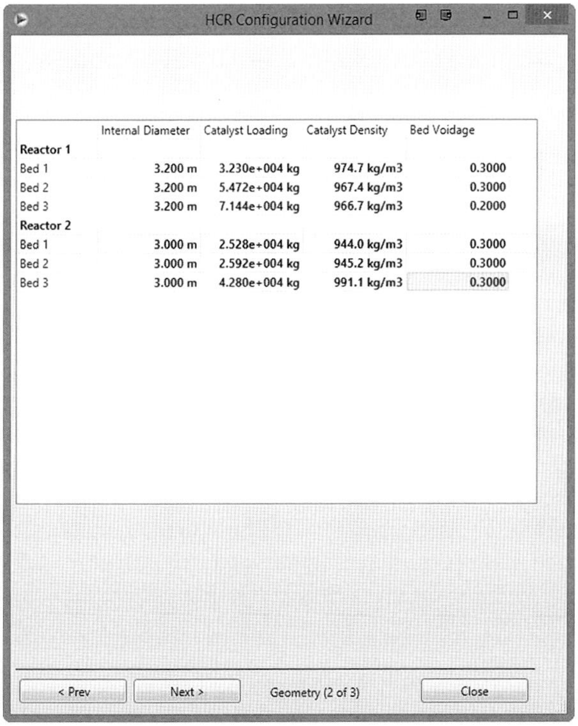

Figure 6.85 Define catalyst bed.

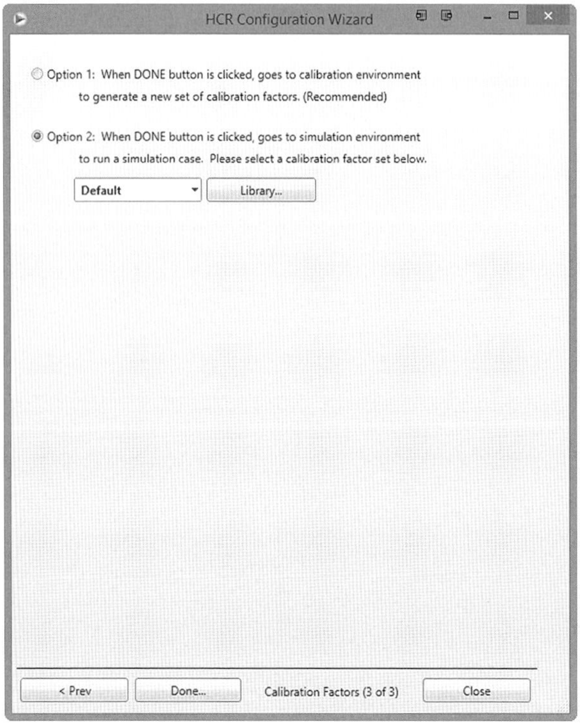

Figure 6.86 Choose set of reaction activity factors.

Step 4. Input the required feed analysis (Figure 6.87).

Step 5. Select an appropriate feed fingerprint (Figure 6.88).

Step 6. Input the conditions of feed streams. The temperature and pressure input here only affect flash calculation of feed stream and have no influence on reactor condition. However, it is important to input correct data of hydrogen flow (Figure 6.89).

Step 7. Input inlet temperature of each reaction bed (Figure 6.90).

Step 8. Input operating data of recycle hydrogen system. It is crucial to ensure that the "outlet pressure of compressor" and "Delta P to Reactor Inlet" are correct because they are used to calculate the inlet pressure of reactor (Figure 6.91).

Step 9. Input the catalyst information provided by vendor. After completing this step, Aspen HYSYS Petroleum Refining will solve the model automatically (Figure 6.92).

Step 10. Increase the number of iterations and reduce the step size of creep step parameters to enhance model convergence (Figure 6.93).

Step 11. Check model results such as product yields and reactor temperature profile. Save the converged simulation file as *Workshop 6.1.hsc* (Figures 6.94 and 6.95).

Figure 6.87 Feed analysis sheet.

Feed Properties

Feed Type	hcrfeed_lvgo
API Gravity	<empty>
Specific Gravity (60F/60F)	0.9103
Distillation Type	D2887
0% Point [C]	279.0
5% Point [C]	323.7
10% Point [C]	359.5
30% Point [C]	411.5
50% Point [C]	418.0
70% Point [C]	426.8
90% Point [C]	468.5
95% Point [C]	505.8
100% Point [C]	531.3
Total Nitrogen [ppmwt]	861.0
Basic Nitrogen [ppmwt]	<empty>
Total/Basic Nitrogen Ratio	4.477
Sulfur Content [%]	1.99
Meas. RI @ Spec. Temp. (Opt.)	<empty>
Meas. RI Temp. (Opt.) [C]	<empty>
RI @ 20C (From Lab Data)	<empty>
RI @ 20C (From Bulk Prop.)	<empty>
Viscosity, cSt@210 F Lab (Opt.)	<empty>
Viscosity, SUS@210 F Lab (Opt.)	<empty>
Viscosity, cSt@210 F (From Bulk Pr(	<empty>
Measured Ca (Opt.)	<empty>
Ca Est. From Total Method	<empty>

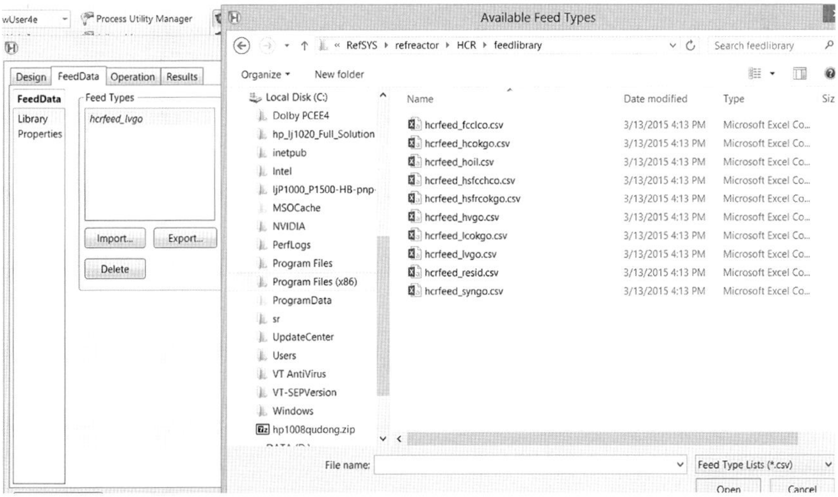

Figure 6.88 Import from the HCR feed library. C → Program Files → AspenTech → Aspen HYSYS 9.0 → RefSYS → refreactor → HCR → feedlibrary → hcrfeed_lvgo.csv.

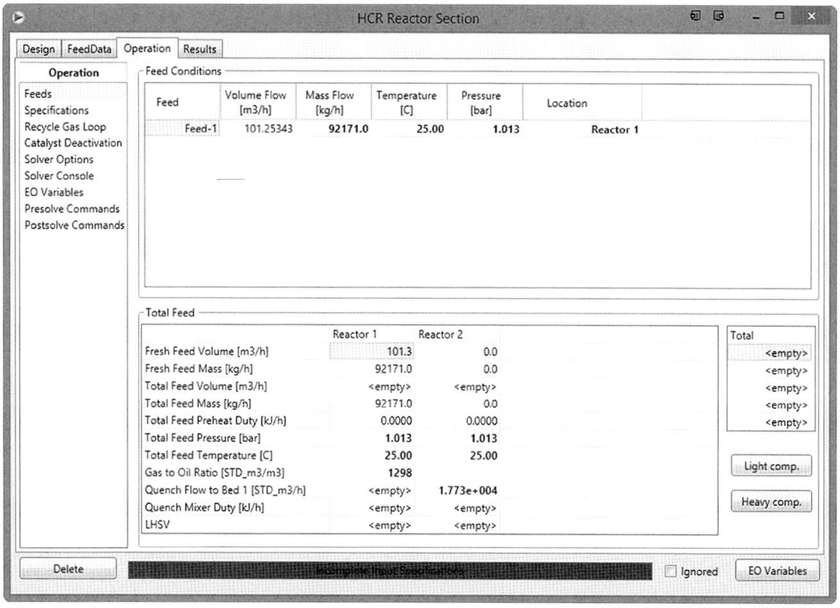

Figure 6.89 Define feed conditions.

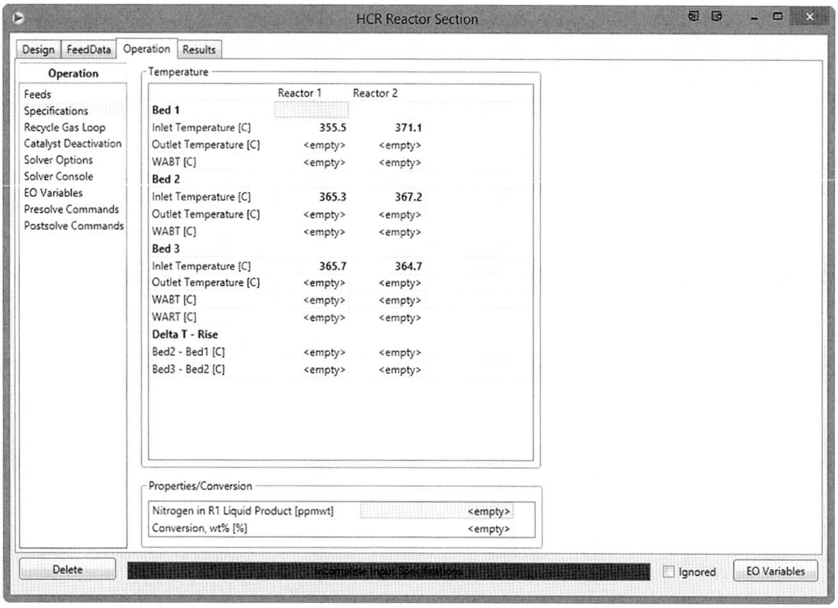

Figure 6.90 Assign reactor temperature.

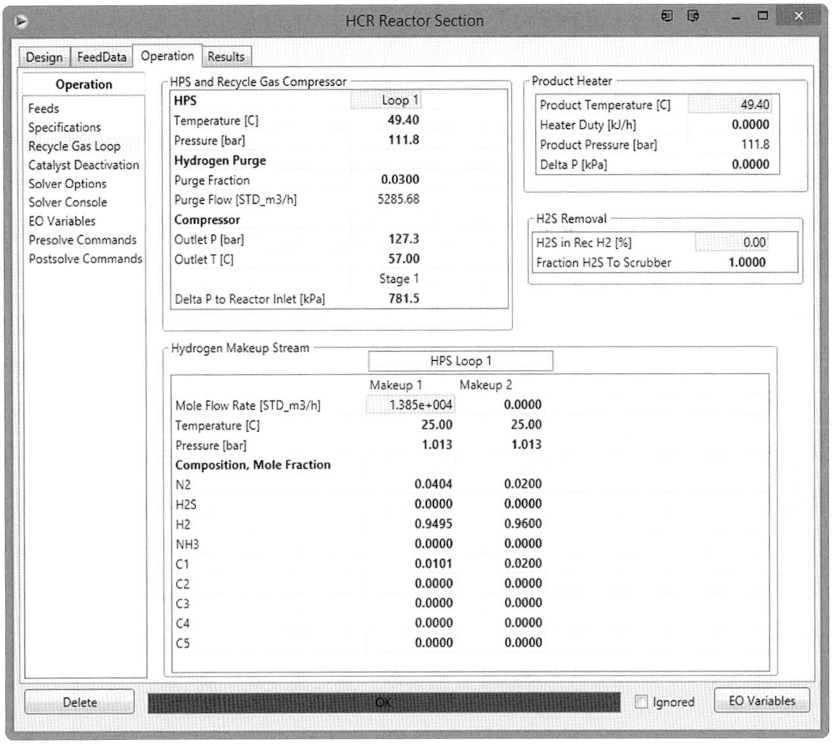

Figure 6.91 Define hydrogen recycle system.

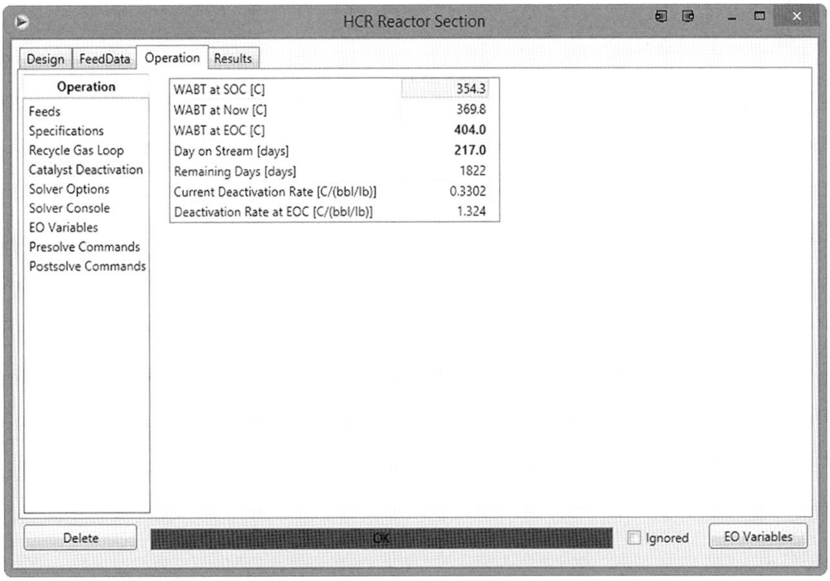

Figure 6.92 Catalyst deactivation information.

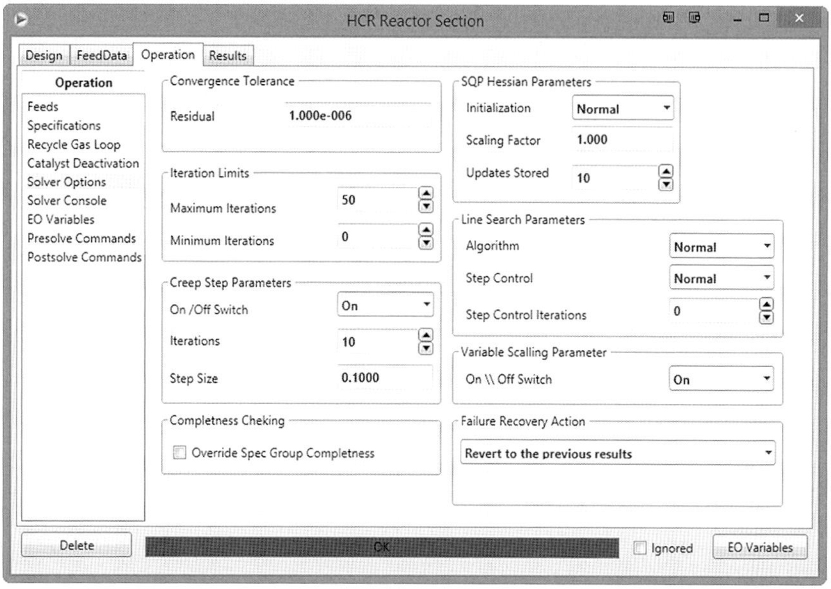

Figure 6.93 Select algorithm for model convergence.

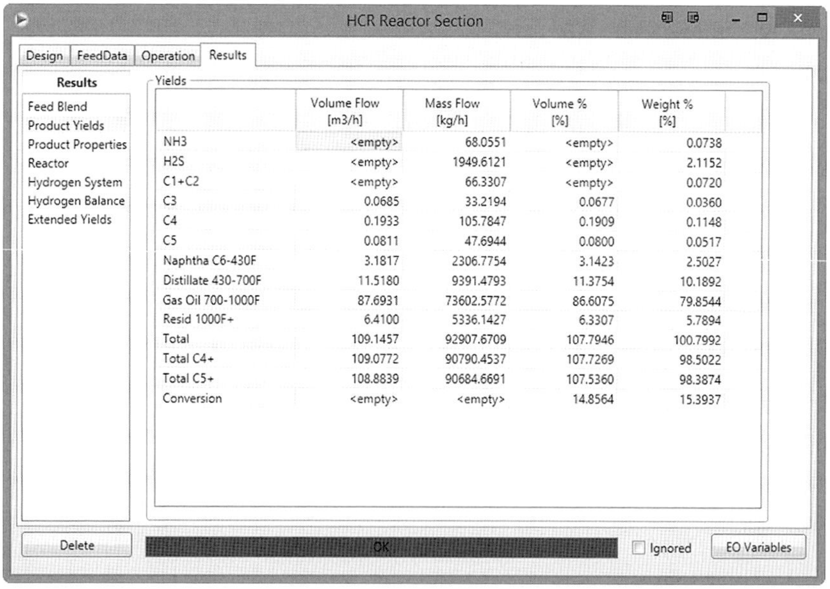

	Volume Flow [m3/h]	Mass Flow [kg/h]	Volume % [%]	Weight % [%]
NH3	<empty>	68.0551	<empty>	0.0738
H2S	<empty>	1949.6121	<empty>	2.1152
C1+C2	<empty>	66.3307	<empty>	0.0720
C3	0.0685	33.2194	0.0677	0.0360
C4	0.1933	105.7847	0.1909	0.1148
C5	0.0811	47.6944	0.0800	0.0517
Naphtha C6-430F	3.1817	2306.7754	3.1423	2.5027
Distillate 430-700F	11.5180	9391.4793	11.3754	10.1892
Gas Oil 700-1000F	87.6931	73602.5772	86.6075	79.8544
Resid 1000F+	6.4100	5336.1427	6.3307	5.7894
Total	109.1457	92907.6709	107.7946	100.7992
Total C4+	109.0772	90790.4537	107.7269	98.5022
Total C5+	108.8839	90684.6691	107.5360	98.3874
Conversion	<empty>	<empty>	14.8564	15.3937

Figure 6.94 Model results – product yield.

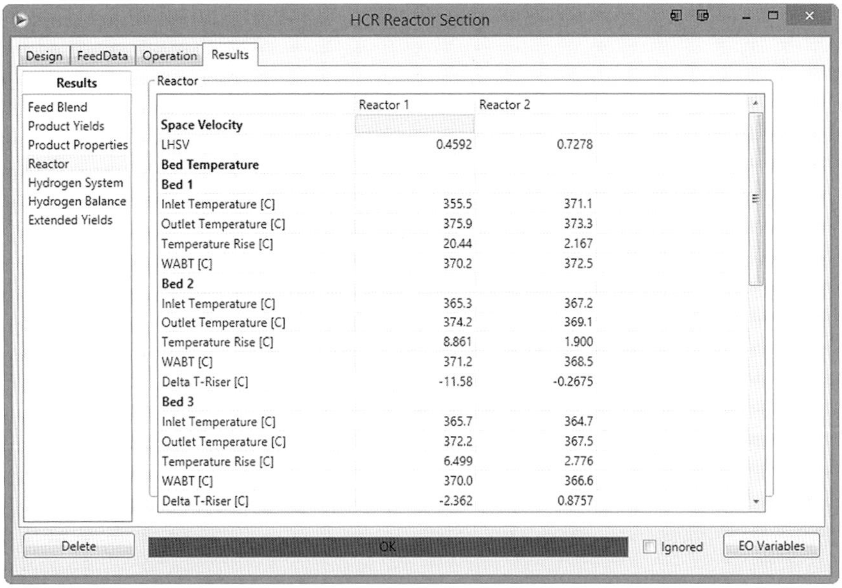

Figure 6.95 Model results – reactor performance.

6.10 Workshop 6.2 – Calibrate Preliminary Reactor Model to Match Plant Data

After completing preliminary model, it is necessary to calibrate the model to match plant measurement. The following section represents a step-by-step guideline to calibrating a preliminary model to match plant measurement of reactor temperature profile and product yields. We continue with simulation file, *Workshop 6.1.hsc,* and save it as *Workshop 6.2-starting.hsc.*

Step 1. Enter the “calibration” environment (Figure 6.96).

Step 2. Click the button of “Pull Data from Simulation” to import the results of the preliminary model (Figure 6.97).

Step 3. Input temperature rise and pressure drop of each reaction bed (Figure 6.98).

Step 4. Input quench flow of each reaction bed, sour gas removal, makeup hydrogen rate, chemical hydrogen consumption, nitrogen content in first reactor's effluent, and composition of purge gas (Figure 6.99).

Step 5. Define the number of cuts in each distillate range (Figure 6.100).

Step 6. Input compositional analyses and flow rates of light ends (Fuel Gas 1, Fuel Gas 2, and LPG1). These are important to calculate the composition of naphtha cuts (Figure 6.101).

Step 7. Input distillation curves, elemental analyses, specific gravities, and flow rates of liquid products. Distillation curves and flow rates are the most important properties and they have to be accurate to ensure that the model

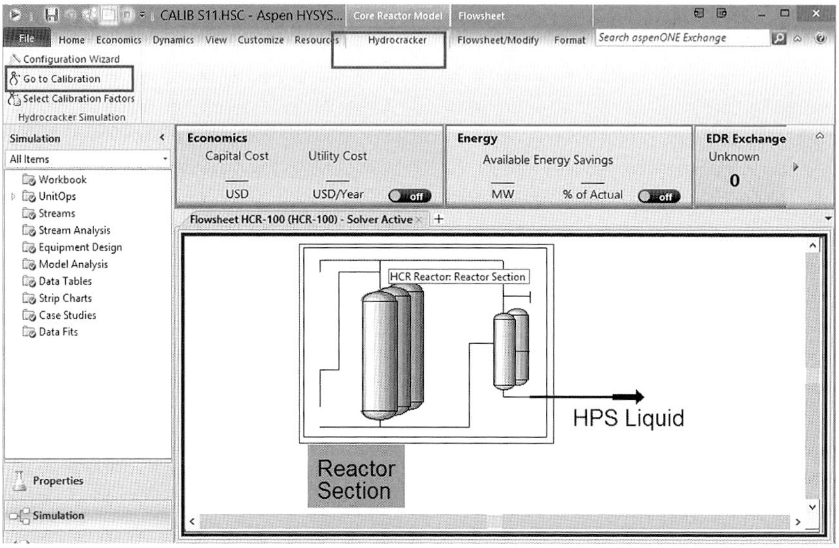

Figure 6.96 Enter calibration environment.

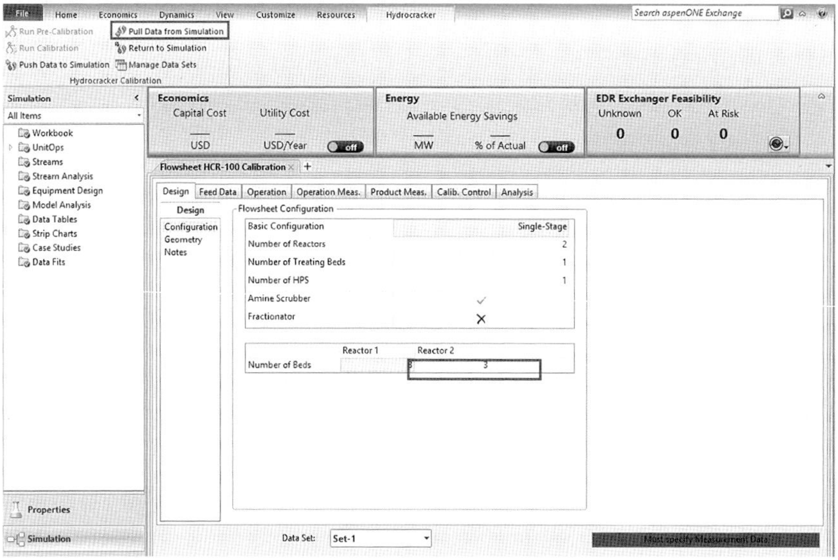

Figure 6.97 Extract data from simulation.

works well. Specific gravity affects the model's prediction on hydrogenation reaction rate. Elemental analysis only affects the severity of HDN and HDS reactions and hydrogen balance, which have little effect on yield predictions of HCR model (Figure 6.102).

Step 8. Change the iteration scheme to enhance model's convergence (Figure 6.103).

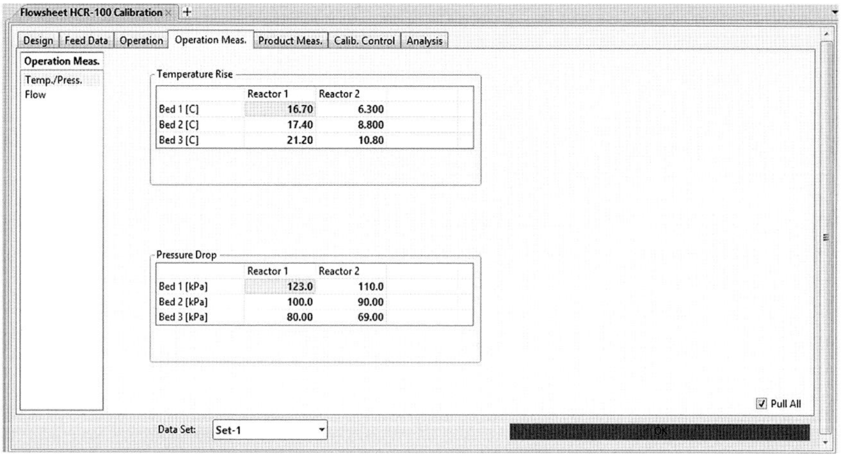

Figure 6.98 Input reactor variables.

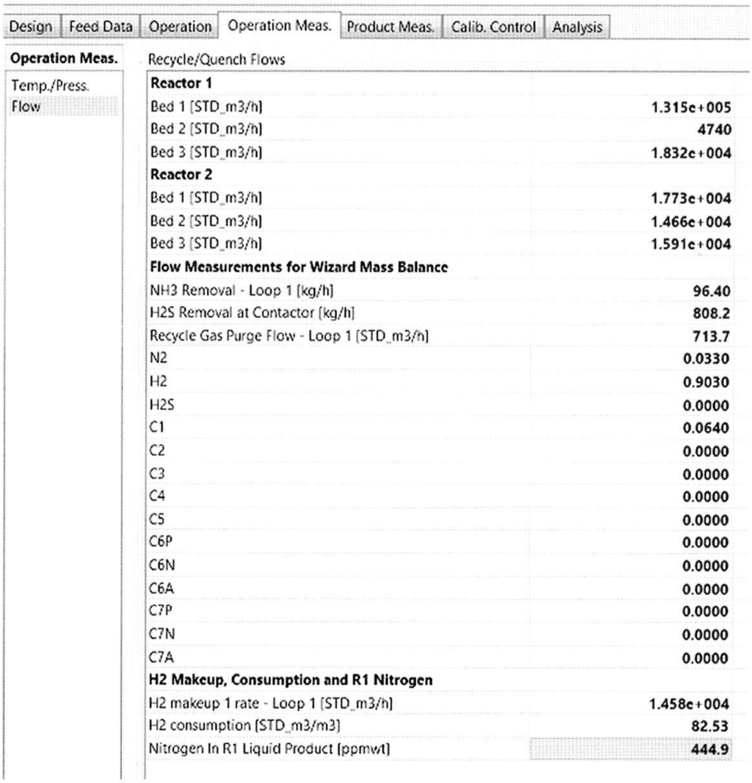

Figure 6.99 Input process data.

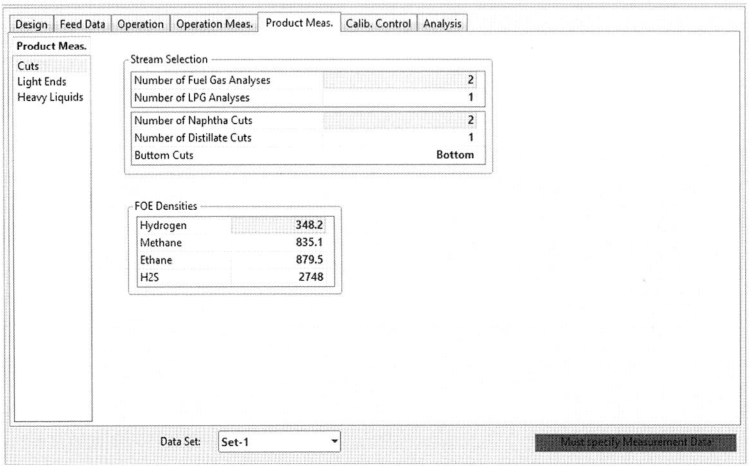

Figure 6.100 Define plant cuts.

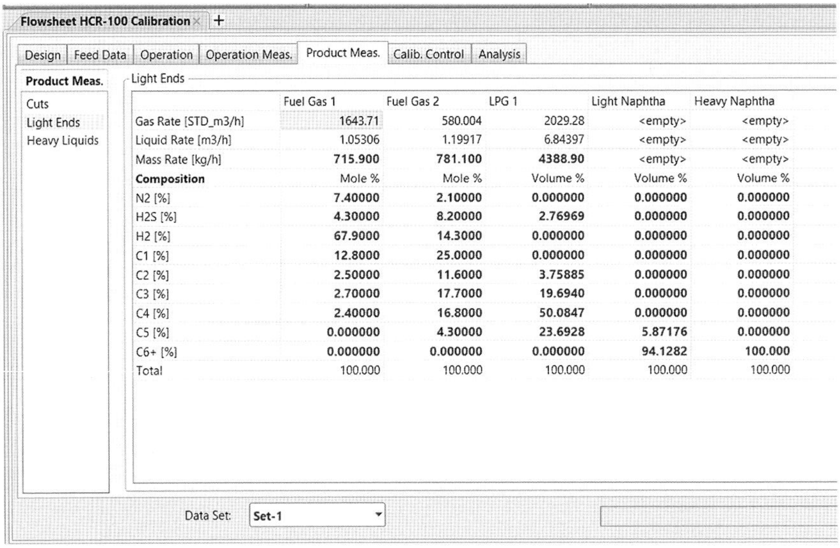

Figure 6.101 Input product yields and analyses (light products).

Step 9. Check *all* of the boxes in "object function" sheet so that we are able to probe into how significantly all of the model results deviate from plant data (Figure 6.104).

Step 10. We can use this sheet to select the reaction activities to be adjusted during automatic calibration by clicking "Run calibration" and change the lower and upper bounds of the selected reaction activities. In this step, we click the button of "Run Pre-Calibration" to run the model with current reaction activities, which are also default values (Figure 6.105). After clicking on "Run Pre-Calibration," Apen HYSYS shows a window of "Validation Wizard for

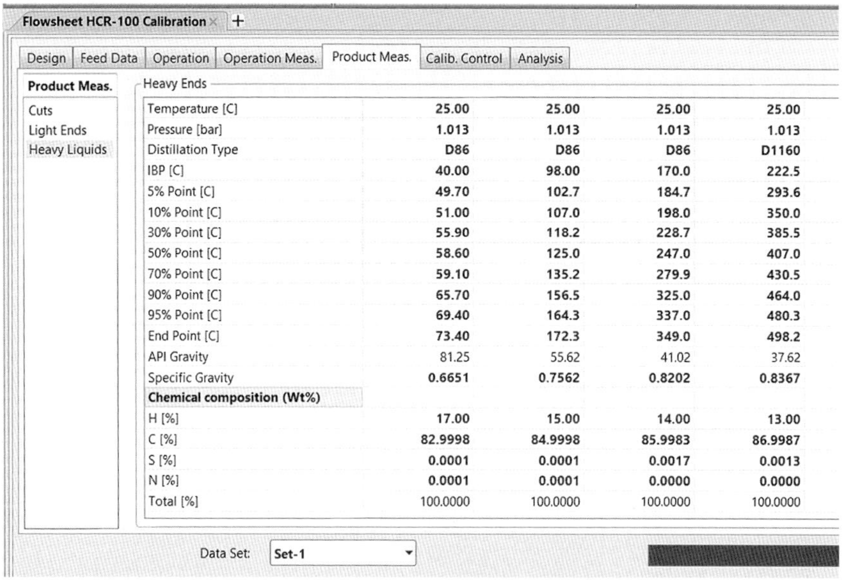

Figure 6.102 Input product yields and analyses (heavy products).

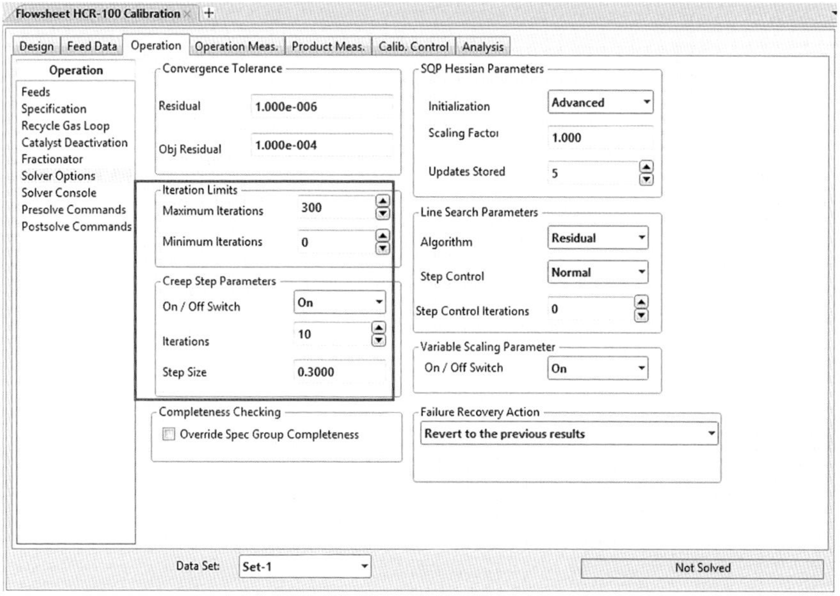

Figure 6.103 Iteration algorithm for model convergence.

Set-1" (Figure 6.106), which compares the measured and adjusted mass flows and mass balance. We click "OK" to continue.

Step 11. Analysis sheet represents the results after running calibration. It also shows the comparisons between model results from current reaction activities and plant data. Save the simulation file as *Workshop 6.2-1.hsc* (Figure 6.107).

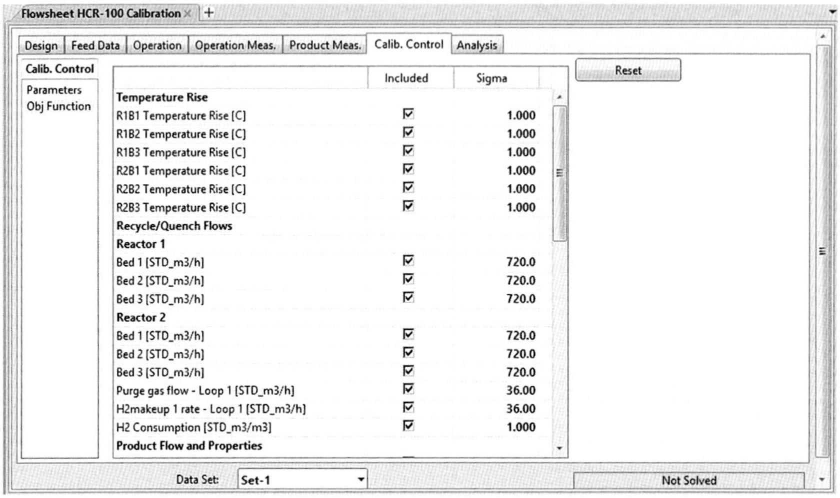

Figure 6.104 Objective function sheet.

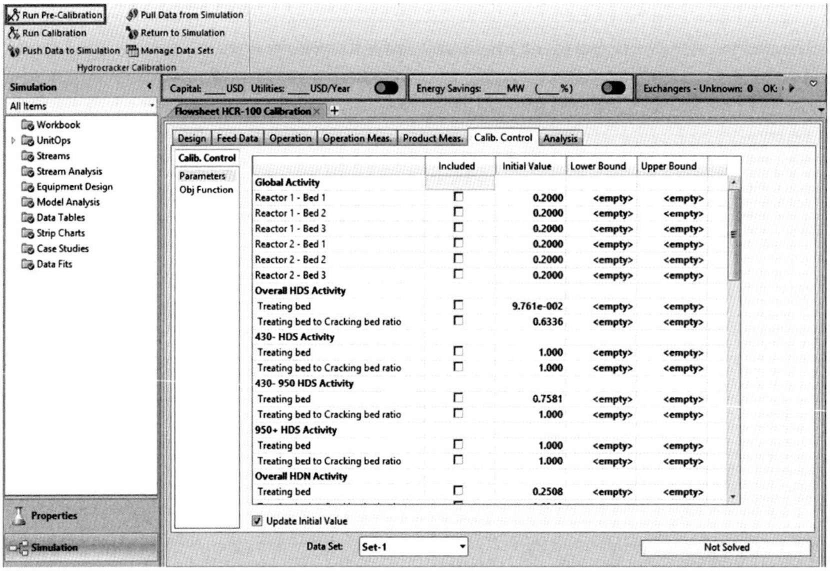

Figure 6.105 Reaction activity factor sheet.

Step 12. Select "R1B1 Temperature Rise" in "Obj. Function" sheet (Figure 6.108).

Step 13. Select "Global activity – Reactor 1 – Bed 1" and assign appropriate lower and upper bounds. We suggest that the lower and upper bounds are ±25% up and down around current value (i.e., initial value). By doing steps 12 and 13, the model will tune "Global Activity – Reactor 1 – Bed 1" within the assigned range to minimize the deviation between model result of "R1B1 Temperature Rise" and plant measurement (Figures 6.109 and 6.110).

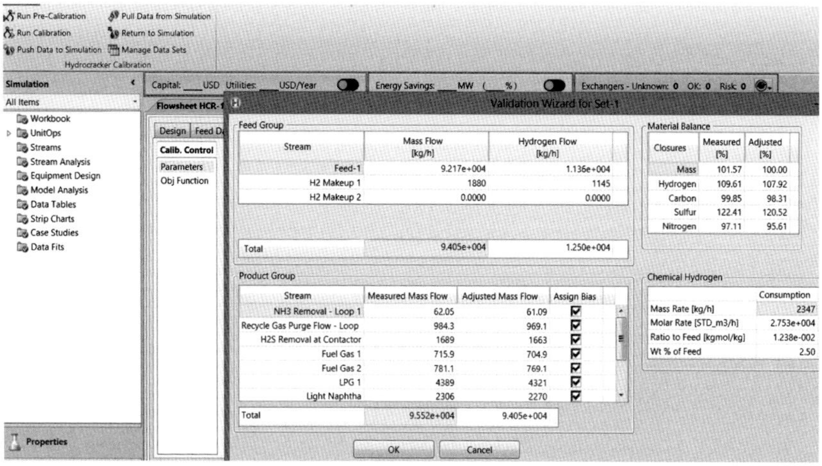

Figure 6.106 Validation wizard for Set-1.

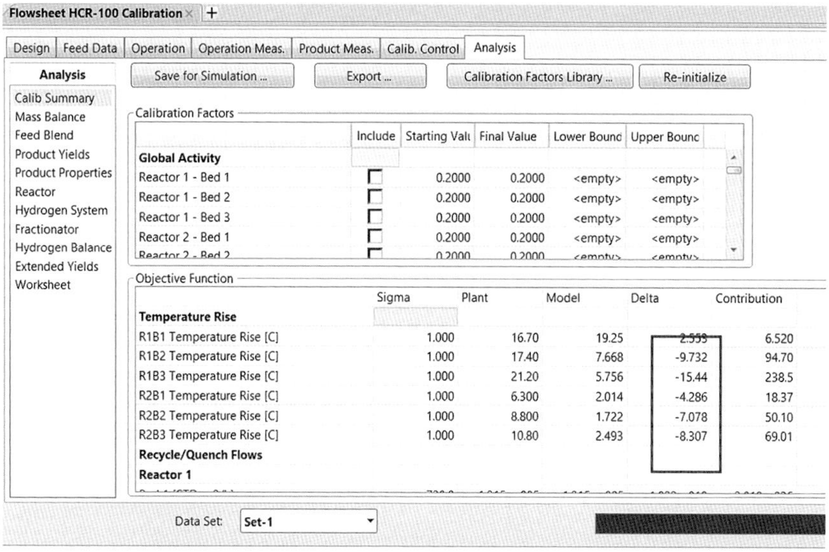

Figure 6.107 Calibration result sheet.

Step 14. Check the results in "Analysis" sheet. Repeat step 14 to assign new lower and upper bounds to calibrate the model again if the model results are not good enough. We save the converged simulation as *Workshop 6.2-2* (Figure 6.111).

Step 15. After obtaining satisfied result of "R1B1 Temperature Rise," uncheck the selections of "R1B1 Temperature Rise" and "Global activity – Reactor 1 – Bed 1." Repeat steps 12–14 for the other reaction bed temperature rises and corresponding global activity reaction activities one by one (R1B2, R1B3, R2B1, R2B2, and R2B3).

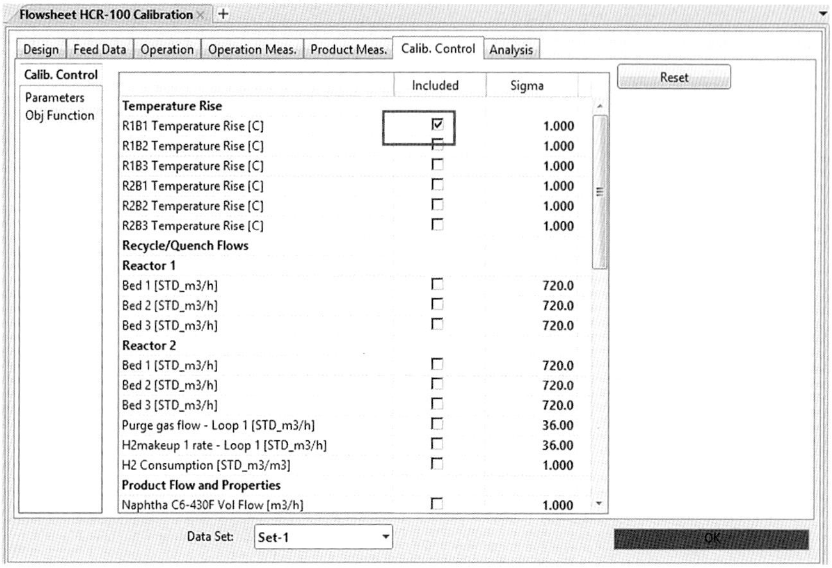

Figure 6.108 Define objective function (first bed).

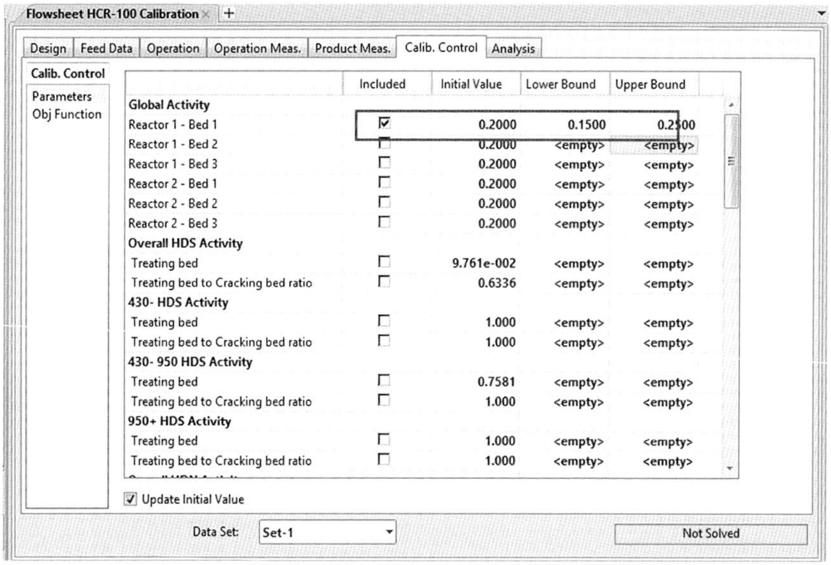

Figure 6.109 Select tuning activity factor (first global activity) before running the calibration.

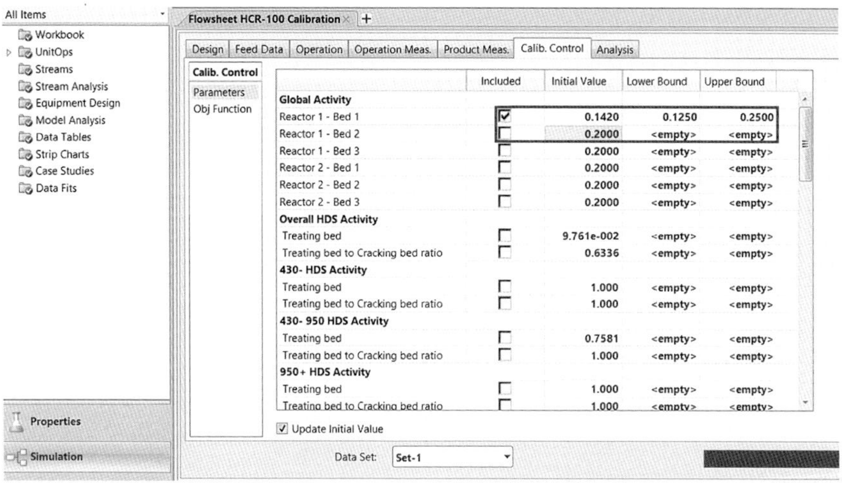

Figure 6.110 Converged tuning activity factor (first global activity) after running the calibration.

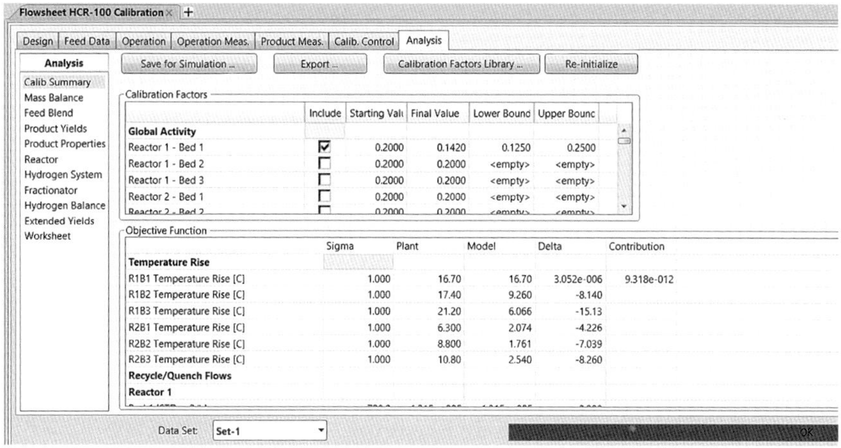

Figure 6.111 Excellent calibration result of temperature rise (first bed).

Step 16. In most cases, the reactor temperature profiles obtained from step 15 will show similar trend as plant measurements rather than perfect agreement. To make model's prediction on reactor temperature profiles match plant measurements well, we select all of the "Temperature Rise" variables as objective functions and assign new initial values and lower and upper bounds to all of the "global reaction activities" (Figures 6.112–6.115).

Step 17. Repeat step 16 until model's predictions on reactor temperature profiles agree well with plant measurements (Figures 6.116 and 6.117).

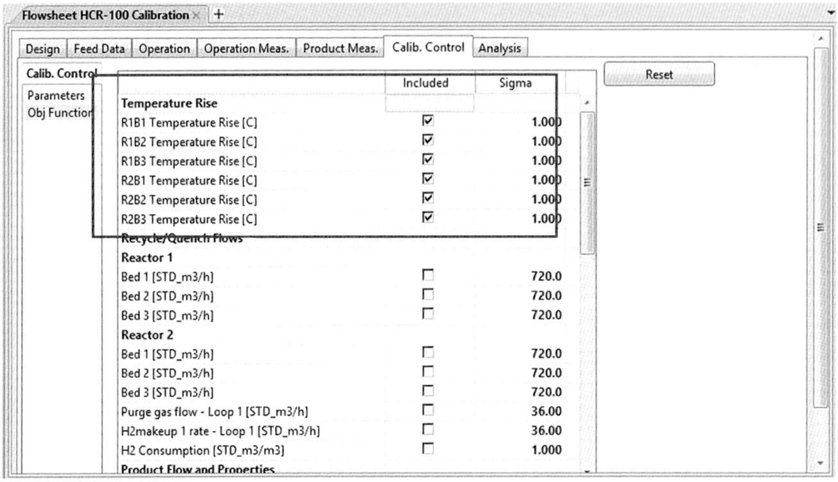

Figure 6.112 Define objective function (all beds).

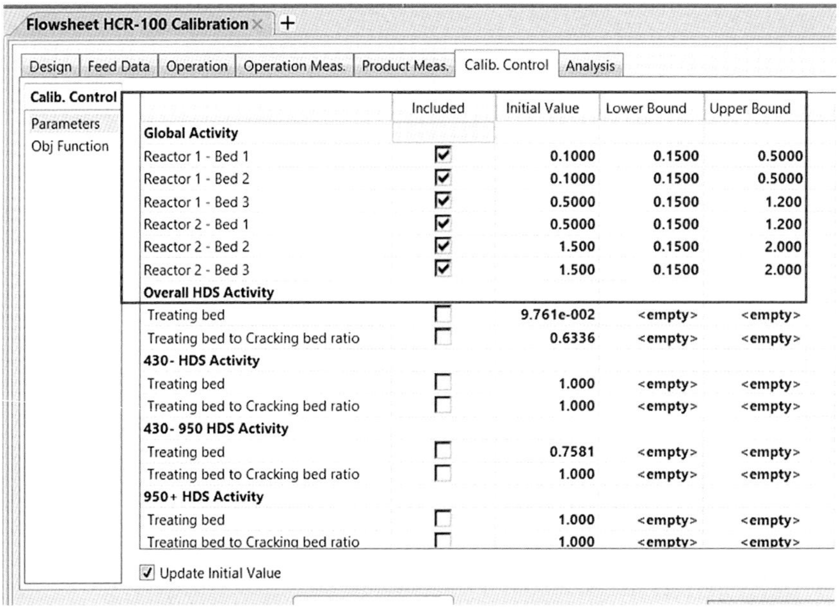

Figure 6.113 Select tuning activity factor (all global activities).

Step 18. Even though the reactor temperature profiles from the model agree well with plant measurements, model's prediction on product yields still significantly deviates from plant data (see Figure 6.118). Save the simulation file as *Workshop 6.2-3.hsc* and then resave the file as *Workshop 6.2-4.hsc* before continuing with step 19.

Step 19. Select the following objective functions and reaction activities to calibrate the model on temperature rise and on recycle/quench flows (by

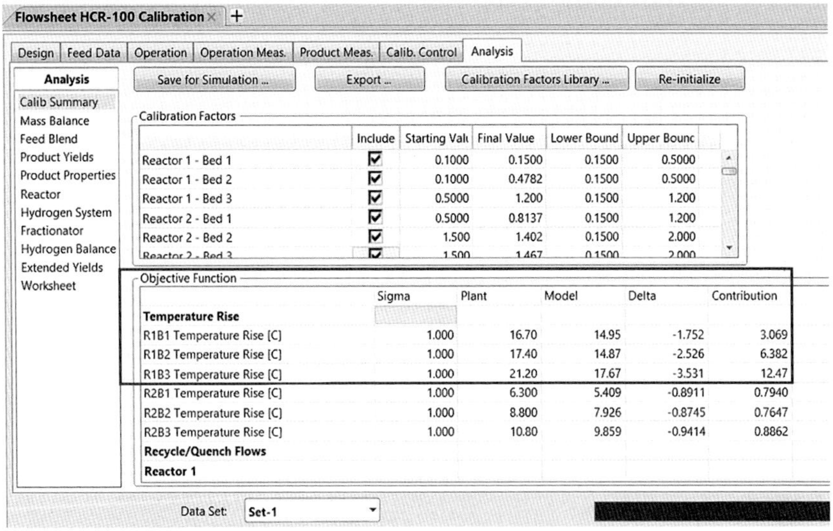

Figure 6.114 Calibration results corresponding to the parameters of Figure 6.113 – further tuning of R1B1, R1B2, and R1B3 temperature rises suggested.

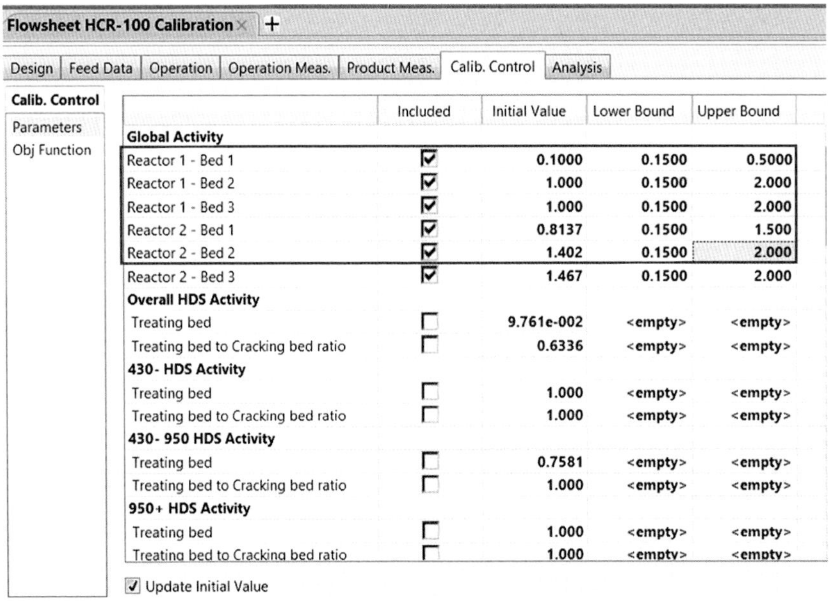

Figure 6.115 Calibration parameter changes from those in Figure 6.113.

clicking the button of "run calib") (see Figures 6.119–6.121). Repeat this step until the model matches plant measurements on reactor temperature profiles and product yields.

Figure 6.122 shows the results of calibrated tuning activity factors.

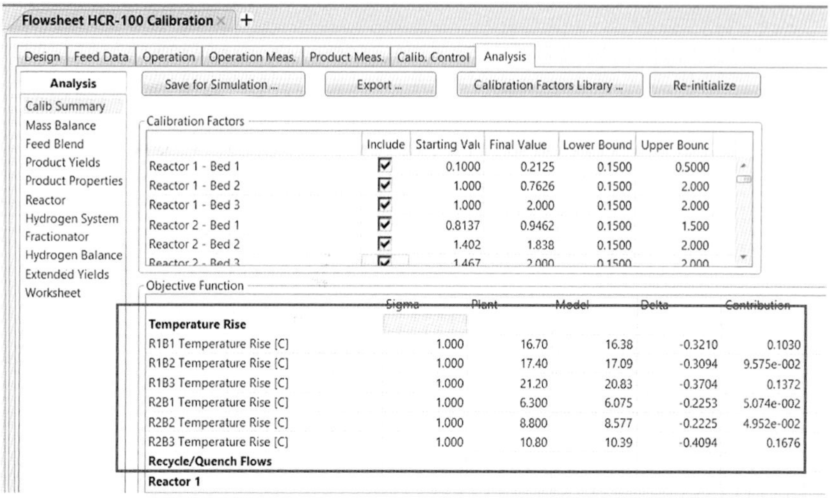

Figure 6.116 Calibration result corresponding to the parameters of Figure 6.115 – increasing the upper bonds of R1B3 and R2B3 suggested.

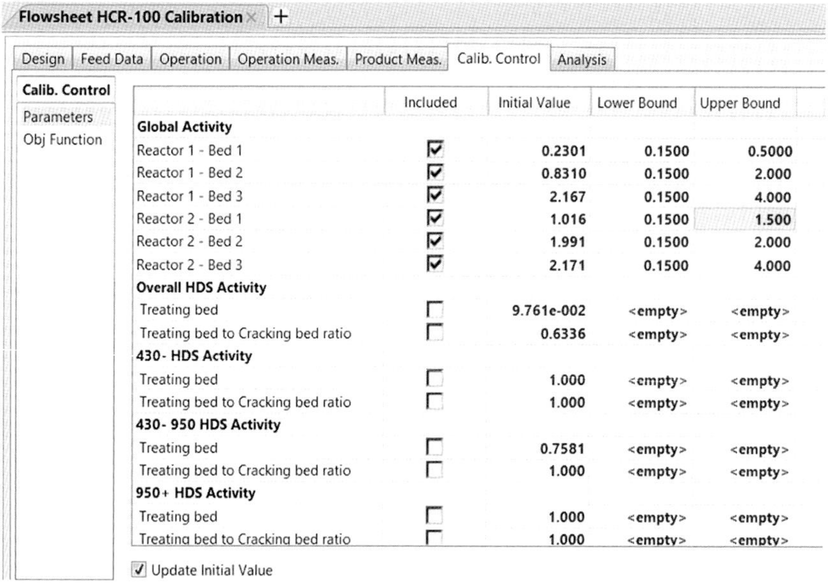

Figure 6.117 Increase the upper bounds for R1B3 and R2B2 in Figure 6.115 to 4.0.

Step 20. In some cases, the model's predictions match most of the plant measurements except for one or two process variables. It is suggested not to run automatic calibration but manually reconcile the model that allows "creep" moving by small step in each run. For example, Figure 6.123 shows that the model only fails to predict the third bed temperature of first reactor (R1B3).

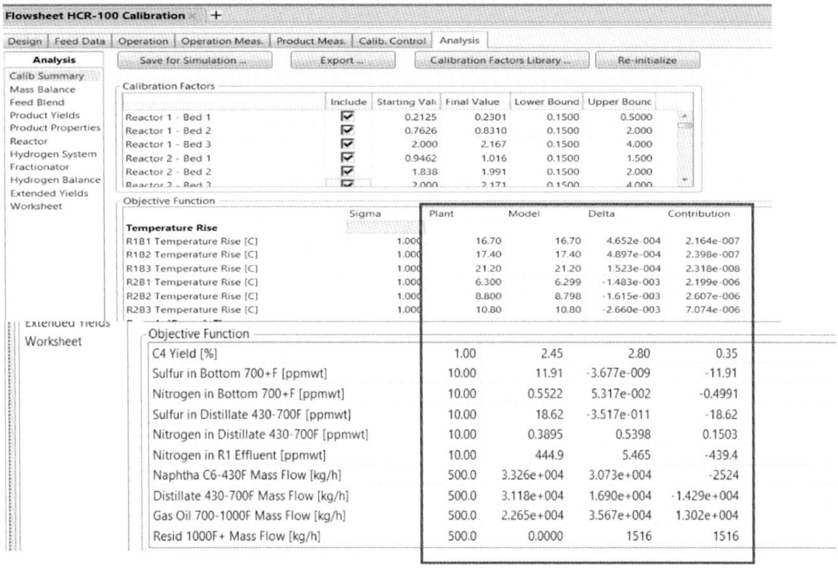

Figure 6.118 Improved calibration results for temperature rise correspond to parameters of Figure 6.117 and notable deviations between plant data and model predictions for naphtha, distillate, gas oil, and resid mass flows.

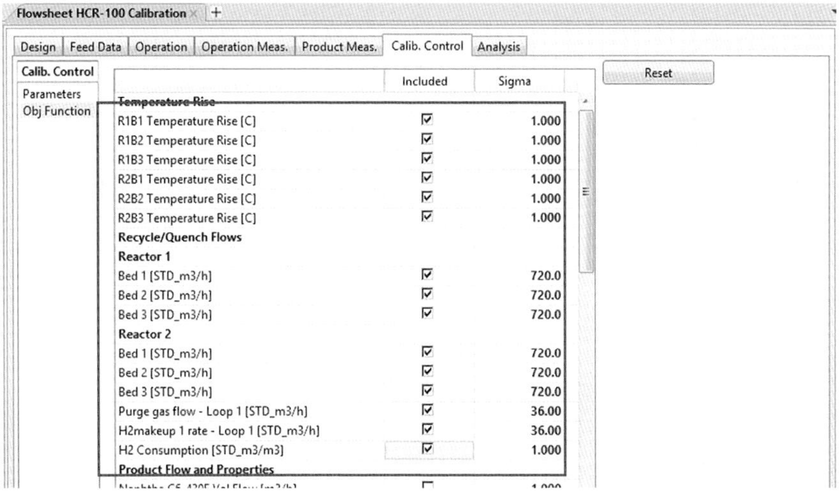

Figure 6.119 Define objective function (all beds).

Step 21. To reconcile the model manually, assign a new value to the related reaction activity. In this case, the predicted temperature is lower than plant measurement and a bigger value of the related reaction reactivity is expected. Thus, we change "Global Activity Reactor 1 – Bed 3" from the current value of 1.944 to 2.0 and click the button "pre-calib" to run the model with current values of reaction activities (Figure 6.124).

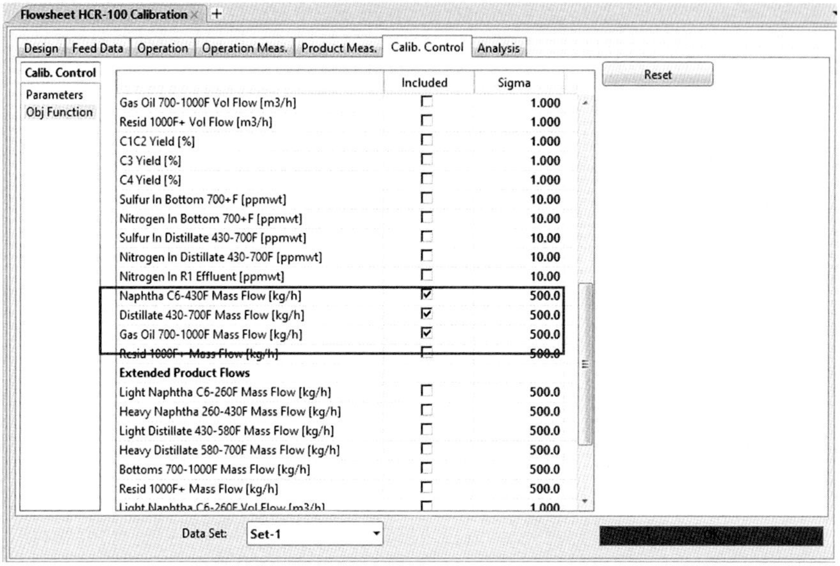

Figure 6.120 Define objective function (all mass yields except for resid).

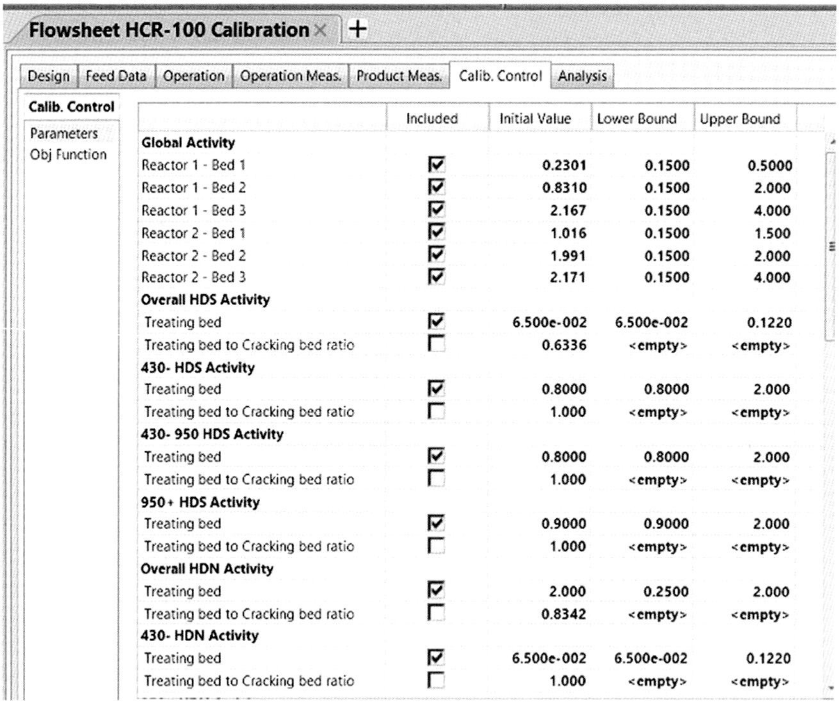

Figure 6.121 Select tuning activity factor (all global activities and all cracking activities on cracking beds).

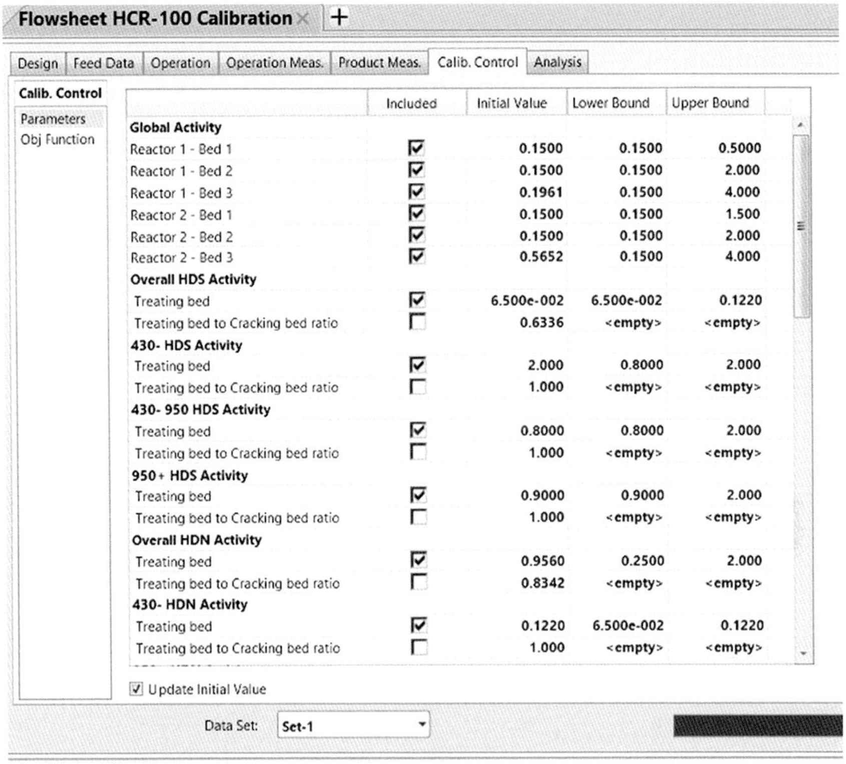

Figure 6.122 Results of calibrated activity factors.

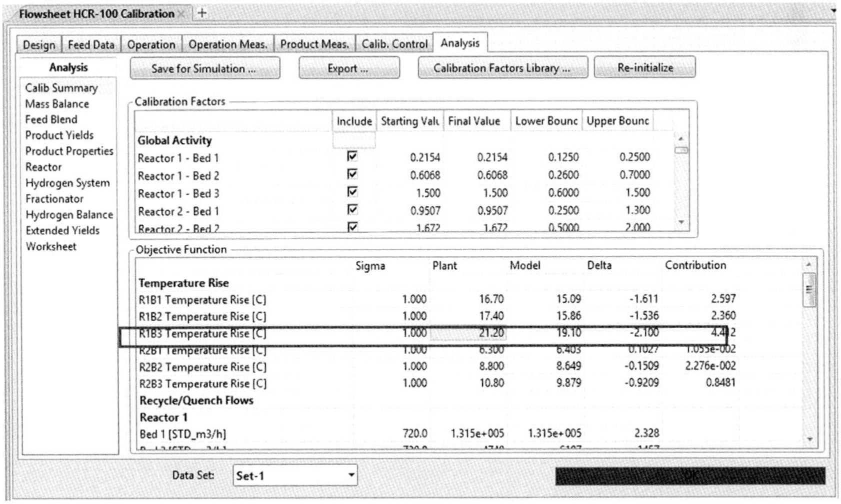

Figure 6.123 Calibration results indicating a significant deviation of R1B3 temperature rise.

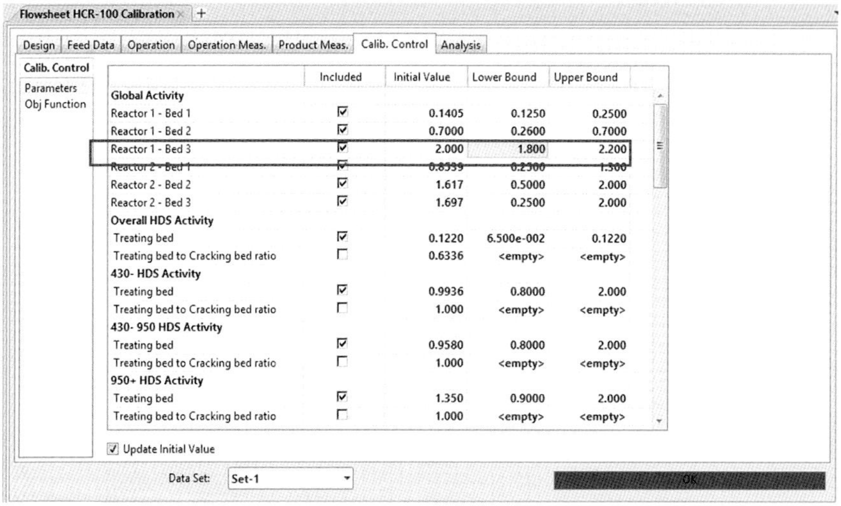

Figure 6.124 Manual calibration of R1B3 temperature rise tuning factor.

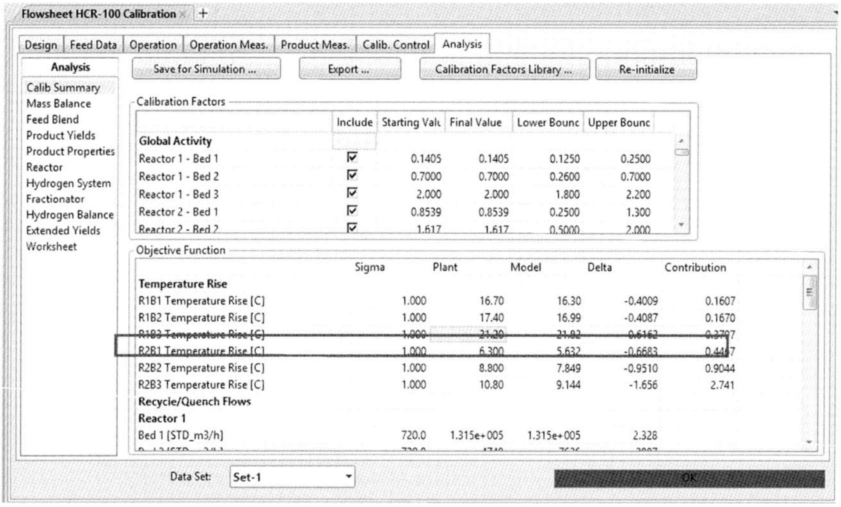

Figure 6.125 Improved calibration results after manual calibration.

Step 22. By observing the results, the temperature of R1B3 is closer to the plant measurement. To obtain better result, it is necessary to repeat step 21 until the temperature profile is within the tolerance. Meanwhile, it is also important to watch out for all of the other objective process variables – other reaction bed temperatures and product yields. It may be necessary to repeat steps 16–21 if the model's predictions fail in other objective process variables during manual calibration (Figure 6.125).

Step 23. Figure 6.126 gives the calibration results in this workshop.

Step 24. After completing model calibration, click the button "push data to simulation" to export updated reaction activities into HCR simulation environment (Figure 6.127).

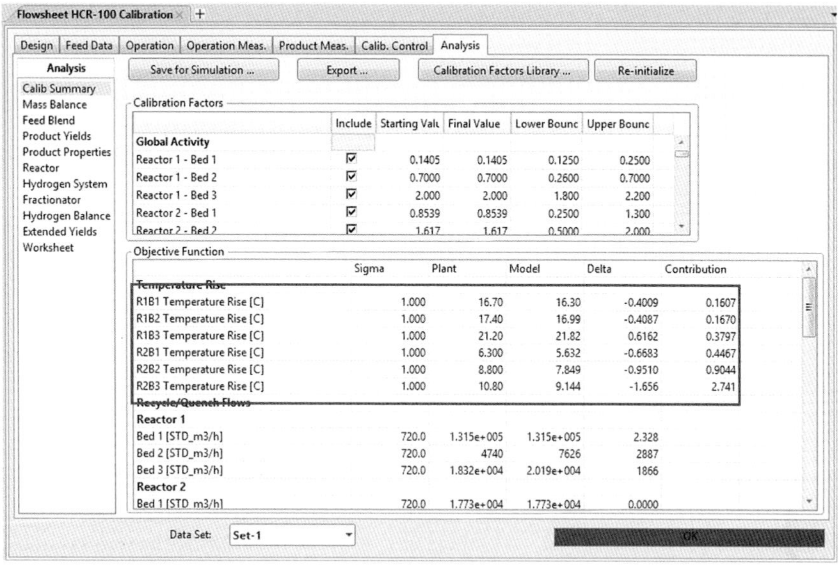

Figure 6.126 Calibration results of this workshop.

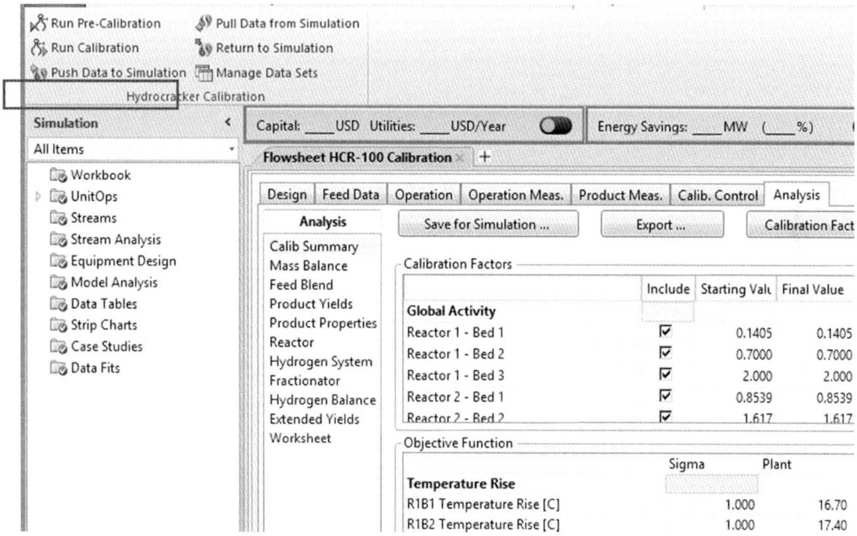

Figure 6.127 Export calibrated activity factors and results into simulation.

6.11 Workshop 6.3 – Case Studies

One of HCR model applications is to investigate different operating scenarios and help answering what-if question by running case studies. This workshop demonstrates how to use developed Aspen HYSYS Petroleum Refining HCR model to investigate the effect of WART of HCR reactor and feed flow rate on product distribution. In real operation, the only way to tune WART is to change the inlet temperature of reaction bed. In this workshop, we will change inlet temperatures of the three HCR beds at the same time to perform case study. We begin with simulation file, *Workshop 6.3-starting.hsc.*

Step 1. Hold the model to avoid automatic calculation while defining variables for the simulation experiment (Figure 6.128).

Step 2. We need to add a "spreadsheet" in Aspen HYSYS Petroleum Refining to make tuning three inlet temperatures possible. Click "Model Palette (F4)" → "add spreadsheet" (Figure 6.129).

Open the reactor model and pay attention to the inlet temperatures of three beds of reactor 2, R2B1, R2B2, and R2B3 (see Figure 6.130).

Step 3. Open the spreadsheet tab and input current values of the three HCR beds' inlet temperatures. In cells A1–A3, enter the three reactor bed names, R2B1, R2B2, and R2B3 (see Figure 6.131). Export the three temperatures to the spreadsheet (see Figures 6.131–6.133).

Step 4. Add a cell called "temp increment," which will be used as an operator to allow a step change of inlet temperature during the simulation experiment. Specify an initial increment value of 0 °C (Figure 6.134).

Step 5. Enter "Feed Mass Flow" to cell D1 and send the current value of feed mass flow to cell E1 of the spreadsheet. Enter "Feed Increment" in cell D2, and enter an initial value of 0 kg/h in cell E2 (Figure 6.135).

Step 6. Add equations to cell C1, C2, C3, and F1 to calculate the new process variables (inlet temperature of HCR bed and feed mass flow rate) while running case study: C1 = B1 + B4, C2 = B2 + B4, C3 = B3 + B4, F1 = E1 + E2 (Figure 6.136).

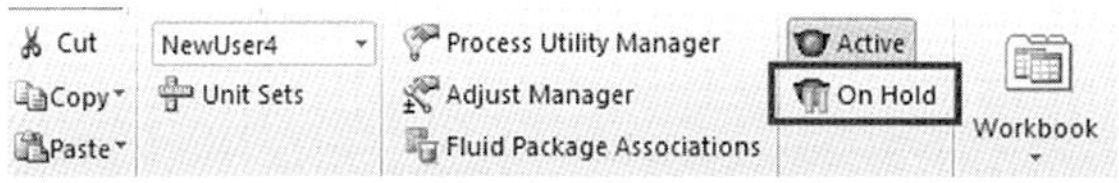

Figure 6.128 Deactivation button.

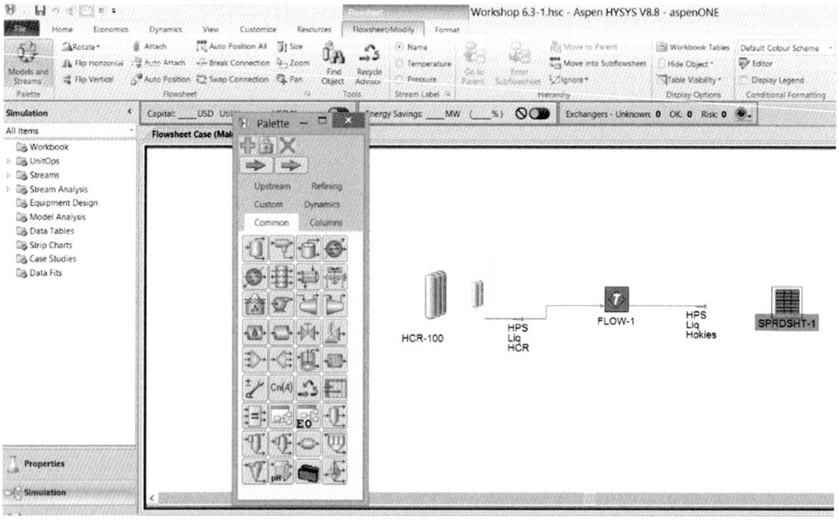

Figure 6.129 Add spreadsheet in Aspen HYSYS.

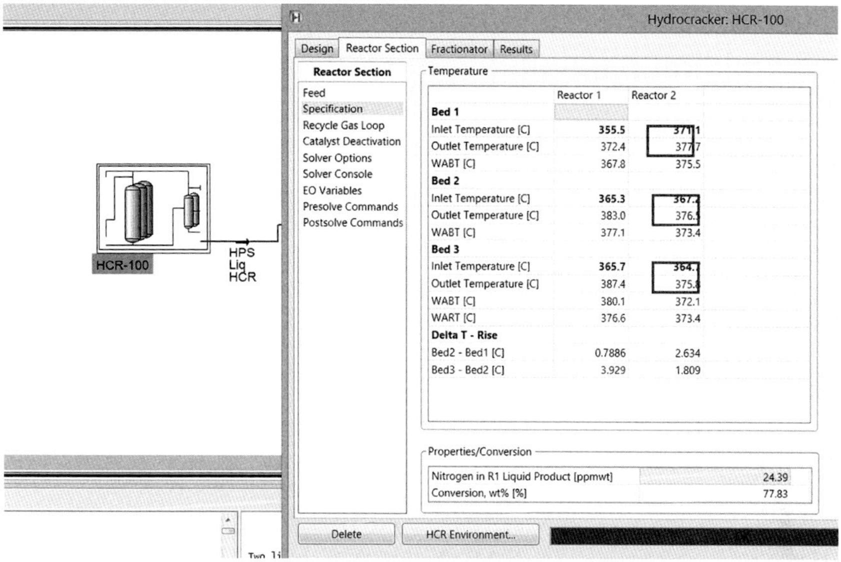

Figure 6.130 Inlet temperatures of three beds of reactor two.

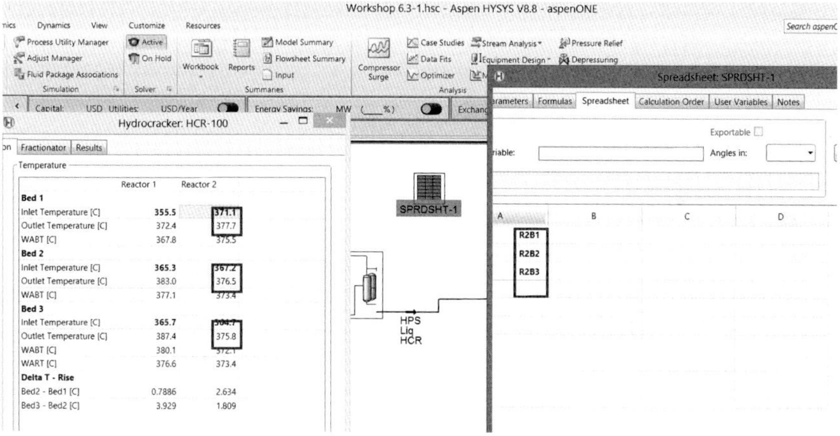

Figure 6.131 Enter the three bed names of reactor two, R2B1–R2B3.

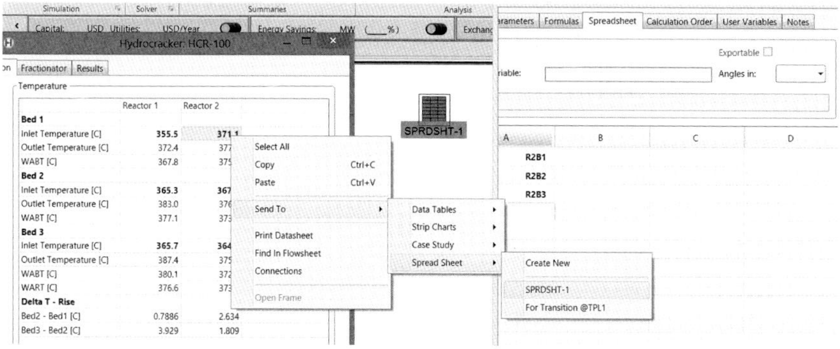

Figure 6.132 Right-click "send to" to export the R2B1, R2B2, and R2B3 temperatures of 371.1 °C, 367.2 °C, and 364.7 °C from the reactor input to cells B1, B2, and B3.

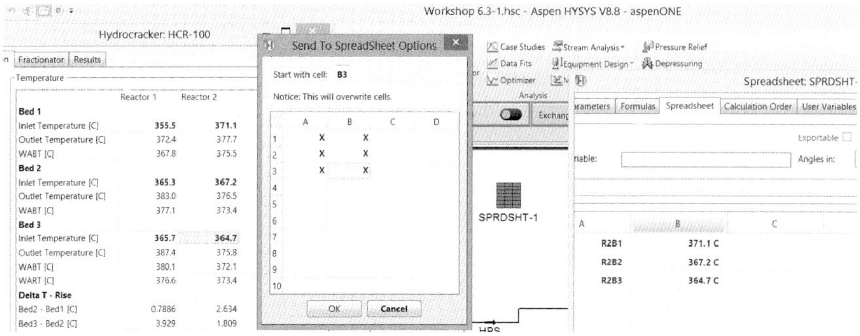

Figure 6.133 Send the reactor bed temperatures to the spreadsheet, cells B1–B3.

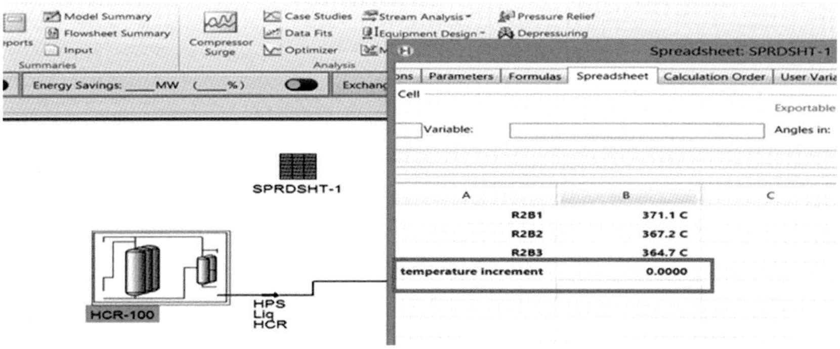

Figure 6.134 Add a temperature increment with an initial value of 0 °C.

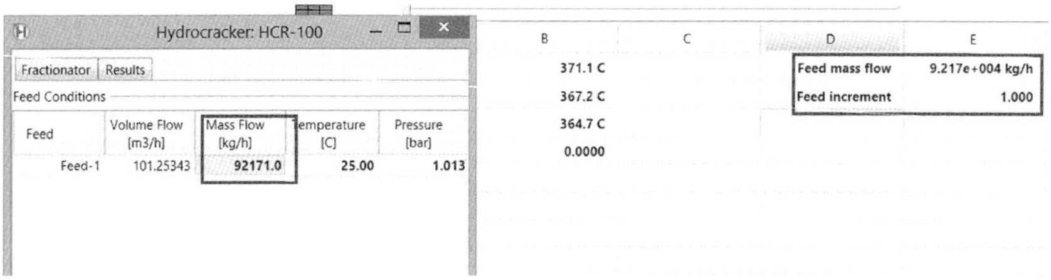

Figure 6.135 Export the feed flow into the spreadsheet.

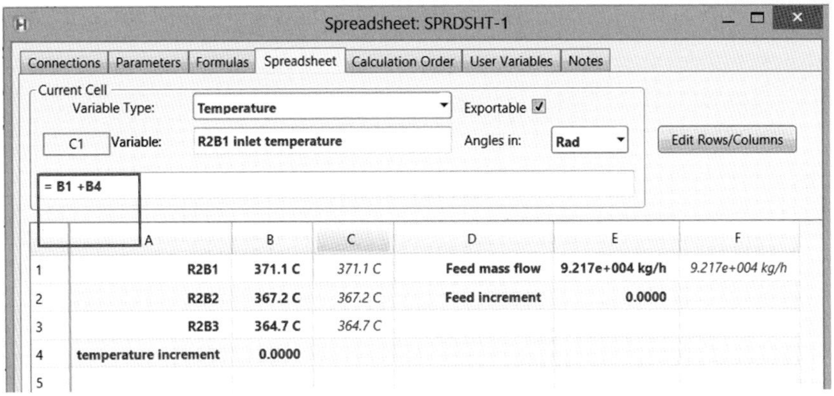

Figure 6.136 Add equations to allow the three reactor temperatures to be tuned at once.

Step 7. In order to link the calculated cell results in the spreadsheet with our case studies later, we right-click on the selected cell and click "export formula result." We do this to export the calculated temperatures for beds 1–3 of reactor 2 (cells C1–C3 for R2B1, R2B2, and R2B3) and for feed mass flow rate (cell F1) [see Figures 6.125–6.127 (Figure 6.137)].

Step 8. Select the inlet temperature of R2B1 to export the calculated temperature (Figures 6.138 and 6.139).

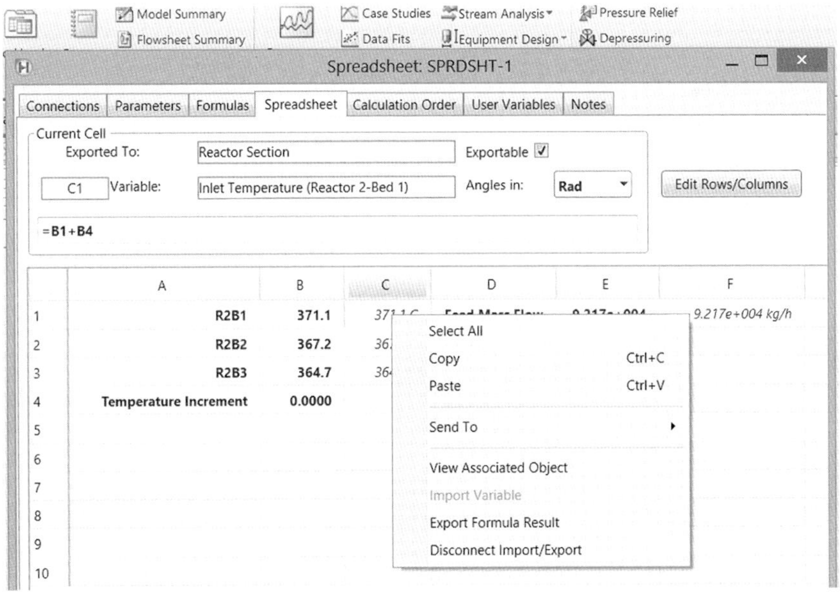

Figure 6.137 Export formula results.

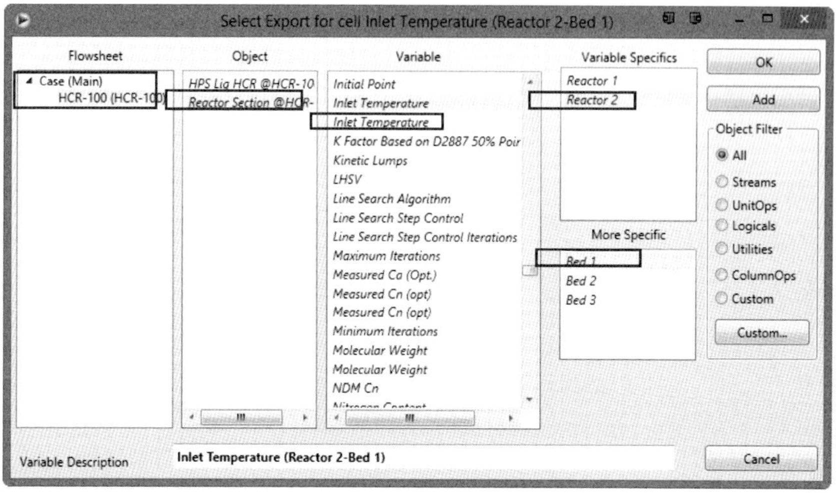

Figure 6.138 Export formula results for R2B1 inlet temperature.

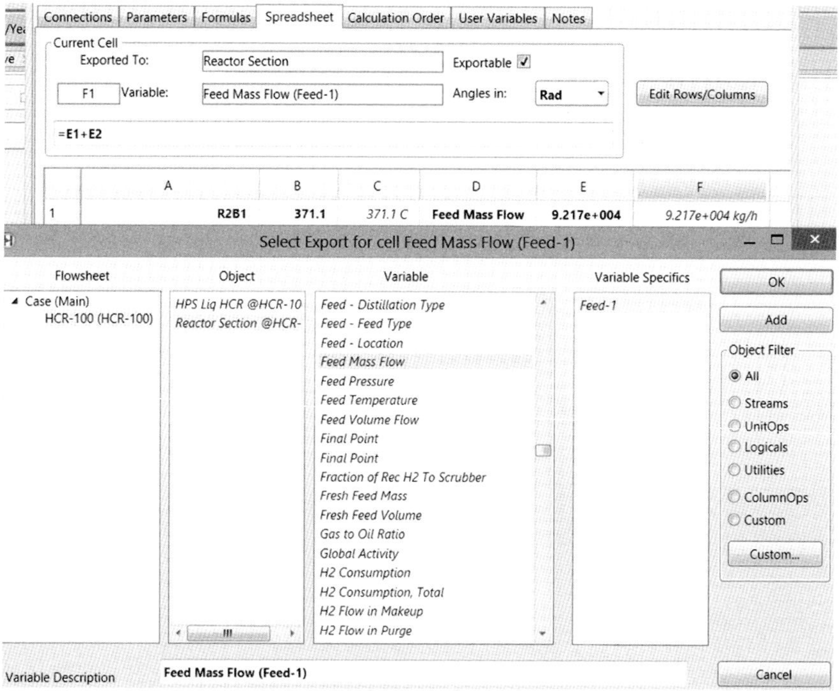

Figure 6.139 Export formula results for feed mass flow.

Step 9. Change the temperature increment to 7 °C and the feed mass flow increment to 2.3E4 kg/h (Figure 6.140).

Step 10. Activate case studies (Figure 6.141).

Step 11. We follow the procedure for case studies demonstrated in Figures 2.69–2.73, Section 2.10.3; and in Workshop 5.4, Section 5.17. Insert feed mass flow, spreadsheet cells B4 (temperature increment) and E2 (feed mass flow

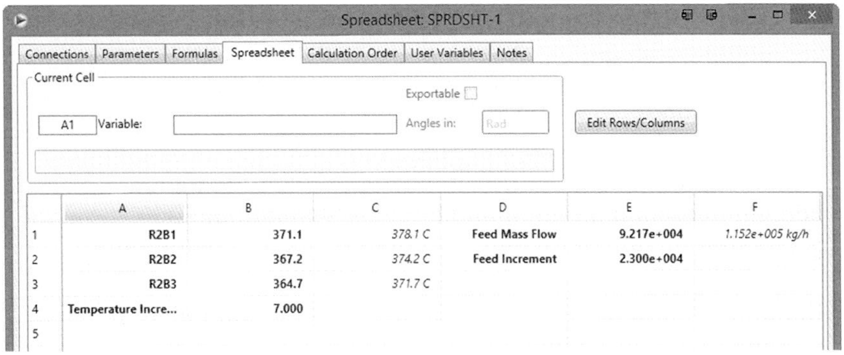

Figure 6.140 Exported formula results in spreadsheet.

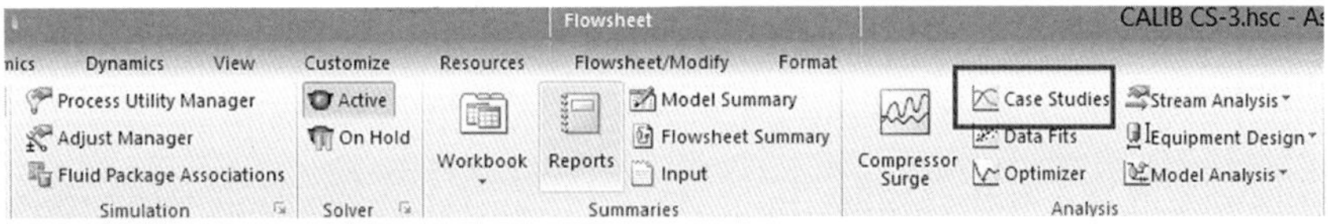

Figure 6.141 New case studies.

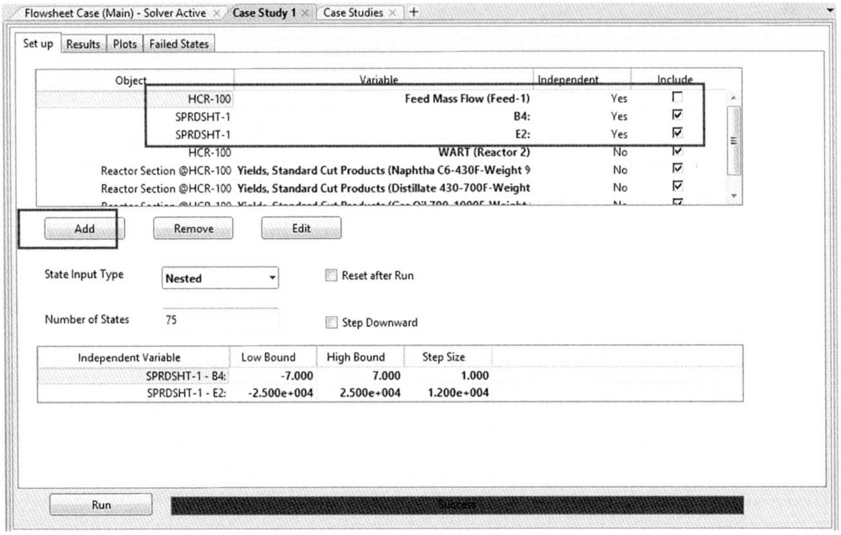

Figure 6.142 Insert independent process variables into the case study.

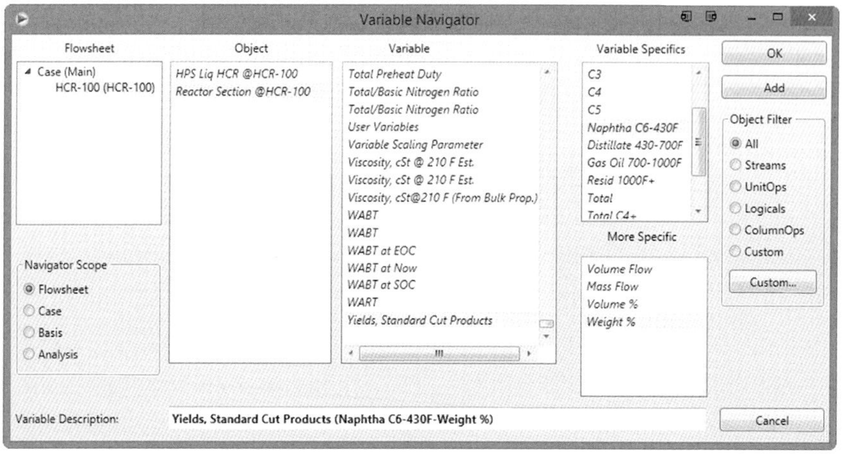

Figure 6.143 Insert product yields into case study.

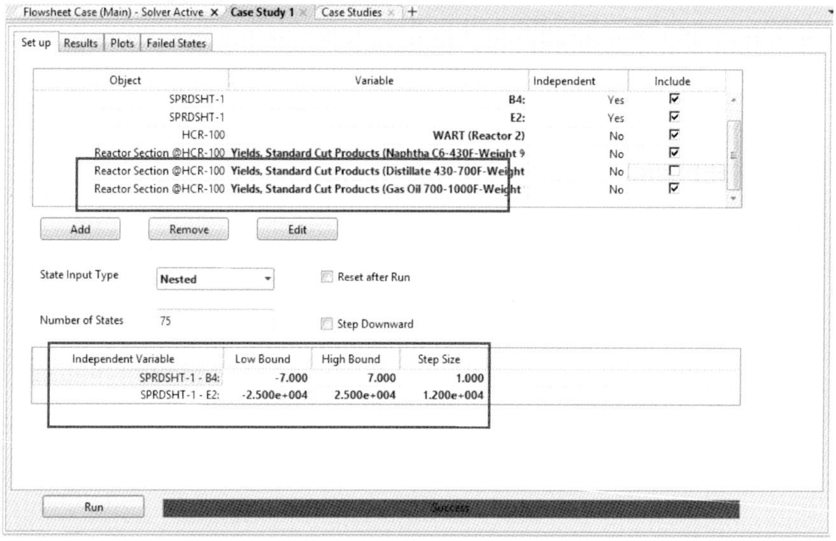

Figure 6.144 Insert dependent process variables in the case study.

increment), WART of reactor 2 (HCR reactor), and standard cut product yields for naphtha, distillate, and gas oil. We also specify the lower and upper bounds of spreadsheet cells B4 (temperature increment) and E2 (feed mass flow increment) and their step sizes (see Figures 6.142–6.144).

Step 12. Click "view" to open a new window to assign lower and upper bounds that allow WART and feed mass flow to change while running the simulation experiment. Click "start" to run a case study.

Step 13. Click "Start" to run the case study and click "Results" to check the results of the case study (see Figure 6.145). Save the simulation file as *Workshop 6.3-done.hsc*.

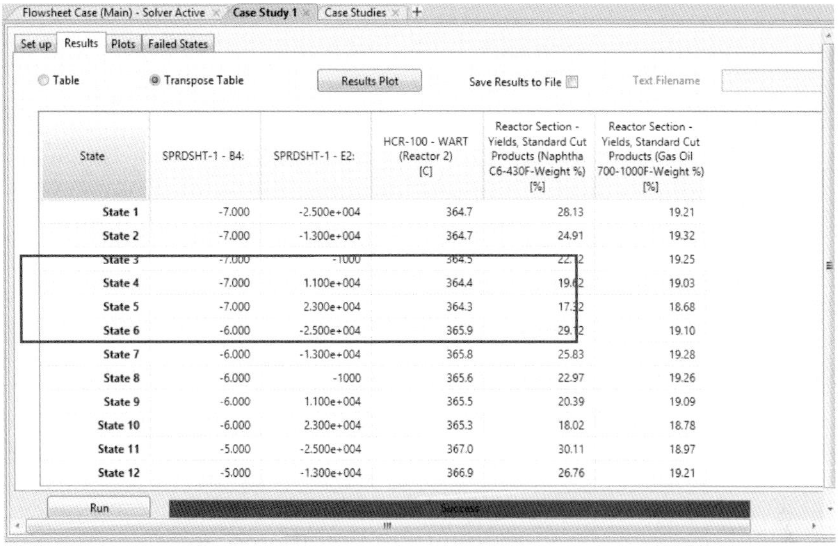

State	SPRDSHT-1 - B4:	SPRDSHT-1 - E2:	HCR-100 - WART (Reactor 2) [C]	Reactor Section - Yields, Standard Cut Products (Naphtha C6-430F-Weight %) [%]	Reactor Section - Yields, Standard Cut Products (Gas Oil 700-1000F-Weight %) [%]
State 1	-7.000	-2.500e+004	364.7	28.13	19.21
State 2	-7.000	-1.300e+004	364.7	24.91	19.32
State 3	-7.000	-1000	364.5	22.[illegible]2	19.25
State 4	-7.000	1.100e+004	364.4	19.62	19.03
State 5	-7.000	2.300e+004	364.3	17.32	18.68
State 6	-6.000	-2.500e+004	365.9	29.[illegible]2	19.10
State 7	-6.000	-1.300e+004	365.8	25.83	19.28
State 8	-6.000	-1000	365.6	22.97	19.26
State 9	-6.000	1.100e+004	365.5	20.39	19.09
State 10	-6.000	2.300e+004	365.3	18.02	18.78
State 11	-5.000	-2.500e+004	367.0	30.11	18.97
State 12	-5.000	-1.300e+004	366.9	26.76	19.21

Figure 6.145 Results of case study.

6.12 Workshop 6.4 – Fractionation System for HCR Reactor

We open the reactor model, *HCR-Reactor.hsc,* and save it as *Workshop 6.4-fractionation.hsc.*

We add exchanger E-100 (see Figure 6.146).

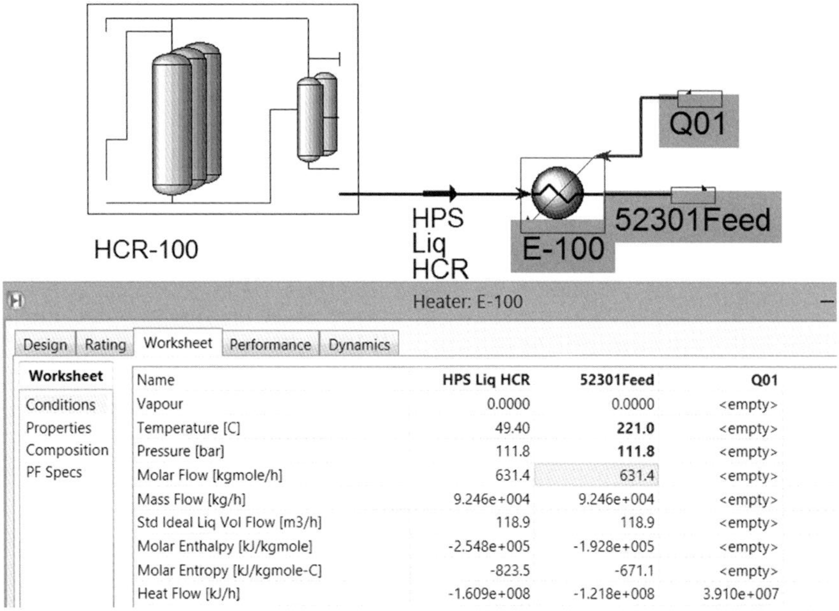

Name	HPS Liq HCR	52301Feed	Q01
Vapour	0.0000	0.0000	<empty>
Temperature [C]	49.40	**221.0**	<empty>
Pressure [bar]	111.8	**111.8**	<empty>
Molar Flow [kgmole/h]	631.4	631.4	<empty>
Mass Flow [kg/h]	9.246e+004	9.246e+004	<empty>
Std Ideal Liq Vol Flow [m3/h]	118.9	118.9	<empty>
Molar Enthalpy [kJ/kgmole]	-2.548e+005	-1.928e+005	<empty>
Molar Entropy [kJ/kgmole-C]	-823.5	-671.1	<empty>
Heat Flow [kJ/h]	-1.609e+008	-1.218e+008	3.910e+007

Figure 6.146 Add a heater to HPS Liq HCR stream.

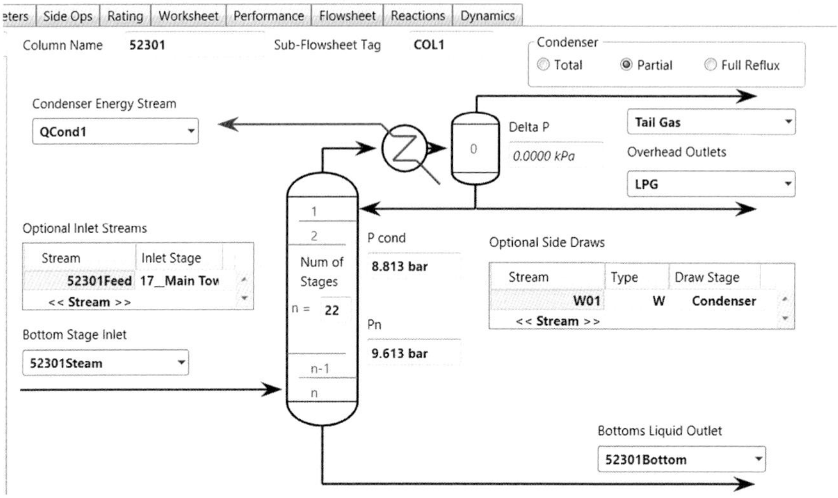

Figure 6.147 Column 52301 definitions with 52301 steam at 345 °C, 11.01 bar, and 680 kg/h.

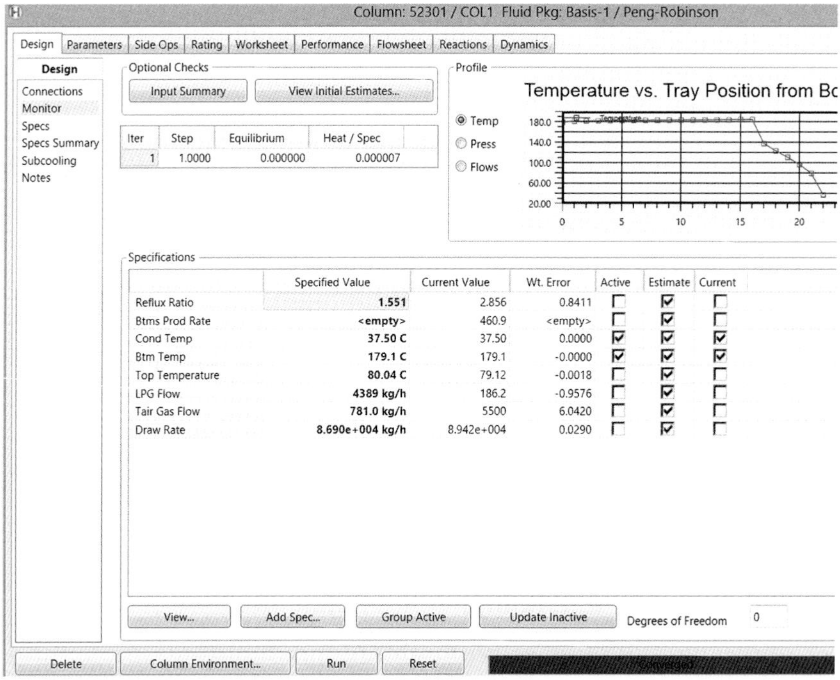

Figure 6.148 Column 52301 specifications for column convergence.

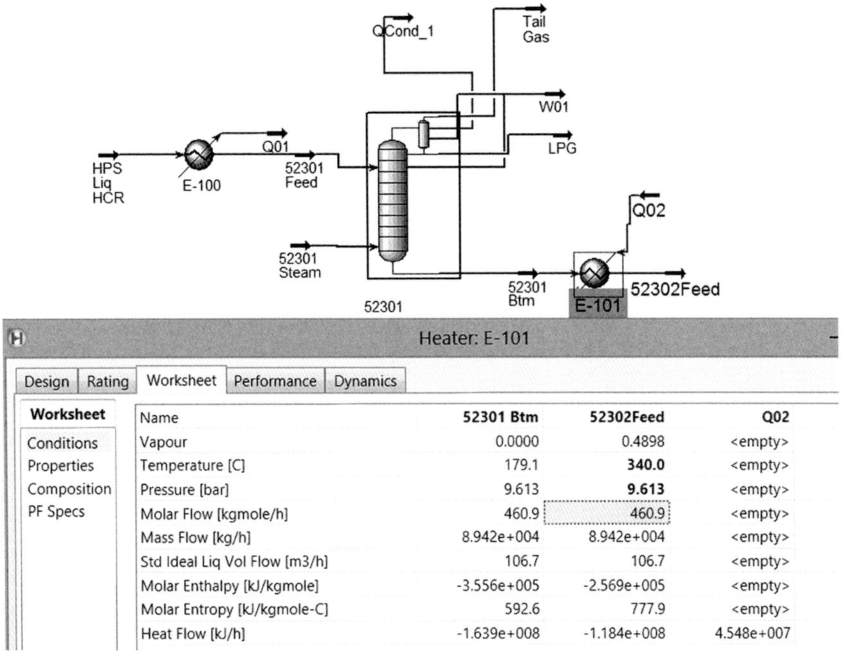

Name	52301 Btm	52302Feed	Q02
Vapour	0.0000	0.4898	<empty>
Temperature [C]	179.1	**340.0**	<empty>
Pressure [bar]	9.613	**9.613**	<empty>
Molar Flow [kgmole/h]	460.9	460.9	<empty>
Mass Flow [kg/h]	8.942e+004	8.942e+004	<empty>
Std Ideal Liq Vol Flow [m3/h]	106.7	106.7	<empty>
Molar Enthalpy [kJ/kgmole]	-3.556e+005	-2.569e+005	<empty>
Molar Entropy [kJ/kgmole-C]	592.6	777.9	<empty>
Heat Flow [kJ/h]	-1.639e+008	-1.184e+008	4.548e+007

Figure 6.149 Add a heater to 52301Btm stream.

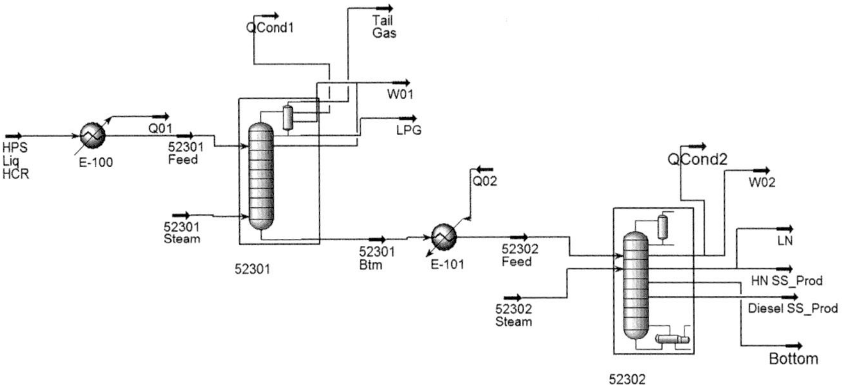

Figure 6.150 Add column 52302.

We then add column 52301, following Figures 6.147 and 6.148.

We continue to add exchanger E-101 (see Figure 6.149).

We then add column 52302, a reboiler absorber with 43 equilibrium stages. This column has two side strippers with reboilers for diesel and heavy naphtha (HN), each having eight equilibrium stages. The column also has a pumparound PA-1 (see Figures 6.150–6.158).

Save the converged simulation as *Workshop 6.4-fractionation.hsc*. Figure 6.159 shows the resulting stream table for column 52302.

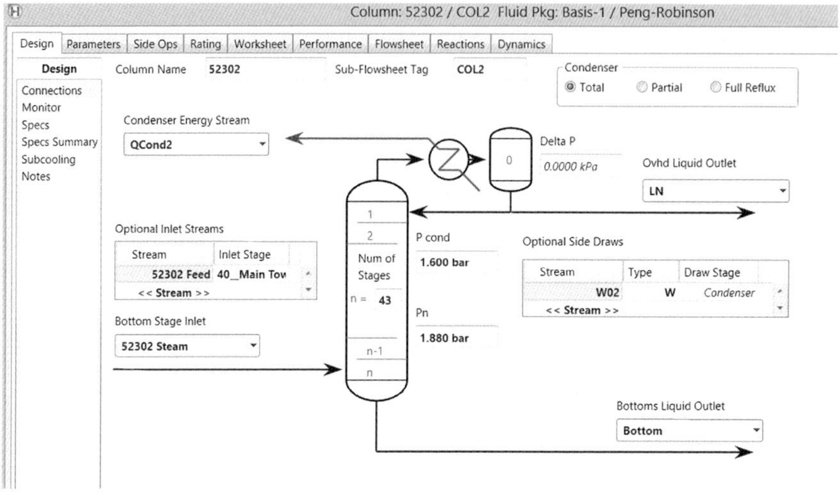

Figure 6.151 Column definitions with 52302 steam at 345 °C, 11.01 bar, and 200 kg/h.

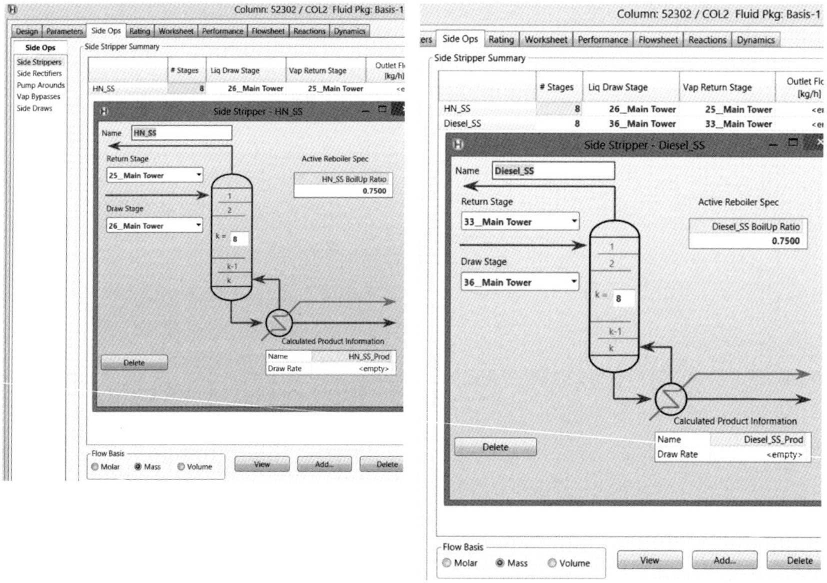

Figure 6.152 Add heavy naphtha (HN) and diesel side strippers.

Figure 6.153 Add a pumparound P-1.

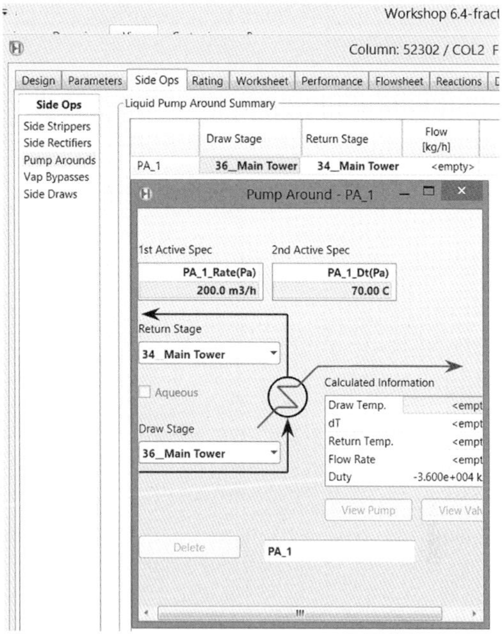

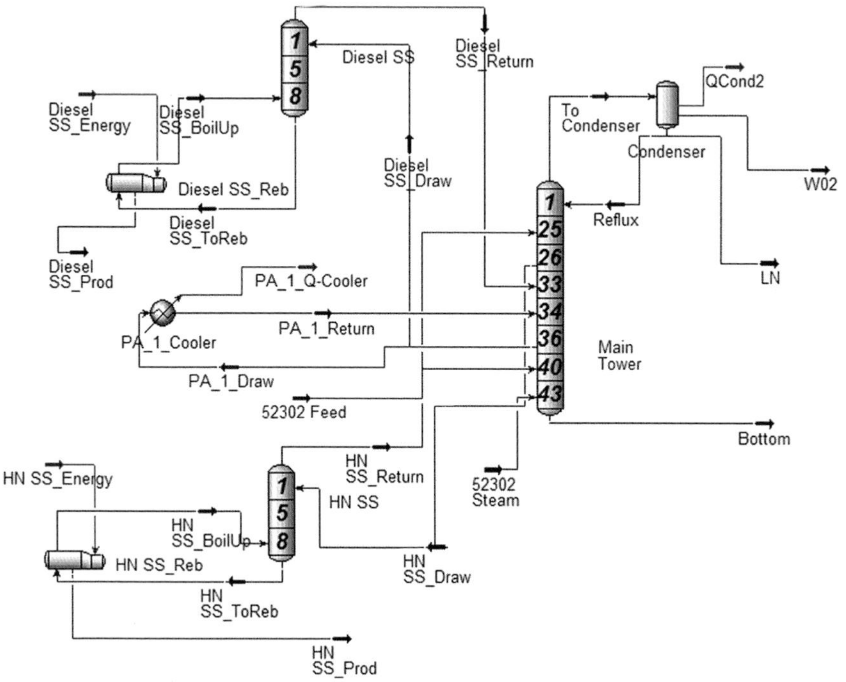

Figure 6.154 Column 52302 with naphtha and diesel side strippers and a pumparound.

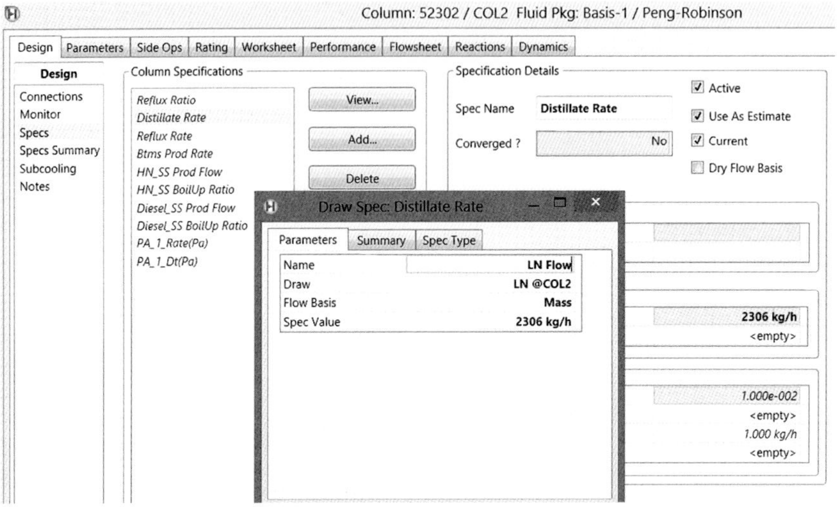

Figure 6.155 Rename "Distillate Rate" as "HN Flow" and specify its value of 2306 kg/h; rename "Btms Prod Rate" as "Bottom Flow" and speciffy its value of 3.047E4 kg/h.

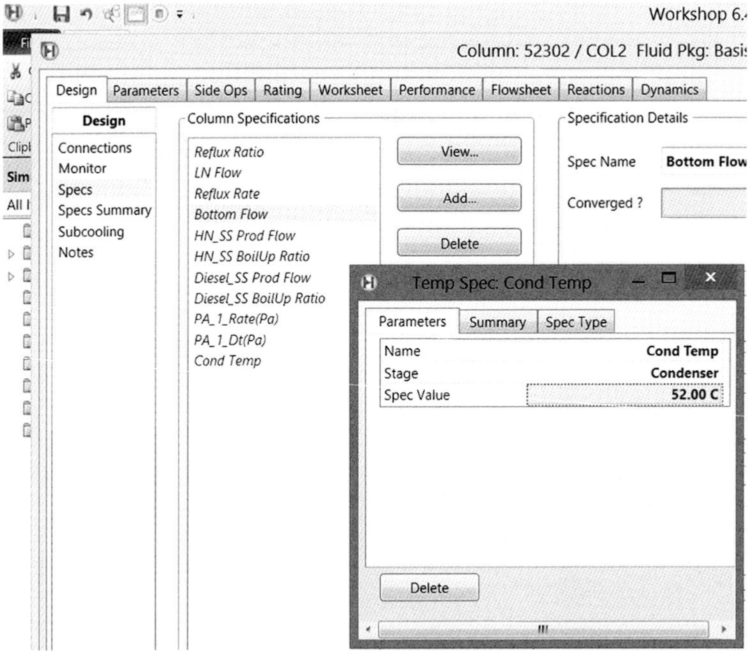

Figure 6.156 Specify condenser, stage 1, and bottom temperatures of 52 °C, 64 °C, and 316 °C.

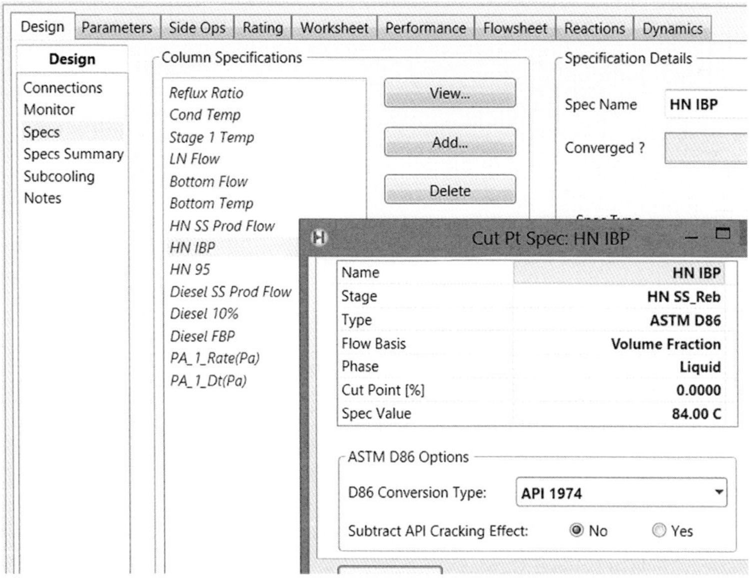

Figure 6.157 Specify "End Point-Based Column Cut Point Spec" for HN IBP (cut point 0%) of 84 °C, HN 95 (95%) of 170 °C, Diesel 10% of 200 °C, and Diesel FBP (100%) of 338 °C.

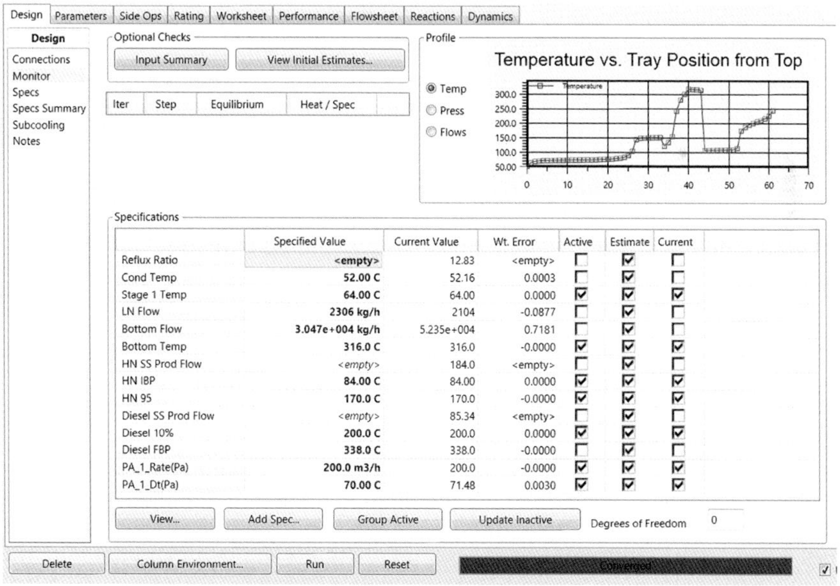

Figure 6.158 Column specifications for simulation convergence.

Column: 52302 / COL2 Fluid Pkg: Basis-1 / Peng-Robinson

Design | Parameters | Side Ops | Rating | Worksheet | Performance | Flowsheet | Reactions | Dynamics

Worksheet: Conditions, Properties, Compositions, PF Specs

Name	52302 Feed @COL2	52302 Steam @COL2	LN @COL2	W02 @COL2	Btm @COL2	HN SS_Prod @COL2	Diesel SS_Prod @COL2
Vapour	0.4898	1.0000	0.0000	0.0000	0.0000	0.0000	0.0000
Temperature [C]	340.0	345.0	52.16	52.16	316.0	115.5	244.9
Pressure [bar]	9.613	11.01	1.600	1.600	1.880	1.767	1.833
Molar Flow [kgmole/h]	460.9	11.10	27.71	24.71	150.2	184.0	85.34
Mass Flow [kg/h]	8.942e+004	200.0	2104	445.2	5.235e+004	1.917e+004	1.555e+004
Std Ideal Liq Vol Flow [m3/h]	106.7	0.2004	3.210	0.4461	57.67	26.46	19.06
Molar Enthalpy [kJ/kgmole]	-2.569e+005	-2.310e+005	-1.694e+005	-2.841e+005	-4.972e+005	-2.060e+005	-2.983e+005
Molar Entropy [kJ/kgmole-C]	777.9	178.8	147.0	60.48	1282	285.3	601.7
Heat Flow [kJ/h]	-1.184e+008	-2.564e+006	-4.694e+006	-7.021e+006	-7.468e+007	-3.790e+007	-2.545e+007

Figure 6.159 Stream table for column 52302.

6.13 Conclusion

HCR process models are usually very sophisticated because their complex feedstocks and highly coupled reaction mechanisms make it difficult to build a plantwide HCR process simulation. However, refiners are mostly concerned about maximizing profit under steady operations, which imply small changes of the process operations and feedstock varieties. Therefore, a good operating model of the refining process only needs to match key product yields, qualities, and process operations under small process changes.

We summarize the key achievements of this work as follows:

1) We develop two integrated HCR process models, which include reactors, fractionators, and hydrogen recycle systems.
2) We provide the step-by-step guideline of model development that has not been reported in the literature.
3) We apply the Gauss–Legendre quadrature to convert kinetic lumps into pseudocomponents based on boiling point ranges (delumping) for rigorous fractionator simulation.
4) Our delumping method gives a continuous response to changes in fractionator specification such as distillate rate.
5) We update API correlations of flash point and freezing point to plant operation and production.
6) The integrated HCR process models are able to predict accurately the product yields, distillation curves of liquid products, and temperature profiles of reactors and fractionators.
7) The integrated HCR process models also gives good estimations on liquid product qualities – density, flash point, and freezing point of diesel fuel (MP HCR) and jet fuel (HP HCR) – by using updated API correlations.
8) We apply the integrated MP HCR process model to conduct simulation experiments to quantify the effects of operating variables on product yields.
9) We apply the integrated HP HCR process model to generate a delta-base vector for LP-based production planning.

This work represents the workflow to build an integrated HCR process model using Aspen HYSYS Petroleum Refining and routine measurement in refinery. We only use routine measurement of feedstock (ASTM D86, specific gravity, total sulfur, and nitrogen content) to build the preliminary model. Furthermore, we also use routine measurements of products (compositional analysis of gas products and distillation curve and specific gravity of liquid product) to calibrate the model. Although the resulting model provides good predictions for 2 months of process and production data, there are several aspects worthy of further studies:

1) Apply SimDist analysis whenever it is available.
2) Currently, the feed lump distribution is developed by routine measurements and it is unachievable for any modeling technique to accurately estimate the molecular information of oil fraction (such as PNA content, multiring aromatics distribution, and hindered and nonhindered sulfur content) by using routine measurement only. Therefore, the resulting model is sensitive to feedstock and needs to be recalibrated when the feedstock is changed from the base case. If detail molecular information of the feedstock is available, the feed lump distribution can be customized to better characterize the feedstock.
3) Users are also allowed to customize the calibration environment to include product property and product composition as objective functions if detail molecular information of product is available.

Nomenclature

CONV	Conversion (–)
F	Molar flow (mol/h)
F_{vi}	The i cut point while applying Gauss–Legendre quadrature into delump
I_m	Inhibition factor in LHHW mechanism (m = NH_3, H_2S, and organic nitrogen compounds) (–)
K_{avg}	Watson K factor (–)
K_{ADS}	LHHW adsorption constants of hydrocarbon (kPa^{-1})
K_{eq}	Equilibrium constant of reversible reaction (–)
K_{total}	Overall activity of reaction group (–)
K_{global_i}	Global activity for the i catalyst bed (–)
$K_{sul\_i\_j}$	Hydrodesulfurization activity of the j distillate cut (j = whole fraction, 430 °F–, 430–950 °F, and 950 °F+) in the i reactor (i = hydrotreating and hydrocracking reactor) (–)
$K_{crc\_i\_j}$	Hydrocracking activity of the j distillate cut (j = whole fraction, 430 °F–, 430–950 °F, and 950 °F+) in the i reactor (i = hydrotreating and hydrocracking reactor) (–)
$K_{hdg\_i\_j}$	Aromatic hydrogenation activity of the j distillate cut (j = whole fraction, 430 °F–, 430–950 °F, and 950 °F+) in the i reactor (i = hydrotreating and hydrocracking reactor) (–)
$K_{ro\_i\_j}$	Ring-opening activity of the j distillate cut (j = whole fraction, 430 (–)°F–, 430–950 (–)°F, and 950(–)°F+) in the i reactor (i = hydrotreating and hydrocracking reactor) (–)
K_{light_i}	Light gas distributing factor (i = C1, C2, C3, and C4) (–)
k	Intrinsic rate constant of reaction (h^{-1})
MeABP	Mean average boiling point (Rankine)

MW	Molecular weight (–)
OBJ_{TR}	The predicting error of temperature rise of catalyst bed (°C)
OBJ_{HQ}	The predicting error of hydrogen quench of catalyst bed (STD m^3/h)
OBJ_{PGF}	The predicting error of flow rate of purge gas (STD m^3/h)
OBJ_{MHF}	The predicting error of flow rate of makeup H_2 (STD m^3/h)
OBJ_{HC}	The predicting error H_2 consumption (STD m^3/m^3)
OBJ_{NVF}	The predicting error of C6 to 430 °F cut (naphtha) volume flow (m^3/h)
OBJ_{DVF}	The predicting error of 430–700 °F cut (diesel) volume flow (m^3/h)
OBJ_{BVF}	The predicting error of 700–1000 °F cut (bottom) volume flow (m^3/h)
OBJ_{RVF}	The predicting error of 1000 °F+ cut (resid) volume flow (m^3/h)
OBJ_{NMF}	The predicting error of C6 to 430 °F cut (naphtha) mass flow (kg/h)
OBJ_{DMF}	The predicting error of 430–700 °F cut (diesel) mass flow (kg/h)
OBJ_{BMF}	The predicting error of 700–1000 °F cut (bottom) mass flow (kg/h)
OBJ_{RMF}	The predicting error of 1000 °F+ cut (resid) mass flow (kg/h)
OBJ_{C1C2}	The predicting error of C1C2 mass yield (wt%)
OBJ_{C3}	The predicting error of C3 mass yield (wt%)
OBJ_{C4}	The predicting error of C4 mass yield (wt%)
OBJ_{SD}	The predicting error of sulfur content of 430–700 °F cut (wt%)
OBJ_{SB}	The predicting error of sulfur content of 700–1000 °F cut (wt%)
OBJ_{ND}	The predicting error of nitrogen content of 430–700 °F cut (ppmwt)
OBJ_{NB}	The predicting error of nitrogen content of 700–1000 °F cut (ppmwt)
OBJ_{NR1}	The predicting error of nitrogen content in reactor 1 effluent (ppmwt)
P_c	Critical pressure (kPa)
P_{H_2}	Partial pressure of hydrogen (kPa)
q_i	The zeros of the Legendre polynomial (–)
SG	Specific gravity 60 °F/60 °F (–)
SV	Space velocity (1/h)
T	Temperature (°C)
T_b	Normal boiling point (°C)
T_c	Critical temperature (°C)
T_r	Reduced temperature (–)
V	Volume (m^3)
V_c	Critical volume (m^3)
w_i	Weight factor of Gauss–Legendre quadrature (–)
ω	Acentric factor (–)
θ	Feed ratio of feed 1 to feed 2 (–)
τ	Residence time (h)

Bibliography

1 Aye, M.M.S. and Zhang, N. (2005) *Chemical Engineering Science*, **60**, 6702.
2 Qader, S.A. and Hill, G.R. (1969) *Industrial & Engineering Chemistry Process Design and Development*, **8**, 98.
3 Quann, R.J. and Jaffe, S.B. (1992) *Industrial & Engineering Chemistry Research*, **31**, 2483.
4 Quann, R.J. and Jaffe, S.B. (1996) *Chemical Engineering Science*, **51**, 1615.

5 Quann, R.R. (1998) *Environmental Health Perspectives Supplements*, **106**, 1501.
6 Froment, G.F. (2005) *Catalysis Reviews – Science and Engineering*, **47**, 83.
7 Ghosh, P., Andrews, A.T., Quann, R.J., and Halbert, T.R. (2009) *Energy & Fuels*, **23**, 5743.
8 Christensen, G., Apelian, M.R., Karlton, J.H., and Jaffe, S.B. (1999) *Chemical Engineering Science*, **54**, 2753.
9 Kumar, H. and Froment, G.F. (2007) *Industrial & Engineering Chemistry Research*, **46**, 5881.
10 Ancheyta, J., Sánchez, S., and Rodríguez, M.A. (2005) *Catalysis Today*, **109**, 76.
11 Ho, T.C. (2008) *Catalysis Reviews: Science and Engineering*, **50**, 287.
12 Valavarasu, G., Bhaskar, M., and Sairam, B. (2005) *Petroleum Science and Technology*, **23**, 1323.
13 Sánchez, S., Rodríguez, M.A., and Ancheyta, J. (2005) *Industrial & Engineering Chemistry Research*, **44**, 9409.
14 Verstraete, J.J., Le Lannic, K., and Guibard, I. (2007) *Chemical Engineering Science*, **62**, 5402.
15 Stangeland, B.E. (1974) *Industrial & Engineering Chemistry Process Design and Development*, **13**, 71.
16 Mohanty, S., Saraf, D.N., and Kunzru, D. (1991) *Fuel Processing Technology*, **29**, 1.
17 Pacheco, M.A. and Dassori, C.G. (2002) *Chemical Engineering Communications*, **189**, 1684.
18 Bhutani, N., Ray, A.K., and Rangaiah, G.P. (2006) *Industrial & Engineering Chemistry Research*, **45**, 1354.
19 Laxminarasimhan, C.S., Verma, R.P., and Ramachandran, P.A. (1996) *AIChE Journal*, **42**, 2645.
20 Basak, K., Sau, M., Manna, U., and Verma, R.P. (2004) *Catalysis Today*, **98**, 253.
21 Fukuyama, H. and Terai, S. (2007) *Petroleum Science and Technology*, **25**, 277.
22 Aspen HYSYS Petroleum Refining Option Guide (2006) AspenTech, Cambridge, MA.
23 Korre, S.C., Klein, M.T., and Quann, R. (1997) *Industrial & Engineering Chemistry Research*, **36**, 2041.
24 Jacob, S.M., Quann, R.J., Sanchez, E., and Wells, M.E. (1998, July) *Oil & Gas Journal*, **6**, 51.
25 Filimonov, V.A., Popov, A.A., Khavkin, V.A., Perezhigina, I.Y., Osipov, L.N., Rogov, S.P., and Agafonov, A.V. (1972) *International Chemical Engineering*, **12**, 21.
26 Jacobs, P.A. (1997) *Industrial & Engineering Chemistry Research*, **36**, 3242.
27 Brown, J.M., Sundaram, A., Saeger, R.B., Wellons, H.S., Kennedy, H.S., and Jaffe, S.B. (2009) WO2009051742.
28 Gomez-Prado, J., Zhang, N., and Theodoropoulos, C. (2008) *Energy*, **33**, 974.
29 Aspen Plus Hydrocracker User's Guide (2006) , AspenTech, Cambridge, MA.
30 Mudt, D.R., Pedersen, C.C., Jett, M.D., Karur, S., McIntyre, B., and Robinson, P.R. (2006) Refinery-wide optimization with rigorous models, in *Practical*

Advances in Petroleum Processing (eds C.S. Hsu and P.R. Robinson), Springer, New York, NY.

31 Satterfield, C.N. (1975) *AIChE Journal*, **21**, 209.

32 Kaes, G.L. (2000) *Refinery Process Modeling: A Practical Guide to Steady State Modeling of Petroleum Processes*, The Athens Printing Company, Athens, GA.

33 Fogler, H.S. (2005) *Elements of Chemical Reaction Engineering*, 4th edn, Prentice Hall, Upper Saddle River, NJ.

34 Aspen HYSYS Simulation Basis (2006) , AspenTech, Cambridge, MA.

35 Daubert, T.E. and Danner, R.P. (1997) *API Technical Data Book – Petroleum Refining*, 6th edn, American Petroleum Institute, Washington, DC.

36 Bollas, G.M., Vasalos, I.A., Lappas, A.A., Iatridis, D.K., and Tsioni, G.K. (2004) *Industrial & Engineering Chemistry Research*, **43**, 3270.

37 Riazi, M.R. (2005) *Characterization and Properties of Petroleum Fractions*, 1st edn, American Society for Testing and Materials, West Conshohocken, PA.

38 Haynes, H.W. Jr. and Matthews, M.A. (1991) *Industrial & Engineering Chemistry Research*, **30**, 1911.

39 Cotterman, R.L., Bender, R., and Prausnitz, J.M. (1985) *Industrial & Engineering Chemistry Process Design and Development*, **24**, 194.

40 Mani, K.C., Mathews, M.A., and Haynes, H.W. Jr. (1993, Feb) *Oil & Gas Journal*, **15**, 76.

41 Riazi, M.R. and Daubert, T.E. (1980) *Hydrocarbon Processing*, **59** (3), 115.

42 Lee, B.I. and Kesler, M.A. (1985) *AIChE Journal*, **31**, 1136.

43 Kister, H.Z. (1992) *Distillation Design*, McGraw-Hill, Inc., New York, NY.

44 Roussel, M., Norsica, S., Lemberton, J.L., Guinet, M., Cseri, T., and Benazzi, E. (2005) *Applied Catalysis*, **279**, 53.

45 Dufresne, P., Bigeard, P.H., and Bilon, A. (1987) *Catalysis Today*, **1**, 367.

46 Scherzer, J. and Gruia, A.J. (1996) *Hydrocracking Science and Technology*, Marcel Dekker, New York, NY.

47 Hu, Z.H., Xiong, Z.L., Shi, Y.H., and Li, D.D. (2005) *Petroleum Processing and Petrochemicals*, **36**, 35.

48 Tippett, T.W. and Ward, J.W. (1985) National Petroleum Refiners Association (NPRA) Annual Meeting, 24 Mar 1985, AM-85-43.

49 Rossi, V.J., Mayer, J.F., and Powell, B.E. (1978, October) *Hydrocarbon Processing*, **15**, 123.

50 Bodington, C.E. and Baker, T.E. (1990) *Interfaces*, **20**, 117.

51 El-Kady, F.Y. (1979) *Indian Journal of Technology*, **17**, 176.

52 Hu, M.C., Powell, R.T., and Kidd, N.F. (1997) *Hydrocarbon Processing*, **76** (6), 81.

53 Chang, A.F. and Liu, Y.A. (2011) Predictive modeling of large-scale integrated refinery reaction and fractionation systems from plant data: 1. Hydrocracking (HCR) processes. *Energy and Fuels*, **25**, 5264–5397.

54 Briggs, B. January 2012 "Hydrocracking Model to Support Crude Selection Process", BP Refining Technology, AspenTech Global Conference: OPTIMIZE 2011, Washington, DC, May 2011; AspenTech Webinar: Improve Refinery Margins with Hydroprocessing Models.

7

Alkylation, Delayed Coking, and Refinery-Wide Simulation

This chapter presents three new topics that are of growing importance in the integrated process modeling and optimization of petroleum refineries. Section 7.1 discusses the alkylation process to produce high-octane blending components for gasoline from reacting isobutane with light olefins. Section 7.2 covers the delayed coking process to upgrade and convert petroleum residual or "bottom of the barrel" materials (e.g., vacuum residue) to valuable liquid and gas product streams (fuel gas, LPG, naphtha, and coke gas oil) and petroleum coke. Section 7.3 demonstrates how to improve profit margins through a refinery-wide process simulation model. Section 7.4 presents the conclusions. Finally, the Bibliography is listed.

7.1 Alkylation

7.1.1 Process Description

Kaes [2], Gary *et al.* [1], and Kranz [4] gave good reviews on the alkylation process, its chemistry, product separation, and technology economics. We describe the key features from these references, focusing mainly on those aspects that are relevant to process simulation and optimization.

Basically, alkylation reactions combine light C3–C5 olefins with isobutane in the presence of a strong acid catalyst. Alkylation can occur at high temperatures without a catalyst, but all commercial processes involve low-to-moderate temperatures using either sulfuric or hydrofluoric acid as a catalyst. Reactions are complete with both catalysts, achieving 100% conversion of the feedstock to isoparaffins and by-products.

Alkylation reactions typically produce 75–150 different isoparaffin isomers. With appropriate operating conditions, the product (alkylate) will fall into the gasoline boiling range with motor octane numbers (MON) of 85–95 and research octane number (RON) of 90–98 [4]. Thus, alkylation operation is important to refineries in producing high-octane blending components for gasoline from reacting isobutane with light olefins.

Figure 7.1 shows a schematic diagram of an alkylation process [2]. A feed stream containing light C3–C5 olefins is combined with an isobutane-rich recycle and

Petroleum Refinery Process Modeling: Integrated Optimization Tools and Applications,
First Edition. Y. A. Liu, Ai-Fu Chang, and Kiran Pashikanti.

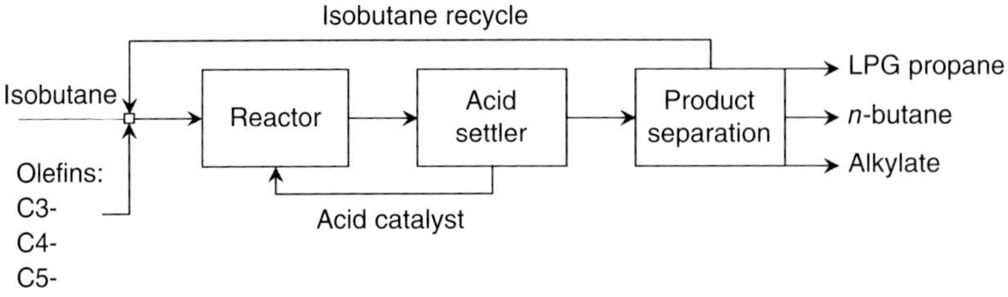

Figure 7.1 A schematic diagram of an alkylation process.

makeup streams and is sent to a reactor, where either HF or H_2SO_4 is used as a catalyst. The reacted mixture passes into an acid settler, where the acid is separated and recycled back to the reactor. We send the hydrocarbon phase from the settler to product separation, where LPG-grade propane, *n*-butane, and alkylate are recovered. A large circulating stream of isobutane is also recovered and sent back to the process as recycle isobutane.

We refer the reader to Refs. [1] and [2] for more detailed process flowsheets using HF or H_2SO_4 as a catalyst, and note the following differences in operating conditions: (1) reaction temperature – 18–45 °C (HF) versus 2–15 °C (H_2SO_4); (2) acid strength – 80–95 wt% (HF) versus 88–95 wt% (H_2SO_4); (3) isobutene concentration – 30–80 vol% (HF) versus 40–80 vol% (H_2SO_4); and (4) hydrocarbon contact time – 8–20 min (HF) versus 20–30 min (H_2SO_4).

7.1.2 Feed Components and Alkylation Kinetics

Aspen HYSYS alkylation model uses 45 pure components and 57 reactions. The Excel spreadsheet *Alkyation components and reactions.xlxs*, in the supplementary material of this book, within this chapter, lists these components and reactions.

Specifically, there are three types of reactions in the alkylation process. First, *the primary alkylation reactions* include all of the C3–C5 olefins reacting with isobutane to form *branched paraffins (BP)* with carbon number from C7 to C13, that is, C7BP to C13BP. Next, *the hydrogen transfer reactions* react C3–C5 olefins with isobutane to form C3–C5 paraffins and isobutene. Third, *the secondary alkylation reactions* are the reactions between the C7–C9 BP formed from primary alkylation and the C3–C5 olefins to form C10–C14 BP. The kinetics of each reaction is represented by a first-order reaction, with its reaction rate constant being represented by the standard Arrhenius equation. On the "Advanced" page of the "Calibration" tab, we can specify advanced kinetic and property parameters to fine-tune the kinetic model to better match plant data. Figure 7.2 is a part of the "kinetic factors" folder, in which E_a is the activation energy, R is the ideal gas constant, and A is the preexponential factor for the Arrhenius equation.

The alkylation model also includes the tuning factors for different classes of reactions (see Figure 7.3). To understand more about tuning these parameters, use functional key F1 to access the Aspen HYSYS V9 help. For example, using functional key F1 teaches us the meanings of activity factors and split factors for C6–C9 (see Figures 7.4 and 7.5).

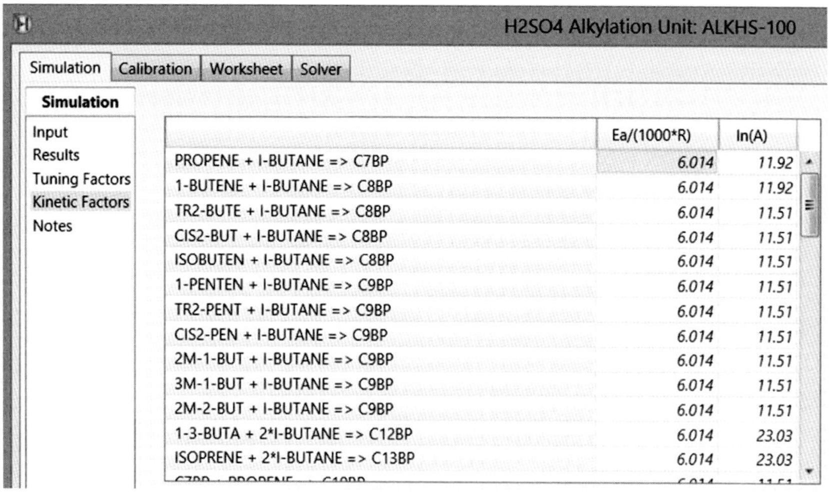

Figure 7.2 A part of the "kinetic factors" folder showing the alkylation reactions.

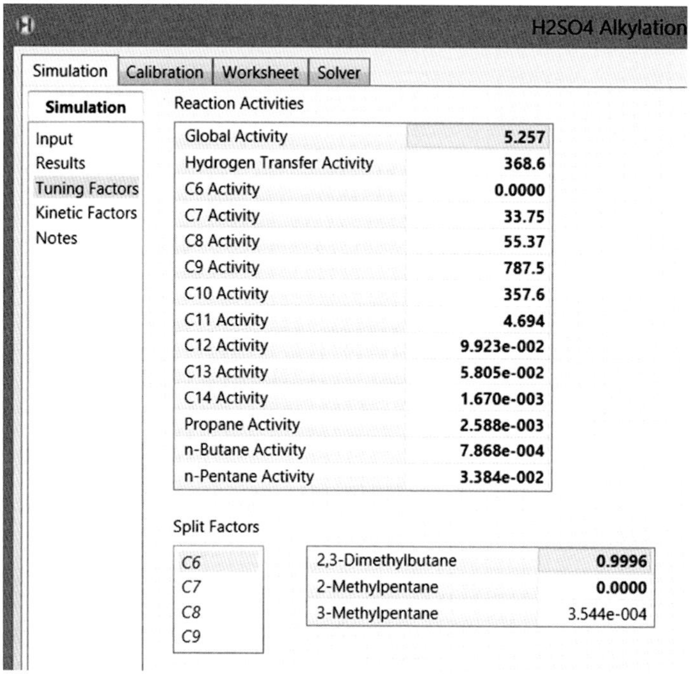

Figure 7.3 Tuning factors for different classes of reactions in the alkylation model.

7.1.3 Workshop 7.1 – Hydrofluoric Acid Alkylation Process Simulation

We start by opening a new case in Aspen HYSYS, go to Properties Environment, and import the component list, *Assay Components Celsius to 850C.cml*. Go to the Chapter 7 folder in the supplementary material of this book and open the file *Workshop 7.1-1 Input Data.xlxs*. Add the additional components relevant

In the **Tuning Factors** page on the **Simulation** tab, you can specify **reaction activities**:

The reaction activities multiply the pre-exponential factors for any reaction in that class.

- C7 to C14 activities affect the primary and secondary alkylation reactions that form C7 through C14 paraffins.
- The hydrogen transfer activity affects all hydrogen transfer reactions.
- The n-Butane activity affects the hydrogen transfer reactions that form normal butane.
- The n-Pentane activity affects the hydrogen transfer reactions that form normal pentane.
- The Propane activity affects the hydrogen transfer reactions that form propane.
- The global activity affects all of the reactions.

Figure 7.4 Online help for alkylation simulation model – manually increase or decrease the tuning factors for different reaction activities to match the plant data.

Split Factors:

Factor		Description
C6	2-3-Dimethylbutane	The split faction of 2-3-Dimethylbutane in C6.
	2-Methylpentane	The split faction of 2-Methylpentane in C6.
	3-Methylpentane	The split faction of 3-Methylpentane in C6.
C7	2-2-Dimethylpentane	The split faction of 2-2-Dimethylpentane in C7.
	2-3-Dimethylpentane	The split faction of 2-3-Dimethylpentane in C7.
	2-4-Dimethylpentane	The split faction of 2-4-Dimethylpentane in C7.
	2-Methylhexane	The split faction of 2-Methylpentane in C7.
	3-Methylhexane	The split faction of 3-Methylpentane in C7.

Figure 7.5 Online help for alkylation simulation model – specification of split fraction of C6 and C7 branched paraffins to a specific component.

to alkylation noted in the Excel to our component list, Component List-1 (see Figure 7.6). Use Peng–Robinson equation of state for the fluid package.

We continue to import a petroleum assay, *Arab Light –1983*, from the Aspen assay library to represent our feeds (see Figures 7.7 and 7.8).

Next, we draw the initial flowsheet of the feed system to the alkylation reactor (see Figure 7.9). We then enter the input data from the above Excel spreadsheet, *Workshop 7.1-1 Alkylation Input Data.xlsx*, to define the two feed streams, Fresh_*i*Butane and ALKY_Feed, and an assumed *i*-butane recycle stream, *i*Butane_RCY.

We use a "SET" operator to define the relationship that the volumetric flow rate of Mixed_*i*Butane is eight times that of ALKF_Feed. Based on the assumed flow rate of *i*-butane recycle stream, *i*Butane_RCY, the "SET" operator can determine the required flow rate of fresh butane, Fresh_*i*Butane (see Figure 7.10).

Next, we add an alkylation reactor to the flowsheet and input the alkylation specifications (see Figures 7.11 and 7.12). Figure 7.13 shows the calculated results of the alkylation reactor.

We now add a petroleum distillation column to simulate the alkylation product separation. Figure 7.14 shows the addition of a petroleum distillation column.

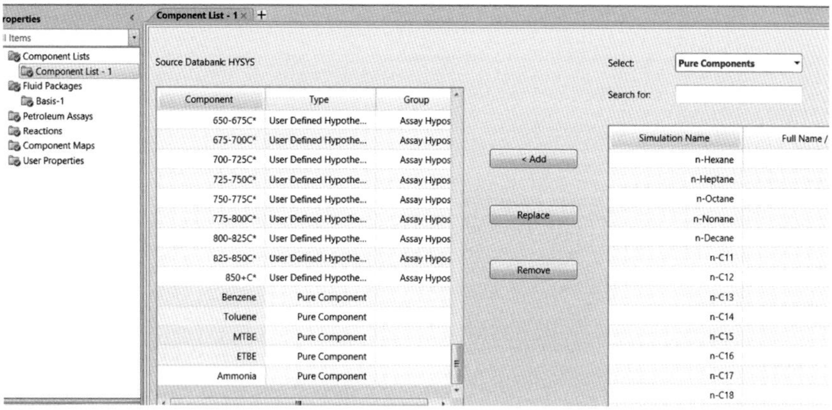

Figure 7.6 Add five more components after importing the component list, *Assay Components Celsius to 850C.cml* to define Component List-1.

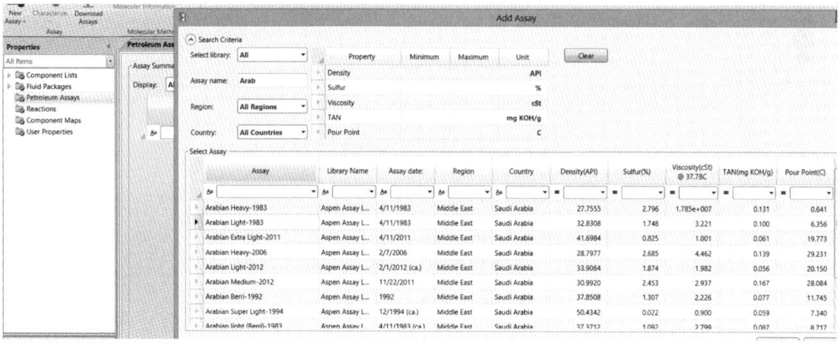

Figure 7.7 Import a petroleum assay, Arab Light – 1983.

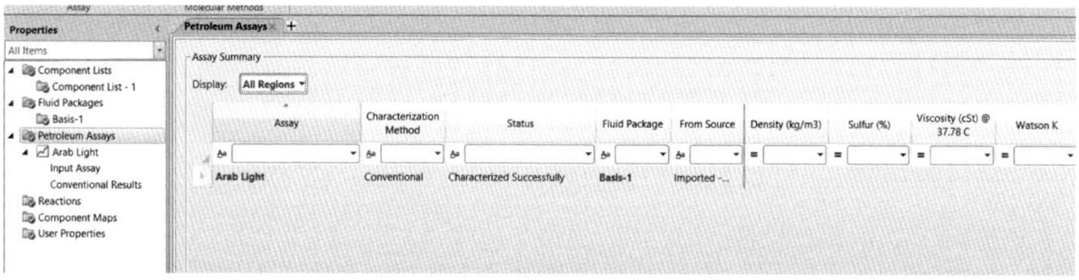

Figure 7.8 The Arab Light petroleum assay added.

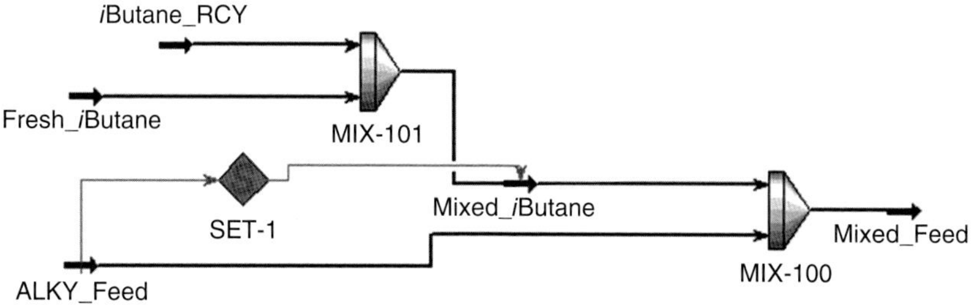

Figure 7.9 Initial flowsheet for the feed system to the alkylation reactor.

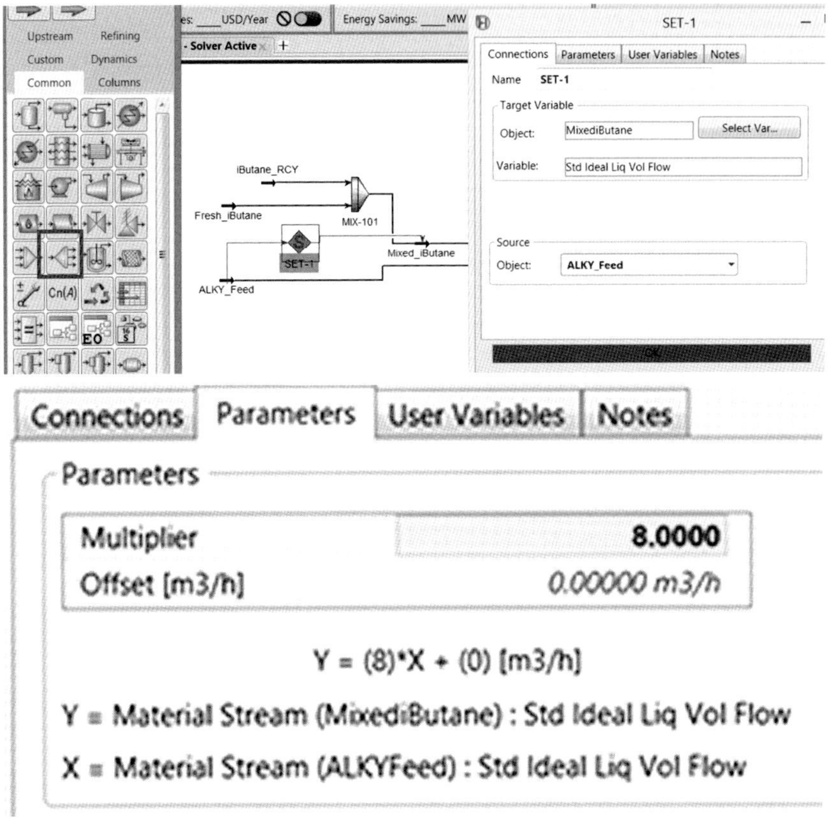

Figure 7.10 Use a "SET" operator to define the flow rate relationship.

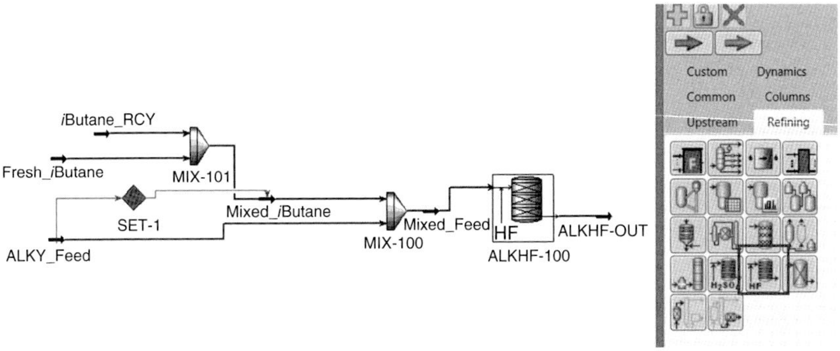

Figure 7.11 Add an alkylation reactor.

Figure 7.15 shows the tuning parameters for the petroleum distillation column. We refer the reader to Section 2.15, Workshop 2.6, Application of the Petroleum Distillation Column, for detailed explanations of the tuning parameters ECP (effective cut point), SI TOP (fractionation index, top section), and SI BOT (fractionation index, bottom section). Figures 7.16 and 7.17 show the predicted product streams by the initial model.

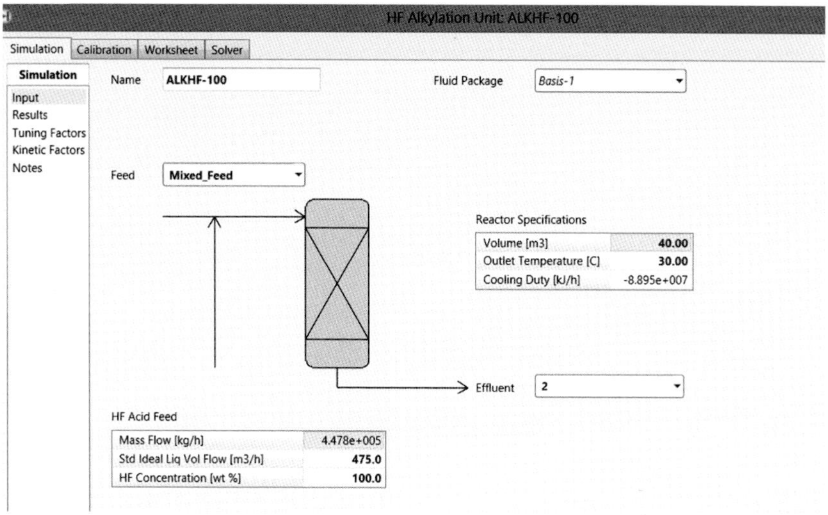

Figure 7.12 Specifications of the alkylation reactor.

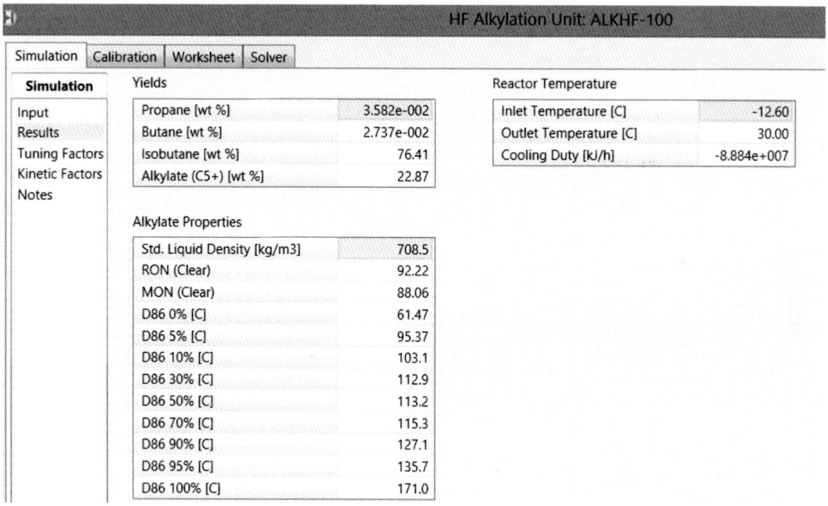

Figure 7.13 Calculated results of the alkylation reactor.

We now close the recycle loop for *i*Butane recycle. See Figure 7.18. First, we delete the stream *i*Butane_RCY to mixer MIX-101 and add a recycle block, RCY-1 with a new recycle stream, *i*Butane_RCY1.

Next, we return *i*Butane_RCY1 to MIX-101 to close the recycle loop, and the simulation converges quickly (see Figures 7.19 and 7.20).

We check the resulting volumetric flow rates: Mixed_*i*Butane has a flow rate of 424 m^3/h, which is exactly eight times that of ALKY_Feed, 53 m^3/h; iButane_RCY1 has a flow rate of 368.9 m^3/h. This requires Fresh-*i*Butane stream to have a flow rate of 424–368.9 or 55.1 m^3/h, which is exactly what the simulation results show. Therefore, the RCY-1 and SET blocks work well. Figures 7.21 and 7.22 show the resulting product streams and their properties.

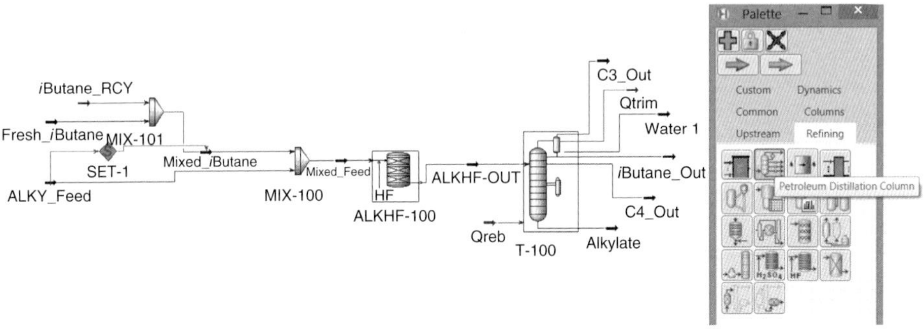

Figure 7.14 Add a petroleum distillation column.

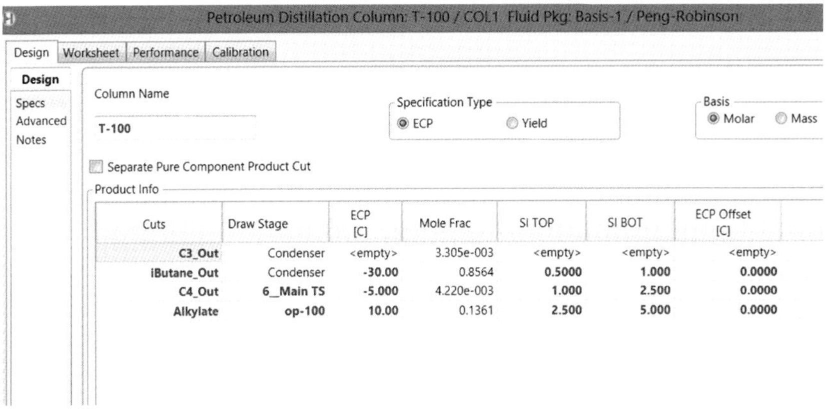

Cuts	Draw Stage	ECP [C]	Mole Frac	SI TOP	SI BOT	ECP Offset [C]
C3_Out	Condenser	<empty>	3.305e-003	<empty>	<empty>	<empty>
iButane_Out	Condenser	-30.00	0.8564	0.5000	1.000	0.0000
C4_Out	6__Main TS	-5.000	4.220e-003	1.000	2.500	0.0000
Alkylate	op-100	10.00	0.1361	2.500	5.000	0.0000

Figure 7.15 Specification of tuning parameters for the petroleum distillation column.

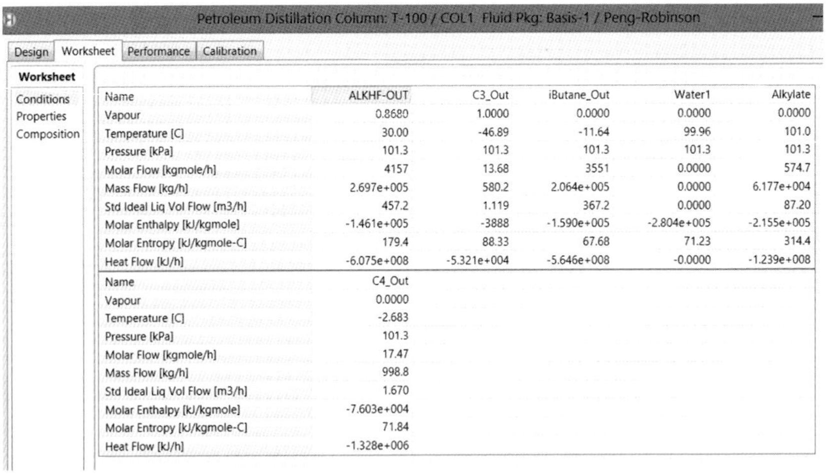

Name	ALKHF-OUT	C3_Out	iButane_Out	Water1	Alkylate
Vapour	0.8680	1.0000	0.0000	0.0000	0.0000
Temperature [C]	30.00	-46.89	-11.64	99.96	101.0
Pressure [kPa]	101.3	101.3	101.3	101.3	101.3
Molar Flow [kgmole/h]	4157	13.68	3551	0.0000	574.7
Mass Flow [kg/h]	2.697e+005	580.2	2.064e+005	0.0000	6.177e+004
Std Ideal Liq Vol Flow [m3/h]	457.2	1.119	367.2	0.0000	87.20
Molar Enthalpy [kJ/kgmole]	-1.461e+005	-3888	-1.590e+005	-2.804e+005	-2.155e+005
Molar Entropy [kJ/kgmole-C]	179.4	88.33	67.68	71.23	314.4
Heat Flow [kJ/h]	-6.075e+008	-5.321e+004	-5.646e+008	-0.0000	-1.239e+008

Name	C4_Out
Vapour	0.0000
Temperature [C]	-2.683
Pressure [kPa]	101.3
Molar Flow [kgmole/h]	17.47
Mass Flow [kg/h]	998.8
Std Ideal Liq Vol Flow [m3/h]	1.670
Molar Enthalpy [kJ/kgmole]	-7.603e+004
Molar Entropy [kJ/kgmole-C]	71.84
Heat Flow [kJ/h]	-1.328e+006

Figure 7.16 Product streams predicted by the initial model.

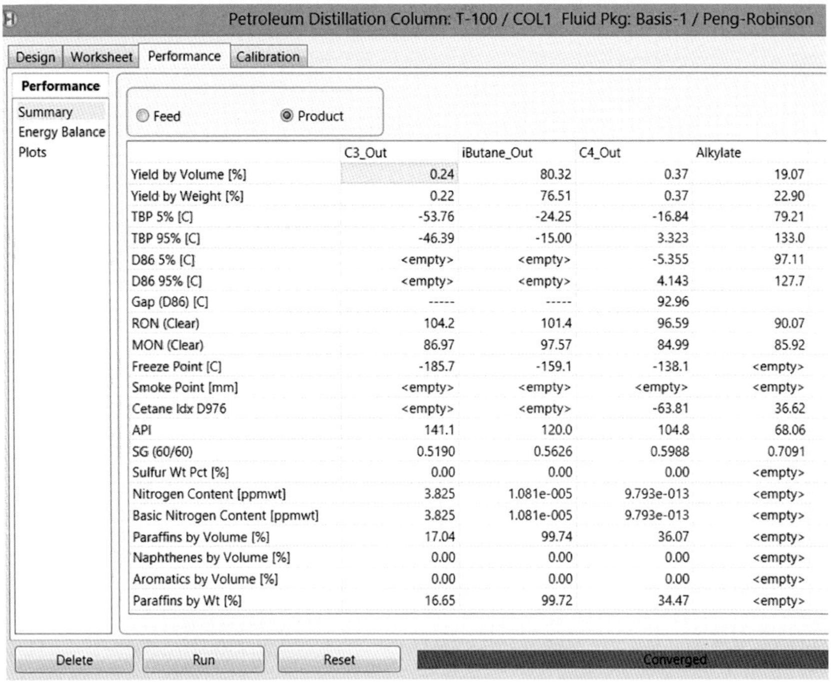

Petroleum Distillation Column: T-100 / COL1 Fluid Pkg: Basis-1 / Peng-Robinson

Design | Worksheet | Performance | Calibration

Performance: Summary, Energy Balance, Plots

Feed / Product

	C3_Out	iButane_Out	C4_Out	Alkylate
Yield by Volume [%]	0.24	80.32	0.37	19.07
Yield by Weight [%]	0.22	76.51	0.37	22.90
TBP 5% [C]	-53.76	-24.25	-16.84	79.21
TBP 95% [C]	-46.39	-15.00	3.323	133.0
D86 5% [C]	<empty>	<empty>	-5.355	97.11
D86 95% [C]	<empty>	<empty>	4.143	127.7
Gap (D86) [C]	-----	-----	92.96	
RON (Clear)	104.2	101.4	96.59	90.07
MON (Clear)	86.97	97.57	84.99	85.92
Freeze Point [C]	-185.7	-159.1	-138.1	<empty>
Smoke Point [mm]	<empty>	<empty>	<empty>	<empty>
Cetane Idx D976	<empty>	<empty>	-63.81	36.62
API	141.1	120.0	104.8	68.06
SG (60/60)	0.5190	0.5626	0.5988	0.7091
Sulfur Wt Pct [%]	0.00	0.00	0.00	<empty>
Nitrogen Content [ppmwt]	3.825	1.081e-005	9.793e-013	<empty>
Basic Nitrogen Content [ppmwt]	3.825	1.081e-005	9.793e-013	<empty>
Paraffins by Volume [%]	17.04	99.74	36.07	<empty>
Naphthenes by Volume [%]	0.00	0.00	0.00	<empty>
Aromatics by Volume [%]	0.00	0.00	0.00	<empty>
Paraffins by Wt [%]	16.65	99.72	34.47	<empty>

Delete | Run | Reset | Converged

Figure 7.17 Predicted properties of product streams.

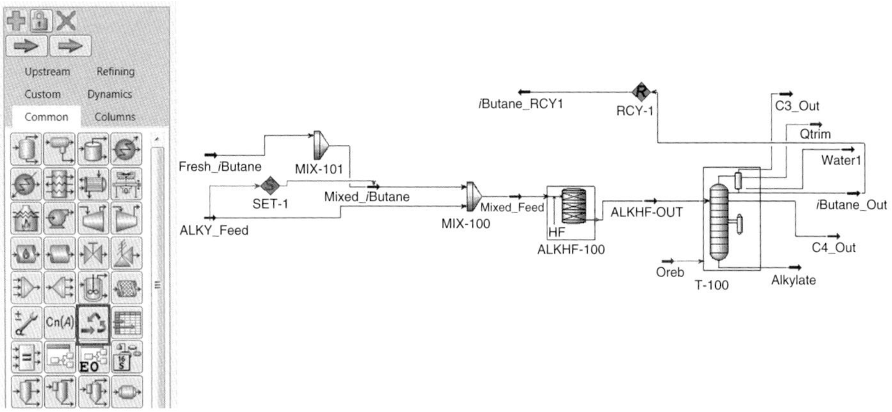

Figure 7.18 Add a recycle block to recycle *i*Butane_Out to MIX-101.

We conclude this workshop by noting that: (1) if plant data are available, we could replace the petroleum distillation column by a rigorous column; (2) the current alkylation model does not include catalyst (HF or H_2SO_4) as a component, but only include its flow rate and concentration in the reaction kinetics calculations; thus, both HF an H_2SO_4 alkylation models are quite similar.

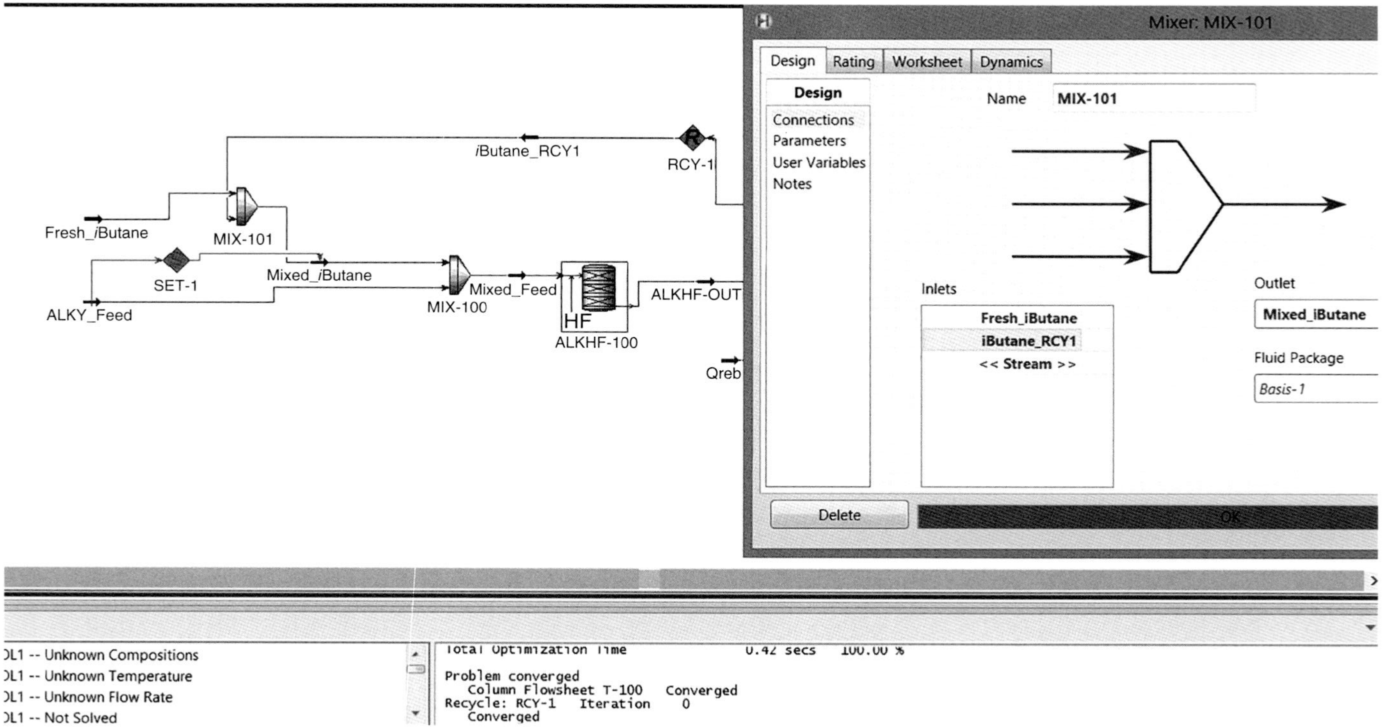

Figure 7.19 Close the *i*Butane recycle loop.

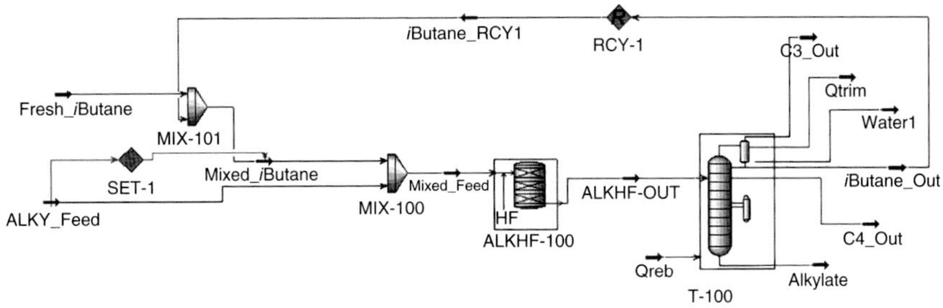

Figure 7.20 The final HF alkylation process flowsheet with *i*Butane recycle.

Design | Worksheet | Performance | Calibration

Worksheet
Conditions
Properties
Composition

Name	ALKHF-OUT	C3_Out	iButane_Out	Water1	Alkylate
Vapour	0.8693	1.0000	0.0000	0.0000	0.0000
Temperature [C]	30.00	-46.89	-11.64	99.96	100.9
Pressure [kPa]	101.3	101.3	101.3	101.3	101.3
Molar Flow [kgmole/h]	4157	13.68	3551	0.0000	574.8
Mass Flow [kg/h]	2.697e+005	580.2	2.064e+005	0.0000	6.178e+004
Std Ideal Liq Vol Flow [m3/h]	457.2	1.119	367.2	0.0000	87.21
Molar Enthalpy [kJ/kgmole]	-1.460e+005	-3888	-1.590e+005	-2.804e+005	-2.150e+005
Molar Entropy [kJ/kgmole-C]	175.1	88.33	67.68	71.23	280.5
Heat Flow [kJ/h]	-6.068e+008	-5.320e+004	-5.646e+008	-0.0000	-1.236e+008

Name	C4_Out
Vapour	0.0000
Temperature [C]	-2.684
Pressure [kPa]	101.3
Molar Flow [kgmole/h]	17.46
Mass Flow [kg/h]	998.7
Std Ideal Liq Vol Flow [m3/h]	1.669
Molar Enthalpy [kJ/kgmole]	-7.602e+004
Molar Entropy [kJ/kgmole-C]	71.85
Heat Flow [kJ/h]	-1.328e+006

Figure 7.21 Product streams from the HF alkylation process.

Design | Worksheet | Performance | Calibration

Worksheet
Conditions
Properties
Composition

Name	ALKHF-OUT	Alkylate	C3_Out	iButane_Out	Water1
Molecular Weight	64.89	107.5	42.40	58.12	18.02
Molar Density [kgmole/m3]	4.746e-002	6.042	5.562e-002	10.22	52.62
Mass Density [kg/m3]	3.080	649.4	2.358	593.9	947.9
Act. Volume Flow [m3/h]	8.759e+004	95.14	246.0	347.5	0.0000
Mass Enthalpy [kJ/kg]	-2250	-2001	-91.70	-2736	-1.556e+004
Mass Entropy [kJ/kg-C]	2.698	2.610	2.083	1.165	3.954
Heat Capacity [kJ/kgmole-C]	116.0	260.0	53.63	124.7	79.22
Mass Heat Capacity [kJ/kg-C]	1.788	2.420	1.265	2.145	4.398
LHV Molar Basis (Std) [kJ/kgmole]	<empty>	<empty>	1.946e+006	2.652e+006	0.0000
HHV Molar Basis (Std) [kJ/kgmole]	<empty>	<empty>	2.076e+006	2.857e+006	4.101e+004
HHV Mass Basis (Std) [kJ/kg]	<empty>	<empty>	4.895e+004	4.917e+004	2276
CO2 Loading	<empty>	<empty>	<empty>	<empty>	<empty>
CO2 Apparent Mole Conc. [kgmole/m3]	<empty>	3.189e-022	<empty>	0.0000	<empty>
CO2 Apparent Wt. Conc. [kgmol/kg]	<empty>	4.911e-025	<empty>	0.0000	<empty>
LHV Mass Basis (Std) [kJ/kg]	<empty>	<empty>	4.590e+004	4.564e+004	0.0000
Phase Fraction [Vol. Basis]	0.8315	<empty>	1.000	<empty>	0.0000
Phase Fraction [Mass Basis]	0.8041	0.0000	1.000	0.0000	0.0000
Phase Fraction [Act. Vol. Basis]	0.9991	0.0000	1.000	0.0000	0.0000
Mass Exergy [kJ/kg]	0.1432	18.97	14.11	54.90	35.22
Partial Pressure of CO2 [kPa]	0.0000	0.0000	4.538e-004	0.0000	0.0000
Cost Based on Flow [Cost/s]	0.0000	0.0000	0.0000	0.0000	0.0000
Act. Gas Flow [ACT_m3/h]	8.751e+004	<empty>	246.0	<empty>	<empty>

Delete | Run | Reset | Ignor

Figure 7.22 Product properties for HF alkylation process.

7.2 Delayed Coking

7.2.1 Process Description

Kaes [2], Gary *et al.* [1], and Ellis and Paul [7] gave good descriptions of the petroleum coke and the delayed coking operation, coking reactions, product separation, and technology economics. We summarize the key features from these references, focusing mainly on those aspects that are relevant to process simulation and optimization.

Basically, delayed coking is a severe thermal cracking operation to upgrade and convert petroleum residual materials (e.g., bottoms from atmospheric and vacuum distillation of crude oil) to liquid and gas product streams (fuel gas, LPG, naphtha, and coke gas oil), leaving behind a large amount of concentrated solid carbon material, petroleum coke [7]. The coker gas oil is a suitable feed to a fluid catalytic cracking unit or to a hydrocracking unit. Figure 7.23 shows a simplified delayed coking process flowsheet.

The feed to a delayed coker is typically a vacuum residue. It combines with a recycle oil from the coker fractionation bottom and with a steam stream and is heated in a charge furnace (fired heater with horizontal tubes) to reach a thermal cracking temperature of 485–505 °C. With a short residence time in the furnace tubes, coking of the feed material is "delayed" until it reaches the bottom of a large coking drum downstream of the furnace [7]. The coker drum has a long residence time, allowing the heavy liquids to polymerize and dehydrogenate to form petroleum coke and to deposit the coke on the sides of the drum. The coke builds in the drum from bottom up. When the accumulated coke reaches a certain height, the drum is taken off-line, and the coke deposit is removed by a hydraulic drill. The removal of coke from a drum consumes a large amount of water, and the contaminated water requires treatment before water reuse [2].

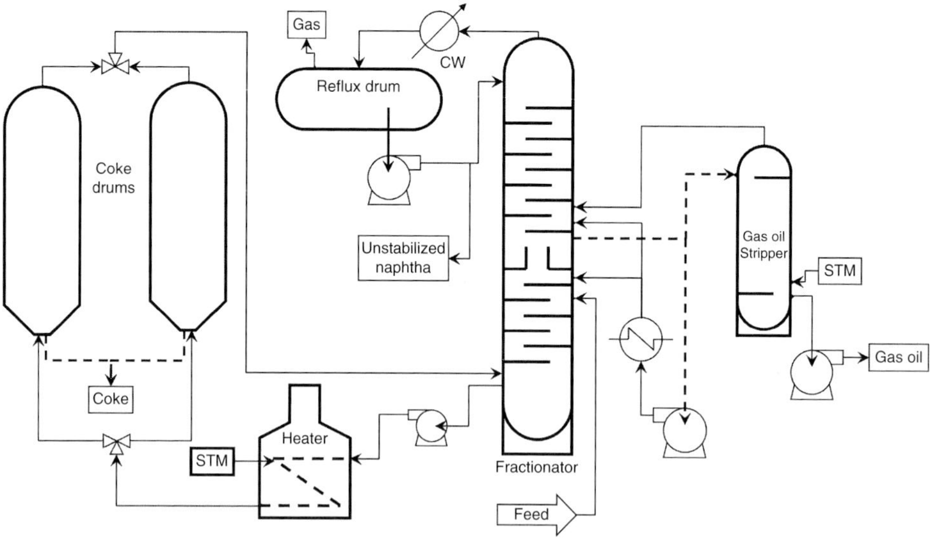

Figure 7.23 A schematic diagram of a typical delayed coking process. (Courtesy of Kaes [2].)

This coke deposit and removal operation are cyclic. At least two coker drums are used. When one drum is filled with coke deposit, it is taken off-line for coke removal, and the other coker drum is brought on-line [2].

The steam and oil vapors that do not coke leave the top of the coker drum and are quenched with cold gas oil to stop the reactions before fractionation. The quenched coke drum vapors are introduced to the wash zone at the bottom of the fractionation column, where they are "washed' with heavy gas oil. The washed zone liquid falls to the bottom of the fractionation column, where it mixes with the fresh residual charge [7].

The main fractionator separates the vapors from the wash zone into fuel gas, LPG, naphtha or unstabilized gasoline, light gas oil (LGO), heavy gas oil (HGO), and a bottom residue stream. The residue stream is recycled to the coker inlet as a coker feed. The coker gas oils may be further processed in hydrotreaters and used as feedstock to other cracking processes.

7.2.2 Feed Characterization, Kinetic Lumps, and Coking Reaction Kinetics

There are three types of reactions occurring in a delayed coker. First, thermal cracking can occur in furnace, drum liquid, and drum vapor phase. Next, coke formation from precipitation or polymerization can occur in the drum liquid phase. Lastly, the asphaltenes can precipitate in the drum to form coke.

To model the reaction kinetics within a delayed coker using Aspen HYSYS, we first characterize the coker feed by *its specific gravity (API gravity), sulfur content, Conradson carbon residue (CCR), nitrogen content, and D2887 9-point distillation curve.* With delayed coking becoming the technology of choice to upgrade the petroleum residue materials ("bottom of the barrel") in a refinery, it is important to be able to predict product yields based on coker feed characterization. Ancheyta [8] and Gary *et al.* [1] have reviewed empirical correlations reported from 1981 to 2006 to predict product yields from a delayed coker, with most of the correlations based on CCR and API gravity. References [9] and [10] present examples of applying these correlations to commercial delayed cokers.

In this chapter, we show the reader how to develop simulation models for the delayed cokers that can predict the effects of operating conditions, in addition to feed properties, using Aspen HYSYS.

Aspen HYSYS delayed coker model uses 37 kinetic lumps and 113 reactions. Aspen HYSYS takes the feed properties listed and calculates properties that are more directly related to the kinetic lumps, particularly the Watson K factor, Eq. (1.7). From the Watson K, Aspen HYSYS estimates paraffins, naphthenes, and aromatics (PNA). We illustrate the feed properties and the estimated PNA contents for the coker feed in the simulation file, *Workshop 7.2-1 Closed Loop_Calibration.hsc.* Open the file and go to COKER-100 → Calibrate → Feed Summary (see Figure 7.24).

Aspen HYSYS uses the PNA contents, along with the distillation curve and CCR to calculate the kinetic lump compositions entering the coker.

To understand the meaning of the acronym of each kinetic lump, we may put the mouse on top of one kinetic lump, say, HP, and click on the functional key F1 to access the Aspen HYSYS V9 Help. We then see the explanation in Figure 7.25.

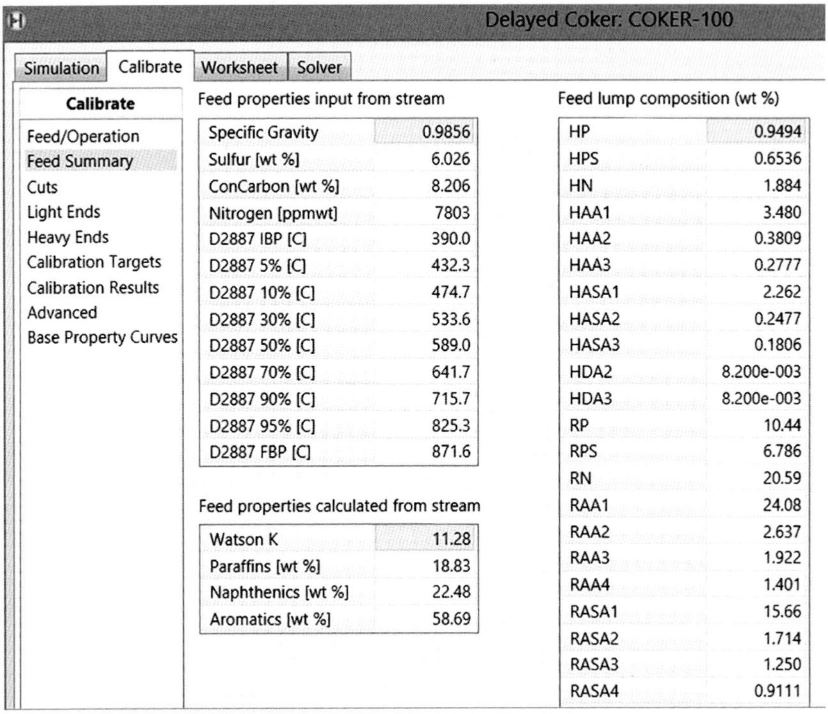

Figure 7.24 Feed properties input from stream, CokerFeed; feed properties calculated from Watson *K* factor – PNA contents; and feed kinetic lump compositions of the coker feed.

In addition to the 24 lumps depicted in Figure 7.24, Aspen HYSYS also include nine more components and kinetic lumps, including H_2S, C (lights), G (gasoline), GS (gasoline S), LP (light paraffins), LPS (light paraffin S), LN (light naphthenes), coke, and water.

The 33 components and kinetic lumps are involved in 133 reactions, with the kinetics of each reaction being represented by a first-order reaction, with its reaction rate constant being represented by the standard Arrhenius equation. On the Advanced page of the Calibration tab, we can specify advanced kinetic and property parameters to fine-tune the kinetic model to better match plant data (see Figure 7.26).

To understand more about tuning these parameters, use functional key F1 to access the Aspen HYSYS V9 help. For example, using functional key F1 teaches us that "fraction of the unconverted resid" represents the resid that did not coke but remains in the coker drum at the end of the drum cycle. This corresponds to the volatile matter in the coke. Increasing this value decreases the coke production and drum temperature.

7.2.3 Workshop 7.2 – Simulation and Calibration of a Delayed Coking Process

In this workshop, we teach the reader how to duplicate a completed delayed coker model available from the installation package of Aspen HYSYS V9 under the file name *delayedcoker_rigorouscolumn.hsc.*

HP	Gas Oil Paraffin
HPS	Gas Oil Sulfides
HN	Gas Oil Naphthenes
HAA1	Gas Oil 1-Ring Alkyl Aromatics
HAA2	Gas Oil 2-Ring Alkyl Aromatics
HAA3	Gas Oil 3-Ring Alkyl Aromatics
HASA1	Gas Oil 1-Ring Sulfide Aromatics
HASA2	Gas Oil 2-Ring Sulfide Aromatics
HASA3	Gas Oil 3-Ring Sulfide Aromatics
HDA2	Gas Oil 2-Ring Denuded Aromatics (No more crackable sidechains)
HDA3	Gas Oil 3-Ring Denuded Aromatics
RP	Resid Paraffin
RPS	Resid Sulfides
RN	Resid Naphthenes
RAA1	Resid 1-Ring Alkyl Aromatics
RAA2	Resid 2-Ring Alkyl Aromatics
RAA3	Resid 3-Ring Alkyl Aromatics
RAA4	Resid 4-Ring Alkyl Aromatics
RASA1	Resid 1-Ring Sulfide Aromatics
RASA2	Resid 2-Ring Sulfide Aromatics
RASA3	Resid 3-Ring Sulfide Aromatics
RASA4	Resid 4-Ring Sulfide Aromatics
RDA3	Resid 3-Ring Denuded Aromatics
RDA4	Resid 4-Ring Denuded Aromatics

Figure 7.25 Aspen HYSYS V9 Help – explanation of the meaning of 24 kinetic lumps.

We start by opening a new case in Aspen HYSYS, go to Properties Environment, and click on Petroleum Assay. Note the three options for adding an assay – import from library, import from file, and manually enter (see Figure 7.27). We save the simulation file as *Workshop 7.2-1.hsc.*

The traditional feed to a delayed coker is a vacuum residue. For this workshop, we choose to import an existing assay in Aspen assay library, Hondo Monterrey – 1983 to define this vacuum residue. After clicking on "Import from library" button on the previous figure, HYSYS asks us to select assay components from a list. We choose component list (cml), *"Assay Component Celsius to 850C.cml".* We then search for the assay *Hondo Monterey – 1983* and import it (see Figures 7.28–7.31).

Next, we enter the Simulation Environment to define a Petroleum Feeder to the flowsheet (see Figure 7.32). We enter the temperature, pressure, and mass flow

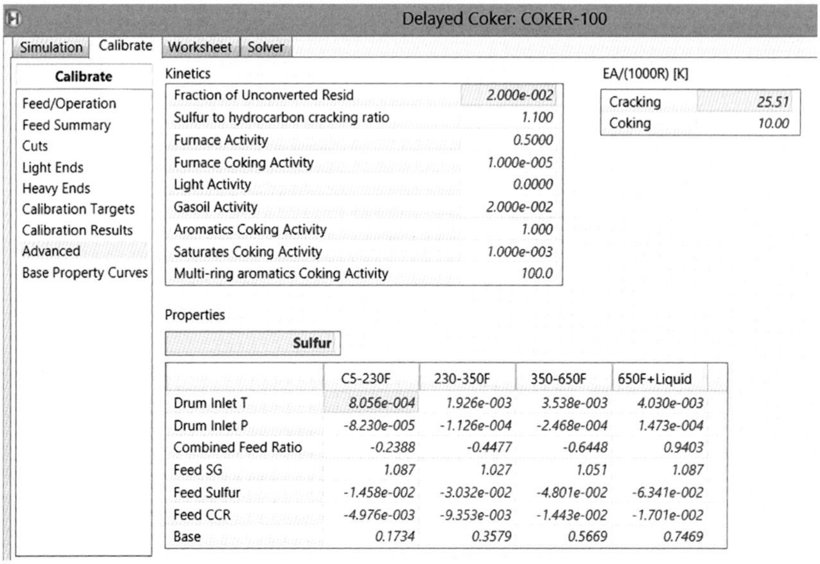

Figure 7.26 Advanced kinetic and property parameters for fine-tuning the kinetic model response to match plant data.

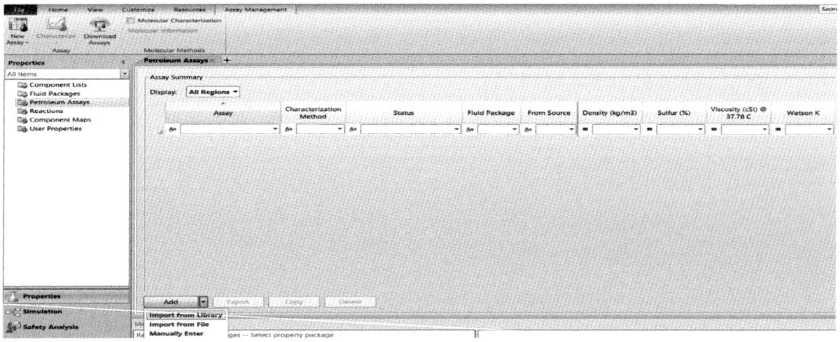

Figure 7.27 Open a new case, enter the Properties environment, click on Petroleum Assays, and choose "Import from Library" to add an assay.

rate of CokerFeed based on the plant data, and we assume initially CokerFeed to have the same composition and properties as the petroleum assay, Honod Monterey – 1983 (see Figure 7.33).

Next, we add the delayed coker into the flowsheet and specify the connecting streams (Figure 7.34).

We enter the required input to the delayed coker. Note that *the combined feed ratio*, CFR, is the flow rate ratio of the coker feed from the coker fractionator to the feed entering the coke fractionator. Generally, this ratio varies between 1.05 and 1.15 (see Figure 7.35).

Next, we add another petroleum feeder based on the same petroleum assay, Honod Monterey – 1983, with a product stream, VR Feed, which is the vacuum

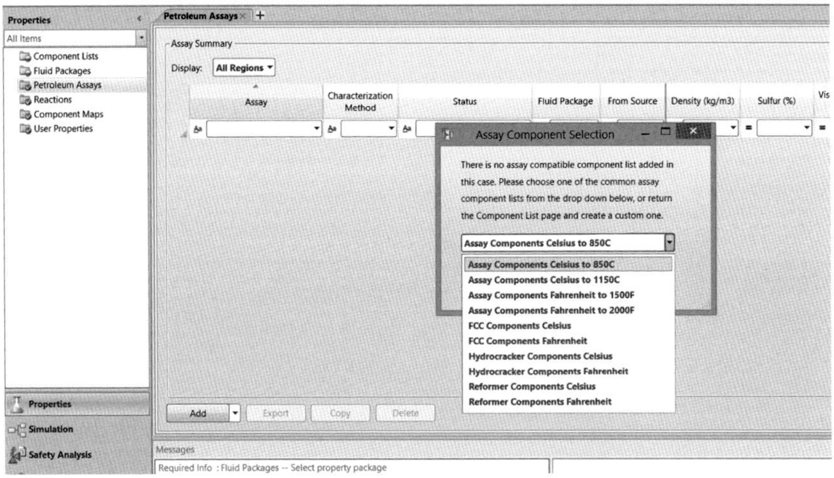

Figure 7.28 Choose a predefined component list from Aspen HYSYS. (Courtesy of ASPEN Technology, Inc.)

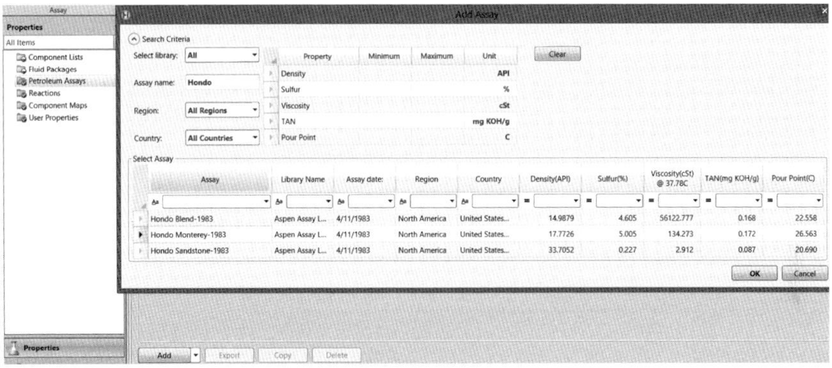

Figure 7.29 Search for the assay, Hondo Monterey – 1983 from Aspen Assay Library. (Courtesy of ASPEN Technology, Inc.)

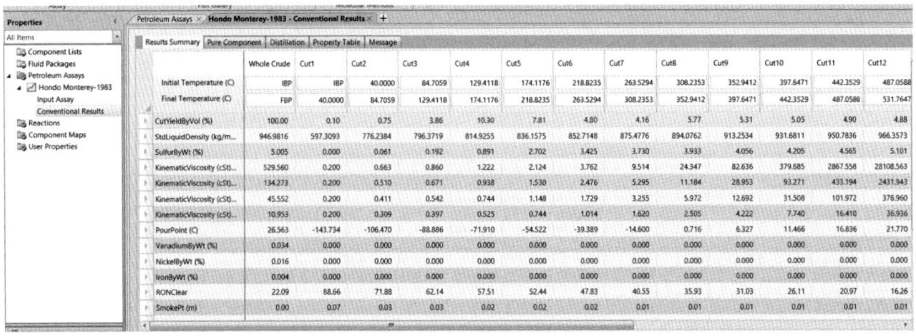

Figure 7.30 The imported petroleum assay, Hondo Monterey – 1983.

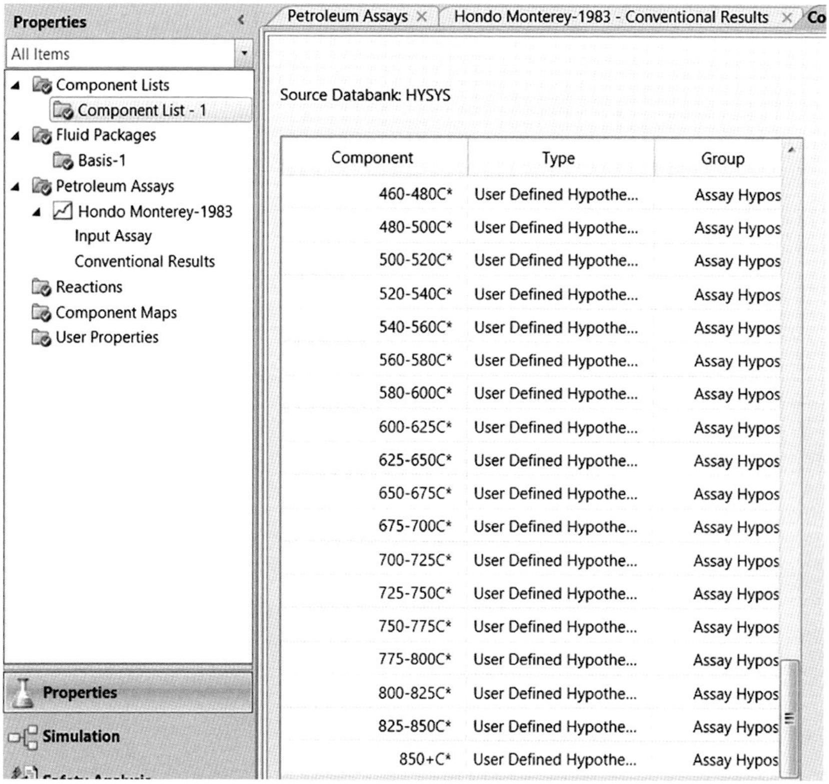

Figure 7.31 The Component List-1, with hypocomponents up to a boiling range above 850 °C.

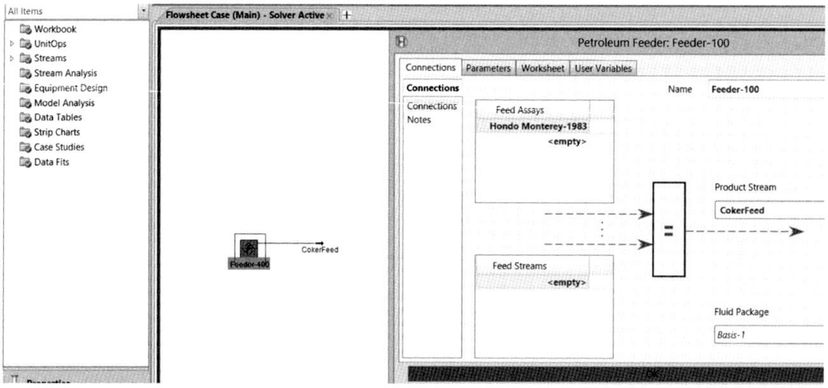

Figure 7.32 Define petroleum feeder showing feed assay, Hondo Monterey – 1983, and product stream, CokerFeed.

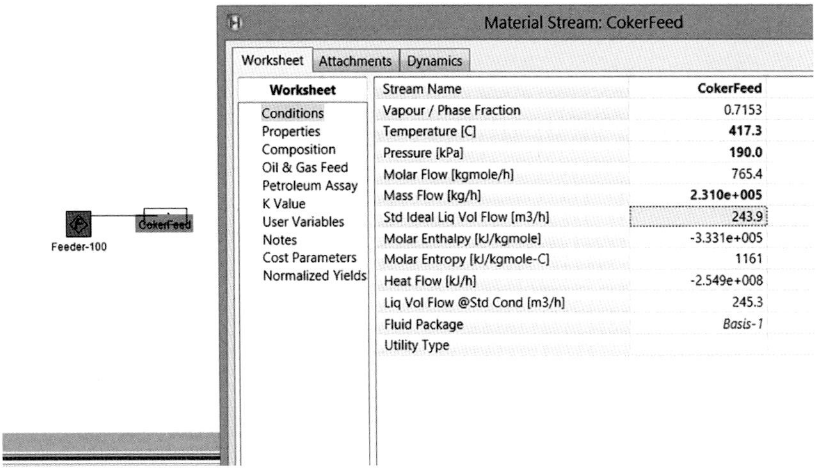

Figure 7.33 Enter the conditions of CokerFeed.

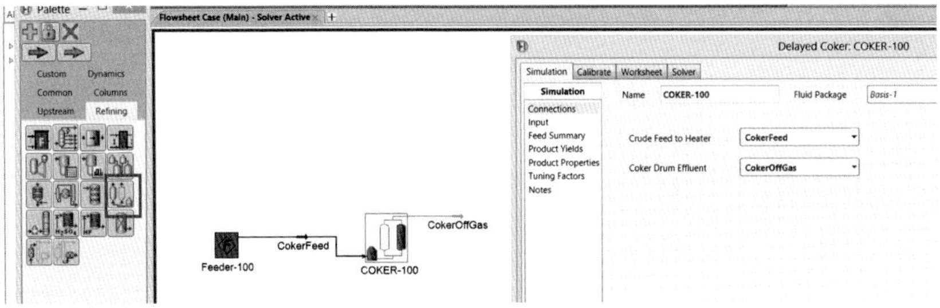

Figure 7.34 Add a delayed coker to the flowsheet.

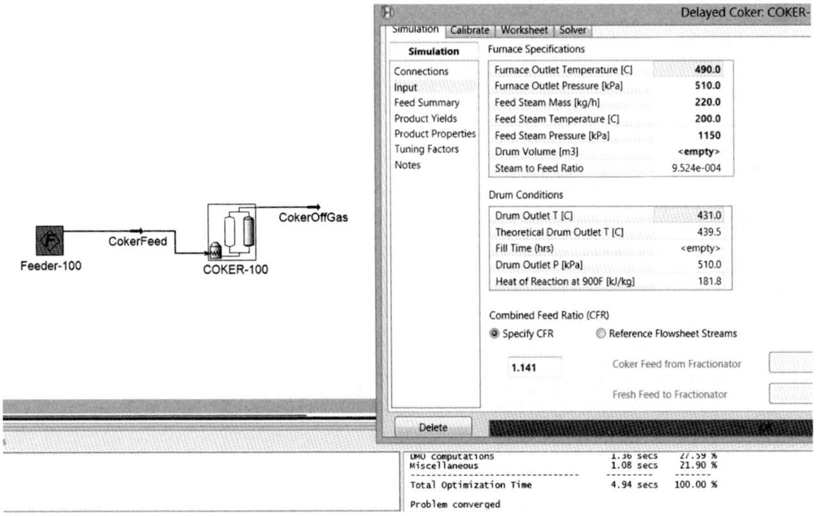

Figure 7.35 Enter input to the delayed coker model and achieve simulation convergence.

reside to the fractionation column. We specify the temperature, pressure, and flow rate of VR Feed (see Figure 7.36).

We continue to add a fractionation column with VR Feed and CokerOffGas as feeds and with fuel gas, LPG, naphtha, LGO, HGO, and coker feed as products. The column has a side stripper and a pumparound. We use a refluxed absorber, with a condenser but no reboiler, as our column model (see details in Figures 7.37–7.40).

Figure 7.41 shows the eight specifications that enable the convergence of the coker fractionation column. We save the converged open-loop simulation as *Workshop 7.2-1 Open Loop_Converged.hsc.*

Next, we delete the petroleum feeder F-100. We add a recycle block RCY-1 and connect the streams as Figure 7.41. We see the converged specifications for the

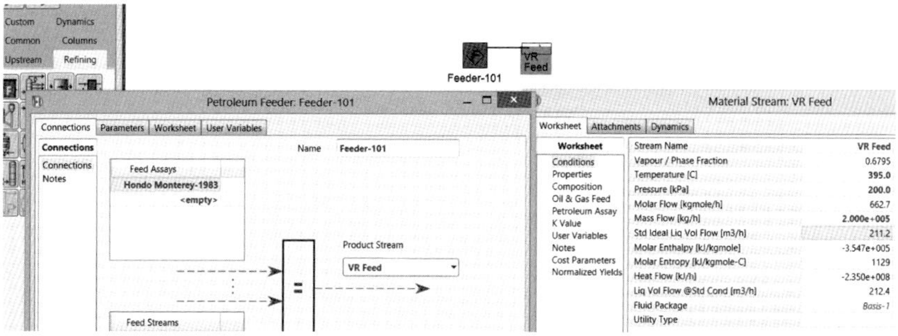

Figure 7.36 Define a petroleum feed and its product stream, VR Feed.

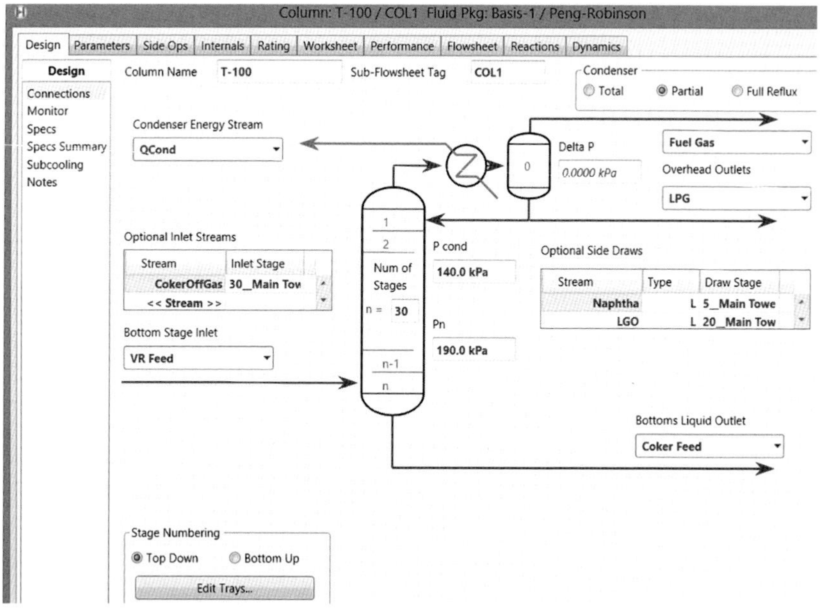

Figure 7.37 Specifications of the coker fractionation column.

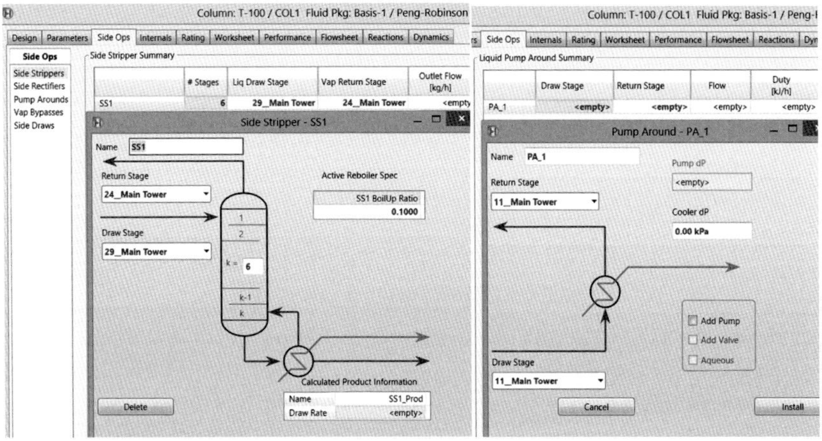

Figure 7.38 Specifications of side stripper and pumparound for the coker fractionation column.

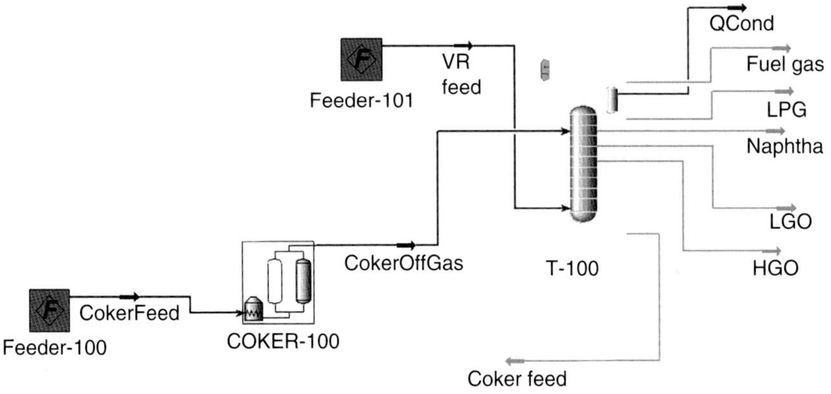

Figure 7.39 Flowsheet with the coker fractionation column before recycle.

coker fractionation column in Figure 7.42. We save the simulation as *Workshop 7.2-1 Closed Loop_Converged.hsc.*

We continue to do calibration of the delayed coker model using plant data. First, we resave the file *Workshop 7.2-1 Closed Loop_Converged.hsc* as *Workshop 7.2-1 Closed Loop_Calibration.hsc.* We open the delayed coker model, go to "Calibration," and click on "Pull Data from Simulation." Figure 7.43 shows the resulting data.

We then enter the available plant data for calibration (see Figures 7.44–7.46). For convenience, we can copy these calibration data from the supplement to this book, under the file *Workshop 7.2-1 Initial and Calibration Data.xlsx.*

The calibration involves three steps: (1) initialize calibration; (2) calibrate; and (3) transfer the resulting calibrated model parameters back to the simulation file (see Figure 7.47).

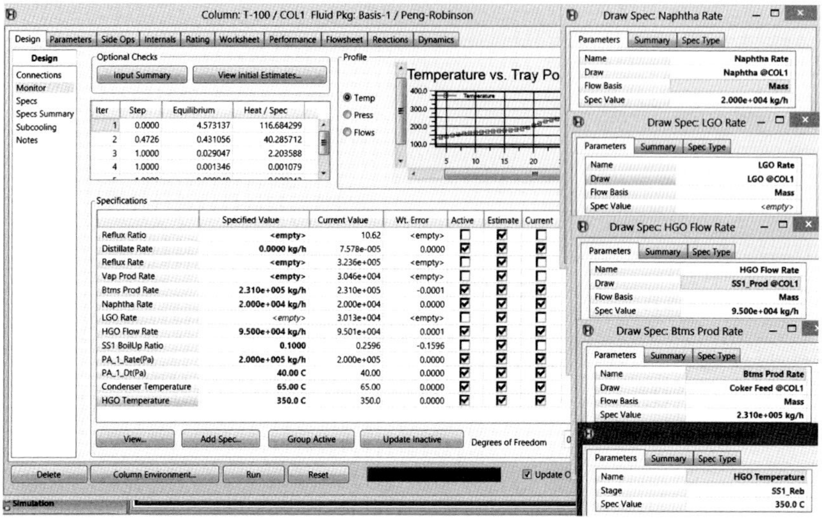

Figure 7.40 Eight specifications to achieve coker fractionation column convergence in the open-loop flowsheet.

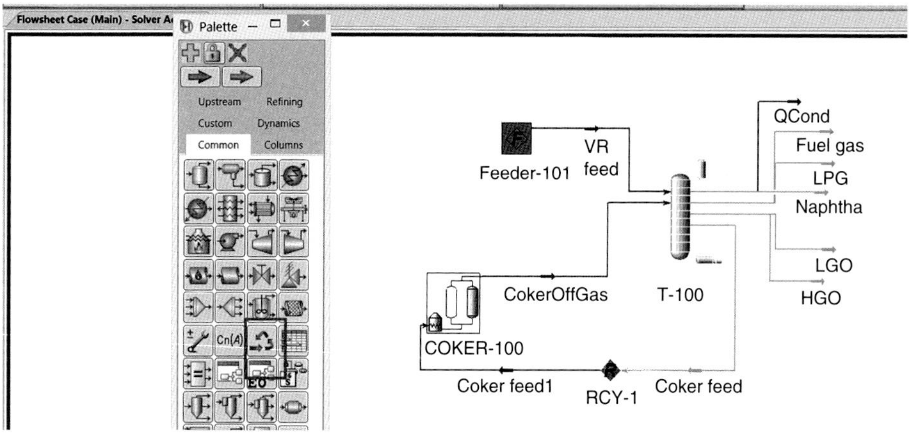

Figure 7.41 Convert the open-loop converged flowsheet to a closed loop.

To conclude this workshop, we note that when plant data are available, the reader can continue to add more fractionation units (such as the stabilizer for the naphtha or unstabilized gasoline product) to our current delayed coking process flowsheet. We can do this fairly easily by referring to Workshop 4.3, Sections 4.14.1–4.15.3, where we present a detailed, step-by-step demonstration of how to build a simulation model for the FCC main fractionator and gas plant system (including the stabilizer from Figures 4.106–4.110). We can apply the same procedure to build the simulation models for the complete fractionation systems for reformers, hydrocrackers, delayed cokers, and so on.

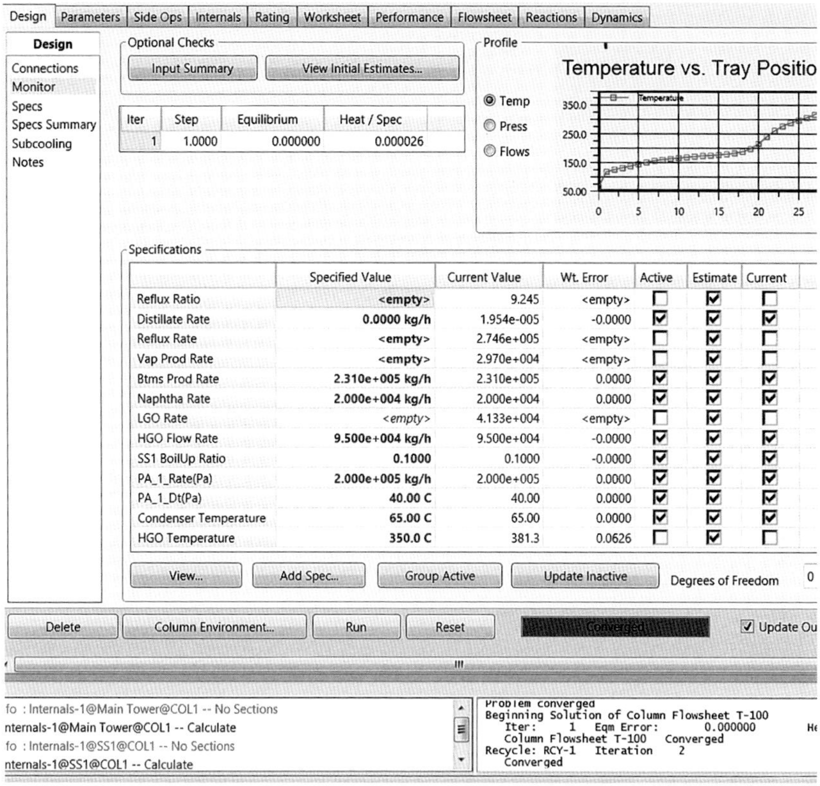

Figure 7.42 Converged fraction column specifications in the closed-loop flowsheet.

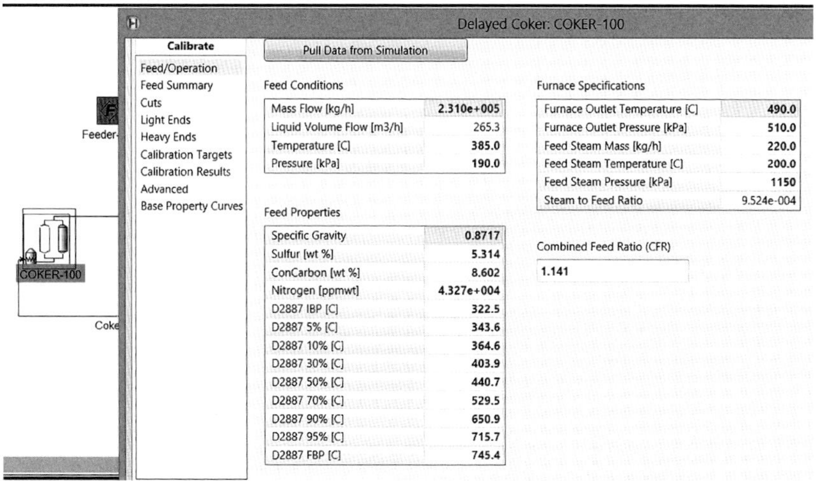

Figure 7.43 The first step in calibration is to pull data from a converged simulation.

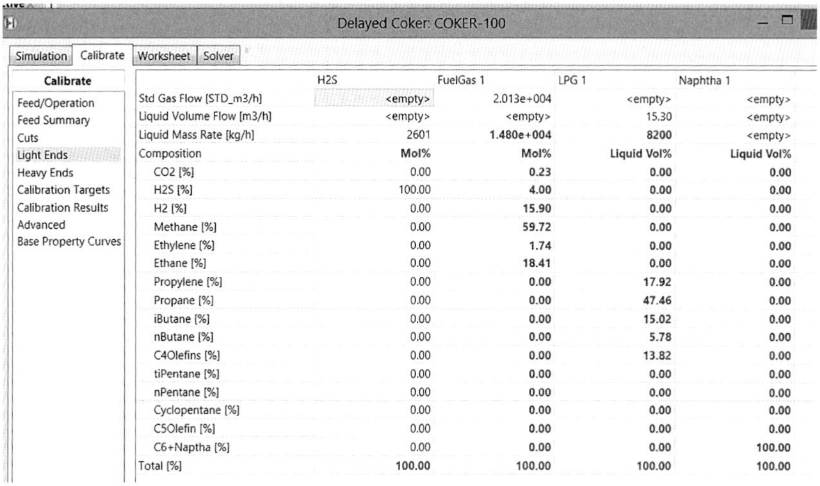

	H2S	FuelGas 1	LPG 1	Naphtha 1
Std Gas Flow [STD_m3/h]	<empty>	2.013e+004	<empty>	<empty>
Liquid Volume Flow [m3/h]	<empty>	<empty>	15.30	<empty>
Liquid Mass Rate [kg/h]	2601	1.480e+004	8200	<empty>
Composition	Mol%	Mol%	Liquid Vol%	Liquid Vol%
CO2 [%]	0.00	0.23	0.00	0.00
H2S [%]	100.00	4.00	0.00	0.00
H2 [%]	0.00	15.90	0.00	0.00
Methane [%]	0.00	59.72	0.00	0.00
Ethylene [%]	0.00	1.74	0.00	0.00
Ethane [%]	0.00	18.41	0.00	0.00
Propylene [%]	0.00	0.00	17.92	0.00
Propane [%]	0.00	0.00	47.46	0.00
iButane [%]	0.00	0.00	15.02	0.00
nButane [%]	0.00	0.00	5.78	0.00
C4Olefins [%]	0.00	0.00	13.82	0.00
tiPentane [%]	0.00	0.00	0.00	0.00
nPentane [%]	0.00	0.00	0.00	0.00
Cyclopentane [%]	0.00	0.00	0.00	0.00
C5Olefin [%]	0.00	0.00	0.00	0.00
C6+Naptha [%]	0.00	0.00	0.00	100.00
Total [%]	100.00	100.00	100.00	100.00

Figure 7.44 Plant data for light ends for calibration.

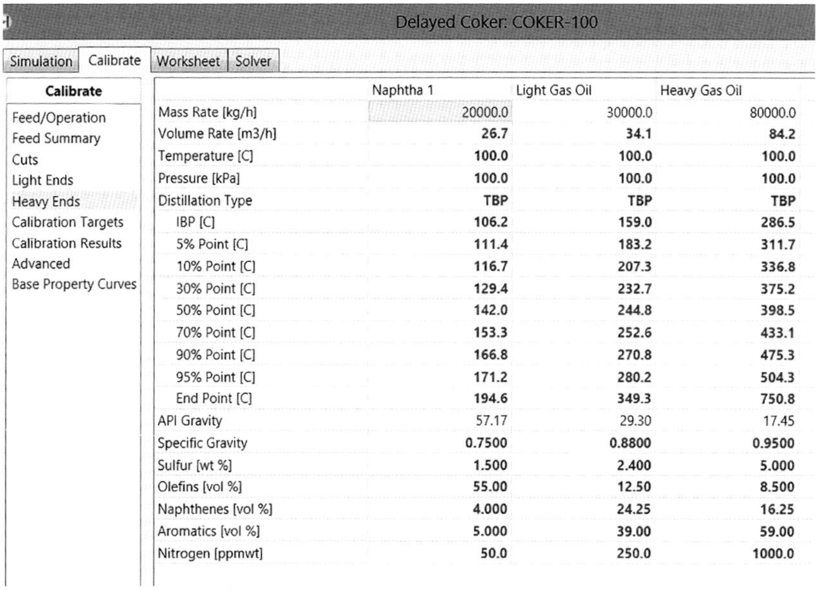

	Naphtha 1	Light Gas Oil	Heavy Gas Oil
Mass Rate [kg/h]	20000.0	30000.0	80000.0
Volume Rate [m3/h]	26.7	34.1	84.2
Temperature [C]	100.0	100.0	100.0
Pressure [kPa]	100.0	100.0	100.0
Distillation Type	TBP	TBP	TBP
IBP [C]	106.2	159.0	286.5
5% Point [C]	111.4	183.2	311.7
10% Point [C]	116.7	207.3	336.8
30% Point [C]	129.4	232.7	375.2
50% Point [C]	142.0	244.8	398.5
70% Point [C]	153.3	252.6	433.1
90% Point [C]	166.8	270.8	475.3
95% Point [C]	171.2	280.2	504.3
End Point [C]	194.6	349.3	750.8
API Gravity	57.17	29.30	17.45
Specific Gravity	0.7500	0.8800	0.9500
Sulfur [wt %]	1.500	2.400	5.000
Olefins [vol %]	55.00	12.50	8.500
Naphthenes [vol %]	4.000	24.25	16.25
Aromatics [vol %]	5.000	39.00	59.00
Nitrogen [ppmwt]	50.0	250.0	1000.0

Figure 7.45 Plant data for heavy ends for calibration.

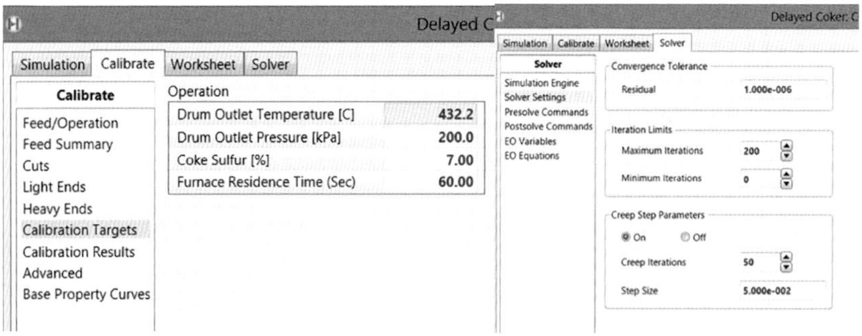

Figure 7.46 Calibration targets and solver settings.

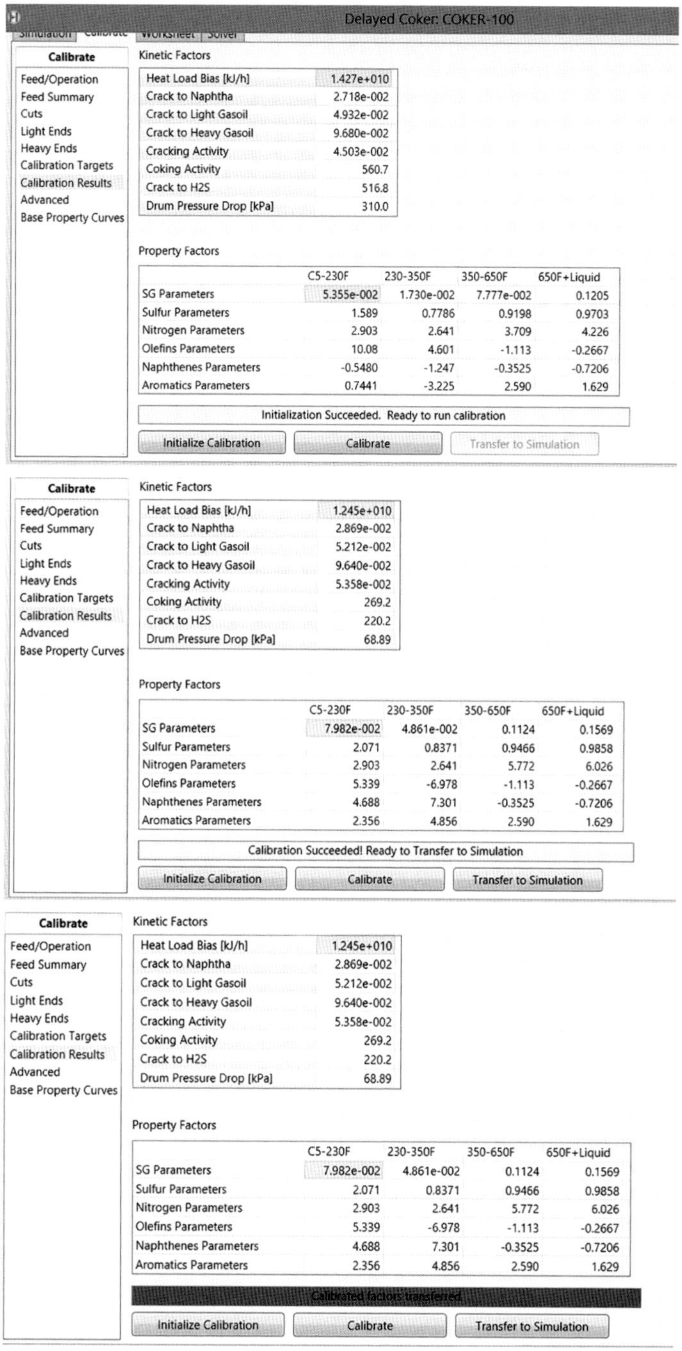

Figure 7.47 Three steps of calibration of the delayed coker model.

7.2.4 Workshop 7.3 – Simplified Model of Delayed Coker by Petroleum Shift Reactor for Production Planning Applications

We use this workshop to introduce the petroleum shift reactor that is widely used in the simplified modeling of refinery reactors for production planning purposes.

The petroleum shift reactor is developed based on the same principle that we have previously discussed in Section 4.12, Production Planning, and in Section 4.17, Workshop 4.5, Generate Delta-Base Vectors for Linear Programming (LP)-Based Planning. Basically, the petroleum shift reactor quantifies the effects of "shifts" (changes) in independent variables on the "shifts" of product yields and properties and on the "shifts" of utility consumptions according to our previous relationship in defining the delta-base vector, Eq. (4.13), now renumbered as Eq. (7.1):

$$\begin{bmatrix} y_1 \\ y_2 \\ \vdots \\ y_m \end{bmatrix} \text{(Prediction)} = \begin{bmatrix} \overline{y_1} \\ \overline{y_2} \\ \vdots \\ \overline{y_m} \end{bmatrix} \text{(Base)} + \begin{bmatrix} \frac{\Delta y_1}{\Delta x_1} & \cdots & \frac{\Delta y_1}{\Delta x_n} \\ \vdots & \cdots & \vdots \\ \frac{\Delta y_m}{\Delta x_1} & \cdots & \frac{\Delta y_m}{\Delta x_n} \end{bmatrix} \times \text{(Delta-Base)} \cdot \begin{bmatrix} \Delta x_1 \\ \vdots \\ \Delta x_n \end{bmatrix} \text{(Delta)} \tag{7.1}$$

In the context of the petroleum shift reactor,

- x_i is the value of the ith independent variable, such as the API gravity, CCR, or sulfur content (wt%) of the vacuum residue feeding to the delayed coker.
- y_i is the value of the ith dependent variable relating to products, properties, and utilities, such as the mass yield (%) of coker product (light ends, naphtha, distillate, gas oil, or coke), the liquid density (a property) of coker distillate, or the mass flow rate of utility steam to the coker.
- The delta vector defines the "shifts" (changes) in the independent variables, x_1 to x_n.
- The base vector gives the base-case values of dependent variables, y_1 to y_m, representing product yields or properties or utility consumptions.
- The delta-base matrix, or Jacobian, consists of element, $\Delta y_m/\Delta x_n$, representing the "shift" of the product yield or property, or "shift" in utility consumption, y_m, per one-unit "shift" in the independent variable x_n.

The petroleum shift reactor enables computational speed with some compromise in the rigorousness required for building large refinery-wide flowsheets. The delta-base concept is also the main approach by which production planning and scheduling software tools, Aspen PIMS and Aspen Petroleum Scheduler, use in modeling. Thus, the petroleum shift reactor makes it easier to integrate Aspen HYSYS for process modeling and optimization with production and scheduling tools into a single framework.

To run the petroleum shift reactor, we must make sure *to attach a petroleum assay to the fluid package of the flowsheet*. This follows that the model requires some petroleum properties of the feed and also needs to calculate some petroleum properties for products.

For this workshop, we open a starting file that contains the models for CDU/VDU discussed in detail in Chapters 2 and 3. This file is available in the supplementary material of this book for this chapter, *Workshop 7.3 – petroleum shift reactor for delayed coker_starting file.hsc*. Figure 7.48 shows the flowsheet for CDU and VDU of the starting file, and we are to send the vacuum residue from the bottom of the VDU to a petroleum shift reactor.

We add a petroleum shift reactor (see Figure 7.49).

We choose the energy stream, *Q*-steam, as medium-pressure (MP) steam (see Figure 7.50).

Based on the Excel spreadsheet, *Workshop 7.3 – petroleum shift reactor for delayed coker_input data.xlsx*, available in the supplement to this book, under Chapter 7, we complete the required input. To complete the "Model Data" under "Design" in Figure 7.51, we do as follows: Design → Model Data → Design Vars (variables) → Independent Vars → Petroleum Shift Reactor: Delayed Coker → Specify Independent Vars → Insert → Case(Main) → Vacuum Tower(COL2) → Vacuum Residue @COL2 → Calculator → Select: (1) *Sulfur Wt Pct* (petrol); (2) *Conradson Carbon Content* (petrol); (3) *API* (petrol). Then, copy and paste the input data from the spreadsheet. (*Note:* Currently, Aspen HYSYS lists the last chosen independent variable, i.e., API (petrol), first; and the first chosen independent variable, i.e., Sulfur Wt Pct (petrol), is listed last. Pay attention to matching the order of independent variables with that of the spreadsheet input data.)

The "Base Yield Fractions" column in Figure 7.51 represents the BASE shift vector in Eq. (7.1); the (5 row × 3 column) matrix with first row [−5.000E−4, 5.5E3, 2.1E−3], … and fifth row [−5E−4, 5.5E−3, 2.1E−3], is the delta base, or Jacobian, matrix in Eq. (7.1), representing the shift (change) of the product yield, y_m, per one-unit shift (change) in the independent variable x_n.

We compare the shifted values of three independent variables (API = 1.709, Conradson carbon = 21.20%, and sulfur = 3.558%) with the corresponding values in the vacuum residue feed (see Figure 7.52).

We continue to complete other input forms (see Figures 7.53–7.56).

Figure 7.57 shows the simulation results for product streams. We save the resulting simulation file as *Workshop 7.3 – petroleum shift reactor for delayed coker-end file.hsc.*

To summarize, this section demonstrates how to do efficient refinery reactor modeling using data tables with petroleum shift reactor, to implement a delta-base vector approach to quantify the effects of shifts (changes) in independent variables on shifts of product yields and properties and on the shift of utility consumptions, and to manipulate product yields and properties using a linear shift relationship.

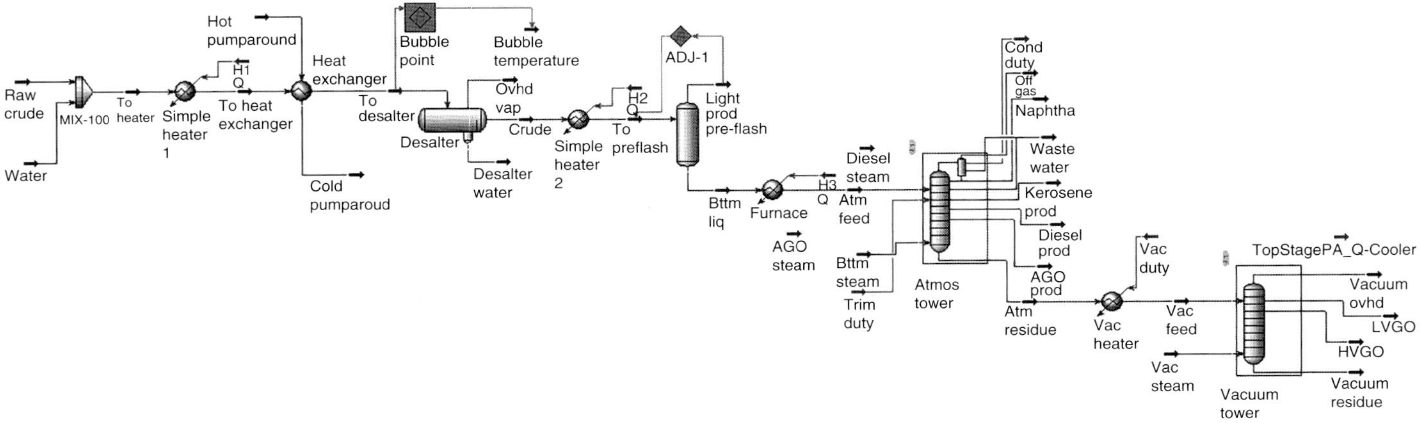

Figure 7.48 The starting flowsheet of the CDU and VDU from which we are to send the vacuum residue to a delayed coker represented by a petroleum shift reactor.

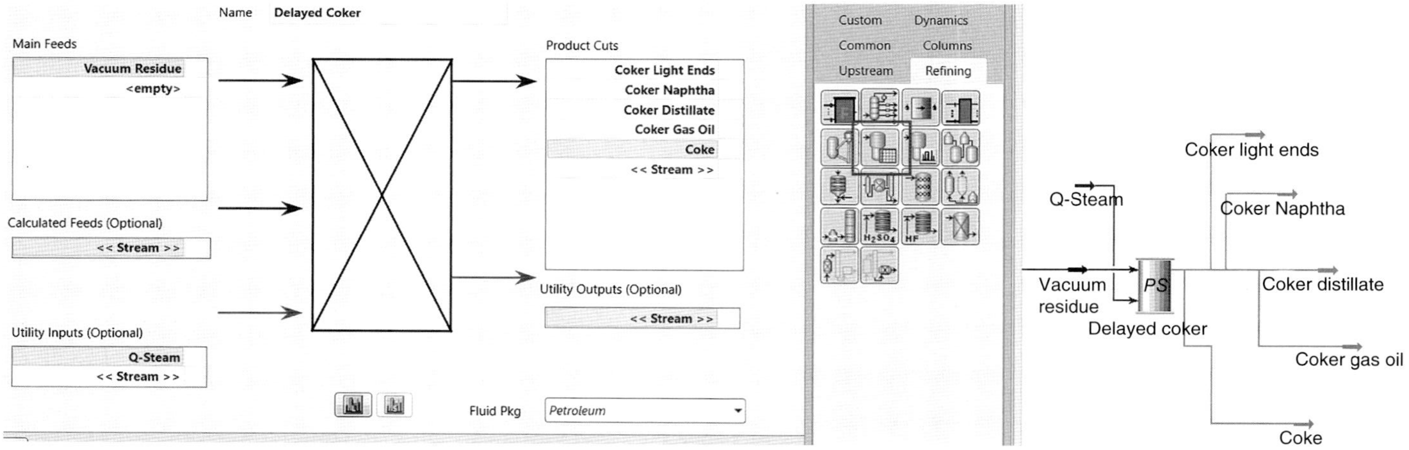

Figure 7.49 Add a petroleum shift reactor to represent the delayed coker.

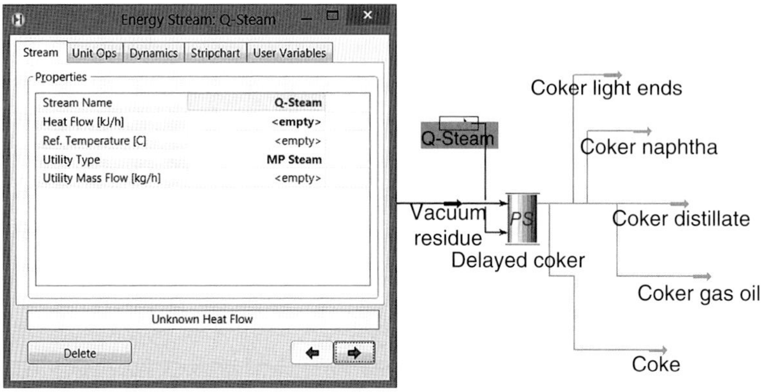

Figure 7.50 Specify the energy stream as MP steam.

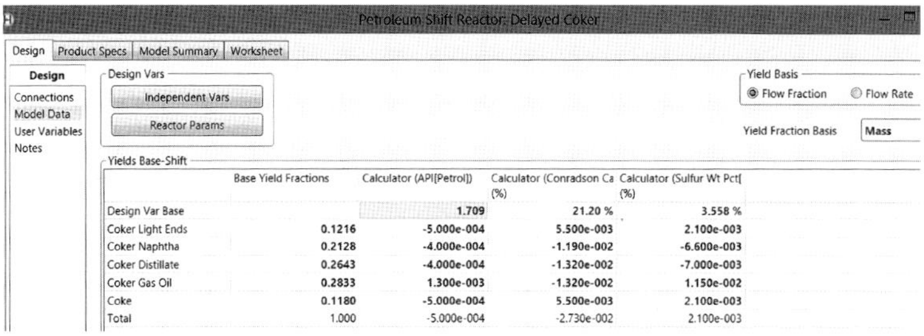

Figure 7.51 Select independent variables and enter model data for the petroleum shift reactor.

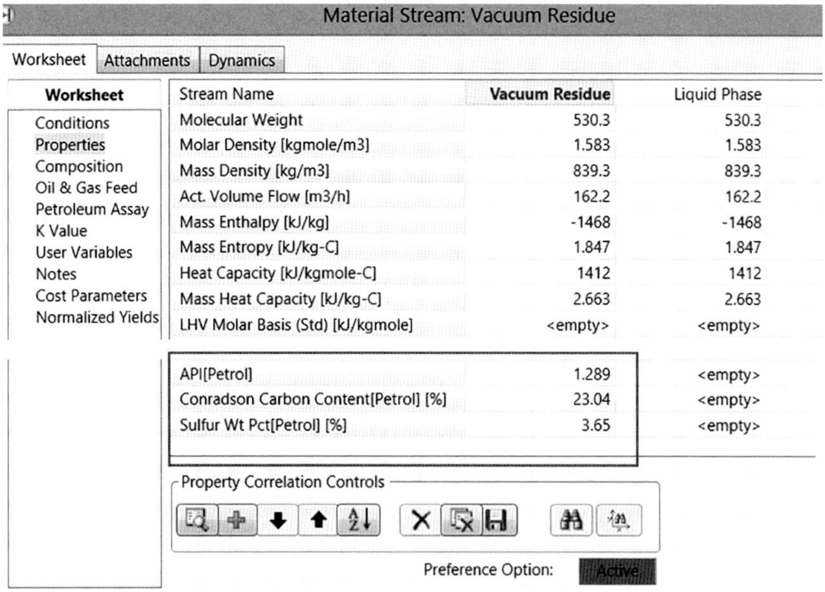

Figure 7.52 Values of three independent variables in the vacuum residue feed to the delayed coker.

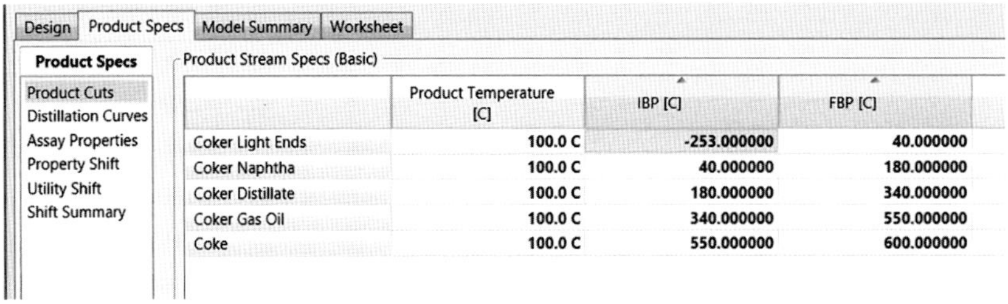

Figure 7.53 Define product stream cuts.

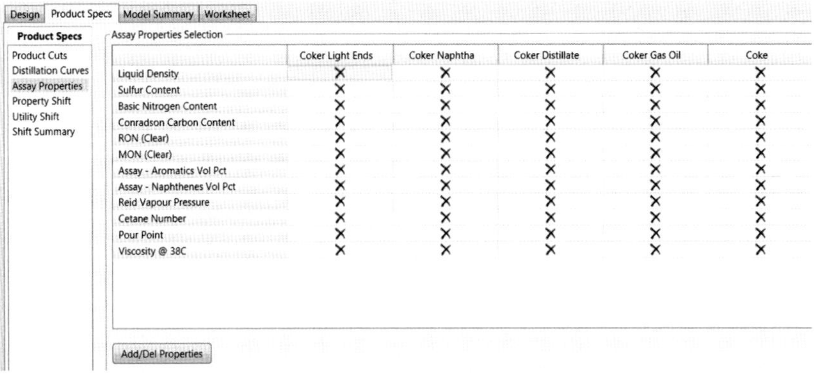

Figure 7.54 Select assay properties.

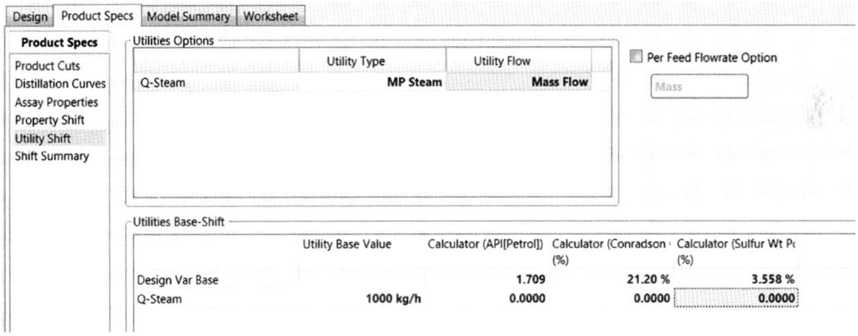

Figure 7.55 Complete utilities base-shift.

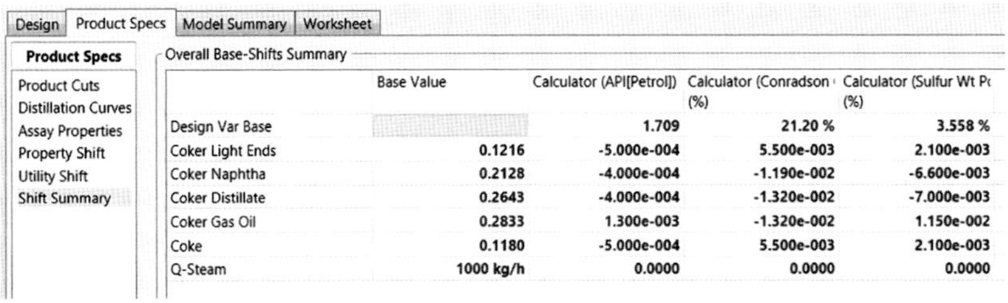

Figure 7.56 Overall base-shifts summary

Design | Product Specs | Model Summary | Worksheet

Worksheet
Conditions
Properties
Composition

Name	Vacuum Residue	Coker Light Ends	Coker Naphtha	Coker Distillate	Coker Gas Oil	Coke	Q-Steam
Vapour	0.0000	1.0000	1.0000	0.0000	0.0000	0.0000	<empty>
Temperature [C]	341.6	100.0	100.0	100.0	100.0	100.0	<empty>
Pressure [kPa]	13.00	13.00	13.00	13.00	13.00	13.00	<empty>
Molar Flow [kgmole/h]	256.7	256.5	247.5	157.8	89.46	31.50	<empty>
Mass Flow [kg/h]	1.361e+005	1.798e+004	2.593e+004	3.262e+004	3.533e+004	1.749e+004	1000
Std Ideal Liq Vol Flow [m3/h]	127.9	23.74	30.51	35.30	35.91	16.53	<empty>
Molar Enthalpy [kJ/kgmole]	-7.784e+005	-1.205e+005	-1.814e+005	-4.154e+005	-7.876e+005	-1.114e+006	<empty>
Molar Entropy [kJ/kgmole-C]	979.2	85.12	46.71	71.57	479.1	478.2	<empty>
Heat Flow [kJ/h]	-1.998e+008	-3.092e+007	-4.490e+007	-6.555e+007	-7.046e+007	-3.510e+007	1.981e+006

Figure 7.57 Simulation results of the petroleum shift reactor for the delayed coker.

7.3 Refinery-Wide Process Simulation

This section benefits from references written by Dziuk and Mohan [11–13]. In particular, we greatly appreciate the kind help by a co-author of these references, Sandeep Mohan, in the preparation of this section.

7.3.1 Refinery-Wide Process Model: A Key to Integrating Process Modeling and Production Planning

Profit margin analysis is a crucial exercise for refineries operating on tight profit margins. Process engineering software tools that enable refinery-wide process modeling can greatly improve and facilitate the profit margin analysis for both process engineers and production planners in the refinery. By developing *a refinery-wide process simulation model,* process engineers can evaluate the economic impact of operational improvements and unexpected events, and they can help production planners in achieving a more accurate assessment of profit margins. The process engineer can use the rigorous model data to easily evaluate why profit margins are low, suggest remedial actions, and predict the effect on profit margins.

A key challenge to developing fully rigorous refinery-wide process simulation models is their cumbersome nature. These models can take a long lead time to develop and run. These models often require a high level of expertise to operate, keeping refineries dependent on expensive third-party service providers to develop and use the models.

A practical solution is to develop a refinery-wide process simulation model by employing a mixture of shortcut and rigorous submodels [11–13]. In particular, there are three recent advances in process simulation technology that are making it easier to develop refinery-wide process simulation models for integrated process engineering and production planning applications. Let us consider, for example, the integration between the process simulation tool (Aspen HYSYS) and the production planning tool (Aspen PIMS) below.

The first advance is the petroleum assay manager, discussed in Sections 1.5 and 1.6. The *same* petroleum assay manager is used in both Aspen HYSYS and Aspen PIMS. Thus, both process engineering and production planning models use an identical set of pure and hypothetical components and characterize the assay by the same set of petroleum properties. This shared assay management tool enables the easy transfer of crude oil information between engineers and planners and facilitates greater accuracy in process simulation and production planning models.

The second advance is the fractionation model. Specifically, both Aspen HYSYS and Aspen PIMS use: (1) the *same* rigorous fractionation model, such as the crude distillation unit (CDU) and the vacuum distillation unit (VDU), discussed in Chapters 2 and 3, and (2) the *same* shortcut petroleum distillation column model discussed in Section 2.6. A benefit of this integration is that we can, for example, better calibrate the Aspen PIMS CDU model to match plant data by using the petroleum distillation model that Aspen PIMS shares with Aspen HYSYS.

The third advance is the availability of the *same* shortcut reactor models, such as the petroleum shift reactors, in both Aspen HYSYS and Aspen PIMS. This means that engineers can use the simulation results from rigorous refinery reactor models (e.g., FCC, reformer, hydrotreating unit, hydrocracker, delayed coker, and alkylation unit) to update the PIMS reactor submodels using petroleum shift reactors. This integration creates a streamlined and more efficient workflow for planning model updates.

With these integrations, process engineers can quickly develop a "clone" of their PIMS refinery model in Aspen HYSYS. This provides the process engineer with a simple refinery-wide process model that has the same credibility of a PIMS planning model. With a newly expanded, complete suite of rigorous reactor models available in the process simulation environment, process engineers can enhance the rigor of the refinery-wide process model by selectively upgrading specific shortcut reactor submodels to rigorous models in Aspen HYSYS. This capability allows process engineers to easily manage and maintain the model, while ensuring the rigor required for accurate refinery margin analysis.

7.3.2 An Example of a Refinery-Wide Process Simulation Model

Figure 7.58 illustrates a refinery-wide process simulation model that consists of nine submodels (subflowsheets) [11]. We include either rigorous or shortcut simulation models in each submodel, depending on the purposes of model applications. These applications include, but not limited to, the following: operational improvements, refinery reconfigurations, response to unexpected events, new unu startups, turnaround planning, planning support, and emission utility analyses [11].

Table 7.1 summarizes the rigorous and shortcut simulation models that are included in each submodel (subflowsheet).

For examples of larger and more complex, refinery-wide process simulation models, open up a new case in Aspen HYSYS V9.0. Go to: Examples → refinery cases → (1) *Refinery-wide model.hsc* and (2) *RefineryWideModel_GulfCoast.hsc.*

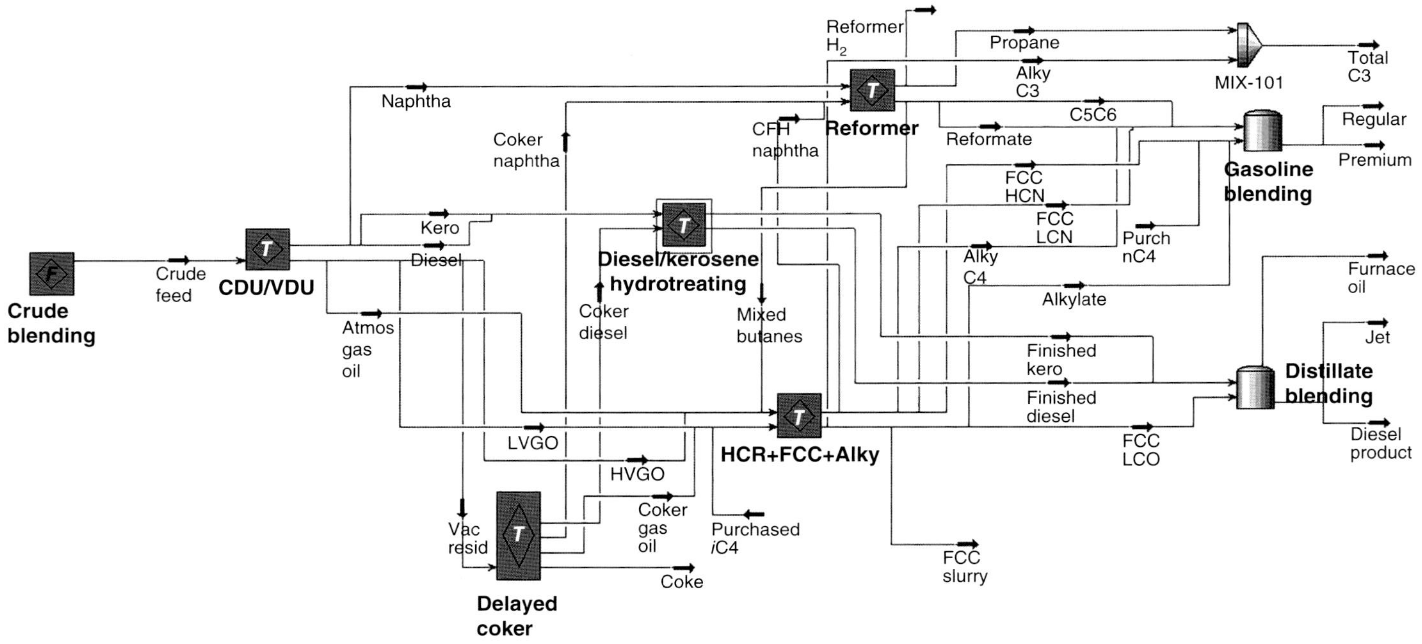

Figure 7.58 An example of a refinery-wide process simulation model. (Adapted from Dziuk and Mohan 2016 [11].)

Table 7.1 A summary of rigorous and shortcut simulation models included in the refinery-wide simulation model of Figure 7.58.

Submodel	Rigorous models used	Shortcut models used
1. Crude blending	Petroleum feeder	
2. CDU/VDU (crude distillation unit/vacuum distillation unit)		Petroleum distillation columns for both CDU and VDU
3. Delayed coker		Petroleum shift reactor for coker; petroleum distillation column for coker fractionation
4. Diesel/kerosene hydrotreating		Petroleum shift reactors for both hydrotreaters
5. HCR/FCC/Alky (hydrocracking/fluid catalytic/alkylation)	Hydrocracking and fluid catalytic cracking reactor models	Conversion reactor for alkylation; petroleum distillation columns for product fractionations
6. Reformer	Naphtha hydrotreating and catalytic reactor models	Petroleum distillation columns for product fractionations
7. Gasoline blending	Product blending for optimization	
8. Distillate blending	Product blender for optimization	

7.3.3 Tools for Developing Refinery-Wide Process Models

The preceding example illustrates the following essential tools to develop refinery-wide process simulation models:

- Petroleum assay manager (Sections 1.5 and 1.6);
- Rigorous CDU (Chapter 2), VDU (Chapter 3), product fractionation unit, and gas plant (Section 4.15);
- Rigorous refinery reactor models (Chapters 3–7): Except for vis breaker and isomerization units, this text has covered all of the other reactor models displayed in Figure 7.59 [11];
- Shortcut petroleum shift reactor (Section 7.2.4) and shortcut petroleum distillation column (Section 2.6);
- Petroleum product blender [14];
- Excel spreadsheet within Aspen HYSYS for easy display of the values of key independent and dependent variables and the profit function.

7.3.4 Deployment and Applications of the Refinery-Wide Process Models for Process Engineering and Production Planning

There are three steps to deploy a refinery-wide process simulation model [11]:

Step 1. Match the Aspen HYSYS model for process engineering with Aspen PIMS model for production planning. This implies the use of shortcut reactor and fractionation models in Aspen HYSYS, particularly the petroleum shift reactor and the petroleum distillation column. Define the scope for Step 2 model upgrade.

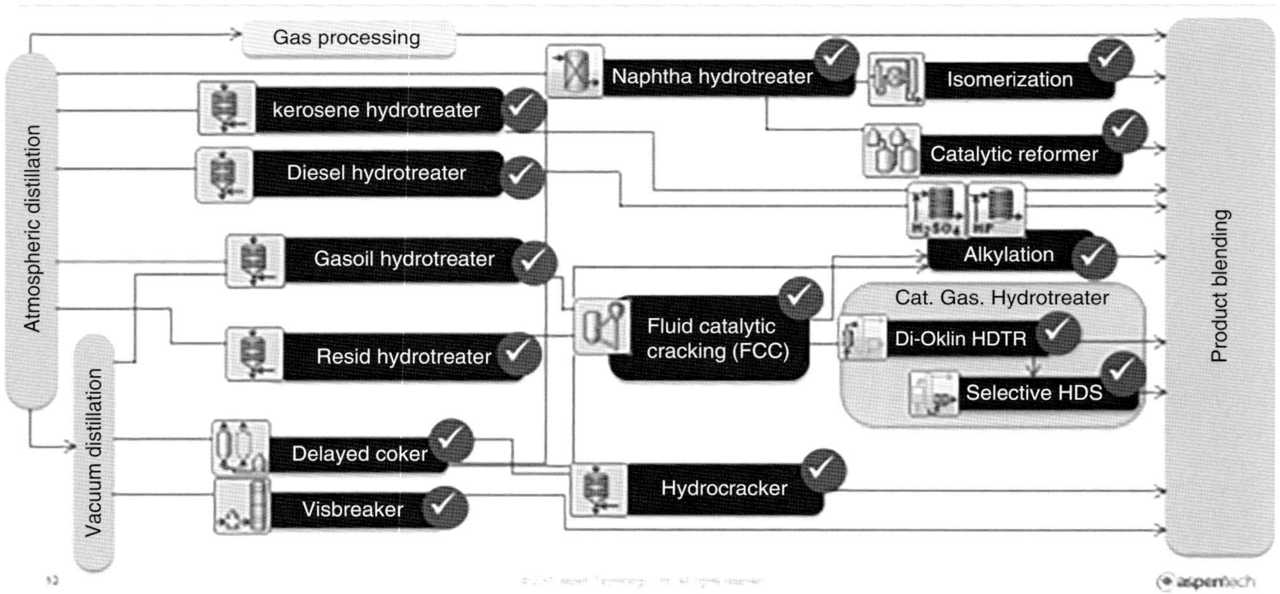

Figure 7.59 Current rigorous refinery reactor models in Aspen HYSYS [11]. (Courtesy of ASPEN Technology, Inc.)

Step 2. Upgrade the Aspen HYSYS model with rigorous reactor and fractionation submodels. This includes the use of improved stream property methods, convert shortcut petroleum shift reactor models to rigorous refinery reactor models (such as those for FCC and for hydrocrackers), and use rigorous refinery fractionation models to replace the shortcut petroleum distillation models.

Step 3. Apply the refinery-wide simulation model to: (1) *refinery reconfigurations* – reconfiguring refineries for changes in crude oil mixtures, for changes in product slates and for strategic planning for capital projects; (2) *operational improvements* – examples include the reevaluation of optimal cut points and reactor set points across the refinery, the evaluation of changes of catalyst for reactor units, perform deeper cuts at the VDU to increase conversion feed, and balance hydrogen production from reformer against hydrogen demand from hydroprocessors; (3) *operational response for unexpected events* – when a key equipment (e.g., a major feed pump) is down, when the crude feed rates decrease due to supply interruptions, and when the production rates decrease due to shipment problems; (4) *improved startup strategy* – examples include the planning of startup for new refinery units and the response to unexpected shutdowns during commissioning; (5) *turnaround planning* – how to rebalance the refinery when large parts of it are shut down for maintenance once every 3–5 years; (6) *improved planning model support* – improved collaboration between process engineers and production planners; improved validation of planning model results; more flexible and powerful modeling tools to support production planning, such as the identical representation of petroleum assays and their properties, and rigorous refinery reactor models to generate the delta-base vectors for linear programming (LP)-based Aspen PIMS planning model (Section 4.17); and (7) *evaluation of refinery-wide emissions and utilities* – use Aspen HYSYS refinery process model to evaluate greenhouse gas emissions for air quality reports and to generate refinery-wide utility balances.

7.4 Conclusions

This chapter covers three important aspects of integrated refinery modeling and production planning: (1) the alkylation process to produce high-octane blending components for gasoline; (2) the delayed coking process to upgrade and convert the "bottom of the barrel" residue materials to valuable liquid and gas product streams; and (3) refinery-wide process simulation model for process engineers and production planners.

As demonstrated in this text, modern refinery simulation models provide a power tool to predict product yields and properties, and to guide both process engineers and production planners how to optimize the process operations and to sustain refinery margins.

Bibliography

1 Gary, J.H., Handwerk, G.E., and Kaiser, M.J. (2007) *Petroleum Refining. Technology and Economics*, 5th edn, CRC Press, Boca Raton, FL.

2 Kaes, G.L. (2000) *Refinery Process Modeling A Practical Guide to Steady State Modeling of Petroleum Processes*, The Athens Printing Company, Athens, GA.

3 Luyben, W.L. (2009) Design and control of an auto refrigerated alkylation process. *Industrial & Engineering Chemistry Research*, **48**, 11081–11093.

4 Kranz, K. (2008) Introduction to alkylation chemistry: mechanisms, operating variables, and olefin interactions. *DuPont STRATCO Clean Fuel Technology*, http://www.dupont.com/content/dam/dupont/products-and-services/consulting-services-and-process-technologies/consulting-services-and-process-technologies-landing/documents/AlkylationChemistry_RU.pdf.

5 Sun, W., Shi, Y., Chen, J., Xi, Z., and Zhao, L. (2013) Alkylation kinetics of isobutane by C4 olefins using sulfuric acid as catalyst. *Industrial & Engineering Chemistry Research*, **52**, 15262–15269.

6 Esteves, P.M., Araujo, C.L., Horta, B.A.C., Alvarez, L.J., Zicovich-Wilson, C.M., and Ramirez-Solis, A. (2005) The isobutylene-isobutane alkylation processing liquid HF revisited. *The Journal of Physical Chemistry*, **109**, 12946–12955.

7 Ellis, P.J. and Paul, C.A. (1998) Delayed Coking Fundamentals. AIChE Spring National Meeting in New Orleans, LA, http://inside.mines.edu/~jjechura/Refining/DECOKTUT.pdf.

8 Ancheyta, J. (2013) *Modeling of Processes and Reactors for Upgrading of Heavy Petroleum*, CRC Press, Boca Raton, FL.

9 Akpabio, E.J. and Ekott, E.J. (2012) Integrating delayed coking process into Nigeria's refinery configuration. *Indian Journal of Science and Technology*, **5**, 2923–2927.

10 Alfeel, A.M.M., Mohamed, A.A.A., Ali, A., Lo-Lujo, E.O.Y. and Mhmound, L.A.M. (2016) *Simulation of Delayed Coking Unit in KRC*, http://repository.sustech.edu/handle/123456789/15098.

11 Dziuk, S. and Mohan, S. (2016) Rapidly Deploy a Refinery-Wide Process Model for Improved Profit Margin Analysis, AspenTech webinar, February 16.

12 Dziuk, S. and Mohan, S. (2016) Improve Profit Margins through a Refinery-wide Process Model. 114th American Fuel and Petrochemical Manufacturers Annual Meeting. AFAM_AM_2016<Day3<AspenTec.pdf, http://www.aspentech.com/resource-library/ → white papers → refining → Aspen HYSYS.

13 Dziuk, S. and Mohan, S. *Secure Your Refinery Profit Margins by Keeping Planning Models Up to Date*, http://www.aspentech.com/resource-library/ → white papers → refining → Aspen PIMS.

14 Aspen Technology, Inc. (2016) Aspen HYSYS Petroleum Refining Unit Operations and Reactor Models V9: Reference Guide, May, pp. 59–80, Chapter 5, "Product Blender".

A

List of Computer Files

Accessible through Wiley-VCH website, wiley-vch.de/en/→shop→bookfinder→ petroleum refinery process modeling→student materials

Chapter 1

- WS1.1 ASTMConvert.xls
  Convert one ASTM distillation curve into another
- WS1.2 Beta.xls
  Extrapolate incomplete distillation curve by beta distribution
- WS1.3 MeABP Iteration.xls
  Calculate MeABP for a given oil fraction
- WS1.4 Oil Manager.hsc
  Answer file of workshop 1.4
- WS1.4 Distillation Curve and Light End Compostition.xlsx
- WS1.5 Petroleum Assay Manager.hsc
- WS1.6 Conversion from Oil Manager to Petroleum Assay Manager.hsc
- Critical_Property_Correlations.xls
  Spreadsheet containing various methods to estimate critical properties of pseudocomponents (Section 1.10.1)

Chapter 2

- Beta_Data_Fit.xls
  Spreadsheet to fit distillation and density/specific gravity curves to beta distributions
- ASTM_Convert.xls
  Spreadsheet to convert distillation curves from ASTM type to another
- Crude Assay Only.hsc
  Define assays for crude column model (Section 2.8.1)
- CDU EX-1.hsc
  Add Feed-Preheat system to column model (Section 2.8.2)
- CDU EX-2.hsc
  Initial Column model before solving (no side ops) (Sections 2.8.3 and 2.8.4)

Petroleum Refinery Process Modeling: Integrated Optimization Tools and Applications,
First Edition. Y. A. Liu, Ai-Fu Chang, and Kiran Pashikanti.

- CDU EX-3.hsc
  Initial Column model (after adding the side stripper for heavy naphtha) (Section 2.8.5)
- CDU EX-4.hsc
  Initial Column model (after adding all three side strippers) (Section 2.8.5)
- CDU EX-5.hsc
  Column simulation model with preheat train and all side strippers and all pumparounds (Section 2.8.6)
- CDU EX-6.hsc
  Converged simulation model with three side strippers and three pumparounds (Section 2.8.7)
- CDU EX-7.hsc
  Converged simulation model after adding custom stream properties (Section 2.8.7)
- CDU EX-8_GS.hsc
  Solve column model with preheat train and all side strippers and all pumparounds with Grayson–Streed thermodynamics (Section 2.9)
- CDU-Case Study.hsc
  Case study of the effects of independent variables on the CDU product petroleum properties (Section 2.10.3)
- CDU-Backblending-1.hsc
  Define the CDU product assays in the petroleum assay manager (Section 2.11.1)
- CDU-Backblending-2.hsc
  Define a new blend of the backblended crude feed (Section 2.11.2)
- CDU-Backblending-3.hsc
  Simulation model for the Heated_FEED to the CDU developed from backblending (Section 2.11.3)
- CDU-Backblending-4.hsc
  CDU simulation model based on the backblended feed (Section 2.11.3)
- CDU-Backblending-5.hsc
  CDU simulation model based on the backblended feed with water draw to speed up column convergence when two liquid phases might be present (Section 2.11.4)
- Workshop 2.3.hsc (Section 2.12)
- Workshop 2.4.hsc (Section 2.13)
- CDU-backblending-internals.hsc (Section 2.14, column hydraulic analysis)
- Workshop 2.5: CDU-blending-internals.hsc
- Workshop 2.6-starting.hsc; Workshop 2.6-end.hsc

Chapter 3

- Workshop 3.1-1.hsc to Workshop 3.1-6.hsc: Step-by-step illustration of the development of the simplified VDU model; VDU-Simplified.hsc – the final converged simulation file (Section 3.3.2)
- Workshop 3.2-1.hsc and VDU-Rigorous.hsc: Step-by-step illustration of the development of the rigorous VDU model (Section 3.3.2)

- VDU Deep Cut-Start.hsc, Workshop 3.3-1.hsc to Workshop 3.3-3.hsc (Section 3.5)

Chapter 4

- Composition_PNA.xls
  PNA correlation for various feed types
- FCC_HeatBalance_US.xls
  Perform heat balance and calculate delta-coke for FCC units
- FCC_DecantOil.xls
  Estimating distillation curves of FCC decant oil
- Fuel_Properties.xls
  Various fuel properties (Cloud, Freeze, etc.) for petroleum fractions
- Lognormal.xls
  Fitting distillation or specific gravity data to lognormal distributions
- FCC Components and Properties.hsc (Section 4.13.4), Workshop 4.1-1.hsc, and Workshop 4.1-done.hsc (Build the basic FCC model, Sections 4.13.5–4.13.10)
- Workshop 4.2-1.hsc and Workshop 4.2-done.hsc (Calibrate a basic FCC model, Section 4.14)
- Workshop 4.3-1.hsc to Workshop 4.3-8.hsc (Section 4.15, Workshop 4.3 – Build a fractionation section for the FCC model)
- Workshop 4.4-1.hsc and Workshop 4.4-2.hsc (Section 4.16 – Case studies)
- Workshop 4.5.hsc (Section 4.17 – Application to production planning)

Chapter 5

- Alternate_Feed_Lumping.xlsm
  Alternate method for feed lumping for reformer
- Feed_AspenHYSYS_Transform.xlxs
  Remap measured component feed into Aspen HYSYS components
- Hydrogen_Balance.xls
  Calculate hydrogen balance of reformer unit with external feeds
- Workshop 5.1.hsc (Section 5.14)
- Workshop 5.1-1.hsc (Section 5.14, using bulk properties), Workshop 5.1-2.hsc (Section 5.14, using measured molecular information)
- Workshop 5.2.hsc (Section 5.15, Model calibration)
- Workshop 5.3.hsc (Section 5.16, Fractionation system)
- Workshop 5.4-1.hsc and Workshop 5.4-2.hsc (Section 5.17, two case studies)

Chapter 6

- GL_Quad Pt.xls
  Allocation points of Gaussian quadrature
- Mass Balance.xls
  Check the mass balance of HCR unit

- Workshop 2 (Folder)
  The simulation file of each step described in workshop 2
- Workshop 3 (Folder)
  The simulation file and results of workshop 3
- Workshop 4 (Folder)
  The simulation files of workshop 4 and the spreadsheet for delumping (Delump.xls)

Chapter 7

- Workshop 7.1-1 Input Data.xlsx
- Workshop 7.1-1 alkylation_open loop_petroleum distillation column.hsc
- Workshop 7.1-1 alkylation_closed loop_petroleum distillation column.hsc
- Workshop 7.2-1 Initial and Calibration Data.xlsx
- Workshop 7.2-1 Open Loop_Converged.hsc
- Workshop 7.2-1 Closed Loop_Converged.hsc
- Workshop 7.2-1 Closed Loop_Calibration.hsc
- Workshop 7.3 Petroleum shift reactor for delayed coker_starting file.hsc
- Workshop 7.3 Petroleum shift reactor for delayed coker_end file.hsc

Index

Petroleum Refinery Process Modeling: Integrated Optimization Tools and Applications,
First Edition. Y. A. Liu, Ai-Fu Chang, and Kiran Pashikanti.

b